DATE DUE

OCT. 28 1994	
MAY 2 2 1995	
APR 1 0 1996	
MAY 01 1996	
JUN 2 8 1996	
MAR 11 1997	
MAR 1 2 1997	
OCT 1 6 1998	

BRODART   Cat. No. 23-221

DIFFERENTIAL EQUATIONS

Third Edition

Shepley L. Ross
University of New Hampshire

John Wiley & Sons
New York • Chichester • Brisbane • Toronto • Singapore

Library of Congress Cataloging in Publication Data:

Ross, Shepley L.
  Differential equations.

  Bibliography: p.
  Includes index.
  1. Differential equations.   I. Title.
QA371.R6 1984   515.3′5   83-21643
ISBN 0-471-03294-8

Printed in the United States of America

10 9 8 7 6 5 4

PREFACE

This third edition, like the first two, is an introduction to the basic methods, theory, and applications of differential equations. A knowledge of elementary calculus is presupposed.

The detailed style of presentation that characterized the previous editions of the text has been retained. Many sections have been taken verbatim from the second edition, while others have been rewritten or rearranged with the sole intention of making them clearer and smoother. As in the earlier editions, the text contains many thoroughly worked out examples. Also, a number of new exercises have been added, and assorted exercise sets rearranged to make them more useful in teaching and learning.

The book is divided into two main parts. The first part (Chapters 1 through 9) deals with the material usually found in a one-semester introductory course in ordinary differential equations. This part is also available separately as *Introduction to Ordinary Differential Equations*, Third Edition (John Wiley & Sons, New York, 1980). The second part of the present text (Chapters 10 through 14) introduces the reader to certain specialized and more advanced methods and provides an introduction to fundamental theory. The table of contents indicates just what topics are treated.

The following additions and modifications are specifically noted.

1. Material emphasizing the second-order linear equation has been inserted at appropriate places in Section 4.1.
2. New illustrative examples, including an especially detailed introductory one, have been written to clarify the Method of Undetermined Coefficients in Section 4.3, and a useful table has also been supplied.
3. Matrix multiplication and inversion have been added to the introductory material on linear algebra in Section 7.5.
4. Additional applications now appear in the text in Sections 3.3 and 7.2.
5. Section 7.6 is a completely new section on the application of matrix algebra to the solution of linear systems with constant coefficients in the special case of two equations in two unknown functions. The theory that occupied this section in the

second edition now appears in Chapter 11 (see note 9 following). We believe that this change represents a major improvement for both introductory and intermediate courses.

6. Section 7.7 extends the matrix method of Section 7.6 to the case of linear systems with constant coefficients involving n equations in n unknown functions. Several detailed examples illustrate the method for the case $n = 3$.

7. Both revised and new material on the Laplace Transform of step functions, translated functions, and periodic functions now appears in Section 9.1.

8. The basic existence theory for systems and higher-order equations, formerly located at the beginning of Chapter 11, has now been placed at the end of Chapter 10. This minor change has resulted in better overall organization.

9. Chapter 11, the Theory of Linear Differential Equations, has been changed considerably. Sections 11.1 through 11.4 present the fundamental theory of linear systems. Much of this material was found in Section 7.6 in the second edition, and some additional results are also included here. Sections 11.5 through 11.7 now present the basic theory of the single nth-order equation, making considerable use of the material of the preceding sections. Section 11.8 introduces second-order self-adjoint equations and proceeds through the fundamentals of classical Sturm Theory. We believe that the linear theory is now presented more coherently than in the previous edition.

10. An appendix presents, without proof, the fundamentals of second and third order determinants.

The book can be used as a text in several different types of courses. The more or less traditional one-semester introductory course could be based on Chapter 1 through Section 7.4 of Chapter 7 if elementary applications are to be included. An alternative one-semester version omitting applications but including numerical methods and Laplace transforms could be based on Chapters 1, 2, 4, 6, 7, 8, and 9. An introductory course designed to lead quickly to the methods of partial differential equations could be based on Chapters 1, 2 (in part), 4, 6, 12, and 14.

The book can also be used as a text in various intermediate courses for juniors and seniors who have already had a one-semester introduction to the subject. An intermediate course emphasizing further methods could be based on Chapters 8, 9, 12, 13, and 14. An intermediate course designed as an introduction to fundamental theory could be based on Chapters 10 through 14. We also note that Chapters 13 and 14 can be interchanged advantageously.

I am grateful to several anonymous reviewers who made useful comments and suggestions. I thank my colleagues William Bonnice and Robert O. Kimball for helpful advice. I also thank my son, Shepley L. Ross, II, graduate student in mathematics, University of Rochester, Rochester, New York, for his careful reviewing and helpful suggestions.

I am grateful to Solange Abbott for her excellent typing. I am pleased to record my appreciation to Editor Gary Ostedt and the Wiley staff for their constant helpfulness and cooperation.

As on several previous occasions, the most thanks goes to my wife who offered encouragement, understanding, patience, and help in many different ways. Thanks, Gin.

Shepley L. Ross

CONTENTS

PART ONE   FUNDAMENTAL METHODS AND APPLICATIONS

One   Differential Equations and Their Solutions   3

1.1   Classification of Differential Equations; Their Origin and Application   3
1.2   Solutions   7
1.3   Initial-Value Problems, Boundary-Value Problems, and Existence of Solutions   15

Two   First-Order Equations for Which Exact Solutions Are Obtainable   25

2.1   Exact Differential Equations and Integrating Factors   25
2.2   Separable Equations and Equations Reducible to This Form   39
2.3   Linear Equations and Bernoulli Equations   49
2.4   Special Integrating Factors and Transformations   61

Three   Applications of First-Order Equations   70

3.1   Orthogonal and Oblique Trajectories   70
3.2   Problems in Mechanics   77
3.3   Rate Problems   89

Four   Explicit Methods of Solving Higher-Order Linear Differential Equations   102

4.1   Basic Theory of Linear Differential Equations   102
4.2   The Homogeneous Linear Equation with Constant Coefficients   125

v

4.3 The Method of Undetermined Coefficients   137
4.4 Variation of Parameters   155
4.5 The Cauchy–Euler Equation   164
4.6 Statements and Proofs of Theorems on the Second-Order Homogeneous Linear
     Equation   170

**Five Applications of Second-Order Linear Differential Equations
with Constant Coefficients**   **179**

5.1 The Differential Equation of the Vibrations of a Mass on a Spring   179
5.2 Free, Undamped Motion   182
5.3 Free, Damped Motion   189
5.4 Forced Motion   199
5.5 Resonance Phenomena   206
5.6 Electric Circuit Problems   211

Six Series Solutions of Linear Differential Equations   **221**

6.1 Power Series Solutions About an Ordinary Point   221
6.2 Solutions About Singular Points; The Method of Frobenius   233
6.3 Bessel's Equation and Bessel Functions   252

Seven Systems of Linear Differential Equations   **264**

7.1 Differential Operators and an Operator Method   264
7.2 Applications   278
7.3 Basic Theory of Linear Systems in Normal Form: Two Equations in Two Unknown
     Functions   290
7.4 Homogeneous Linear Systems with Constant Coefficients: Two Equations in Two
     Unknown Functions   301
7.5 Matrices and Vectors   312
7.6 The Matrix Method for Homogeneous Linear Systems with Constant Coefficients:
     Two Equations in Two Unknown Functions   346
7.7 The Matrix Method for Homogeneous Linear Systems with Constant Coefficients:
     n Equations in n Unknown Functions   355

Eight Approximate Methods of Solving First-Order Equations   **377**

8.1 Graphical Methods   377
8.2 Power Series Methods   384
8.3 The Method of Successive Approximations   390
8.4 Numerical Methods   394

Nine The Laplace Transform   **411**

9.1 Definition, Existence, and Basic Properties of the Laplace Transform   411
9.2 The Inverse Transform and the Convolution   431
9.3 Laplace Transform Solution of Linear Differential Equations with Constant
     Coefficients   441
9.4 Laplace Transform Solution of Linear Systems   453

PART TWO   FUNDAMENTAL THEORY AND FURTHER METHODS

Ten   Existence and Uniqueness Theory   461

10.1   Some Concepts from Real Function Theory   461
10.2   The Fundamental Existence and Uniqueness Theorem   473
10.3   Dependence of Solutions on Initial Conditions and on the Function f 488
10.4   Existence and Uniqueness Theorems for Systems and Higher-Order Equations   495

Eleven   The Theory of Linear Differential Equations   505

11.1   Introduction   505
11.2   Basic Theory of the Homogeneous Linear System   510
11.3   Further Theory of the Homogeneous Linear System   522
11.4   The Nonhomogeneous Linear System   533
11.5   Basic Theory of the nth-Order Homogeneous Linear Differential Equation   543
11.6   Further Properties of the nth-Order Homogeneous Linear Differential Equation   558
11.7   The nth-Order Nonhomogeneous Linear Equation   569
11.8   Sturm Theory   573

Twelve   Sturm–Liouville Boundary-Value Problems and Fourier Series   588

12.1   Sturm–Liouville Problems   588
12.2   Orthogonality of Characteristic Functions   597
12.3   The Expansion of a Function in a Series of Orthonormal Functions   601
12.4   Trigonometric Fourier Series   608

Thirteen   Nonlinear Differential Equations   632

13.1   Phase Plane, Paths, and Critical Points   632
13.2   Critical Points and Paths of Linear Systems   644
13.3   Critical Points and Paths of Nonlinear Systems   658
13.4   Limit Cycles and Periodic Solutions   692
13.5   The Method of Kryloff and Bogoliuboff   707

Fourteen   Partial Differential Equations   715

14.1   Some Basic Concepts and Examples   715
14.2   The Method of Separation of Variables   722
14.3   Canonical Forms of Second-Order Linear Equations with Constant Coefficients   743
14.4   An Initial-Value Problem; Characteristics   757

Appendices   771
Answers   777
Suggested Reading   801
Index   803

PART ONE
FUNDAMENTAL METHODS AND APPLICATIONS

CHAPTER ONE

Differential Equations and Their Solutions

The subject of differential equations constitutes a large and very important branch of modern mathematics. From the early days of the calculus the subject has been an area of great theoretical research and practical applications, and it continues to be so in our day. This much stated, several questions naturally arise. Just what is a differential equation and what does it signify? Where and how do differential equations originate and of what use are they? Confronted with a differential equation, what does one do with it, how does one do it, and what are the results of such activity? These questions indicate three major aspects of the subject: theory, method, and application. The purpose of this chapter is to introduce the reader to the basic aspects of the subject and at the same time give a brief survey of the three aspects just mentioned. In the course of the chapter, we shall find answers to the general questions raised above, answers that will become more and more meaningful as we proceed with the study of differential equations in the following chapters.

1.1 CLASSIFICATION OF DIFFERENTIAL EQUATIONS; THEIR ORIGIN AND APPLICATION

A. Differential Equations and Their Classification

DEFINITION

An equation involving derivatives of one or more dependent variables with respect to one or more independent variables is called a differential equation.*

---

* In connection with this basic definition, we do *not* include in the class of differential equations those equations that are actually derivative identities. For example, we exclude such expressions as

$$\frac{d}{dx}(e^{ax}) = ae^{ax}, \qquad \frac{d}{dx}(uv) = u\frac{dv}{dx} + v\frac{du}{dx}, \qquad \text{and so forth.}$$

3

▶ **Example 1.1**

For examples of differential equations we list the following:

$$\frac{d^2y}{dx^2} + xy\left(\frac{dy}{dx}\right)^2 = 0, \tag{1.1}$$

$$\frac{d^4x}{dt^4} + 5\frac{d^2x}{dt^2} + 3x = \sin t, \tag{1.2}$$

$$\frac{\partial v}{\partial s} + \frac{\partial v}{\partial t} = v, \tag{1.3}$$

$$\frac{\partial^2 u}{\partial x^2} + \frac{\partial^2 u}{\partial y^2} + \frac{\partial^2 u}{\partial z^2} = 0. \tag{1.4}$$

From the brief list of differential equations in Example 1.1 it is clear that the various variables and derivatives involved in a differential equation can occur in a variety of ways. Clearly some kind of classification must be made. To begin with, we classify differential equations according to whether there is one or more than one independent variable involved.

DEFINITION

A differential equation involving ordinary derivatives of one or more dependent variables with respect to a single independent variable is called an ordinary differential equation.

▶ **Example 1.2**

Equations (1.1) and (1.2) are ordinary differential equations. In Equation (1.1) the variable x is the single independent variable, and y is a dependent variable. In Equation (1.2) the independent variable is t, whereas x is dependent.

DEFINITION

A differential equation involving partial derivatives of one or more dependent variables with respect to more than one independent variable is called a partial differential equation.

▶ **Example 1.3**

Equations (1.3) and (1.4) are partial differential equations. In Equation (1.3) the variables s and t are independent variables and v is a dependent variable. In Equation (1.4) there are three independent variables: x, y, and z; in this equation u is dependent.

We further classify differential equations, both ordinary and partial, according to the order of the highest derivative appearing in the equation. For this purpose we give the following definition.

DEFINITION

The order of the highest ordered derivative involved in a differential equation is called the order *of the differential equation.*

▶ **Example 1.4**

The ordinary differential equation (1.1) is of the second order, since the highest derivative involved is a second derivative. Equation (1.2) is an ordinary differential equation of the fourth order. The partial differential equations (1.3) and (1.4) are of the first and second orders, respectively.

Proceeding with our study of ordinary differential equations, we now introduce the important concept of *linearity* applied to such equations. This concept will enable us to classify these equations still further.

DEFINITION

A linear ordinary differential equation *of order n, in the dependent variable y and the independent variable x, is an equation that is in, or can be expressed in, the form*

$$a_0(x)\frac{d^n y}{dx^n} + a_1(x)\frac{d^{n-1} y}{dx^{n-1}} + \cdots + a_{n-1}(x)\frac{dy}{dx} + a_n(x)y = b(x),$$

where a_0 is not identically zero.

Observe (1) that the dependent variable y and its various derivatives occur to the first degree only, (2) that no products of y and/or any of its derivatives are present, and (3) that no transcendental functions of y and/or its derivatives occur.

▶ **Example 1.5**

The following ordinary differential equations are both linear. In each case y is the dependent variable. Observe that y and its various derivatives occur to the first degree only and that no products of y and/or any of its derivatives are present.

$$\frac{d^2 y}{dx^2} + 5\frac{dy}{dx} + 6y = 0, \tag{1.5}$$

$$\frac{d^4 y}{dx^4} + x^2\frac{d^3 y}{dx^3} + x^3\frac{dy}{dx} = xe^x. \tag{1.6}$$

DEFINITION

A nonlinear ordinary differential equation *is an ordinary differential equation that is not linear.*

▶ **Example 1.6**

The following ordinary differential equations are all nonlinear:

$$\frac{d^2y}{dx^2} + 5\frac{dy}{dx} + 6y^2 = 0, \tag{1.7}$$

$$\frac{d^2y}{dx^2} + 5\left(\frac{dy}{dx}\right)^3 + 6y = 0, \tag{1.8}$$

$$\frac{d^2y}{dx^2} + 5y\frac{dy}{dx} + 6y = 0. \tag{1.9}$$

Equation (1.7) is nonlinear because the dependent variable y appears to the second degree in the term $6y^2$. Equation (1.8) owes its nonlinearity to the presence of the term $5(dy/dx)^3$, which involves the third power of the first derivative. Finally, Equation (1.9) is nonlinear because of the term $5y(dy/dx)$, which involves the product of the dependent variable and its first derivative.

Linear ordinary differential equations are further classified according to the nature of the coefficients of the dependent variables and their derivatives. For example, Equation (1.5) is said to be linear with *constant coefficients*, while Equation (1.6) is linear with *variable coefficients*.

B. Origin and Application of Differential Equations

Having classified differential equations in various ways, let us now consider briefly where, and how, such equations actually originate. In this way we shall obtain some indication of the great variety of subjects to which the theory and methods of differential equations may be applied.

Differential equations occur in connection with numerous problems that are encountered in the various branches of science and engineering. We indicate a few such problems in the following list, which could easily be extended to fill many pages.

1. The problem of determining the motion of a projectile, rocket, satellite, or planet.
2. The problem of determining the charge or current in an electric circuit.
3. The problem of the conduction of heat in a rod or in a slab.
4. The problem of determining the vibrations of a wire or a membrane.
5. The study of the rate of decomposition of a radioactive substance or the rate of growth of a population.
6. The study of the reactions of chemicals.
7. The problem of the determination of curves that have certain geometrical properties.

The mathematical formulation of such problems give rise to differential equations. But just how does this occur? In the situations under consideration in each of the above problems the objects involved obey certain scientific laws. These laws involve various rates of change of one or more quantities with respect to other quantities. Let us recall that such rates of change are expressed mathematically by derivatives. In the mathematical formulation of each of the above situations, the various rates of change are thus expressed by various derivatives and the scientific laws themselves become mathematical equations involving derivatives, that is, differential equations.

In this process of mathematical formulation, certain simplifying assumptions generally have to be made in order that the resulting differential equations be tractable. For example, if the actual situation in a certain aspect of the problem is of a relatively complicated nature, we are often forced to modify this by assuming instead an approximate situation that is of a comparatively simple nature. Indeed, certain relatively unimportant aspects of the problem must often be entirely eliminated. The result of such changes from the actual nature of things means that the resulting differential equation is actually that of an idealized situation. Nonetheless, the information obtained from such an equation is of the greatest value to the scientist.

A natural question now is the following: How does one obtain useful information from a differential equation? The answer is essentially that if it is possible to do so, one solves the differential equation to obtain a solution; if this is not possible, one uses the theory of differential equations to obtain information *about* the solution. To understand the meaning of this answer, we must discuss what is meant by a solution of a differential equation; this is done in the next section.

Exercises

Classify each of the following differential equations as ordinary or partial differential equations; state the order of each equation; and determine whether the equation under consideration is linear or nonlinear.

1. $\dfrac{dy}{dx} + x^2 y = xe^x$.

2. $\dfrac{d^3 y}{dx^3} + 4\dfrac{d^2 y}{dx^2} - 5\dfrac{dy}{dx} + 3y = \sin x$.

3. $\dfrac{\partial^2 u}{\partial x^2} + \dfrac{\partial^2 u}{\partial y^2} = 0$.

4. $x^2\, dy + y^2\, dx = 0$.

5. $\dfrac{d^4 y}{dx^4} + 3\left(\dfrac{d^2 y}{dx^2}\right)^5 + 5y = 0$.

6. $\dfrac{\partial^4 u}{\partial x^2\, \partial y^2} + \dfrac{\partial^2 u}{\partial x^2} + \dfrac{\partial^2 u}{\partial y^2} + u = 0$.

7. $\dfrac{d^2 y}{dx^2} + y \sin x = 0$.

8. $\dfrac{d^2 y}{dx^2} + x \sin y = 0$.

9. $\dfrac{d^6 x}{dt^6} + \left(\dfrac{d^4 x}{dt^4}\right)\left(\dfrac{d^3 x}{dt^3}\right) + x = t$.

10. $\left(\dfrac{dr}{ds}\right)^3 = \sqrt{\dfrac{d^2 r}{ds^2} + 1}$.

1.2   SOLUTIONS

A. Nature of Solutions

We now consider the concept of a solution of the *n*th-order ordinary differential equation.

DEFINITION

Consider the nth-order ordinary differential equation

$$F\left[x, y, \frac{dy}{dx}, \ldots, \frac{d^n y}{dx^n}\right] = 0, \tag{1.10}$$

where F is a real function of its (n + 2) arguments $x, y, \dfrac{dy}{dx}, \ldots, \dfrac{d^n y}{dx^n}$.

1. Let f be a real function defined for all x in a real interval I and having an nth derivative (and hence also all lower ordered derivatives) for all $x \in I$. *The function f is called an* explicit solution *of the differential equation* (1.10) *on I if it fulfills the following two requirements:*

$$F[x, f(x), f'(x), \ldots, f^{(n)}(x)] \tag{A}$$

is defined for all $x \in I$, *and*

$$F[x, f(x), f'(x), \ldots, f^{(n)}(x)] = 0 \tag{B}$$

for all $x \in I$. *That is, the substitution of f(x) and its various derivations for y and its corresponding derivatives, respectively, in* (1.10) *reduces* (1.10) *to an identity on I.*

2. A relation $g(x, y) = 0$ *is called an* implicit solution *of* (1.10) *if this relation defines at least one real function f of the variable x on an interval I such that this function is an explicit solution of* (1.10) *on this interval.*

3. Both explicit solutions and implicit solutions will usually be called simply solutions.

Roughly speaking, then, we may say that a solution of the differential equation (1.10) is a relation—explicit or implicit—between x and y, not containing derivatives, which identically satisfies (1.10).

▶ **Example 1.7**

The function f defined for all real x by

$$f(x) = 2 \sin x + 3 \cos x \tag{1.11}$$

is an explicit solution of the differential equation

$$\frac{d^2 y}{dx^2} + y = 0 \tag{1.12}$$

for all real x. First note that f is defined and has a second derivative for all real x. Next observe that

$$f'(x) = 2 \cos x - 3 \sin x,$$

$$f''(x) = -2 \sin x - 3 \cos x.$$

Upon substituting $f''(x)$ for $d^2 y/dx^2$ and $f(x)$ for y in the differential equation (1.12), it reduces to the identity

$$(-2 \sin x - 3 \cos x) + (2 \sin x + 3 \cos x) = 0,$$

which holds for all real x. Thus the function f defined by (1.11) is an explicit solution of the differential equation (1.12) for all real x.

▶ **Example 1.8**

The relation

$$x^2 + y^2 - 25 = 0 \qquad (1.13)$$

is an implicit solution of the differential equation

$$x + y\frac{dy}{dx} = 0 \qquad (1.14)$$

on the interval I defined by $-5 < x < 5$. For the relation (1.13) defines the two real functions f_1 and f_2 given by

$$f_1(x) = \sqrt{25 - x^2}$$

and

$$f_2(x) = -\sqrt{25 - x^2},$$

respectively, for all real $x \in I$, and both of these functions are explicit solutions of the differential equations (1.14) on I.

Let us illustrate this for the function f_1. Since

$$f_1(x) = \sqrt{25 - x^2},$$

we see that

$$f'_1(x) = \frac{-x}{\sqrt{25 - x^2}}$$

for all real $x \in I$. Substituting $f_1(x)$ for y and $f'_1(x)$ for dy/dx in (1.14), we obtain the identity

$$x + (\sqrt{25 - x^2})\left(\frac{-x}{\sqrt{25 - x^2}}\right) = 0 \quad \text{or} \quad x - x = 0,$$

which holds for all real $x \in I$. Thus the function f_1 is an explicit solution of (1.14) on the interval I.

Now consider the relation

$$x^2 + y^2 + 25 = 0. \qquad (1.15)$$

Is this also an implicit solution of Equation (1.14)? Let us differentiate the relation (1.15) implicitly with respect to x. We obtain

$$2x + 2y\frac{dy}{dx} = 0 \quad \text{or} \quad \frac{dy}{dx} = -\frac{x}{y}.$$

Substituting this into the differential equation (1.14), we obtain the *formal* identity

$$x + y\left(-\frac{x}{y}\right) = 0.$$

Thus the relation (1.15) *formally* satisfies the differential equation (1.14). Can we conclude from this alone that (1.15) is an implicit solution of (1.14)? The answer to this question is "no," for we have no assurance from this that the relation (1.15) defines any function that is an explicit solution of (1.14) on any real interval *I*. All that we have shown is that (1.15) is a relation between *x* and *y* that, upon implicit differentiation and substitution, *formally* reduces the differential equation (1.14) to a *formal* identity. It is called a *formal* solution; it has the *appearance* of a solution; but that is all that we know about it at this stage of our investigation.

Let us investigate a little further. Solving (1.15) for *y*, we find that

$$y = \pm\sqrt{-25 - x^2}.$$

Since this expression yields nonreal values of *y* for all real values of *x*, we conclude that the relation (1.15) does not define any real function on any interval. Thus the relation (1.15) is not truly an implicit solution but merely a *formal solution* of the differential equation (1.14).

In applying the methods of the following chapters we shall often obtain relations that we can readily verify are at least formal solutions. Our main objective will be to gain familiarity with the methods themselves and we shall often be content to refer to the relations so obtained as "solutions," although we have no assurance that these relations are actually true implicit solutions. If a critical examination of the situation is required, one must undertake to determine whether or not these formal solutions so obtained are actually true implicit solutions which define explicit solutions.

In order to gain further insight into the significance of differential equations and their solutions, we now examine the simple equation of the following example.

▶ **Example 1.9**

Consider the first-order differential equation

$$\frac{dy}{dx} = 2x. \tag{1.16}$$

The function f_0 defined for all real *x* by $f_0(x) = x^2$ is a solution of this equation. So also are the functions f_1, f_2, and f_3 defined for all real *x* by $f_1(x) = x^2 + 1$, $f_2(x) = x^2 + 2$, and $f_3(x) = x^2 + 3$, respectively. In fact, for each real number *c*, the function f_c defined for all real *x* by

$$f_c(x) = x^2 + c \tag{1.17}$$

is a solution of the differential equation (1.16). In other words, the formula (1.17) defines an infinite family of functions, one for each real constant *c*, and every function of this family is a solution of (1.16). We call the constant *c* in (1.17) an *arbitrary constant* or *parameter* and refer to the family of functions defined by (1.17) as a *one-parameter faimily of solutions* of the differential equation (1.16). We write this one-parameter family of solutions as

$$y = x^2 + c. \tag{1.18}$$

Although it is clear that every function of the family defined by (1.18) is a solution of (1.16), we have not shown that the family of functions defined by (1.18) includes *all* of the solutions of (1.16). However, we point out (without proof) that this is indeed the case here; that is, every solution of (1.16) is actually of the form (1.18) for some appropriate real number c.

Note. We must not conclude from the last sentence of Example 1.9 that *every* first-order ordinary differential equation has a so-called one-parameter family of solutions which contains *all* solutions of the differential equation, for this is by no means the case. Indeed, some first-order differential equations have no solution at all (see Exercise 7(a) at the end of this section), while others have a one-parameter family of solutions plus one or more "extra" solutions which appear to be "different" from all those of the family (see Exercise 7(b) at the end of this section).

The differential equation of Example 1.9 enables us to obtain a better understanding of the analytic significance of differential equations. Briefly stated, the differential equation of that example *defines functions*, namely, its solutions. We shall see that this is the case with many other differential equations of both first and higher order. Thus we may say that a differential equation is merely an expression involving derivatives which may serve as a means of defining a certain set of functions: its solutions. Indeed, many of the now familiar functions originally appeared in the form of differential equations that define them.

We now consider the geometric significance of differential equations and their solutions. We first recall that a real function F may be represented geometrically by a curve $y = F(x)$ in the xy plane and that the value of the derivative of F at x, $F'(x)$, may be interpreted as the slope of the curve $y = F(x)$ at x. Thus the general first-order differential equation

$$\frac{dy}{dx} = f(x, y), \tag{1.19}$$

where f is a real function, may be interpreted geometrically as defining a slope $f(x, y)$ at every point (x, y) at which the function f is defined. Now assume that the differential equation (1.19) has a so-called one-parameter family of solutions that can be written in the form

$$y = F(x, c), \tag{1.20}$$

where c is the arbitrary constant or parameter of the family. The one-parameter family of functions defined by (1.20) is represented geometrically by a so-called *one-parameter family of curves* in the xy plane, the slopes of which are given by the differential equation (1.19). These curves, the graphs of the solutions of the differential equation (1.19), are called the *integral curves* of the differential equation (1.19).

▶ **Example 1.10**

Consider again the first-order differential equation

$$\frac{dy}{dx} = 2x \tag{1.16}$$

of Example 1.9. This differential equation may be interpreted as defining the slope $2x$ at the point with coordinates (x, y) for every real x. Now, we observed in Example 1.9 that the differential equation (1.16) has a one-parameter family of solutions of the form

$$y = x^2 + c, \tag{1.18}$$

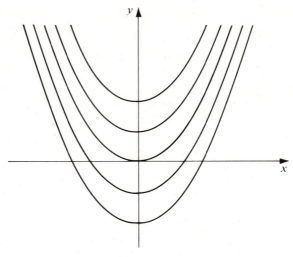

Figure 1.1

where c is the arbitrary constant or parameter of the family. The one-parameter family of functions defined by (1.18) is represented geometrically by a one-parameter family of curves in the xy plane, namely, the family of *parabolas* with Equation (1.18). The slope of each of these parabolas is given by the differential equation (1.16) of the family. Thus we see that the family of parabolas (1.18) defined by differential equation (1.16) is that family of parabolas, each of which has slope $2x$ at the point (x, y) for every real x, and all of which have the y axis as axis. These parabolas are the integral curves of the differential equation (1.16). See Figure 1.1.

B. Methods of Solution

When we say that we shall solve a differential equation we mean that we shall find one or more of its solutions. How is this done and what does it really mean? The greater part of this text is concerned with various methods of solving differential equations. The method to be employed depends upon the type of differential equation under consideration, and we shall not enter into the details of specific methods here.

But suppose we solve a differential equation, using one or another of the various methods. Does this necessarily mean that we have found an explicit solution f expressed in the so-called closed form of a finite sum of known elementary functions? That is, roughly speaking, when we have solved a differential equation, does this necessarily mean that we have found a "formula" for the solution? The answer is "no." Comparatively few differential equations have solutions so expressible; in fact, a closed-form solution is really a luxury in differential equations. In Chapters 2 and 4 we shall consider certain types of differential equations that do have such closed-form solutions and study the exact methods available for finding these desirable solutions. But, as we have just noted, such equations are actually in the minority and we must consider what it means to "solve" equations for which exact methods are unavailable. Such equations are solved approximately by various methods, some of which are considered in Chapters 6 and 8. Among such methods are series methods, numerical methods, and graphical methods. What do such approximate methods actually yield? The answer to this depends upon the method under consideration.

Series methods yield solutions in the form of infinite series; numerical methods give approximate values of the solution functions corresponding to selected values of the independent variables; and graphical methods produce approximately the graphs of solutions (the integral curves). These methods are not so desirable as exact methods because of the amount of work involved in them and because the results obtained from them are only approximate; but if exact methods are not applicable, one has no choice but to turn to approximate methods. Modern science and engineering problems continue to give rise to differential equations to which exact methods do not apply, and approximate methods are becoming increasingly more important.

Exercises

1. Show that each of the functions defined in Column I is a solution of the corresponding differential equation in Column II on every interval $a < x < b$ of the x axis.

I

II

(a) $f(x) = x + 3e^{-x}$

$\dfrac{dy}{dx} + y = x + 1$

(b) $f(x) = 2e^{3x} - 5e^{4x}$

$\dfrac{d^2y}{dx^2} - 7\dfrac{dy}{dx} + 12y = 0$

(c) $f(x) = e^x + 2x^2 + 6x + 7$

$\dfrac{d^2y}{dx^2} - 3\dfrac{dy}{dx} + 2y = 4x^2$

(d) $f(x) = \dfrac{1}{1 + x^2}$

$(1 + x^2)\dfrac{d^2y}{dx^2} + 4x\dfrac{dy}{dx} + 2y = 0$

2. (a) Show that $x^3 + 3xy^2 = 1$ is an implicit solution of the differential equation $2xy(dy/dx) + x^2 + y^2 = 0$ on the interval $0 < x < 1$.
   (b) Show that $5x^2y^2 - 2x^3y^2 = 1$ is an implicit solution of the differential equation $x(dy/dx) + y = x^3y^3$ on the interval $0 < x < \frac{5}{2}$.

3. (a) Show that every function f defined by

$$f(x) = (x^3 + c)e^{-3x},$$

where c is an arbitrary constant, is a solution of the differential equation

$$\dfrac{dy}{dx} + 3y = 3x^2e^{-3x}.$$

   (b) Show that every function f defined by

$$f(x) = 2 + ce^{-2x^2},$$

where c is an arbitrary constant, is a solution of the differential equation

$$\dfrac{dy}{dx} + 4xy = 8x.$$

4. (a) Show that every function f defined by $f(x) = c_1 e^{4x} + c_2 e^{-2x}$, where c_1 and c_2 are arbitrary constants, is a solution of the differential equation

$$\frac{d^2 y}{dx^2} - 2\frac{dy}{dx} - 8y = 0.$$

(b) Show that every function g defined by $g(x) = c_1 e^{2x} + c_2 x e^{2x} + c_3 e^{-2x}$, where c_1, c_2, and c_3 are arbitrary constants, is a solution of the differential equation

$$\frac{d^3 y}{dx^3} - 2\frac{d^2 y}{dx^2} - 4\frac{dy}{dx} + 8y = 0.$$

5. (a) For certain values of the constant m the function f defined by $f(x) = e^{mx}$ is a solution of the differential equation

$$\frac{d^3 y}{dx^3} - 3\frac{d^2 y}{dx^2} - 4\frac{dy}{dx} + 12y = 0.$$

Determine all such values of m.

(b) For certain values of the constant n the function g defined by $g(x) = x^n$ is a solution of the differential equation

$$x^3 \frac{d^3 y}{dx^3} + 2x^2 \frac{d^2 y}{dx^2} - 10x\frac{dy}{dx} - 8y = 0.$$

Determine all such values of n.

6. (a) Show that the function f defined by $f(x) = (2x^2 + 2e^{3x} + 3)e^{-2x}$ satisfies the differential equation

$$\frac{dy}{dx} + 2y = 6e^x + 4xe^{-2x}$$

and also the condition $f(0) = 5$.

(b) Show that the function f defined by $f(x) = 3e^{2x} - 2xe^{2x} - \cos 2x$ satisfies the differential equation

$$\frac{d^2 y}{dx^2} - 4\frac{dy}{dx} + 4y = -8 \sin 2x$$

and also the conditions that $f(0) = 2$ and $f'(0) = 4$.

7. (a) Show that the first-order differential equation

$$\left|\frac{dy}{dx}\right| + |y| + 1 = 0$$

has *no* (real) solutions.

(b) Show that the first-order differential equation

$$\left(\frac{dy}{dx}\right)^2 - 4y = 0$$

has a one-parameter family of solutions of the form $f(x) = (x + c)^2$, where c is an arbitrary constant, plus the "extra" solution $g(x) = 0$ that is *not* a member of this family $f(x) = (x + c)^2$ for *any* choice of the constant c.

1.3   INITIAL-VALUE PROBLEMS, BOUNDARY-VALUE PROBLEMS, AND EXISTENCE OF SOLUTIONS

A. Initial-Value Problems and Boundary-Value Problems

We shall begin this section by considering the rather simple problem of the following example.

▶ **Example 1.11**

Problem.   Find a solution f of the differential equation

$$\frac{dy}{dx} = 2x \tag{1.21}$$

such that at $x = 1$ this solution f has the value 4.

Explanation.   First let us be certain that we thoroughly understand this problem. We seek a real function f which fulfills the two following requirements:

1. The function f must satisfy the differential equation (1.21). That is, the function f must be such that $f'(x) = 2x$ for all real x in a real interval I.
2. The function f must have the value 4 at $x = 1$. That is, the function f must be such that $f(1) = 4$.

Notation.   The stated problem may be expressed in the following somewhat abbreviated notation:

$$\frac{dy}{dx} = 2x,$$

$$y(1) = 4.$$

In this notation we may regard y as representing the desired solution. Then the differential equation itself obviously represents requirement 1, and the statement $y(1) = 4$ stands for requirement 2. More specifically, the notation $y(1) = 4$ states that the desired solution y must have the value 4 at $x = 1$; that is, $y = 4$ at $x = 1$.

Solution.   We observed in Example 1.9 that the differential equation (1.21) has a one-parameter family of solutions which we write as

$$y = x^2 + c, \tag{1.22}$$

where c is an arbitrary constant, and that each of these solutions satisfies requirement 1. Let us now attempt to determine the constant c so that (1.22) satisfies requirement 2, that is, $y = 4$ at $x = 1$. Substituting $x = 1$, $y = 4$ into (1.22), we obtain $4 = 1 + c$, and hence $c = 3$. Now substituting the value $c = 3$ thus determined back into (1.22), we obtain

$$y = x^2 + 3,$$

which is indeed a solution of the differential equation (1.21), which has the value 4 at $x = 1$. In other words, the function f defined by

$$f(x) = x^2 + 3,$$

satisfies both of the requirements set forth in the problem.

Comment on Requirement 2 and Its Notation.   In a problem of this type, requirement 2 is regarded as a *supplementary condition* that the solution of the differential equation must also satisfy. The abbreviated notation $y(1) = 4$, which we used to express this condition, is in some way undesirable, but it has the advantages of being both customary and convenient.

In the application of both first- and higher-order differential equations the problems most frequently encountered are similar to the above introductory problem in that they involve *both* a differential equation *and* one or more supplementary conditions which the solution of the given differential equation must satisfy. If all of the associated supplementary conditions relate to *one* x value, the problem is called an *initial-value problem* (or one-point boundary-value problem). If the conditions relate to *two* different x values, the problem is called a *two-point boundary-value problem* (or simply a boundary-value problem). We shall illustrate these concepts with examples and then consider one such type of problem in detail. Concerning notation, we generally employ abbreviated notations for the supplementary conditions that are similar to the abbreviated notation introduced in Example 1.11.

▶ **Example 1.12**

$$\frac{d^2 y}{dx^2} + y = 0,$$

$$y(1) = 3,$$

$$y'(1) = -4.$$

This problem consists in finding a solution of the differential equation

$$\frac{d^2 y}{dx^2} + y = 0,$$

which assumes the value 3 at $x = 1$ and whose first derivative assumes the value -4 at $x = 1$. Both of these conditions relate to one x value, namely, $x = 1$. Thus this is an initial-value problem. We shall see later that this problem has a unique solution.

▶ **Example 1.13**

$$\frac{d^2 y}{dx^2} + y = 0,$$

$$y(0) = 1,$$

$$y\left(\frac{\pi}{2}\right) = 5.$$

In this problem we again seek a solution of the same differential equation, but this time the solution must assume the value 1 at $x = 0$ and the value 5 at $x = \pi/2$. That is, the conditions relate to the *two* different x values, 0 and $\pi/2$. This is a (two-point) boundary-value problem. This problem also has a unique solution; but the boundary-value problem

$$\frac{d^2y}{dx^2} + y = 0,$$

$$y(0) = 1, \qquad y(\pi) = 5,$$

has no solution at all! This simple fact may lead one to the correct conclusion that boundary-value problems are not to be taken lightly!

We now turn to a more detailed consideration of the initial-value problem for a first-order differential equation.

DEFINITION

Consider the first-order differential equation

$$\frac{dy}{dx} = f(x, y), \tag{1.23}$$

where f is a continuous function of x and y in some domain* D of the xy plane; and let (x_0, y_0) be a point of D. The initial-value *problem associated with (1.23) is to find a solution ϕ of the differential equation (1.23), defined on some real interval containing x_0, and satisfying the* initial condition

$$\phi(x_0) = y_0.$$

In the customary abbreviated notation, this initial-value problem may be written

$$\frac{dy}{dx} = f(x, y),$$

$$y(x_0) = y_0.$$

To solve this problem, we must find a function ϕ that not only satisfies the differential equation (1.23) but that also satisfies the initial condition that it has the value y_0 when x has value x_0. The geometric interpretation of the initial condition, and hence of the entire initial-value problem, is easily understood. The graph of the desired solution ϕ must pass through the point with coordinates (x_0, y_0). That is, interpreted geometrically, the initial-value problem is to find an integral curve of the differential equation (1.23) that passes through the point (x_0, y_0).

The method of actually finding the desired solution ϕ depends upon the nature of the differential equation of the problem, that is, upon the form of $f(x, y)$. Certain special types of differential equations have a one-parameter family of solutions whose equation may be found exactly by following definite procedures (see Chapter 2). If the differential equation of the problem is of some such special type, one first obtains the equation of its one-parameter family of solutions and then applies the initial condition

---

* A *domain* is an open, connected set. For those unfamiliar with such concepts, D may be regarded as the interior of some simple closed curve in the plane.

to this equation in an attempt to obtain a "particular" solution ϕ that satisfies the entire initial-value problem. We shall explain this situation more precisely in the next paragraph. Before doing so, however, we point out that in general one cannot find the equation of a one-parameter family of solutions of the differential equation; approximate methods must then be used (see Chapter 8).

Now suppose one can determine the equation

$$g(x, y, c) = 0 \tag{1.24}$$

of a one-parameter family of solutions of the differential equation of the problem. Then, since the initial condition requires that $y = y_0$ at $x = x_0$, we let $x = x_0$ and $y = y_0$ in (1.24) and thereby obtain

$$g(x_0, y_0, c) = 0.$$

Solving this for c, in general we obtain a particular value of c which we denote here by c_0. We now replace the arbitrary constant c by the particular constant c_0 in (1.24), thus obtaining the particular solution

$$g(x, y, c_0) = 0.$$

The particular explicit solution satisfying the two conditions (differential equation and initial condition) of the problem is then determined from this, if possible.

We have already solved one initial-value problem in Example 1.11. We now give another example in order to illustrate the concepts and procedures more thoroughly.

▶ **Example 1.14**

Solve the initial-value problem

$$\frac{dy}{dx} = -\frac{x}{y}, \tag{1.25}$$

$$y(3) = 4, \tag{1.26}$$

given that the differential equation (1.25) has a one-parameter family of solutions which may be written in the form

$$x^2 + y^2 = c^2. \tag{1.27}$$

The condition (1.26) means that we seek the solution of (1.25) such that $y = 4$ at $x = 3$. Thus the pair of values $(3, 4)$ must satisfy the relation (1.27). Substituting $x = 3$ and $y = 4$ into (1.27), we find

$$9 + 16 = c^2 \quad \text{or} \quad c^2 = 25.$$

Now substituting this value of c^2 into (1.27), we have

$$x^2 + y^2 = 25.$$

Solving this for y, we obtain

$$y = \pm\sqrt{25 - x^2}.$$

Obviously the positive sign must be chosen to give y the value $+4$ at $x = 3$. Thus the function f defined by

$$f(x) = \sqrt{25 - x^2}, \qquad -5 < x < 5,$$

is the solution of the problem. In the usual abbreviated notation, we write this solution as $y = \sqrt{25 - x^2}$.

B. Existence of Solutions

In Example 1.14 we were able to find a solution of the initial-value problem under consideration. But do all initial-value and boundary-value problems have solutions? We have already answered this question in the negative, for we have pointed out that the boundary-value problem

$$\frac{d^2 y}{dx^2} + y = 0,$$

$$y(0) = 1,$$

$$y(\pi) = 5,$$

mentioned at the end of Example 1.13, has no solution! Thus arises the question of *existence* of solutions: given an initial-value or boundary-value problem, does it actually have a solution? Let us consider the question for the initial-value problem defined on page 17. Here we can give a definite answer. Every initial-value problem that satisfies the definition on page 17 has *at least one* solution.

But now another question is suggested, the question of *uniqueness*. Does such a problem ever have *more than one* solution? Let us consider the initial-value problem

$$\frac{dy}{dx} = y^{1/3},$$

$$y(0) = 0.$$

One may verify that the functions f_1 and f_2 defined, respectively, by

$$f_1(x) = 0 \quad \text{for all real } x;$$

and

$$f_2(x) = \left(\tfrac{2}{3}x\right)^{3/2}, \quad x \geq 0; \qquad f_2(x) = 0, \quad x \leq 0;$$

are *both* solutions of this initial-value problem! In fact, this problem has infinitely many solutions! The answer to the uniqueness question is clear: the initial-value problem, as stated, need not have a *unique* solution. In order tb ensure uniqueness, some additional requirement must certainly be imposed. We shall see what this is in Theorem 1.1, which we shall now state.

THEOREM 1.1.   BASIC EXISTENCE AND UNIQUENESS THEOREM

Hypothesis. *Consider the differential equation*

$$\frac{dy}{dx} = f(x, y), \tag{1.28}$$

where

   1. *The function f is a continuous function of x and y in some domain D of the xy plane,* and
   2. *The partial derivative* $\partial f/\partial y$ *is also a continuous function of x and y in D; and let* (x_0, y_0) *be a point in D.*

   Conclusion.   *There exists a unique solution* ϕ *of the differential equation* (*1.28*), *defined on some interval* $|x - x_0| \leq h$, *where h is sufficiently small, that satisfies the condition*

$$\phi(x_0) = y_0. \tag{1.29}$$

   Explanatory Remarks.   This basic theorem is the first theorem from the theory of differential equations which we have encountered. We shall therefore attempt to explain its meaning in detail.

   1. It is an *existence and uniqueness theorem.* This means that it is a theorem which tells us that under certain conditions (stated in the hypothesis) something *exists* (the solution described in the conclusion) and is *unique* (there is *only one* such solution). It gives no hint whatsoever concerning *how* to find this solution but merely tells us that the problem *has* a solution.
   2. The *hypothesis* tells us what conditions are required of the quantities involved. It deals with two objects: the differential equation (1.28) and the point (x_0, y_0). As far as the differential equation (1.28) is concerned, the hypothesis requires that *both* the function f *and* the function $\partial f/\partial y$ (obtained by differentiating $f(x, y)$ partially with respect to y) must be continuous in some domain D of the xy plane. As far as the point (x_0, y_0) is concerned, it must be a point in this same domain D, where f and $\partial f/\partial y$ are so well behaved (that is, continuous).
   3. The *conclusion* tells us of what we can be assured when the stated hypothesis is satisfied. It tells us that we are assured that there exists one and only one solution ϕ of the differential equation, which is defined on some interval $|x - x_0| \leq h$ centered about x_0 and which assumes the value y_0 when x takes on the value x_0. That is, it tells us that, under the given hypothesis on $f(x, y)$, the *initial-value problem*

$$\frac{dy}{dx} = f(x, y),$$

$$y(x_0) = y_0,$$

has a *unique solution* that is valid in some interval about the initial point x_0.
   4. The *proof* of this theorem is omitted. It is proved under somewhat less restrictive hypotheses in Chapter 10.
   5. The *value* of an existence theorem may be worth a bit of attention. What good is it, one might ask, if it does not tell us how to obtain the solution? The answer to this question is quite simple: an existence theorem will assure us that there *is* a solution to look for! It would be rather pointless to spend time, energy, and even money in trying to find a solution when there was actually no solution to be found! As for the value of the uniqueness, it would be equally pointless to waste time and energy finding one particular solution only to learn later that there were others and that the one found was not the one wanted!

We have included this rather lengthy discussion in the hope that the student, who has probably never before encountered a theorem of this type, will obtain a clearer idea of what this important theorem really means. We further hope that this discussion will help him to analyze theorems which he will encounter in the future, both in this book and elsewhere. We now consider two simple examples which illustrate Theorem 1.1.

▶ **Example 1.15**

Consider the initial-value problem

$$\frac{dy}{dx} = x^2 + y^2,$$

$$y(1) = 3.$$

Let us apply Theorem 1.1. We first check the hypothesis. Here $f(x, y) = x^2 + y^2$ and $\dfrac{\partial f(x, y)}{\partial y} = 2y$. Both of the functions f and $\partial f/\partial y$ are continuous in every domain D of the xy plane. The initial condition $y(1) = 3$ means that $x_0 = 1$ and $y_0 = 3$, and the point $(1, 3)$ certainly lies in some such domain D. Thus all hypotheses are satisfied and the conclusion holds. That is, there is a unique solution ϕ of the differential equation $dy/dx = x^2 + y^2$, defined on some interval $|x - 1| \leq h$ about $x_0 = 1$, which satisfies that initial condition, that is, which is such that $\phi(1) = 3$.

▶ **Example 1.16**

Consider the two problems:

1. $\dfrac{dy}{dx} = \dfrac{y}{\sqrt{x}},$ $y(1) = 2,$

2. $\dfrac{dy}{dx} = \dfrac{y}{\sqrt{x}},$ $y(0) = 2.$

Here

$$f(x, y) = \frac{y}{x^{1/2}} \quad \text{and} \quad \frac{\partial f(x, y)}{\partial y} = \frac{1}{x^{1/2}}.$$

These functions are both continuous *except* for $x = 0$ (that is, along the y axis). In problem 1, $x_0 = 1$, $y_0 = 2$. The square of side 1 centered about $(1, 2)$ does *not* contain the y axis, and so both f and $\partial f/\partial y$ satisfy the required hypotheses in this square. Its interior may thus be taken to be the domain D of Theorem 1.1; and $(1, 2)$ certainly lies within it. Thus the conclusion of Theorem 1.1 applies to problem 1 and we know the problem has a unique solution defined in some sufficiently small interval about $x_0 = 1$.

Now let us turn to problem 2. Here $x_0 = 0$, $y_0 = 2$. At this point neither f nor $\partial f/\partial y$ are continuous. In other words, the point $(0, 2)$ cannot be included in a domain D where the required hypotheses are satisfied. Thus we can *not* conclude from Theorem 1.1 that problem 2 has a solution. We are *not* saying that it does *not* have one. Theorem 1.1 simply gives no information one way or the other.

Exercises

1.   Show that

$$y = 4e^{2x} + 2e^{-3x}$$

is a solution of the initial-value problem

$$\frac{d^2y}{dx^2} + \frac{dy}{dx} - 6y = 0,$$

$$y(0) = 6,$$

$$y'(0) = 2.$$

Is $y = 2e^{2x} + 4e^{-3x}$ also a solution of this problem? Explain why or why not.

2.   Given that every solution of

$$\frac{dy}{dx} + y = 2xe^{-x}$$

may be written in the form $y = (x^2 + c)e^{-x}$, for some choice of the arbitrary constant c, solve the following initial-value problems:

(a) $\dfrac{dy}{dx} + y = 2xe^{-x}$,

$\quad\quad y(0) = 2.$

(b) $\dfrac{dy}{dx} + y = 2xe^{-x}$,

$\quad\quad y(-1) = e + 3.$

3.   Given that every solution of

$$\frac{d^2y}{dx^2} - \frac{dy}{dx} - 12y = 0$$

may be written in the form

$$y = c_1 e^{4x} + c_2 e^{-3x},$$

for some choice of the arbitrary constants c_1 and c_2, solve the following initial-value problems:

(a) $\dfrac{d^2y}{dx^2} - \dfrac{dy}{dx} - 12y = 0,$

$\quad\quad y(0) = 5,$

$\quad\quad y'(0) = 6.$

(b) $\dfrac{d^2y}{dx^2} - \dfrac{dy}{dx} - 12y = 0,$

$\quad\quad y(0) = -2,$

$\quad\quad y'(0) = 6.$

4.   Every solution of the differential equation

$$\frac{d^2y}{dx^2} + y = 0$$

may be written in the form $y = c_1 \sin x + c_2 \cos x$, for some choice of the arbitrary constants c_1 and c_2. Using this information, show that boundary problems (a) and (b) possess solutions but that (c) does not.

(a) $\dfrac{d^2 y}{dx^2} + y = 0,$

      $y(0) = 0,$

      $y(\pi/2) = 1.$

(b) $\dfrac{d^2 y}{dx^2} + y = 0,$

      $y(0) = 1,$

      $y'(\pi/2) = -1.$

(c) $\dfrac{d^2 y}{dx^2} + y = 0,$

      $y(0) = 0,$

      $y(\pi) = 1.$

5. Given that every solution of

$$x^3 \frac{d^3 y}{dx^3} - 3x^2 \frac{d^2 y}{dx^2} + 6x \frac{dy}{dx} - 6y = 0$$

may be written in the form $y = c_1 x + c_2 x^2 + c_3 x^3$ for some choice of the arbitrary constants c_1, c_2, and c_3, solve the initial-value problem consisting of the above differential equation plus the three conditions

$$y(2) = 0, \qquad y'(2) = 2, \qquad y''(2) = 6.$$

6. Apply Theorem 1.1 to show that each of the following initial-value problems has a unique solution defined on some sufficiently small inverval $|x - 1| \le h$ about $x_0 = 1$:

(a) $\dfrac{dy}{dx} = x^2 \sin y,$

      $y(1) = -2.$

(b) $\dfrac{dy}{dx} = \dfrac{y^2}{x - 2},$

      $y(1) = 0.$

7. Consider the initial-value problem

$$\frac{dy}{dx} = P(x) y^2 + Q(x) y,$$

$$y(2) = 5,$$

where $P(x)$ and $Q(x)$ are both third-degree polynomials in x. Has this problem a unique solution on some interval $|x - 2| \le h$ about $x_0 = 2$? Explain why or why not.

8. On page 19 we stated that the initial-value problem

$$\frac{dy}{dx} = y^{1/3},$$

$$y(0) = 0,$$

has infinitely many solutions.

(a)   Verify that this is indeed the case by showing that

$$y = \begin{cases} 0, & x \le c, \\ [\frac{2}{3}(x - c)]^{3/2}, & x \ge c, \end{cases}$$

is a solution of the stated problem for every real number $c \ge 0$.

(b)   Carefully graph the solution for which $c = 0$. Then, using this particular graph, also graph the solutions for which $c = 1$, $c = 2$, and $c = 3$.

CHAPTER TWO

First-Order Equations for Which Exact Solutions Are Obtainable

In this chapter we consider certain basic types of first-order equations for which exact solutions may be obtained by definite procedures. The purpose of this chapter is to gain ability to recognize these various types and to apply the corresponding methods of solutions. Of the types considered here, the so-called exact equations considered in Section 2.1 are in a sense the most basic, while the separable equations of Section 2.2 are in a sense the "easiest." The most important, from the point of view of applications, are the separable equations of Section 2.2 and the linear equations of Section 2.3. The remaining types are of various very special forms, and the correponding methods of solution involve various devices. In short, we might describe this chapter as a collection of special "methods," "devices," "tricks," or "recipes," in descending order of kindness!

2.1 EXACT DIFFERENTIAL EQUATIONS AND INTEGRATING FACTORS

A. Standard Forms of First-Order Differential Equations

The first-order differential equations to be studied in this chapter may be expressed in either the derivative form

$$\frac{dy}{dx} = f(x, y) \tag{2.1}$$

or the differential form

$$M(x, y)\, dx + N(x, y)\, dy = 0. \tag{2.2}$$

25

An equation in one of these forms may readily be written in the other form. For example, the equation

$$\frac{dy}{dx} = \frac{x^2 + y^2}{x - y}$$

is of the form (2.1). It may be written

$$(x^2 + y^2)\,dx + (y - x)\,dy = 0,$$

which is of the form (2.2). The equation

$$(\sin x + y)\,dx + (x + 3y)\,dy = 0,$$

which is of the form (2.2), may be written in the form (2.1) as

$$\frac{dy}{dx} = -\frac{\sin x + y}{x + 3y}.$$

In the form (2.1) it is clear from the notation itself that y is regarded as the dependent variable and x as the independent one; but in the form (2.2) we may actually regard either variable as the dependent one and the other as the independent. However, in this text, in all differential equations of the form (2.2) in x and y, we shall regard y as dependent and x as independent, unless the contrary is specifically stated.

B. Exact Differential Equations

DEFINITION

Let F be a function of two real variables such that F has continuous first partial derivatives in a domain D. The total differential dF *of the function F is defined by the formula*

$$dF(x, y) = \frac{\partial F(x, y)}{\partial x}\,dx + \frac{\partial F(x, y)}{\partial y}\,dy$$

for all $(x, y) \in D$.

▶ **Example 2.1**

Let F be the function of two real variables defined by

$$F(x, y) = xy^2 + 2x^3 y$$

for all real (x, y). Then

$$\frac{\partial F(x, y)}{\partial x} = y^2 + 6x^2 y, \qquad \frac{\partial F(x, y)}{\partial y} = 2xy + 2x^3,$$

and the total differential dF is defined by

$$dF(x, y) = (y^2 + 6x^2 y)\,dx + (2xy + 2x^3)\,dy$$

for all real (x, y).

DEFINITION

The expression

$$M(x, y)\, dx + N(x, y)\, dy \tag{2.3}$$

is called an exact differential *in a domain D if there exists a function F of two real variables such that this expression equals the total differential dF(x, y) for all (x, y) ∈ D. That is, expression (2.3) is an exact differential in D if there exists a function F such that*

$$\frac{\partial F(x, y)}{\partial x} = M(x, y) \quad and \quad \frac{\partial F(x, y)}{\partial y} = N(x, y)$$

for all (x, y) ∈ D.

If M(x, y) dx + N(x, y) dy is an exact differential, then the differential equation

$$M(x, y)\, dx + N(x, y)\, dy = 0$$

is called an exact differential equation.

▶ **Example 2.2**

The differential equation

$$y^2\, dx + 2xy\, dy = 0 \tag{2.4}$$

is an exact differential equation, since the expression $y^2\, dx + 2xy\, dy$ is an exact differential. Indeed, it is the total differential of the function F defined for all (x, y) by $F(x, y) = xy^2$, since the coefficient of dx is $\partial F(x, y)/(\partial x) = y^2$ and that of dy is $\partial F(x, y)/(\partial y) = 2xy$. On the other hand, the more simple appearing equation

$$y\, dx + 2x\, dy = 0, \tag{2.5}$$

obtained from (2.4) by dividing through by y, is *not* exact.

In Example 2.2 we stated without hesitation that the differential equation (2.4) is exact but the differential equation (2.5) is not. In the case of Equation (2.4), we verified our assertion by actually exhibiting the function F of which the expression $y^2\, dx + 2xy\, dy$ is the total differential. But in the case of Equation (2.5), we did not back up our statement by showing that there is no fucntion F such that $y\, dx + 2x\, dy$ is its total differential. It is clear that we need a simple test to determine whether or not a given differential equation is exact. This is given by the following theorem.

THEOREM 2.1

Consider the differential equation

$$M(x, y)\, dx + N(x, y)\, dy = 0, \tag{2.6}$$

where M and N have continuous first partial derivatives at all points (x, y) in a rectangular domain D.

1. If *the differential equation (2.6) is exact in D*, then

$$\frac{\partial M(x, y)}{\partial y} = \frac{\partial N(x, y)}{\partial x} \tag{2.7}$$

for all $(x, y) \in D.$

2. Conversely, if

$$\frac{\partial M(x, y)}{\partial y} = \frac{\partial N(x, y)}{\partial x}$$

for all $(x, y) \in D$, then *the differential equation (2.6) is exact in D.*

Proof.   Part 1.   If the differential equation (2.6) is exact in D, then $M\,dx + N\,dy$ is an exact differential in D. By definition of an exact differential, there exists a function F such that

$$\frac{\partial F(x, y)}{\partial x} = M(x, y) \quad \text{and} \quad \frac{\partial F(x, y)}{\partial y} = N(x, y)$$

for all $(x, y) \in D$. Then

$$\frac{\partial^2 F(x, y)}{\partial y\,\partial x} = \frac{\partial M(x, y)}{\partial y} \quad \text{and} \quad \frac{\partial^2 F(x, y)}{\partial x\,\partial y} = \frac{\partial N(x, y)}{\partial x}$$

for all $(x, y) \in D$. But, using the continuity of the first partial derivatives of M and N, we have

$$\frac{\partial^2 F(x, y)}{\partial y\,\partial x} = \frac{\partial^2 F(x, y)}{\partial x\,\partial y}$$

and therefore

$$\frac{\partial M(x, y)}{\partial y} = \frac{\partial N(x, y)}{\partial x}$$

for all $(x, y) \in D$.

Part 2. This being the converse of Part 1, we start with the hypothesis that

$$\frac{\partial M(x, y)}{\partial y} = \frac{\partial N(x, y)}{\partial x}$$

for all $(x, y) \in D$, and set out to show that $M\,dx + N\,dy = 0$ is exact in D. This means that we must prove that there exists a function F such that

$$\frac{\partial F(x, y)}{\partial x} = M(x, y) \tag{2.8}$$

and

$$\frac{\partial F(x, y)}{\partial y} = N(x, y) \tag{2.9}$$

for all $(x, y) \in D$. We can certainly find some $F(x, y)$ satisfying either (2.8) or (2.9), but what about both? Let us assume that F satisfies (2.8) and proceed. Then

$$F(x, y) = \int M(x, y)\,\partial x + \phi(y), \tag{2.10}$$

where $\int M(x, y)\, \partial x$ indicates a partial integration with respect to x, holding y constant, and ϕ is an arbitrary function of y only. This $\phi(y)$ is needed in (2.10) so that $F(x, y)$ given by (2.10) will represent *all* solutions of (2.8). It corresponds to a constant of integration in the "one-variable" case. Differentiating (2.10) partially with respect to y, we obtain

$$\frac{\partial F(x, y)}{\partial y} = \frac{\partial}{\partial y} \int M(x, y)\, \partial x + \frac{d\phi(y)}{dy}.$$

Now if (2.9) is to be satisfied, we must have

$$N(x, y) = \frac{\partial}{\partial y} \int M(x, y)\, \partial x + \frac{d\phi(y)}{dy} \qquad (2.11)$$

and hence

$$\frac{d\phi(y)}{dy} = N(x, y) - \frac{\partial}{\partial y} \int M(x, y)\, \partial x.$$

Since ϕ is a function of y only, the derivative $d\phi/dy$ must also be independent of x. That is, in order for (2.11) to hold,

$$N(x, y) - \frac{\partial}{\partial y} \int M(x, y)\, \partial x \qquad (2.12)$$

must be independent of x.

We shall show that

$$\frac{\partial}{\partial x} \left[N(x, y) - \frac{\partial}{\partial y} \int M(x, y)\, \partial x \right] = 0.$$

We at once have

$$\frac{\partial}{\partial x} \left[N(x, y) - \frac{\partial}{\partial y} \int M(x, y)\, \partial x \right] = \frac{\partial N(x, y)}{\partial x} - \frac{\partial^2}{\partial x\, \partial y} \int M(x, y)\, \partial x.$$

If (2.8) and (2.9) are to be satisfied, then using the hypothesis (2.7), we must have

$$\frac{\partial^2}{\partial x\, \partial y} \int M(x, y)\, \partial x = \frac{\partial^2 F(x, y)}{\partial x\, \partial y} = \frac{\partial^2 F(x, y)}{\partial y\, \partial x} = \frac{\partial^2}{\partial y\, \partial x} \int M(x, y)\, \partial x.$$

Thus we obtain

$$\frac{\partial}{\partial x} \left[N(x, y) - \frac{\partial}{\partial y} \int M(x, y)\, \partial x \right] = \frac{\partial N(x, y)}{\partial x} - \frac{\partial^2}{\partial y\, \partial x} \int M(x, y)\, \partial x$$

and hence

$$\frac{\partial}{\partial x} \left[N(x, y) - \frac{\partial}{\partial y} \int M(x, y)\, \partial x \right] = \frac{\partial N(x, y)}{\partial x} - \frac{\partial M(x, y)}{\partial y}.$$

But by hypothesis (2.7),

$$\frac{\partial M(x, y)}{\partial y} = \frac{\partial N(x, y)}{\partial x}$$

for all $(x, y) \in D$. Thus

$$\frac{\partial}{\partial x}\left[N(x, y) - \frac{\partial}{\partial y}\int M(x, y)\, \partial x \right] = 0$$

for all $(x, y) \in D$, and so (2.12) *is* independent of x. Thus we may write

$$\phi(y) = \int \left[N(x, y) - \int \frac{\partial M(x, y)}{\partial y}\, \partial x \right] dy.$$

Substituting this into Equation (2.10), we have

$$F(x, y) = \int M(x, y)\, \partial x + \int \left[N(x, y) - \int \frac{\partial M(x, y)}{\partial y}\, \partial x \right] dy. \tag{2.13}$$

This $F(x, y)$ thus satisfies both (2.8) and (2.9) for all $(x, y) \in D$, and so $M\, dx + N\, dy = 0$ is exact in D.   Q.E.D.

Students well versed in the terminology of higher mathematics will recognize that Theorem 2.1 may be stated in the following words: A necessary and sufficient condition that Equation (2.6) be exact in D is that condition (2.7) hold for all $(x, y) \in D$. For students not so well versed, let us emphasize that condition (2.7),

$$\frac{\partial M(x, y)}{\partial y} = \frac{\partial N(x, y)}{\partial x},$$

is the criterion for exactness. If (2.7) holds, then (2.6) is exact; if (2.7) does *not* hold, then (2.6) is *not* exact.

▶ **Example 2.3**

We apply the exactness criterion (2.7) to Equations (2.4) and (2.5), introduced in Example 2.2. For the equation

$$y^2\, dx + 2xy\, dy = 0 \tag{2.4}$$

we have

$$M(x, y) = y^2, \qquad N(x, y) = 2xy,$$

$$\frac{\partial M(x, y)}{\partial y} = 2y = \frac{\partial N(x, y)}{\partial x}$$

for all (x, y). Thus Equation (2.4) is exact in every rectangular domain D. On the other hand, for the equation

$$y\, dx + 2x\, dy = 0, \tag{2.5}$$

we have

$$M(x, y) = y, \qquad N(x, y) = 2x,$$

$$\frac{\partial M(x, y)}{\partial y} = 1 \neq 2 = \frac{\partial N(x, y)}{\partial x}$$

for all (x, y). Thus Equation (2.5) is not exact in any rectangular domain D.

► **Example 2.4**

Consider the differential equation

$$(2x \sin y + y^3 e^x) \, dx + (x^2 \cos y + 3y^2 e^x) \, dy = 0.$$

Here

$$M(x, y) = 2x \sin y + y^3 e^x,$$

$$N(x, y) = x^2 \cos y + 3y^2 e^x,$$

$$\frac{\partial M(x, y)}{\partial y} = 2x \cos y + 3y^2 e^x = \frac{\partial N(x, y)}{\partial x}$$

in every rectangular domain D. Thus this differential equation is exact in every such domain.

These examples illustrate the use of the test given by (2.7) for determining whether or not an equation of the form $M(x, y) \, dx + N(x, y) \, dy = 0$ is exact. It should be observed that the equation *must* be in the standard form $M(x, y) \, dx + N(x, y) \, dy = 0$ in order to use the exactness test (2.7). Note this carefully: an equation may be encountered in the *nonstandard* form $M(x, y) \, dx = N(x, y) \, dy$, and in this form the test (2.7) does *not* apply.

C. The Solution of Exact Differential Equations

Now that we have a test with which to determine exactness, let us proceed to solve exact differential equations. If the equation $M(x, y) \, dx + N(x, y) \, dy = 0$ is exact in a rectangular domain D, then there exists a function F such that

$$\frac{\partial F(x, y)}{\partial x} = M(x, y) \quad \text{and} \quad \frac{\partial F(x, y)}{\partial y} = N(x, y) \quad \text{for all } (x, y) \in D.$$

Then the equation may be written

$$\frac{\partial F(x, y)}{\partial x} \, dx + \frac{\partial F(x, y)}{\partial y} \, dy = 0 \quad \text{or simply} \quad dF(x, y) = 0.$$

The relation $F(x, y) = c$ is obviously a solution of this, where c is an arbitrary constant. We summarize this observation in the following theorem.

THEOREM 2.2

Suppose the differential equation $M(x, y) \, dx + N(x, y) \, dy = 0$ satisfies the differentiability requirements of Theorem 2.1 and is exact in a rectangular domain D. Then a one-parameter family of solutions of this differential equation is given by $F(x, y) = c$, where F is a function such that

$$\frac{\partial F(x, y)}{\partial x} = M(x, y) \quad \text{and} \quad \frac{\partial F(x, y)}{\partial y} = N(x, y) \quad \text{for all } (x, y) \in D.$$

and c is an arbitrary constant.

Referring to Theorem 2.1, we observe that $F(x, y)$ is given by formula (2.13). However, in solving exact differential equations it is neither necessary nor desirable to use this formula. Instead one obtains $F(x, y)$ either by proceeding as in the proof of Theorem 2.1, Part 2, or by the so-called "method of grouping," which will be explained in the following examples.

▶ **Example 2.5**

Solve the equation

$$(3x^2 + 4xy)\, dx + (2x^2 + 2y)\, dy = 0.$$

Our first duty is to determine whether or not the equation is exact. Here

$$M(x, y) = 3x^2 + 4xy, \qquad N(x, y) = 2x^2 + 2y,$$

$$\frac{\partial M(x, y)}{\partial y} = 4x, \qquad \frac{\partial N(x, y)}{\partial x} = 4x,$$

for all real (x, y), and so the equation is exact in every rectangular domain D. Thus we must find F such that

$$\frac{\partial F(x, y)}{\partial x} = M(x, y) = 3x^2 + 4xy \quad \text{and} \quad \frac{\partial F(x, y)}{\partial y} = N(x, y) = 2x^2 + 2y.$$

From the first of these,

$$F(x, y) = \int M(x, y)\, \partial x + \phi(y) = \int (3x^2 + 4xy)\, \partial x + \phi(y)$$

$$= x^3 + 2x^2 y + \phi(y).$$

Then

$$\frac{\partial F(x, y)}{\partial y} = 2x^2 + \frac{d\phi(y)}{dy}.$$

But we must have

$$\frac{\partial F(x, y)}{\partial y} = N(x, y) = 2x^2 + 2y.$$

Thus

$$2x^2 + 2y = 2x^2 + \frac{d\phi(y)}{dy}$$

or

$$\frac{d\phi(y)}{dy} = 2y.$$

Thus $\phi(y) = y^2 + c_0$, where c_0 is an arbitrary constant, and so

$$F(x, y) = x^3 + 2x^2 y + y^2 + c_0.$$

Hence a one-parameter family of solution is $F(x, y) = c_1$, or

$$x^3 + 2x^2y + y^2 + c_0 = c_1.$$

Combining the constants c_0 and c_1 we may write this solution as

$$x^3 + 2x^2y + y^2 = c,$$

where $c = c_1 - c_0$ is an arbitrary constant. The student will observe that there is no loss in generality by taking $c_0 = 0$ and writing $\phi(y) = y^2$. We now consider an alternative procedure.

Method of Grouping. We shall now solve the differential equation of this example by grouping the terms in such a way that its left member appears as the sum of certain exact differentials. We write the differential equation

$$(3x^2 + 4xy)\, dx + (2x^2 + 2y)\, dy = 0$$

in the form

$$3x^2\, dx + (4xy\, dx + 2x^2\, dy) + 2y\, dy = 0.$$

We now recognize this as

$$d(x^3) + d(2x^2y) + d(y^2) = d(c),$$

where c is an arbitrary constant, or

$$d(x^3 + 2x^2y + y^2) = d(c).$$

From this we have at once

$$x^3 + 2x^2y + y^2 = c.$$

Clearly this procedure is much quicker, but it requires a good "working knowledge" of differentials and a certain amount of ingenuity to determine just how the terms should be grouped. The standard method may require more "work" and take longer, but it is perfectly straightforward. It is recommended for those who like to follow a pattern and for those who have a tendency to jump at conculsions.

Just to make certain that we have both procedures well in hand, we shall consider an initial-value problem involving an exact differential equation.

▶ **Example 2.6**

Solve the initial-value problem

$$(2x \cos y + 3x^2y)\, dx + (x^3 - x^2 \sin y - y)\, dy = 0,$$
$$y(0) = 2.$$

We first observe that the equation is exact in every rectangular domain D, since

$$\frac{\partial M(x, y)}{\partial y} = -2x \sin y + 3x^2 = \frac{\partial N(x, y)}{\partial x}$$

for all real (x, y).

Standard Method. We must find F such that

$$\frac{\partial F(x, y)}{\partial x} = M(x, y) = 2x \cos y + 3x^2 y$$

and

$$\frac{\partial F(x, y)}{\partial y} = N(x, y) = x^3 - x^2 \sin y - y.$$

Then

$$F(x, y) = \int M(x, y)\, \partial x + \phi(y)$$

$$= \int (2x \cos y + 3x^2 y)\, \partial x + \phi(y)$$

$$= x^2 \cos y + x^3 y + \phi(y),$$

$$\frac{\partial F(x, y)}{\partial y} = -x^2 \sin y + x^3 + \frac{d\phi(y)}{dy}.$$

But also

$$\frac{\partial F(x, y)}{\partial y} = N(x, y) = x^3 - x^2 \sin y - y$$

and so

$$\frac{d\phi(y)}{dy} = -y$$

and hence

$$\phi(y) = -\frac{y^2}{2} + c_0.$$

Thus

$$F(x, y) = x^2 \cos y + x^3 y - \frac{y^2}{2} + c_0.$$

Hence a one-parameter family of solutions is $F(x, y) = c_1$, which may be expressed as

$$x^2 \cos y + x^3 y - \frac{y^2}{2} = c.$$

Applying the initial condition $y = 2$ when $x = 0$, we find $c = -2$. Thus the solution of the given initial-value problem is

$$x^2 \cos y + x^3 y - \frac{y^2}{2} = -2.$$

Method of Grouping. We group the terms as follows:

$$(2x \cos y\, dx - x^2 \sin y\, dy) + (3x^2 y\, dx + x^3\, dy) - y\, dy = 0.$$

Thus we have

$$d(x^2 \cos y) + d(x^3 y) - d\left(\frac{y^2}{2}\right) = d(c);$$

and so

$$x^2 \cos y + x^3 y - \frac{y^2}{2} = c$$

is a one-parameter family of solutions of the differential equation. Of course the initial condition $y(0) = 2$ again yields the particular solution already obtained.

D. Integrating Factors

Given the differential equation

$$M(x, y) \, dx + N(x, y) \, dy = 0,$$

if

$$\frac{\partial M(x, y)}{\partial y} = \frac{\partial N(x, y)}{\partial x},$$

then the equation is exact and we can obtain a one-parameter family of solutions by one of the procedures explained above. But if

$$\frac{\partial M(x, y)}{\partial y} \neq \frac{\partial N(x, y)}{\partial x},$$

then the equation is *not* exact and the above procedures do not apply. What shall we do in such a case? Perhaps we can multiply the nonexact equation by some expression that will transform it into an essentially equivalent exact equation. If so, we can proceed to solve the resulting exact equation by one of the above procedures. Let us consider again the equation

$$y \, dx + 2x \, dy = 0, \tag{2.5}$$

which was introduced in Example 2.2. In that example we observed that this equation is *not* exact. However, if we multiply Equation (2.5) by y, it is transformed into the essentially equivalent equation

$$y^2 \, dx + 2xy \, dy = 0, \tag{2.4}$$

which *is* exact (see Example 2.2). Since this resulting exact equation (2.4) is integrable, we call y an *integrating factor* of Equation (2.5). In general, we have the following definition:

DEFINITION

If the differential equation

$$M(x, y) \, dx + N(x, y) \, dy = 0 \tag{2.14}$$

is not *exact in a domain D but the differential equation*

$$\mu(x, y)M(x, y)\, dx + \mu(x, y)N(x, y)\, dy = 0 \qquad (2.15)$$

is *exact in D, then* $\mu(x, y)$ *is called an* integrating factor *of the differential equation* (2.14).

▶ **Example 2.7**

Consider the differential equation

$$(3y + 4xy^2)\, dx + (2x + 3x^2y)\, dy = 0. \qquad (2.16)$$

This equation is of the form (2.14), where

$$M(x, y) = 3y + 4xy^2, \qquad N(x, y) = 2x + 3x^2y,$$

$$\frac{\partial M(x, y)}{\partial y} = 3 + 8xy, \qquad \frac{\partial N(x, y)}{\partial x} = 2 + 6xy.$$

Since

$$\frac{\partial M(x, y)}{\partial y} \neq \frac{\partial N(x, y)}{\partial x}$$

except for (x, y) such that $2xy + 1 = 0$, Equation (2.16) is *not* exact in any rectangular domain D.

Let $\mu(x, y) = x^2y$. Then the corresponding differential equation of the form (2.15) is

$$(3x^2y^2 + 4x^3y^3)\, dx + (2x^3y + 3x^4y^2)\, dy = 0.$$

This equation is exact in every rectangular domain D, since

$$\frac{\partial [\mu(x, y)M(x, y)]}{\partial y} = 6x^2y + 12x^3y^2 = \frac{\partial [\mu(x, y)N(x, y)]}{\partial x}$$

for all real (x, y). Hence $\mu(x, y) = x^2y$ is an integrating factor of Equation (2.16).

Multiplication of a nonexact differential equation by an integrating factor thus transforms the nonexact equation into an exact one. We have referred to this resulting exact equation as "essentially equivalent" to the original. This so-called essentially equivalent exact equation has the same one-parameter family of solutions as the nonexact original. However, the multiplication of the original equation by the integrating factor may result in either (1) the loss of (one or more) solutions of the original, or (2) the gain of (one or more) functions which are solutions of the "new" equation but *not* of the original, or (3) both of these phenomena. Hence, whenever we transform a nonexact equation into an exact one by multiplication by an integrating factor, we should check carefully to determine whether any solutions may have been lost or gained. We shall illustrate an important special case of these phenomena when we consider separable equations in Section 2.2. See also Exercise 22 at the end of this section.

The question now arises: How is an integrating factor found? We shall not attempt to answer this question at this time. Instead we shall proceed to a study of the important class of separable equations in Section 2.2 and linear equations in Section 2.3. We shall see that separable equations always possess integrating factors that are perfectly obvious, while linear equations always have integrating factors of a certain special form. We shall return to the question raised above in Section 2.4. Our object here has been merely to introduce the concept of an integrating factor.

Exercises

In Exercises 1–10 determine whether or not each of the given equations is exact; solve those that are exact.

1. $(3x + 2y) dx + (2x + y) dy = 0.$

2. $(y^2 + 3) dx + (2xy - 4) dy = 0.$

3. $(2xy + 1) dx + (x^2 + 4y) dy = 0.$

4. $(3x^2y + 2) dx - (x^3 + y) dy = 0.$

5. $(6xy + 2y^2 - 5) dx + (3x^2 + 4xy - 6) dy = 0.$

6. $(\theta^2 + 1)\cos r \, dr + 2\theta \sin r \, d\theta = 0.$

7. $(y \sec^2 x + \sec x \tan x) dx + (\tan x + 2y) dy = 0.$

8. $\left(\dfrac{x}{y^2} + x\right) dx + \left(\dfrac{x^2}{y^3} + y\right) dy = 0.$

9. $\left(\dfrac{2s - 1}{t}\right) ds + \left(\dfrac{s - s^2}{t^2}\right) dt = 0.$

10. $\dfrac{2y^{3/2} + 1}{x^{1/2}} dx + (3x^{1/2}y^{1/2} - 1) dy = 0.$

Solve the initial-value problems in Exercises 11–16.

11. $(2xy - 3) dx + (x^2 + 4y) dy = 0, \qquad y(1) = 2.$

12. $(3x^2y^2 - y^3 + 2x) dx + (2x^3y - 3xy^2 + 1) dy = 0, \qquad y(-2) = 1.$

13. $(2y \sin x \cos x + y^2 \sin x) dx + (\sin^2 x - 2y \cos x) dy = 0, \qquad y(0) = 3.$

14. $(ye^x + 2e^x + y^2) dx + (e^x + 2xy) dy = 0, \qquad y(0) = 6.$

15. $\left(\dfrac{3 - y}{x^2}\right) dx + \left(\dfrac{y^2 - 2x}{xy^2}\right) dy = 0, \qquad y(-1) = 2.$

16. $\dfrac{1 + 8xy^{2/3}}{x^{2/3}y^{1/3}} dx + \dfrac{2x^{4/3}y^{2/3} - x^{1/3}}{y^{4/3}} dy = 0, \qquad y(1) = 8.$

17. In each of the following equations determine the constant A such that the equation is exact, and solve the resulting exact equation:

    (a) $(x^2 + 3xy) dx + (Ax^2 + 4y) dy = 0.$

    (b) $\left(\dfrac{1}{x^2} + \dfrac{1}{y^2}\right) dx + \left(\dfrac{Ax + 1}{y^3}\right) dy = 0.$

18. In each of the following equations determine the constant A such that the equation is exact, and solve the resulting exact equation:

(a) $(Ax^2y + 2y^2) \, dx + (x^3 + 4xy) \, dy = 0.$

(b) $\left(\dfrac{Ay}{x^3} + \dfrac{y}{x^2}\right) dx + \left(\dfrac{1}{x^2} - \dfrac{1}{x}\right) dy = 0.$

19. In each of the following equations determine the most general function $N(x, y)$ such that the equation is exact:

(a) $(x^3 + xy^2) \, dx + N(x, y) \, dy = 0.$

(b) $(x^{-2}y^{-2} + xy^{-3}) \, dx + N(x, y) \, dy = 0.$

20. In each of the following equations determine the most general function $M(x, y)$ such that the equation is exact:

(a) $M(x, y) \, dx + (2x^2y^3 + x^4y) \, dy = 0.$

(b) $M(x, y) \, dx + (2ye^x + y^2e^{3x}) \, dy = 0.$

21. Consider the differential equation

$$(4x + 3y^2) \, dx + 2xy \, dy = 0.$$

(a) Show that this equation is not exact.

(b) Find an integrating factor of the form x^n, where n is a positive integer.

(c) Multiply the given equation through by the integrating factor found in (b) and solve the resulting exact equation.

22. Consider the differential equation

$$(y^2 + 2xy) \, dx - x^2 \, dy = 0.$$

(a) Show that this equation is not exact.

(b) Multiply the given equation through by y^n, where n is an integer, and then determine n so that y^n is an integrating factor of the given equation.

(c) Multiply the given equation through by the integrating factor found in (b) and solve the resulting exact equation.

(d) Show that $y = 0$ is a solution of the original nonexact equation but is not a solution of the essentially equivalent exact equation found in step (c).

(e) Graph several integral curves of the original equation, including all those whose equations are (or can be written) in some "special" form.

23. Consider a differential equation of the form

$$[y + xf(x^2 + y^2)] \, dx + [yf(x^2 + y^2) - x] \, dy = 0.$$

(a) Show that an equation of this form is not exact.

(b) Show that $1/(x^2 + y^2)$ is an integrating factor of an equation of this form.

24. Use the result of Exercise 23(b) to solve the equation

$$[y + x(x^2 + y^2)^2] \, dx + [y(x^2 + y^2)^2 - x] \, dy = 0.$$

2.2 SEPARABLE EQUATIONS AND EQUATIONS REDUCIBLE TO THIS FORM

A. Separable Equations

DEFINITION

An equation of the form

$$F(x)G(y)\,dx + f(x)g(y)\,dy = 0 \tag{2.17}$$

is called an equation with variables separable *or simply a* separable equation.

For example, the equation $(x - 4)y^4\,dx - x^3(y^2 - 3)\,dy = 0$ is a separable equation.

In general the separable equation (2.17) is not exact, but it possesses an obvious integrating factor, namely $1/f(x)G(y)$. For if we multiply Equation (2.17) by this expression, we separate the variables, reducing (2.17) to the essentially equivalent equation

$$\frac{F(x)}{f(x)}\,dx + \frac{g(y)}{G(y)}\,dy = 0. \tag{2.18}$$

This equation is exact, since

$$\frac{\partial}{\partial y}\left[\frac{F(x)}{f(x)}\right] = 0 = \frac{\partial}{\partial x}\left[\frac{g(y)}{G(y)}\right].$$

Denoting $F(x)/f(x)$ by $M(x)$ and $g(y)/G(y)$ by $N(y)$, Equation (2.18) takes the form $M(x)\,dx + N(y)\,dy = 0$. Since M is a function of x only and N is a function of y only, we see at once that a one-parameter family of solutions is

$$\int M(x)\,dx + \int N(y)\,dy = c, \tag{2.19}$$

where c is the arbitrary constant. Thus the problem of finding such a family of solutions of the separable equation (2.17) has reduced to that of performing the integrations indicated in Equation (2.19). It is in this sense that separable equations are the simplest first-order differential equations.

Since we obtained the separated exact equation (2.18) from the nonexact equation (2.17) by multiplying (2.17) by the integrating factor $1/f(x)G(y)$, solutions may have been lost or gained in this process. We now consider this more carefully. In formally multiplying by the integrating factor $1/f(x)G(y)$, we actually divided by $f(x)G(y)$. We did this under the tacit assumption that neither $f(x)$ nor $G(y)$ is zero; and, under this assumption, we proceeded to obtain the one-parameter family of solutions given by (2.19). Now, we should investigate the possible loss or gain of solutions that may have occurred in this formal process. In particular, regarding y as the dependent variable as usual, we consider the situation that occurs if $G(y)$ is zero. Writing the original differential equation (2.17) in the derivative form

$$f(x)g(y)\frac{dy}{dx} + F(x)G(y) = 0,$$

we immediately note the following: If y_0 is any real number such that $G(y_0) = 0$, then $y = y_0$ is a (constant) solution of the original differential equation; and this solution may (or may not) have been lost in the formal separation process.

In finding a one-parameter family of solutions a separable equation, we shall always make the assumption that any factors by which we divide in the formal separation process are not zero. Then we must find the solutions $y = y_0$ of the equation $G(y) = 0$ and determine whether any of these are solutions of the original equation which were lost in the formal separation process.

▶ **Example 2.8**

Solve the equation

$$(x - 4)y^4 \, dx - x^3(y^2 - 3) \, dy = 0.$$

The equation is separable; separating the variables by dividing by $x^3 y^4$, we obtain

$$\frac{(x - 4) \, dx}{x^3} - \frac{(y^2 - 3) \, dy}{y^4} = 0$$

or

$$(x^{-2} - 4x^{-3}) \, dx - (y^{-2} - 3y^{-4}) \, dy = 0.$$

Integrating, we have the one-parameter family of solutions

$$-\frac{1}{x} + \frac{2}{x^2} + \frac{1}{y} - \frac{1}{y^3} = c,$$

where c is the arbitrary constant.

In dividing by $x^3 y^4$ in the separation process, we assumed that $x^3 \neq 0$ and $y^4 \neq 0$. We now consider the solution $y = 0$ of $y^4 = 0$. It is not a member of the one-parameter family of solutions which we obtained. However, writing the original differential equation of the problem in the derivative form

$$\frac{dy}{dx} = \frac{(x - 4)y^4}{x^3(y^2 - 3)},$$

it is obvious that $y = 0$ *is* a solution of the original equation. We conclude that it is a solution which was lost in the separation process.

▶ **Example 2.9**

Solve the initial-value problem that consists of the differential equation

$$x \sin y \, dx + (x^2 + 1) \cos y \, dy = 0 \qquad (2.20)$$

and the initial condition

$$y(1) = \frac{\pi}{2}. \qquad (2.21)$$

We first obtain a one-parameter family of solutions of the differential equation (2.20).

Separating the variables by dividing by $(x^2 + 1)\sin y$, we obtain

$$\frac{x}{x^2 + 1}\, dx + \frac{\cos y}{\sin y}\, dy = 0.$$

Thus

$$\int \frac{x\, dx}{x^2 + 1} + \int \frac{\cos y}{\sin y}\, dy = c_0,$$

where c_0 is an arbitrary constant. Recall that

$$\int \frac{du}{u} = \ln|u| + C \quad \text{and} \quad |u| = \begin{cases} u & \text{if} \quad u \geq 0, \\ -u & \text{if} \quad u \leq 0. \end{cases}$$

Then, carrying out the integrations, we find

$$\tfrac{1}{2}\ln(x^2 + 1) + \ln|\sin y| = c_0. \tag{2.22}$$

We could leave the family of solutions in this form, but we can put it in a neater form in the following way. Since each term of the left member of this equation involves the logarithm of a function, it would seem reasonable that something might be accomplished by writing the arbitrary constant c_0 in the form $\ln|c_1|$. This we do, obtaining

$$\tfrac{1}{2}\ln(x^2 + 1) + \ln|\sin y| = \ln|c_1|.$$

Multiplying by 2, we have

$$\ln(x^2 + 1) + 2\ln|\sin y| = 2\ln|c_1|.$$

Since

$$2\ln|\sin y| = \ln(\sin y)^2,$$

and

$$2\ln|c_1| = \ln c_1^2 = \ln c,$$

where

$$c = c_1^2 \geq 0,$$

we now have

$$\ln(x^2 + 1) + \ln\sin^2 y = \ln c.$$

Since $\ln A + \ln B = \ln AB$, this equation may be written

$$\ln(x^2 + 1)\sin^2 y = \ln c.$$

From this we have at once

$$(x^2 + 1)\sin^2 y = c. \tag{2.23}$$

Clearly (2.23) is of a neater form than (2.22).

In dividing by $(x^2 + 1)\sin y$ in the separation process, we assumed that $\sin y \neq 0$. Now consider the solutions of $\sin y = 0$. These are given by $y = n\pi$ ($n = 0, \pm 1, \pm 2, \dots$). Writing the original differential equation (2.20) in the derivative form, it is clear that each of these solutions $y = n\pi (n = 0, \pm 1, \pm 2, \dots)$ of $\sin y = 0$ is a constant solution of the original differential equation. Now, each of these constant solutions

$y = n\pi$ is a member of the one-parameter family (2.23) of solutions of (2.20) for $c = 0$. Thus none of these solutions was lost in the separation process.

We now apply the initial condition (2.21) to the family of solutions (2.23). We have

$$(1^2 + 1)\sin^2 \frac{\pi}{2} = c$$

and so $c = 2$. Therefore the solution of the initial-value problem under consideration is

$$(x^2 + 1)\sin^2 y = 2.$$

B. Homogeneous Equations

We now consider a class of differential equations that can be reduced to separable equations by a change of variables.

DEFINITION

The first-order differential equation $M(x, y)\,dx + N(x, y)\,dy = 0$ is said to be homogeneous if, when written in the derivative form $(dy/dx) = f(x, y)$, there exists a function g such that $f(x, y)$ can be expressed in the form $g(y/x)$.

► **Example 2.10**

The differential equation $(x^2 - 3y^2)\,dx + 2xy\,dy = 0$ is homogeneous. To see this, we first write this equation in the derivative form

$$\frac{dy}{dx} = \frac{3y^2 - x^2}{2xy}.$$

Now observing that

$$\frac{3y^2 - x^2}{2xy} = \frac{3y}{2x} - \frac{x}{2y} = \frac{3}{2}\left(\frac{y}{x}\right) - \frac{1}{2}\left(\frac{1}{y/x}\right),$$

we see that the differential equation under consideration may be written as

$$\frac{dy}{dx} = \frac{3}{2}\left(\frac{y}{x}\right) - \frac{1}{2}\left(\frac{1}{y/x}\right),$$

in which the right member is of the form $g(y/x)$ for a certain function g.

► **Example 2.11**

The equation

$$(y + \sqrt{x^2 + y^2})\,dx - x\,dy = 0$$

is homogeneous. When written in the form

$$\frac{dy}{dx} = \frac{y + \sqrt{x^2 + y^2}}{x},$$

the right member may be expressed as

$$\frac{y}{x} \pm \frac{\sqrt{x^2 + y^2}}{\sqrt{x^2}}$$

or

$$\frac{y}{x} \pm \sqrt{1 + \left(\frac{y}{x}\right)^2},$$

depending on the sign of x. This is obviously of the form $g(y/x)$.

Before proceeding to the actual solution of homogeneous equations we shall consider slightly different procedure for recognizing such equations. A function F is called *homogeneous of degree n* if $F(tx, ty) = t^n F(x, y)$. This means that if tx and ty are substituted for x and y, respectively, in $F(x, y)$, and if t^n is then factored out, the other factor that remains is the original expression $F(x, y)$ itself. For example, the function F given by $F(x, y) = x^2 + y^2$ is homogeneous of degree 2, since

$$F(tx, ty) = (tx)^2 + (ty)^2 = t^2(x^2 + y^2) = t^2 F(x, y).$$

Now suppose the functions M and N in the differential equation $M(x, y)\, dx + N(x, y)\, dy = 0$ are both homogeneous of the *same* degree n. Then since $M(tx, ty) = t^n M(x, y)$, if we let $t = 1/x$, we have

$$M\left(\frac{1}{x} \cdot x, \frac{1}{x} \cdot y\right) = \left(\frac{1}{x}\right)^n M(x, y).$$

Clearly this may be written more simply as

$$M\left(1, \frac{y}{x}\right) = \left(\frac{1}{x}\right)^n M(x, y);$$

and from this we at once obtain

$$M(x, y) = \left(\frac{1}{x}\right)^{-n} M\left(1, \frac{y}{x}\right).$$

Likewise, we find

$$N(x, y) = \left(\frac{1}{x}\right)^{-n} N\left(1, \frac{y}{x}\right).$$

Now writing the differential equation $M(x, y)\, dx + N(x, y)\, dy = 0$ in the form

$$\frac{dy}{dx} = -\frac{M(x, y)}{N(x, y)},$$

we find

$$\frac{dy}{dx} = -\frac{\left(\dfrac{1}{x}\right)^{-n} M\left(1, \dfrac{y}{x}\right)}{\left(\dfrac{1}{x}\right)^{-n} N\left(1, \dfrac{y}{x}\right)} = -\frac{M\left(1, \dfrac{y}{x}\right)}{N\left(1, \dfrac{y}{x}\right)}.$$

Clearly the expression on the right is of the form $g(y/x)$, and so the equation $M(x, y)\, dx + N(x, y)\, dy = 0$ is homogeneous in the sense of the original definition of homogeneity. Thus we conclude that if M and N in $M(x, y)\,dx + N(x, y)\,dy = 0$ are both homogeneous functions of the same degree n, then the differential equation is a homogeneous differential equation.

Let us now look back at Examples 2.10 and 2.11 in this light. In Example 2.10, $M(x, y) = x^2 - 3y^2$ and $N(x, y) = 2xy$. Both M and N are homogeneous of degree 2. Thus we know at once that the equation $(x^2 - 3y^2)\,dx + 2xy\,dy = 0$ is a homogeneous equation. In Example 2.11, $M(x, y) = y + \sqrt{x^2 + y^2}$ and $N(x, y) = -x$. Clearly N is homogeneous of degree 1. Since

$$M(tx, ty) = ty + \sqrt{(tx)^2 + (ty)^2} = t(y + \sqrt{x^2 + y^2}) = t^1 M(x, y),$$

we see that M is also homogeneous of degree 1. Thus we conclude that the equation

$$(y + \sqrt{x^2 + y^2})\,dx - x\,dy = 0$$

is indeed homogeneous.

We now show that every homogeneous equation can be reduced to a separable equation by proving the following theorem.

THEOREM 2.3

If

$$M(x, y)\,dx + N(x, y)\,dy = 0 \qquad (2.24)$$

is a homogeneous equation, then the change of variables $y = vx$ transforms (2.24) into a separable equation in the variables v and x.

Proof. Since $M(x, y)\,dx + N(x, y)\,dy = 0$ is homogeneous, it may be written in the form

$$\frac{dy}{dx} = g\left(\frac{y}{x}\right).$$

Let $y = vx$. Then

$$\frac{dy}{dx} = v + x\frac{dv}{dx}$$

and (2.24) becomes

$$v + x\frac{dv}{dx} = g(v)$$

or

$$[v - g(v)]\,dx + x\,dv = 0.$$

This equation is separable. Separating the variables we obtain

$$\frac{dv}{v - g(v)} + \frac{dx}{x} = 0. \qquad (2.25)$$

Q.E.D.

Thus to solve a homogeneous differential equation of the form (2.24), we let $y = vx$ and transform the homogeneous equation into a separable equation of the form (2.25). From this, we have

$$\int \frac{dv}{v - g(v)} + \int \frac{dx}{x} = c,$$

where c is an arbitrary constant. Letting $F(v)$ denote

$$\int \frac{dv}{v - g(v)}$$

and returning to the original dependent variable y, the solution takes the form

$$F\left(\frac{y}{x}\right) + \ln|x| = c.$$

▶   **Example 2.12**

Solve the equation

$$(x^2 - 3y^2)\, dx + 2xy\, dy = 0.$$

We have already observed that this equation is homogeneous. Writing it in the form

$$\frac{dy}{dx} = -\frac{x}{2y} + \frac{3y}{2x}$$

and letting $y = vx$, we obtain

$$v + x\frac{dv}{dx} = -\frac{1}{2v} + \frac{3v}{2},$$

or

$$x\frac{dv}{dx} = -\frac{1}{2v} + \frac{v}{2},$$

or, finally,

$$x\frac{dv}{dx} = \frac{v^2 - 1}{2v}.$$

This equation is separable. Separating the variables, we obtain

$$\frac{2v\, dv}{v^2 - 1} = \frac{dx}{x}.$$

Integrating, we find

$$\ln|v^2 - 1| = \ln|x| + \ln|c|,$$

and hence

$$|v^2 - 1| = |cx|,$$

where c is an arbitrary constant. The reader should observe that no solutions were lost in the separation process. Now, replacing v by y/x we obtain the solutions in the form

$$\left| \frac{y^2}{x^2} - 1 \right| = |cx|$$

or

$$|y^2 - x^2| = |cx|x^2.$$

If $y \geq x \geq 0$, then this may be expressed somewhat more simply as

$$y^2 - x^2 = cx^3.$$

▶ **Example 2.13**

Solve the initial-value problem

$$(y + \sqrt{x^2 + y^2})\, dx - x\, dy = 0,$$
$$y(1) = 0.$$

We have seen that the differential equation is homogeneous. As before, we write it in the form

$$\frac{dy}{dx} = \frac{y + \sqrt{x^2 + y^2}}{x}.$$

Since the initial x value is 1, we consider $x > 0$ and take $x = \sqrt{x^2}$ and obtain

$$\frac{dy}{dx} = \frac{y}{x} + \sqrt{1 + \left(\frac{y}{x}\right)^2}.$$

We let $y = vx$ and obtain

$$v + x\frac{dv}{dx} = v + \sqrt{1 + v^2}$$

or

$$x\frac{dv}{dx} = \sqrt{1 + v^2}.$$

Separating variables, we find

$$\frac{dv}{\sqrt{v^2 + 1}} = \frac{dx}{x}.$$

Using tables, we perform the required integrations to obtain

$$\ln|v + \sqrt{v^2 + 1}| = \ln|x| + \ln|c|,$$

or

$$v + \sqrt{v^2 + 1} = cx.$$

Now replacing v by y/x, we obtain the general solution of the differential equation in the form

$$\frac{y}{x} + \sqrt{\frac{y^2}{x^2} + 1} = cx$$

or

$$y + \sqrt{x^2 + y^2} = cx^2.$$

The initial condition requires that $y = 0$ when $x = 1$. This gives $c = 1$ and hence

$$y + \sqrt{x^2 + y^2} = x^2,$$

from which it follows that

$$y = \tfrac{1}{2}(x^2 - 1).$$

Exercises

Solve each of the differential equations in Exercises 1–14.

1. $4xy\, dx + (x^2 + 1)\, dy = 0.$

2. $(xy + 2x + y + 2)\, dx + (x^2 + 2x)\, dy = 0.$

3. $2r(s^2 + 1)\, dr + (r^4 + 1)\, ds = 0.$

4. $\csc y\, dx + \sec x\, dy = 0.$

5. $\tan \theta\, dr + 2r\, d\theta = 0.$

6. $(e^v + 1)\cos u\, du + e^v(\sin u + 1)\, dv = 0.$

7. $(x + 4)(y^2 + 1)\, dx + y(x^2 + 3x + 2)\, dy = 0.$

8. $(x + y)\, dx - x\, dy = 0.$

9. $(2xy + 3y^2)\, dx - (2xy + x^2)\, dy = 0.$

10. $v^3\, du + (u^3 - uv^2)\, dv = 0.$

11. $\left(x \tan \dfrac{y}{x} + y\right) dx - x\, dy = 0.$

12. $(2s^2 + 2st + t^2)\, ds + (s^2 + 2st - t^2)\, dt = 0.$

13. $(x^3 + y^2 \sqrt{x^2 + y^2})\, dx - xy \sqrt{x^2 + y^2}\, dy = 0.$

14. $(\sqrt{x + y} + \sqrt{x - y})\, dx + (\sqrt{x - y} - \sqrt{x + y})\, dy = 0.$

Solve the initial-value problems in Exercises 15–20.

15. $(y + 2)\, dx + y(x + 4)\, dy = 0, \qquad y(-3) = -1.$

16. $8 \cos^2 y\, dx + \csc^2 x\, dy = 0, \qquad y\!\left(\dfrac{\pi}{12}\right) = \dfrac{\pi}{4}.$

17. $(3x + 8)(y^2 + 4)\, dx - 4y(x^2 + 5x + 6)\, dy = 0, \qquad y(1) = 2.$

18. $(x^2 + 3y^2) dx - 2xy \, dy = 0,$ $y(2) = 6.$

19. $(2x - 5y) dx + (4x - y) dy = 0,$ $y(1) = 4.$

20. $(3x^2 + 9xy + 5y^2) dx - (6x^2 + 4xy) dy = 0,$ $y(2) = -6.$

21. (a) Show that the homogeneous equation

$$(Ax + By) dx + (Cx + Dy) dy = 0$$

   is exact if and only if $B = C$.

   (b) Show that the homogeneous equation

$$(Ax^2 + Bxy + Cy^2) dx + (Dx^2 + Exy + Fy^2) dy = 0$$

   is exact if and only if $B = 2D$ and $E = 2C$.

22. Solve each of the following by two methods (see Exercise 21(a)):

   (a) $(x + 2y) dx + (2x - y) dy = 0.$

   (b) $(3x - y) dx - (x + y) dy = 0.$

23. Solve each of the following by two methods (see Exercise 21(b)):

   (a) $(x^2 + 2y^2) dx + (4xy - y^2) dy = 0.$

   (b) $(2x^2 + 2xy + y^2) dx + (x^2 + 2xy) dy = 0.$

24. (a) Prove that if $M \, dx + N \, dy = 0$ is a homogeneous equation, then the change of variables $x = uy$ transforms this equation into a separable equation in the variables u and x.

   (b) Use the result of (a) to solve the equation of Example 2.12 of the text.

   (c) Use the result of (a) to solve the equation of Example 2.13 of the text.

25. Suppose the equation $M \, dx + N \, dy = 0$ is homogeneous. Show that the transformation $x = r \cos \theta$, $y = r \sin \theta$ reduces this equation to a separable equation in the variables r and θ.

26. (a) Use the method of Exercise 25 to solve Exercise 8.

   (b) Use the method of Exercise 25 to solve Exercise 9.

27. Suppose the equation

$$M \, dx + N \, dy = 0 \tag{A}$$

   is homogeneous.

   (a) Show that Equation (A) is invariant under the transformation

$$x = k\xi, \qquad y = k\eta, \tag{B}$$

   where k is a constant.

   (b) Show that the general solution of Equation (A) can be written in the form

$$x = c\phi\left(\frac{y}{x}\right), \tag{C}$$

   where c is an arbitrary constant.

(c) Use the result of (b) to show that the solution (C) is also invariant under the transformation (B).

(d) Interpret geometrically the results proved in (a) and (c).

2.3 LINEAR EQUATIONS AND BERNOULLI EQUATIONS

A. Linear Equations

In Chapter 1 we gave the definition of the linear ordinary differential equation of order n; we now consider the linear ordinary differential equation of the first order.

DEFINITION

A first-order ordinary differential equation is linear *in the dependent variable y and the independent variable x if it is, or can be, written in the form*

$$\frac{dy}{dx} + P(x)y = Q(x). \tag{2.26}$$

For example, the equation

$$x\frac{dy}{dx} + (x + 1)y = x^3$$

is a first-order linear differential equation, for it can be written as

$$\frac{dy}{dx} + \left(1 + \frac{1}{x}\right)y = x^2,$$

which is of the form (2.26) with $P(x) = 1 + (1/x)$ and $Q(x) = x^2$.

Let us write Equation (2.26) in the form

$$[P(x)y - Q(x)]\, dx + dy = 0. \tag{2.27}$$

Equation (2.27) is of the form

$$M(x, y)\, dx + N(x, y)\, dy = 0,$$

where

$$M(x, y) = P(x)y - Q(x) \quad \text{and} \quad N(x, y) = 1.$$

Since

$$\frac{\partial M(x, y)}{\partial y} = P(x) \quad \text{and} \quad \frac{\partial N(x, y)}{\partial x} = 0,$$

Equation (2.27) is *not* exact unless $P(x) = 0$, in which case Equation (2.26) degenerates into a simple separable equation. However, Equation (2.27) possesses an integrating factor that depends on x only and may easily be found. Let us proceed to find it. Let us multiply Equation (2.27) by $\mu(x)$, obtaining

$$[\mu(x)P(x)y - \mu(x)Q(x)]\, dx + \mu(x)\, dy = 0. \tag{2.28}$$

By definition, $\mu(x)$ is an integrating factor of Equation (2.28) if and only if Equation (2.28) is exact; that is, if and only if

$$\frac{\partial}{\partial y}\left[\mu(x)P(x)y - \mu(x)Q(x)\right] = \frac{\partial}{\partial x}\left[\mu(x)\right].$$

This condition reduces to

$$\mu(x)P(x) = \frac{d}{dx}\left[\mu(x)\right]. \tag{2.29}$$

In (2.29), P is a known function of the independent variable x, but μ is an unknown function of x that we are trying to determine. Thus we write (2.29) as the differential equation

$$\mu P(x) = \frac{d\mu}{dx},$$

in the dependent variable μ and the independent variable x, where P is a known function of x. This differential equation is separable; separating the variables, we have

$$\frac{d\mu}{\mu} = P(x)\,dx.$$

Integrating, we obtain the particular solution

$$\ln|\mu| = \int P(x)\,dx$$

or

$$\mu = e^{\int P(x)\,dx}, \tag{2.30}$$

where it is clear that $\mu > 0$. Thus the linear equation (2.26) possesses an integrating factor of the form (2.30). Multiplying (2.26) by (2.30) gives

$$e^{\int P(x)\,dx}\frac{dy}{dx} + e^{\int P(x)\,dx}P(x)y = Q(x)e^{\int P(x)\,dx},$$

which is precisely

$$\frac{d}{dx}\left[e^{\int P(x)\,dx}y\right] = Q(x)e^{\int P(x)\,dx}.$$

Integrating this we obtain the solution of Equation (2.26) in the form

$$e^{\int P(x)\,dx}y = \int e^{\int P(x)\,dx}Q(x)\,dx + c,$$

where c is an arbitrary constant.

Summarizing this discussion, we have the following theorem:

THEOREM 2.4

The linear differential equation

$$\frac{dy}{dx} + P(x)y = Q(x) \tag{2.26}$$

has an integrating factor of the form

$$e^{\int P(x)\,dx}.$$

A one-parameter family of solutions of this equation is

$$ye^{\int P(x)\,dx} = \int e^{\int P(x)\,dx}Q(x)\,dx + c;$$

that is,

$$y = e^{-\int P(x)\,dx}\left[\int e^{\int P(x)\,dx}Q(x)\,dx + c\right].$$

Furthermore, it can be shown that this one-parameter family of solutions of the linear equation (2.26) includes all solutions of (2.26).

We consider several examples.

▶ **Example 2.14**

$$\frac{dy}{dx} + \left(\frac{2x+1}{x}\right)y = e^{-2x}. \tag{2.31}$$

Here

$$P(x) = \frac{2x+1}{x}$$

and hence an integrating factor is

$$\exp\left[\int P(x)\,dx\right] = \exp\left[\int\left(\frac{2x+1}{x}\right)dx\right] = \exp(2x + \ln|x|)$$

$$= \exp(2x)\exp(\ln|x|) = x\exp(2x).*$$

Multiplying Equation (2.31) through by this integrating factor, we obtain

$$xe^{2x}\frac{dy}{dx} + e^{2x}(2x+1)y = x$$

or

product rule

$$\frac{d}{dx}(xe^{2x}y) = x.$$

Integrating, we obtain the solutions

$$xe^{2x}y = \frac{x^2}{2} + c$$

or

$$y = \tfrac{1}{2}xe^{-2x} + \frac{c}{x}e^{-2x},$$

where c is an arbitrary constant.

---

* The expressions e^x and exp x are identical.

▶ **Example 2.15**

Solve the initial-value problem that consists of the differential equation

$$(x^2 + 1)\frac{dy}{dx} + 4xy = x \tag{2.32}$$

and the initial condition

$$y(2) = 1. \tag{2.33}$$

The differential equation (2.32) is not in the form (2.26). We therefore divide by $x^2 + 1$ to obtain

$$\frac{dy}{dx} + \frac{4x}{x^2 + 1}y = \frac{x}{x^2 + 1}. \tag{2.34}$$

Equation (2.34) is in the standard form (2.26), where

$$P(x) = \frac{4x}{x^2 + 1}.$$

An integrating factor is

$$\exp\left[\int P(x)\,dx\right] = \exp\left(\int \frac{4x\,dx}{x^2 + 1}\right) = \exp[\ln(x^2 + 1)^2] = (x^2 + 1)^2.$$

Multiplying Equation (2.34) through by this integrating factor, we have

$$(x^2 + 1)^2 \frac{dy}{dx} + 4x(x^2 + 1)y = x(x^2 + 1)$$

or

$$\frac{d}{dx}[(x^2 + 1)^2 y] = x^3 + x.$$

We now integrate to obtain a one-parameter family of solutions of Equation (2.23) in the form

$$(x^2 + 1)^2 y = \frac{x^4}{4} + \frac{x^2}{2} + c.$$

Applying the initial condition (2.33), we have

$$25 = 6 + c.$$

Thus $c = 19$ and the solution of the initial-value problem under consideration is

$$(x^2 + 1)^2 y = \frac{x^4}{4} + \frac{x^2}{2} + 19.$$

▶ **Example 2.16**

Consider the differential equation

$$y^2\,dx + (3xy - 1)\,dy = 0. \tag{2.35}$$

Solving for dy/dx, this becomes

$$\frac{dy}{dx} = \frac{y^2}{1 - 3xy},$$

which is clearly *not* linear in y. Also, Equation (2.35) is *not* exact, separable, or homogeneous. It appears to be of a type that we have not yet encountered; but let us look a little closer. In Section 2.1, we pointed out that in the differential form of a first-order differential equation the roles of x and y are interchangeable, in the sense that either variable may be regarded as the dependent variable and the other as the independent variable. Considering differential equation (2.35) with this in mind, let us now regard x as the dependent variable and y as the independent variable. With this interpretation, we now write (2.35) in the derivative form

$$\frac{dx}{dy} = \frac{1 - 3xy}{y^2}$$

or

$$\frac{dx}{dy} + \frac{3}{y}x = \frac{1}{y^2}. \tag{2.36}$$

Now observe that Equation (2.36) is of the form

$$\frac{dx}{dy} + P(y)x = Q(y)$$

and so is *linear in x*. Thus the theory developed in this section may be applied to Equation (2.36) merely by interchanging the roles played by x and y. Thus an integrating factor is

$$\exp\left[\int P(y)\,dy\right] = \exp\left(\int \frac{3}{y}\,dy\right) = \exp(\ln|y|^3) = y^3.$$

Multiplying (2.36) by y^3 we obtain

$$y^3 \frac{dx}{dy} + 3y^2 x = y$$

or

$$\frac{d}{dy}[y^3 x] = y.$$

Integrating, we find the solutions in the form

$$y^3 x = \frac{y^2}{2} + c$$

or

$$x = \frac{1}{2y} + \frac{c}{y^3},$$

where c is an arbitrary constant.

B. Bernoulli Equations

We now consider a rather special type of equation that can be reduced to a linear equation by an appropriate transformation. This is the so-called Bernoulli equation.

DEFINITION

An equation of the form

$$\frac{dy}{dx} + P(x)y = Q(x)y^n \tag{2.37}$$

is called a Bernoulli differential equation.

We observe that if $n = 0$ or 1, then the Bernoulli equation (2.37) is actually a linear equation and is therefore readily solvable as such. However, in the general case in which $n \neq 0$ or 1, this simple situation does not hold and we must proceed in a different manner. We now state and prove Theorem 2.5, which gives a method of solution in the general case.

THEOREM 2.5

Suppose $n \neq 0$ or 1. Then the transformation $v = y^{1-n}$ reduces the Bernoulli equation

$$\frac{dy}{dx} + P(x)y = Q(x)y^n \tag{2.37}$$

to a linear equation in v.

Proof. We first multiply Equation (2.37) by y^{-n}, thereby expressing it in the equivalent form

$$y^{-n}\frac{dy}{dx} + P(x)y^{1-n} = Q(x). \tag{2.38}$$

If we let $v = y^{1-n}$, then

$$\text{directly} \longrightarrow \quad \frac{dv}{dx} = (1-n)y^{-n}\frac{dy}{dx}$$

and Equation (2.38) transforms into

$$\frac{1}{1-n}\frac{dv}{dx} + P(x)v = Q(x)$$

or, equivalently,

$$\frac{dv}{dx} + (1-n)P(x)v = (1-n)Q(x).$$

Letting

$$P_1(x) = (1-n)P(x)$$

and

$$Q_1(x) = (1 - n)Q(x),$$

this may be written

$$\frac{dv}{dx} + P_1(x)v = Q_1(x).$$

which is linear in v.

Q.E.D.

▶ **Example 2.17**

$$\frac{dy}{dx} + y = xy^3. \tag{2.39}$$

This is a Bernoulli differential equation, where $n = 3$. We first multiply the equation through by y^{-3}, thereby expressing it in the equivalent form

$$y^{-3}\frac{dy}{dx} + y^{-2} = x.$$

If we let $v = y^{1-n} = y^{-2}$, then $dv/dx = -2y^{-3}(dy/dx)$ and the preceding differential equation transforms into the linear equation

$$-\frac{1}{2}\frac{dv}{dx} + v = x.$$

Writing this linear equation in the standard form

$$\frac{dv}{dx} - 2v = -2x, \tag{2.40}$$

we see that an integrating factor for this equation is

$$e^{\int P(x)\,dx} = e^{-\int 2\,dx} = e^{-2x}.$$

Multiplying (2.40) by e^{-2x}, we find

$$e^{-2x}\frac{dv}{dx} - 2e^{-2x}v = -2xe^{-2x}$$

or

$$\frac{d}{dx}(e^{-2x}v) = -2xe^{-2x}.$$

Integrating, we find

$$e^{-2x}v = \tfrac{1}{2}e^{-2x}(2x + 1) + c,$$
$$v = x + \tfrac{1}{2} + ce^{2x},$$

where c is an arbitrary constant. But

$$v = \frac{1}{y^2}.$$

Thus we obtain the solutions of (2.39) in the form

$$\frac{1}{y^2} = x + \tfrac{1}{2} + ce^{2x}.$$

Note. Consider the equation

$$\frac{df(y)dy}{dy\,dx} + P(x)f(y) = Q(x), \tag{2.41}$$

where f is a known function of y. Letting $v = f(y)$, we have

$$\frac{dv}{dx} = \frac{dv}{dy}\frac{dy}{dx} = \frac{df(y)}{dy}\frac{dy}{dx},$$

and Equation (2.41) becomes

$$\frac{dv}{dx} + P(x)v = Q(x),$$

which is linear in v. We now observe that the Bernoulli differential equation (2.37) is a special case of Equation (2.41). Writing (2.37) in the form

$$y^{-n}\frac{dy}{dx} + P(x)y^{1-n} = Q(x)$$

and then multiplying through by $(1 - n)$, we have

$$(1 - n)y^{-n}\frac{dy}{dx} + P_1(x)y^{1-n} = Q_1(x),$$

where $P_1(x) = (1 - n)P(x)$ and $Q_1(x) = (1 - n)Q(x)$. This is of the form (2.41), where $f(y) = y^{1-n}$; letting $v = y^{1-n}$, it becomes

$$\frac{dv}{dx} + P_1(x)v = Q_1(x),$$

which is linear in v. For other special cases of (2.41), see Exercise 37.

Exercises

Solve the given differential equations in Exercises 1–18.

1. $\dfrac{dy}{dx} + \dfrac{3y}{x} = 6x^2.$

2. $x^4\dfrac{dy}{dx} + 2x^3y = 1.$

3. $\dfrac{dy}{dx} + 3y = 3x^2e^{-3x}.$

4. $\dfrac{dy}{dx} + 4xy = 8x.$

5. $\dfrac{dx}{dt} + \dfrac{x}{t^2} = \dfrac{1}{t^2}.$

6. $(u^2 + 1)\dfrac{dv}{du} + 4uv = 3u.$

7. $x\dfrac{dy}{dx} + \dfrac{2x + 1}{x + 1}y = x - 1.$

8. $(x^2 + x - 2)\dfrac{dy}{dx} + 3(x + 1)y = x - 1.$

9. $x\,dy + (xy + y - 1)\,dx = 0.$

10. $y\,dx + (xy^2 + x - y)\,dy = 0.$

11. $\dfrac{dr}{d\theta} + r\tan\theta = \cos\theta.$

12. $\cos\theta\,dr + (r\sin\theta - \cos^4\theta)\,d\theta = 0.$

13. $(\cos^2 x - y\cos x)\,dx - (1 + \sin x)\,dy = 0.$

14. $(y\sin 2x - \cos x)\,dx + (1 + \sin^2 x)\,dy = 0.$

15. $\dfrac{dy}{dx} - \dfrac{y}{x} = -\dfrac{y^2}{x}.$

16. $x\dfrac{dy}{dx} + y = -2x^6 y^4.$

17. $dy + (4y - 8y^{-3})x\,dx = 0.$

18. $\dfrac{dx}{dt} + \dfrac{t + 1}{2t}x = \dfrac{t + 1}{xt}.$

Solve the initial-value problems in Exercises 19–30.

19. $x\dfrac{dy}{dx} - 2y = 2x^4, \qquad y(2) = 8.$

20. $\dfrac{dy}{dx} + 3x^2 y = x^2, \qquad y(0) = 2.$

21. $e^x[y - 3(e^x + 1)^2]\,dx + (e^x + 1)\,dy = 0, \qquad y(0) = 4.$

22. $2x(y + 1)\,dx - (x^2 + 1)\,dy = 0, \qquad y(1) = -5.$

23. $\dfrac{dr}{d\theta} + r\tan\theta = \cos^2\theta, \qquad r\left(\dfrac{\pi}{4}\right) = 1.$

24. $\dfrac{dx}{dt} - x = \sin 2t, \qquad x(0) = 0.$

25. $\dfrac{dy}{dx} + \dfrac{y}{2x} = \dfrac{x}{y^3}, \qquad y(1) = 2.$

26. $x\dfrac{dy}{dx} + y = (xy)^{3/2}, \qquad y(1) = 4.$

27. $\dfrac{dy}{dx} + y = f(x), \quad$ where $\quad f(x) = \begin{cases} 2, & 0 \le x < 1, \\ 0, & x \ge 1, \end{cases} \qquad y(0) = 0.$

28. $\dfrac{dy}{dx} + y = f(x), \quad$ where $\quad f(x) = \begin{cases} 5, & 0 \le x < 10, \\ 1, & x \ge 10, \end{cases} \qquad y(0) = 6.$

29. $\dfrac{dy}{dx} + y = f(x), \quad$ where $\quad f(x) = \begin{cases} e^{-x}, & 0 \le x < 2, \\ e^{-2}, & x \ge 2, \end{cases} \qquad y(0) = 1.$

30. $(x + 2)\dfrac{dy}{dx} + y = f(x), \quad$ where $\quad f(x) = \begin{cases} 2x, & 0 \le x < 2, \\ 4, & x \ge 2, \end{cases} \qquad y(0) = 4.$

31. Consider the equation $a(dy/dx) + by = ke^{-\lambda x}$, where a, b, and k are positive constants and λ is a nonnegative constant.

    (a) Solve this equation.

    (b) Show that if $\lambda = 0$ every solution approaches k/b as $x \to \infty$, but if $\lambda > 0$ every solution approaches 0 as $x \to \infty$.

32. Consider the differential equation

$$\frac{dy}{dx} + P(x)y = 0.$$

    (a) Show that if f and g are two solutions of this equation and c_1 and c_2 are arbitrary constants, then $c_1 f + c_2 g$ is also a solution of this equation.

    (b) Extending the result of (a), show that if $f_1, f_2, \ldots, f_n$ are n solutions of this equation and $c_1, c_2, \ldots, c_n$ are n arbitrary constants, then

$$\sum_{k=1}^{n} c_k f_k$$

    is also a solution of this equation..

33. Consider the differential equation

$$\frac{dy}{dx} + P(x)y = 0, \qquad\qquad \text{(A)}$$

    where P is continuous on a real interval I.

    (a) Show that the function f such that $f(x) = 0$ for all $x \in I$ is a solution of this equation.

    (b) Show that if f is a solution of (A) such that $f(x_0) = 0$ for some $x_0 \in I$, then $f(x) = 0$ for all $x \in I$.

    (c) Show that if f and g are two solutions of (A) such that $f(x_0) = g(x_0)$ for some $x_0 \in I$, then $f(x) = g(x)$ for all $x \in I$.

34. (a) Prove that if f and g are two different solutions of

$$\frac{dy}{dx} + P(x)y = Q(x), \qquad\qquad \text{(A)}$$

    then $f - g$ is a solution of the equation

$$\frac{dy}{dx} + P(x)y = 0.$$

    (b) Thus show that if f and g are two different solutions of Equation (A) and c is an arbitrary constant, then

$$c(f - g) + f$$

    is a one-parameter family of solutions of (A).

35. (a) Let f_1 be a solution of

$$\frac{dy}{dx} + P(x)y = Q_1(x)$$

and f_2 be a solution of

$$\frac{dy}{dx} + P(x)y = Q_2(x),$$

where P, Q_1, and Q_2 are all defined on the same real interval I. Prove that $f_1 + f_2$ is a solution of

$$\frac{dy}{dx} + P(x)y = Q_1(x) + Q_2(x)$$

on I.

(b) Use the result of (a) to solve the equation

$$\frac{dy}{dx} + y = 2 \sin x + 5 \sin 2x.$$

36. (a) Extend the result of Exercise 35(a) to cover the case of the equation

$$\frac{dy}{dx} + P(x)y = \sum_{k=1}^{n} Q_k(x),$$

where P, $Q_k(k = 1, 2, \ldots, n)$ are all defined on the same real interval I.

(b) Use the result obtained in (a) to solve the equation

$$\frac{dy}{dx} + y = \sum_{k=1}^{5} \sin kx.$$

37. Solve each of the following equations of the form (2.41):

(a) $\cos y \dfrac{dy}{dx} + \dfrac{1}{x} \sin y = 1.$

(b) $(y + 1) \dfrac{dy}{dx} + x(y^2 + 2y) = x.$

38. The equation

$$\frac{dy}{dx} = A(x)y^2 + B(x)y + C(x) \tag{A}$$

is called *Riccati's equation.*

(a) Show that if $A(x) = 0$ for all x, then Equation (A) is a linear equation, whereas if $C(x) = 0$ for all x, then Equation (A) is a Bernoulli equation.

(b) Show that if f is any solution of Equation (A), then the transformation

$$y = f + \frac{1}{v}$$

reduces (A) to a linear equation in v.

In each of Exercises 39–41, use the result of Exercise 38(b) and the given solution to find a one-parameter family of solutions of the given Riccati equation:

39. $\dfrac{dy}{dx} = (1 - x)y^2 + (2x - 1)y - x$; given solution $f(x) = 1.$

40. $\dfrac{dy}{dx} = -y^2 + xy + 1$; given solution $f(x) = x$.

41. $\dfrac{dy}{dx} = -8xy^2 + 4x(4x + 1)y - (8x^3 + 4x^2 - 1)$; given solution $f(x) = x$.

Exercises: Miscellaneous Review

Solve each of the differential equations in Exercises 1–14. Several can be solved by at least two different methods.

1. $6x^2y\, dx - (x^3 + 1)\, dy = 0.$

2. $(3x^2y^2 - x)\, dy + (2xy^3 - y)\, dx = 0.$

3. $(y - 1)\, dx + x(x + 1)\, dy = 0.$

4. $(x^2 - 2y)\, dx - x\, dy = 0.$

5. $(3x - 5y)\, dx + (x + y)\, dy = 0.$

6. $e^{2x}y^2\, dx + (e^{2x}y - 2y)\, dy = 0.$

7. $(8x^3y - 12x^3)\, dx + (x^4 + 1)\, dy = 0.$

8. $(2x^2 + xy + y^2)\, dx + 2x^2\, dy = 0.$

9. $\dfrac{dy}{dx} = \dfrac{4x^3y^2 - 3x^2y}{x^3 - 2x^4y}.$

10. $(x + 1)\dfrac{dy}{dx} + xy = e^{-x}.$

11. $\dfrac{dy}{dx} = \dfrac{2x - 7y}{3y - 8x}.$

12. $x^2\dfrac{dy}{dx} + xy = xy^3.$

13. $(x^3 + 1)\dfrac{dy}{dx} + 6x^2y = 6x^2.$

14. $\dfrac{dy}{dx} = \dfrac{2x^2 + y^2}{2xy - x^2}.$

Solve the initial value problems in Exercises 15–24.

15. $(x^2 + y^2)\, dx - 2xy\, dy = 0,$ $\qquad y(1) = 2.$

16. $2(y^2 + 4)\, dx + (1 - x^2)y\, dy = 0,$ $\qquad y(3) = 0.$

17. $(e^{2x}y^2 - 2x)\, dx + e^{2x}y\, dy = 0,$ $\qquad y(0) = 2.$

18. $(3x^2 + 2xy^2)\, dx + (2x^2y + 6y^2)\, dy = 0,$ $\qquad y(1) = 2.$

19. $4xy\dfrac{dy}{dx} = y^2 + 1,$ $\qquad y(2) = 1.$

20. $\dfrac{dy}{dx} = \dfrac{2x + 7y}{2x - 2y}$, $y(1) = 2$.

21. $\dfrac{dy}{dx} = \dfrac{xy}{x^2 + 1}$, $y(\sqrt{15}) = 2$.

22. $\dfrac{dy}{dx} + y = f(x)$,   where $f(x) = \begin{cases} 1, & 0 \le x < 2, \\ 0, & x > 2, \end{cases}$ $y(0) = 0$.

23. $(x + 2)\dfrac{dy}{dx} + y = f(x)$,   where $f(x) = \begin{cases} 2x, & 0 \le x \le 2, \\ 4, & x > 2, \end{cases}$ $y(0) = 4$.

24. $x^2 \dfrac{dy}{dx} + xy = \dfrac{y^3}{x}$, $y(1) = 1$.

2.4 SPECIAL INTEGRATING FACTORS AND TRANSFORMATIONS

We have thus far encountered five distinct types of first-order equations for which solutions may be obtained by exact methods, namely, exact, separable, homogeneous, linear, and Bernoulli equations. In the case of exact equations, we follow a definite procedure to directly obtain solutions. For the other four types definite procedures for solution are also available, but in these cases the procedures are actually not quite so direct. In the cases of both separable and linear equations we actually multiply by appropriate integrating factors that reduce the given equations to equations that are of the more basic exact type. For both homogeneous and Bernoulli equations we make appropriate transformations that reduce such equations to equations that are of the more basic separable and linear types, respectively.

This suggests two general plans of attack to be used in solving a differential equation that is *not* of one of the five types mentioned. Either (1) we might multiply the given equation by an appropriate integrating factor and directly reduce it to an exact equation, or (2) we might make an appropriate transformation that will reduce the given equation to an equation of some more basic type (say, one of the five types already studied). Unfortunately no general directions can be given for finding an appropriate integrating factor or transformation in all cases. However, there is a variety of special types of equations that either possess special types of integrating factors or to which special transformations may be applied. We shall consider a few of these in this section. Since these types are relatively unimportant, in most cases we shall simply state the relevant theorem and leave the proof to the exercises.

A. Finding Integrating Factors

The so-called separable equations considered in Section 2.2 always possess integrating factors that may be determined by immediate inspection. While it is true that some nonseparable equations also possess integrating factors that may be determined "by inspection," such equations are rarely encountered except in differential equations texts on pages devoted to an exposition of this dubious "method." Even then a considerable amount of knowledge and skill are often required.

Let us attempt to attack the problem more systematically. Suppose the equation

$$M(x, y)\, dx + N(x, y)\, dy = 0 \tag{2.42}$$

is *not* exact and that $\mu(x, y)$ is an integrating factor of it. Then the equation

$$\mu(x, y)M(x, y)\, dx + \mu(x, y)N(x, y)\, dy = 0 \tag{2.43}$$

is exact. Now using the criterion (2.7) for exactness, Equation (2.43) is exact if and only if

$$\frac{\partial}{\partial y}\left[\mu(x, y)M(x, y)\right] = \frac{\partial}{\partial x}\left[\mu(x, y)N(x, y)\right].$$

This condition reduces to

$$N(x, y)\frac{\partial \mu(x, y)}{\partial x} - M(x, y)\frac{\partial \mu(x, y)}{\partial y} = \left[\frac{\partial M(x, y)}{\partial y} - \frac{\partial N(x, y)}{\partial x}\right]\mu(x, y).$$

Here M and N are known functions of x and y, but μ is an unknown function of x and y that we are trying to determine. Thus we write the preceding condition in the form

$$N(x, y)\frac{\partial \mu}{\partial x} - M(x, y)\frac{\partial \mu}{\partial y} = \left[\frac{\partial M(x, y)}{\partial y} - \frac{\partial N(x, y)}{\partial x}\right]\mu. \tag{2.44}$$

Hence μ is an integrating factor of the differential equation (2.42) if and only if it is a solution of the differential equation (2.44). Equation (2.44) is a partial differential equation for the general integrating factor μ, and we are in no position to attempt to solve such an equation. Let us instead attempt to determine integrating factors of certain special types. But what special types might we consider? Let us recall that the linear differential equation

$$\frac{dy}{dx} + P(x)y = Q(x)$$

always possesses the integrating factor $e^{\int P(x)\,dx}$, which depends only upon x. Perhaps other equations also have integrating factors that depend only upon x. We therefore multiply Equation (2.42) by $\mu(x)$, where μ depends upon x alone. We obtain

$$\mu(x)M(x, y)\, dx + \mu(x)N(x, y)\, dy = 0.$$

This is exact if and only if

$$\frac{\partial}{\partial y}\left[\mu(x)M(x, y)\right] = \frac{\partial}{\partial x}\left[\mu(x)N(x, y)\right].$$

Now M and N are known functions of both x and y, but here the intergrating factor μ depends only upon x. Thus the above condition reduces to

$$\mu(x)\frac{\partial M(x, y)}{\partial y} = \mu(x)\frac{\partial N(x, y)}{\partial x} + N(x, y)\frac{d\mu(x)}{dx}$$

or

$$\frac{d\mu(x)}{\mu(x)} = \frac{1}{N(x, y)}\left[\frac{\partial M(x, y)}{\partial y} - \frac{\partial N(x, y)}{\partial x}\right] dx. \tag{2.45}$$

If

$$\frac{1}{N(x, y)} \left[\frac{\partial M(x, y)}{\partial y} - \frac{\partial N(x, y)}{\partial x} \right]$$

involves the variable y, this equation then involves two dependent variables and we again have difficulties. However, if

$$\frac{1}{N(x, y)} \left[\frac{\partial M(x, y)}{\partial y} - \frac{\partial N(x, y)}{\partial x} \right]$$

depends upon x only, Equation (2.45) is a separated ordinary equation in the single independent variable x and the single dependent variable μ. In this case we may integrate to obtain the integrating factor

$$\mu(x) = \exp\left\{ \int \frac{1}{N(x, y)} \left[\frac{\partial M(x, y)}{\partial y} - \frac{\partial N(x, y)}{\partial x} \right] dx \right\}.$$

In like manner, if

$$\frac{1}{M(x, y)} \left[\frac{\partial N(x, y)}{\partial x} - \frac{\partial M(x, y)}{\partial y} \right]$$

depends upon y only, then we may obtain an integrating factor that depends only on y. We summarize these observations in the following theorem.

THEOREM 2.6

Consider the differential equation

$$M(x, y) \, dx + N(x, y) \, dy = 0. \tag{2.42}$$

If

$$\frac{1}{N(x, y)} \left[\frac{\partial M(x, y)}{\partial y} - \frac{\partial N(x, y)}{\partial x} \right] \tag{2.46}$$

depends upon x only, then

$$\exp\left\{ \int \frac{1}{N(x, y)} \left[\frac{\partial M(x, y)}{\partial y} - \frac{\partial N(x, y)}{\partial x} \right] dx \right\} \tag{2.47}$$

is an integrating factor of Equation (2.42). If

$$\frac{1}{M(x, y)} \left[\frac{\partial N(x, y)}{\partial x} - \frac{\partial M(x, y)}{\partial y} \right] \tag{2.48}$$

depends upon y only, then

$$\exp\left\{ \int \frac{1}{M(x, y)} \left[\frac{\partial N(x, y)}{\partial x} - \frac{\partial M(x, y)}{\partial y} \right] dy \right\} \tag{2.49}$$

is an integrating factor of Equation (2.42).

We emphasize that, given a differential equation, we have no assurance in general that either of these procedures will apply. It may well turn out that (2.46) involves y

and (2.48) involves x for the differential equation under consideration. Then we must seek other procedures. However, since the calculation of the expressions (2.46) and (2.48) is generally quite simple, it is often worthwhile to calculate them before trying something more complicated.

▶ **Example 2.18**

Consider the differential equation

$$(2x^2 + y)\,dx + (x^2 y - x)\,dy = 0. \tag{2.50}$$

Let us first observe that this equation is *not* exact, separable, homogeneous, linear, or Bernoulli. Let us then see if Theorem 2.6 applies. Here $M(x, y) = 2x^2 + y$, and $N(x, y) = x^2 y - x$, and the expression (2.46) becomes

$$\frac{1}{x^2 y - x}\,[1 - (2xy - 1)] = \frac{2(1 - xy)}{x(xy - 1)} = -\frac{2}{x}.$$

This depends upon x only, and so

$$\exp\left(-\int \frac{2}{x}\,dx\right) = \exp(-2 \ln|x|) = \frac{1}{x^2}$$

is an integrating factor of Equation (2.50). Multiplying (2.50) by this integrating factor, we obtain the equation

$$\left(2 + \frac{y}{x^2}\right) dx + \left(y - \frac{1}{x}\right) dy = 0. \tag{2.51}$$

The student may readily verify that Equation (2.51) is indeed exact and that the solution is

$$2x + \frac{y^2}{2} - \frac{y}{x} = c.$$

More and more specialized results concerning particular types of integrating factors corresponding to particular types of equations are known. However, instead of going into such special cases we shall now proceed to investigate certain useful transformations.

B. A Special Transformation

We have already made use of transformations in reducing both homogeneous and Bernoulli equations to more tractable types. Another type of equation that can be reduced to a more basic type by means of a suitable transformation is an equation of the form

$$(a_1 x + b_1 y + c_1)\,dx + (a_2 x + b_2 y + c_2)\,dy = 0.$$

We state the following theorem concerning this equation.

THEOREM 2.7

Consider the equation

$$(a_1 x + b_1 y + c_1)\, dx + (a_2 x + b_2 y + c_2)\, dy = 0, \tag{2.52}$$

where $a_1, b_1, c_1, a_2, b_2,$ *and* c_2 *are constants.*

    Case 1. *If* $a_2/a_1 \neq b_2/b_1$, *then the transformation*

$$x = X + h,$$
$$y = Y + k,$$

where (h, k) *is the solution of the system*

$$a_1 h + b_1 k + c_1 = 0,$$
$$a_2 h + b_2 k + c_2 = 0,$$

reduces Equation (2.52) to the homogeneous equation

$$(a_1 X + b_1 Y)\, dX + (a_2 X + b_2 Y)\, dY = 0$$

in the variables X and Y.

    Case 2. *If* $a_2/a_1 = b_2/b_1 = k$, *then the transformation* $z = a_1 x + b_1 y$ *reduces the equation (2.52) to a separable equation in the variables x and z.*

Examples 2.19 and 2.20 illustrate the two cases of this theorem.

▶ **Example 2.19**

$$(x - 2y + 1)\, dx + (4x - 3y - 6)\, dy = 0. \tag{2.53}$$

Here $a_1 = 1$, $b_1 = -2$, $a_2 = 4$, $b_2 = -3$, and so

$$\frac{a_2}{a_1} = 4 \quad \text{but} \quad \frac{b_2}{b_1} = \frac{3}{2} \neq \frac{a_2}{a_1}.$$

Therefore this is Case 1 of Theorem 2.7. We make the transformation

$$x = X + h,$$
$$y = Y + k,$$

where (h, k) is the solution of the system

$$h - 2k + 1 = 0,$$
$$4h - 3k - 6 = 0.$$

The solution of this system is $h = 3$, $k = 2$, and so the transformation is

$$x = X + 3,$$
$$y = Y + 2.$$

This reduces Equation (2.53) to the homogeneous equation

$$(X - 2Y)\,dX + (4X - 3Y)\,dY = 0. \tag{2.54}$$

Now following the procedure in Section 2.2 we first put this homogeneous equation in the form

$$\frac{dY}{dX} = \frac{1 - 2(Y/X)}{3(Y/X) - 4}$$

and let $Y = vX$ to obtain

$$v + X\frac{dv}{dX} = \frac{1 - 2v}{3v - 4}.$$

This reduces to

$$\frac{(3v - 4)\,dv}{3v^2 - 2v - 1} = -\frac{dX}{X}. \tag{2.55}$$

Integrating (we recommend the use of tables here), we obtain

$$\tfrac{1}{2}\ln|3v^2 - 2v - 1| - \tfrac{3}{4}\ln\left|\frac{3v - 3}{3v + 1}\right| = -\ln|X| + \ln|c_1|,$$

or

$$\ln(3v^2 - 2v - 1)^2 - \ln\left|\frac{3v - 3}{3v + 1}\right|^3 = \ln\left(\frac{c_1^4}{X^4}\right),$$

or

$$\ln\left|\frac{(3v + 1)^5}{v - 1}\right| = \ln\left(\frac{c_1^4}{X^4}\right),$$

or, finally,

$$X^4|(3v + 1)^5| = c|v - 1|,$$

where $c = c_1^4$. These are the solutions of the separable equation (2.55). Now replacing v by Y/X, we obtain the solutions of the homogeneous equation (2.54) in the form

$$|3Y + X|^5 = c|Y - X|.$$

Finally, replacing X by $x - 3$ and Y by $y - 2$ from the original transformation, we obtain the solutions of the differential equation (2.53) in the form

$$|3(y - 2) + (x - 3)|^5 = c|y - 2 - x + 3|$$

or

$$|x + 3y - 9|^5 = c|y - x + 1|.$$

▶ **Example 2.20**

$$(x + 2y + 3)\,dx + (2x + 4y - 1)\,dy = 0. \tag{2.56}$$

Here $a_1 = 1, b_1 = 2, a_2 = 2, b_2 = 4$, and $a_2/a_1 = b_2/b_1 = 2$. Therefore, this is Case 2 of Theorem 2.7. We therefore let

$$z = x + 2y,$$

and Equation (2.56) transforms into

$$(z + 3)\, dx + (2z - 1)\left(\frac{dz - dx}{2}\right) = 0$$

or

$$7\, dx + (2z - 1)\, dz = 0,$$

which is separable. Integrating, we have

$$7x + z^2 - z = c.$$

Replacing z by $x + 2y$ we obtain the solution of Equation (2.56) in the form

$$7x + (x + 2y)^2 - (x + 2y) = c$$

or

$$x^2 + 4xy + 4y^2 + 6x - 2y = c.$$

C. Other Special Types and Methods; An Important Reference

Many other special types of first-order equations exist for which corresponding special methods of solution are known. We shall not go into such highly specialized types in this book. Instead we refer the reader to the book *Differentialgleichungen: Losungsmethoden und Losungen*, by E. Kamke (Chelsea, New York, 1948). This remarkable volume contains discussions of a large number of special types of equations and their solutions. We strongly suggest that the reader consult this book whenever he encounters an unfamiliar type of equation. Of course one may encounter an equation for which no exact method of solution is known. In such a case one must resort to various methods of approximation. We shall consider some of these general methods in Chapter 8.

Exercises

Solve each differential equation in Exercises 1–4 by first finding an integrating factor.

1. $(5xy + 4y^2 + 1)\, dx + (x^2 + 2xy)\, dy = 0.$

2. $(2x + \tan y)\, dx + (x - x^2 \tan y)\, dy = 0.$

3. $[y^2(x + 1) + y]\, dx + (2xy + 1)\, dy = 0.$

4. $(2xy^2 + y)\, dx + (2y^3 - x)\, dy = 0.$

In each of Exercises 5 and 6 find an integrating factor of the form $x^p y^q$ and solve.

5. $(4xy^2 + 6y)\, dx + (5x^2 y + 8x)\, dy = 0.$

6. $(8x^2 y^3 - 2y^4)\, dx + (5x^3 y^2 - 8xy^3)\, dy = 0.$

Solve each differential equation in Exercises 7–10 by making a suitable transformation.

7. $(5x + 2y + 1) \, dx + (2x + y + 1) \, dy = 0.$

8. $(3x - y + 1) \, dx - (6x - 2y - 3) \, dy = 0.$

9. $(x - 2y - 3) \, dx + (2x + y - 1) \, dy = 0.$

10. $(10x - 4y + 12) \, dx - (x + 5y + 3) \, dy = 0.$

Solve the initial-value problems in Exercises 11–14.

11. $(6x + 4y + 1) \, dx + (4x + 2y + 2) \, dy = 0, \qquad y(\tfrac{1}{2}) = 3.$

12. $(3x - y - 6) \, dx + (x + y + 2) \, dy = 0, \qquad y(2) = -2.$

13. $(2x + 3y + 1) \, dx + (4x + 6y + 1) \, dy = 0, \qquad y(-2) = 2.$

14. $(4x + 3y + 1) \, dx + (x + y + 1) \, dy = 0, \qquad y(3) = -4.$

15. Prove Theorem 2.6.

16. Prove Theorem 2.7.

17. Show that if $\mu(x, y)$ and $v(x, y)$ are integrating factors of
$$M(x, y) \, dx + N(x, y) \, dy = 0 \qquad \text{(A)}$$
such that $\mu(x, y)/v(x, y)$ is not constant, then
$$\mu(x, y) = cv(x, y)$$
is a solution of Equation (A) for every constant c.

18. Show that if the equation
$$M(x, y) \, dx + N(x, y) \, dy = 0 \qquad \text{(A)}$$
is homogeneous and $M(x, y)x + N(x, y)y \neq 0$, then $1/[M(x, y)x + N(x, y)y]$ is an integrating factor of (A).

19. Show that if the equation $M(x, y) \, dx + N(x, y) \, dy = 0$ is both homogeneous and exact and if $M(x, y)x + N(x, y)y$ is not a constant, then the solution of this equation is $M(x, y)x + N(x, y)y = c$, where c is an arbitrary constant.

20. An equation that is of the form
$$y = px + f(p), \qquad \text{(A)}$$
where $p \equiv dy/dx$ and f is a given function, is called a *Clairaut equation*. Given such an equation, proceed as follows:

   1. Differentiate (A) with respect to x and simplify to obtain
$$[x + f'(p)]\frac{dp}{dx} = 0. \qquad \text{(B)}$$

Observe that (B) is a first-order differential equation in x and p.

   2. Assume $x + f'(p) \neq 0$, divide through by this factor, and solve the resulting equation to obtain
$$p = c, \qquad \text{(C)}$$

where c is an arbitrary constant.

3. Eliminate p between (A) and (C) to obtain

$$y = cx + f(c). \tag{D}$$

Note that (D) is a one-parameter family of solutions of (A) and compare the *form* of differential equation (A) with the *form* of the family of solutions (D).

4. *Remark.* Assuming $x + f'(p) = 0$ and then eliminating p between (A) and $x + f'(p) = 0$ may lead to an "extra" solution that is *not* a member of the one-parameter family of solutions of the form (D). Such an extra solution is usually called a *singular solution*. For a specific example, see Exercise 21.

21. Consider the *Clairaut equation*

$$y = px + p^2, \quad \text{where} \quad p \equiv \frac{dy}{dx}.$$

   (a) Find a one-parameter family of solutions of this equation.

   (b) Proceed as in the Remark of Exercise 20 and find an "extra" solution that is not a member of the one-parameter family found in part (a).

   (c) Graph the integral curves corresponding to several members of the one-parameter family of part (a); graph the integral curve corresponding to the "extra" solution of part (b); and describe the geometric relationship between the graphs of the members of the one-parameter family and the graph of the "extra" solution.

CHAPTER THREE

Applications of First-Order Equations

In Chapter 1 we pointed out that differential equations originate from the mathematical formulation of a great variety of problems in science and engineering. In this chapter we consider problems that give rise to some of the types of first-order ordinary differential equations studied in Chapter 2. First, we formulate the problem mathematically, thereby obtaining a differential equation. Then we solve the equation and attempt to interpret the solution in terms of the quantities involved in the original problem.

3.1 ORTHOGONAL AND OBLIQUE TRAJECTORIES

A. Orthogonal Trajectories

DEFINITION

Let

$$F(x, y, c) = 0 \qquad (3.1)$$

be a given one-parameter family of curves in the xy plane. A curve that intersects the curves of the family (3.1) at right angles is called an orthogonal trajectory *of the given family.*

▶ **Example 3.1**

Consider the family of circles

$$x^2 + y^2 = c^2 \qquad (3.2)$$

70

with center at the origin and radius c. Each straight line through the origin,

$$y = kx, \qquad (3.3)$$

is an orthogonal trajectory of the family of circles (3.2). Conversely, each circle of the family (3.2) is an orthogonal trajectory of the family of straight lines (3.3). The families (3.2) and (3.3) are orthogonal trajectories of each other. In Figure 3.1 several members of the family of circles (3.2), drawn solidly, and several members of the family of straight lines (3.3), drawn with dashes, are shown.

The problem of finding the orthogonal trajectories of a given family of curves arises in many physical situations. For example, in a two-dimensional electric field the lines of force (flux lines) and the equipotential curves are orthogonal trajectories of each other.

We now proceed to find the orthogonal trajectories of a family of curves

$$F(x, y, c) = 0. \qquad (3.1)$$

We obtain the differential equation of the family (3.1) by first differentiating Equation (3.1) implicitly with respect to x and then eliminating the parameter c between the derived equation so obtained and the given equation (3.1) itself. We assume that the resulting differential equation of the family (3.1) can be expressed in the form

$$\frac{dy}{dx} = f(x, y). \qquad (3.4)$$

Thus the curve C of the given family (3.1) which passes through the point (x, y) has the slope $f(x, y)$ there. Since an orthogonal trajectory of the given family intersects each curve of the family at right angles, the slope of the orthogonal trajectory to C at (x, y) is

$$-\frac{1}{f(x, y)}.$$

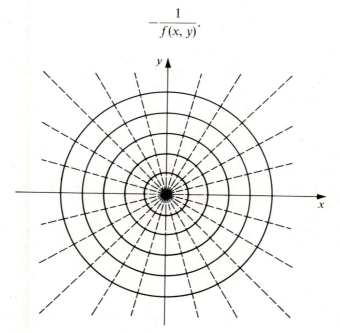

Figure 3.1

Thus the differential equation of the family of orthogonal trajectories is

$$\frac{dy}{dx} = -\frac{1}{f(x, y)}. \tag{3.5}$$

A one-parameter family

$$G(x, y, c) = 0$$

or

$$y = F(x, c)$$

of solutions of the differential equation (3.5) represents the family of orthogonal trajectories of the original family (3.1), except possibly for certain trajectories that are vertical lines.

We summarize this procedure as follows:

Procedure for Finding the Orthogonal Trajectories of a Given Family of Curves

Step 1. From the equation

$$F(x, y, c) = 0 \tag{3.1}$$

of the given family of curves, find the differential equation

$$\frac{dy}{dx} = f(x, y) \tag{3.4}$$

of this family.

Step 2. In the differential equation $dy/dx = f(x, y)$ so found in Step 1, replace $f(x, y)$ by its negative reciprocal $-1/f(x, y)$. This gives the differential equation

$$\frac{dy}{dx} = -\frac{1}{f(x, y)} \tag{3.5}$$

of the orthogonal trajectories.

Step 3. Obtain a one-parameter family

$$G(x, y, c) = 0 \quad \text{or} \quad y = F(x, c)$$

of solutions of the differential equation (3.5), thus obtaining the desired family of orthogonal trajectories (except possibly for certain trajectories that are vertical lines and must be determined separately).

Caution. In Step 1, in finding the differential equation (3.4) of the given family, be sure to eliminate the parameter c during the process.

▶ Example 3.2

In Example 3.1 we stated that the set of orthogonal trajectories of the family of circles

$$x^2 + y^2 = c^2 \tag{3.2}$$

is the family of straight lines

$$y = kx. \tag{3.3}$$

Let us verify this using the procedure outlined above.

Step 1. Differentiating the equation

$$x^2 + y^2 = c^2 \tag{3.2}$$

of the given family, we obtain

$$x + y\frac{dy}{dx} = 0.$$

From this we obtain the differential equation

$$\frac{dy}{dx} = -\frac{x}{y} \tag{3.6}$$

of the given family (3.2). (Note that the parameter c was automatically eliminated in this case.)

Step 2. We replace $-x/y$ by its negative reciprocal y/x in the differential equation (3.6) to obtain the differential equation

$$\frac{dy}{dx} = \frac{y}{x} \tag{3.7}$$

of the orthogonal trajectories.

Step 3. We now solve the differential equation (3.7). Separating variables, we have

$$\frac{dy}{y} = \frac{dx}{x};$$

integrating, we obtain

$$y = kx. \tag{3.3}$$

This is a one-parameter family of solutions of the differential equation (3.7) and thus represents the family of orthogonal trajectories of the given family of circles (3.2) (except for the single trajectory that is the vertical line $x = 0$ and this may be determined by inspection).

▶ **Example 3.3**

Find the orthogonal trajectories of the family of parabolas $y = cx^2$.

Step 1. We first find the differential equation of the given family

$$y = cx^2. \tag{3.8}$$

Differentiating, we obtain

$$\frac{dy}{dx} = 2cx. \tag{3.9}$$

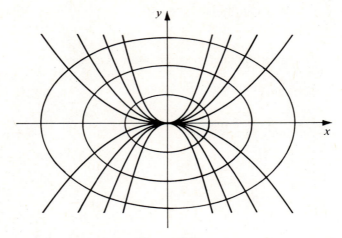

Figure 3.2

Eliminating the parameter c between Equations (3.8) and (3.9), we obtain the differential equation of the family (3.8) in the form

$$\frac{dy}{dx} = \frac{2y}{x}. \tag{3.10}$$

Step 2. We now find the differential equation of the orthogonal trajectories by replacing $2y/x$ in (3.10) by its negative reciprocal, obtaining

$$\frac{dy}{dx} = -\frac{x}{2y}. \tag{3.11}$$

Step 3. We now solve the differential equation (3.11). Separating variables, we have

$$2y\, dy = -x\, dx.$$

Integrating, we obtain the one-parameter family of solutions of (3.11) in the form

$$x^2 + 2y^2 = k^2,$$

where k is an arbitrary constant. This is the family of orthogonal trajectories of (3.8); it is clearly a family of ellipses with centers at the origin and major axes along the x axis. Some members of the original family of parabolas and some of the orthogonal trajectories (the ellipses) are shown in Figure 3.2.

B. Oblique Trajectories

DEFINITION

Let

$$F(x, y, c) = 0 \tag{3.12}$$

be a one-parameter family of curves. A curve that intersects the curves of the family (3.12) at a constant angle $\alpha \neq 90°$ is called an oblique trajectory *of the given family.*

Suppose the differential equation of a family is

$$\frac{dy}{dx} = f(x, y). \tag{3.13}$$

Then the curve of the family (3.13) through the point (x, y) has slope $f(x, y)$ at (x, y) and hence its tangent line has angle of inclination $\tan^{-1}[f(x, y)]$ there. The tangent line of an oblique trajectory that intersects this curve at the angle α will thus have angle of inclination

$$\tan^{-1}[f(x, y)] + \alpha$$

at the point (x, y). Hence the slope of this oblique trajectory is given by

$$\tan\left\{\tan^{-1}[f(x, y)] + \alpha\right\} = \frac{f(x, y) + \tan \alpha}{1 - f(x, y)\tan \alpha}.$$

Thus the differential equation of such a family of oblique trajectories is given by

$$\frac{dy}{dx} = \frac{f(x, y) + \tan \alpha}{1 - f(x, y)\tan \alpha}.$$

Thus to obtain a family of oblique trajectories intersecting a given family of curves at the constant angle $\alpha \neq 90°$, we may follow the three steps in the above procedure (page 72) for finding the orthogonal trajectories, except that we replace Step 2 by the following step:

Step 2′. In the differential equation $dy/dx = f(x, y)$ of the given family, replace $f(x, y)$ by the expression

$$\frac{f(x, y) + \tan \alpha}{1 - f(x, y)\tan \alpha}. \tag{3.14}$$

▶ **Example 3.4**

Find a family of oblique trajectories that intersect the family of straight lines $y = cx$ at angle $45°$.

Step 1. From $y = cx$, we find $dy/dx = c$. Eliminating c, we obtain the differential equation

$$\frac{dy}{dx} = \frac{y}{x} \tag{3.15}$$

of the given family of straight lines.

Step 2′. We replace $f(x, y) = y/x$ in Equation (3.15) by

$$\frac{f(x, y) + \tan \alpha}{1 - f(x, y)\tan \alpha} = \frac{y/x + 1}{1 - y/x} = \frac{x + y}{x - y}$$

($\tan \alpha = \tan 45° = 1$ here). Thus the differential equation of the desired oblique trajectories is

$$\frac{dy}{dx} = \frac{x + y}{x - y}.$$ (3.16)

Step 3. We now solve the differential equation (3.16). Observing that it is a homogeneous differential equation, we let $y = vx$ to obtain

$$v + x\frac{dv}{dx} = \frac{1 + v}{1 - v}.$$

After simplifications this becomes

$$\frac{(v - 1)\,dv}{v^2 + 1} = -\frac{dx}{x}.$$

Integrating we obtain

$$\tfrac{1}{2}\ln(v^2 + 1) - \arctan v = -\ln|x| - \ln|c|$$

or

$$\ln c^2 x^2 (v^2 + 1) - 2\arctan v = 0.$$

Replacing v by y/x, we obtain the family of oblique trajectories in the form

$$\ln c^2(x^2 + y^2) - 2\arctan\frac{y}{x} = 0.$$

Exercises

In Exercises 1–9 find the orthogonal trajectories of each given family of curves. In each case sketch several members of the family and several of the orthogonal trajectories on the same set of axes.

1. $y = cx^3$.

2. $y^2 = cx$.

3. $cx^2 + y^2 = 1$.

4. $y = e^{cx}$.

5. $y = x - 1 + ce^{-x}$.

6. $x - y = cx^2$.

7. $x^2 + y^2 = cx^3$.

8. $x^2 = 2y - 1 + ce^{-2y}$.

9. $x = \dfrac{y^2}{4} + \dfrac{c}{y^2}$.

10. Find the orthogonal trajectories of the family of ellipses having center at the origin, a focus at the point $(c, 0)$, and semimajor axis of length $2c$.

11. Find the orthogonal trajectories of the family of circles which are tangent to the y axis at the origin.

12. Find the value of K such that the parabolas $y = c_1 x^2 + K$ are the orthogonal trajectories of the family of ellipses $x^2 + 2y^2 - y = c_2$.

13. Find the value of n such that the curves $x^n + y^n = c_1$ are the orthogonal trajectories of the family

$$y = \frac{x}{1 - c_2 x}.$$

14. A given family of curves is said to be *self-orthogonal* if its family of orthogonal trajectories is the same as the given family. Show that the family of parabolas $y^2 = 2cx + c^2$ is self-orthogonal.

15. Find a family of oblique trajectories that intersect the family of circles $x^2 + y^2 = c^2$ at angle $45°$.

16. Find a family of oblique trajectories that intersect the family of parabolas $y^2 = cx$ at angle $60°$.

17. Find a family of oblique trajectories that intersect the family of curves $x + y = cx^2$ at angle α such that $\tan \alpha = 2$.

3.2 PROBLEMS IN MECHANICS

A. Introduction

Before we apply our knowledge of differential equations to certain problems in mechanics, let us briefly recall certain principles of that subject. The *momentum* of a body is defined to be the product mv of its mass m and its velocity v. The velocity v and hence the momentum are vector quantities. We now state the following basic law of mechanics:

Newton's Second Law. The time rate of change of momentum of a body is proportional to the resultant force acting on the body and is in the direction of this resultant force.

In mathematical language, this law states that

$$\frac{d}{dt}(mv) = KF,$$

where m is the mass of the body, v is its velocity, F is the resultant force acting upon it, and K is a constant of proportionality. If the mass m is considered constant, this reduces to

$$m\frac{dv}{dt} = KF,$$

or

$$a = K\frac{F}{m}, \tag{3.17}$$

or

$$F = kma, \tag{3.18}$$

where $k = 1/K$ and $a = dv/dt$ is the acceleration of the body. The form (3.17) is a direct mathematical statement of the manner in which Newton's second law is usually expressed in words, the mass being considered constant. However, we shall make use of the equivalent form (3.18). The magnitude of the constant of proportionality k depends upon the units employed for force, mass, and acceleration. Obviously the simplest systems of units are those for which $k = 1$. When such a system is used (3.18) reduces to

$$F = ma. \tag{3.19}$$

It is in this form that we shall use Newton's second law. Observe that Equation (3.19) is a vector equation.

Several systems of units for which $k = 1$ are in use. In this text we shall use only three: the British gravitational system (British), the centimeter-gram-second system (cgs), and the meter-kilogram-second system (mks). We summarize the various units of these three systems in Table 3.1.

Recall that the force of gravitational attraction that the earth exerts on a body is called the weight of the body. The weight, being a force, is expressed in force units. Thus in the British system the weight is measured in pounds; in the cgs system, in dynes; and in the mks system, in newtons.

Let us now apply Newtons's second law to a freely falling body (a body falling toward the earth in the absence of air resistance). Let the mass of the body be m and let w denote its weight. The only force acting on the body is its weight and so this is the resultant force. The acceleration is that due to gravity, denoted by g, which is approximately 32 ft/sec² in the British system, 980 cm/sec² in the cgs system, and 9.8 m/sec² in the mks system (for points near the earth's surface). Newton's second law $F = ma$ thus reduces to $w = mg$. Thus

$$m = \frac{w}{g}, \tag{3.20}$$

a relation that we shall frequently employ.

Let us now consider a body B in rectilinear motion, that is, in motion along a straight line L. On L we choose a fixed reference point as origin O, a fixed direction as positive, and a unit of distance. Then the coordinate x of the position of B from the origin O tells us the distance or displacement of B. (See Figure 3.3.) The *instantaneous velocity* of B is the time rate of change of x:

$$v = \frac{dx}{dt};$$

TABLE 3.1

	British System	cgs System	mks System
force	pound	dyne	newton
mass	slug	gram	kilogram
distance	foot	centimeter	meter
time	second	second	second
acceleration	ft/sec²	cm/sec²	m/sec²

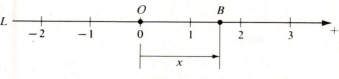

Figure 3.3

and the *instantaneous acceleration* of B is the time rate of change of v:

$$a = \frac{dv}{dt} = \frac{d^2x}{dt^2}.$$

Note that x, v, and a are vector quantities. All forces, displacements, velocities, and accelerations in the positive direction on L are positive quantities; while those in the negative direction are negative quantities.

If we now apply Newton's second law $F = ma$ to the motion of B along L, noting that

$$\frac{dv}{dt} = \frac{dv}{dx}\frac{dx}{dt} = v\frac{dv}{dx},$$

we may express the law in any of the following three forms:

$$m\frac{dv}{dt} = F, \tag{3.21}$$

$$m\frac{d^2x}{dt^2} = F, \tag{3.22}$$

$$mv\frac{dv}{dx} = F, \tag{3.23}$$

where F is the resultant force acting on the body. The form to use depends upon the way in which F is expressed. For example, if F is a function of time t only and we desire to obtain the velocity v as a function of t, we would use (3.21); whereas if F is expressed as a function of the displacement x and we wish to find v as a function of x, we would employ (3.23).

B. Falling Body Problems

We shall now consider some examples of a body falling through air toward the earth. In such a circumstance the body encounters air resistance as it falls. The amount of air resistance depends upon the velocity of the body, but no general law exactly expressing this dependence is known. In some instances the law $R = kv$ appears to be quite satisfactory, while in others $R = kv^2$ appears to be more exact. In any case, the constant of proportionality k in turn depends on several circumstances. In the examples that follow we shall assume certain reasonable resistance laws in each case. Thus we shall actually be dealing with idealized problems in which the true resistance law is approximated and in which certain comparatively negligible factors are disregarded.

▶ **Example 3.5**

A body weighing 8 lb falls from rest toward the earth from a great height. As it falls, air resistance act upon it, and we shall assume that this resistence (in pounds) is numerically equal to $2v$, where v is the velocity (in feet per second). Find the velocity and distance fallen at time t seconds.

   Formulation.   We choose the positive x axis vertically downward along the path of the body B and the origin at the point from which the body fell. The forces acting on the body are:

1. F_1, its weight, 8 lb, which acts downward and hence is positive.
2. F_2, the air resistance, numerically equal to $2v$, which acts upward and hence is the negative quantity $-2v$.

See Figure 3.4, where these forces are indicated.

   Newton's second law, $F = ma$, becomes

$$m \frac{dv}{dt} = F_1 + F_2$$

or, taking $g = 32$ and using $m = w/g = \frac{8}{32} = \frac{1}{4}$,

$$\frac{1}{4} \frac{dv}{dt} = 8 - 2v. \qquad (3.24)$$

Since the body was initially at rest, we have the initial condition

$$v(0) = 0. \qquad (3.25)$$

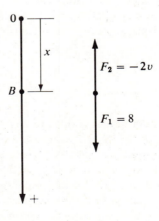

Earth

Figure 3.4

Solution.   Equation (3.24) is separable. Separating variables, we have

$$\frac{dv}{8 - 2v} = 4\, dt.$$

Integrating we find

$$-\tfrac{1}{2} \ln |8 - 2v| = 4t + c_0,$$

which reduces to

$$8 - 2v = c_1 e^{-8t}.$$

Applying the condition (3.25) we find $c_1 = 8$. Thus the velocity at time t is given by

$$v = 4(1 - e^{-8t}). \tag{3.26}$$

Now to determine the distance fallen at time t, we write (3.26) in the form

$$\frac{dx}{dt} = 4(1 - e^{-8t})$$

and note that $x(0) = 0$. Integrating the above equation, we obtain

$$x = 4(t + \tfrac{1}{8}e^{-8t}) + c_2.$$

Since $x = 0$ when $t = 0$, we find $c_2 = -\tfrac{1}{2}$ and hence the distance fallen is given by

$$x = 4(t + \tfrac{1}{8}e^{-8t} - \tfrac{1}{8}). \tag{3.27}$$

Interpretation of Results.   Equation (3.26) shows us that as $t \to \infty$, the velocity v approaches the *limiting velocity* 4(ft/sec). We also observe that this limiting velocity is approximately attained in a very short time. Equation (3.27) states that as $t \to \infty$, x also $\to \infty$. Does this imply that the body will plow through the earth and continue forever? Of course not; for when the body reaches the earth's surface its motion will certainly cease. How then do we reconcile this obvious end to the motion with the statement of Equation (3.27)? It is simple; when the body reaches the earth's surface, the differential equation (3.24) and hence Equation (3.27) no longer apply!

▶ **Example 3.6**

A skydiver equipped with parachute and other essential equipment falls from rest toward the earth. The total weight of the man plus the equipment is 160 lb. Before the parachute opens, the air resistance (in pounds) is numerically equal to $\tfrac{1}{2}v$, where v is the velocity (in feet per second). The parachute opens 5 sec after the fall begins; after it opens, the air resistance (in pounds) is numerically equal to $\tfrac{5}{8}v^2$, where v is the velocity (in feet per second). Find the velocity of the skydiver (A) before the parachute opens, and (B) after the parachute opens.

Formulation.   We again choose the positive x axis vertically downward with the origin at the point where the fall began. The statement of the problem suggests that we break it into two parts: (A) *before* the parachute opens; (B) *after* it opens.

We first consider problem (A). Before the parachute opens, the forces acting upon the skydiver are:

1. F_1, the weight, 160 lb, which acts downward and hence is positive.

2. F_2, the air resistance, numerically equal to $\frac{1}{2}v$, which acts upward and hence is the negative quantity $-\frac{1}{2}v$.

We use Newton's second law $F = ma$, where $F = F_1 + F_2$, let $m = w/g$, and take $g = 32$. We obtain

$$5\frac{dv}{dt} = 160 - \frac{1}{2}v.$$

Since the skydiver was initially at rest, $v = 0$ when $t = 0$. Thus, problem (A), concerned with the time *before* the parachute opens, is formulated as follows:

$$5\frac{dv}{dt} = 160 - \frac{1}{2}v. \tag{3.28}$$

$$v(0) = 0. \tag{3.29}$$

We now turn to the formulation of problem (B). Reasoning as before, we see that after the parachute opens, the forces acting upon the skydiver are:

1. $F_1 = 160$, exactly as before.
2. $F_2 = -\frac{5}{8}v^2$ (instead of $-\frac{1}{2}v$).

Thus, proceeding as above, we obtain the differential equation

$$5\frac{dv}{dt} = 160 - \frac{5}{8}v^2.$$

Since the parachute opens 5 sec after the fall begins, we have $v = v_1$ when $t = 5$, where v_1 is the velocity attained when the parachute opened. Thus, problem (B), concerned with the time *after* the parachute opens, is formulated as follows:

$$5\frac{dv}{dt} = 160 - \frac{5}{8}v^2, \tag{3.30}$$

$$v(5) = v_1. \tag{3.31}$$

Solution.   We shall first consider problem (A). We find a one-parameter family of solution of

$$5\frac{dv}{dt} = 160 - \frac{1}{2}v. \tag{3.28}$$

Separating variables, we obtain

$$\frac{dv}{v - 320} = -\frac{1}{10}\,dt.$$

Integration yields

$$\ln(v - 320) = -\frac{1}{10}t + c_0,$$

which readily simplifies to the form

$$v = 320 + ce^{-t/10}.$$

Applying the initial condition (3.29) that $v = 0$ at $t = 0$, we find that $c = -320$. Hence the solution to problem (A) is

$$v = 320(1 - e^{-t/10}), \tag{3.32}$$

which is valid for $0 \le t \le 5$. In particular, where $t = 5$, we obtain

$$v_1 = 320(1 - e^{-1/2}) \approx 126, \tag{3.33}$$

which is the velocity when the parachute opens.

Now let us consider problem (B). We first find a one-parameter family of solutions of the differential equation

$$5 \frac{dv}{dt} = 160 - \tfrac{5}{8}v^2 \tag{3.30}$$

Simplifying and separating variables, we obtain

$$\frac{dv}{v^2 - 256} = -\frac{dt}{8}.$$

Integration yields

$$\frac{1}{32} \ln \frac{v - 16}{v + 16} = -\frac{t}{8} + c_2$$

or

$$\ln \frac{v - 16}{v + 16} = -4t + c_1.$$

This readily simplifies to the form

$$\frac{v - 16}{v + 16} = ce^{-4t}, \tag{3.34}$$

and solving this for v we obtain

$$v = \frac{16(ce^{-4t} + 1)}{1 - ce^{-4t}}. \tag{3.35}$$

Applying the initial condition (3.31) that $v = v_1$ at $t = 5$, where v_1 is given by (3.33) and is approximately 126, to (3.34), we obtain

$$c = \tfrac{110}{142}e^{20}.$$

Substituting this into (3.35) we obtain

$$v = \frac{16(\tfrac{110}{142}e^{20-4t} + 1)}{1 - \tfrac{110}{142}e^{20-4t}}, \tag{3.36}$$

which is valid for $t \ge 5$.

Interpretation of Results. Let us first consider the solution of problem (A), given by Equation (3.32). According to this, as $t \to \infty$, v approaches the limiting velocity 320 ft/sec. Thus if the parachute never opened, the velocity would have been approximately 320 ft/sec at the time when the unfortunate skydiver would have struck the earth! But, according to the statement of the problem, the parachute *does* open 5 sec after the fall begins (we tacitly and thoughfully assume $5 \ll T$, where T is the time when the earth *is* reached!). Then, referring to the solution of problem (B), Equation (3.36), we see that as $t \to \infty$, v approaches the limiting velocity 16 ft/sec. Thus, assuming that the parachute opens at a considerable distance above the earth, the velocity is approximately 16 ft/sec when the earth is finally reached. We thus obtain the well-known fact that the velocity of impact with the open parachute is a small fraction of the impact velocity that would have occured if the parachute had not opened. The calculations in this problem are somewhat complicated, but the moral is clear: Make certain that the parachute opens!

C. Frictional Forces

If a body moves on a rough surface, it will encounter not only air resistance but also another resistance force due to the roughness of the surface. This additional force is called *friction*. It is shown in physics that the friction is given by μN, where

1. μ is a constant of proportionality called the *coefficient of friction*, which depends upon the roughness of the given surface; and
2. N is the normal (that is, perpendicular) force which the surface exerts on the body.

We now apply Newton's second law to a problem in which friction is involved.

▶ **Example 3.7**

An object weighing 48 lb is released from rest at the top of a plane metal slide that is inclined 30° to the horizontal. Air resistance (in pounds) is numerically equal to one-half the velocity (in feet per second), and the coefficient of friction is one-quarter.

A. What is the velocity of the object 2 sec after it is released?
B. If the slide is 24 ft long, what is the velocity when the object reaches the bottom?

Formulation. The line of motion is along the slide. We choose the origin at the top and the positive x direction down the slide. If we temporarily neglect the friction and air resistance, the forces acting upon the object A are:

1. Its weight, 48 lb, which acts vertically downward; and
2. The normal force, N, exerted by the slide which acts in an upward direction perpendicular to the slide. (See Figure 3.5.)

The components of the weight parallel and perpendicular to the slide have magnitude

$$48 \sin 30° = 24$$

and

$$48 \cos 30° = 24\sqrt{3},$$

respectively, The components perpendicular to the slide are in equilibrium and hence the normal force N has magnitude $24\sqrt{3}$.

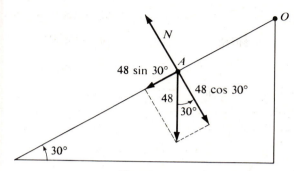

Figure 3.5

Now, taking into consideration the friction and air resistance, we see that the forces acting on the object as it moves along the slide are the following:

1. F_1, the component on the weight parallel to the plane, having numerical value 24. Since this force acts in the positive (downward) direction along the slide, we have

$$F_1 = 24.$$

2. F_2, the frictional force, having numerical value $\mu N = \frac{1}{4}(24\sqrt{3})$. Since this acts in the negative (upward) direction along the side, we have

$$F_2 = -6\sqrt{3}.$$

3. F_3, the air resistance, having numerical value $\frac{1}{2}v$. Since $v > 0$ and this also acts in the negative direction, we have

$$F_3 = -\tfrac{1}{2}v.$$

We apply Newton's second law $F = ma$. Here $F = F_1 + F_2 + F_3 = 24 - 6\sqrt{3} - \frac{1}{2}v$ and $m = w/g = \frac{48}{32} = \frac{3}{2}$. Thus we have the differential equation

$$\frac{3}{2}\frac{dv}{dt} = 24 - 6\sqrt{3} - \tfrac{1}{2}v. \tag{3.37}$$

Since the object is released from rest, the initial condition is

$$v(0) = 0. \tag{3.38}$$

Solution.   Equation (3.37) is separable; separating variables we have

$$\frac{dv}{48 - 12\sqrt{3} - v} = \frac{dt}{3}.$$

Integrating and simplifying, we find

$$v = 48 - 12\sqrt{3} - c_1 e^{-t/3}.$$

The condition (3.38) gives $c_1 = 48 - 12\sqrt{3}$. Thus we obtain

$$v = (48 - 12\sqrt{3})(1 - e^{-t/3}). \tag{3.39}$$

Question A is thus answered by letting $t = 2$ in Equation (3.39). We find

$$v(2) = (48 - 12\sqrt{3})(1 - e^{-2/3}) \approx 10.2 (\text{ft/sec}).$$

In order to answer question B, we integrate (3.39) to obtain

$$x = (48 - 12\sqrt{3})(t + 3e^{-t/3}) + c_2.$$

Since $x(0) = 0$, $c_2 = -(48 - 12\sqrt{3})(3)$. Thus the distance covered at time t is given by

$$x = (48 - 12\sqrt{3})(t + 3e^{-t/3} - 3).$$

Since the slide is 24 ft long, the object reaches the bottom at the time T determined from the transcendental equation

$$24 = (48 - 12\sqrt{3})(T + 3e^{-T/3} - 3),$$

which may be written as

$$3e^{-T/3} = \frac{47 + 2\sqrt{3}}{13} - T.$$

The value of T that satisfies this equation is approximately 2.6. Thus from Equation (3.39) the velocity of the object when it reaches the bottom is given approximately by

$$(48 - 12\sqrt{3})(1 - e^{-0.9}) \approx 12.3 \text{ (ft/sec)}.$$

Exercises

1. A stone weighing 4 lb falls from rest toward the earth from a great height. As it falls it is acted upon by air resistance that is numercially equal to $\frac{1}{2}v$ (in pounds), where v is the velocity (in feet per second).

    (a) Find the velocity and distance fallen at time t sec.

    (b) Find the velocity and distance fallen at the end of 5 sec.

2. A ball weighing 6 lb is thrown vertically downward toward the earth from a height of 1000 ft with an initial velocity of 6 ft/sec. As it falls it is acted upon by air resistance that is numerically equal to $\frac{2}{3}v$ (in pounds), where v is the velocity (in feet per second).

    (a) What is the velocity and distance fallen at the end of one minute?

    (b) With what velocity does the ball strike the earth?

3. A ball weighing $\frac{3}{4}$ lb is thrown vertically upward from a point 6 ft above the surface of the earth with an initial velocity of 20 ft/sec. As it rises it is acted upon by air resistance that is numerically equal to $\frac{1}{64}v$ (in pounds), where v is the velocity (in feet per second). How high will the ball rise?

4. A ship which weighs 32,000 tons starts from rest under the force of a constant propeller thrust of 100,000 lb. The resistance in pounds is numerically equal to $8000v$, where v is in feet per second.

    (a) Find the velocity of the ship as a function of the time.

    (b) Find the limiting velocity (that is, the limit of v as $t \to +\infty$).

    (c) Find how long it takes the ship to attain a velocity of 80% of the limiting velocity.

5. A body of mass 100 g is dropped from rest toward the earth from a height of 1000 m. As it falls, air resistance acts upon it, and this resistance (in newtons) is

proportional to the velocity v (in meters per second). Suppose the limiting velocity is 245 m/sec.

(a)   Find the velocity and distance fallen at time t secs.

(b)   Find the time at which the velocity is one-fifth of the limiting velocity.

6.   An object of mass 100 g is thrown vertically upward from a point 60 cm above the earth's surface with an initial velocity of 150 cm/sec. It rises briefly and then falls vertically to the earth, all of which time it is acted on by air resistance that is numerically equal to $200v$ (in dynes), where v is the velocity (in cm/sec).

(a)   Find the velocity 0.1 sec after the object is thrown.

(b)   Find the velocity 0.1 sec after the object stops rising and starts falling.

7.   Two people are riding in a motorboat and the combined weight of individuals, motor, boat, and equipment is 640 lb. The motor exerts a constant force of 20 lb on the boat in the direction of motion, while the resistance (in pounds) is numerically equal to one and one-half times the velocity (in feet per second). If the boat started from rest, find the velocity of the boat after (a) 20 sec, (b) 1 min.

8.   A boat weighing 150 lb with a single rider weighing 170 lb is being towed in a certain direction at the rate of 20 mph. At time $t = 0$ the tow rope is suddenly cast off and the rider begins to row in the same direction, exerting a force equivalent to a constant force of 12 lb in this direction. The resistance (in pounds) is numerically equal to twice the velocity (in feet per second).

(a)   Find the velocity of the boat 15 sec after the tow rope was cast off.

(b)   How many seconds after the tow rope is cast off will the velocity be one-half that at which the boat was being towed?

9.   A bullet weighing 1 oz is fired vertically downward from a stationary helicopter with a muzzle velocity of 1200 ft/sec. The air resistance (in pounds) is numerically equal to $16^{-5}v^2$, where v is the velocity (in feet per second). Find the velocity of the bullet as a function of the time.

10.   A shell weighing 1 lb is fired vertically upward from the earth's surface with a muzzle velocity of 1000 ft/sec. The air resistance (in pounds) is numerically equal to $10^{-4}v^2$, where v is the velocity (in feet per second).

(a)   Find the velocity of the rising shell as a function of the time.

(b)   How long will the shell rise?

11.   An object weighing 16 lb is dropped from rest on the surface of a calm lake and thereafter starts to sink. While its weight tends to force it downward, the buoyancy of the object tends to force it back upward. If this buoyancy force is one of 6 lb and the resistance of the water (in pounds) is numerically equal to twice the square of the velocity (in feet per second), find the formula for the velocity of the sinking object as a function of the time.

12.   An object weighing 12 lb is placed beneath the surface of a calm lake. The buoyancy of the object is 30 lb; because of this the object begins to rise. If the resistance of the water (in pounds) is numerically equal to the square of the velocity (in feet per second) and the object surfaces in 5 sec, find the velocity of the object at the instant when it reaches the surface.

13. A man is pushing a loaded sled across a level field of ice at the constant speed of 10 ft/sec. When the man is halfway across the ice field, he stops pushing and lets the loaded sled continue on. The combined weight of the sled and its load is 80 lb; the air resistance (in pounds) is numerically equal to $\frac{3}{4}v$, where v is the velocity of the sled (in feet per second); and the coefficient of friction of the runners on the ice is 0.04. How far will the sled continue to move after the man stops pushing?

14. A girl on her sled has just slid down a hill onto a level field of ice and is starting to slow down. At the instant when their speed is 5 ft/sec, the girl's father runs up and begins to push the sled forward, exerting a constant force of 15 lb in the direction of motion. The combined weight of the girl and the sled is 96 lb, the air resistance (in pounds) is numerically equal to one-half the velocity (in feet per second), and the coefficient of friction of the runners on the ice is 0.05. How fast is the sled moving 10 sec after the father begins pushing?

15. A case of canned milk weighing 24 lb is released from rest at the top of a plane metal slide which is 30 ft long and inclined 45° to the horizontal. Air resistance (in pounds) is numerically equal to one-third the velocity (in feet per second) and the coefficent of friction is 0.4.

  (a) What is the velocity of the moving case 1 sec after it is released?

  (b) What is the velocity when the case reaches the bottom of the slide?

16. A boy goes sledding down a long 30° slope. The combined weight of the boy and his sled is 72 lb and the air resistance (in pounds) is numerically equal to twice their velocity (in feet per second). If they started from rest and their velocity at the end of 5 sec is 10 ft/sec, what is the coefficient of friction of the sled runners on the snow?

17. An object weighing 32 lb is released from rest 50 ft above the surface of a calm lake. Before the object reaches the surface of the lake, the air resistance (in pounds) is given by $2v$, where v is the velocity (in feet per second). After the object passes beneath the surface, the water resistance (in pounds) is given by $6v$. Further, the object is then buoyed up by a buoyancy force of 8 lb. Find the velocity of the object 2 sec after it passes beneath the surface of the lake.

18. A rocket of mass m is fired vertically upward from the surface of the earth with initial velocity $v = v_0$. The only force on the rocket that we consider is the gravitational attraction of the earth. Then, according to Newton's law of gravitation, the acceleration a of the rocket is given by $a = -k/x^2$, where $k > 0$ is a constant of proportionality and x is the distance "upward" from the center of the earth along the line of motion. At time $t = 0$, $x = R$ (where R is the radius of the earth), $a = -g$ (where g is the acceleration due to gravity), and $v = v_0$. Express $a = dv/dt$ as in Equation (3.23), apply the appropriate initial data, and note that v satisfies the differential equation

$$v\frac{dv}{dx} = -\frac{gR^2}{x^2}.$$

Solve this differential equation, apply the appropriate initial condition, and thus express v as a function of x. In particular, show that the minimum value of v_0 for which the rocket will escape from the earth is $\sqrt{2gR}$. This is the so-called *veloc-*

ity of escape; and using $R = 4000$ miles, $g = 32$ ft/sec^2, one finds that this is approximately 25,000 mph (or 7 mi/sec).

19.   A body of mass m is in rectilinear motion along a horizontal axis. The resultant force acting on the body is given by $-kx$, where $k > 0$ is a constant of proportionality and x is the distance along the axis from a fixed point O. The body has initial velocity $v = v_0$ when $x = x_0$. Apply Newton's second law in the form (3.23) and thus write the differential equation of motion in the form

$$mv \frac{dv}{dx} = -kx.$$

Solve the differential equation, apply the initial condition, and thus express the square of the velocity v as a function of the distance x. Recalling that $v = dx/dt$, show that the relation between v and x thus obtained is satisfied for all time t by

$$x = \sqrt{x_0^2 + \frac{mv_0^2}{k}} \sin\left(\sqrt{\frac{k}{m}}\, t + \phi\right),$$

where ϕ is a constant.

3.3 RATE PROBLEMS

In certain problems the rate at which a quantity changes is a known function of the amount present and/or the time, and it is desired to find the quantity itself. If x denotes the amount of the quantity present at time t, then dx/dt denotes the rate at which the quantity changes and we are at once led to a differential equation. In this section we consider certain problems of this type.

A. Rate of Growth and Decay

▶ **Example 3.8**

The rate at which radioactive nuclei decay is proportional to the number of such nuclei that are present in a given sample. Half of the original number of radioactive nuclei have undergone disintegration in a period of 1500 years.

1.   What percentage of the original radioactive nuclei will remain after 4500 years?
2.   In how many years will only one-tenth of the original number remain?

Mathematical Formulation.   Let x be the amount of radioactive nuclei present after t years. Then dx/dt represents the rate at which the nuclei decay. Since the nuclei decay at a rate proportional to the amount present, we have

$$\frac{dx}{dt} = Kx, \tag{3.40}$$

where K is a constant of proportionality. The amount x is clearly positive; further, since x is decreasing, $dx/dt < 0$. Thus, from Equation (3.40), we must have $K < 0$. In order to emphasize that x is decreasing, we prefer to replace K by a positive constant preceded by a minus sign. Thus we let $k = -K > 0$ and write the differential equation (3.40) in the form

$$\frac{dx}{dt} = -kx. \tag{3.41}$$

Letting x_0 denote the amount initially present, we also have the initial condition

$$x(0) = x_0. \tag{3.42}$$

We know that we shall need such a condition in order to determine the arbitrary constant that will appear in a one-parameter family of solutions of the differential equation (3.41). However, we shall apparently need something else, for Equation (3.41) contains an unknown constant of proportionality k. This "something else" appears in the statement of the problem, for we are told that half of the original number disintegrate in 1500 years. Thus half also remain at that time, and this at once gives the condition

$$x(1500) = \tfrac{1}{2}x_0. \tag{3.43}$$

Solution. The differential equation (3.41) is clearly separable; separating variables, integrating, and simplifying, we have at once

$$x = ce^{-kt}.$$

Applying the initial condition (3.42), $x = x_0$ when $t = 0$, we find that $c = x_0$ and hence we obtain

$$x = x_0 e^{-kt}. \tag{3.44}$$

We have not yet determined k. Thus we now apply condition (3.43), $x = \tfrac{1}{2}x_0$ when $t = 1500$, to Equation (3.44). We find

$$\tfrac{1}{2}x_0 = x_0 e^{-1500k},$$

or

$$(e^{-k})^{1500} = \tfrac{1}{2},$$

or finally

$$e^{-k} = \left(\tfrac{1}{2}\right)^{1/1500}. \tag{3.45}$$

From this equation we could determine k explicitly and substitute the result into Equation (3.44). However, we see from Equation (3.44) that we actually do not need k itself but rather only e^{-k}, which we have just obtained in Equation (3.45). Thus we substitute e^{-k} from (3.45) into (3.44) to obtain

$$x = x_0(e^{-k})^t = x_0\left[\left(\tfrac{1}{2}\right)^{1/1500}\right]^t$$

or

$$x = x_0\left(\tfrac{1}{2}\right)^{t/1500}. \tag{3.46}$$

Equation (3.46) gives the number x of radioactive nuclei that are present at time t. Question 1 asks us what percentage of the original number will remain after 4500 years. We thus let $t = 4500$ in Equation (3.46) and find

$$x = x_0\left(\tfrac{1}{2}\right)^3 = \tfrac{1}{8}x_0.$$

Thus, one-eighth or 12.5% of the original number remain after 4500 years. Question 2 asks us when only one-tenth will remain. Thus we let $x = \frac{1}{10}x_0$ in Equation (3.46) and solve for t. We have

$$\tfrac{1}{10} = \left(\tfrac{1}{2}\right)^{t/1500}.$$

Using logarithms, we then obtain

$$\ln\left(\tfrac{1}{10}\right) = \ln\left(\tfrac{1}{2}\right)^{t/1500} = \frac{t}{1500}\ln\left(\tfrac{1}{2}\right).$$

From this it follows at once that

$$\frac{t}{1500} = \frac{\ln\frac{1}{10}}{\ln\frac{1}{2}}$$

or

$$t = \frac{1500 \ln 10}{\ln 2} \approx 4985 \text{ (years)}.$$

B. Population Growth

We next consider the growth of a population (for example, human, an animal species, or a bacteria colony) as a function of time. Note that a population actually increases discontinuously by whole number amounts. However, if the population is very large, such individual increases in it are essentially negligible compared to the entire population itself. In other words, the population increase is approximately continuous. We shall therefore assume that this increase is indeed continuous and in fact that the population is a continuous and differentiable fucntion of time.

Given a population, we let x be the number of individuals in it at time t. If we assume that the rate of change of the population is proportional to the number of individuals in it at any time, we are led to the differential equation

$$\frac{dx}{dt} = kx, \tag{3.47}$$

where k is a constant of proportionality. The population x is positive and is increasing and hence $dx/dx > 0$. Therefore, from (3.47), we must have $k > 0$. Now suppose that at time t_0 the population is x_0. Then, in addition to the differential equation (3.47), we have the initial condition

$$x(t_0) = x_0. \tag{3.48}$$

The differential equation (3.47) is separable. Separating variables, integrating, and simplifying, we obtain

$$x = ce^{kt}.$$

Applying the initial condition (3.48), $x = x_0$ at $t = t_0$, to this, we have $x_0 = ce^{kt_0}$. From this we at once find $c = x_0 e^{-kt_0}$ and hence obtain the unique solution

$$x = x_0 e^{k(t - t_0)} \tag{3.49}$$

of the differential equation (3.47), which satisfies the initial condition (3.48).

From (3.49) we see that a population governed by the differential equation (3.47) with $k > 0$ and initial condition (3.48) is one that increases exponentially with time. This law of population growth is called the *Malthusian law*. We should now inquire whether or not there are cases in which such a model for population growth is indeed realistic. In answer to this, it can be shown that this model, with a suitable value of k, is remarkably accurate in the case of the human population of the earth during the last several decades (see Problem 8(b)). It is also known to be outstandingly accurate for certain mammalian species, with suitable k, under certain realizable conditions and for certain time periods. On the other hand, turning back to the case of the human population of the earth, it can be shown that the *Malthusian law* turns out to be quite unreasonable when applied to the distant future (see Problem 8(e)). It is also completely unrealistic for other populations (for example, bacteria colonies) when applied over sufficiently long periods of time. The reason for this is not hard to see. For, according to (3.49), a population modeled by this law always increases and indeed does so at an ever increasing rate; whereas observation shows that a given population simply does not grow indefinitely.

Population growth is represented more realistically in many cases by assuming that the number of individuals x in the population at time t is described by a differential equation of the form

$$\frac{dx}{dt} = kx - \lambda x^2, \tag{3.50}$$

where $k > 0$ and $\lambda > 0$ are constants. The additional term $-\lambda x^2$ is the result of some cause that tends to limit the ultimate growth of the population. For example, such a cause could be insufficient living space or food supply, when the population becomes sufficiently large. Concerning the choice of $-\lambda x^2$ for the term representing the effect of the cause, one can argue as follows: Assuming the cause affects the entire population of x members, then the effect on any one individual is proportional to x. Thus the effect on all x individuals in the population would be proportional to $x \cdot x = x^2$.

We thus assume that a population is described by a differential equation of the form (3.50), with constants $k > 0$ and $\lambda > 0$, and an initial condition of the form (3.48). In most such cases, it turns out that the constant λ is very small compared to the constant k. Thus for sufficiently small x, the term kx predominates, and so the population grows very rapidly for a time. However, when x becomes sufficiently large, the term $-\lambda x^2$ is of comparatively greater influence, and the result of this is a decrease in the rapid growth rate. We note that the differential equation (3.50) is both a separable equation and a Bernoulli equation. The law of population growth so described is called the *logistic law* of growth. We now consider a specific example of this type of growth.

▶ **Example 3.9**

The population x of a certain city satisfies the logistic law

$$\frac{dx}{dt} = \frac{1}{100} x - \frac{1}{(10)^8} x^2 \tag{3.51}$$

where time t is measured in years. Given that the population of this city is 100,000 in 1980, determine the population as a function of time for $t > 1980$. In particular, answer the following questions:

(a) What will be the population in 2000?
(b) In what year does the 1980 population double?
(c) Assuming the differential equation (3.51) applies for all $t > 1980$, how large will the population ultimately be?

   Solution. We must solve the separable differential equation (3.51) subject to the initial solution

$$x(1980) = 100,000. \tag{3.52}$$

Separating variables in (3.51), we obtain

$$\frac{dx}{(10)^{-2}x - (10)^{-8}x^2} = dt$$

and hence

$$\frac{dx}{(10)^{-2}x[1 - (10)^{-6}x]} = dt.$$

Using partial fractions, this becomes

$$100 \left[\frac{1}{x} + \frac{(10)^{-6}}{1 - (10)^{-6}x} \right] dx = dt.$$

Integrating, assuming $0 < x < 10^6$, we obtain

$$100 \{ \ln x - \ln[1 - (10)^{-6}x] \} = t + c_1$$

and hence

$$\ln \left[\frac{x}{1 - (10)^{-6}x} \right] = \tfrac{1}{100}t + c_2.$$

Thus we find

$$\frac{x}{1 - (10)^{-6}x} = ce^{t/100}.$$

Solving this for x, we finally obtain

$$x = \frac{ce^{t/100}}{1 + (10)^{-6}ce^{t/100}}. \tag{3.53}$$

Now applying the initial condition (3.52) to this, we have

$$(10)^5 = \frac{ce^{19.8}}{1 + (10)^{-6}ce^{19.8}},$$

from which we obtain

$$c = \frac{(10)^5}{e^{19.8}[1 - (10)^5(10)^{-6}]} = \frac{(10)^6}{9e^{19.8}}.$$

Substituting this value for c back into (3.53) and simplifying, we obtain the solution in the form

$$x = \frac{(10)^6}{1 + 9e^{19.8 - t/100}}. \tag{3.54}$$

This gives the population x as a function of time for $t > 1980$.

We now consider the questions (a), (b), and (c) of the problem. Question (a) asks for the population in the year 2000. Thus we let $t = 2000$ in (3.54) and obtain

$$x = \frac{(10)^6}{1 + 9e^{-0.2}} \approx 119{,}495.$$

Question (b) asks for the year in which the population doubles. Thus we let $x = 200{,}000 = 2(10)^5$ in (3.54) and solve for t. We have

$$2(10)^5 = \frac{(10)^6}{1 + 9e^{19.8 - t/100}},$$

from which

$$e^{19.8 - t/100} = \tfrac{4}{9},$$

and hence

$$t \approx 2061.$$

Question (c) asks how large the population will ultimately be, assuming the differential equation (3.51) applies for all $t > 1980$. To answer this, we evaluate $\lim x$ as $t \to \infty$ using the solution (3.54) of (3.51). We find

$$\lim_{t \to \infty} x = \lim_{t \to \infty} \frac{(10)^6}{1 + 9e^{19.8 - t/100}} = (10)^6 = 1{,}000{,}000.$$

C. Mixture Problems

We now consider rate problems involving mixtures. A substance S is allowed to flow into a certain mixture in a container at a certain rate, and the mixture is kept uniform by stirring. Further, in one such situation, this uniform mixture simultaneously flows out of the container at another (generally different) rate; in another situation this may not be the case. In either case we seek to determine the quantity of the substance S present in the mixture at time t.

Letting x denote the amount of S present at time t, the derivative dx/dt denotes the rate of change of x with respect to t. If IN denotes the rate at which S enters the mixture and OUT the rate at which it leaves, we have at once the basic equation

$$\frac{dx}{dt} = \text{IN} - \text{OUT} \tag{3.55}$$

from which to determine the amount x of S at time t. We now consider examples.

▶ **Example 3.10**

A tank initially contains 50 gal of pure water. Starting at time $t = 0$ a brine containing 2 lb of dissolved salt per gallon flows into the tank at the rate of 3 gal/min. The mixture is kept uniform by stirring and the well-stirred mixture simultaneously flows out of the tank at the same rate.

1. How much salt is in the tank at any time $t > 0$?
2. How much salt is present at the end of 25 min?
3. How much salt is present after a long time?

Mathematical Formulation.   Let x denote the amount of salt in the tank at time t. We apply the basic equation (3.55),

$$\frac{dx}{dt} = \text{IN} - \text{OUT}.$$

The brine flows in at the rate of 3 gal/min, and each gallon contains 2 lb of salt. Thus

$$\text{IN} = (2 \text{ lb/gal})(3 \text{ gal/min}) = 6 \text{ lb/min}.$$

Since the rate of outflow equals the rate of inflow, the tank contains 50 gal of the mixture at any time t. This 50 gal contains x lb of salt at time t, and so the concentration of salt at time t is $\frac{1}{50}x$ lb/gal. Thus, since the mixture flows out at the rate of 3 gal/min, we have

$$\text{OUT} = \left(\frac{x}{50} \text{ lb/gal}\right)(3 \text{ gal/min}) = \frac{3x}{50} \text{ lb/min}.$$

Thus the differential equation for x as a function of t is

$$\frac{dx}{dt} = 6 - \frac{3x}{50}. \tag{3.56}$$

Since initially there was no salt in the tank, we also have the initial condition

$$x(0) = 0. \tag{3.57}$$

Solution.   Equation (3.56) is both linear and separable. Separating variables, we have

$$\frac{dx}{100 - x} = \frac{3}{50} dt.$$

Integrating and simplifying, we obtain

$$x = 100 + ce^{-3t/50}.$$

Applying the condition (3.57), $x = 0$ at $t = 0$, we find that $c = -100$. Thus we have

$$x = 100(1 - e^{-3t/50}). \tag{3.58}$$

This is the answer to question 1. As for question 2, at the end of 25 min, $t = 25$, and Equation (3.58) gives

$$x(25) = 100(1 - e^{-1.5}) \approx 78 \text{(lb)}.$$

Question 3 essentially asks us how much salt is present as $t \to \infty$. To answer this we let $t \to \infty$ in Equation (3.58) and observe that $x \to 100$.

▶ **Example 3.11**

A large tank initially contains 50 gal of brine in which there is dissolved 10 lb of salt. Brine containing 2 lb of dissolved salt per gallon flows into the tank at the rate of 5 gal/min. The mixture is kept uniform by stirring, and the stirred mixture simultaneously flows out at the slower rate of 3 gal/min. How much salt is in the tank at any time $t > 0$?

Mathematical Formulation.   Let $x =$ the amount of salt at time t. Again we shall use Equation (3.55):

$$\frac{dx}{dt} = \text{IN} - \text{OUT}.$$

Proceeding as in Example 3.10,

$$\text{IN} = (2 \text{ lb/gal})(5 \text{ gal/min}) = 10 \text{ lb/min};$$

also, once again

$$\text{OUT} = (C \text{ lb/gal})(3 \text{ gal/min}),$$

where C lb/gal denotes the concentration. But here, since the rate of outflow is different from that of inflow, the concentration is not quite so simple. At time $t = 0$, the tank contains 50 gal of brine. Since brine flows in at the rate of 5 gal/min but flows out at the slower rate of 3 gal/min, there is a net gain of $5 - 3 = 2$ gal/min of brine in the tank. Thus at the end of t minutes the amount of brine in the tank is

$$50 + 2t \text{ gal.}$$

Hence the concentration at time t minutes is

$$\frac{x}{50 + 2t} \text{ lb/gal,}$$

and so

$$\text{OUT} = \frac{3x}{50 + 2t} \text{ lb/min.}$$

Thus the differential equation becomes

$$\frac{dx}{dt} = 10 - \frac{3x}{50 + 2t}. \tag{3.59}$$

Since there was initially 10 lb of salt in the tank, we have the initial condition

$$x(0) = 10. \tag{3.60}$$

Solution.   The differential equation (3.59) is *not* separable but it *is* linear. Putting it in standard form,

$$\frac{dx}{dt} + \frac{3}{2t + 50} x = 10,$$

we find the integrating factor

$$\exp\left(\int \frac{3}{2t + 50} \, dt\right) = (2t + 50)^{3/2}.$$

Multiplying through by this, we have

$$(2t + 50)^{3/2} \frac{dx}{dt} + 3(2t + 50)^{1/2}x = 10(2t + 50)^{3/2}$$

or

$$\frac{d}{dt}[(2t + 50)^{3/2}x] = 10(2t + 50)^{3/2}.$$

Thus

$$(2t + 50)^{3/2}x = 2(2t + 50)^{5/2} + c$$

or

$$x = 4(t + 25) + \frac{c}{(2t + 50)^{3/2}}.$$

Applying condition (3.60), $x = 10$ at $t = 0$, we find

$$10 = 100 + \frac{c}{(50)^{3/2}}$$

or

$$c = -(90)(50)^{3/2} = -22,500\sqrt{2}.$$

Thus the amount of salt at any time $t > 0$ is given by

$$x = 4t + 100 - \frac{22,500\sqrt{2}}{(2t + 50)^{3/2}}.$$

Exercises

1. Assume that the rate at which radioactive nuclei decay is proportional to the number of such nuclei that are present in a given sample. In a certain sample 10% of the original number of radioactive nuclei have undergone disintegration in a period of 100 years.

    (a) What percentage of the original radioactive nuclei will remain after 1000 years?

    (b) In how many years will only one-fourth of the original number remain?

2. A certain chemical is converted into another chemical by a chemical reaction. The rate at which the first chemical is converted is proportional to the amount of this chemical present at any instant. Ten percent of the original amount of the first chemical has been converted in 5 min.

    (a) What percent of the first chemical will have been converted in 20 min?

    (b) In how many minutes will 60% of the first chemical have been converted?

3. A chemical reaction converts a certain chemical into another chemical, and the rate at which the first chemical is converted is proportional to the amount of this chemical present at any time. At the end of one hour, 50 gm of the first chemical remain; while at the end of three hours, only 25 gm remain.

    (a) How many grams of the first chemical were present initially?

    (b) How many grams of the first chemical will remain at the end of five hours?

    (c) In how many hours will only 2 gm of the first chemical remain?

4. A chemical reaction converts a certain chemical into another chemical, and the rate at which the first chemical is converted is proportional to the amount of this chemical present at any time. At the end of one hour, two-thirds kg of the first chemical remains, while at the end of four hours, only one-third kg remains.

    (a) What fraction of the first chemical remains at the end of seven hours?

    (b) When will only one-tenth of the first chemical remain?

5. Assume that the population of a certain city increases at a rate proportional to the number of inhabitants at any time. If the population doubles in 40 years, in how many years will it triple?

6. The population of the city of Bingville increases at a rate proportional to the number of its inhabitants present at any time t. If the population of Bingville was 30,000 in 1970 and 35,000 in 1980, what will be the population of Bingville in 1990?

7. In a certain bacteria culture the rate of increase in the number of bacteria is proportional to the number present.

    (a) If the number triples in 5 hr, how many will be present in 10 hr?

    (b) When will the number present be 10 times the number initially present?

8. Assume that the rate of change of the human population of the earth is proportional to the number of people on earth at any time, and suppose that this population is increasing at the rate of 2% per year. The 1979 *World Almanac* gives the 1978 world population estimate as 4,219 million; assume this figure is in fact correct.

    (a) Using this data, express the human population of the earth as a function of time.

    (b) According to the formula of part (a), what was the population of the earth in 1950? The 1979 *World Almanac* gives the 1950 world population estimate as 2,510 million. Assuming this estimate is very nearly correct, comment on the accuracy of the formula of part (a) in checking such past populations.

    (c) According to the formula of part (a), what will be the population of the earth in 2000? Does this seem reasonable?

    (d) According to the formula of part (a), what was the population of the earth in 1900? The 1979 *World Almanac* gives the 1900 world population estimate as 1,600 million. Assuming this estimate is very nearly correct, comment on the accuracy of the formula of part (a) in checking such past populations.

    (e) According to the formula of part (a), what will be the population of the earth in 2100? Does this seem reasonable?

9. The human population of a certain island satisfies the logistic law (3.50) with $k = 0.03$, $\lambda = 3(10)^{-8}$, and time t measured in years.

   (a) If the population in 1980 is 200,000, find a formula for the population in future years.

   (b) According to the formula of part (a), what will be the population in the year 2000?

   (c) What is the limiting value of the population as $t \to \infty$?

10. This is a general problem about the logistic law of growth. A population satisfies the logistic law (3.50) and has x_0 members at time t_0.

   (a) Solve the differential equation (3.50) and thus express the population x as a function of t.

   (b) Show that as $t \to \infty$, the population x approaches the limiting value k/λ.

   (c) Show that dx/dt is increasing if $x < k/2\lambda$ and decreasing if $x > k/2\lambda$.

   (d) Graph x as a function of t for $t > t_0$.

   (e) Interpret the results of parts (b), (c), and (d).

11. The human population of a certain small island would satisfy the logistic law (3.50), with $k = \frac{1}{400}$, $\lambda = (10)^{-8}$, and t measured in years, provided the annual emigration from the island is neglected. However, the fact is that every year 100 people become disenchanted with island life and move from the island to the mainland. Modify the logistic differential equation (3.50) with the given k and λ so as to include the stated annual emigration. Assuming that the population in 1980 is 20,000, solve the resulting initial-value problem and thus find the population of the island as a function of time.

12. Under natural circumstances the population of mice on a certain island would increase at a rate proportional to the number of mice present at any time, provided the island had no cats. There were no cats on the island from the beginning of 1970 to the beginning of 1980, and during this time the mouse population doubled, reaching an all-time high of 100,000 at the beginning of 1980. At this time the people of the island, alarmed by the increasing number of mice, imported a number of cats to kill the mice. If the indicated natural rate of increase of mice was thereafter offset by the work of the cats, who killed 1000 mice a month, how many mice remained at the beginning of 1981?

13. An amount of invested money is said to draw interest *compounded continuously* if the amount of money increases at a rate proportional to the amount present. Suppose \$1000 is invested and draws interest compounded continuously, where the annual interest rate is 6%.

   (a) How much money will be present 10 years after the original amount was invested?

   (b) How long will it take the original amount of money to double?

14. Suppose a certain amount of money is invested and draws interest compounded continuously.

   (a) If the original amount doubles in two years, then what is the annual interest rate?

(b) If the original amount increases 50% in six months, then how long will it take the original amount to double?

15. A tank initially contains 100 gal of brine in which there is dissolved 20 lb of salt. Starting at time $t = 0$, brine containing 3 lb of dissolved salt per gallon flows into the tank at the rate of 4 gal/min. The mixture is kept uniform by stirring and the well-stirred mixture simultaneously flows out of the tank at the same rate.

(a) How much salt is in the tank at the end of 10 min?

(b) When is there 160 lb of salt in the tank?

16. A large tank initially contains 100 gal of brine in which 10 lb of salt is dissolved. Starting at $t = 0$, pure water flows into the tank at the rate of 5 gal/min. The mixture is kept uniform by stirring and the well-stirred mixture simultaneously flows out at the slower rate of 2 gal/min.

(a) How much salt is in the tank at the end of 15 min and what is the concentration at that time?

(b) If the capacity of the tank is 250 gal, what is the concentration at the instant the tank overflows?

17. A tank initially contains 100 gal of pure water. Starting at $t = 0$, a brine containing 4 lb of salt per gallon flows into the tank at the rate of 5 gal/min. The mixture is kept uniform by stirring and the well-stirred mixture flows out at the slower rate of 3 gal/min.

(a) How much salt is in the tank at the end of 20 min?

(b) When is there 50 lb of salt in the tank?

18. A large tank initially contains 200 gal of brine in which 15 lb of salt is dissolved. Starting at $t = 0$, brine containing 4 lb of salt per gallon flows into the tank at the rate of 3.5 gal/min. The mixture is kept uniform by stirring and the well-stirred mixture leaves the tank at the rate of 4 gal/min.

(a) How much salt is in the tank at the end of one hour?

(b) How much salt is in the tank when the tank contains only 50 gal of brine?

19. A 500 liter tank initially contains 300 liters of fluid in which there is dissolved 50 gm of a certain chemical. Fluid containing 30 gm per liter of the dissolved chemical flows into the tank at the rate of 4 liters/min. The mixture is kept uniform by stirring, and the stirred mixture simultaneously flows out at the rate of 2.5 liters/min. How much of the chemical is in the tank at the instant it overflows?

20. A 200 liter tank is initially full of fluid in which there is dissolved 40 gm of a certain chemical. Fluid containing 50 gm per liter of this chemical flows into the tank at the rate of 5 liters/min. The mixture is kept uniform by stirring, and the stirred mixture simultaneously flows out at the rate of 7 liters/min. How much of the chemical is in the tank when it is only half full?

21. The air in a room whose volume is 10,000 cu ft tests 0.15% carbon dioxide. Starting at $t = 0$, outside air testing 0.05% carbon dioxide is admitted at the rate of 5000 cu ft/min.

(a) What is the percentage of carbon dioxide in the air in the room after 3 min?

(b) When does the air in the room test 0.1% carbon dioxide?

22. The air in a room 50 ft by 20 ft by 8 ft tests 0.2% carbon dioxide. Starting at $t = 0$, outside air testing 0.05% carbon dioxide is admitted to the room. How many cubic feet of this outside air must be admitted per minute in order that the air in the room test 0.1% at the end of 30 min?

23. Newton's law of cooling states that the rate at which a body cools is proportional to the difference between the temperature of the body and that of the medium in which it is situated. A body of temperature 80 °F is placed at time $t = 0$ in a medium the temperature of which is maintained at 50 °F. At the end of 5 min, the body has cooled to a temperature of 70 °F.

    (a)   What is the temperature of the body at the end of 10 min?

    (b)   When will the temperature of the body be 60 °F?

24. A body cools from 60 °C to 50 °C in 15 min in air which is maintained to 30 °C. How long will it take this body to cool from 100 °C to 80 °C in air that is maintained at 50 °C? Assume Newton's law of cooling (Exercise 23).

25. The rate at which a certain substance dissolves in water is proportional at the product of the amount undissolved and the difference $c_1 - c_2$, where c_1 is the concentration in the saturated solution and c_2 is the concentration in the actual solution. If saturated, 50 gm of water would dissolve 20 gm of the substance. If 10 gm of the substance is placed in 50 gm of water and half of the substance is then dissolved in 90 min, how much will be dissolved in 3 hr?

CHAPTER FOUR

Explicit Methods of Solving Higher-Order
Linear Differential Equations

The subject of ordinary linear differential equations is one of great theoretical and practical importance. Theoretically, the subject is one of simplicity and elegance. Practically, linear differential equations originate in a variety of applications to science and engineering. Fortunately many of the linear differential equations that thus occur are of a special type, linear with constant coefficients, for which explicit methods of solution are available. The main purpose of this chapter is to study certain of these methods. First, however, we need to consider certain basic theorems that will be used throughout the chapter. These theorems are stated and illustrated in Section 4.1, but proofs are omitted in this introductory section. By far the most important case is that of the *second*-order linear differential equation, and we shall explicitly consider and illustrate this case for each important concept and result presented. In the final section of the chapter we return to this fundamental theory and present theorems *and* proofs in this important special case. Proofs in the general case are given in Chapter 11.

4.1 BASIC THEORY OF LINEAR DIFFERENTIAL EQUATIONS

A. Definition and Basic Existence Theorem

DEFINITION

A linear ordinary differential equation of order n in the dependent variable y and the independent variable x is an equation that is in, or can be expressed in, the form

$$a_0(x)\frac{d^n y}{dx^n} + a_1(x)\frac{d^{n-1} y}{dx^{n-1}} + \cdots + a_{n-1}(x)\frac{dy}{dx} + a_n(x)y = F(x), \qquad (4.1)$$

where a_0 is not identically zero. We shall assume that $a_0, a_1, \ldots, a_n$ and F are continuous real functions on a real interval $a \le x \le b$ and that $a_0(x) \ne 0$ for any x on $a \le x \le b$. The right-hand member $F(x)$ is called the nonhomogeneous *term. If F is identically zero, Equation (4.1) reduces to*

$$a_0(x)\frac{d^n y}{dx^n} + a_1(x)\frac{d^{n-1}y}{dx^{n-1}} + \cdots + a_{n-1}(x)\frac{dy}{dx} + a_n(x)y = 0 \qquad (4.2)$$

and is then called homogeneous.

For $n = 2$, Equation (4.1) reduces to the *second*-order nonhomogeneous linear differential equation

$$a_0(x)\frac{d^2 y}{dx^2} + a_1(x)\frac{dy}{dx} + a_2(x)y = F(x) \qquad (4.3)$$

and (4.2) reduces to the corresponding second-order *homogeneous* equation

$$a_0(x)\frac{d^2 y}{dx^2} + a_1(x)\frac{dy}{dx} + a_2(x)y = 0. \qquad (4.4)$$

Here we assume that a_0, a_1, a_2, and F are continuous real functions on a real interval $a \le x \le b$ and that $a_0(x) \ne 0$ for any x on $a \le x \le b$.

▶ **Example 4.1**

The equation

$$\frac{d^2 y}{dx^2} + 3x\frac{dy}{dx} + x^3 y = e^x$$

is a linear ordinary differential equation of the second order.

▶ **Example 4.2**

The equation

$$\frac{d^3 y}{dx^3} + x\frac{d^2 y}{dx^2} + 3x^2\frac{dy}{dx} - 5y = \sin x$$

is a linear ordinary differential equation of the third order.

We now state the basic existence theorem for initial-value problems associated with and *n*th-order linear ordinary differential equation:

THEOREM 4.1

Hypothesis

1. Consider the nth-order linear differential equation

$$a_0(x)\frac{d^n y}{dx^n} + a_1(x)\frac{d^{n-1}y}{dx^{n-1}} + \cdots + a_{n-1}(x)\frac{dy}{dx} + a_n(x)y = F(x), \qquad (4.1)$$

where $a_0, a_1, \ldots, a_n$ and F are continuous real functions on a real interval $a \le x \le b$ and $a_0(x) \ne 0$ for any x on $a \le x \le b$.

2. Let x_0 be any point of the interval $a \le x \le b$, and let $c_0, c_1, \ldots, c_{n-1}$ be n arbitrary real constants.

Conclusion. *There exists a unique solution f of (4.1) such that*

$$f(x_0) = c_0, f'(x_0) = c_1, \ldots, f^{(n-1)}(x_0) = c_{n-1},$$

and this solution is defined over the entire interval $a \le x \le b$.

Suppose that we are considering an nth-order linear differential equation (4.1), the coefficients and nonhomogeneous term of which all possess the continuity requirements set forth in Hypothesis 1 of Theorem 4.1 on a certain interval of the x axis. Then, given *any* point x_0 of this interval and *any* n real numbers $c_0, c_1, \ldots, c_{n-1}$, the theorem assures us that there is *precisely one* solution of the differential equation that assumes the value c_0 at $x = x_0$ and whose kth derivative assumes the value c_k for each $k = 1, 2, \ldots, n - 1$ at $x = x_0$. Further, the theorem asserts that this unique solution is defined for *all* x in the above-mentioned interval.

For the *second*-order linear differential equation,

$$a_0(x)\frac{d^2 y}{dx^2} + a_1(x)\frac{dy}{dx} + a_2(x)y = F(x), \tag{4.3}$$

the requirements of Hypothesis 1 of Theorem 4.1 are that a_0, a_1, a_2, and F be continuous on a real interval $a \le x \le b$ and that $a_0(x) \ne 0$ for any x on this interval. Then, if x_0 is *any* point of the interval $a \le x \le b$ and c_0 and c_1 are *any two* real numbers, the theorem assures us that there is *precisely one* solution f of the second-order differential equation (4.3) which assumes the value c_0 at $x = x_0$ and whose first derivative assumes the value c_1 at $x = x_0$:

$$f(x_0) = c_0, \qquad f'(x_0) = c_1. \tag{4.5}$$

Moreover, the theorem asserts that this unique solution f of Equation (4.3) which satisfies conditions (4.5) is defined for *all* x on the interval $a \le x \le b$.

▶ **Example 4.3**

Consider the initial-value problem

$$\frac{d^2 y}{dx^2} + 3x\frac{dy}{dx} + x^3 y = e^x,$$

$$y(1) = 2,$$

$$y'(1) = -5.$$

The coefficients $1, 3x$, and x^3, as well as the nonhomogeneous term e^x, in this second-order differential equation are all continuous for all values of x, $-\infty < x < \infty$. The point x_0 here is the point 1, which certainly belongs to this interval; and the real numbers c_0 and c_1 are 2 and -5, respectively. Thus Theorem 4.1 assures us that a solution of the given problem exists, is unique, and is defined for all x, $-\infty < x < \infty$.

▶ **Example 4.4**

Consider the initial-value problem

$$2\frac{d^3y}{dx^3} + x\frac{d^2y}{dx^2} + 3x^2\frac{dy}{dx} - 5y = \sin x,$$

$$y(4) = 3,$$

$$y'(4) = 5,$$

$$y''(4) = -\tfrac{7}{2}.$$

Here we have a third-order problem. The coefficients $2, x, 3x^2$, and -5, as well as the nonhomogeneous term $\sin x$, are all continuous for all x, $-\infty < x < \infty$. The point $x_0 = 4$ certainly belongs to this interval; the real numbers c_0, c_1, and c_2 in this problem are 3, 5, and $-\tfrac{7}{2}$, respectively. Theorem 4.1 assures us that this problem also has a unique solution which is defined for all x, $-\infty < x < \infty$.

A useful corollary to Theorem 4.1 is the following:

COROLLARY

Hypothesis. *Let f be a solution of the nth-order* homogeneous *linear differential equation*

$$a_0(x)\frac{d^ny}{dx^n} + a_1(x)\frac{d^{n-1}y}{dx^{n-1}} + \cdots + a_{n-1}(x)\frac{dy}{dx} + a_n(x)y = 0 \tag{4.2}$$

such that

$$f(x_0) = 0, f'(x_0) = 0, \ldots, f^{(n-1)}(x_0) = 0,$$

where x_0 is a point of the interval $a \le x \le b$ in which the coefficients $a_0, a_1, \ldots, a_n$ are all continuous and $a_0(x) \ne 0$.

Conclusion. *Then $f(x) = 0$ for all x on $a \le x \le b$.*

Let us suppose that we are considering a homogeneous equation of the form (4.2), all the coefficients of which are continuous on a certain interval of the x axis. Suppose further that we have a solution f of this equation which is such that f and its first $n - 1$ derivatives all equal zero at a point x_0 of this interval. Then this corollary states that this solution is the "trivial" solution f such that $f(x) = 0$ for *all* x on the above-mentioned interval.

▶ **Example 4.5**

The unique solution f of the third-order homogeneous equation

$$\frac{d^3y}{dx^3} + 2\frac{d^2y}{dx^2} + 4x\frac{dy}{dx} + x^2y = 0,$$

which is such that

$$f(2) = f'(2) = f''_{(2)} = 0,$$

is the trivial solution f such that $f(x) = 0$ for *all* x.

B. The Homogeneous Equation

We now consider the fundamental results concerning the homogeneous equation

$$a_0(x)\frac{d^n y}{dx^n} + a_1(x)\frac{d^{n-1}y}{dx^{n-1}} + \cdots + a_{n-1}(x)\frac{dy}{dx} + a_n(x)y = 0. \tag{4.2}$$

We first state the following basic theorem:

THEOREM 4.2 BASIC THEOREM ON LINEAR HOMOGENEOUS DIFFERENTIAL EQUATIONS

Hypothesis. *Let* $f_1, f_2, \ldots, f_m$ *be any m solutions of the homogeneous linear differential equation (4.2).*

Conclusion. *Then* $c_1 f_1 + c_2 f_2 + \cdots + c_m f_m$ *is also a solution of (4.2), where* $c_1, c_2, \ldots, c_m$ *are m arbitrary constants.*

Theorem 4.2 states that if m known solutions of (4.2) are each multiplied by an arbitrary constant and the resulting products are then added together, the resulting sum is also a solution of (4.2). We may put this theorem in a very simple form by means of the concept of linear combination, which we now introduce.

DEFINITION

If $f_1, f_2, \ldots, f_m$ *are m given functions, and* $c_1, c_2, \ldots, c_m$ *are m constants, then the expression*

$$c_1 f_1 + c_2 f_2 + \cdots + c_m f_m$$

is called a linear combination *of* $f_1, f_2, \ldots, f_m$.

In terms of this concept, Theorem 4.2 may be stated as follows:

THEOREM 4.2 (RESTATED)

Any linear combination of solutions of the homogeneous linear differential equation (4.2) is also a solution of (4.2).

In particular, any linear combination

$$c_1 f_1 + c_2 f_2 + \cdots + c_m f_m$$

of m solutions $f_1, f_2, \ldots, f_m$ of the *second*-order homogeneous linear differential equation

$$a_0(x) \frac{d^2 y}{dx^2} + a_1(x) \frac{dy}{dx} + a_2(x)y = 0 \tag{4.4}$$

is also a solution of (4.4).

▶ **Example 4.6**

The student will readily verify that $\sin x$ and $\cos x$ are solutions of

$$\frac{d^2 y}{dx^2} + y = 0.$$

Theorem 4.2 states that the linear combination $c_1 \sin x + c_2 \cos x$ is also a solution for any constants c_1 and c_2. For example, the particular linear combination

$$5 \sin x + 6 \cos x$$

is a solution.

▶ **Example 4.7**

The student may verify that e^x, e^{-x}, and e^{2x} are solutions of

$$\frac{d^3 y}{dx^3} - 2\frac{d^2 y}{dx^2} - \frac{dy}{dx} + 2y = 0.$$

Theorem 4.2 states that the linear combination $c_1 e^x + c_2 e^{-x} + c_3 e^{2x}$ is also a solution for any constants c_1, c_2, and c_3. For example, the particular linear combination

$$2e^x - 3e^{-x} + \tfrac{2}{3}e^{2x}$$

is a solution.

We now consider what constitutes the so-called general solution of (4.2). To understand this we first introduce the concepts of *linear dependence* and *linear independence*.

DEFINITION

The n functions $f_1, f_2, \ldots, f_n$ are called linearly dependent *on $a \le x \le b$ if there exist constants $c_1, c_2, \ldots, c_n$, not all zero, such that*

$$c_1 f_1(x) + c_2 f_2(x) + \cdots + c_n f_n(x) = 0$$

for all x such that $a \le x \le b$.

In particular, two functions f_1 and f_2 are linearly dependent *on $a \le x \le b$ if there exist constants c_1, c_2, not both zero, such that*

$$c_1 f_1(x) + c_2 f_2(x) = 0$$

for all x such that $a \le x \le b$.

▶ **Example 4.8**

We observe that x and $2x$ are linearly dependent on the interval $0 \leq x \leq 1$. For there exist constants c_1 and c_2, *not both zero*, such that

$$c_1 x + c_2(2x) = 0$$

for all x on the interval $0 \leq x \leq 1$. For example, let $c_1 = 2, c_2 = -1$.

▶ **Example 4.9**

We observe that $\sin x$, $3 \sin x$, and $-\sin x$ are linearly dependent on the interval $-1 \leq x \leq 2$. For there exist constants c_1, c_2, c_3, *not all zero*, such that

$$c_1 \sin x + c_2(3 \sin x) + c_3(-\sin x) = 0$$

for all x on the interval $-1 \leq x \leq 2$. For example, let $c_1 = 1, c_2 = 1, c_3 = 4$.

DEFINITION

The n functions $f_1, f_2, \ldots, f_n$ are called linearly independent *on the interval $a \leq x \leq b$ if they are* not *linearly dependent there. That is, the functions $f_1, f_2, \ldots, f_n$ are linearly independent on $a \leq x \leq b$ if the relation*

$$c_1 f_1(x) + c_2 f_2(x) + \cdots + c_n f_n(x) = 0$$

for all x such that $a \leq x \leq b$ implies that

$$c_1 = c_2 = \cdots = c_n = 0.$$

In other words, the only linear combination of $f_1, f_2, \ldots, f_n$ that is identically zero on $a \leq x \leq b$ is the trivial linear combination

$$0 \cdot f_1 + 0 \cdot f_2 + \cdots + 0 \cdot f_n.$$

In particular, two functions f_1 and f_2 are linearly independent *on $a \leq x \leq b$ if the relation*

$$c_1 f_1(x) + c_2 f_2(x) = 0$$

for all x on $a \leq x \leq b$ implies that

$$c_1 = c_2 = 0.$$

▶ **Example 4.10**

We assert that x and x^2 are linearly independent on $0 \leq x \leq 1$, since $c_1 x + c_2 x^2 = 0$ for *all* x on $0 \leq x \leq 1$ implies that both $c_1 = 0$ and $c_2 = 0$. We may verify this in the following way. We differentiate both sides of $c_1 x + c_2 x^2 = 0$ to obtain $c_1 + 2c_2 x = 0$, which must also hold for all x on $0 \leq x \leq 1$. Then from this we also have $c_1 x + 2c_2 x^2 = 0$ for all such x. Thus we have both

$$c_1 x + c_2 x^2 = 0 \quad \text{and} \quad c_1 x + 2c_2 x^2 = 0 \tag{4.6}$$

for *all* x on $0 \leq x \leq 1$. Subtracting the first from the second gives $c_2 x^2 = 0$ for all x on $0 \leq x \leq 1$, which at once implies $c_2 = 0$. Then either of (4.6) show similarly that $c_1 = 0$.

The next theorem is concerned with the existence of sets of linearly independent solutions of an nth-order homogeneous linear differential equation and with the significance of such linearly independent sets.

THEOREM 4.3

The nth-order homogeneous linear differential equation

$$a_0(x)\frac{d^n y}{dx^n} + a_1(x)\frac{d^{n-1} y}{dx^{n-1}} + \cdots + a_{n-1}(x)\frac{dy}{dx} + a_n(x)y = 0 \qquad (4.2)$$

always possesses n solutions that are linearly independent. Further, if $f_1, f_2, \ldots, f_n$ are n linearly independent solutions of (4.2), then every solution f of (4.2) can be expressed as a linear combination

$$c_1 f_1 + c_2 f_2 + \cdots + c_n f_n$$

of these n linearly independent solutions by proper choice of the constants $c_1, c_2, \ldots, c_n$.

Given an nth-order homogeneous linear differential equation, this theorem assures us first that a set of n linearly independent solutions actually exists. The existence of such a linearly independent set assured, the theorem goes on to tell us that *any solution* whatsoever of (4.2) can be written as a linear combination of such a linearly independent set of n solutions by suitable choice of the constants $c_1, c_2, \ldots, c_n$.

For the *second*-order homogeneous linear differential equation

$$a_0(x)\frac{d^2 y}{dx^2} + a_1(x)\frac{dy}{dx} + a_2(x)y = 0, \qquad (4.4)$$

Theorem 4.3 first assures us that a set of *two* linearly independent solutions exists. The existence of such a linearly independent set assured, let f_1 and f_2 be a set of two linearly independent solutions. Then if f is *any* solution of (4.4), the theorem also assures us that f can be expressed as a linear combination $c_1 f_1 + c_2 f_2$ of the two linearly independent solutions f_1 and f_2 by proper choice of the constants c_1 and c_2.

▶ **Example 4.11**

We have observed that $\sin x$ and $\cos x$ are solutions of

$$\frac{d^2 y}{dx^2} + y = 0 \qquad (4.7)$$

for all x, $-\infty < x < \infty$. Further, one can show that these two solutions are linearly independent. Now suppose f is *any* solution of (4.7). Then by Theorem 4.3 f can be expressed as a certain linear combination $c_1 \sin x + c_2 \cos x$ of the two linearly independent solutions $\sin x$ and $\cos x$ by proper choice of c_1 and c_2. That is, there exist two particular constants c_1 and c_2 such that

$$f(x) = c_1 \sin x + c_2 \cos x \qquad (4.8)$$

for all x, $-\infty < x < \infty$. For example, one can easily verify that $f(x) = \sin(x + \pi/6)$ is a solution of Equation (4.7). Since

$$\sin\left(x + \frac{\pi}{6}\right) = \sin x \cos\frac{\pi}{6} + \cos x \sin\frac{\pi}{6} = \frac{\sqrt{3}}{2}\sin x + \frac{1}{2}\cos x,$$

we see that the solution $\sin(x + \pi/6)$ can be expressed as the linear combination

$$\frac{\sqrt{3}}{2}\sin x + \tfrac{1}{2}\cos x$$

of the two linearly independent solutions $\sin x$ and $\cos x$. Note that this is of the form in the right member of (4.8) with $c_1 = \sqrt{3/2}$ and $c_2 = \tfrac{1}{2}$.

Now let $f_1, f_2, \ldots, f_n$ be a set of n linearly independent solutions of (4.2). Then by Theorem 4.2 we know that the linear combination

$$c_1 f_1 + c_2 f_2 + \cdots + c_n f_n, \tag{4.9}$$

where $c_1, c_2, \ldots, c_n$ are n arbitrary constants, is also a solution of (4.2). On the other hand, by Theorem 4.3 we know that if f is any solution of (4.2), then it can be expressed as a linear combination (4.9) of the n linearly independent solutions $f_1, f_2, \ldots, f_n$, by a suitable choice of the constants $c_1, c_2, \ldots, c_n$. Thus a linear combination (4.9) of the n linearly independent solutions $f_1, f_2, \ldots, f_n$ in which $c_1, c_2, \ldots, c_n$ are arbitrary constants must include all solutions of (4.2). For this reason, we refer to a set of n linearly independent solutions of (4.2) as a "fundamental set" of (4.2) and call a "general" linear combination of n linearly independent solutions a "general solution" of (4.2), in accordance with the following definition:

DEFINITION

If $f_1, f_2, \ldots, f_n$ are n linearly independent solutions of the nth-order homogeneous linear differential equation

$$a_0(x)\frac{d^n y}{dx^n} + a_1(x)\frac{d^{n-1}y}{dx^{n-1}} + \cdots + a_{n-1}(x)\frac{dy}{dx} + a_n(x)y = 0 \tag{4.2}$$

on $a \le x \le b$, then the set $f_1, f_2, \ldots, f_n$ is called a fundamental set *of solutions of (4.2) and the function f defined by*

$$f(x) = c_1 f_1(x) + c_2 f_2(x) + \cdots + c_n f_n(x), \qquad a \le x \le b,$$

where $c_1, c_2, \ldots, c_n$ are arbitrary constants, is called a general solution *of (4.2) on $a \le x \le b$.*

Therefore, if we can find n linearly independent solutions of (4.2), we can at once write the general solution of (4.2) as a general linear combination of these n solutions.

For the *second*-order homogeneous linear differential equation

$$a_0(x)\frac{d^2 y}{dx^2} + a_1(x)\frac{dy}{dx} + a_2(x)y = 0, \tag{4.4}$$

a *fundamental set* consists of *two* linearly independent solutions. If f_1 and f_2 are a fundamental set of (4.4) on $a \leq x \leq b$, then a *general solution* of (4.4) on $a \leq x \leq b$ is defined by

$$c_1 f_1(x) + c_2 f_2(x), \qquad a \leq x \leq b,$$

where c_1 and c_2 are arbitrary constants.

▶ **Example 4.12**

We have observed that $\sin x$ and $\cos x$ are solutions of

$$\frac{d^2 y}{dx^2} + y = 0$$

for all x, $-\infty < x < \infty$. Further, one can show that these two solutions are linearly independent. Thus, they constitute a fundamental set of solutions of the given differential equation, and its general solution may be expressed as the linear combination

$$c_1 \sin x + c_2 \cos x,$$

where c_1 and c_2 are arbitrary constants. We write this as $y = c_1 \sin x + c_2 \cos x$.

▶ **Example 4.13**

The solutions e^x, e^{-x}, and e^{2x} of

$$\frac{d^3 y}{dx^3} - 2\frac{d^2 y}{dx^2} - \frac{dy}{dx} + 2y = 0$$

may be shown to be linearly independent for all x, $-\infty < x < \infty$. Thus, e^x, e^{-x}, and e^{2x} constitute a fundamental set of the given differential equation, and its general solution may be expressed as the linear combination

$$c_1 e^x + c_2 e^{-x} + c_3 e^{2x},$$

where c_1, c_2, and c_3 are arbitrary constants. We write this as

$$y = c_1 e^x + c_2 e^{-x} + c_3 e^{2x}.$$

The next theorem gives a simple criterion for determining whether or not n solutions of (4.2) are linearly independent. We first introduce another concept.

DEFINITION

Let $f_1, f_2, \ldots, f_n$ be n real functions each of which has an $(n-1)$st derivative on a real interval $a \leq x \leq b$. The determinant

$$W(f_1, f_2, \ldots, f_n) = \begin{vmatrix} f_1 & f_2 & \cdots & f_n \\ f'_1 & f'_2 & \cdots & f'_n \\ \vdots & \vdots & & \vdots \\ f_1^{(n-1)} & f_2^{(n-1)} & \cdots & f_n^{(n-1)} \end{vmatrix},$$

in which primes denote derivatives, is called the Wronskian *of these n functions. We observe that $W(f_1, f_2, \ldots, f_n)$ is itself a real function defined on $a \le x \le b$. Its value at x is denoted by $W(f_1, f_2, \ldots, f_n)(x)$ or by $W[f_1(x), f_2(x), \ldots, f_n(x)]$.*

THEOREM 4.4

The n solutions $f_1, f_2, \ldots, f_n$ of the nth-order homogeneous linear differential equation (4.2) are linearly independent on $a \le x \le b$ if and only if the Wronskian of $f_1, f_2, \ldots, f_n$ is different from zero for some x on the interval $a \le x \le b$.

We have further:

THEOREM 4.5

The Wronskian of n solutions $f_1, f_2, \ldots, f_n$ of (4.2) is either identically zero on $a \le x \le b$ or else is never zero on $a \le x \le b$.

Thus if we can find n solutions of (4.2), we can apply the Theorems 4.4. and 4.5 to determine whether or not they are linearly independent. If they are linearly independent, then we can form the general solution as a linear combination of these n linearly independent solutions.

In the case of the general *second*-order homogeneous linear differential equation

$$a_0(x)\frac{d^2y}{dx^2} + a_1(x)\frac{dy}{dx} + a_2(x)y = 0, \tag{4.4}$$

the Wronskian of two solutions f_1 and f_2 is the second-order determinant

$$W(f_1, f_2) = \begin{vmatrix} f_1 & f_2 \\ f'_1 & f'_2 \end{vmatrix} = f_1 f'_2 - f'_1 f_2.$$

By Theorem 4.4, two solutions f_1 and f_2 of (4.4) are linearly independent on $a \le x \le b$ if and only if their Wronskian is different from zero for some x on $a \le x \le b$; and by Theorem 4.5, this Wronskian is either always zero or never zero on $a \le x \le b$. Thus if $W[f_1(x), f_2(x)] \ne 0$ on $a \le x \le b$, solutions f_1 and f_2 of (4.4) are linearly independent on $a \le x \le b$ and the general solution of (4.4) can be written as the linear combination

$$c_1 f_1(x) + c_2 f_2(x),$$

where c_1 and c_2 are arbitrary constants.

▶ **Example 4.14**

We apply Theorem 4.4 to show that the solutions $\sin x$ and $\cos x$ of

$$\frac{d^2y}{dx^2} + y = 0$$

are linearly independent. We find that

$$W(\sin x, \cos x) = \begin{vmatrix} \sin x & \cos x \\ \cos x & -\sin x \end{vmatrix} = -\sin^2 x - \cos^2 x = -1 \neq 0$$

for all real x. Thus, since $W(\sin x, \cos x) \neq 0$ for all real x, we conclude that $\sin x$ and $\cos x$ are indeed linearly independent solutions of the given differential equation on every real interval.

▶ **Example 4.15**

The solutions e^x, e^{-x}, and e^{2x} of

$$\frac{d^3y}{dx^3} - 2\frac{d^2y}{dx^2} - \frac{dy}{dx} + 2y = 0$$

are linearly independent on every real interval, for

$$W(e^x, e^{-x}, e^{2x}) = \begin{vmatrix} e^x & e^{-x} & e^{2x} \\ e^x & -e^{-x} & 2e^{2x} \\ e^x & e^{-x} & 4e^{2x} \end{vmatrix} = e^{2x} \begin{vmatrix} 1 & 1 & 1 \\ 1 & -1 & 2 \\ 1 & 1 & 4 \end{vmatrix} = -6e^{2x} \neq 0$$

for all real x.

Exercises

1. Theorem 4.1 applies to one of the following problems but not to the other. Determine to which of the problems the theorem applies and state precisely the conclusion which can be drawn in this case. Explain why the theorem does not apply to the remaining problem.

   (a) $\dfrac{d^2y}{dx^2} + 5\dfrac{dy}{dx} + 6y = e^x,$ $y(0) = 5,$ $y'(0) = 7.$

   (b) $\dfrac{d^2y}{dx^2} + 5\dfrac{dy}{dx} + 6y = e^x,$ $y(0) = 5,$ $y'(1) = 7.$

2. Answer orally: What is the solution of the following initial-value problem? Why?

   $$\frac{d^2y}{dx^2} + x\frac{dy}{dx} + x^2y = 0, \qquad y(1) = 0, \qquad y'(1) = 0.$$

3. Prove Theorem 4.2 for the case $m = n = 2$. That is, prove that if $f_1(x)$ and $f_2(x)$ are two solutions of

   $$a_0(x)\frac{d^2y}{dx^2} + a_1(x)\frac{dy}{dx} + a_2(x)y = 0,$$

   then $c_1 f_1(x) + c_2 f_2(x)$ is also a solution of this equation, where c_1 and c_2 are arbitrary constants.

4. Consider the differential equation

$$\frac{d^2 y}{dx^2} - 4\frac{dy}{dx} + 3y = 0. \qquad (A)$$

(a) Show that each of the functions e^x and e^{3x} is a solution of differential equation (A) on the interval $a \le x \le b$, where a and b are arbitrary real numbers such that $a < b$.

(b) What theorem enables us to conclude at once that each of the functions

$$5e^x + 2e^{3x}, \qquad 6e^x - 4e^{3x}, \quad \text{and} \quad -7e^x + 5e^{3x}$$

is also a solution of differential equation (A) on $a \le x \le b$?

(c) Each of the functions

$$3e^x, \qquad -4e^x, \qquad 5e^x, \quad \text{and} \quad 6e^x$$

is also a solution of differential equation (A) on $a \le x \le b$. Why?

5. Again consider the differential equation (A) of Exercise 4.

(a) Use the definition of linear dependence to show that the four functions of part (c) of Exercise 4 are linearly dependent on $a \le x \le b$.

(b) Use Theorem 4.4 to show that each pair of the four solutions of differential equation (A) listed in part (c) of Exercise 4 are linearly dependent on $a \le x \le b$.

6. Again consider the differential equation (A) of Exercise 4.

(a) Use the definition of linear independence to show that the two functions e^x and e^{3x} are linearly independent on $a \le x \le b$.

(b) Use Theorem 4.4 to show that the two solutions e^x and e^{3x} of differential equation (A) are linearly independent on $a \le x \le b$.

7. Consider the differential equation

$$\frac{d^2 y}{dx^2} - 5\frac{dy}{dx} + 6y = 0.$$

(a) Show that e^{2x} and e^{3x} are linearly independent solutions of this equation on the interval $-\infty < x < \infty$.

(b) Write the general solution of the given equation.

(c) Find the solution that satisfies the conditions $y(0) = 2$, $y'(0) = 3$. Explain why this solution is unique. Over what interval is it defined?

8. Consider the differential equation

$$\frac{d^2 y}{dx^2} - 2\frac{dy}{dx} + y = 0.$$

(a) Show that e^x and xe^x are linearly independent solutions of this equation on the interval $-\infty < x < \infty$.

(b) Write the general solution of the given equation.

(c) Find the solution that satisfies the condition $y(0) = 1$, $y'(0) = 4$. Explain why this solution is unique. Over what interval is it defined?

9. Consider the differential equation

$$x^2 \frac{d^2 y}{dx^2} - 2x \frac{dy}{dx} + 2y = 0.$$

(a) Show that x and x^2 are linearly independent solutions of this equation on the interval $0 < x < \infty$.

(b) Write the general solution of the given equation.

(c) Find the solution that satisfies the conditions $y(1) = 3$, $y'(1) = 2$. Explain why this solution is unique. Over what interval is this solution defined?

10. Consider the differential equation

$$x^2 \frac{d^2 y}{dx^2} + x \frac{dy}{dx} - 4y = 0.$$

(a) Show that x^2 and $1/x^2$ are linearly independent solutions of this equation on the interval $0 < x < \infty$.

(b) Write the general solution of the given equation.

(c) Find the solution that satisfies the conditions $y(2) = 3$, $y'(2) = -1$. Explain why this solution is unique. Over what interval is this solution defined?

11. Consider the differential equation

$$\frac{d^2 y}{dx^2} - 5 \frac{dy}{dx} + 4y = 0.$$

(a) Show that each of the functions e^x, e^{4x}, and $2e^x - 3e^{4x}$ is a solution of this equation on the interval $-\infty < x < \infty$.

(b) Show that the solutions e^x and e^{4x} are linearly independent on $-\infty < x < \infty$.

(c) Show that the solutions e^x and $2e^x - 3e^{4x}$ are also linearly independent on $-\infty < x < \infty$.

(d) Are the solutions e^{4x} and $2e^x - 3e^{4x}$ still another pair of linearly independent solutions on $-\infty < x < \infty$? Justify your answer.

12. Given that e^{-x}, e^{3x}, and e^{4x} are all solutions of

$$\frac{d^3 y}{dx^3} - 6 \frac{d^2 y}{dx^2} + 5 \frac{dy}{dx} + 12y = 0,$$

show that they are linearly independent on the interval $-\infty < x < \infty$ and write the general solution.

13. Given that x, x^2, and x^4 are all solutions of

$$x^3 \frac{d^3 y}{dx^3} - 4x^2 \frac{d^2 y}{dx^2} + 8x \frac{dy}{dx} - 8y = 0,$$

show that they are linearly independent on the interval $0 < x < \infty$ and write the general solution.

C. Reduction of Order

In Section 4.2 we shall begin to study methods for obtaining explicit solutions of higher-order linear differential equations. There and in later sections we shall find that the following theorem on reduction of order is often quite useful.

THEOREM 4.6

Hypothesis.   *Let f be a nontrivial solution of the nth-order homogeneous linear differential equation*

$$a_0(x)\frac{d^n y}{dx^n} + a_1(x)\frac{d^{n-1}y}{dx^{n-1}} + \cdots + a_{n-1}(x)\frac{dy}{dx} + a_n(x)y = 0. \tag{4.2}$$

Conclusion.   *The transformation $y = f(x)v$ reduces Equation (4.2) to an $(n-1)$st-order homogeneous linear differential equation in the dependent variable $w = dv/dx$.*

This theorem states that if one nonzero solution of the nth-order homogeneous linear differential equation (4.2) is known, then by making the appropriate transformation we may reduce the given equation to another homogeneous linear equation that is one order lower than the original. Since this theorem will be most useful for us in connection with second-order homogeneous linear equations (the case where $n = 2$), we shall now investigate the second-order case in detail. Suppose f is a *known* nontrivial solution of the second-order homogeneous linear equation

$$a_0(x)\frac{d^2 y}{dx^2} + a_1(x)\frac{dy}{dx} + a_2(x)y = 0. \tag{4.10}$$

Let us make the transformation

$$y = f(x)v, \tag{4.11}$$

where f is the *known* solution of (4.10) and v is a function of x that will be determined. Then, differentiating, we obtain

$$\frac{dy}{dx} = f(x)\frac{dv}{dx} + f'(x)v, \tag{4.12}$$

$$\frac{d^2 y}{dx^2} = f(x)\frac{d^2 v}{dx^2} + 2f'(x)\frac{dv}{dx} + f''(x)v. \tag{4.13}$$

Substituting (4.11), (4.12), and (4.13) into (4.10), we obtain

$$a_0(x)\left[f(x)\frac{d^2 v}{dx^2} + 2f'(x)\frac{dv}{dx} + f''(x)v\right] + a_1(x)\left[f(x)\frac{dv}{dx} + f'(x)v\right] + a_2(x)f(x)v = 0$$

or

$$a_0(x)f(x)\frac{d^2 v}{dx^2} + [2a_0(x)f'(x) + a_1(x)f(x)]\frac{dv}{dx}$$
$$+ [a_0(x)f''(x) + a_1(x)f'(x) + a_2(x)f(x)]v = 0.$$

Since f is a solution of (4.10), the coefficient of v is zero, and so the last equation reduces to

$$a_0(x)f(x)\frac{d^2v}{dx^2} + [2a_0(x)f'(x) + a_1(x)f(x)]\frac{dv}{dx} = 0.$$

Letting $w = dv/dx$, this becomes

$$a_0(x)f(x)\frac{dw}{dx} + [2a_0(x)f'(x) + a_1(x)f(x)]w = 0. \tag{4.14}$$

This is a *first*-order homogeneous linear differential equation in the dependent variable w. The equation is separable; thus assuming $f(x) \neq 0$ and $a_0(x) \neq 0$, we may write

$$\frac{dw}{w} = -\left[2\frac{f'(x)}{f(x)} + \frac{a_1(x)}{a_0(x)}\right]dx.$$

Thus integrating, we obtain

$$\ln|w| = -\ln[f(x)]^2 - \int\frac{a_1(x)}{a_0(x)}dx + \ln|c|$$

or

$$w = \frac{c\exp\left[-\int\frac{a_1(x)}{a_0(x)}dx\right]}{[f(x)]^2}.$$

This is the general solution of Equation (4.14); choosing the particular solution for which $c = 1$, recalling that $dv/dx = w$, and integrating again, we now obtain

$$v = \int\frac{\exp\left[-\int\frac{a_1(x)}{a_0(x)}dx\right]}{[f(x)]^2}dx.$$

Finally, from (4.11), we obtain

$$y = f(x)\int\frac{\exp\left[-\int\frac{a_1(x)}{a_0(x)}dx\right]}{[f(x)]^2}dx. \tag{4.15}$$

The function defined in the right member of (4.15), which we shall henceforth denote by g, is actually a solution of the original second-order equation (4.10). Furthermore, this new solution g and the original known solution f are linearly independent, since

$$W(f, g)(x) = \begin{vmatrix} f(x) & g(x) \\ f'(x) & g'(x) \end{vmatrix} = \begin{vmatrix} f(x) & f(x)v \\ f'(x) & f(x)v' + f'(x)v \end{vmatrix}$$

$$= [f(x)]^2v' = \exp\left[-\int\frac{a_1(x)}{a_0(x)}dx\right] \neq 0.$$

Thus the linear combination

$$c_1f + c_2g$$

is the general solution of Equation (4.10). We now summarize this discussion in the following theorem.

THEOREM 4.7

Hypothesis. *Let f be a nontrivial solution of the second-order homogeneous linear differential equation*

$$a_0(x)\frac{d^2 y}{dx^2} + a_1(x)\frac{dy}{dx} + a_2(x)y = 0. \tag{4.10}$$

Conclusion 1. *The transformation $y = f(x)v$ reduces Equation (4.10) to the first-order homogeneous linear differential equation*

$$a_0(x)f(x)\frac{dw}{dx} + [2a_0(x)f'(x) + a_1(x)f(x)]w = 0 \tag{4.14}$$

in the dependent variable w, where $w = dv/dx$.

Conclusion 2. *The particular solution*

$$w = \frac{\exp\left[-\int \frac{a_1(x)}{a_0(x)}\,dx\right]}{[f(x)]^2}$$

of Equation (4.14) gives rise to the function v, where

$$v(x) = \int \frac{\exp\left[-\int \frac{a_1(x)}{a_0(x)}\,dx\right]}{[f(x)]^2}\,dx.$$

The function g defined by $g(x) = f(x)v(x)$ is then a solution of the second-order equation (4.10).

Conclusion 3. *The original known solution f and the "new" solution g are linearly independent solutions of (4.10), and hence the general solution of (4.10) may be expressed as the linear combination*

$$c_1 f + c_2 g.$$

Let us emphasize the utility of this theorem and at the same time clearly recognize its limitations. Certainly its utility is by now obvious. It tells us that if one solution of the second-order equation (4.10) is known, then we can reduce the order to obtain a linearly independent solution and thereby obtain the general solution of (4.10). But the limitations of the theorem are equally obvious. One solution of Equation (4.10) must already be known to us in order to apply the theorem. How does one "already know" a solution? In general one does not. In some cases the form of the equation itself or related physical considerations suggest that there may be a solution of a certain special form: for example, an exponential solution or a linear solution. However, such cases are not too common and if no solution at all can be so ascertained, then the theorem will not aid us.

We now illustrate the method of reduction of order by means of the following example.

▶ **Example 4.16**

Given that $y = x$ is a solution of

$$(x^2 + 1)\frac{d^2y}{dx^2} - 2x\frac{dy}{dx} + 2y = 0, \qquad (4.16)$$

find a linearly independent solution by reducing the order.

Solution. First observe that $y = x$ *does* satisfy Equation (4.16). Then let

$$y = xv.$$

Then

$$\frac{dy}{dx} = x\frac{dv}{dx} + v \quad \text{and} \quad \frac{d^2y}{dx^2} = x\frac{d^2v}{dx^2} + 2\frac{dv}{dx}.$$

Substituting the expressions for y, dy/dx, and d^2y/dx^2 into Equation (4.16), we obtain

$$(x^2 + 1)\left(x\frac{d^2v}{dx^2} + 2\frac{dv}{dx}\right) - 2x\left(x\frac{dv}{dx} + v\right) + 2xv = 0$$

or

$$x(x^2 + 1)\frac{d^2v}{dx^2} + 2\frac{dv}{dx} = 0.$$

Letting $w = dv/dx$ we obtain the *first*-order homogeneous linear equation

$$x(x^2 + 1)\frac{dw}{dx} + 2w = 0.$$

Treating this as a separable equation, we obtain

$$\frac{dw}{w} = -\frac{2\,dx}{x(x^2 + 1)}$$

or

$$\frac{dw}{w} = \left(-\frac{2}{x} + \frac{2x}{x^2 + 1}\right)dx.$$

Integrating, we obtain the general solution

$$w = \frac{c(x^2 + 1)}{x^2}.$$

Choosing $c = 1$, we recall that $dv/dx = w$ and integrate to obtain the function v given by

$$v(x) = x - \frac{1}{x}.$$

Now forming $g = fv$, where $f(x)$ denotes the *known* solution x, we obtain the function g defined by

$$g(x) = x\left(x - \frac{1}{x}\right) = x^2 - 1.$$

By Theorem 4.7 we know that this is the desired linearly independent solution. The general solution of Equation (4.16) may thus be expressed as the linear combination $c_1 x + c_2(x^2 - 1)$ of the linearly independent solutions f and g. We thus write the general solution of Equation (4.16) as

$$y = c_1 x + c_2(x^2 - 1).$$

D. The Nonhomogeneous Equation

We now return briefly to the nonhomogeneous equation

$$a_0(x)\frac{d^n y}{dx^n} + a_1(x)\frac{d^{n-1} y}{dx^{n-1}} + \cdots + a_{n-1}(x)\frac{dy}{dx} + a_n(x)y = F(x). \tag{4.1}$$

The basic theorem dealing with this equation is the following.

THEOREM 4.8

Hypothesis

(1) Let v be any solution of the given (nonhomogeneous) nth-order linear differential equation (4.1). (2) Let u be any solution of the corresponding homogeneous equation

$$a_0(x)\frac{d^n y}{dx^n} + a_1(x)\frac{d^{n-1} y}{dx^{n-1}} + \cdots + a_{n-1}(x)\frac{dy}{dx} + a_n(x)y = 0. \tag{4.2}$$

Conclusion. *Then $u + v$ is also a solution of the given (nonhomogeneous) equation (4.1).*

▶ **Example 4.17**

Observe that $y = x$ is a solution of the nonhomogeneous equation

$$\frac{d^2 y}{dx^2} + y = x.$$

and that $y = \sin x$ is a solution of the corresponding homogeneous equation

$$\frac{d^2 y}{dx^2} + y = 0.$$

Then by Theorem 4.8 the sum

$$\sin x + x$$

is also a solution of the given nonhomogeneous equation

$$\frac{d^2 y}{dx^2} + y = x.$$

The student should check that this is indeed true.

Now let us apply Theorem 4.8 in the case where v is a given solution y_p of the nonhomogeneous equation (4.1) involving no arbitrary constants, and u is the general solution

$$y_c = c_1 y_1 + c_2 y_2 + \cdots + c_n y_n$$

of the corresponding homogeneous equation (4.2). Then by this theorem,

$$y_c + y_p$$

is also a solution of the nonhomogeneous equation (4.1), and it is a solution involving n arbitrary constants $c_1, c_2, \ldots, c_n$. Concerning the significance of such a solution, we now state the following result.

THEOREM 4.9

Hypothesis

(1) Let y_p be a given solution of the nth-order nonhomogeneous linear equation (4.1) involving no arbitrary constants. (2) Let

$$y_c = c_1 y_1 + c_2 y_2 + \cdots + c_n y_n$$

be the general solution of the corresponding homogeneous equation (4.2).

Conclusion. *Then every solution ϕ of the nth-order nonhomogeneous equation (4.1) can be expressed in the form*

$$y_c + y_p,$$

that is,

$$c_1 y_1 + c_2 y_2 + \cdots + c_n y_n + y_p$$

for suitable choice of the n arbitrary constants $c_1, c_2, \ldots, c_n$.

This result suggests that we call a solution of Equation (4.1) of the form $y_c + y_p$, a *general solution* of (4.1), in accordance with the following definition:

DEFINITION

Consider the nth-order (nonhomogeneous) linear differential equation

$$a_0(x)\frac{d^n y}{dx^n} + a_1(x)\frac{d^{n-1} y}{dx^{n-1}} + \cdots + a_{n-1}(x)\frac{dy}{dx} + a_n(x)y = F(x) \qquad (4.1)$$

and the corresponding homogeneous equation

$$a_0(x)\frac{d^n y}{dx^n} + a_1(x)\frac{d^{n-1} y}{dx^{n-1}} + \cdots + a_{n-1}(x)\frac{dy}{dx} + a_n(x)y = 0. \qquad (4.2)$$

1. *The general solution of (4.2) is called the* complementary function *of Equation (4.1). We shall denote this by y_c.*

2. *Any particular solution of (4.1) involving no arbitrary constants is called a* particular integral *of (4.1). We shall denote this by* y_p.
3. *The solution* $y_c + y_p$ *of (4.1), where* y_c *is the complementary function and* y_p *is a particular integral of (4.1), is called the* general solution *of (4.1).*

Thus to find the general solution of (4.1), we need merely find:

1. *The complementary function*, that is, a "general" linear combination of n linearly independent solutions of the corresponding homogeneous equation (4.2); and
2. *A particular integral*, that is, any particular solution of (4.1) involving no arbitrary constants.

▶ **Example 4.18**

Consider the differential equation

$$\frac{d^2 y}{dx^2} + y = x.$$

The complementary function is the general solution

$$y_c = c_1 \sin x + c_2 \cos x$$

of the corresponding homogeneous equation

$$\frac{d^2 y}{dx^2} + y = 0.$$

A particular integral is given by

$$y_p = x.$$

Thus the general solution of the given equation may be written

$$y = y_c + y_p = c_1 \sin x + c_2 \cos x + x.$$

In the remaining sections of this chapter we shall proceed to study methods of obtaining the two constituent parts of the general solution.

We point out that if the nonhomogeneous member $F(x)$ of the linear differential equation (4.1) is expressed as a linear combination of two or more functions, then the following theorem may often be used to advantage in finding a particular integral.

THEOREM 4.10

Hypothesis

1. Let f_1 *be a particular integral of*

$$a_0(x)\frac{d^n y}{dx^n} + a_1(x)\frac{d^{n-1} y}{dx^{n-1}} + \cdots + a_{n-1}(x)\frac{dy}{dx} + a_n(x)y = F_1(x). \qquad (4.17)$$

2. Let f_2 be a particular integral of

$$a_o(x)\frac{d^n y}{dx^n} + a_1(x)\frac{d^{n-1} y}{dx^{n-1}} + \cdots + a_{n-1}(x)\frac{dy}{dx} + a_n(x)y = F_2(x). \qquad (4.18)$$

Conclusion. Then $k_1 f_1 + k_2 f_2$ is a particular integral of

$$a_0(x)\frac{d^n y}{dx^n} + a_1(x)\frac{d^{n-1} y}{dx^{n-1}} + \cdots + a_{n-1}(x)\frac{dy}{dx} + a_n(x)y = k_1 F_1(x) + k_2 F_2(x),$$

$$(4.19)$$

where k_1 and k_2 are constants.

▶ **Example 4.19**

Suppose we seek a particular integral of

$$\frac{d^2 y}{dx^2} + y = 3x + 5 \tan x. \qquad (4.20)$$

We may then consider the two equations

$$\frac{d^2 y}{dx^2} + y = x \qquad (4.21)$$

and

$$\frac{d^2 y}{dx^2} + y = \tan x. \qquad (4.22)$$

We have already noted in Example 4.18 that a particular integral of Equation (4.21) is given by

$$y = x.$$

Further, we can verify (by direct substitution) that a particular integral of Equation (4.22) is given by

$$y = -(\cos x)\ln |\sec x + \tan x|.$$

Therefore, applying Theorem 4.10, a particular integral of Equation (4.22) is

$$y = 3x - 5(\cos x)\ln |\sec x + \tan x|.$$

This example makes the utility of Theorem 4.10 apparent. The particular integral $y = x$ of (4.21) can be quickly determined by the method of Section 4.3 (or by direct inspection!), whereas the particular integral

$$y = -(\cos x)\ln |\sec x + \tan x|$$

of (4.22) must be determined by the method of Section 4.4, and this requires considerably greater computation.

Exercises

1. Given that $y = x$ is a solution of

    $$x^2 \frac{d^2 y}{dx^2} - 4x \frac{dy}{dx} + 4y = 0,$$

    find a linearly independent solution by reducing the order. Write the general solution.

2. Given that $y = x + 1$ is a solution of

    $$(x + 1)^2 \frac{d^2 y}{dx^2} - 3(x + 1) \frac{dy}{dx} + 3y = 0,$$

    find a linearly independent solution by reducing the order. Write the general solution.

3. Given that $y = x$ is a solution of

    $$(x^2 - 1) \frac{d^2 y}{dx^2} - 2x \frac{dy}{dx} + 2y = 0,$$

    find a linearly independent solution by reducing the order. Write the general solution.

4. Given that $y = x$ is a solution of

    $$(x^2 - x + 1) \frac{d^2 y}{dx^2} - (x^2 + x) \frac{dy}{dx} + (x + 1)y = 0,$$

    find a linearly independent solution by reducing the order. Write the general solution.

5. Given that $y = e^{2x}$ is a solution of

    $$(2x + 1) \frac{d^2 y}{dx^2} - 4(x + 1) \frac{dy}{dx} + 4y = 0,$$

    find a linearly independent solution by reducing the order. Write the general solution.

6. Given that $y = x^2$ is a solution of

    $$(x^3 - x^2) \frac{d^2 y}{dx^2} - (x^3 + 2x^2 - 2x) \frac{dy}{dx} + (2x^2 + 2x - 2)y = 0,$$

    find a linearly independent solution by reducing the order. Write the general solution.

7. Prove Theorem 4.8 for the case $n = 2$. That is, prove that if u is any solution of

    $$a_0(x) \frac{d^2 y}{dx^2} + a_1(x) \frac{dy}{dx} + a_2(x)y = 0$$

    and v is any solution of

    $$a_0(x) \frac{d^2 y}{dx^2} + a_1(x) \frac{dy}{dx} + a_2(x)y = F(x),$$

    then $u + v$ is also a solution of this latter nonhomogeneous equation.

8. Consider the nonhomogeneous differential equation

$$\frac{d^2y}{dx^2} - 3\frac{dy}{dx} + 2y = 4x^2.$$

(a) Show that e^x and e^{2x} are linearly independent solutions of the corresponding homogeneous equation

$$\frac{d^2y}{dx^2} - 3\frac{dy}{dx} + 2y = 0.$$

(b) What is the complementary function of the given nonhomogeneous equation?

(c) Show that $2x^2 + 6x + 7$ is a particular integral of the given equation.

(d) What is the general solution of the given equation?

9. Given that a particular integral of

$$\frac{d^2y}{dx^2} - 5\frac{dy}{dx} + 6y = 1 \quad \text{is} \quad y = \frac{1}{6},$$

a particular integral of

$$\frac{d^2y}{dx^2} - 5\frac{dy}{dx} + 6y = x \quad \text{is} \quad y = \frac{x}{6} + \frac{5}{36},$$

and a particular integral of

$$\frac{d^2y}{dx^2} - 5\frac{dy}{dx} + 6y = e^x \quad \text{is} \quad y = \frac{e^x}{2},$$

use Theorem 4.10 to find a particular integral of

$$\frac{d^2y}{dx^2} - 5\frac{dy}{dx} + 6y = 2 - 12x + 6e^x.$$

4.2 THE HOMOGENEOUS LINEAR EQUATION WITH CONSTANT COEFFICIENTS

A. Introduction

In this section we consider the special case of the nth-order homogeneous linear differential equation in which all of the coefficients are real constants. That is, we shall be concerned with the equation

$$a_0\frac{d^ny}{dx^n} + a_1\frac{d^{n-1}y}{dx^{n-1}} + \cdots + a_{n-1}\frac{dy}{dx} + a_ny = 0 \tag{4.23}$$

where $a_0, a_1, \ldots, a_{n-1}, a_n$ are real constants. We shall show that the general solution of this equation can be found explicitly.

In an attempt to find solutions of a differential equation we would naturally inquire whether or not any familiar type of function might possibly have the properties that would enable it to be a solution. The differential equation (4.23) requires a function f

having the property such that if it and its various derivatives are each multiplied by certain constants, the a_i, and the resulting products, $a_i f^{(n-i)}$, are then added, the result will equal zero for all values of x for which this result is defined. For this to be the case we need a function such that its derivatives are constant multiples of itself. Do we know of functions f having this property that

$$\frac{d^k}{dx^k}[f(x)] = cf(x)$$

for all x? The answer is "yes," for the exponential function f such that $f(x) = e^{mx}$, where m is a constant, is such that

$$\frac{d^k}{dx^k}(e^{mx}) = m^k e^{mx}.$$

Thus we shall seek solutions of (4.23) of the form $y = e^{mx}$, where the constant m will be chosen such that e^{mx} *does* satisfy the equation. Assuming then that $y = e^{mx}$ is a solution for certain m, we have:

$$\frac{dy}{dx} = me^{mx},$$

$$\frac{d^2 y}{dx^2} = m^2 e^{mx},$$

$$\vdots$$

$$\frac{d^n y}{dx^n} = m^n e^{mx}.$$

Substituting in (4.23), we obtain

$$a_0 m^n e^{mx} + a_1 m^{n-1} e^{mx} + \cdots + a_{n-1} m e^{mx} + a_n e^{mx} = 0$$

or

$$e^{mx}(a_0 m^n + a_1 m^{n-1} + \cdots + a_{n-1} m + a_n) = 0.$$

Since $e^{mx} \neq 0$, we obtain the polynomial equation in the unknown m:

$$a_0 m^n + a_1 m^{n-1} + \cdots + a_{n-1} m + a_n = 0. \tag{4.24}$$

This equation is called the *auxiliary equation* or the *characteristic equation* of the given differential equation (4.23). If $y = e^{mx}$ is a solution of (4.23) then we see that the constant m must satisfy (4.24). Hence, to solve (4.23), we write the auxiliary equation (4.24) and solve it for m. Observe that (4.24) is formally obtained from (4.23) by merely replacing the kth derivative in (4.23) by $m^k (k = 0, 1, 2, \ldots, n)$. Three cases arise, according as the roots of (4.24) are real and distinct, real and repeated, or complex.

B. Case 1. Distinct Real Roots

Suppose the roots of (4.24) are the n distinct real numbers

$$m_1, m_2, \ldots, m_n.$$

Then

$$e^{m_1 x}, e^{m_2 x}, \ldots, e^{m_n x}$$

are n distinct solutions of (4.23). Further, using the Wronskian determinant one may show that these n solutions are linearly independent. Thus we have the following result.

THEOREM 4.11

Consider the nth-order homogeneous linear differential equation (4.23) with constant coefficients. If the auxiliary equation (4.24) has the n distinct real roots $m_1, m_2, \ldots, m_n$, then the general solution of (4.23) is

$$y = c_1 e^{m_1 x} + c_2 e^{m_2 x} + \cdots + c_n e^{m_n x},$$

where $c_1, c_2, \ldots, c_n$ are arbitrary constants.

▶ **Example 4.20**

Consider the differential equation

$$\frac{d^2 y}{dx^2} - 3 \frac{dy}{dx} + 2y = 0.$$

The auxiliary equation is

$$m^2 - 3m + 2 = 0.$$

Hence

$$(m - 1)(m - 2) = 0, \qquad m_1 = 1, \qquad m_2 = 2.$$

The roots are real and distinct. Thus e^x and e^{2x} are solutions and the general solution may be written

$$y = c_1 e^x + c_2 e^{2x}.$$

We verify that e^x and e^{2x} are indeed linearly independent. Their Wronskian is

$$W(e^x, e^{2x}) = \begin{vmatrix} e^x & e^{2x} \\ e^x & 2e^{2x} \end{vmatrix} = e^{3x} \neq 0.$$

Thus by Theorem 4.4 we are assured of their linear independence.

▶ **Example 4.21**

Consider the differential equation

$$\frac{d^3 y}{dx^3} - 4 \frac{d^2 y}{dx^2} + \frac{dy}{dx} + 6y = 0.$$

The auxiliary equation is

$$m^3 - 4m^2 + m + 6 = 0.$$

We observe that $m = -1$ is a root of this equation. By synthetic division we obtain the factorization

$$(m + 1)(m^2 - 5m + 6) = 0$$

or

$$(m + 1)(m - 2)(m - 3) = 0.$$

Thus the roots are the distinct real numbers

$$m_1 = -1, \qquad m_2 = 2, \qquad m_3 = 3,$$

and the general solution is

$$y = c_1 e^{-x} + c_2 e^{2x} + c_3 e^{3x}.$$

C. Case 2. Repeated Real Roots

We shall begin our study of this case by considering a simple example.

▶ **Example 4.22: Introductory Example**

Consider the differential equation

$$\frac{d^2 y}{dx^2} - 6\frac{dy}{dx} + 9y = 0. \tag{4.25}$$

The auxiliary equation is

$$m^2 - 6m + 9 = 0$$

or

$$(m - 3)^2 = 0.$$

The roots of this equation are

$$m_1 = 3, \qquad m_2 = 3$$

(real but *not* distinct).

Corresponding to the root m_1 we have the solution e^{3x}, and corresponding to m_2 we have the *same* solution e^{3x}. The linear combination $c_1 e^{3x} + c_2 e^{3x}$ of these "two" solutions is clearly *not* the general solution of the differential equation (4.25), for it is *not* a linear combination of *two linearly independent* solutions. Indeed we may write the combination $c_1 e^{3x} + c_2 e^{3x}$ as simply $c_0 e^{3x}$, where $c_0 = c_1 + c_2$; and clearly $y = c_0 e^{3x}$, involving *one* arbitrary constant, is not the general solution of the given *second*-order equation.

We must find a linearly independent solution; but how shall we proceed to do so? Since we already know the one solution e^{3x}, we may apply Theorem 4.7 and reduce the order. We let

$$y = e^{3x}v,$$

where v is to be determined. Then

$$\frac{dy}{dx} = e^{3x}\frac{dv}{dx} + 3e^{3x}v,$$

$$\frac{d^2y}{dx^2} = e^{3x}\frac{d^2v}{dx^2} + 6e^{3x}\frac{dv}{dx} + 9e^{3x}v.$$

Substituting into Equation (4.25) we have

$$\left(e^{3x}\frac{d^2v}{dx^2} + 6e^{3x}\frac{dv}{dx} + 9e^{3x}v\right) - 6\left(e^{3x}\frac{dv}{dx} + 3e^{3x}v\right) + 9e^{3x}v = 0$$

or

$$e^{3x}\frac{d^2v}{dx^2} = 0.$$

Letting $w = dv/dx$, we have the first-order equation

$$e^{3x}\frac{dw}{dx} = 0$$

or simply

$$\frac{dw}{dx} = 0.$$

The solutions of this first-order equation are simply $w = c$, where c is an arbitrary constant. Choosing the particular solution $w = 1$ and recalling that $dv/dx = w$, we find

$$v(x) = x + c_0,$$

where c_0 is an arbitrary constant. By Theorem 4.7 we know that for any choice of the constant c_0, $v(x)e^{3x} = (x + c_0)e^{3x}$ is a solution of the given second-order equation (4.25). Further, by Theorem 4.7, we know that this solution and the previously known solution e^{3x} are linearly independent. Choosing $c_0 = 0$ we obtain the solution

$$y = xe^{3x},$$

and thus corresponding to the *double* root 3 we find the linearly independent solutions

$$e^{3x} \quad \text{and} \quad xe^{3x}$$

of Equation (4.25).

Thus the general solution of Equation (4.25) may be written

$$y = c_1 e^{3x} + c_2 xe^{3x} \tag{4.26}$$

or

$$y = (c_1 + c_2 x)e^{3x}. \tag{4.27}$$

With this example as a guide, let us return to the general nth-order equation (4.23). If the auxiliary equation (4.24) has the *double* real root m, we would surely expect that e^{mx} and xe^{mx} would be the corresponding linearly independent solutions. This is indeed the case. Specifically, suppose the roots of (4.24) are the double real root m and the $(n-2)$ distinct real roots

$$m_1, m_2, \ldots, m_{n-2}.$$

Then linearly independent solutions of (4.23) are

$$e^{mx}, xe^{mx}, e^{m_1 x}, e^{m_2 x}, \ldots, e^{m_{n-2} x},$$

and the general solution may be written

$$y = c_1 e^{mx} + c_2 x e^{mx} + c_3 e^{m_1 x} + c_4 e^{m_2 x} + \cdots + c_n e^{m_{n-2} x}$$

or

$$y = (c_1 + c_2 x) e^{mx} + c_3 e^{m_1 x} + c_4 e^{m_2 x} + \cdots + c_n e^{m_{n-2} x}.$$

In like manner, if the auxiliary equation (4.24) has the triple real root m, corresponding linearly independent solutions are

$$e^{mx}, xe^{mx}, \quad \text{and} \quad x^2 e^{mx}.$$

The corresponding part of the general solution may be written

$$(c_1 + c_2 x + c_3 x^2) e^{mx}.$$

Proceeding further in like manner, we summarize Case 2 in the following theorem:

THEOREM 4.12

1. Consider the nth-order homogeneous linear differential equation (4.23) with constant coefficients. If the auxiliary equation (4.24) has the real root m occurring k times, then the part of the general solution of (4.23) corresponding to this k-fold repeated root is

$$(c_1 + c_2 x + c_3 x^2 + \cdots + c_k x^{k-1}) e^{mx}.$$

2. If, further, the remaining roots of the auxiliary equation (4.24) are the distinct real numbers $m_{k+1}, \ldots, m_n$, then the general solution of (4.23) is

$$y = (c_1 + c_2 x + c_3 x^2 + \cdots + c_k x^{k-1}) e^{mx} + c_{k+1} e^{m_{k+1} x} + \cdots + c_n e^{m_n x}.$$

3. If, however, any of the remaining roots are also repeated, then the parts of the general solution of (4.23) corresponding to each of these other repeated roots are expressions similar to that corresponding to m in part 1.

We now consider several examples.

▶ **Example 4.23**

Find the general solution of

$$\frac{d^3 y}{dx^3} - 4 \frac{d^2 y}{dx^2} - 3 \frac{dy}{dx} + 18y = 0.$$

The auxiliary equation

$$m^3 - 4m^2 - 3m + 18 = 0$$

has the roots, 3, 3, -2. The general solution is

$$y = c_1 e^{3x} + c_2 x e^{3x} + c_3 e^{-2x}$$

or

$$y = (c_1 + c_2 x)e^{3x} + c_3 e^{-2x}.$$

▶ **Example 4.24**

Find the general solution of

$$\frac{d^4 y}{dx^4} - 5\frac{d^3 y}{dx^3} + 6\frac{d^2 y}{dx^2} + 4\frac{dy}{dx} - 8y = 0.$$

The auxiliary equation is

$$m^4 - 5m^3 + 6m^2 + 4m - 8 = 0,$$

with roots $2, 2, 2, -1$. The part of the general solution corresponding to the three-fold root 2 is

$$y_1 = (c_1 + c_2 x + c_3 x^2)e^{2x}$$

and that corresponding to the simple root -1 is simply

$$y_2 = c_4 e^{-x}.$$

Thus the general solution is $y = y_1 + y_2$, that is,

$$y = (c_1 + c_2 x + c_3 x^2)e^{2x} + c_4 e^{-x}.$$

D. Case 3. Conjugate Complex Roots

Now suppose that the auxiliary equation has the complex number $a + bi$ (a, b real, $i^2 = -1$, $b \neq 0$) as a nonrepeated root. Then, since the coefficients are real, the conjugate complex number $a - bi$ is also a nonrepeated root. The corresponding part of the general solution is

$$k_1 e^{(a+bi)x} + k_2 e^{(a-bi)x},$$

where k_1 and k_2 are arbitrary constants. The solutions defined by $e^{(a+bi)x}$ and $e^{(a-bi)x}$ are complex functions of the real variable x. It is desirable to replace these by two *real* linearly independent solutions. This can be accomplished by using Euler's formula,

$$e^{i\theta} = \cos\theta + i\sin\theta,*$$

which holds for all real θ. Using this we have:

$$\begin{aligned}
k_1 e^{(a+bi)x} + k_2 e^{(a-bi)x} &= k_1 e^{ax}e^{bix} + k_2 e^{ax}e^{-bix} \\
&= e^{ax}[k_1 e^{ibx} + k_2 e^{-ibx}] \\
&= e^{ax}[k_1(\cos bx + i\sin bx) + k_2(\cos bx - i\sin bx)] \\
&= e^{ax}[(k_1 + k_2)\cos bx + i(k_1 - k_2)\sin bx] \\
&= e^{ax}[c_1 \sin bx + c_2 \cos bx],
\end{aligned}$$

---

* We borrow this basic identity from complex variable theory, as well as the fact that $e^{ax+bix} = e^{ax}e^{ibx}$ holds for complex exponents.

where $c_1 = i(k_1 - k_2)$, $c_2 = k_1 + k_2$ are two new arbitrary constants. Thus the part of the general solution corresponding to the nonrepeated conjugate complex roots $a \pm bi$ is

$$e^{ax}[c_1 \sin bx + c_2 \cos bx].$$

Combining this with the results of Case 2, we have the following theorem covering Case 3.

THEOREM 4.13

1. Consider the nth-order homogeneous linear differential equation (4.23) with constant coefficients. If the auxiliary equation (4.24) has the conjugate complex roots $a + bi$ and $a - bi$, neither repeated, then the corresponding part of the general solution of (4.23) may be written

$$y = e^{ax}(c_1 \sin bx + c_2 \cos bx).$$

2. If, however, $a + bi$ and $a - bi$ are each k-fold roots of the auxiliary equation (4.24), then the corresponding part of the general solution of (4.23) may be written

$$y = e^{ax}[(c_1 + c_2 x + c_3 x^2 + \cdots + c_k x^{k-1}) \sin bx$$
$$+ (c_{k+1} + c_{k+2} x + c_{k+3} x^2 + \cdots + c_{2k} x^{k-1}) \cos bx].$$

We now give several examples.

▶ **Example 4.25**

Find the general solution of

$$\frac{d^2 y}{dx^2} + y = 0.$$

We have already used this equation to illustrate the theorems of Section 4.1. Let us now obtain its solution using Theorem 4.13. The auxiliary equation $m^2 + 1 = 0$ has the roots $m = \pm i$. These are the pure imaginary complex numbers $a \pm bi$, where $a = 0$, $b = 1$. The general solution is thus

$$y = e^{0x}(c_1 \sin 1 \cdot x + c_2 \cos 1 \cdot x),$$

which is simply

$$y = c_1 \sin x + c_2 \cos x.$$

▶ **Example 4.26**

Find the general solution of

$$\frac{d^2 y}{dx^2} - 6 \frac{dy}{dx} + 25y = 0.$$

The auxiliary equation is $m^2 - 6m + 25 = 0$. Solving it, we find

$$m = \frac{6 \pm \sqrt{36 - 100}}{2} = \frac{6 \pm 8i}{2} = 3 \pm 4i.$$

Here the roots are the conjugate complex numbers $a \pm bi$, where $a = 3$, $b = 4$. The general solution may be written

$$y = e^{3x}(c_1 \sin 4x + c_2 \cos 4x).$$

▶ **Example 4.27**

Find the general solution of

$$\frac{d^4y}{dx^4} - 4\frac{d^3y}{dx^3} + 14\frac{d^2y}{dx^2} - 20\frac{dy}{dx} + 25y = 0.$$

The auxiliary equation is

$$m^4 - 4m^3 + 14m^2 - 20m + 25 = 0.$$

The solution of this equation presents some ingenuity and labor. Since our purpose in this example is not to display our mastery of the solution of algebraic equations but rather to illustrate the above principles of determining the general solution of differential equations, we unblushingly list the roots without further apologies.
   They are

$$1 + 2i, \qquad 1 - 2i, \qquad 1 + 2i, \qquad 1 - 2i.$$

Since each pair of conjugate complex roots is double, the general solution is

$$y = e^x[(c_1 + c_2 x)\sin 2x + (c_3 + c_4 x)\cos 2x]$$

or

$$y = c_1 e^x \sin 2x + c_2 x e^x \sin 2x + c_3 e^x \cos 2x + c_4 x e^x \cos 2x.$$

E. An Initial-Value Problem

We now apply the results concerning the general solution of a homogeneous linear equation with constant coefficients to an initial-value problem involving such an equation.

▶ **Example 4.28**

Solve the initial-value problem

$$\frac{d^2y}{dx^2} - 6\frac{dy}{dx} + 25y = 0, \tag{4.28}$$

$$y(0) = -3, \tag{4.29}$$

$$y'(0) = -1. \tag{4.30}$$

First let us note that by Theorem 4.1 this problem has a unique solution defined for all x, $-\infty < x < \infty$. We now proceed to find this solution; that is, we seek the particular solution of the differential equation (4.28) that satisfies the two initial conditions (4.29) and (4.30). We have already found the general solution of the differential equation (4.28) in Example 4.26. It is

$$y = e^{3x}(c_1 \sin 4x + c_2 \cos 4x). \tag{4.31}$$

From this, we find

$$\frac{dy}{dx} = e^{3x}[(3c_1 - 4c_2)\sin 4x + (4c_1 + 3c_2)\cos 4x]. \tag{4.32}$$

We now apply the initial conditions. Applying condition (4.29), $y(0) = -3$, to Equation (4.31), we find

$$-3 = e^0(c_1 \sin 0 + c_2 \cos 0),$$

which reduces at once to

$$c_2 = -3. \tag{4.33}$$

Applying condition (4.30), $y'(0) = -1$, to Equation (4.32), we obtain

$$-1 = e^0[(3c_1 - 4c_2)\sin 0 + (4c_1 + 3c_2)\cos 0],$$

which reduces to

$$4c_1 + 3c_2 = -1. \tag{4.34}$$

Solving Equations (4.33) and (4.34) for the unknowns c_1 and c_2, we find

$$c_1 = 2, \qquad c_2 = -3.$$

Replacing c_1 and c_2 in Equation (4.31) by these values, we obtain the unique solution of the given initial-value problem in the form

$$y = e^{3x}(2 \sin 4x - 3 \cos 4x).$$

Recall from trigonometry that a linear combination of a sine term and a cosine term having a common argument cx may be expressed as an appropriate constant multiple of the sine of the sum of this common argument cx and an appropriate constant angle ϕ. Thus the preceding solution can be reexpressed in an alternative form involving the factor $\sin(4x + \phi)$ for some suitable ϕ. To do this we first multiply and divide by $\sqrt{(2^2) + (-3)^2} = \sqrt{13}$, thereby obtaining

$$y = \sqrt{13}e^{3x}\left[\frac{2}{\sqrt{13}} \sin 4x - \frac{3}{\sqrt{13}} \cos 4x\right].$$

From this we may express the solution in the alternative form

$$y = \sqrt{13}e^{3x} \sin(4x + \phi),$$

where the angle ϕ is defined by the equations

$$\sin \phi = -\frac{3}{\sqrt{13}}, \qquad \cos \phi = \frac{2}{\sqrt{13}}.$$

Exercises

Find the general solution of each of the differential equations in Exercises 1–24.

1. $\dfrac{d^2 y}{dx^2} - 5\dfrac{dy}{dx} + 6y = 0.$

2. $\dfrac{d^2 y}{dx^2} - 2\dfrac{dy}{dx} - 3y = 0.$

3. $4\dfrac{d^2 y}{dx^2} - 12\dfrac{dy}{dx} + 5y = 0.$

4. $3\dfrac{d^2 y}{dx^2} - 14\dfrac{dy}{dx} - 5y = 0.$

5. $\dfrac{d^3 y}{dx^3} - 3\dfrac{d^2 y}{dx^2} - \dfrac{dy}{dx} + 3y = 0.$

6. $\dfrac{d^3 y}{dx^3} - 6\dfrac{d^2 y}{dx^2} + 5\dfrac{dy}{dx} + 12y = 0.$

7. $\dfrac{d^2 y}{dx^2} - 8\dfrac{dy}{dx} + 16y = 0.$

8. $4\dfrac{d^2 y}{dx^2} + 4\dfrac{dy}{dx} + y = 0.$

9. $\dfrac{d^2 y}{dx^2} - 4\dfrac{dy}{dx} + 13y = 0.$

10. $\dfrac{d^2 y}{dx^2} + 6\dfrac{dy}{dx} + 25y = 0.$

11. $\dfrac{d^2 y}{dx^2} + 9y = 0.$

12. $4\dfrac{d^2 y}{dx^2} + y = 0.$

13. $\dfrac{d^3 y}{dx^3} - 5\dfrac{d^2 y}{dx^2} + 7\dfrac{dy}{dx} - 3y = 0.$

14. $4\dfrac{d^3 y}{dx^3} + 4\dfrac{d^2 y}{dx^2} - 7\dfrac{dy}{dx} + 2y = 0.$

15. $\dfrac{d^3 y}{dx^3} - 6\dfrac{d^2 y}{dx^2} + 12\dfrac{dy}{dx} - 8y = 0.$

16. $\dfrac{d^3 y}{dx^3} + 4\dfrac{d^2 y}{dx^2} + 5\dfrac{dy}{dx} + 6y = 0.$

17. $\dfrac{d^3 y}{dx^3} - \dfrac{d^2 y}{dx^2} + \dfrac{dy}{dx} - y = 0.$

18. $\dfrac{d^4 y}{dx^4} + 8\dfrac{d^2 y}{dx^2} + 16y = 0.$

19. $\dfrac{d^5 y}{dx^5} - 2\dfrac{d^4 y}{dx^4} + \dfrac{d^3 y}{dx^3} = 0.$

20. $\dfrac{d^4 y}{dx^4} - \dfrac{d^3 y}{dx^3} - 3\dfrac{d^2 y}{dx^2} + \dfrac{dy}{dx} + 2y = 0.$

21. $\dfrac{d^4 y}{dx^4} - 3\dfrac{d^3 y}{dx^3} - 2\dfrac{d^2 y}{dx^2} + 2\dfrac{dy}{dx} + 12y = 0.$

22. $\dfrac{d^4 y}{dx^4} + 6\dfrac{d^3 y}{dx^3} + 15\dfrac{d^2 y}{dx^2} + 20\dfrac{dy}{dx} + 12y = 0.$

23. $\dfrac{d^4 y}{dx^4} + y = 0.$

24. $\dfrac{d^5 y}{dx^5} = 0.$

Solve the initial-value problems in Exercises 25–42:

25. $\dfrac{d^2 y}{dx^2} - \dfrac{dy}{dx} - 12y = 0,$ $y(0) = 3,$ $y'(0) = 5.$

26. $\dfrac{d^2 y}{dx^2} + 7\dfrac{dy}{dx} + 10y = 0,$ $y(0) = -4,$ $y'(0) = 2.$

27. $\dfrac{d^2 y}{dx^2} - 6\dfrac{dy}{dx} + 8y = 0,$ $y(0) = 1,$ $y'(0) = 6.$

28. $3\dfrac{d^2 y}{dx^2} + 4\dfrac{dy}{dx} - 4y = 0,$ $y(0) = 2,$ $y'(0) = -4.$

29. $\dfrac{d^2 y}{dx^2} + 6\dfrac{dy}{dx} + 9y = 0,$ $y(0) = 2,$ $y'(0) = -3.$

30. $4\dfrac{d^2 y}{dx^2} - 12\dfrac{dy}{dx} + 9y = 0,$ $y(0) = 4,$ $y'(0) = 9.$

31. $\dfrac{d^2 y}{dx^2} + 4\dfrac{dy}{dx} + 4y = 0,$ $y(0) = 3,$ $y'(0) = 7.$

32. $9\dfrac{d^2 y}{dx^2} - 6\dfrac{dy}{dx} + y = 0,$ $y(0) = 3,$ $y'(0) = -1.$

33. $\dfrac{d^2 y}{dx^2} - 4\dfrac{dy}{dx} + 29y = 0,$ $y(0) = 0,$ $y'(0) = 5.$

34. $\dfrac{d^2 y}{dx^2} + 6\dfrac{dy}{dx} + 58y = 0,$ $y(0) = -1,$ $y'(0) = 5.$

35. $\dfrac{d^2 y}{dx^2} + 6\dfrac{dy}{dx} + 13y = 0,$ $y(0) = 3,$ $y'(0) = -1.$

36. $\dfrac{d^2 y}{dx^2} + 2\dfrac{dy}{dx} + 5y = 0,$ $y(0) = 2,$ $y'(0) = 6.$

37. $9\dfrac{d^2 y}{dx^2} + 6\dfrac{dy}{dx} + 5y = 0,$ $y(0) = 6,$ $y'(0) = 0.$

38. $4\dfrac{d^2 y}{dx^2} + 4\dfrac{dy}{dx} + 37y = 0,$ $y(0) = 2,$ $y'(0) = -4.$

39. $\dfrac{d^3 y}{dx^3} - 6\dfrac{d^2 y}{dx^2} + 11\dfrac{dy}{dx} - 6y = 0,$ $y(0) = 0,$ $y'(0) = 0,$ $y''(0) = 2.$

40. $\dfrac{d^3 y}{dx^3} - 2\dfrac{d^2 y}{dx^2} + 4\dfrac{dy}{dx} - 8y = 0,$ $y(0) = 2,$ $y'(0) = 0,$ $y''(0) = 0.$

41. $\dfrac{d^3 y}{dx^3} - 3\dfrac{d^2 y}{dx^2} + 4y = 0,$ $y(0) = 1,$ $y'(0) = -8,$ $y''(0) = -4.$

42. $\dfrac{d^3 y}{dx^3} - 5\dfrac{d^2 y}{dx^2} + 9\dfrac{dy}{dx} - 5y = 0,$ $y(0) = 0,$ $y'(0) = 1,$ $y''(0) = 6.$

43. The roots of the auxiliary equation, corresponding to a certain 10th-order homogeneous linear differential equation with constant coefficients, are

$$4, \quad 4, \quad 4, \quad 4, \quad 2 + 3i, \quad 2 - 3i, \quad 2 + 3i, \quad 2 - 3i, \quad 2 + 3i, \quad 2 - 3i.$$

    Write the general solution.

44. The roots of the auxiliary equation, corresponding to a certain 12th-order homogeneous linear differential equation with constant coefficients, are

$$2, \quad 2, \quad 2, \quad 2, \quad 2, \quad 2, \quad 3 + 4i, \quad 3 - 4i, \quad 3 + 4i, \quad 3 - 4i, \quad 3 + 4i, \quad 3 - 4i.$$

    Write the general solution.

45. Given that $\sin x$ is a solution of

$$\frac{d^4 y}{dx^4} + 2\frac{d^3 y}{dx^3} + 6\frac{d^2 y}{dx^2} + 2\frac{dy}{dx} + 5y = 0,$$

    find the general solution.

46. Given that $e^x \sin 2x$ is a solution of

$$\frac{d^4 y}{dx^4} + 3\frac{d^3 y}{dx^3} + \frac{d^2 y}{dx^2} + 13\frac{dy}{dx} + 30y = 0,$$

    find the general solution.

4.3 THE METHOD OF UNDETERMINED COEFFICIENTS

A. Introduction; An Illustrative Example

We now consider the (nonhomogeneous) differential equation

$$a_0 \frac{d^n y}{dx^n} + a_1 \frac{d^{n-1} y}{dx^{n-1}} + \cdots + a_{n-1}\frac{dy}{dx} + a_n y = F(x), \tag{4.35}$$

where the coefficients $a_0, a_1, \ldots, a_n$ are constants but where the nonhomogeneous term F is (in general) a nonconstant function of x. Recall that the general solution of (4.35) may be written

$$y = y_c + y_p,$$

where y_c is the *complementary function*, that is, the general solution of the corresponding homogeneous equation (Equation (4.35) with F replaced by 0), and y_p is a *particular integral*, that is, any solution of (4.35) containing no arbitrary constants. In Section 4.2 we learned how to find the complementary function; now we consider methods of determining a particular integral.

We consider first the method of *undetermined coefficients*. Mathematically speaking, the class of functions F to which this method applies is actually quite restricted; but this mathematically narrow class includes functions of frequent occurrence and considerable importance in various physical applications. And this method has one distinct advantage—when it *does apply*, it is relatively simple!

▶ **Example 4.29: Introductory Example**

$$\frac{d^2y}{dx^2} - 2\frac{dy}{dx} - 3y = 2e^{4x} \tag{4.36}$$

We proceed to seek a particular solution y_p; but what type of function might be a possible candidate for such a particular solution? The differential equation (4.36) requires a solution which is such that its second derivative, minus twice its first derivative, minus three times the solution itself, add up to twice the exponential function e^{4x}. Since the derivatives of e^{4x} are constant multiples of e^{4x}, it seems reasonable that the desired particular solution might also be a constant multiple of e^{4x}. Thus we assume a particular solution of the form

$$y_p = Ae^{4x}, \tag{4.37}$$

where A is a constant (undetermined coefficient) to be determined such that (4.37) *is a* solution of (4.36). Differentiating (4.37), we obtain

$$y_p' = 4Ae^{4x} \quad \text{and} \quad y_p'' = 16Ae^{4x}.$$

Then substituting into (4.36), we obtain

$$16Ae^{4x} - 2(4Ae^{4x}) - 3Ae^{4x} = 2e^{4x}$$

or

$$5Ae^{4x} = 2e^{4x}. \tag{4.38}$$

Since the solution (4.37) is to satisfy the differential equation identically for *all* x on some real interval, the relation (4.38) must be an identity for all such x and hence the coefficients of e^{4x} on both sides of (4.38) must be respectively equal. Equating these coefficients, we obtain the equation

$$5A = 2,$$

from which we determine the previously undetermined coefficient

$$A = \tfrac{2}{5}.$$

Substituting this back into (4.37), we obtain the particular solution

$$y_p = \tfrac{2}{5}e^{4x}.$$

Now consider the differential equation

$$\frac{d^2y}{dx^2} - 2\frac{dy}{dx} - 3y = 2e^{3x} \tag{4.39}$$

which is exactly the same as Equation (4.36) except that e^{4x} in the right member has been replaced by e^{3x}. Reasoning as in the case of differential equation (4.36), we would now assume a particular solution of the form

$$y_p = Ae^{3x}. \tag{4.40}$$

Then differentiating (4.40), we obtain

$$y_p' = 3Ae^{3x} \quad \text{and} \quad y_p'' = 9Ae^{3x}.$$

Then substituting into (4.39), we obtain

$$9Ae^{3x} - 2(3Ae^{3x}) - 3(Ae^{3x}) = 2e^{3x}$$

or

$$0 \cdot Ae^{3x} = 2e^{3x}.$$

or simply

$$0 = 2e^{3x},$$

which does not hold for any real x. This impossible situation tells us that there is no particular solution of the assumed form (4.40).

As noted, Equations (4.36) and (4.39) are almost the same, the only difference between them being the constant multiple of x in the exponents of their respective nonhomogeneous terms $2e^{4x}$ and $2e^{3x}$. The equation (4.36) involving $2e^{4x}$ had a particular solution of the assumed form Ae^{4x}, whereas Equation (4.39) involving $2e^{3x}$ did *not* have one of the assumed form Ae^{3x}. What is the difference in these two so apparently similar cases?

The answer to this is found by examining the solutions of the differential equation

$$\frac{d^2y}{dx^2} - 2\frac{dy}{dx} - 3y = 0 \tag{4.41}$$

which is the homogeneous equation corresponding to both (4.36) and (4.39). The auxiliary equation is $m^2 - 2m - 3 = 0$ with roots 3 and -1; and so

$$e^{3x} \quad \text{and} \quad e^{-x}$$

are (linearly independent) solutions of (4.41). This suggests that the failure to obtain a solution of the form $y_p = Ae^{3x}$ for Equation (4.39) is due to the fact that the function e^{3x} in this assumed solution is a solution of the homogeneous equation (4.41) corresponding to (4.39); and this is indeed the case. For, since Ae^{3x} satisfies the *homogeneous* equation (4.41), it reduces the common left member

$$\frac{d^2y}{dx^2} - 2\frac{dy}{dx} - 3y$$

of both (4.41) and (4.39) to 0, *not* $2e^{3x}$, which a particular solution of Equation (4.39) would have to do.

Now that we have considered what caused the difficulty in attempting to obtain a particular solution of the form Ae^{3x} for (4.39), we naturally ask what form of solution should we seek? Recall that in the case of a double root m for an auxiliary equation, a solution linearly independent of the basic solution e^{mx} was xe^{mx}. While this in itself tells us nothing about the situation at hand, it might suggest that we seek a particular solution of (4.39) of the form

$$y_p = Axe^{3x}. \tag{4.42}$$

Differentiating (4.42), we obtain

$$y_p' = 3Axe^{3x} + Ae^{3x}, \qquad y_p'' = 9Axe^{3x} + 6Ae^{3x}.$$

Then substituting into (4.39), we obtain

$$(9Axe^{3x} + 6Ae^{3x}) - 2(3Axe^{3x} + Ae^{3x}) - 3Axe^{3x} = 2e^{3x}$$

or

$$(9A - 6A - 3A)xe^{3x} + 4Ae^{3x} = 2e^{3x}.$$

or simply

$$0xe^{3x} + 4Ae^{3x} = 2e^{3x}. \tag{4.43}$$

Since the (assumed) solution (4.42) is to satisfy the differential equation identically for *all* x on some real interval, the relation (4.43) must be an identity for all such x and hence the coefficients of e^{3x} on both sides of (4.43) must be respectively equal. Equating coefficients, we obtain the equation

$$4A = 2,$$

from which we determine the previously undetermined coefficient

$$A = \tfrac{1}{2}.$$

Substituting this back into (4.42), we obtain the particular solution

$$y_p = \tfrac{1}{2}xe^{3x}.$$

We summarize the results of this example. The differential equations

$$\frac{d^2y}{dx^2} - 2\frac{dy}{dx} - 3y = 2e^{4x} \tag{4.36}$$

and

$$\frac{d^2y}{dx^2} - 2\frac{dy}{dx} - 3y = 2e^{3x} \tag{4.39}$$

each have the same corresponding homogeneous equation

$$\frac{d^2y}{dx^2} - 2\frac{dy}{dx} - 3y = 0. \tag{4.41}$$

This homogeneous equation has linearly independent solutions

$$e^{3x} \quad \text{and} \quad e^{-x},$$

and so the complementary function of both (4.36) and (4.39) is

$$y_c = c_1 e^{3x} + c_2 e^{-x}.$$

The right member $2e^{4x}$ of (4.36) is *not* a solution of the corresponding homogeneous equation (4.41), and the attempted particular solution

$$y_p = Ae^{4x} \tag{4.37}$$

suggested by this right member did indeed lead to a particular solution of this assumed form, namely, $y_p = \tfrac{2}{5}e^{4x}$. On the other hand, the right member $2e^{3x}$ of (4.39) *is a* solution of the corresponding homogeneous equation (4.41) [with $c_1 = 2$ and $c_2 = 0$], and the attempted particular solution

$$y_p = Ae^{3x} \tag{4.40}$$

suggested by this right member *failed* to lead to a particular solution of this form. However, in this case, the revised attempted particular solution,

$$y_p = Axe^{3x}, \tag{4.42}$$

obtained from (4.40) by multiplying by x, led to a particular solution of this assumed form, namely, $y_p = \frac{1}{2}xe^{3x}$.

The general solutions of (4.36) and (4.39) are, respectively,

$$y = c_1 e^{3x} + c_2 e^{-x} + \tfrac{2}{5}e^{4x}$$

and

$$y = c_1 e^{3x} + c_2 e^{-x} + \tfrac{1}{2}xe^{3x}.$$

The preceding example illustrates a particular case of the method of undetermined coefficients. It suggests that in some cases the assumed particular solution y_p corresponding to a nonhomogeneous term in the differential equation is of the same type as that nonhomogeneous term, whereas in other cases the assumed y_p ought to be some sort of modification of that nonhomogeneous term. It turns out that this is essentially the case. We now proceed to present the method systematically.

B. The Method

We begin by introducing certain preliminary definitions.

DEFINITION

We shall call a function a UC *function if it is either (1) a function defined by one of the following:*

(i) x^n, *where n is a positive integer or zero,*
(ii) e^{ax}, *where a is a constant $\neq 0$,*
(iii) $\sin(bx + c)$, *where b and c are constants, $b \neq 0$,*
(iv) $\cos(bx + c)$, *where b and c are constants, $b \neq 0$,*

or (2) a function defined as a finite product of two or more functions of these four types.

▶ **Example 4.20**

Examples of UC functions of the four basic types (*i*), (*ii*), (*iii*), (*iv*) of the preceeding definition are those defined respectively by

$$x^3, \qquad e^{-2x}, \qquad \sin(3x/2), \qquad \cos(2x + \pi/4).$$

Examples of UC functions defined as finite products of two or more of these four basic types are those defined respectively by

$$x^2 e^{3x}, \qquad x\cos 2x, \qquad e^{5x}\sin 3x,$$

$$\sin 2x \cos 3x, \qquad x^3 e^{4x}\sin 5x.$$

The method of undetermined coefficients applies when the nonhomogeneous function F in the differential equation is a finite linear combination of UC functions.

Observe that given a UC function f, each successive derivative of f is either itself a constant multiple of a UC function or else a linear combination of UC functions.

DEFINITION

Consider a UC function f. The set of functions consisting of f itself and all linearly independent UC functions of which the successive derivatives of f are either constant multiples or linear combinations will be called the UC set *of f.*

▶ **Example 4.31**

The function f defined for all real x by $f(x) = x^3$ is a UC function. Computing derivatives of f, we find

$$f'(x) = 3x^2, \qquad f''(x) = 6x, \qquad f'''(x) = 6 = 6 \cdot 1, \qquad f^{(n)}(x) = 0 \quad \text{for} \quad n > 3.$$

The linearly independent UC functions of which the successive derivatives of f are either constant multiples or linear combinations are those given by

$$x^2, \qquad x, \qquad 1.$$

Thus the *UC set* of x^3 is the set $S = \{x^3, x^2, x, 1\}$.

▶ **Example 4.32**

The function f defined for all real x by $f(x) = \sin 2x$ is a UC function. Computing derivatives of f, we find

$$f'(x) = 2 \cos 2x, \qquad f''(x) = -4 \sin 2x, \qquad \ldots.$$

The only linearly independent UC function of which the successive derivatives of f are constant multiples or linear combinations is that given by $\cos 2x$. Thus the *UC set* of $\sin 2x$ is the set $S = \{\sin 2x, \cos 2x\}$.

These and similar examples of the four basic types of UC functions lead to the results listed as numbers 1, 2, and 3 of Table 4.1.

▶ **Example 4.33**

The function f defined for all real x by $f(x) = x^2 \sin x$ is the product of the two UC functions defined by x^2 and $\sin x$. Hence f is itself a UC function. Computing derivatives of f, we find

$$f'(x) = 2x \sin x + x^2 \cos x,$$

$$f''(x) = 2 \sin x + 4x \cos x - x^2 \sin x,$$

$$f'''(x) = 6 \cos x - 6x \sin x - x^2 \cos x, \qquad \cdots.$$

No "new" types of functions will occur from further differentiation. Each derivative of f is a linear combination of certain of the six UC functions given by $x^2 \sin x$, $x^2 \cos x$,

TABLE 4.1

	UC function	UC set
1	x^n	$\{x^n, x^{n-1}, x^{n-2}, \ldots, x, 1\}$
2	e^{ax}	$\{e^{ax}\}$
3	$\sin(bx + c)$ or $\cos(bx + c)$	$\{\sin(bx + c), \cos(bx + c)\}$
4	$x^n e^{ax}$	$\{x^n e^{ax}, x^{n-1} e^{ax}, x^{n-2} e^{ax}, \ldots, x e^{ax}, e^{ax}\}$
5	$x^n \sin(bx + c)$ or $x^n \cos(bx + c)$	$\{x^n \sin(bx + c), x^n \cos(bx + c),$ $x^{n-1} \sin(bx + c), x^{n-1} \cos(bx + c),$ $\ldots, x \sin(bx + c), x \cos(bx + c),$ $\sin(bx + c), \cos(bx + c)\}$
6	$e^{ax} \sin(bx + c)$ or $e^{ax} \cos(bx + c)$	$\{e^{ax} \sin(bx + c), e^{ax} \cos(bx + c)\}$
7	$x^n e^{ax} \sin(bx + c)$ or $x^n e^{ax} \cos(bx + c)$	$\{x^n e^{ax} \sin(bx + c), x^n e^{ax} \cos(bx + c),$ $x^{n-1} e^{ax} \sin(bx + c), x^{n-1} e^{ax} \cos(bx + c), \ldots,$ $x e^{ax} \sin(bx + c), x e^{ax} \cos(bx + c),$ $e^{ax} \sin(bx + c), e^{ax} \cos(bx + c)\}$

$x \sin x, x \cos x, \sin x,$ and $\cos x$. Thus the set

$$S = \{x^2 \sin x, x^2 \cos x, x \sin x, x \cos x, \sin x, \cos x\}$$

is the *UC set* of $x^2 \sin x$. Note carefully that x^2, x, and 1 are *not* members of this UC set.

Observe that the UC set of the product $x^2 \sin x$ is the set of all products obtained by multiplying the various members of the UC set $\{x^2, x, 1\}$ of x^2 by the various members of the UC set $\{\sin x, \cos x\}$ of $\sin x$. This observation illustrates the general situation regarding the UC set of a UC function defined as a finite product of two or more UC functions of the four basic types. In particular, suppose h is a UC function defined as the product fg of two basic UC functions f and g. Then the UC set of the product function h is the set of all the products obtained by multiplying the various members of the UC set of f by the various members of the UC set of g. Results of this type are listed as numbers 4, 5, and 6 of Table 4.1 and a specific illustration is presented in Example 4.34.

► **Example 4.34**

The function defined for all real x by $f(x) = x^3 \cos 2x$ is the product of the two UC functions defined by x^3 and $\cos 2x$. Using the result stated in the preceding paragraph, the UC set of this product $x^3 \cos 2x$ is the set of all products obtained by multiplying the various members of the UC set of x^3 by the various members of the UC set of $\cos 2x$. Using the definition of UC set or the appropriate numbers of Table 4.1, we find that the UC set of x^3 is

$$\{x^3, x^2, x, 1\}$$

and that of $\cos 2x$ is

$$\{\sin 2x, \cos 2x\}.$$

Thus the UC set of the product $x^3 \cos 2x$ is the set of all products of each of x^3, x^2, x, and 1 by each of $\sin 2x$ and $\cos 2x$, and so it is

$$\{x^3 \sin 2x, x^3 \cos 2x, x^2 \sin 2x, x^2 \cos 2x, x \sin 2x, x \cos 2x, \sin 2x, \cos 2x\}.$$

Observe that this can be found directly from Table 4.1, number 5, with $n = 3$, $b = 2$, and $c = 0$.

We now outline the method of undetermined coefficients for finding a particular integral y_p of

$$a_0 \frac{d^n y}{dx^n} + a_1 \frac{d^{n-1} y}{dx^{n-1}} + \cdots + a_{n-1} \frac{dy}{dx} + a_n y = F(x),$$

where F is a finite linear combination

$$F = A_1 u_1 + A_2 u_2 + \cdots + A_m u_m$$

of UC functions $u_1, u_2, \ldots, u_m$, the A_i being known constants. Assuming the complementary function y_c has already been obtained, we proceed as follows:

1. For *each* of the UC functions

$$u_1, \ldots, u_m$$

of which F is a linear combination, form the corresponding UC set, thus obtaining the respective sets

$$S_1, S_2, \ldots, S_m.$$

2. Suppose that one of the UC sets so formed, say S_j, is identical with or completely included in another, say S_k. In this case, we omit the (identical or smaller) set S_j from further consideration (retaining the set S_k).

3. We now consider in turn each of the UC sets which still remain after Step 2. Suppose now that one of these UC sets, say S_l, includes one or more members which are solutions of the corresponding homogeneous differential equation. If this is the case, we multiply *each* member of S_l by the lowest positive integral power of x so that the resulting revised set will contain no members that are solutions of the corresponding homogeneous differential equation. We now replace S_l by this revised set, so obtained. Note that here we consider one UC set at a time and perform the indicated multiplication, if needed, only upon the members of the one UC set under consideration at the moment.

4. In general there now remains:
   (i) certain of the original UC sets, which were neither omitted in Step 2 nor needed revision in Step 3, and
   (ii) certain revised sets resulting from the needed revision in Step 3.

Now form a linear combination of *all* of the sets of these two categories, with unknown constant coefficients (*undetermined coefficients*).

5. Determine these unknown coefficients by substituting the linear combination formed in Step 4 into the differential equation and demanding that it identically satisfy the differential equation (that is, that it be a particular solution).

This outline of procedure at once covers all of the various special cases to which the method of undetermined coefficients applies, thereby freeing one from the need of considering separately each of these special cases.

Before going on to the illustrative examples of Part C following, let us look back and observe that we actually followed this procedure in solving the differential equations (4.36) and (4.39) of the Introductory Example 4.29. In each of those equations, the nonhomogeneous member consisted of a single term that was a constant multiple of a UC function; and in each case we followed the outline procedure step by step, as far as it applied.

For the differential equation (4.36), the UC function involved was e^{4x}; and we formed its UC set, which was simply $\{e^{4x}\}$ (Step 1). Step 2 obviously did not apply. Nor did Step 3, for as we noted later, e^{4x} was not a solution of the corresponding homogeneous equation (4.41). Thus we assumed $y_p = Ae^{4x}$ (Step 4) substituted in differential equation (4.36), and found A and hence y_p (Step 5).

For the differential equation (4.39), the UC function involved was e^{3x}; and we formed its UC set, which was simply $\{e^{3x}\}$ (Step 1). Step 2 did not apply here either. But Step 3 was very much needed, for e^{3x} *was* a solution of the corresponding homogeneous equation (4.41). Thus we applied Step 3 and multiplied e^{3x} in the UC set $\{e^{3x}\}$ by x, obtaining the revised UC set $\{xe^{3x}\}$, whose single member was *not* a solution of (4.41). Thus we assumed $y_p = Axe^{3x}$ (Step 4), substituted in the differential equation (4.39), and found A and hence y_p (Step 5).

The outline generalizes what the procedure for the differential equation of Introductory Example 4.29 suggested. Equation (4.39) of that example has already brought out the necessity for the revision described in Step 3 when it applies. We give here a brief illustration involving this critical step.

▶ **Example 4.35**

Consider the two equations

$$\frac{d^2y}{dx^2} - 3\frac{dy}{dx} + 2y = x^2e^x \qquad (4.44)$$

and

$$\frac{d^2y}{dx^2} - 2\frac{dy}{dx} + y = x^2e^x \qquad (4.45)$$

The UC set of x^2e^x is

$$S = \{x^2e^x, xe^x, e^x\}.$$

The homogeneous equation corresponding to (4.44) has linearly independent solutions e^x and e^{2x}, and so the complementary function of (4.44) is $y_c = c_1e^x + c_2e^{2x}$. Since member e^x of UC set S is a solution of the homogeneous equation corresponding to (4.44), we multiply each member of UC set S by the lowest positive integral power of x so that the resulting revised set will contain no members that are solutions of the homogeneous equation corresponding to (4.44). This turns out to be x itself; for the revised set

$$S' = \{x^3e^x, x^2e^x, xe^x\}$$

has no members that satisfy the homogeneous equation corresponding to (4.44).

The homogeneous equation corresponding to (4.45) has linearly independent solutions e^x and xe^x, and so the complementary function of (4.45) is $y_c = c_1 e^x + c_2 xe^x$. Since the two members e^x and xe^x of UC set S are solutions of the homogeneous equation corresponding to (4.45), we must modify S here also. But now x itself will not do, for we would get S', which still contains xe^x. Thus we must here multiply each member of S by x^2 to obtain the revised set

$$S'' = \{x^4 e^x, x^3 e^x, x^2 e^x\},$$

which has no member that satisfies the homogeneous equation corresponding to (4.45).

C. Examples

A few illustrative examples, with reference to the above outline, should make the procedure clear. Our first example will be a simple one in which the situations of Steps 2 and 3 do not occur.

► **Example 4.36**

$$\frac{d^2 y}{dx^2} - 2\frac{dy}{dx} - 3y = 2e^x - 10 \sin x.$$

The corresponding homogeneous equation is

$$\frac{d^2 y}{dx^2} - 2\frac{dy}{dx} - 3y = 0$$

and the complementary function is

$$y_c = c_1 e^{3x} + c_2 e^{-x}.$$

The nonhomogenous term is the linear combination $2e^x - 10 \sin x$ of the two UC functions given by e^x and $\sin x$.

1. Form the UC set for each of these two functions. We find

$$S_1 = \{e^x\},$$

$$S_2 = \{\sin x, \cos x\}.$$

2. Note that neither of these sets is identical with nor included in the other; hence both are retained.

3. Furthermore, by examining the complementary function, we see that none of the functions e^x, $\sin x$, $\cos x$ in either of these sets is a solution of the corresponding homogeneous equation. Hence neither set needs to be revised.

4. Thus the original sets S_1 and S_2 remain intact in this problem, and we form the linear combination

$$Ae^x + B \sin x + C \cos x$$

of the three elements e^x, $\sin x$, $\cos x$ of S_1 and S_2, with the undetermined coefficients A, B, C.

5. We determine these unknown coefficients by substituting the linear combination formed in Step 4 into the differential equation and demanding that it satisfy the differential equation identically. That is, we take

$$y_p = Ae^x + B \sin x + C \cos x$$

as a particular solution. Then

$$y'_p = Ae^x + B \cos x - C \sin x,$$
$$y''_p = Ae^x - B \sin x - C \cos x.$$

Actually substituting, we find

$$(Ae^x - B \sin x - C \cos x) - 2(Ae^x + B \cos x - C \sin x)$$
$$- 3(Ae^x + B \sin x + C \cos x) = 2e^x - 10 \sin x$$

or

$$-4Ae^x + (-4B + 2C)\sin x + (-4C - 2B)\cos x = 2e^x - 10 \sin x.$$

Since the solution is to satisfy the differential equation identically for *all* x on some real interval, this relation must be an identity for all such x and hence the coefficients of like terms on both sides must be respectively equal. Equating coefficients of these like terms, we obtain the equations

$$-4A = 2, \qquad -4B + 2C = -10, \qquad -4C - 2B = 0.$$

From these equations, we find that

$$A = -\tfrac{1}{2}, \qquad B = 2, \qquad C = -1,$$

and hence we obtain the particular integral

$$y_p = -\tfrac{1}{2}e^x + 2 \sin x - \cos x.$$

Thus the general solution of the differential equation under consideration is

$$y = y_c + y_p = c_1 e^{3x} + c_2 e^{-x} - \tfrac{1}{2}e^x + 2 \sin x - \cos x.$$

▶ **Example 4.37**

$$\frac{d^2y}{dx^2} - 3\frac{dy}{dx} + 2y = 2x^2 + e^x + 2xe^x + 4e^{3x}.$$

The corresponding homogeneous equation is

$$\frac{d^2y}{dx^2} - 3\frac{dy}{dx} + 2y = 0$$

and the complementary function is

$$y_c = c_1 e^x + c_2 e^{2x}.$$

The nonhomogeneous term is the linear combination

$$2x^2 + e^x + 2xe^x + 4e^{3x}$$

of the four UC functions given by x^2, e^x, xe^x, and e^{3x}.

1. Form the UC set for each of these functions. We have

$$S_1 = \{x^2, x, 1\},$$
$$S_2 = \{e^x\},$$
$$S_3 = \{xe^x, e^x\},$$
$$S_4 = \{e^{3x}\}.$$

2. We note that S_2 is completely included in S_3, so S_2 is omitted from further consideration, leaving the three sets

$$S_1 = \{x^2, x, 1\} \qquad S_3 = \{xe^x, e^x\}, \qquad S_4 = \{e^{3x}\}.$$

3. We now observe that $S_3 = \{xe^x, e^x\}$ includes e^x, which is included in the complementary function and so is a solution of the corresponding homogeneous differential equation. Thus we multiply *each* member of S_3 by x to obtain the revised family

$$S_3' = \{x^2 e^x, xe^x\},$$

which contains no members that are solutions of the corresponding homogeneous equation.

4. Thus there remain the original UC sets

$$S_1 = \{x^2, x, 1\}$$

and

$$S_4 = \{e^{3x}\}$$

and the revised set

$$S_3' = \{x^2 e^x, xe^x\}.$$

These contain the six elements

$$x^2, \quad x, \quad 1, \quad e^{3x}, \quad x^2 e^x, \quad xe^x.$$

We form the linear combination

$$Ax^2 + Bx + C + De^{3x} + Ex^2 e^x + Fxe^x$$

of these six elements.

5. Thus we take as our particular solution,

$$y_p = Ax^2 + Bx + C + De^{3x} + Ex^2 e^x + Fxe^x.$$

From this, we have

$$y_p' = 2Ax + B + 3De^{3x} + Ex^2 e^x + 2Exe^x + Fxe^x + Fe^x,$$

$$y_p'' = 2A + 9De^{3x} + Ex^2 e^x + 4Exe^x + 2Ee^x + Fxe^x + 2Fe^x.$$

We substitute y_p, y_p', y_p'' into the differential equation for $y, dy/dx, d^2y/dx^2$, respectively, to obtain:

$$2A + 9De^{3x} + Ex^2 e^x + (4E + F)xe^x + (2E + 2F)e^x$$
$$- 3[2Ax + B + 3De^{3x} + Ex^2 e^x + (2E + F)xe^x + Fe^x]$$
$$+ 2(Ax^2 + Bx + C + De^{3x} + Ex^2 e^x + Fxe^x)$$
$$= 2x^2 + e^x + 2xe^x + 4e^{3x},$$

or

$$(2A - 3B + 2C) + (2B - 6A)x + 2Ax^2 + 2De^{3x} + (-2E)xe^x + (2E - F)e^x$$
$$= 2x^2 + e^x + 2xe^x + 4e^{3x}.$$

Equating coefficients of like terms, we have:

$$2A - 3B + 2C = 0,$$
$$2B - 6A = 0,$$
$$2A = 2,$$
$$2D = 4,$$
$$-2E = 2,$$
$$2E - F = 1.$$

From this $A = 1$, $B = 3$, $C = \frac{7}{2}$, $D = 2$, $E = -1$, $F = -3$, and so the particular integral is

$$y_p = x^2 + 3x + \tfrac{7}{2} + 2e^{3x} - x^2e^x - 3xe^x.$$

The general solution is therefore

$$y = y_c + y_p = c_1 e^x + c_2 e^{2x} + x^2 + 3x + \tfrac{7}{2} + 2e^{3x} - x^2e^x - 3xe^x.$$

▶ **Example 4.38**

$$\frac{d^4y}{dx^4} + \frac{d^2y}{dx^2} = 3x^2 + 4 \sin x - 2 \cos x.$$

The corresponding homogeneous equation is

$$\frac{d^4y}{dx^4} + \frac{d^2y}{dx^2} = 0,$$

and the complementary function is

$$y_c = c_1 + c_2x + c_3 \sin x + c_4 \cos x.$$

The nonhomogeneous term is the linear combination

$$3x^2 + 4 \sin x - 2 \cos x$$

of the three UC functions given by

$$x^2, \quad \sin x, \quad \text{and} \quad \cos x.$$

1. Form the UC set for each of these three functions. These sets are, respectively,

$$S_1 = \{x^2, x, 1\},$$
$$S_2 = \{\sin x, \cos x\},$$
$$S_3 = \{\cos x, \sin x\}.$$

2. Observe that S_2 and S_3 are identical and so we retain only one of them, leaving the two sets

$$S_1 = \{x^2, x, 1\}, \qquad S_2 = \{\sin x, \cos x\}.$$

3. Now observe that $S_1 = \{x^2, x, 1\}$ includes 1 and x, which, as the complementary function shows, are both solutions of the corresponding homogeneous differential equation. Thus we multiply each member of the set S_1 by x^2 to obtain the revised set

$$S_1' = \{x^4, x^3, x^2\},$$

none of whose members are solutions of the homogeneous differential equation. We observe that multiplication by x instead of x^2 would not be sufficient, since the resulting set would be (x^3, x^2, x), which still includes the homogeneous solution x. Turning to the set S_2, observe that both of its members, $\sin x$ and $\cos x$, are also solutions of the homogeneous differential equation. Hence we replace S_2 by the revised set

$$S_2' = \{x \sin x, x \cos x\}.$$

4. None of the original UC sets remain here. They have been replaced by the revised sets S_1' and S_2' containing the five elements

$$x^4, \qquad x^3, \qquad x^2, \qquad x \sin x, \qquad x \cos x.$$

We form a linear combination of these,

$$Ax^4 + Bx^3 + Cx^2 + Dx \sin x + Ex \cos x,$$

with undetermined coefficients A, B, C, D, E.

5. We now take this as our particular solution

$$y_p = Ax^4 + Bx^3 + Cx^2 + Dx \sin x + Ex \cos x.$$

Then

$$y_p' = 4Ax^3 + 3Bx^2 + 2Cx + Dx \cos x + D \sin x - Ex \sin x + E \cos x,$$

$$y_p'' = 12Ax^2 + 6Bx + 2C - Dx \sin x + 2D \cos x - Ex \cos x - 2E \sin x,$$

$$y_p''' = 24Ax + 6B - Dx \cos x - 3D \sin x + Ex \sin x - 3E \cos x,$$

$$y_p^{(iv)} = 24A + Dx \sin x - 4D \cos x + Ex \cos x + 4E \sin x.$$

Substituting into the differential equation, we obtain

$$24A + Dx \sin x - 4D \cos x + Ex \cos x + 4E \sin x + 12Ax^2 + 6Bx + 2C$$
$$- Dx \sin x + 2D \cos x - Ex \cos x - 2E \sin x$$
$$= 3x^2 + 4 \sin x - 2 \cos x.$$

Equating coefficients, we find

$$24A + 2C = 0$$
$$6B = 0$$
$$12A = 3$$
$$-2D = -2$$
$$2E = 4.$$

Hence $A = \frac{1}{4}$, $B = 0$, $C = -3$, $D = 1$, $E = 2$, and the particular integral is

$$y_p = \tfrac{1}{4}x^4 - 3x^2 + x \sin x + 2x \cos x.$$

The general solution is

$$y = y_c + y_p$$
$$= c_1 + c_2 x + c_3 \sin x + c_4 \cos x + \tfrac{1}{4}x^4 - 3x^2 + x \sin x + 2x \cos x.$$

► **Example 4.39 An Initial-Value Problem**

We close this section by applying our results to the solution of the initial-value problem

$$\frac{d^2y}{dx^2} - 2\frac{dy}{dx} - 3y = 2e^x - 10 \sin x, \tag{4.46}$$

$$y(0) = 2, \tag{4.47}$$

$$y'(0) = 4. \tag{4.48}$$

By Theorem 4.1, this problem has a unique solution, defined for all x, $-\infty < x < \infty$; let us proceed to find it. In Example 4.36 we found that the general solution of the differential equation (4.46) is

$$y = c_1 e^{3x} + c_2 e^{-x} - \tfrac{1}{2}e^x + 2 \sin x - \cos x. \tag{4.49}$$

From this, we have

$$\frac{dy}{dx} = 3c_1 e^{3x} - c_2 e^{-x} - \tfrac{1}{2}e^x + 2 \cos x + \sin x. \tag{4.50}$$

Applying the initial conditions (4.47) and (4.48) to Equations (4.49) and (4.50), respectively, we have

$$2 = c_1 e^0 + c_2 e^0 - \tfrac{1}{2}e^0 + 2 \sin 0 - \cos 0,$$
$$4 = 3c_1 e^0 - c_2 e^0 - \tfrac{1}{2}e^0 + 2 \cos 0 + \sin 0.$$

These equations simplify at once to the following:

$$c_1 + c_2 = \tfrac{7}{2}, \qquad 3c_1 - c_2 = \tfrac{5}{2}.$$

From these two equations we obtain

$$c_1 = \tfrac{3}{2}, \qquad c_2 = 2.$$

Substituting these values for c_1 and c_2 into Equation (4.49) we obtain the unique solution of the given initial-value problem in the form

$$y = \tfrac{3}{2}e^{3x} + 2e^{-x} - \tfrac{1}{2}e^x + 2 \sin x - \cos x.$$

Exercises

Find the general solution of each of the differential equations in Exercises 1–24.

1. $\dfrac{d^2y}{dx^2} - 3\dfrac{dy}{dx} + 2y = 4x^2.$

2. $\dfrac{d^2y}{dx^2} - 2\dfrac{dy}{dx} - 8y = 4e^{2x} - 21e^{-3x}.$

3. $\dfrac{d^2y}{dx^2} + 2\dfrac{dy}{dx} + 5y = 6\sin 2x + 7\cos 2x.$

4. $\dfrac{d^2y}{dx^2} + 2\dfrac{dy}{dx} + 2y = 10\sin 4x.$

5. $\dfrac{d^2y}{dx^2} + 2\dfrac{dy}{dx} + 4y = \cos 4x.$

6. $\dfrac{d^2y}{dx^2} - 3\dfrac{dy}{dx} - 4y = 16x - 12e^{2x}.$

7. $\dfrac{d^2y}{dx^2} + 6\dfrac{dy}{dx} + 5y = 2e^x + 10e^{5x}.$

8. $\dfrac{d^2y}{dx^2} + 2\dfrac{dy}{dx} + 10y = 5xe^{-2x}.$

9. $\dfrac{d^3y}{dx^3} + 4\dfrac{d^2y}{dx^2} + \dfrac{dy}{dx} - 6y = -18x^2 + 1.$

10. $\dfrac{d^3y}{dx^3} + 2\dfrac{d^2y}{dx^2} - 3\dfrac{dy}{dx} - 10y = 8xe^{-2x}.$

11. $\dfrac{d^3y}{dx^3} + \dfrac{d^2y}{dx^2} + 3\dfrac{dy}{dx} - 5y = 5\sin 2x + 10x^2 + 3x + 7.$

12. $4\dfrac{d^3y}{dx^3} - 4\dfrac{d^2y}{dx^2} - 5\dfrac{dy}{dx} + 3y = 3x^3 - 8x.$

13. $\dfrac{d^2y}{dx^2} + \dfrac{dy}{dx} - 6y = 10e^{2x} - 18e^{3x} - 6x - 11.$

14. $\dfrac{d^2y}{dx^2} + \dfrac{dy}{dx} - 2y = 6e^{-2x} + 3e^x - 4x^2.$

15. $\dfrac{d^3y}{dx^3} - 3\dfrac{d^2y}{dx^2} + 4y = 4e^x - 18e^{-x}.$

16. $\dfrac{d^3y}{dx^3} - 2\dfrac{d^2y}{dx^2} - \dfrac{dy}{dx} + 2y = 9e^{2x} - 8e^{3x}.$

17. $\dfrac{d^3y}{dx^3} + \dfrac{dy}{dx} = 2x^2 + 4\sin x.$

18. $\dfrac{d^4y}{dx^4} - 3\dfrac{d^3y}{dx^3} + 2\dfrac{d^2y}{dx^2} = 3e^{-x} + 6e^{2x} - 6x.$

19. $\dfrac{d^3y}{dx^3} - 6\dfrac{d^2y}{dx^2} + 11\dfrac{dy}{dx} - 6y = xe^x - 4e^{2x} + 6e^{4x}.$

20. $\dfrac{d^3y}{dx^3} - 4\dfrac{d^2y}{dx^2} + 5\dfrac{dy}{dx} - 2y = 3x^2e^x - 7e^x.$

21. $\dfrac{d^2 y}{dx^2} + y = x \sin x.$

22. $\dfrac{d^2 y}{dx^2} + 4y = 12x^2 - 16x \cos 2x.$

23. $\dfrac{d^4 y}{dx^4} + 2\dfrac{d^3 y}{dx^3} - 3\dfrac{d^2 y}{dx^2} = 18x^2 + 16xe^x + 4e^{3x} - 9.$

24. $\dfrac{d^4 y}{dx^4} - 5\dfrac{d^3 y}{dx^3} + 7\dfrac{d^2 y}{dx^2} - 5\dfrac{dy}{dx} + 6y = 5 \sin x - 12 \sin 2x.$

Solve the initial-value problems in Exercises 25–40.

25. $\dfrac{d^2 y}{dx^2} - 4\dfrac{dy}{dx} + 3y = 9x^2 + 4, \qquad y(0) = 6, \qquad y'(0) = 8.$

26. $\dfrac{d^2 y}{dx^2} + 5\dfrac{dy}{dx} + 4y = 16x + 20e^x, \qquad y(0) = 0, \qquad y'(0) = 3.$

27. $\dfrac{d^2 y}{dx^2} - 8\dfrac{dy}{dx} + 15y = 9xe^{2x}, \qquad y(0) = 5, \qquad y'(0) = 10.$

28. $\dfrac{d^2 y}{dx^2} + 7\dfrac{dy}{dx} + 10y = 4xe^{-3x}, \qquad y(0) = 0, \qquad y'(0) = -1.$

29. $\dfrac{d^2 y}{dx^2} + 8\dfrac{dy}{dx} + 16y = 8e^{-2x}, \qquad y(0) = 2, \qquad y'(0) = 0.$

30. $\dfrac{d^2 y}{dx^2} + 6\dfrac{dy}{dx} + 9y = 27e^{-6x}, \qquad y(0) = -2, \qquad y'(0) = 0.$

31. $\dfrac{d^2 y}{dx^2} + 4\dfrac{dy}{dx} + 13y = 18e^{-2x}, \qquad y(0) = 0, \qquad y'(0) = 4.$

32. $\dfrac{d^2 y}{dx^2} - 10\dfrac{dy}{dx} + 29y = 8e^{5x}, \qquad y(0) = 0, \qquad y'(0) = 8.$

33. $\dfrac{d^2 y}{dx^2} - 4\dfrac{dy}{dx} + 13y = 8 \sin 3x, \qquad y(0) = 1, \qquad y'(0) = 2.$

34. $\dfrac{d^2 y}{dx^2} - \dfrac{dy}{dx} - 6y = 8e^{2x} - 5e^{3x}, \qquad y(0) = 3, \qquad y'(0) = 5.$

35. $\dfrac{d^2 y}{dx^2} - 2\dfrac{dy}{dx} + y = 2xe^{2x} + 6e^x, \qquad y(0) = 1, \qquad y'(0) = 0.$

36. $\dfrac{d^2 y}{dx^2} - y = 3x^2 e^x, \qquad y(0) = 1, \qquad y'(0) = 2.$

37. $\dfrac{d^2 y}{dx^2} + y = 3x^2 - 4 \sin x, \qquad y(0) = 0, \qquad y'(0) = 1.$

38. $\dfrac{d^2 y}{dx^2} + 4y = 8 \sin 2x, \qquad y(0) = 6, \qquad y'(0) = 8.$

39. $\dfrac{d^3y}{dx^3} - 4\dfrac{d^2y}{dx^2} + \dfrac{dy}{dx} + 6y = 3xe^x + 2e^x - \sin x,$

$$y(0) = \frac{33}{40}, \qquad y'(0) = 0, \qquad y''(0) = 0.$$

40. $\dfrac{d^3y}{dx^3} - 6\dfrac{d^2y}{dx^2} + 9\dfrac{dy}{dx} - 4y = 8x^2 + 3 - 6e^{2x},$

$$y(0) = 1, \qquad y'(0) = 7, \qquad y''(0) = 10.$$

For each of the differential equations in Exercises 41–54 *set up* the correct linear combination of functions with undetermined literal coefficients to use in finding a particular integral by the method of undetermined coefficients. (Do not actually find the particular integrals.)

41. $\dfrac{d^2y}{dx^2} - 6\dfrac{dy}{dx} + 8y = x^3 + x + e^{-2x}.$

42. $\dfrac{d^2y}{dx^2} + 9y = e^{3x} + e^{-3x} + e^{3x}\sin 3x.$

43. $\dfrac{d^2y}{dx^2} + 4\dfrac{dy}{dx} + 5y = e^{-2x}(1 + \cos x).$

44. $\dfrac{d^2y}{dx^2} - 6\dfrac{dy}{dx} + 9y = x^4e^x + x^3e^{2x} + x^2e^{3x}.$

45. $\dfrac{d^2y}{dx^2} + 6\dfrac{dy}{dx} + 13y = xe^{-3x}\sin 2x + x^2e^{-2x}\sin 3x.$

46. $\dfrac{d^3y}{dx^3} - 3\dfrac{d^2y}{dx^2} + 2\dfrac{dy}{dx} = x^2e^x + 3xe^{2x} + 5x^2.$

47. $\dfrac{d^3y}{dx^3} - 6\dfrac{d^2y}{dx^2} + 12\dfrac{dy}{dx} - 8y = xe^{2x} + x^2e^{3x}.$

48. $\dfrac{d^4y}{dx^4} + 3\dfrac{d^3y}{dx^3} + 4\dfrac{d^2y}{dx^2} + 3\dfrac{dy}{dx} + y = x^2e^{-x} + 3e^{-x/2}\cos\dfrac{\sqrt{3}}{2}x.$

49. $\dfrac{d^4y}{dx^4} - 16y = x^2\sin 2x + x^4e^{2x}.$

50. $\dfrac{d^6y}{dx^6} + 2\dfrac{d^5y}{dx^5} + 5\dfrac{d^4y}{dx^4} = x^3 + x^2e^{-x} + e^{-x}\sin 2x.$

51. $\dfrac{d^4y}{dx^4} + 2\dfrac{d^2y}{dx^2} + y = x^2\cos x.$

52. $\dfrac{d^4y}{dx^4} + 16y = xe^{\sqrt{2}x}\sin\sqrt{2}x + e^{-\sqrt{2}x}\cos\sqrt{2}x.$

53. $\dfrac{d^4y}{dx^4} + 3\dfrac{d^2y}{dx^2} - 4y = \cos^2 x - \cosh x.$

54. $\dfrac{d^4y}{dx^4} + 10\dfrac{d^2y}{dx^2} + 9y = \sin x \sin 2x.$

4.4 VARIATION OF PARAMETERS

A. The Method

While the process of carrying out the method of undetermined coefficients is actually quite straightforward (involving only techniques of college algebra and differentiation), the method applies in general to a rather small class of problems. For example, it would not apply to the apparently simply equation

$$\frac{d^2 y}{dx^2} + y = \tan x.$$

We thus seek a method of finding a particular integral that applies in all cases (including variable coefficients) in which the complementary function is known. Such a method is the method of *variation of parameters*, which we now consider.

We shall develop this method in connection with the general second-order linear differential equation with variable coefficients

$$a_0(x)\frac{d^2 y}{dx^2} + a_1(x)\frac{dy}{dx} + a_2(x)y = F(x). \tag{4.51}$$

Suppose that y_1 and y_2 are linearly independent solutions of the corresponding homogeneous equation

$$a_0(x)\frac{d^2 y}{dx^2} + a_1(x)\frac{dy}{dx} + a_2(x)y = 0. \tag{4.52}$$

Then the complementary function of Equation (4.51) is

$$c_1 y_1(x) + c_2 y_2(x),$$

where y_1 and y_2 are linearly independent solutions of (4.52) and c_1 and c_2 are arbitrary constants. The procedure in the method of variation of parameters is to replace the arbitrary constants c_1 and c_2 in the complementary function by respective *functions* v_1 and v_2 which will be determined so that the resulting function, which is defined by

$$v_1(x)y_1(x) + v_2(x)y_2(x), \tag{4.53}$$

will be a particular integral of Equation (4.51) (hence the name, *variation of parameters*).

We have at our disposal the *two functions* v_1 and v_2 with which to satisfy the *one condition* that (4.53) be a solution of (4.51). Since we have *two* functions but only *one* condition on them, we are thus free to impose a second condition, provided this second condition does not violate the first one. We shall see when and how to impose this additional condition as we proceed.

We thus assume a solution of the form (4.53) and write

$$y_p(x) = v_1(x)y_1(x) + v_2(x)y_2(x). \tag{4.54}$$

Differentiating (4.54), we have

$$y_p'(x) = v_1(x)y_1'(x) + v_2(x)y_2'(x) + v_1'(x)y_1(x) + v_2'(x)y_2(x), \tag{4.55}$$

where we use primes to denote differentiations. At this point we impose the aforementioned second condition; we simplify y'_p by demanding that

$$v'_1(x)y_1(x) + v'_2(x)y_2(x) = 0. \tag{4.56}$$

With this condition imposed, (4.55) reduces to

$$y'_p(x) = v_1(x)y'_1(x) + v_2(x)y'_2(x). \tag{4.57}$$

Now differentiating (4.57), we obtain

$$y''_p(x) = v_1(x)y''_1(x) + v_2(x)y''_2(x) + v'_1(x)y'_1(x) + v'_2(x)y'_2(x). \tag{4.58}$$

We now impose the basic condition that (4.54) be a solution of Equation (4.51). Thus we substitute (4.54), (4.57), and (4.58) for y, dy/dx, and d^2y/dx^2, respectively, in Equation (4.51) and obtain the identity

$$a_0(x)[v_1(x)y''_1(x) + v_2(x)y''_2(x) + v'_1(x)y'_1(x) + v'_2(x)y'_2(x)]$$
$$+ a_1(x)[v_1(x)y'_1(x) + v_2(x)y'_2(x)] + a_2(x)[v_1(x)y_1(x) + v_2(x)y_2(x)] = F(x).$$

This can be written as

$$v_1(x)[a_0(x)y''_1(x) + a_1(x)y'_1(x) + a_2(x)y_1(x)]$$
$$+ v_2(x)[a_0(x)y''_2(x) + a_1(x)y'_2(x) + a_2(x)y_2(x)]$$
$$+ a_0(x)[v'_1(x)y'_1(x) + v'_2(x)y'_2(x)] = F(x). \tag{4.59}$$

Since y_1 and y_2 are solutions of the corresponding homogeneous differential equation (4.52), the expressions in the first two brackets in (4.59) are identically zero. This leaves merely

$$v'_1(x)y'_1(x) + v'_2(x)y'_2(x) = \frac{F(x)}{a_0(x)}. \tag{4.60}$$

This is actually what the basic condition demands. Thus the two imposed conditions require that the functions v_1 and v_2 be chosen such that the system of equations

$$y_1(x)v'_1(x) + y_2(x)v'_2(x) = 0,$$
$$y'_1(x)v'_1(x) + y'_2(x)v'_2(x) = \frac{F(x)}{a_0(x)}, \tag{4.61}$$

is satisfied. The determinant of coefficients of this system is precisely

$$W[y_1(x), y_2(x)] = \begin{vmatrix} y_1(x) & y_2(x) \\ y'_1(x) & y'_2(x) \end{vmatrix}.$$

Since y_1 and y_2 are linearly independent solutions of the corresponding homogeneous differential equation (4.52), we know that $W[y_1(x), y_2(x)] \neq 0$. Hence the system (4.61) has a unique solution. Actually solving this system, we obtain

$$v'_1(x) = \frac{\begin{vmatrix} 0 & y_2(x) \\ \dfrac{F(x)}{a_0(x)} & y'_2(x) \end{vmatrix}}{\begin{vmatrix} y_1(x) & y_2(x) \\ y'_1(x) & y'_2(x) \end{vmatrix}} = -\frac{F(x)y_2(x)}{a_0(x)W[y_1(x), y_2(x)]},$$

$$v_2'(x) = \frac{\begin{vmatrix} y_1(x) & 0 \\ y_1'(x) & \dfrac{F(x)}{a_0(x)} \end{vmatrix}}{\begin{vmatrix} y_1(x) & y_2(x) \\ y_1'(x) & y_2'(x) \end{vmatrix}} = \frac{F(x)y_1(x)}{a_0(x)W[y_1(x), y_2(x)]}.$$

Thus we obtain the functions v_1 and v_2 defined by

$$v_1(x) = -\int^x \frac{F(t)y_2(t)\,dt}{a_0(t)W[y_1(t), y_2(t)]},$$

$$\tag{4.62}$$

$$v_2(x) = \int^x \frac{F(t)y_1(t)\,dt}{a_0(t)W[y_1(t), y_2(t)]}.$$

Therefore a particular integral y_p of Equation (4.51) is defined by

$$y_p(x) = v_1(x)y_1(x) + v_2(x)y_2(x),$$

where v_1 and v_2 are defined by (4.62).

B. Examples

▶ Example 4.40

Consider the differential equation

$$\frac{d^2y}{dx^2} + y = \tan x. \tag{4.63}$$

The complementary function is defined by

$$y_c(x) = c_1 \sin x + c_2 \cos x.$$

We assume
$$y_p(x) = v_1(x)\sin x + v_2(x)\cos x, \tag{4.64}$$

where the functions v_1 and v_2 will be determined such that this is a particular integral of the differential equation (4.63). Then

$$y_p'(x) = v_1(x)\cos x - v_2(x)\sin x + v_1'(x)\sin x + v_2'(x)\cos x.$$

We impose the condition

$$v_1'(x)\sin x + v_2'(x)\cos x = 0, \tag{4.65}$$

leaving
$$y_p'(x) = v_1(x)\cos x - v_2(x)\sin x.$$

From this

$$y_p''(x) = -v_1(x)\sin x - v_2(x)\cos x + v_1'(x)\cos x - v_2'(x)\sin x. \tag{4.66}$$

Substituting (4.64) and (4.66) into (4.63) we obtain

$$v_1'(x)\cos x - v_2'(x)\sin x = \tan x. \tag{4.67}$$

Thus we have the two equations (4.65) and (4.67) from which to determine $v'_1(x)$, $v'_2(x)$:

$$v'_1(x)\sin x + v'_2(x)\cos x = 0,$$

$$v'_1(x)\cos x - v'_2(x)\sin x = \tan x.$$

Solving we find:

$$v'_1(x) = \frac{\begin{vmatrix} 0 & \cos x \\ \tan x & -\sin x \end{vmatrix}}{\begin{vmatrix} \sin x & \cos x \\ \cos x & -\sin x \end{vmatrix}} = \frac{-\cos x \tan x}{-1} = \sin x,$$

$$v'_2(x) = \frac{\begin{vmatrix} \sin x & 0 \\ \cos x & \tan x \end{vmatrix}}{\begin{vmatrix} \sin x & \cos x \\ \cos x & -\sin x \end{vmatrix}} = \frac{\sin x \tan x}{-1} = \frac{-\sin^2 x}{\cos x}$$

$$= \frac{\cos^2 x - 1}{\cos x} = \cos x - \sec x.$$

Integrating we find:

$$v_1(x) = -\cos x + c_3, \qquad v_2(x) = \sin x - \ln|\sec x + \tan x| + c_4. \qquad (4.68)$$

Substituting (4.68) into (4.64) we have

$$y_p(x) = (-\cos x + c_3)\sin x + (\sin x - \ln|\sec x + \tan x| + c_4)\cos x$$

$$= -\sin x \cos x + c_3 \sin x + \sin x \cos x$$

$$\quad - \ln|\sec x + \tan x|(\cos x) + c_4 \cos x$$

$$= c_3 \sin x + c_4 \cos x - (\cos x)(\ln|\sec x + \tan x|).$$

Since a particular integral is a solution free of arbitrary constants, we may assign any particular values A and B to c_3 and c_4, respectively, and the result will be the particular integral

$$A \sin x + B \cos x - (\cos x)(\ln|\sec x + \tan x|).$$

Thus $y = y_c + y_p$ becomes

$$y = c_1 \sin x + c_2 \cos x + A \sin x + B \cos x - (\cos x)(\ln|\sec x + \tan x|),$$

which we may write as

$$y = C_1 \sin x + C_2 \cos x - (\cos x)(\ln|\sec x + \tan x|),$$

where $C_1 = c_1 + A$, $C_2 = c_2 + B$.

Thus we see that we might as well have chosen the constants c_3 and c_4 both equal to 0 in (4.68), for essentially the same result,

$$y = c_1 \sin x + c_2 \cos x - (\cos x)(\ln|\sec x + \tan x|),$$

would have been obtained. This is the general solution of the differential equation (4.63).

The method of variation of parameters extends to higher-order linear equations. We now illustrate the extension to a third-order equation in Example 4.41, although we hasten to point out that the equation of this example can be solved more readily by the method of undetermined coefficients.

▶ **Example 4.41**

Consider the differential equation

$$\frac{d^3y}{dx^3} - 6\frac{d^2y}{dx^2} + 11\frac{dy}{dx} - 6y = e^x. \tag{4.69}$$

The complementary function is

$$y_c(x) = c_1 e^x + c_2 e^{2x} + c_3 e^{3x}.$$

We assume as a particular integral

$$y_p(x) = v_1(x)e^x + v_2(x)e^{2x} + v_3(x)e^{3x}. \tag{4.70}$$

Since we have *three* functions v_1, v_2, v_3 at our disposal in this case, we can apply three conditions. We have:

$$y'_p(x) = v_1(x)e^x + 2v_2(x)e^{2x} + 3v_3(x)e^{3x} + v'_1(x)e^x + v'_2(x)e^{2x} + v'_3(x)e^{3x}.$$

Proceeding in a manner analogous to that of the second-order case, we impose the condition

$$v'_1(x)e^x + v'_2(x)e^{2x} + v'_3(x)e^{3x} = 0, \tag{4.71}$$

leaving

$$y'_p(x) = v_1(x)e^x + 2v_2(x)e^{2x} + 3v_3(x)e^{3x}. \tag{4.72}$$

Then

$$y''_p(x) = v_1(x)e^x + 4v_2(x)e^{2x} + 9v_3(x)e^{3x} + v'_1(x)e^x + 2v'_2(x)e^{2x} + 3v'_3(x)e^{3x}.$$

We now impose the condition

$$v'_1(x)e^x + 2v'_2(x)e^{2x} + 3v'_3(x)e^{3x} = 0, \tag{4.73}$$

leaving

$$y''_p(x) = v_1(x)e^x + 4v_2(x)e^{2x} + 9v_3(x)e^{3x}. \tag{4.74}$$

From this,

$$y'''_p(x) = v_1(x)e^x + 8v_2(x)e^{2x} + 27v_3(x)e^{3x} + v'_1(x)e^x + 4v'_2(x)e^{2x} + 9v'_3(x)e^{3x}. \tag{4.75}$$

We substitute (4.70), (4.72), (4.74), and (4.75) into the differential equation (4.69), obtaining:

$$v_1(x)e^x + 8v_2(x)e^{2x} + 27v_3(x)e^{3x} + v'_1(x)e^x + 4v'_2(x)e^{2x} + 9v'_3(x)e^{3x}$$
$$- 6v_1(x)e^x - 24v_2(x)e^{2x} - 54v_3(x)e^{3x} + 11v_1(x)e^x + 22v_2(x)e^{2x} + 33v_3(x)e^{3x}$$
$$- 6v_1(x)e^x - 6v_2(x)e^{2x} - 6v_3(x)e^{3x} = e^x$$

or

$$v'_1(x)e^x + 4v'_2(x)e^{2x} + 9v'_3(x)e^{3x} = e^x. \tag{4.76}$$

Thus we have the three equations (4.71), (4.73), (4.76) from which to determine $v'_1(x)$, $v'_2(x)$, $v'_3(x)$:

$$v'_1(x)e^x + v'_2(x)e^{2x} + v'_3(x)e^{3x} = 0,$$
$$v'_1(x)e^x + 2v'_2(x)e^{2x} + 3v'_3(x)e^{3x} = 0,$$
$$v'_1(x)e^x + 4v'_2(x)e^{2x} + 9v'_3(x)e^{3x} = e^x.$$

Solving, we find

$$v'_1(x) = \frac{\begin{vmatrix} 0 & e^{2x} & e^{3x} \\ 0 & 2e^{2x} & 3e^{3x} \\ e^x & 4e^{2x} & 9e^{3x} \end{vmatrix}}{\begin{vmatrix} e^x & e^{2x} & e^{3x} \\ e^x & 2e^{2x} & 3e^{3x} \\ e^x & 4e^{2x} & 9e^{3x} \end{vmatrix}} = \frac{e^{6x}\begin{vmatrix} 1 & 1 \\ 2 & 3 \end{vmatrix}}{e^{6x}\begin{vmatrix} 1 & 1 & 1 \\ 1 & 2 & 3 \\ 1 & 4 & 9 \end{vmatrix}} = \frac{1}{2},$$

$$v'_2(x) = \frac{\begin{vmatrix} e^x & 0 & e^{3x} \\ e^x & 0 & 3e^{3x} \\ e^x & e^x & 9e^{3x} \end{vmatrix}}{\begin{vmatrix} e^x & e^{2x} & e^{3x} \\ e^x & 2e^{2x} & 3e^{3x} \\ e^x & 4e^{2x} & 9e^{3x} \end{vmatrix}} = \frac{-e^{5x}\begin{vmatrix} 1 & 1 \\ 1 & 3 \end{vmatrix}}{2e^{6x}} = -e^{-x},$$

$$v'_3(x) = \frac{\begin{vmatrix} e^x & e^{2x} & 0 \\ e^x & 2e^{2x} & 0 \\ e^x & 4e^{2x} & e^x \end{vmatrix}}{\begin{vmatrix} e^x & e^{2x} & e^{3x} \\ e^x & 2e^{2x} & 3e^{3x} \\ e^x & 4e^{2x} & 9e^{3x} \end{vmatrix}} = \frac{e^{4x}\begin{vmatrix} 1 & 1 \\ 1 & 2 \end{vmatrix}}{2e^{6x}} = \frac{1}{2}e^{-2x}.$$

We now integrate, choosing all the constants of integration to be zero (as the previous example showed was possible). We find:

$$v_1(x) = \tfrac{1}{2}x, \qquad v_2(x) = e^{-x}, \qquad v_3(x) = -\tfrac{1}{4}e^{-2x}.$$

Thus

$$y_p(x) = \tfrac{1}{2}xe^x + e^{-x}e^{2x} - \tfrac{1}{4}e^{-2x}e^{3x} = \tfrac{1}{2}xe^x + \tfrac{3}{4}e^x.$$

Thus the general solution of Equation (4.53) is

$$y = y_c + y_p = c_1e^x + c_2e^{2x} + c_3e^{3x} + \tfrac{1}{2}xe^x + \tfrac{3}{4}e^x$$

or

$$y = c'_1e^x + c_2e^{2x} + c_3e^{3x} + \tfrac{1}{2}xe^x,$$

where $c'_1 = c_1 + \tfrac{3}{4}$.

In Examples 4.40 and 4.41 the coefficients in the differential equation were constants. The general discussion at the beginning of this section shows that the method applies equally well to linear differential equations with variable coefficients, once the complementary function y_c is known. We now illustrate its application to such an equation in Example 4.42.

▶ **Example 4.42**

Consider the differential equation

$$(x^2 + 1)\frac{d^2 y}{dx^2} - 2x\frac{dy}{dx} + 2y = 6(x^2 + 1)^2. \tag{4.77}$$

In Example 4.16 we solved the corresponding homogeneous equation

$$(x^2 + 1)\frac{d^2 y}{dx^2} - 2x\frac{dy}{dx} + 2y = 0.$$

From the results of that example, we see that the complementary function of equation (4.77) is

$$y_c(x) = c_1 x + c_2(x^2 - 1).$$

To find a particular integral of Equation (4.77), we therefore let

$$y_p(x) = v_1(x)x + v_2(x)(x^2 - 1). \tag{4.78}$$

Then

$$y_p'(x) = v_1(x) \cdot 1 + v_2(x) \cdot 2x + v_1'(x)x + v_2'(x)(x^2 - 1).$$

We impose the condition

$$v_1'(x)x + v_2'(x)(x^2 - 1) = 0, \tag{4.79}$$

leaving

$$y_p'(x) = v_1(x) \cdot 1 + v_2(x) \cdot 2x. \tag{4.80}$$

From this, we find

$$y_p''(x) = v_1'(x) + 2v_2(x) + v_2'(x) \cdot 2x. \tag{4.81}$$

Substituting (4.78), (4.80), and (4.81) into (4.77) we obtain

$$(x^2 + 1)[v_1'(x) + 2v_2(x) + 2xv_2'(x)] - 2x[v_1(x) + 2xv_2(x)]$$
$$+ 2[v_1(x)x + v_2(x)(x^2 - 1)] = 6(x^2 + 1)^2$$

or

$$(x^2 + 1)[v_1'(x) + 2xv_2'(x)] = 6(x^2 + 1)^2. \tag{4.82}$$

Thus we have the two equations (4.79) and (4.82) from which to determine $v_1'(x)$ and $v_2'(x)$; that is, $v_1'(x)$ and $v_2'(x)$ satisfy the system

$$v_1'(x)x + v_2'(x)[x^2 - 1] = 0,$$
$$v_1'(x) + v_2'(x)[2x] = 6(x^2 + 1).$$

Solving this system, we find

$$v_1'(x) = \frac{\begin{vmatrix} 0 & x^2 - 1 \\ 6(x^2 + 1) & 2x \end{vmatrix}}{\begin{vmatrix} x & x^2 - 1 \\ 1 & 2x \end{vmatrix}} = \frac{-6(x^2 + 1)(x^2 - 1)}{x^2 + 1} = -6(x^2 - 1),$$

$$v_2'(x) = \frac{\begin{vmatrix} x & 0 \\ 1 & 6(x^2 + 1) \end{vmatrix}}{\begin{vmatrix} x & x^2 - 1 \\ 1 & 2x \end{vmatrix}} = \frac{6x(x^2 + 1)}{x^2 + 1} = 6x.$$

Integrating, we obtain

$$v_1(x) = -2x^3 + 6x, \qquad v_2(x) = 3x^2, \tag{4.83}$$

where we have chosen both constants of integration to be zero. Substituting (4.83) into (4.78), we have

$$y_p(x) = (-2x^3 + 6x)x + 3x^2(x^2 - 1)$$
$$= x^4 + 3x^2.$$

Therefore the general solution of Equation (4.77) may be expressed in the form

$$y = y_c + y_p$$
$$= c_1 x + c_2(x^2 - 1) + x^4 + 3x^2.$$

Exercises

Find the general solution of each of the differential equations in Exercises 1–18.

1. $\dfrac{d^2y}{dx^2} + y = \cot x.$
2. $\dfrac{d^2y}{dx^2} + y = \tan^2 x.$

3. $\dfrac{d^2y}{dx^2} + y = \sec x.$
4. $\dfrac{d^2y}{dx^2} + y = \sec^3 x.$

5. $\dfrac{d^2y}{dx^2} + 4y = \sec^2 2x.$

6. $\dfrac{d^2y}{dx^2} + y = \tan x \sec x.$

7. $\dfrac{d^2y}{dx^2} + 4\dfrac{dy}{dx} + 5y = e^{-2x} \sec x.$

8. $\dfrac{d^2y}{dx^2} - 2\dfrac{dy}{dx} + 5y = e^x \tan 2x.$

9. $\dfrac{d^2y}{dx^2} + 6\dfrac{dy}{dx} + 9y = \dfrac{e^{-3x}}{x^3}.$

10. $\dfrac{d^2y}{dx^2} - 2\dfrac{dy}{dx} + y = xe^x \ln x \; (x > 0).$

11. $\dfrac{d^2 y}{dx^2} + y = \sec x \csc x.$

12. $\dfrac{d^2 y}{dx^2} + y = \tan^3 x.$

13. $\dfrac{d^2 y}{dx^2} + 3\dfrac{dy}{dx} + 2y = \dfrac{1}{1 + e^x}.$

14. $\dfrac{d^2 y}{dx^2} + 3\dfrac{dy}{dx} + 2y = \dfrac{1}{1 + e^{2x}}.$

15. $\dfrac{d^2 y}{dx^2} + y = \dfrac{1}{1 + \sin x}.$

16. $\dfrac{d^2 y}{dx^2} - 2\dfrac{dy}{dx} + y = e^x \sin^{-1} x.$

17. $\dfrac{d^2 y}{dx^2} + 3\dfrac{dy}{dx} + 2y = \dfrac{e^{-x}}{x}.$

18. $\dfrac{d^2 y}{dx^2} - 2\dfrac{dy}{dx} + y = x \ln x \quad (x > 0).$

19. Find the general solution of

$$x^2 \frac{d^2 y}{dx^2} - 6x\frac{dy}{dx} + 10y = 3x^4 + 6x^3,$$

given that $y = x^2$ and $y = x^5$ are linearly independent solutions of the corresponding homogeneous equation.

20. Find the general solution of

$$(x + 1)^2 \frac{d^2 y}{dx^2} - 2(x + 1)\frac{dy}{dx} + 2y = 1,$$

given that $y = x + 1$ and $y = (x + 1)^2$ are linearly independent solutions of the corresponding homogeneous equation.

21. Find the general solution of

$$(x^2 + 2x) \frac{d^2 y}{dx^2} - 2(x + 1)\frac{dy}{dx} + 2y = (x + 2)^2,$$

given that $y = x + 1$ and $y = x^2$ are linearly independent solutions of the corresponding homogeneous equation.

22. Find the general solution of

$$x^2 \frac{d^2 y}{dx^2} - x(x + 2)\frac{dy}{dx} + (x + 2)y = x^3,$$

given that $y = x$ and $y = xe^x$ are linearly independent solutions of the corresponding homogeneous equation.

23. Find the general solution of

$$x(x-2)\frac{d^2 y}{dx^2} - (x^2 - 2)\frac{dy}{dx} + 2(x-1)y = 3x^2(x-2)^2 e^x,$$

given that $y = e^x$ and $y = x^2$ are linearly independent solutions of the corresponding homogeneous equation.

24. Find the general solution of

$$(2x+1)(x+1)\frac{d^2 y}{dx^2} + 2x\frac{dy}{dx} - 2y = (2x+1)^2,$$

given that $y = x$ and $y = (x+1)^{-1}$ are linearly independent solutions of the corresponding homogeneous equation.

25. Find the general solution of

$$(\sin^2 x)\frac{d^2 y}{dx^2} - 2\sin x \cos x \frac{dy}{dx} + (\cos^2 x + 1)y = \sin^3 x,$$

given that $y = \sin x$ and $y = x \sin x$ are linearly independent solutions of the corresponding homogeneous equation.

26. Find the general solution by two methods:

$$\frac{d^3 y}{dx^3} - 3\frac{d^2 y}{dx^2} - \frac{dy}{dx} + 3y = x^2 e^x.$$

4.5   THE CAUCHY–EULER EQUATION

A. The Equation and the Method of Solution

In the preceding sections we have seen how to obtain the general solution of the *nth*-order linear differential equation with *constant* coefficients. We have seen that in such cases the form of the complementary function may be readily determined. The general *nth*-order linear equation with *variable* coefficients is quite a different matter, however, and only in certain special cases can the complementary function be obtained explicity in closed form. One special case of considerable practical importance for which it is fortunate that this can be done is the so-called *Cauchy-Euler equation* (or *equidimensional* equation). This is an equation of the form

$$a_0 x^n \frac{d^n y}{dx^n} + a_1 x^{n-1}\frac{d^{n-1} y}{dx^{n-1}} + \cdots + a_{n-1} x \frac{dy}{dx} + a_n y = F(x), \qquad (4.84)$$

where $a_0, a_1, \ldots, a_{n-1}, a_n$ are constants. Note the characteristic feature of this equation: each term in the left member is a constant multiple of an expression of the form

$$x^k \frac{d^k y}{dx^k}.$$

How should one proceed to solve such an equation? About the only hopeful thought that comes to mind at this stage of our study is to attempt a transformation. But what transformation should we attempt and where will it lead us? While it is certainly worthwhile to stop for a moment and consider what sort of transformation we might use in solving a "new" type of equation when we first encounter it, it is certainly not worthwhile to spend a great deal of time looking for clever devices which mathematicians have known about for many years. The facts are stated in the following theorem.

THEOREM 4.14

The transformation $x = e^t$ reduces the equation

$$a_0 x^n \frac{d^n y}{dx^n} + a_1 x^{n-1} \frac{d^{n-1} y}{dx^{n-1}} + \cdots + a_{n-1} x \frac{dy}{dx} + a_n y = F(x) \qquad (4.84)$$

to a linear differential equation with constant coefficients.

This is what we need! We shall prove this theorem for the case of the *second*-order Cauchy–Euler differential equation

$$a_0 x^2 \frac{d^2 y}{dx^2} + a_1 x \frac{dy}{dx} + a_2 y = F(x). \qquad (4.85)$$

The proof in the general nth-order case proceeds in a similar fashion. Letting $x = e^t$, assuming $x > 0$, we have $t = \ln x$. Then

$$\frac{dy}{dx} = \frac{dy}{dt} \frac{dt}{dx} = \frac{1}{x} \frac{dy}{dt}$$

and

$$\frac{d^2 y}{dx^2} = \frac{1}{x} \frac{d}{dx} \left(\frac{dy}{dt} \right) + \frac{dy}{dt} \frac{d}{dx} \left(\frac{1}{x} \right) = \frac{1}{x} \left(\frac{d^2 y}{dt^2} \frac{dt}{dx} \right) - \frac{1}{x^2} \frac{dy}{dt}$$

$$\frac{1}{x} \left(\frac{d^2 y}{dt^2} \frac{1}{x} \right) - \frac{1}{x^2} \frac{dy}{dt} = \frac{1}{x^2} \left(\frac{d^2 y}{dt^2} - \frac{dy}{dt} \right).$$

Thus

$$x \frac{dy}{dx} = \frac{dy}{dt} \quad \text{and} \quad x^2 \frac{d^2 y}{dx^2} = \frac{d^2 y}{dt^2} - \frac{dy}{dt}.$$

Substituting into Equation (4.85) we obtain

$$a_0 \left(\frac{d^2 y}{dt^2} - \frac{dy}{dt} \right) + a_1 \frac{dy}{dt} + a_2 y = F(e^t)$$

or

$$A_0 \frac{d^2 y}{dt^2} + A_1 \frac{dy}{dt} + A_2 y = G(t), \qquad (4.86)$$

where

$$A_0 = a_0, \quad A_1 = a_1 - a_0, \quad A_2 = a_2, \quad G(t) = F(e^t).$$

This is a second-order linear differential equation with *constant* coefficients, which was what we wished to show.

Remarks.   1. Note that the leading coefficient $a_0 x^n$ in Equation (4.84) is zero for $x = 0$. Thus the basic interval $a \le x \le b$, referred to in the general theorems of Section 4.1, does *not* include $x = 0$.

2. Observe that in the above proof we assumed that $x > 0$. If $x < 0$, the substitution $x = -e^t$ is actually the correct one. Unless the contrary is explicitly stated, we shall assume $x > 0$ when finding the general solution of a Cauchy–Euler differential equation.

B. Examples

▶ **Example 4.43**

$$x^2 \frac{d^2 y}{dx^2} - 2x \frac{dy}{dx} + 2y = x^3. \tag{4.87}$$

Let $x = e^t$. Then, assuming $x > 0$, we have $t = \ln x$, and

$$\frac{dy}{dx} = \frac{dy}{dt} \frac{dt}{dx} = \frac{1}{x} \frac{dy}{dt},$$

$$\frac{d^2 y}{dx^2} = \frac{1}{x}\left(\frac{d^2 y}{dt^2} \frac{dt}{dx} \right) - \frac{1}{x^2} \frac{dy}{dt} = \frac{1}{x^2}\left(\frac{d^2 y}{dt^2} - \frac{dy}{dt} \right).$$

Thus Equation (4.87) becomes

$$\frac{d^2 y}{dt^2} - \frac{dy}{dt} - 2 \frac{dy}{dt} + 2y = e^{3t}$$

or

$$\frac{d^2 y}{dt^2} - 3 \frac{dy}{dt} + 2y = e^{3t}. \tag{4.88}$$

The complementary function of this equation is $y_c = c_1 e^t + c_2 e^{2t}$. We find a particular integral by the method of undetermined coefficients. We assume $y_p = A e^{3t}$. Then $y'_p = 3A e^{3t}$, $y''_p = 9A e^{3t}$, and substituting into Equation (4.88) we obtain

$$2A e^{3t} = e^{3t}.$$

Thus $A = \frac{1}{2}$ and we have $y_p = \frac{1}{2} e^{3t}$. The general solution of Equation (4.88) is then

$$y = c_1 e^t + c_2 e^{2t} + \tfrac{1}{2} e^{3t}.$$

But we are not yet finished! We must return to the original independent variable x. Since $e^t = x$, we find

$$y = c_1 x + c_2 x^2 + \tfrac{1}{2} x^3.$$

This is the general solution of Equation (4.87).

Remarks.   1. Note carefully that under the transformation $x = e^t$ the right member of (4.87), x^3, transforms into e^{3t}. The student should be careful to transform *both* sides of the equation if he intends to obtain a particular integral of the given equation by finding a particular integral of the transformed equation, as we have done here.

2. We hasten to point out that the following alternative procedure may be used. After finding the complementary function of the transformed equation one can immediately write the complementary function of the original given equation and then proceed to obtain a particular integral of the original equation by variation of parameters. In Example 4.43, upon finding the complementary function $c_1 e^t + c_2 e^{2t}$ of Equation (4.88), one can immediately write the complementary function $c_1 x + c_2 x^2$ of Equation (4.87), then assume the particular integral $y_p(x) = v_1(x)x + v_2(x)x^2$, and from here proceed by the method of variation of parameters. However, when the nonhomogeneous function F transforms into a linear combination of UC functions, as it does in this example, the procedure illustrated is generally simpler.

▶ **Example 4.44**

$$x^3 \frac{d^3 y}{dx^3} - 4x^2 \frac{d^2 y}{dx^2} + 8x \frac{dy}{dx} - 8y = 4 \ln x. \tag{4.89}$$

Assuming $x > 0$, we let $x = e^t$. Then $t = \ln x$, and

$$\frac{dy}{dx} = \frac{1}{x} \frac{dy}{dt},$$

$$\frac{d^2 y}{dx^2} = \frac{1}{x^2} \left(\frac{d^2 y}{dt^2} - \frac{dy}{dt} \right).$$

Now we must consider $\dfrac{d^3 y}{dx^3}$.

$$\frac{d^3 y}{dx^3} = \frac{1}{x^2} \frac{d}{dx} \left(\frac{d^2 y}{dt^2} - \frac{dy}{dt} \right) - \frac{2}{x^3} \left(\frac{d^2 y}{dt^2} - \frac{dy}{dt} \right)$$

$$= \frac{1}{x^2} \left(\frac{d^3 y}{dt^3} \frac{dt}{dx} - \frac{d^2 y}{dt^2} \frac{dt}{dx} \right) - \frac{2}{x^3} \left(\frac{d^2 y}{dt^2} - \frac{dy}{dt} \right)$$

$$= \frac{1}{x^3} \left(\frac{d^3 y}{dt^3} - \frac{d^2 y}{dt^2} \right) - \frac{2}{x^3} \left(\frac{d^2 y}{dt^2} - \frac{dy}{dt} \right)$$

$$= \frac{1}{x^3} \left(\frac{d^3 y}{dt^3} - 3 \frac{d^2 y}{dt^2} + 2 \frac{dy}{dt} \right).$$

Thus, substituting into Equation (4.89), we obtain

$$\left(\frac{d^3 y}{dt^3} - 3 \frac{d^2 y}{dt^2} + 2 \frac{dy}{dt} \right) - 4 \left(\frac{d^2 y}{dt^2} - \frac{dy}{dt} \right) + 8 \left(\frac{dy}{dt} \right) - 8y = 4t$$

or

$$\frac{d^3 y}{dt^3} - 7 \frac{d^2 y}{dt^2} + 14 \frac{dy}{dt} - 8y = 4t. \tag{4.90}$$

The complementary function of the transformed equation (4.90) is

$$y_c = c_1 e^t + c_2 e^{2t} + c_3 e^{4t}.$$

We procced to obtain a particular integral of Equation (4.90) by the method of undetermined coefficients. We assume $y_p = At + B$. Then $y_p' = A$, $y_p'' = y_p''' = 0$. Substituting into Equation (4.90), we find

$$14A - 8At - 8B = 4t.$$

Thus

$$-8A = 4, \qquad 14A - 8B = 0,$$

and so $A = -\frac{1}{2}$, $B = -\frac{7}{8}$. Thus the general solution of Equation (4.90) is

$$y = c_1 e^t + c_2 e^{2t} + c_3 e^{4t} - \tfrac{1}{2}t - \tfrac{7}{8},$$

and so the general solution of Equation (4.89) is

$$y = c_1 x + c_2 x^2 + c_3 x^4 - \tfrac{1}{2}\ln x - \tfrac{7}{8}.$$

Remarks. In solving the Cauchy–Euler equations of the preceding examples, we observe that the transformation $x = e^t$ reduces

$$x\frac{dy}{dx} \quad \text{to} \quad \frac{dy}{dt}, \qquad x^2\frac{d^2y}{dx^2} \quad \text{to} \quad \frac{d^2y}{dt^2} - \frac{dy}{dt},$$

and

$$x^3\frac{d^3y}{dx^3} \quad \text{to} \quad \frac{d^3y}{dt^3} - 3\frac{d^2y}{dt^2} + 2\frac{dy}{dt}.$$

We now show (without proof) how to find the expression into which the general term

$$x^n\frac{d^n y}{dx^n},$$

where n is an *arbitrary* positive integer, reduces under the transformation $x = e^t$. We present this as the following formal four-step procedure.

1. For the given positive integer n, determine
$$r(r - 1)(r - 2)\cdots[r - (n - 1)].$$

2. Expand the preceding as a polynomial of degree n in r.

3. Replace r^k by $\dfrac{d^k y}{dt^k}$, for each $k = 1, 2, 3, \ldots, n$.

4. Equate $x^n\dfrac{d^n y}{dx^n}$ to the result in Step 3.

For example, when $n = 3$, we have the following illustration.

1. Since $n = 3$, $n - 1 = 2$ and we determine $r(r - 1)(r - 2)$.

2. Expanding the preceding, we obtain $r^3 - 3r^2 + 2r$.

3. Replacing r^3 by $\dfrac{d^3 y}{dt^3}$, r^2 by $\dfrac{d^2 y}{dt^2}$, and r by $\dfrac{dy}{dt}$, we have

$$\frac{d^3y}{dt^3} - 3\frac{d^2y}{dt^2} + 2\frac{dy}{dt}.$$

4. Equating $x^3 \dfrac{d^3 y}{dx^3}$ to this, we have the relation

$$x^3 \frac{d^3 y}{dx^3} = \frac{d^3 y}{dt^3} - 3 \frac{d^2 y}{dt^2} + 2 \frac{dy}{dt}.$$

Note that this is precisely the relation we found in Example 4.44 and stated above.

Exercises

Find the general solution of each of the differential equations in Exercises 1–19. In each case assume $x > 0$.

1. $x^2 \dfrac{d^2 y}{dx^2} - 3x \dfrac{dy}{dx} + 3y = 0.$

2. $x^2 \dfrac{d^2 y}{dx^2} + x \dfrac{dy}{dx} - 4y = 0.$

3. $4x^2 \dfrac{d^2 y}{dx^2} - 4x \dfrac{dy}{dx} + 3y = 0.$

4. $x^2 \dfrac{d^2 y}{dx^2} - 3x \dfrac{dy}{dx} + 4y = 0.$

5. $x^2 \dfrac{d^2 y}{dx^2} + x \dfrac{dy}{dx} + 4y = 0.$

6. $x^2 \dfrac{d^2 y}{dx^2} - 3x \dfrac{dy}{dx} + 13y = 0.$

7. $3x^2 \dfrac{d^2 y}{dx^2} - 4x \dfrac{dy}{dx} + 2y = 0.$

8. $x^2 \dfrac{d^2 y}{dx^2} + x \dfrac{dy}{dx} + 9y = 0.$

9. $9x^2 \dfrac{d^2 y}{dx^2} + 3x \dfrac{dy}{dx} + y = 0.$

10. $x^2 \dfrac{d^2 y}{dx^2} - 5x \dfrac{dy}{dx} + 10y = 0.$

11. $x^3 \dfrac{d^3 y}{dx^3} - 3x^2 \dfrac{d^2 y}{dx^2} + 6x \dfrac{dy}{dx} - 6y = 0.$

12. $x^3 \dfrac{d^3 y}{dx^3} + 2x^2 \dfrac{d^2 y}{dx^2} - 10x \dfrac{dy}{dx} - 8y = 0.$

13. $x^3 \dfrac{d^3 y}{dx^3} - x^2 \dfrac{d^2 y}{dx^2} - 6x \dfrac{dy}{dx} + 18y = 0.$

14. $x^2 \dfrac{d^2 y}{dx^2} - 4x \dfrac{dy}{dx} + 6y = 4x - 6.$

15. $x^2 \dfrac{d^2 y}{dx^2} - 5x \dfrac{dy}{dx} + 8y = 2x^3.$

16. $x^2 \dfrac{d^2 y}{dx^2} + 4x \dfrac{dy}{dx} + 2y = 4 \ln x.$

17. $x^2 \dfrac{d^2 y}{dx^2} + x \dfrac{dy}{dx} + 4y = 2x \ln x.$

18. $x^2 \dfrac{d^2 y}{dx^2} + x \dfrac{dy}{dx} + y = 4 \sin \ln x.$

19. $x^3 \dfrac{d^3 y}{dx^3} - x^2 \dfrac{d^2 y}{dx^2} + 2x \dfrac{dy}{dx} - 2y = x^3.$

Solve the initial-value problem in each of Exercises 20–27. In each case assume $x > 0$.

20. $x^2 \dfrac{d^2 y}{dx^2} - 2x \dfrac{dy}{dx} - 10y = 0, \qquad y(1) = 5, \qquad y'(1) = 4.$

21. $x^2 \dfrac{d^2 y}{dx^2} - 4x \dfrac{dy}{dx} + 6y = 0, \qquad y(2) = 0, \qquad y'(2) = 4.$

22. $x^2 \dfrac{d^2 y}{dx^2} + 5x \dfrac{dy}{dx} + 3y = 0, \qquad y(1) = 1, \qquad y'(1) = -5.$

23. $x^2 \dfrac{d^2 y}{dx^2} - 2y = 4x - 8, \qquad y(1) = 4, \qquad y'(1) = -1.$

24. $x^2 \dfrac{d^2 y}{dx^2} - 4x \dfrac{dy}{dx} + 4y = 4x^2 - 6x^3, \qquad y(2) = 4, \qquad y'(2) = -1.$

25. $x^2 \dfrac{d^2 y}{dx^2} + 2x \dfrac{dy}{dx} - 6y = 10x^2, \qquad y(1) = 1, \qquad y'(1) = -6.$

26. $x^2 \dfrac{d^2 y}{dx^2} - 5x \dfrac{dy}{dx} + 8y = 2x^3, \qquad y(2) = 0, \qquad y'(2) = -8.$

27. $x^2 \dfrac{d^2 y}{dx^2} - 6y = \ln x, \qquad y(1) = \frac{1}{6}, \qquad y'(1) = -\frac{1}{6}.$

28. Solve:
$$(x + 2)^2 \dfrac{d^2 y}{dx^2} - (x + 2) \dfrac{dy}{dx} - 3y = 0.$$

29. Solve:
$$(2x - 3)^2 \dfrac{d^2 y}{dx^2} - 6(2x - 3) \dfrac{dy}{dx} + 12y = 0.$$

4.6 STATEMENTS AND PROOFS OF THEOREMS ON THE SECOND-ORDER HOMOGENEOUS LINEAR EQUATION

Having considered the most fundamental methods of solving higher-order linear differential equations, we now return briefly to the theoretical side of the subject and present detailed statements and proofs of the basic theorems concerning the second-order homogeneous equation. The corresponding results for both the general nth-order equation and the special second-order equation were introduced in Section 4.1B and employed frequently thereafter. By restricting attention here to the second-order case we shall be able to present proofs which are completely explicit in every detail. However, we point out that each of these proofs may be extended in a straight-forward manner to provide a proof of the corresponding theorem for the general nth-order case. For general proofs, we again refer to Chapter 11.

We thus consider the second-order homogeneous linear differential equation

$$a_0(x)\frac{d^2y}{dx^2} + a_1(x)\frac{dy}{dx} + a_2(x)y = 0. \tag{4.91}$$

where $a_0, a_1,$ and a_2 are continuous real functions on a real interval $a \leq x \leq b$ and $a_0(x) \neq 0$ for any x on $a \leq x \leq b$.

In order to obtain the basic results concerning this equation, we shall need to make use of the following special case of Theorem 4.1 and its corollary.

THEOREM A

Hypothesis. *Consider the second-order homogeneous linear equation (4.91), where* $a_0, a_1,$ *and* a_2 *are continuous real functions on a real interval* $a \leq x \leq b$ *and* $a_0(x) \neq 0$ *for any* x *on* $a \leq x \leq b$. *Let* x_0 *be any point of* $a \leq x \leq b$; *and let* c_0 *and* c_1 *be any two real constants.*

Conclusion 1. *Then there exists a unique solution* f *of Equation (4.91) such that* $f(x_0) = c_0$ *and* $f'(x_0) = c_1$, *and this solution* f *is defined over the entire interval* $a \leq x \leq b$.

Conclusion 2. *In particular, the unique solution* f *of Equation (4.91), which is such that* $f(x_0) = 0$ *and* $f'(x_0) = 0$, *is the function* f *such that* $f(x) = 0$ *for all* x *on* $a \leq x \leq b$.

Besides this result, we shall also need the following two theorems from algebra.

THEOREM B

Two homogeneous linear algebraic equations in two unknowns have a nontrivial solution if and only if the determinant of coefficients of the system is equal to zero.

THEOREM C

Two linear algebraic equations in two unknowns have a unique solution if and only if the determinant of coefficients of the system is unequal to zero.

We shall now proceed to obtain the basic results concerning Equation (4.91). Since each of the concepts involved has already been introduced and illustrated in Section 4.1, we shall state and prove the various theorems without further comments or examples.

THEOREM 4.15

Hypothesis. *Let the functions* f_1 *and* f_2 *be any two solutions of the homogeneous linear differential equation (4.91) on* $a \leq x \leq b$, *and let* c_1 *and* c_2 *be any two arbitrary constants.*

Conclusion. *Then the linear combination $c_1 f_1 + c_2 f_2$ of f_1 and f_2 is also a solution of Equation (4.91) on $a \leq x \leq b$.*

Proof. We must show that the function f defined by

$$f(x) = c_1 f_1(x) + c_2 f_2(x), \qquad a \leq x \leq b, \tag{4.92}$$

satisfies the differential equation (4.91) on $a \leq x \leq b$. From (4.92), we see that

$$f'(x) = c_1 f'_1(x) + c_2 f'_2(x), \qquad a \leq x \leq b, \tag{4.93}$$

and

$$f''(x) = c_1 f''_1(x) + c_2 f''_2(x), \qquad a \leq x \leq b. \tag{4.94}$$

Substituting $f(x)$ given by (4.92), $f'(x)$ given by (4.93), and $f''(x)$ given by (4.94) for y, dy/dx, and d^2y/dx^2, respectively, in the left member of differential equation (4.91), we obtain

$$a_0(x)[c_1 f''_1(x) + c_2 f''_2(x)] + a_1(x)[c_1 f'_1(x) + c_2 f'_2(x)]$$
$$+ a_2(x)[c_1 f_1(x) + c_2 f_2(x)]. \tag{4.95}$$

By rearranging terms, we express this as

$$c_1[a_0(x)f''_1(x) + a_1(x)f'_1(x) + a_2(x)f_1(x)]$$
$$+ c_2[a_0(x)f''_2(x) + a_1(x)f'_2(x) + a_2(x)f_2(x)]. \tag{4.96}$$

Since by hypothesis, f_1 and f_2 are solutions of differential equation (4.91) on $a \leq x \leq b$, we have, respectively,

$$a_0(x)f''_1(x) + a_1(x)f'_1(x) + a_2(x)f_1(x) = 0$$

and

$$a_0(x)f''_2(x) + a_1(x)f'_2(x) + a_2(x)f_2(x) = 0$$

for all x on $a \leq x \leq b$.

Thus the expression (4.96) is equal to zero for all x on $a \leq x \leq b$, and therefore so is the expression (4.95). That is, we have

$$a_0(x)[c_1 f''_1(x) + c_2 f''_2(x)] + a_1(x)[c_1 f'_1(x) + c_2 f'_2(x)]$$
$$+ a_2(x)[c_1 f_1(x) + c_2 f_2(x)] = 0$$

for all x on $a \leq x \leq b$, and so the function $c_1 f_1 + c_2 f_2$ is also a solution of differential equation (4.91) on this interval. Q.E.D.

THEOREM 4.16

Hypothesis. *Consider the second-order homogeneous linear differential equation (4.91), where a_0, a_1, and a_2 are continuous on $a \leq x \leq b$ and $a_0(x) \neq 0$ on $a \leq x \leq b$.*

Conclusion. *There exists a set of two solutions of Equation (4.91) that are linearly independent on $a \leq x \leq b$.*

Proof. We prove this theorem by actually exhibiting such a set of solutions. Let x_0 be a point of the interval $a \leq x \leq b$. Then by Theorem A, Conclusion 1, there exists a unique solution f_1 of Equation (4.91) such that

$$f_1(x_0) = 1 \quad \text{and} \quad f'_1(x_0) = 0 \tag{4.97}$$

and a unique solution f_2 of Equation (4.91) such that

$$f_2(x_0) = 0 \quad \text{and} \quad f'_2(x_0) = 1. \tag{4.98}$$

We now show that these two solutions f_1 and f_2 are indeed linearly independent. Suppose they were not. Then they would be linear *dependent*; and so by the definition of linear dependence, there would exist constants c_1 and c_2, not both zero, such that

$$c_1 f_1(x) + c_2 f_2(x) = 0 \tag{4.99}$$

for all x such that $a \leq x \leq b$. Then also

$$c_1 f'_1(x) + c_2 f'_2(x) = 0 \tag{4.100}$$

for all x such that $a \leq x \leq b$. The identities (4.99) and (4.100) hold at $x = x_0$, giving

$$c_1 f_1(x_0) + c_2 f_2(x_0) = 0, \qquad c_1 f'_1(x_0) + c_2 f'_2(x_0) = 0.$$

Now apply conditions (4.97) and (4.98) to this set of equations. They reduce to

$$c_1(1) + c_2(0) = 0, \qquad c_1(0) + c_2(1) = 0.$$

or simply $c_1 = c_2 = 0$, which is a contradiction (since c_1 and c_2 are not both zero). Thus the solutions f_1 and f_2 defined respectively by (4.97) and (4.98) are linearly independent on $a \leq x \leq b$.

Q.E.D.

THEOREM 4.17

Two solutions f_1 and f_2 of the second-order homogeneous linear differential equation (4.91) are linear independent on $a \leq x \leq b$ if and only if the value of the Wronskian of f_1 and f_2 is different from zero for some x on the interval $a \leq x \leq b$.

Method of Proof. We prove this theorem by proving the following equivalent theorem.

THEOREM 4.18

Two solutions f_1 and f_2 of the second-order homogeneous linear differential equation (4.91) are linearly dependent on $a \leq x \leq b$ if and only if the value of the Wronskian of f_1 and f_2 is zero for all x on $a \leq x \leq b$:

$$\begin{vmatrix} f_1(x) & f_2(x) \\ f'_1(x) & f'_2(x) \end{vmatrix} = 0 \quad \text{for all } x \text{ on } a \leq x \leq b.$$

Proof. Part 1. We must show that if the value of the Wronskian of f_1 and f_2 is zero for all x on $a \leq x \leq b$, then f_1 and f_2 are linearly dependent on $a \leq x \leq b$. We thus assume that

$$\begin{vmatrix} f_1(x) & f_2(x) \\ f_1'(x) & f_2'(x) \end{vmatrix} = 0$$

for all x such that $a \le x \le b$. Then at any particular x_0 such that $a \le x_0 \le b$, we have

$$\begin{vmatrix} f_1(x_0) & f_2(x_0) \\ f_1'(x_0) & f_2'(x_0) \end{vmatrix} = 0.$$

Thus, by Theorem B, there exist constants c_1 and c_2, not both zero, such that

$$c_1 f_1(x_0) + c_2 f_2(x_0) = 0,$$
$$c_1 f_1'(x_0) + c_2 f_2'(x_0) = 0.$$

(4.101)

Now consider the function f defined by

$$f(x) = c_1 f_1(x) + c_2 f_2(x), \qquad a \le x \le b.$$

By Theorem 4.15, since f_1 and f_2 are solutions of differential equation (4.91), this function f is also a solution of Equation (4.91). From (4.101), we have

$$f(x_0) = 0 \quad \text{and} \quad f'(x_0) = 0.$$

Thus by Theorem A, Conclusion 2, we know that

$$f(x) = 0 \quad \text{for all } x \text{ on } a \le x \le b.$$

That is,

$$c_1 f_1(x) + c_2 f_2(x) = 0$$

for all x on $a \le x \le b$, where c_1 and c_2 are not both zero. Therefore the solutions f_1 and f_2 are linearly dependent on $a \le x \le b$.

Part 2. We must now show that if f_1 and f_2 are linearly dependent on $a \le x \le b$, then their Wronskian has the value zero for all x on this interval. We thus assume that f_1 and f_2 are linearly dependent on $a \le x \le b$. Then there exist constants c_1 and c_2, not both zero, such that

$$c_1 f_1(x) + c_2 f_2(x) = 0$$

(4.102)

for all x on $a \le x \le b$. From (4.102), we also have

$$c_1 f_1'(x) + c_2 f_2'(x) = 0$$

(4.103)

for all x on $a \le x \le b$. Now let $x = x_0$ be an arbitrary point of the interval $a \le x \le b$. Then (4.102) and (4.103) hold at $x = x_0$. That is,

$$c_1 f_1(x_0) + c_2 f_2(x_0) = 0,$$
$$c_1 f_1'(x_0) + c_2 f_2'(x_0) = 0,$$

where c_1 and c_2 are not both zero. Thus, by Theorem B, we have

$$\begin{vmatrix} f_1(x_0) & f_2(x_0) \\ f_1'(x_0) & f_2'(x_0) \end{vmatrix} = 0.$$

But this determinant is the value of the Wronskian of f_1 and f_2 at $x = x_0$, and x_0 is an *arbitrary* point of $a \leq x \leq b$. Thus we have

$$\begin{vmatrix} f_1(x) & f_2(x) \\ f_1'(x) & f_2'(x) \end{vmatrix} = 0$$

for *all* x on $a \leq x \leq b$.

Q.E.D.

THEOREM 4.19

The value of the Wronskian of two solutions f_1 and f_2 of differential equation (4.91) either is zero for all x on $a \leq x \leq b$ or is zero for no x on $a \leq x \leq b$.

Proof. If f_1 and f_2 are linearly dependent on $a \leq x \leq b$, then by Theorem 4.18, the value of the Wronskian of f_1 and f_2 is zero for all x on $a \leq x \leq b$.

Now let f_1 and f_2 be linearly independent on $a \leq x \leq b$; and let W denote the Wronskian of f_1 and f_2, so that

$$W(x) = \begin{vmatrix} f_1(x) & f_2(x) \\ f_1'(x) & f_2'(x) \end{vmatrix}.$$

Differentiating this, we obtain

$$W'(x) = \begin{vmatrix} f_1'(x) & f_2'(x) \\ f_1'(x) & f_2'(x) \end{vmatrix} + \begin{vmatrix} f_1(x) & f_2(x) \\ f_1''(x) & f_2''(x) \end{vmatrix},$$

and this reduces at once to

$$W'(x) = \begin{vmatrix} f_1(x) & f_2(x) \\ f_1''(x) & f_2''(x) \end{vmatrix}. \tag{4.104}$$

Since f_1 and f_2 are solutions of differential equation (4.91), we have, respectively,

$$a_0(x)f_1''(x) + a_1(x)f_1'(x) + a_2(x)f_1(x) = 0,$$

$$a_0(x)f_2''(x) + a_1(x)f_2'(x) + a_2(x)f_2(x) = 0,$$

and hence

$$f_1''(x) = -\frac{a_1(x)}{a_0(x)}f_1'(x) - \frac{a_2(x)}{a_0(x)}f_1(x),$$

$$f_2''(x) = -\frac{a_1(x)}{a_0(x)}f_2'(x) - \frac{a_2(x)}{a_0(x)}f_2(x)$$

on $a \leq x \leq b$. Substituting these expressions into (4.104), we obtain

$$W'(x) = \begin{vmatrix} f_1(x) & f_2(x) \\ -\dfrac{a_1(x)}{a_0(x)}f_1'(x) - \dfrac{a_2(x)}{a_0(x)}f_1(x) & -\dfrac{a_1(x)}{a_0(x)}f_2'(x) - \dfrac{a_2(x)}{a_0(x)}f_2(x) \end{vmatrix}.$$

This reduces at once to

$$W'(x) = \begin{vmatrix} f_1(x) & f_2(x) \\ -\dfrac{a_1(x)}{a_0(x)}f_1'(x) & -\dfrac{a_1(x)}{a_0(x)}f_2'(x) \end{vmatrix} + \begin{vmatrix} f_1(x) & f_2(x) \\ -\dfrac{a_2(x)}{a_0(x)}f_1(x) & -\dfrac{a_2(x)}{a_0(x)}f_2(x) \end{vmatrix},$$

and since the last determinant has two proportional rows, this in turn reduces to

$$W'(x) = -\frac{a_1(x)}{a_0(x)} \begin{vmatrix} f_1(x) & f_2(x) \\ f'_1(x) & f'_2(x) \end{vmatrix},$$

which is simply

$$W'(x) = -\frac{a_1(x)}{a_0(x)} W(x).$$

Thus the Wronskian W satisfies the first-order homogeneous linear differential equation

$$\frac{dW}{dx} + \frac{a_1(x)}{a_0(x)} W = 0.$$

Integrating this from x_0 to x, where x_0 is an arbitrary point of $a \le x \le b$, we obtain

$$W(x) = c \exp\left[-\int_{x_0}^{x} \frac{a_1(t)}{a_0(t)} \, dt \right].$$

Letting $x = x_0$, we find that $c = W(x_0)$. Hence we obtain the identity

$$W(x) = W(x_0)\exp\left[-\int_{x_0}^{x} \frac{a_1(t)}{a_0(t)} \, dt \right], \tag{4.105}$$

valid for all x on $a \le x \le b$, where x_0 is an arbitrary point of this interval.

Now assume that $W(x_0) = 0$. Then by identity (4.105), we have $W(x) = 0$ for all x on $a \le x \le b$. Thus by Theorem 4.18, the solutions f_1 and f_2 must be linearly dependent on $a \le x \le b$. This is a contradiction, since f_1 and f_2 are linearly independent. Therefore the assumption that $W(x_0) = 0$ is false, and so $W(x_0) \ne 0$. But x_0 is an *arbitrary* point of $a \le x \le b$. Thus $W(x)$ is zero for no x on $a \le x \le b$.

Q.E.D.

THEOREM 4.20

Hypothesis.   Let f_1 and f_2 be any two linearly independent solutions of differential equation (4.91) on $a \le x \le b$.

Conclusion.   Then every solution f of differential equation (4.91) can be expressed as a suitable linear combination

$$c_1 f_1 + c_2 f_2$$

of these two linear independent solutions.

Proof.   Let x_0 be an arbitrary point of the interval $a \le x \le b$, and consider the following system of two linear algebraic equations in the two unknowns k_1 and k_2:

$$k_1 f_1(x_0) + k_2 f_2(x_0) = f(x_0),$$
$$k_1 f'_1(x_0) + k_2 f'_2(x_0) = f'(x_0). \tag{4.106}$$

Since f_1 and f_2 are linearly independent on $a \le x \le b$, we know by Theorem 4.17 that the value of the Wronskian of f_1 and f_2 is different from zero at some point of this interval. Then by Theorem 4.19 the value of the Wronskian is zero for no x on $a \le x \le b$ and hence its value at x_0 is not zero. That is,

$$\begin{vmatrix} f_1(x_0) & f_2(x_0) \\ f'_1(x_0) & f'_2(x_0) \end{vmatrix} \neq 0.$$

Thus by Theorem C, the algebraic system (4.106) has a unique solution $k_1 = c_1$ and $k_2 = c_2$. Thus for $k_1 = c_1$ and $k_2 = c_2$, each left member of system (4.106) is the same number as the corresponding right member of (4.106). That is, the number $c_1 f_1(x_0) + c_2 f_2(x_0)$ is equal to the number $f(x_0)$, and the number $c_1 f'_1(x_0) + c_2 f'_2(x_0)$ is equal to the number $f'(x_0)$. But the numbers $c_1 f_1(x_0) + c_2 f_2(x_0)$ and $c_1 f'_1(x_0) + c_2 f'_2(x_0)$ are the values of the solution $c_1 f_1 + c_2 f_2$ and its first derivative, respectively, at x_0; and the numbers $f(x_0)$ and $f'(x_0)$ are the values of the solution f and its first derivative, respectively, at x_0. Thus the two solutions $c_1 f_1 + c_2 f_2$ and f have equal values and their first derivative also have equal values at x_0. Hence by Theorem A, Conclusion 1, we know that these two solutions are identical throughout the interval $a \le x \le b$. That is,

$$f(x) = c_1 f_1(x) + c_2 f_2(x)$$

for all x on $a \le x \le b$, and so f is expressed as a linear combination of f_1 and f_2.

Q.E.D.

Exercises

1. Consider the second-order homogenous linear differential equation

$$\frac{d^2 y}{dx^2} - 3\frac{dy}{dx} + 2y = 0.$$

(a) Find the two linearly independent solutions f_1 and f_2 of this equation which are such that

$$f_1(0) = 1 \quad \text{and} \quad f'_1(0) = 0$$

and

$$f_2(0) = 0 \quad \text{and} \quad f'_2(0) = 1.$$

(b) Express the solution

$$3e^x + 2e^{2x}$$

as a linear combination of the two linearly independent solutions f_1 and f_2 defined in part (a).

2. Consider the second-order homogeneous linear differential equation

$$a_0(x)\frac{d^2 y}{dx^2} + a_1(x)\frac{dy}{dx} + a_2(x)y = 0, \tag{A}$$

where a_0, a_1, and a_2 are continuous on a real interval $a \le x \le b$, and $a_0(x) \neq 0$ for all x on this interval. Let f_1 and f_2 be two distinct solutions of differential equation (A) on $a \le x \le b$, and suppose $f_2(x) \neq 0$ for all x on this interval. Let $W[f_1(x), f_2(x)]$ be the value of the Wronskian of f_1 and f_2 at x.

(a) Show that

$$\frac{d}{dx}\left[\frac{f_1(x)}{f_2(x)}\right] = -\frac{W[f_1(x), f_2(x)]}{[f_2(x)]^2}$$

for all x on $a \leq x \leq b$.

(b) Use the result of part (a) to show that if $W[f_1(x), f_2(x)] = 0$ for all x such that $a \leq x \leq b$, then the solutions f_1 and f_2 are linearly dependent on this interval.

(c) Suppose the solutions f_1 and f_2 are linearly independent on $a \leq x \leq b$, and let f be the function defined by $f(x) = f_1(x)/f_2(x)$, $a \leq x \leq b$. Show that f is a monotonic function on $a \leq x \leq b$.

3. Let f_1 and f_2 be two solutions of the second-order homogeneous linear differential equation (A) of Exercise 2.

(a) Show that if f_1 and f_2 have a common zero at a point x_0 of the interval $a \leq x \leq b$, then f_1 and f_2 are linearly dependent on $a \leq x \leq b$.

(b) Show that if f_1 and f_2 have relative maxima at a common point x_0 of the interval $a \leq x \leq b$, then f_1 and f_2 are linearly dependent on $a \leq x \leq b$.

4. Consider the second-order homogeneous linear differential equation (A) of Exercise 2.

(a) Let f_1 and f_2 be two solutions of this equation. Show that if f_1 and f_2 are linearly independent on $a \leq x \leq b$ and A_1, A_2, B_1, and B_2 are constants such that $A_1 B_2 - A_2 B_1 \neq 0$, then the solutions $A_1 f_1 + A_2 f_2$ and $B_1 f_1 + B_2 f_2$ of Equation (A) are also linearly independent on $a \leq x \leq b$.

(b) Let $\{f_1, f_2\}$ be one set of two linearly independent solutions of Equation (A) on $a \leq x \leq b$, and let $\{g_1, g_2\}$ be another set of two linearly independent solutions of Equation (A) on this interval. Let $W[f_1(x), f_2(x)]$ denote the value of the Wronskian of f_1 and f_2 at x, and let $W[g_1(x), g_2(x)]$ denote the value of the Wronskian of g_1 and g_2 at x. Show that there exists a constant $c \neq 0$ such that

$$W[f_1(x), f_2(x)] = cW[g_1(x), g_2(x)]$$

for all x on $a \leq x \leq b$.

5. Let f_1 and f_2 be two solutions of the second-order homogeneous linear differential equation (A) of Exercise 2. Show that if f_1 and f_2 are linearly independent on $a \leq x \leq b$ and are such that $f_1''(x_0) = f_2''(x_0) = 0$ at some point x_0 of this interval, then $a_1(x_0) = a_2(x_0) = 0$.

CHAPTER FIVE

Applications of Second-Order Linear Differential Equations with Constant Coefficients

Higher-order linear differential equations, which were introduced in the previous chapter, are equations having a great variety of important applications. In particular, second-order linear differential equations with constant coefficients have numerous applications in physics and in electrical and mechanical engineering. Two of these applications will be considered in the present chapter. In Sections 5.1–5.5 we shall discuss the motion of a mass vibrating up and down at the end of a spring, while in Section 5.6 we shall consider problems in electric circuit theory.

5.1 THE DIFFERENTIAL EQUATION OF THE VIBRATIONS OF A MASS ON A SPRING

The Basic Problem

A coil spring is suspended vertically from a fixed point on a ceiling, beam, or other similar object. A mass is attached to its lower end and allowed to come to rest in an equilibrium position. The system is then set in motion either (1) by pulling the mass down a distance below its equilibrium position (or pushing it up a distance above it) and subsequently releasing it with an initial velocity (zero or nonzero, downward or upward) at $t = 0$; or (2) by forcing the mass out of its equilibrium position by giving it a nonzero initial velocity (downward or upward) at $t = 0$. Our problem is to determine the resulting motion of the mass on the spring. In order to do so we must also consider certain other phenomena that may be present. For one thing, assuming the system is located in some sort of medium (say "ordinary" air or perhaps water), this medium produces a resistance force that tends to retard the motion. Also, certain external forces may be present. For example, a magnetic force from outside the system may be acting

179

upon the mass. Let us then attempt to determine the motion of the mass on the spring, taking into account both the resistance of the medium and possible external forces. We shall do this by first obtaining and then solving the differential equation for the motion.

In order to set up the differential equation for this problem we shall need two laws of physics: Newton's second law and Hooke's law. Newton's second law was encountered in Chapter 3, and we shall not go into a further discussion of it here. Let us then recall the other law that we shall need.

Hooke's Law

The magnitude of the force needed to produce a certain elongation of a spring is directly proportional to the amount of this elongation, provided this elongation is not too great. In mathematical form,

$$|F| = ks,$$

where F is the magnitude of the force, s is the amount of elongation, and k is a constant of proportionality which we shall call the *spring constant*.

The spring constant k depends upon the spring under consideration and is a measure of its stiffness. For example, if a 30-lb weight stretches a spring 2 ft, then Hooke's law gives $30 = (k)(2)$; thus for this spring $k = 15$ lb/ft.

When a mass is hung upon a spring of spring constant k and thus produces an elongation of amount s, the force F of the mass upon the spring therefore has magnitude ks. The spring at the same time exerts a force upon the mass called the *restoring force* of the spring. This force is equal in magnitude but opposite in sign to F and hence has magnitude $-ks$.

Let us formulate the problem systematically. Let the coil spring have natural (unstretched) length L. The mass m is attached to its lower end and comes to rest in its equilibrium position, thereby stretching the spring an amount l so that its stretched length is $L + l$. We choose the axis along the line of the spring, with the origin O at the equilibrium position and the positive direction downward. Thus, letting x denote the displacement of the mass from O along this line, we see that x is positive, zero, or negative according to whether the mass is below, at, or above its equilibrium position. (See Figure 5.1.)

Forces Acting Upon the Mass

We now enumerate the various forces that act upon the mass. Forces tending to pull the mass downward are positive, while those tending to pull it upward are negative. The forces are:

1. F_1, the *force of gravity*, of magnitude mg, where g is the acceleration due to gravity. Since this acts in the downward direction, it is positive, and so

$$F_1 = mg. \tag{5.1}$$

2. F_2, the *restoring force* of the spring. Since $x + l$ is the total amount of elongation, by Hooke's law the magnitude of this force is $k(x + l)$. When the mass is *below* the end of the unstretched spring, this force acts in the *upward* direction and so is *negative*. Also,

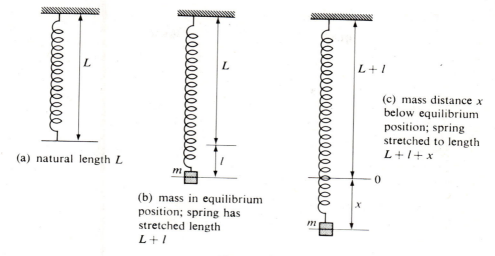

(a) natural length L

(b) mass in equilibrium position; spring has stretched length $L + l$

(c) mass distance x below equilibrium position; spring stretched to length $L + l + x$

Figure 5.1

for the mass in such a position, $x + l$ is *positive*. Thus, when the mass is *below* the end of the unstretched spring, the restoring force is given by

$$F_2 = -k(x + l). \tag{5.2}$$

This also gives the restoring force when the mass is *above* the end of the unstretched spring, as one can see by replacing each italicized word in the three preceding sentences by its opposite. When the mass is at rest in its equilibrium position the restoring force F_2 is equal in magnitude but opposite in direction to the force of gravity and so is given by $-mg$. Since in this position $x = 0$, Equation (5.2) gives

$$-mg = -k(0 + l)$$

or

$$mg = kl.$$

Replacing kl by mg in Equation (5.2) we see that the restoring force can thus be written as

$$F_2 = -kx - mg. \tag{5.3}$$

3. F_3, the *resisting force* of the medium, called the *damping force*. Although the magnitude of this force is not known *exactly*, it is known that for small velocities it is *approximately* proportional to the magnitude of the velocity:

$$|F_3| = a \left| \frac{dx}{dt} \right|, \tag{5.4}$$

where $a > 0$ is called the *damping constant*. When the mass is moving *downward*, F_3 acts in the *upward* direction (opposite to that of the motion) and so $F_3 < 0$. Also, since m is moving *downward*, x is *increasing* and dx/dt is *positive*. Thus, assuming Equation (5.4) to hold, when the mass is moving *downward*, the damping force is given by

$$F_3 = -a \frac{dx}{dt} \qquad (a > 0). \tag{5.5}$$

This also gives the damping force when the mass is moving *upward*, as one may see by replacing each italicized word in the three preceding sentences by its opposite.

4. F_4, any *external impressed forces* that act upon the mass. Let us denote the resultant of all such external forces at time t simply by $F(t)$ and write

$$F_4 = F(t). \tag{5.6}$$

We now apply Newton's second law, $F = ma$, where $F = F_1 + F_2 + F_3 + F_4$. Using (5.1), (5.3), (5.5), and (5.6), we find

$$m\frac{d^2x}{dt^2} = mg - kx - mg - a\frac{dx}{dt} + F(t)$$

or

$$m\frac{d^2x}{dt^2} + a\frac{dx}{dt} + kx = F(t). \tag{5.7}$$

This we take as the differential equation for the motion of the mass on the spring. Observe that it is a nonhomogeneous second-order linear differential equation with constant coefficients. If $a = 0$ the motion is called *undamped*; otherwise it is called *damped*. If there are no external impressed forces, $F(t) = 0$ for all t and the motion is called *free*; otherwise it is called *forced*. In the following sections we consider the solution of (5.7) in each of these cases.

5.2   FREE, UNDAMPED MOTION

We now consider the special case of *free, undamped motion*, that is, the case in which both $a = 0$ and $F(t) = 0$ for all t. The differential equation (5.7) then reduces to

$$m\frac{d^2x}{dt^2} + kx = 0, \tag{5.8}$$

where $m(>0)$ is the mass and $k(>0)$ is the spring constant. Dividing through by m and letting $k/m = \lambda^2$, we write (5.8) in the form

$$\frac{d^2x}{dt^2} + \lambda^2 x = 0. \tag{5.9}$$

The auxiliary equation

$$r^2 + \lambda^2 = 0$$

has roots $r = \pm \lambda i$ and hence the general solution of (5.8) can be written

$$x = c_1 \sin \lambda t + c_2 \cos \lambda t, \tag{5.10}$$

where c_1 and c_2 are arbitrary constants.

Let us now assume that the mass was initially displaced a distance x_0 from its equilibrium position and released from that point with initial velocity v_0. Then, in addition to the differential equation (5.8) [or (5.9)], we have the initial conditions

$$x(0) = x_0, \tag{5.11}$$

$$x'(0) = v_0. \tag{5.12}$$

Differentiating (5.10) with respect to t, we have

$$\frac{dx}{dt} = c_1 \lambda \cos \lambda t - c_2 \lambda \sin \lambda t. \tag{5.13}$$

Applying conditions (5.11) and (5.12) to Equations (5.10) and (5.13), respectively, we see at once that

$$c_2 = x_0,$$

$$c_1 \lambda = v_0.$$

Substituting the values of c_1 and c_2 so determined into Equation (5.10) gives the particular solution of the differential equation (5.8) satisfying the conditions (5.11) and (5.12) in the form

$$x = \frac{v_0}{\lambda} \sin \lambda t + x_0 \cos \lambda t.$$

We put this in an alternative form by first writing it as

$$x = c\left[\frac{(v_0/\lambda)}{c} \sin \lambda t + \frac{x_0}{c} \cos \lambda t\right], \tag{5.14}$$

where

$$c = \sqrt{\left(\frac{v_0}{\lambda}\right)^2 + x_0^2} > 0. \tag{5.15}$$

Then, letting

$$\frac{(v_0/\lambda)}{c} = -\sin \phi,$$

$$\frac{x_0}{c} = \cos \phi, \tag{5.16}$$

Equation (5.14) reduces at once to

$$x = c \cos(\lambda t + \phi), \tag{5.17}$$

where c is given by Equation (5.15) and ϕ is determined by Equations (5.16). Since $\lambda = \sqrt{k/m}$, we now write the solution (5.17) in the form

$$x = c \cos\left(\sqrt{\frac{k}{m}} t + \phi\right). \tag{5.18}$$

This, then, gives the displacement x of the mass from the equilibrium position O as a function of the time t $(t > 0)$. We see at once that the free, undamped motion of the mass is a *simple harmonic motion*. The constant c is called the *amplitude* of the motion and gives the maximum (positive) displacement of the mass from its equilibrium position. The motion is a *periodic* motion, and the mass oscillates back and forth between $x = c$ and $x = -c$. We have $x = c$ if and only if

$$\sqrt{\frac{k}{m}} t + \phi = \pm 2n\pi,$$

$n = 0, 1, 2, 3, \ldots; t > 0$. Thus the maximum (positive) displacement occurs if and only if

$$t = \sqrt{\frac{m}{k}}\,(\pm 2n\pi - \phi) > 0, \tag{5.19}$$

where $n = 0, 1, 2, 3, \ldots$.

The time interval between two successive maxima is called the *period* of the motion. Using (5.19), we see that it is given by

$$\frac{2\pi}{\sqrt{k/m}} = \frac{2\pi}{\lambda}. \tag{5.20}$$

The reciprocal of the period, which gives the number of oscillations per second, is called the *natural frequency* (or simply *frequency*) of the motion. The number ϕ is called the *phase constant* (or *phase angle*). The graph of this motion is shown in Figure 5.2.

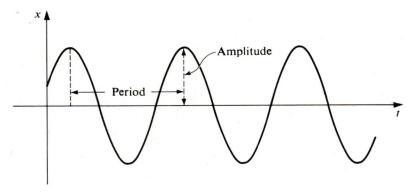

Figure 5.2

▶ **Example 5.1**

An 8-lb weight is placed upon the lower end of a coil spring suspended from the ceiling. The weight comes to rest in its equilibrium position, thereby stretching the spring 6 in. The weight is then pulled down 3 in. below its equilibrium position and released at $t = 0$ with an initial velocity of 1 ft/sec, directed downward. Neglecting the resistance of the medium and assuming that no external forces are present, determine the amplitude, period and frequency of the resulting motion.

Formulation. This is clearly an example of free, undamped motion and hence Equation (5.8) applies. Since the 8-lb weight stretches the spring 6 in. $= \frac{1}{2}$ ft, Hooke's law $F = ks$ gives $8 = k(\frac{1}{2})$ and so $k = 16$ lb/ft. Also, $m = w/g = \frac{8}{32}$ (slugs), and so Equation (5.8) gives

$$\frac{8}{32}\frac{d^2x}{dt^2} + 16x = 0$$

or

$$\frac{d^2x}{dt^2} + 64x = 0. \tag{5.21}$$

Since the weight was released with a downward initial velocity of 1 ft/sec from a point 3 in. ($= \frac{1}{4}$ ft) below its equilibrium position, we also have the initial conditions

$$x(0) = \tfrac{1}{4}, \qquad x'(0) = 1. \tag{5.22}$$

Solution. The auxiliary equation corresponding to Equation (5.21) is $r^2 + 64 = 0$, and hence $r = \pm 8i$. Thus the general solution of the differential equation (5.21) may be written

$$x = c_1 \sin 8t + c_2 \cos 8t, \tag{5.23}$$

where c_1 and c_2 are arbitrary constants. Applying the first of conditions (5.22) to this, we find $c_2 = \frac{1}{4}$. Differentiating (5.23), we have

$$\frac{dx}{dt} = 8c_1 \cos 8t - 8c_2 \sin 8t.$$

Applying the second of conditions (5.22) to this, we have $8c_1 = 1$ and hence $c_1 = \frac{1}{8}$. Thus the solution of the differential equation (5.21) satisfying the conditions (5.22) is

$$x = \tfrac{1}{8} \sin 8t + \tfrac{1}{4} \cos 8t. \tag{5.24}$$

Let us put this in the form (5.18). We find

$$\sqrt{\left(\frac{1}{8}\right)^2 + \left(\frac{1}{4}\right)^2} = \frac{\sqrt{5}}{8}$$

and thus write

$$x = \frac{\sqrt{5}}{8}\left(\frac{\sqrt{5}}{5} \sin 8t + \frac{2\sqrt{5}}{5} \cos 8t\right).$$

Thus, letting

$$\cos \phi = \frac{2\sqrt{5}}{5},$$

$$\sin \phi = -\frac{\sqrt{5}}{5}, \tag{5.25}$$

we write the solution (5.24) in the form

$$x = \frac{\sqrt{5}}{8} \cos(8t + \phi), \tag{5.26}$$

where ϕ is determined by Equations (5.25). From these equations we find that $\phi \approx -0.46$ radians. Taking $\sqrt{5} \approx 2.236$, the solution (5.26) is thus given approximately by

$$x = 0.280 \cos(8t - 0.46).$$

The amplitude of the motion $\sqrt{5}/8 \approx 0.280$ (ft). By formula (5.20), the period is $2\pi/8 = \pi/4$ (sec), and the frequency is $4/\pi$ oscillations/sec. The graph is shown in Figure 5.3.

Before leaving this problem, let us be certain that we can set up initial conditions correctly. Let us replace the third sentence in the statement of the problem by the following: "The weight is then *pushed up* 4 in. *above* its equilibrium position and

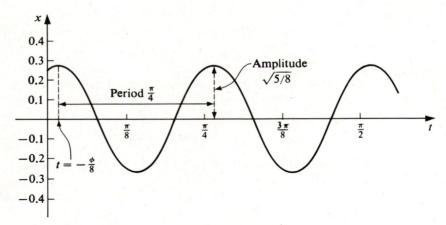

Figure 5.3

released at $t = 0$, with an initial velocity of 2 ft/sec, directed *upward*." The initial conditions (5.22) would then have been replaced by

$$x(0) = -\tfrac{1}{3},$$
$$x'(0) = -2.$$

The minus sign appears before the $\frac{1}{3}$ because the initial position is 4 in. $= \frac{1}{3}$ foot *above* the equilibrium position and hence is *negative*. The minus sign before the 2 is due to the fact that the initial velocity is directed *upward*, that is, in the *negative* direction.

Exercises

Note: In Exercises 1–7 neglect the resistance of the medium and assume that no external forces are present.

1. A 12-lb weight is placed upon the lower end of a coil spring suspended from the ceiling. The weight comes to rest in its equilibrium position, thereby stretching the spring 1.5 in. The weight is then pulled down 2 in. below its equilibrium position and released from rest at $t = 0$. Find the displacement of the weight as a function of the time; determine the amplitude, period, and frequency of the resulting motion; and graph the displacement as a function of the time.

2. A 16-lb weight is placed upon the lower end of a coil spring suspended vertically from a fixed support. The weight comes to rest in its equilibrium position, thereby stretching the spring 6 in. Determine the resulting displacement as a function of time in each of the following cases:

   (a) If the weight is then pulled down 4 in. below its equilibrium position and released at $t = 0$ with an initial velocity of 2 ft/sec, directed downward.

   (b) If the weight is then pulled down 4 in. below its equilibrium position and released at $t = 0$ with an initial velocity of 2 ft/sec, directed upward.

   (c) If the weight is then pushed up 4 in. above its equilibrium position and released at $t = 0$ with an initial velocity of 2 ft/sec, directed downward.

3. A 4-lb weight is attached to the lower end of a coil spring suspended from the ceiling. The weight comes to rest in its equilibrium position, thereby stretching the spring 6 in. At time $t = 0$ the weight is then struck so as to set it into motion with an initial velocity of 2 ft/sec, directed downward.

   (a) Determine the resulting displacement and velocity of the weight as functions of the time.

   (b) Find the amplitude, period, and frequency of the motion.

   (c) Determine the times at which the weight is 1.5 in. below its equilibrium position and moving downward.

   (d) Determine the times at which it is 1.5 in. below its equilibrium position and moving upward.

4. A 64-lb weight is placed upon the lower end of a coil spring suspended from a rigid beam. The weight comes to rest in its equilibrium position, thereby stretching the spring 2 ft. The weight is then pulled down 1 ft below its equilibrium position and released from rest at $t = 0$.

   (a) What is the position of the weight at $t = 5\pi/12$? How fast and which way is it moving at the time?

   (b) At what time is the weight 6 in. above its equilibrium position and moving downward? What is its velocity at such time?

5. A coil spring is such that a 25-lb weight would stretch it 6 in. The spring is suspended from the ceiling, a 16-lb weight is attached to the end of it, and the weight then comes to rest in its equilibrium position. It is then pulled down 4 in. below its equilibrium position and released at $t = 0$ with an initial velocity of 2 ft/sec, directed upward.

   (a) Determine the resulting displacement of the weight as a function of the time.

   (b) Find the amplitude, period, and frequency of the resulting motion.

   (c) At what time does the weight first pass through its equilibrium position and what is its velocity at this instant?

6. An 8-lb weight is attached to the end of a coil spring suspended from a beam and comes to rest in its equilibrium position. The weight is then pulled down A feet below its equilibrium position and released at $t = 0$ with an initial velocity of 3 ft/sec, directed downward. Determine the spring constant k and the constant A if the amplitude of the resulting motion is $\sqrt{\frac{10}{2}}$ and the period is $\pi/2$.

7. An 8-lb weight is placed at the end of a coil spring suspended from the ceiling. After coming to rest in its equilibrium position, the weight is set into vertical motion and the period of the resulting motion is 4 sec. After a time this motion is stopped, and the 8-lb weight is replaced by another weight. After this other weight has come to rest in its equilibrium position, it is set into vertical motion. If the period of this new motion is 6 sec, how heavy is the second weight?

8. A simple pendulum is composed of a mass m (the bob) at the end of a straight wire of negligible mass and length l. It is suspended from a fixed point S (its point of support) and is free to vibrate in a vertical plane (see Figure 5.4). Let SP denote the straight wire; and let θ denote the angle that SP makes with the vertical SP_0 at time t, positive when measured counterclockwise. We neglect air resistance and assume that only two forces act on the mass m: F_1, the tension in the wire; and F_2, the force

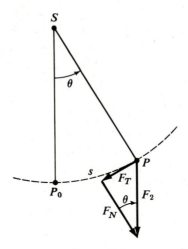

Figure 5.4

due to gravity, which acts vertically downward and is of magnitude mg. We write $F_2 = F_T + F_N$, where F_T is the component of F_2 along the tangent to the path of m and F_N is the component of F_2 normal to F_T. Then $F_N = -F_1$ and $F_T = -mg \sin \theta$, and so the net force acting on m is $F_1 + F_2 = F_1 + F_T + F_N = -mg \sin \theta$, along the arc $P_0 P$. Letting s denote the length of the arc $P_0 P$, the acceleration along this arc is $d^2 s/dt^2$. Hence applying Newton's second law, we have $m d^2 s/dt^2 = -mg \sin \theta$. But since $s = l\theta$, this reduces to the differential equation

$$ml \frac{d^2 \theta}{dt^2} = -mg \sin \theta \quad \text{or} \quad \frac{d^2 \theta}{dt^2} + \frac{g}{l} \sin \theta = 0.$$

(a)   The equation

$$\frac{d^2 \theta}{dt^2} + \frac{g}{l} \sin \theta = 0$$

is a *nonlinear* second-order differential equation. Now recall that

$$\sin \theta = \theta - \frac{\theta^3}{3!} + \frac{\theta^5}{5!} - \cdots.$$

Hence if θ is sufficiently small, we may replace $\sin \theta$ by θ and consider the *approximate linear* equation

$$\frac{d^2 \theta}{dt^2} + \frac{g}{l} \theta = 0.$$

Assume that $\theta = \theta_0$ and $d\theta/dt = 0$ when $t = 0$. Obtain the solution of this approximate equation that satisfies these initial conditions and find the amplitude and period of the resulting solution. Observe that this period is independent of the initial displacement.

(b)   Now return to the nonlinear equation

$$\frac{d^2 \theta}{dt^2} + \frac{g}{l} \sin \theta = 0.$$

Multiply through by $2\,d\theta/dt$, integrate, and apply the initial condition $\theta = \theta_0$, $d\theta/dt = 0$. Then separate variables in the resulting equation to obtain

$$\frac{d\theta}{\sqrt{\cos\theta - \cos\theta_0}} = \pm\sqrt{\frac{2g}{l}}\,dt.$$

From this equation determine the angular velocity $d\theta/dt$ as a function of θ. Note that the left member cannot be integrated in terms of elementary functions to obtain the exact solution $\theta(t)$ of the nonlinear differential equation.

5.3 FREE, DAMPED MOTION

We now consider the effect of the resistance of the medium upon the mass on the spring. Still assuming that no external forces are present, this is then the case of *free, damped motion*. Hence with the damping coefficient $a > 0$ and $F(t) = 0$ for all t, the basic differential equation (5.7) reduces to

$$m\frac{d^2x}{dt^2} + a\frac{dx}{dt} + kx = 0. \tag{5.27}$$

Dividing through by m and putting $k/m = \lambda^2$ and $a/m = 2b$ (for convenience) we have the differential equation (5.27) in the form

$$\frac{d^2x}{dt^2} + 2b\frac{dx}{dt} + \lambda^2 x = 0. \tag{5.28}$$

Observe that since a is positive, b is also positive. The auxiliary equation is

$$r^2 + 2br + \lambda^2 = 0. \tag{5.29}$$

Using the quadratic formula we find that the roots of (5.29) are

$$\frac{-2b \pm \sqrt{4b^2 - 4\lambda^2}}{2} = -b \pm \sqrt{b^2 - \lambda^2}. \tag{5.30}$$

Three distinct cases occur, depending upon the nature of these roots, which in turn depends upon the sign of $b^2 - \lambda^2$.

Case 1. Damped, Oscillatory Motion. Here we consider the case in which $b < \lambda$, which implies that $b^2 - \lambda^2 < 0$. Then the roots (5.30) are the conjugate complex numbers $-b \pm \sqrt{\lambda^2 - b^2}\,i$ and the general solution of Equation (5.28) is thus

$$x = e^{-bt}(c_1 \sin\sqrt{\lambda^2 - b^2}\,t + c_2 \cos\sqrt{\lambda^2 - b^2}\,t), \tag{5.31}$$

where c_1 and c_2 are arbitrary constants. We may write this in the alternative form

$$x = ce^{-bt}\cos(\sqrt{\lambda^2 - b^2}\,t + \phi), \tag{5.32}$$

where $c = \sqrt{c_1^2 + c_2^2} > 0$ and ϕ is determined by the equations

$$\frac{c_1}{\sqrt{c_1^2 + c_2^2}} = -\sin\phi,$$

$$\frac{c_2}{\sqrt{c_1^2 + c_2^2}} = \cos\phi.$$

The right member of Equation (5.32) consists of two factors,

$$ce^{-bt} \quad \text{and} \quad \cos(\sqrt{\lambda^2 - b^2}\, t + \phi).$$

The factor ce^{-bt} is called the *damping factor*, or *time-varying amplitude*. Since $c > 0$, it is positive; and since $b > 0$, it tends to zero monotonically as $t \to \infty$. In other words, as time goes on this positive factor becomes smaller and smaller and eventually becomes negligible. The remaining factor, $\cos(\sqrt{\lambda^2 - b^2}\, t + \phi)$, is, of course, of a periodic, oscillatory character; indeed it represents a simple harmonic motion. The product of these two factors, which is precisely the right member of Equation (5.32), therefore represents an oscillatory motion in which the oscillations become successively smaller and smaller. The oscillations are said to be "damped out," and the motion is described as *damped, oscillatory motion*. Of course, the motion is no longer periodic, but the time interval between two successive (positive) maximum displacements is still referred to as the *period*. This is given by

$$\frac{2\pi}{\sqrt{\lambda^2 - b^2}}.$$

The graph of such a motion is shown in Figure 5.5, in which the damping factor ce^{-bt} and its negative are indicated by dashed curves.

The ratio of the amplitude at any time T to that at time

$$T - \frac{2\pi}{\sqrt{\lambda^2 - b^2}}$$

one period before T is the constant

$$\exp\!\left(-\frac{2\pi b}{\sqrt{\lambda^2 - b^2}}\right).$$

Thus the quantity $2\pi b/\sqrt{\lambda^2 - b^2}$ is the decrease in the logarithm of the amplitude ce^{-bt} over a time interval of one period. It is called the *logarithmic decrement*.

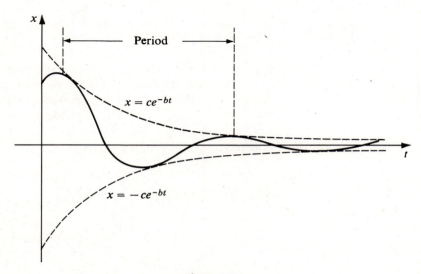

Figure 5.5

If we now return to the original notation of the differential equation (5.27), we see from Equation (5.32) that in terms of the original constants m, a, and k, the general solution of (5.27) is

$$x = ce^{-(a/2m)t} \cos\left(\sqrt{\frac{k}{m} - \frac{a^2}{4m^2}}\, t + \phi\right). \tag{5.33}$$

Since $b < \lambda$ is equivalent to $a/2m < \sqrt{k/m}$, we can say that the general solution of (5.27) is given by (5.33) and that damped, oscillatory motion occurs when $a < 2\sqrt{km}$. The frequency of the oscillations

$$\cos\left(\sqrt{\frac{k}{m} - \frac{a^2}{4m^2}}\, t + \phi\right) \tag{5.34}$$

is

$$\frac{1}{2\pi}\sqrt{\frac{k}{m} - \frac{a^2}{4m^2}}\,.$$

If damping were not present, a would equal zero and the natural frequency of an undamped system would be $(1/2\pi)\sqrt{k/m}$. Thus the frequency of the oscillations (5.34) in the damped oscillatory motion (5.33) is less than the natural frequency of the corresponding undamped system.

Case 2. Critical Damping. This is the case in which $b = \lambda$, which implies that $b^2 - \lambda^2 = 0$. The roots (5.30) are thus both equal to the real negative number $-b$, and the general solution of Equation (5.28) is thus

$$x = (c_1 + c_2 t)e^{-bt}. \tag{5.35}$$

The motion is no longer oscillatory; the damping is just great enough to prevent oscillations. Any slight decrease in the amount of damping, however, will change the situation back to that of Case 1 and damped oscillatory motion will then occur. Case 2 then is a borderline case; the motion is said to be *critically damped.*

From Equation (5.35) we see that

$$\lim_{t \to \infty} x = \lim_{t \to \infty} \frac{c_1 + c_2 t}{e^{bt}} = 0.$$

Hence the mass tends to its equilibrium position as $t \to \infty$. Depending upon the initial conditions present, the following possibilities can occur in this motion:

1. The mass neither passes through its equilibrium position nor attains an extremum (maximum or minimum) displacement from equilibrium for $t > 0$. It simply approaches its equilibrium position monotonically as $t \to \infty$. (See Figure 5.6a.)
2. The mass does not pass through its equilibrium position for $t > 0$, but its displacement from equilibrium attains a single extremum for $t = T_1 > 0$. After this extreme displacement occurs, the mass tends to its equilibrium position monotonically as $t \to \infty$. (See Figure 5.6b.)
3. The mass passes through its equilibrium position once at $t = T_2 > 0$ and then attains an extreme displacement at $t = T_3 > T_2$, following which it tends to its equilibrium position monotonically as $t \to \infty$. (See Figure 5.6c.)

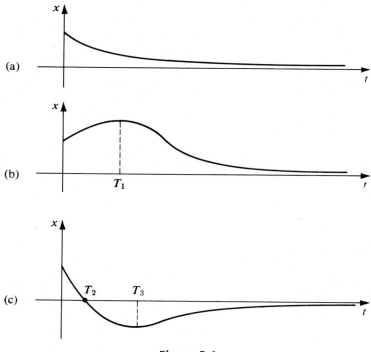

Figure 5.6

Case 3. Overcritical Damping. Finally, we consider here the case in which $b > \lambda$, which implies that $b^2 - \lambda^2 > 0$. Here the roots (5.30) are the distinct, real negative numbers

$$r_1 = -b + \sqrt{b^2 - \lambda^2}$$

and

$$r_2 = -b - \sqrt{b^2 - \lambda^2}.$$

The general solution of (5.28) in this case is

$$x = c_1 e^{r_1 t} + c_2 e^{r_2 t}. \tag{5.36}$$

The damping is now so great that no oscillations can occur. Further, we can no longer say that *every* decrease in the amount of damping will result in oscillations, as we could in Case 2. The motion here is said to be *overcritically damped* (or simply *overdamped*).

Equation (5.36) shows us at once that the displacement x approaches zero as $t \to \infty$. As in Case 2 this approach to zero is monotonic for t sufficiently large. Indeed, the three possible motions in Cases 2 and 3 are qualitatively the same. Thus the three motions illustrated in Figure 5.6 can also serve to illustrate the three types of motion possible in Case 3.

▶ **Example 5.2**

A 32-lb weight is attached to the lower end of a coil spring suspended from the ceiling. The weight comes to rest in its equilibrium position, thereby stretching the spring 2 ft.

The weight is then pulled down 6 in. below its equilibrium position and released at $t = 0$. No external forces are present; but the resistance of the medium in pounds is numerically equal to $4(dx/dt)$, where dx/dt is the instantaneous velocity in feet per second. Determine the resulting motion of the weight on the spring.

Formulation. This is a free, damped motion and Equation (5.27) applies. Since the 32-lb weight stretches the spring 2 ft, Hooke's law, $F = ks$, gives $32 = k(2)$ and so $k = 16$ lb/ft. Thus, since $m = w/g = \frac{32}{32} = 1$ (slug), and the damping constant $a = 4$, Equation (5.27) becomes

$$\frac{d^2x}{dt^2} + 4\frac{dx}{dt} + 16x = 0. \tag{5.37}$$

The initial conditions are

$$x(0) = \tfrac{1}{2},$$
$$x'(0) = 0. \tag{5.38}$$

Solution. The auxiliary equation of Equation (5.37) is

$$r^2 + 4r + 16 = 0.$$

Its roots are the conjugate complex numbers $-2 \pm 2\sqrt{3}i$. Thus the general solution of (5.37) may be written

$$x = e^{-2t}(c_1 \sin 2\sqrt{3}t + c_2 \cos 2\sqrt{3}t), \tag{5.39}$$

where c_1 and c_2 are arbitrary constants. Differentiating (5.39) with respect to t we obtain

$$\frac{dx}{dt} = e^{-2t}[(-2c_1 - 2\sqrt{3}c_2)\sin 2\sqrt{3}t + (2\sqrt{3}c_1 - 2c_2)\cos 2\sqrt{3}t]. \tag{5.40}$$

Applying the initial conditions (5.38) to Equations (5.39) and (5.40), we obtain

$$c_2 = \tfrac{1}{2},$$
$$2\sqrt{3}c_1 - 2c_2 = 0.$$

Thus $c_1 = \sqrt{3}/6$, $c_2 = \tfrac{1}{2}$ and the solution is

$$x = e^{-2t}\left(\frac{\sqrt{3}}{6}\sin 2\sqrt{3}t + \frac{1}{2}\cos 2\sqrt{3}t\right). \tag{5.41}$$

Let us put this in the alternative form (5.32). We have

$$\frac{\sqrt{3}}{6}\sin 2\sqrt{3}t + \frac{1}{2}\cos 2\sqrt{3}t = \frac{\sqrt{3}}{3}\left[\frac{1}{2}\sin 2\sqrt{3}t + \frac{\sqrt{3}}{2}\cos 2\sqrt{3}t\right]$$
$$= \frac{\sqrt{3}}{3}\cos\left(2\sqrt{3}t - \frac{\pi}{6}\right).$$

Thus the solution (5.41) may be written

$$x = \frac{\sqrt{3}}{3}e^{-2t}\cos\left(2\sqrt{3}t - \frac{\pi}{6}\right). \tag{5.42}$$

Interpretation. This is a *damped oscillatory motion* (Case 1). The damping factor is $(\sqrt{3}/3)e^{-2t}$, the "period" is $2\pi/2\sqrt{3} = \sqrt{3}\pi/3$, and the logarithmic decrement is $2\sqrt{3}\pi/3$. The graph of the solution (5.42) is shown in Figure 5.7, where the curves $x = \pm(\sqrt{3}/3)e^{-2t}$ are drawn dashed.

▶ **Example 5.3**

Determine the motion of the weight on the spring described in Example 5.2 if the resistance of the medium in pounds is numerically equal to $8(dx/dt)$ instead of $4(dx/dt)$ (as stated there), all other circumstances being the same as stated in Example 5.2.

Formulation. Once again Equation (5.27) applies, and exactly as in Example 5.2 we find that $m = 1$ (slug) and $k = 16$ lb/ft. But now the damping has increased, and we have $a = 8$. Thus Equation (5.27) now becomes

$$\frac{d^2x}{dt^2} + 8\frac{dx}{dt} + 16x = 0. \tag{5.43}$$

The initial conditions

$$x(0) = \tfrac{1}{2}, \\ x'(0) = 0, \tag{5.44}$$

are, of course, unchanged from Example 5.2.

Solution. The auxiliary equation is now

$$r^2 + 8r + 16 = 0$$

and has the equal roots $r = -4, -4$. The general solution of Equation (5.43) is thus

$$x = (c_1 + c_2 t)e^{-4t}, \tag{5.45}$$

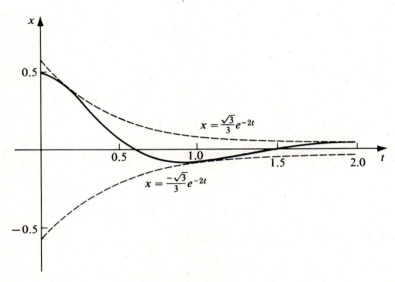

Figure 5.7

where c_1 and c_2 are arbitrary constants. Differentiating (5.45) with respect to t, we have

$$\frac{dx}{dt} = (c_2 - 4c_1 - 4c_2 t)e^{-4t}. \tag{5.46}$$

Applying the initial conditions (5.44) to Equations (5.45) and (5.46) we obtain

$$c_1 = \tfrac{1}{2},$$
$$c_2 - 4c_1 = 0.$$

Thus $c_1 = \tfrac{1}{2}$, $c_2 = 2$ and the soltuion is

$$x = (\tfrac{1}{2} + 2t)e^{-4t}. \tag{5.47}$$

Interpretation. The motion is critically damped. Using (5.47), we see that $x = 0$ if and only if $t = -\tfrac{1}{4}$. Thus $x \neq 0$ for $t > 0$ and the weight does not pass through its equilibrium position. Differentiating (5.47) one finds at once that $dx/dt < 0$ for all $t > 0$. Thus the displacement of the weight from its equilibrium position is a decreasing function of t for all $t > 0$. In other words, the weight starts to move back toward its equilibrium position at once and $x \to 0$ monotonically as $t \to \infty$. The motion is therefore an example of possibility 1 described in the general discussion of Case 2 above. The graph of the solution (5.47) is shown as the solid curve in Figure 5.8.

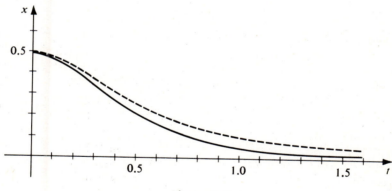

Figure 5.8

▶ **Example 5.4**

Determine the motion of the weight on the spring described in Example 5.2 if the resistance of the medium in pounds is numerically equal to $10(dx/dt)$ instead of $4(dx/dt)$ (as stated there), all other circumstances being the same as stated in Example 5.2.

Formulation. The only difference between this and the two previous examples is in the damping constant. In Example 5.2, $a = 4$; in Example 5.3, $a = 8$; and now we have even greater damping, for here $a = 10$. As before $m = 1$ (slug) and $k = 16$ lb/ft. The differential equation (5.27) thus becomes

$$\frac{d^2 x}{dt^2} + 10\frac{dx}{dt} + 16x = 0. \tag{5.48}$$

The initial conditions (5.44) or (5.38) still hold.

Solution. The auxiliary equation is now

$$r^2 + 10r + 16 = 0$$

and its roots are $r = -2, -8$. Thus the general solution of Equation (5.48) is

$$x = c_1 e^{-2t} + c_2 e^{-8t},$$

where c_1 and c_2 are arbitrary constants. Differentiating this with respect to t to obtain

$$\frac{dx}{dt} = -2c_1 e^{-2t} - 8c_2 e^{-8t}$$

and applying the initial conditions (5.44), we find the following equations for the determination of c_1 and c_2:

$$c_1 + c_2 = \tfrac{1}{2},$$

$$-2c_1 - 8c_2 = 0.$$

The solution of this system is $c_1 = \tfrac{2}{3}, c_2 = -\tfrac{1}{6}$; thus the solution of the problem is

$$x = \tfrac{2}{3} e^{-2t} - \tfrac{1}{6} e^{-8t}. \tag{5.49}$$

Interpretation. Clearly the motion described by Equation (5.49) is an example of the overdamped case (Case 3). Qualitatively the motion is the same as that of the solution (5.47) of Example 5.3. Here, however, due to the increased damping, the weight returns to its equilibrium position at a slower rate. The graph of (5.49) is shown as the dashed curve in Figure 5.8. Note that in each of Examples 5.2, 5.3, and 5.4, all circumstances (the weight, the spring, and the initial conditions) were exactly the same, *except* for the damping. In Example 5.2, the damping constant $a = 4$, and the resulting motion was the damped oscillatory motion shown in Figure 5.7. In Example 5.3 the damping was increased to such an extent ($a = 8$) that oscillations no longer occurred, the motion being shown by the solid curve of Figure 5.8. Finally in Example 5.4 the damping was further increased ($a = 10$) and the resulting motion, indicated by the dashed curve of Figure 5.8, was similar to but slower than that of Example 5.3.

Exercises

1. An 8-lb weight is attached to the lower end of a coil spring suspended from the ceiling and comes to rest in its equilibrium position, thereby stretching the spring 0.4 ft. The weight is then pulled down 6 in. below its equilibrium position and released at $t = 0$. The resistance of the medium in pounds is numerically equal to $2(dx/dt)$, where dx/dt is the instantaneous velocity in feet per second.

   (a) Set up the differential equation for the motion and list the initial conditions.

   (b) Solve the initial-value problem set up in part (a) to determine the displacement of the weight as a function of the time.

   (c) Express the solution found in step (b) in the alternative form (5.32) of the text.

   (d) What is the so-called "period" of the motion?

   (e) Graph the displacement as a function of the time.

2. A 16-lb weight is placed upon the lower end of a coil spring suspended from the ceiling and comes to rest in its equilibrium position, thereby stretching the spring 8 in. At time $t = 0$ the weight is then struck so as to set it into motion with an initial velocity of 2 ft/sec, directed downward. The medium offers a resistance in pounds numerically equal to $6(dx/dt)$, where dx/dt is the instantaneous velocity in feet per second. Determine the resulting displacement of the weight as a function of time and graph this displacement.

3. An 8-lb weight is attached to the lower end of a coil spring suspended from a fixed support. The weight comes to rest in its equilibrium position, thereby stretching the spring 6 in. The weight is then pulled down 9 in. below its equilibrium position and released at $t = 0$. The medium offers a resistance in pounds numerically equal to $4(dx/dt)$, where dx/dt is the instantaneous velocity in feet per second. Determine the displacement of the weight as a function of the time and graph this displacement.

4. A 16-lb weight is attached to the lower end of a coil spring suspended from a fixed support. The weight comes to rest in its equilibrium position, thereby stretching the spring 6 in. The weight is then pulled down 3 in. below its equilibrium position and released at $t = 0$. The medium offers a resistance in pounds numerically equal to $10(dx/dt)$, where dx/dt is the instantaneous velocity in feet per second. Determine the displacement of the weight as a function of the time and graph this displacement.

5. A spring is such that a force of 20 lb would stretch it 6 in. The spring hangs vertically and a 4-lb weight is attached to the end of it. After this 4-lb weight comes to rest in its equilibrium position it is pulled down 8 in. below this position and then released at $t = 0$. The medium offers a resistance in pounds numerically equal to $2(dx/dt)$, where dx/dt is the instantaneous velocity in feet per second.

   (a) Determine the displacement of the weight as a function of the time and express this displacement in the alternative form (5.32) of the text.

   (b) Find the so-called "period" and determine the logarithmic decrement.

   (c) At what time does the weight first pass through its equilibrium position?

6. A 4-lb weight is hung upon the lower end of a coil spring hanging vertically from a fixed support. The weight comes to rest in its equilibrium position, thereby stretching the spring 8 in. The weight is then pulled down a certain distance below this equilibrium position and released at $t = 0$. The medium offers a resistance in pounds numerically equal to $a(dx/dt)$, where $a > 0$ and dx/dt is the instantaneous velocity in feet per second. Show that the motion is oscillatory if $a < \sqrt{3}$, critically damped if $a = \sqrt{3}$, and overdamped if $a > \sqrt{3}$.

7. A 4-lb weight is attached to the lower end of a coil spring that hangs vertically from a fixed support. The weight comes to rest in its equilibrium position, thereby stretching the spring 6 in. The weight is then pulled down 3 in. below this equilibrium position and released at $t = 0$. The medium offers a resistance in pounds numerically equal to $a(dx/dt)$, where $a > 0$ and dx/dt is the instantaneous velocity in feet per second.

   (a) Determine the value of a such that the resulting motion would be critically damped and determine the displacement for this critical value of a.

   (b) Determine the displacement if a is equal to one-half the critical value found in step (a).

   (c) Determine the displacement if a is equal to twice the critical value found in step (a).

8. A 10-lb weight is attached to the lower end of a coil spring suspended from the ceiling, the spring constant being 20 lb/ft. The weight comes to rest in its equilibrium position. It is then pulled down 6 in. below this position and released at $t = 0$ with an initial velocity of 1 ft/sec, directed downward. The resistance of the medium in pounds is numerically equal to $a(dx/dt)$, where $a > 0$ and dx/dt is the instantaneous velocity in feet per second.

   (a) Determine the smallest value of the damping coefficient a for which the motion is not oscillatory.

   (b) Using the value of a found in part (a) find the displacement of the weight as a function of the time.

   (c) Show that the weight attains a single extreme displacement from its equilibrium position at time $t = \frac{1}{40}$, determine this extreme displacement, and show that the weight then tends monotonically to its equilibrium position as $t \to \infty$.

   (d) Graph the displacement found in step (b).

9. A 32-lb weight is attached to the lower end of a coil spring suspended from the ceiling. After the weight comes to rest in its equilibrium position, it is then pulled down a certain distance below this position and released at $t = 0$. If the medium offered no resistance, the natural frequency of the resulting undamped motion would be $4/\pi$ cycles per second. However, the medium does offer a resistance in pounds numerically equal to $a(dx/dt)$, where $a > 0$ and dx/dt is the instantaneous velocity in feet per second; and the frequency of the resulting damped oscillatory motion is only half as great as the natural frequency.

   (a) Determine the spring constant k.

   (b) Find the value of the damping coefficient a.

10. The differential equation for the vertical motion of a mass m on a coil spring of spring constant k in a medium in which the damping is proportional to the instantaneous velocity is given by Equation (5.27). In the case of damped oscillatory motion the solution of this equation is given by (5.33). Show that the displacement x so defined attains an extremum (maximum or minimum) at the times $t_n(n = 0, 1, 2, \ldots)$ given by

$$t_n = \frac{1}{\omega_1}\left[\arctan\left(-\frac{a}{2m\omega_1}\right) + n\pi - \phi\right],$$

   where

$$\omega_1 = \sqrt{\frac{k}{m} - \frac{a^2}{4m^2}}.$$

11. The differential equation for the vertical motion of a unit mass on a certain coil spring in a certain medium is

$$\frac{d^2x}{dt^2} + 2b\frac{dx}{dt} + b^2x = 0,$$

where $b > 0$. The initial displacement of the mass is A feet and its initial velocity is B feet per second.

(a) Show that the motion is critically damped and that the displacement is given by

$$x = (A + Bt + Abt)e^{-bt}.$$

(b) If A and B are such that

$$-\frac{A}{B + Ab} \quad \text{and} \quad \frac{B}{b(B + Ab)}$$

are both negative, show that the mass approaches its equilibrium position monotonically as $t \to \infty$ without either passing through this equilibrium position or attaining an extreme displacement from it for $t > 0$.

(c) If A and B are such that $-A/(B + Ab)$ is negative but $B/b(B + Ab)$ is positive, show that the mass does not pass through its equilibrium position for $t > 0$, that its displacement from this position attains a single extremum at $t = B/b(B + Ab)$, and that thereafter the mass tends to its equilibrium position monotonically as $t \to \infty$.

(d) If A and B are such that $-A/(B + Ab)$ is positive, show that the mass passes through its equilibrium position at $t = -A/(B + Ab)$, attains an extreme displacement at $t = B/b(B + Ab)$, and thereafter tends to its equilibrium position monotonically as $t \to \infty$.

5.4 FORCED MOTION

We now consider an important special case of *forced motion*. That is, we not only consider the effect of damping upon the mass on the spring but also the effect upon it of a periodic external impressed force F defined by $F(t) = F_1 \cos \omega t$ for all $t \geq 0$, where F_1 and ω are constants. Then the basic differential equation (5.7) becomes

$$m\frac{d^2x}{dt^2} + a\frac{dx}{dt} + kx = F_1 \cos \omega t. \tag{5.50}$$

Dividing through by m and letting

$$\frac{a}{m} = 2b, \quad \frac{k}{m} = \lambda^2, \quad \text{and} \quad \frac{F_1}{m} = E_1,$$

this takes the more convenient form

$$\frac{d^2x}{dt^2} + 2b\frac{dx}{dt} + \lambda^2 x = E_1 \cos \omega t. \tag{5.51}$$

We shall assume that the positive damping constant a is small enough so that the damping is less than critical. In other words we assume that $b < \lambda$. Hence by Equation (5.32) the complementary function of Equation (5.51) can be written

$$x_c = ce^{-bt}\cos(\sqrt{\lambda^2 - b^2}\,t + \phi). \tag{5.52}$$

We shall now find a particular integral of (5.51) by the method of undetermined coefficients. We let

$$x_p = A \cos \omega t + B \sin \omega t. \tag{5.53}$$

Then

$$\frac{dx_p}{dt} = -\omega A \sin \omega t + \omega B \cos \omega t,$$

$$\frac{d^2 x_p}{dt^2} = -\omega^2 A \cos \omega t - \omega^2 B \sin \omega t.$$

Substituting into Equation (5.51), we have

$$[-2b\omega A + (\lambda^2 - \omega^2)B]\sin \omega t + [(\lambda^2 - \omega^2)A + 2b\omega B]\cos \omega t = E_1 \cos \omega t.$$

Thus, we have the following two equations from which to determine A and B:

$$-2b\omega A + (\lambda^2 - \omega^2)B = 0,$$

$$(\lambda^2 - \omega^2)A + 2b\omega B = E_1.$$

Solving these, we obtain

$$A = \frac{E_1(\lambda^2 - \omega^2)}{(\lambda^2 - \omega^2)^2 + 4b^2\omega^2},$$

$$B = \frac{2b\omega E_1}{(\lambda^2 - \omega^2)^2 + 4b^2\omega^2}. \tag{5.54}$$

Substituting these values into Equation (5.53), we obtain a particular integral in the form

$$x_p = \frac{E_1}{(\lambda^2 - \omega^2)^2 + 4b^2\omega^2} [(\lambda^2 - \omega^2)\cos \omega t + 2b\omega \sin \omega t].$$

We now put this in the alternative "phase angle" form; we write

$$(\lambda^2 - \omega^2)\cos \omega t + 2b\omega \sin \omega t$$

$$= \sqrt{(\lambda^2 - \omega^2)^2 + 4b^2\omega^2} \left[\frac{\lambda^2 - \omega^2}{\sqrt{(\lambda^2 - \omega^2)^2 + 4b^2\omega^2}} \cos \omega t \right.$$

$$\left. + \frac{2b\omega}{\sqrt{(\lambda^2 - \omega^2)^2 + 4b^2\omega^2}} \sin \omega t \right]$$

$$= \sqrt{(\lambda^2 - \omega^2)^2 + 4b^2\omega^2} [\cos \omega t \cos \theta + \sin \omega t \sin \theta]$$

$$= \sqrt{(\lambda^2 - \omega^2)^2 + 4b^2\omega^2} \cos(\omega t - \theta),$$

where

$$\cos \theta = \frac{\lambda^2 - \omega^2}{\sqrt{(\lambda^2 - \omega^2)^2 + 4b^2\omega^2}},$$

$$\sin \theta = \frac{2b\omega}{\sqrt{(\lambda^2 - \omega^2)^2 + 4b^2\omega^2}}. \tag{5.55}$$

Thus the particular integral appears in the form

$$x_p = \frac{E_1}{\sqrt{(\lambda^2 - \omega^2)^2 + 4b^2\omega^2}} \cos(\omega t - \theta),$$ (5.56)

where θ is determined from Equations (5.55). Thus, using (5.52) and (5.56) the general solution of Equation (5.51) is

$$x = x_c + x_p = ce^{-bt} \cos(\sqrt{\lambda^2 - b^2}t + \phi) + \frac{E_1}{\sqrt{(\lambda^2 - \omega^2)^2 + 4b^2\omega^2}} \cos(\omega t - \theta).$$ (5.57)

Observe that this solution is the sum of two terms. The first term, $ce^{-bt} \cos(\sqrt{\lambda^2 - b^2}t + \phi)$, represents the damped oscillation that would be the entire motion of the system if the external force $F_1 \cos \omega t$ were not present. The second term,

$$\frac{E_1}{\sqrt{(\lambda^2 - \omega^2)^2 + 4b^2\omega^2}} \cos(\omega t - \theta),$$

which results from the presence of the external force, represents a simple harmonic motion of period $2\pi/\omega$. Because of the damping factor ce^{-bt} the contribution of the first term will become smaller and smaller as time goes on and will eventually become negligible. The first term is thus called the *transient* term. The second term, however, being a cosine term of constant amplitude, continues to contribute to the motion in a periodic, oscillatory manner. Eventually, the transient term having become relatively small, the entire motion will consist essentially of that given by this second term. This second term is thus called the *steady-state* term.

▶ **Example 5.5**

A 16-lb weight is attached to the lower end of a coil spring suspended from the ceiling, the spring constant of the spring being 10 lb/ft. The weight comes to rest in its equilibrium position. Beginning at $t = 0$ an external force given by $F(t) = 5 \cos 2t$ is applied to the system. Determine the resulting motion if the damping force in pounds is numerically equal to $2(dx/dt)$, where dx/dt is the instantaneous velocity in feet per second.

Formulation. The basic differential equation for the motion is

$$m\frac{d^2x}{dt^2} + a\frac{dx}{dt} + kx = F(t).$$ (5.58)

Here $m = w/g = \frac{16}{32} = \frac{1}{2}$ (slug), $a = 2$, $k = 10$, and $F(t) = 5 \cos 2t$. Thus Equation (5.58) becomes

$$\frac{1}{2}\frac{d^2x}{dt^2} + 2\frac{dx}{dt} + 10x = 5 \cos 2t$$

or

$$\frac{d^2x}{dt^2} + 4\frac{dx}{dt} + 20x = 10 \cos 2t.$$ (5.59)

The initial conditions are

$$x(0) = 0,$$
$$x'(0) = 0. \tag{5.60}$$

Solution.   The auxiliary equation of the homogeneous equation corresponding to (5.59) is $r^2 + 4r + 20 = 0$; its roots are $-2 \pm 4i$. Thus the complementary function of Equation (5.59) is

$$x_c = e^{-2t}(c_1 \sin 4t + c_2 \cos 4t),$$

where c_1 and c_2 are arbitrary constants. Using the method of undetermined coefficients to obtain a particular integral, we let

$$x_p = A \cos 2t + B \sin 2t.$$

Upon differentiating and substituting into (5.59), we find the following equations for the determination of A and B.

$$-8A + 16B = 0,$$
$$16A + 8B = 10.$$

Solving these, we find

$$A = \tfrac{1}{2}, \qquad B = \tfrac{1}{4}.$$

Thus a particular integral is

$$x_p = \tfrac{1}{2} \cos 2t + \tfrac{1}{4} \sin 2t$$

and the general solution of (5.59) is

$$x = x_c + x_p = e^{-2t}(c_1 \sin 4t + c_2 \cos 4t) + \tfrac{1}{2} \cos 2t + \tfrac{1}{4} \sin 2t. \tag{5.61}$$

Differentiating (5.61) with respect to t, we obtain

$$\frac{dx}{dt} = e^{-2t}[(-2c_1 - 4c_2)\sin 4t + (-2c_2 + 4c_1)\cos 4t] - \sin 2t + \tfrac{1}{2}\cos 2t. \tag{5.62}$$

Applying the initial conditions (5.60) to Equations (5.61) and (5.62), we see that

$$c_2 + \tfrac{1}{2} = 0,$$
$$4c_1 - 2c_2 + \tfrac{1}{2} = 0.$$

From these equations we find that

$$c_1 = -\tfrac{3}{8}, \qquad c_2 = -\tfrac{1}{2}.$$

Hence the solution is

$$x = e^{-2t}(-\tfrac{3}{8} \sin 4t - \tfrac{1}{2} \cos 4t) + \tfrac{1}{2} \cos 2t + \tfrac{1}{4} \sin 2t. \tag{5.63}$$

Let us write this in the "phase angle" form. We have first

$$3 \sin 4t + 4 \cos 4t = 5(\tfrac{3}{5} \sin 4t + \tfrac{4}{5} \cos 4t) = 5 \cos(4t - \phi),$$

where

$$\cos \phi = \tfrac{4}{5}, \qquad \sin \phi = \tfrac{3}{5}. \tag{5.64}$$

Also, we have

$$2 \cos 2t + \sin 2t = \sqrt{5} \left(\frac{2}{\sqrt{5}} \cos 2t + \frac{1}{\sqrt{5}} \sin 2t \right) = \sqrt{5} \cos(2t - \theta),$$

where

$$\cos \theta = \frac{2}{\sqrt{5}}, \qquad \sin \theta = \frac{1}{\sqrt{5}}. \tag{5.65}$$

Thus we may write the solution (5.63) as

$$x = -\frac{5e^{-2t}}{8} \cos(4t - \phi) + \frac{\sqrt{5}}{4} \cos(2t - \theta), \tag{5.66}$$

where ϕ and θ are determined by Equations (5.64) and (5.65), respectively. We find that $\phi \approx 0.64$ (rad) and $\theta \approx 0.46$ (rad). Thus the solution (5.66) is given approximately by

$$x = -0.63e^{-2t} \cos(4t - 0.64) + 0.56 \cos(2t - 0.46).$$

Interpretation.   The term

$$-\frac{5e^{-2t}}{8} \cos(4t - \phi) \approx -0.63e^{-2t} \cos(4t - 0.64)$$

is the *transient* term, representing a damped oscillatory motion. It becomes negligible in a short time; for example, for $t > 3$, its numerical value is less than 0.002. Its graph is shown in Figure 5.9a. The term

$$\frac{\sqrt{5}}{4} \cos(2t - \theta) \approx 0.56 \cos(2t - 0.46)$$

is the *steady-state* term, representing a simple harmonic motion of amplitude

$$\frac{\sqrt{5}}{4} \approx 0.56$$

and period π. Its graph appears in Figure 5.9b. The graph in Figure 5.9c is that of the complete solution (5.66). It is clear from this that the effect of the transient term soon becomes negligible, and that after a short time the contribution of the steady-state term is essentially all that remains.

Exercises

1. A 6 lb weight is attached to the lower end of a coil spring suspended from the ceiling, the spring constant of the spring being 27 lb/ft. The weight comes to rest in its equilibrium position, and beginning at $t = 0$ an external force given by $F(t) = 12 \cos 20t$ is applied to the system. Determine the resulting displacement as a function of the time, assuming damping is negligible.

2. A 16-lb weight is attached to the lower end of a coil spring suspended from the ceiling. The weight comes to rest in its equilibrium position, thereby stretching the spring 0.4 ft. Then, beginning at $t = 0$, an external force given by $F(t) = 40 \cos 16t$

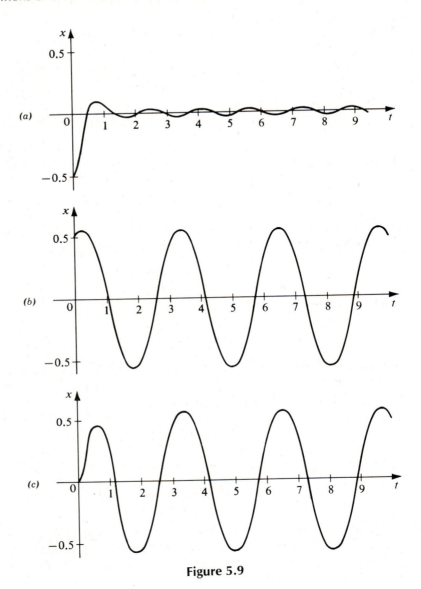

Figure 5.9

is applied to the system. The medium offers a resistance in pounds numerically equal to $4(dx/dt)$, where dx/dt is the instantaneous velocity in feet per second.

(a) Find the displacement of the weight as a function of the time.

(b) Graph separately the transient and steady-state terms of the motion found in step (a) and then use the curves so obtained to graph the entire displacement itself.

3. A 10-lb weight is hung on the lower end of a coil spring suspended from the ceiling, the spring constant of the spring being 20 lb/ft. The weight comes to rest in its equilibrium position, and beginning at $t = 0$ an external force given by $F(t) = 10 \cos 8t$ is applied to the system. The medium offers a resistance in

pounds numerically equal to $5(dx/dt)$, where dx/dt is the instantaneous velocity in feet per second. Find the displacement of the weight as a function of the time.

4. A 4-lb weight is hung on the lower end of a coil spring suspended from a beam. The weight comes to rest in its equilibrium position, thereby stretching the spring 3 in. The weight is then pulled down 6 in. below this position and released at $t = 0$. At this instant an external force given by $F(t) = 13 \sin 4t$ is applied to the system. The resistance of the medium in pounds is numerically equal to twice the instantaneous velocity, measured in feet per second.
    (a) Find the displacement of the weight as a function of the time.
    (b) Observe that the displacement is the sum of a transient term and a steady-state term, and find the amplitude of the steady-state term.

5. A 6-lb weight is hung on the lower end of a coil spring suspended from the ceiling. The weight comes to rest in its equilibrium position, thereby stretching the spring 4 in. Then beginning at $t = 0$ an external force given by $F(t) = 27 \sin 4t - 3 \cos 4t$ is applied to the system. If the medium offers a resistance in pounds numerically equal to three times the instantaneous velocity, measured in feet per second, find the displacement as a function of the time.

6. A certain coil spring having spring constant 10 lb/ft is suspended from the ceiling. A 32-lb weight is attached to the lower end of the spring and comes to rest in its equilibrium position. Beginning at $t = 0$ an external force given by $F(t) = \sin t + \frac{1}{4} \sin 2t + \frac{1}{9} \sin 3t$ is applied to the system. The medium offers a resistance in pounds numerically equal to twice the instantaneous velocity, measured in feet per second. Find the displacement of the weight as a function of the time, using Chapter 4, Theorem 4.10 to obtain the steady-state term.

7. A coil spring having spring constant 20 lb/ft is suspended from the ceiling. A 32-lb weight is attached to the lower end of the spring and comes to rest in its equilibrium position. Beginning at $t = 0$ an external force given by $F(t) = 40 \cos 2t$ is applied to the system. This force then remains in effect until $t = \pi$, at which instant it ceases to be applied. For $t > \pi$, no external forces are present. The medium offers a resistance in pounds numerically equal to $4(dx/dt)$, where dx/dt is the instantaneous velocity in feet per second. Find the displacement of the weight as a function of the time for all $t \geq 0$.

8. Consider the basic differential equation (5.7) for the motion of a mass m vibrating up and down on a coil spring suspended vertically from a fixed support; and suppose the external impressed force F is the periodic function defined by $F(t) = F_2 \sin \omega t$ for all $t \geq 0$, where F_2 and ω are constants. Show that in this case the steady-state term in the solution of Equation (5.7) may be written

$$x_p = \frac{E_2}{\sqrt{(\lambda^2 - \omega^2)^2 + 4b^2\omega^2}} \sin(\omega t - \theta),$$

where $b = a/2m$, $\lambda^2 = k/m$, $E_2 = F_2/m$, and θ is determined from Equations (5.55).

9. A 32-lb weight is attached to the lower end of a coil spring suspended vertically from a fixed support and comes to rest in its equilibrium position, thereby stretching the spring 6 in. Beginning at $t = 0$ an external force given by $F(t) = 15 \cos 7t$ is applied to the system. Assume that the damping is negligible.

(a) Find the displacement of the weight as a function of the time.

(b) Show that this displacement may be expressed as $x = A(t)\sin(15t/2)$, where $A(t) = 2\sin\frac{1}{2}t$. The function $A(t)$ may be regarded as the "slowly varying" amplitude of the more rapid oscillation $\sin(15t/2)$. When a phenomenon involving such fluctuations in maximum amplitude takes place in acoustical applications, *beats* are said to occur.

(c) Carefully graph the slowly varying amplitude $A(t) = 2\sin(t/2)$ and its negative $-A(t)$ and then use these "bounding curves" to graph the displacement $x = A(t)\sin(15t/2)$.

10. A 16-lb weight is attached to the lower end of a coil spring that is suspended vertically from a support and for which the spring constant k is 10 lb/ft. The weight comes to rest in its equilibrium position and is then pulled down 6 in. below this position and released at $t = 0$. At this instant the support of the spring begins a vertical oscillation such that its distance from its initial position is given by $\frac{1}{2}\sin 2t$ for $t \geq 0$. The resistance of the medium in pounds is numerically equal to $2(dx/dt)$, where dx/dt is the instantaneous velocity of the moving weight in feet per second.

(a) Show that the differential equation for the displacement of the weight from its equilibrium position is

$$\frac{1}{2}\frac{d^2x}{dt^2} = -10(x - y) - 2\frac{dx}{dt}, \qquad \text{where} \quad y = \frac{1}{2}\sin 2t,$$

and hence that this differential equation may be written

$$\frac{d^2x}{dt^2} + 4\frac{dx}{dt} + 20x = 10\sin 2t.$$

(b) Solve the differential equation obtained in step (a), apply the relevant initial conditions, and thus obtain the displacement x as a function of time.

5.5 RESONANCE PHENOMENA

We now consider the amplitude of the steady-state vibration that results from the periodic external force defined for all t by $F(t) = F_1 \cos \omega t$, where we assume that $F_1 > 0$. For fixed b, λ, and E_1 we see from Equation (5.56) that this is the function f of ω defined by

$$f(\omega) = \frac{E_1}{\sqrt{(\lambda^2 - \omega^2)^2 + 4b^2\omega^2}}. \qquad (5.67)$$

If $\omega = 0$, the force $F(t)$ is the constant F_1 and the amplitude $f(\omega)$ has the value $E_1/\lambda^2 > 0$. Also, as $\omega \to \infty$, we see from (5.67) that $f(\omega) \to 0$. Let us consider the function f for $0 < \omega < \infty$. Calculating the derivative $f'(\omega)$ we find that this derivative equals zero only if

$$4\omega[2b^2 - (\lambda^2 - \omega^2)] = 0$$

and hence only if $\omega = 0$ or $\omega = \sqrt{\lambda^2 - 2b^2}$. If $\lambda^2 < 2b^2$, $\sqrt{\lambda^2 - 2b^2}$ is a complex number. Hence in this case f has no extremum for $0 < \omega < \infty$, but rather f decreases monotonically for $0 < \omega < \infty$ from the value E_1/λ^2 at $\omega = 0$ and approaches zero as $\omega \to \infty$. Let us assume that $\lambda^2 > 2b^2$. Then the function f has a relative maximum at $\omega_1 = \sqrt{\lambda^2 - 2b^2}$, and this maximum value is given by

$$f(\omega_1) = \frac{E_1}{\sqrt{(2b^2)^2 + 4b^2(\lambda^2 - 2b^2)}} = \frac{E_1}{2b\sqrt{\lambda^2 - b^2}}.$$

When the frequency of the forcing function $F_1 \cos \omega t$ is such that $\omega = \omega_1$, then the forcing function is said to be in *resonance* with the system. In other words, the forcing function defined by $F_1 \cos \omega t$ is in resonance with the system when ω assumes the value ω_1 at which $f(\omega)$ is a maximum. The value $\omega_1/2\pi$ is called the *resonance frequency* of the system. Note carefully that resonance can occur only if $\lambda^2 > 2b^2$. Since then $\lambda^2 > b^2$, the damping must be less than critical in such a case.

We now return to the original notation of Equation (5.50). In terms of the quantities m, a, k, and F_1 of that equation, the function f is given by

$$f(\omega) = \frac{\dfrac{F_1}{m}}{\sqrt{\left(\dfrac{k}{m} - \omega^2\right)^2 + \left(\dfrac{a}{m}\right)^2 \omega^2}} \qquad (5.68)$$

In this original notation the resonance frequency is

$$\frac{1}{2\pi}\sqrt{\frac{k}{m} - \frac{a^2}{2m^2}}. \qquad (5.69)$$

Since the frequency of the corresponding free, damped oscillation is

$$\frac{1}{2\pi}\sqrt{\frac{k}{m} - \frac{a^2}{4m^2}},$$

we see that the resonance frequency is less than that of the corresponding free, damped oscillation.

The graph of $f(\omega)$ is called the *resonance curve* of the system. For a given system with fixed m, k, and F_1, there is a resonance curve corresponding to each value of the damping coefficient $a \geq 0$. Let us choose $m = k = F_1 = 1$, for example, and graph the resonance curves corresponding to certain selected values of a. In this case we have

$$f(\omega) = \frac{1}{\sqrt{(1 - \omega^2)^2 + a^2\omega^2}}$$

and the resonance frequency is given by $(1/2\pi)\sqrt{1 - a^2/2}$. The graphs appear in Figure 5.10.

Observe that resonance occurs in this case only if $a < \sqrt{2}$. As a decreases from $\sqrt{2}$ to 0, the value ω_1 at which resonance occurs increases from 0 to 1 and the corresponding maximum value of $f(\omega)$ becomes larger and larger. In the limiting case $a = 0$, the maximum has disappeared and an infinite discontinuity occurs at $\omega = 1$. In this case our solution actually breaks down, for then

$$f(\omega) = \frac{1}{\sqrt{(1 - \omega^2)^2}} = \frac{1}{1 - \omega^2}$$

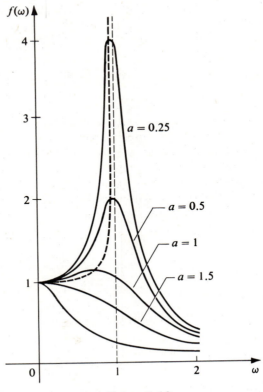

Figure 5.10

and $f(1)$ is undefined. This limiting case is an example of *undamped resonance*, a phenomenon that we shall now investigate.

Undamped resonance occurs when there is no damping and the frequency of the impressed force is equal to the natural frequency of the system. Since in this case $a = 0$ and the frequency $\omega/2\pi$ equals the natural frequency $(1/2\pi)\sqrt{k/m}$, the differential equation (5.50) reduces to

$$m\frac{d^2x}{dt^2} + kx = F_1 \cos\sqrt{\frac{k}{m}}\, t$$

or

$$\frac{d^2x}{dt^2} + \frac{k}{m}x = E_1 \cos\sqrt{\frac{k}{m}}\, t, \tag{5.70}$$

where $E_1 = F_1/m$. Since the complementary function of Equation (5.70) is

$$x_c = c_1 \sin\sqrt{\frac{k}{m}}\, t + c_2 \cos\sqrt{\frac{k}{m}}\, t, \tag{5.71}$$

we cannot assume a particular integral of the form

$$A \sin\sqrt{\frac{k}{m}}\, t + B \cos\sqrt{\frac{k}{m}}\, t.$$

Rather we must assume

$$x_p = At \sin\sqrt{\frac{k}{m}}\,t + Bt \cos\sqrt{\frac{k}{m}}\,t.$$

Differentiating this twice and substituting into Equation (5.70), we find that

$$A = \frac{E_1}{2}\sqrt{\frac{m}{k}} \quad \text{and} \quad B = 0.$$

Thus the particular integral of Equation (5.70) resulting from the forcing function $E_1 \cos\sqrt{k/m}\,t$ is given by

$$x_p = \frac{E_1}{2}\sqrt{\frac{m}{k}}\,t \sin\sqrt{\frac{k}{m}}\,t.$$

Expressing the complementary function (5.71) in the equivalent "phase-angle" form, we see that the general solution of Equation (5.70) is given by

$$x = c \cos\left(\sqrt{\frac{k}{m}}\,t + \phi\right) + \frac{E_1}{2}\sqrt{\frac{m}{k}}\,t \sin\sqrt{\frac{k}{m}}\,t. \qquad (5.72)$$

The motion defined by (5.72) is thus the sum of a periodic term and an oscillatory term whose amplitude $(E_1/2)\sqrt{m/k}\,t$ increases with t. The graph of the function defined by this latter term,

$$\frac{E_1}{2}\sqrt{\frac{m}{k}}\,t \sin\sqrt{\frac{k}{m}}\,t,$$

appears in Figure 5.11. As t increases, this term clearly dominates the entire motion. One might argue that Equation (5.72) informs us that as $t \to \infty$ the oscillations will become infinite. However, common sense intervenes and convinces us that before this exciting phenomenon can occur the system will break down and then Equation (5.72) will no longer apply!

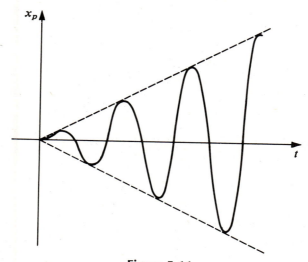

Figure 5.11

▶ **Example 5.6**

A 64-lb weight is attached to the lower end of a coil spring suspended from the ceiling, the spring constant being 18 lb/ft. The weight comes to rest in its equilibrium position. It is then pulled down 6 in. below its equilibrium position and released at $t = 0$. At this instant an external force given by $F(t) = 3 \cos \omega t$ is applied to the system.

1. Assuming the damping force in pounds is numerically equal to $4(dx/dt)$, where dx/dt is the instantaneous velocity in feet per second, determine the resonance frequency of the resulting motion.
2. Assuming there is no damping, determine the value of ω that gives rise to undamped resonance.

Solution. Since $m = w/g = \frac{64}{32} = 2$ (slugs), $k = 18$, and $F(t) = 3 \cos \omega t$, the differential equation is

$$2 \frac{d^2x}{dt^2} + a \frac{dx}{dt} + 18x = 3 \cos \omega t,$$

where a is the damping coefficient. In Part 1, $a = 4$ and so in this case the differential equation reduces to

$$2 \frac{d^2x}{dt^2} + 4 \frac{dx}{dt} + 18x = 3 \cos \omega t.$$

Here we are not asked to solve the differential equation but merely to determine the resonance frequency. Using formula (5.69) we find that this is

$$\frac{1}{2\pi} \sqrt{\frac{18}{2} - \frac{1}{2}\left(\frac{16}{4}\right)} = \frac{1}{2\pi} \sqrt{7} \approx 0.42 \text{ (cycles/sec)}.$$

Thus resonance occurs when $\omega = \sqrt{7} \approx 2.65$.

In Part 2, $a = 0$ and so the differential equation reduces to

$$\frac{d^2x}{dt^2} + 9x = \tfrac{3}{2} \cos \omega t. \tag{5.73}$$

Undamped resonance occurs when the frequency $\omega/2\pi$ of the impressed force is equal to the natural frequency. The complementary function of Equation (5.73) is

$$x_c = c_1 \sin 3t + c_2 \cos 3t,$$

and from this we see that the natural frequency is $3/2\pi$. Thus $\omega = 3$ gives rise to undamped resonance and the differential equation (5.73) in this case becomes

$$\frac{d^2x}{dt^2} + 9x = \tfrac{3}{2} \cos 3t. \tag{5.74}$$

The initial conditions are $x(0) = \frac{1}{2}$, $x'(0) = 0$. The reader should show that the solution of Equation (5.74) satisfying these conditions is

$$x = \tfrac{1}{2} \cos 3t + \tfrac{1}{4} t \sin 3t.$$

Exercises

1. A 12-lb weight is attached to the lower end of a coil spring suspended from the ceiling. The weight comes to rest in its equilibrium position thereby stretching the spring 6 in. Beginning at $t = 0$ an external force given by $F(t) = 2 \cos \omega t$ is applied to the system.

    (a)   If the damping force in pounds is numerically equal to $3(dx/dt)$, where dx/dt is the instantaneous velocity in feet per second, determine the resonance frequency of the resulting motion and find the displacement as a function of the time when the forcing function is in resonance with the system.

    (b)   Assuming there is no damping, determine the value of ω that gives rise to undamped resonance and find the displacement as a function of the time in this case.

2. A 20-lb weight is attached to the lower end of a coil spring suspended from the ceiling. The weight comes to rest in its equilibrium position, thereby stretching the spring 6 in. Various external forces of the form $F(t) = \cos \omega t$ are applied to the system and it is found that the resonance frequency is 0.5 cycles/sec. Assuming that the resistance of the medium in pounds is numerically equal to $a(dx/dt)$, where dx/dt is the instantaneous velocity in feet per second, determine the damping coefficient a.

3. The differential equation for the motion of a unit mass on a certain coil spring under the action of an external force of the form $F(t) = 30 \cos \omega t$ is

    $$\frac{d^2x}{dt^2} + a \frac{dx}{dt} + 24x = 30 \cos \omega t,$$

    where $a \geq 0$ is the damping coefficient.

    (a)   Graph the resonance curves of the system for $a = 0, 2, 4, 6$, and $4\sqrt{3}$.

    (b)   If $a = 4$, find the resonance frequency and determine the amplitude of the steady-state vibration when the forcing function is in resonance with the system.

    (c)   Proceed as in part (b) if $a = 2$.

5.6   ELECTRIC CIRCUIT PROBLEMS

In this section we consider the application of differential equations to series circuits containing (1) an electromotive force, and (2) resistors, inductors, and capacitors. We assume that the reader is somewhat familiar with these items and so we shall avoid an extensive discussion. Let us simply recall that the electromotive force (for example, a battery or generator) produces a flow of current in a closed circuit and that this current produces a so-called *voltage drop* across each resistor, inductor, and capacitor. Further, the following three laws concerning the voltage drops across these various elements are known to hold:

1. The voltage drop across a resistor is given by

$$E_R = Ri, \tag{5.75}$$

where R is a constant of proportionality called the *resistance*, and i is the current.

2. The voltage drop across an inductor is given by

$$E_L = L\frac{di}{dt},$$

(5.76)

where L is a constant of proportionality called the *inductance*, and i again denotes the current.

3. The voltage drop across a capacitor is given by

$$E_C = \frac{1}{C}q,$$

(5.77)

where C is a constant of proportionality called the *capacitance* and q is the instantaneous charge on the capacitor. Since $i = dq/dt$, this is often written as

$$E_C = \frac{1}{C}\int i\,dt.$$

The units in common use are listed in Table 5.1.

TABLE 5.1

Quantity and symbol	Unit
emf or voltage E	volt (V)
current i	ampere
charge q	coulomb
resistance R	ohm (Ω)
inductance L	henry (H)
capacitance C	farad

The fundamental law in the study of electric circuits is the following:

Kirchhoff's Voltage Law (Form 1). The algebraic sum of the instantaneous voltage drops around a close circuit in a specific direction is zero.

Since voltage drops across resistors, inductors, and capacitors have the opposite sign from voltages arising from electromotive forces, we may state this law in the following alternative form:

Kirchhoff's Voltage Law (Form 2). The sum of the voltage drops across resistors, inductors, and capacitors is equal to the total electromotive force in a closed circuit.

We now consider the circuit shown in Figure 5.12.

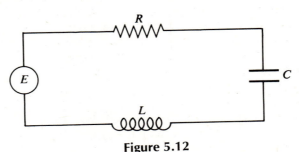

Figure 5.12

Here and in later diagrams the following conventional symbols are employed:

E Electromotive force (battery or generator)

R Resistor

L Inductor

C Capacitor

Let us apply Kirchhoff's law to the circuit of Figure 5.12. Letting E denote the electromotive force, and using the laws 1, 2, and 3 for voltage drops that were given above, we are led at once to the equation

$$L\frac{di}{dt} + Ri + \frac{1}{C}q = E. \tag{5.78}$$

This equation contains *two* dependent variables i and q. However, we recall that these two variables are related to each other by the equation

$$i = \frac{dq}{dt}. \tag{5.79}$$

Using this we may eliminate i from Equation (5.78) and write it in the form

$$L\frac{d^2q}{dt^2} + R\frac{dq}{dt} + \frac{1}{C}q = E. \tag{5.80}$$

Equation (5.80) is a second-order linear differential equation in the single dependent variable q. On the other hand, if we differentiate Equation (5.78) with respect to t and make use of (5.79), we may eliminate q from Equation (5.78) and write

$$L\frac{d^2i}{dt^2} + R\frac{di}{dt} + \frac{1}{C}i = \frac{dE}{dt}. \tag{5.81}$$

This is a second-order linear differential equation in the single dependent variable i.

Thus we have the two second-order linear differential equations (5.80) and (5.81) for the charge q and current i, respectively. Further observe that in two very simple cases the problem reduces to a *first*-order linear differential equation. If the circuit contains no capacitor, Equation (5.78) itself reduces directly to

$$L\frac{di}{dt} + Ri = E;$$

while if no inductor is present, Equation (5.80) reduces to

$$R\frac{dq}{dt} + \frac{1}{C}q = E.$$

Before considering examples, we observe an interesting and useful analogy. The differential equation (5.80) for the charge is exactly the same as the differential equation (5.7) of Section 5.1 for the vibrations of a mass on a coil spring, except for the notations used. That is, the electrical system described by Equation (5.80) is analogous to the mechanical system described by Equation (5.7) of Section 5.1. This analogy is brought out by Table 5.2.

TABLE 5.2

Mechanical system	*Electrical system*
mass m	inductance L
damping constant a	resistance R
spring constant k	reciprocal of capacitance $1/C$
impressed force $F(t)$	impressed voltage or emf E
displacement x	charge q
velocity $v = dx/dt$	current $i = dq/dt$

▶ **Example 5.7**

A circuit has in series an electromotive force given by $E = 100 \sin 40t$ V, a resistor of $10\ \Omega$ and an inductor of 0.5 H. If the initial current is 0, find the current at time $t > 0$.

Formulation. The circuit diagram is shown in Figure 5.13. Let i denote the current in amperes at time t. The total electromotive force is $100 \sin 40t$ V. Using the laws 1 and 2, we find that the voltage drops are as follows:

1. Across the resistor: $E_R = Ri = 10i$.

2. Across the inductor: $E_L = L\dfrac{di}{dt} = \dfrac{1}{2}\dfrac{di}{dt}$.

Applying Kirchhoff's law, have the differential equation

$$\frac{1}{2}\frac{di}{dt} + 10i = 100 \sin 40t,$$

or

$$\frac{di}{dt} + 20i = 200 \sin 40t. \tag{5.82}$$

Since the initial current is 0, the initial condition is

$$i(0) = 0. \tag{5.83}$$

Solution. Equation (5.82) is a first-order linear equation. An integrating factor is

$$e^{\int 20\,dt} = e^{20t}.$$

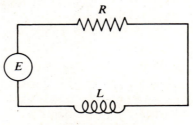

Figure 5.13

Multiplying (5.82) by this, we obtain

$$e^{20t} \frac{di}{dt} + 20e^{20t}i = 200e^{20t} \sin 40t$$

or

$$\frac{d}{dt}(e^{20t}i) = 200e^{20t} \sin 40t.$$

Integrating and simplifying, we find

$$i = 2(\sin 40t - 2 \cos 40t) + Ce^{-20t}.$$

Applying the condition (5.83), $i = 0$ when $t = 0$, we find $C = 4$. Thus the solution is

$$i = 2(\sin 40t - 2 \cos 40t) + 4e^{-20t}.$$

Expressing the trigonometric terms in a "phase-angle" form, we have

$$i = 2\sqrt{5}\left(\frac{1}{\sqrt{5}} \sin 40t - \frac{2}{\sqrt{5}} \cos 40t\right) + 4e^{-20t}$$

or

$$i = 2\sqrt{5} \sin(40t + \phi) + 4e^{-20t}, \tag{5.84}$$

where ϕ is determined by the equation

$$\phi = \arccos \frac{1}{\sqrt{5}} = \arcsin\left(-\frac{2}{\sqrt{5}}\right).$$

We find $\phi \approx -1.11$ rad, and hence the current is given approximately by

$$i = 4.47 \sin(40t - 1.11) + 4e^{-20t}.$$

Interpretation. The current is clearly expressed as the sum of a sinusoidal term and an exponential. The exponential becomes so very small in a short time that its effect is soon practically negligible; it is the *transient* term. Thus, after a short time, essentially all that remains is the sinusoidal term; it is the *steady-state current*. Observe that its *period* $\pi/20$ is the same as that of the electromotive force. However, the *phase angle* $\phi \approx -1.11$ indicates that the electromotive force leads the steady-state current by approximately $\frac{1}{40}$ (1.11). The graph of the current as a function of time appears in Figure 5.14.

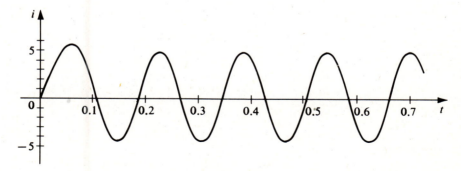

Figure 5.14

▶ **Example 5.8**

A circuit has in series an electromotive force given by $E = 100 \sin 60t$ V, a resistor of
$2\,\Omega$, an inductor of 0.1 H, and a capacitor of $\frac{1}{260}$ farads. (See Figure 5.15.) If the initial
current and the initial charge on the capacitor are both zero, find the charge on the
capacitor at any time $t > 0$.

   Formulation 1, *by directly applying Kirchhoff's law:* Let i denote the current
and q the charge on the capacitor at time t. The total electromotive force is
$100 \sin 60t$ (volts). Using the voltage drop laws 1, 2, and 3 we find that the voltage drops
are as follows:

1. Across the resistor: $E_R = Ri = 2i$.

2. Across the inductor: $E_L = L\dfrac{di}{dt} = \dfrac{1}{10}\dfrac{di}{dt}$.

3. Across the capacitor: $E_c = \dfrac{1}{C}q = 260q$.

Now applying Kirchhoff's law we have at once:

$$\frac{1}{10}\frac{di}{dt} + 2i + 260q = 100 \sin 60t.$$

Since $i = dq/dt$, this reduces to

$$\frac{1}{10}\frac{d^2q}{dt^2} + 2\frac{dq}{dt} + 260q = 100 \sin 60t. \tag{5.85}$$

   Formulation 2, *applying Equation (5.80) for the charge:* We have $L = \frac{1}{10}, R = 2,$
$C = \frac{1}{260}, E = 100 \sin 60t$. Substituting these values directly into Equation (5.80) we
again obtain Equation (5.85) at once.
   Multiplying Equation (5.85) through by 10, we consider the differential equation in
the form

$$\frac{d^2q}{dt^2} + 20\frac{dq}{dt} + 2600q = 1000 \sin 60t. \tag{5.86}$$

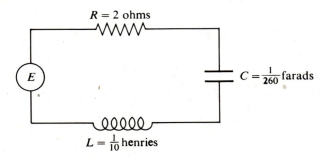

$R = 2$ ohms

$C = \dfrac{1}{260}$ farads

$L = \dfrac{1}{10}$ henries

Figure 5.15

Since the charge q is initially zero, we have as a first initial condition

$$q(0) = 0. \tag{5.87}$$

Since the current i is also initially zero and $i = dq/dt$, we take the second initial condition in the form

$$q'(0) = 0. \tag{5.88}$$

Solution. The homogeneous equation corresponding to (5.86) has the auxiliary equation

$$r^2 + 20r + 2600 = 0.$$

The roots of this equation are $-10 \pm 50i$ and so the complementary function of Equation (5.86) is

$$q_c = e^{-10t}(C_1 \sin 50t + C_2 \cos 50t).$$

Employing the method of undetermined coefficients to find a particular integral of (5.86), we write

$$q_p = A \sin 60t + B \cos 60t.$$

Differentiating twice and substituting into Equation (5.86) we find that

$$A = -\tfrac{25}{61} \quad \text{and} \quad B = -\tfrac{30}{61},$$

and so the general solution of Equation (5.86) is

$$q = e^{-10t}(C_1 \sin 50t + C_2 \cos 50t) - \tfrac{25}{61} \sin 60t - \tfrac{30}{61} \cos 60t. \tag{5.89}$$

Differentiating (5.89), we obtain

$$\frac{dq}{dt} = e^{-10t}[(-10C_1 - 50C_2)\sin 50t + (50C_1 - 10C_2)\cos 50t]$$

$$- \tfrac{1500}{61} \cos 60t + \tfrac{1800}{61} \sin 60t. \tag{5.90}$$

Applying condition (5.87) to Equation (5.89) and condition (5.88) to Equation (5.90), we have

$$C_2 - \tfrac{30}{61} = 0 \quad \text{and} \quad 50C_1 - 10C_2 - \tfrac{1500}{61} = 0.$$

From these equations, we find that

$$C_1 = \tfrac{36}{61} \quad \text{and} \quad C_2 = \tfrac{30}{61}.$$

Thus the solution of the problem is

$$q = \frac{6e^{-10t}}{61} (6 \sin 50t + 5 \cos 50t) - \frac{5}{61} (5 \sin 60t + 6 \cos 60t)$$

or

$$q = \frac{6\sqrt{61}}{61} e^{-10t} \cos(50t - \phi) - \frac{5\sqrt{61}}{61} \cos(60t - \theta),$$

where $\cos \phi = 5/\sqrt{61}$, $\sin \phi = 6/\sqrt{61}$ and $\cos \theta = 6/\sqrt{61}$, $\sin \theta = 5/\sqrt{61}$. From these equations we determine $\phi \approx 0.88$ (radians) and $\theta \approx 0.69$ (radians). Thus our solution is given approximately by

$$q = 0.77e^{-10t}\cos(50t - 0.88) - 0.64\cos(60t - 0.69).$$

Interpretation. The first term in the above solution clearly becomes negligible after a relatively short time; it is the *transient* term. After a sufficient time essentially all that remains is the periodic second term; this is the *steady-state term*. The graphs of these two components and that of their sum (the complete solution) are shown in Figure 5.16.

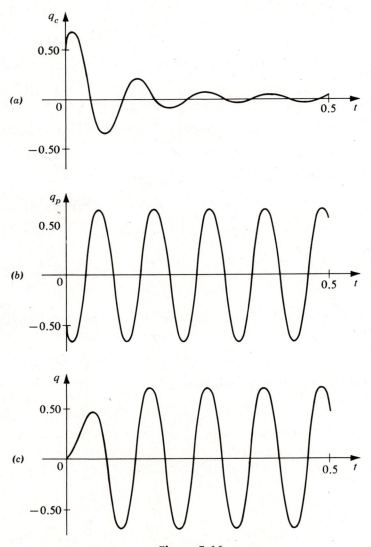

Figure 5.16

Exercises

1. A circuit has in series a constant electromotive force of 40 V, a resistor of 10 Ω, and an inductor of 0.2 H. If the initial current is 0, find the current at time $t > 0$.

2. Solve Exercise 1 if the electromotive force is given by $E(t) = 150 \cos 200t$ V instead of the constant electromotive force given in that problem.

3. A circuit has in series a constant electromotive force of 100 V, a resistor of 10 Ω, and a capacitor of 2×10^{-4} farads. The switch is closed at time $t = 0$, and the charge on the capacitor at this instant is zero. Find the charge and the current at time $t > 0$.

4. A circuit has in series an electromotive force given by $E(t) = 5 \sin 100t$ V, a resistor of 10 Ω, an inductor of 0.05 H, and a capacitor of 2×10^{-4} farads. If the initial current and the initial charge on the capacitor are both zero, find the charge on the capacitor at any time $t > 0$.

5. A circuit has in series an electromotive force given by $E(t) = 100 \sin 200t$ V, a resistor of 40 Ω, an inductor of 0.25 H, and a capacitor of 4×10^{-4} farads. If the initial current is zero, and the initial charge on the capacitor is 0.01 coulombs, find the current at any time $t > 0$.

6. A circuit has in series an electromotive force given by $E(t) = 200e^{-100t}$ V, a resistor of 80 Ω, an inductor of 0.2 H, and a capacitor of 5×10^{-6} farads. If the initial current and the initial charge on the capacitor are zero, find the current at any time $t > 0$.

7. A circuit has in series a resistor R Ω, and inductor of L H, and a capacitor of C farads. The initial current is zero and the initial charge on the capacitor is Q_0 coulombs.

    (a) Show that the charge and the current are damped oscillatory functions of time if and only if $R < 2\sqrt{L/C}$, and find the expressions for the charge and the current in this case.

    (b) If $R \geq 2\sqrt{L/C}$, discuss the nature of the charge and the current as functions of time.

8. A circuit has in series an electromotive force given by $E(t) = E_0 \sin \omega t$ V, a resistor of R Ω, an inductor of L H, and a capacitor of C farads.

    (a) Show that the steady-state current is

    $$ i = \frac{E_0}{Z}\left(\frac{R}{Z}\sin \omega t - \frac{X}{Z}\cos \omega t\right), $$

    where $X = L\omega - 1/C\omega$ and $Z = \sqrt{X^2 + R^2}$. The quantity X is called the *reactance* of the circuit and Z is called the *impedance*.

    (b) Using the result of part (a) show that the steady-state current may be written

    $$ i = \frac{E_0}{Z}\sin(\omega t - \phi), $$

    where ϕ is determined by the equations

    $$ \cos \phi = \frac{R}{Z}, \qquad \sin \phi = \frac{X}{Z}. $$

Thus show that the steady-state current attains its maximum absolute value E_0/Z at times $t_n + \phi/\omega$, where

$$t_n = \frac{1}{\omega}\left[\frac{(2n-1)\pi}{2}\right] \qquad (n = 1, 2, 3, \ldots),$$

are the times at which the electromotive force attains its maximum absolute value E_0.

(c) Show that the amplitude of the steady-state current is a maximum when

$$\omega = \frac{1}{\sqrt{LC}}.$$

For this value of ω *electrical resonance* is said to occur.

(d) If $R = 20$, $L = \frac{1}{4}$, $C = 10^{-4}$, and $E_0 = 100$, find the value of ω that gives rise to electrical resonance and determine the amplitude of the steady-state current in this case.

CHAPTER SIX

Series Solutions of Linear Differential Equations

In Chapter 4 we learned that certain types of higher-order linear differential equations (for example, those with constant coefficients) have solutions that can be expressed as finite linear combinations of known elementary functions. In general, however, higher-order linear equations have no solutions that can be expressed in such a simple manner. Thus we must seek other means of expression for the solutions of these equations. One such means of expression is furnished by infinite series representations, and the present chapter is devoted to methods of obtaining solutions in infinite series form.

6.1 POWER SERIES SOLUTIONS ABOUT AN ORDINARY POINT

A. Basic Concepts and Results

Consider the second-order homogeneous linear differential equation

$$a_0(x)\frac{d^2y}{dx^2} + a_1(x)\frac{dy}{dx} + a_2(x)y = 0, \tag{6.1}$$

and suppose that this equation has no solution that is expressible as a finite linear combination of known elementary functions. Let us assume, however, that it does have a solution that can be expressed in the form of an infinite series. Specifically, we assume that it has a solution expressible in the form

$$c_0 + c_1(x - x_0) + c_2(x - x_0)^2 + \cdots = \sum_{n=0}^{\infty} c_n(x - x_0)^n, \tag{6.2}$$

where $c_0, c_1, c_2, \ldots$ are constants. An expression of the form (6.2) is called a *power series* in $x - x_0$. We have thus assumed that the differential equation (6.1) has a so-called *power series solution* of the form (6.2). Assuming that this assumption is valid, we

221

can proceed to determine the coefficients $c_0, c_1, c_2, \ldots$ in (6.2) in such a manner that the expression (6.2) does indeed satisfy the Equation (6.1).

But under what conditions is this assumption actually valid? That is, under what conditions can we be certain that the differential equation (6.1) actually *does* have a solution of the form (6.2)? This is a question of considerable importance; for it would be quite absurd to actually try to find a "solution" of the form (6.2) if there were really no such solution to be found! In order to answer this important question concerning the existence of a solution of the form (6.2), we shall first introduce certain basic definitions. For this purpose let us write the differential equation (6.1) in the equivalent normalized form

$$\frac{d^2 y}{dx^2} + P_1(x) \frac{dy}{dx} + P_2(x) y = 0, \qquad (6.3)$$

where

$$P_1(x) = \frac{a_1(x)}{a_0(x)} \quad \text{and} \quad P_2(x) = \frac{a_2(x)}{a_0(x)}.$$

DEFINITION

A function f is said to be analytic *at x_0 if its Taylor series about x_0,*

$$\sum_{n=0}^{\infty} \frac{f^{(n)}(x_0)}{n!} (x - x_0)^n,$$

exists and converges to $f(x)$ for all x in some open interval including x_0.

We note that all polynomial functions are analytic everywhere; so also are the functions with values e^x, $\sin x$, and $\cos x$. A rational function is analytic except at those values of x at which its denominator is zero. For example, the rational function defined by $1/(x^2 - 3x + 2)$ is analytic except at $x = 1$ and $x = 2$.

DEFINITION

The point x_0 is called an ordinary point *of the differential equation (6.1) if both of the functions P_1 and P_2 in the equivalent normalized equation (6.3) are analytic at x_0. If either (or both) of these functions is* not *anaytic at x_0, then x_0 is called a* singular point *of the differential equation (6.1).*

▶ **Example 6.1**

Consider the differential equation

$$\frac{d^2 y}{dx^2} + x \frac{dy}{dx} + (x^2 + 2) y = 0. \qquad (6.4)$$

Here $P_1(x) = x$ and $P_2(x) = x^2 + 2$. Both of the functions P_1 and P_2 are polynomial functions and so they are analytic everywhere. Thus all points are ordinary points of this differential equation.

▶ **Example 6.2**

Consider the differential equation

$$(x - 1)\frac{d^2 y}{dx^2} + x\frac{dy}{dx} + \frac{1}{x} y = 0. \tag{6.5}$$

Equation (6.5) has not been written in the normalized form (6.3). We must first express (6.5) in the normalized form, thereby obtaining

$$\frac{d^2 y}{dx^2} + \frac{x}{x - 1}\frac{dy}{dx} + \frac{1}{x(x - 1)} y = 0.$$

Here

$$P_1(x) = \frac{x}{x - 1} \quad \text{and} \quad P_2(x) = \frac{1}{x(x - 1)}.$$

The function P_1 is analytic, except at $x = 1$, and P_2 is analytic except at $x = 0$ and $x = 1$. Thus $x = 0$ and $x = 1$ are singular points of the differential equation under consideration. All other points are ordinary points. Note clearly that $x = 0$ is a singular point, even though P_1 is analytic at $x = 0$. We mention this fact to emphasize that *both* P_1 and P_2 must be analytic at x_0 in order for x_0 to be an ordinary point.

We are now in a position to state a theorem concerning the existence of power series solutions of the form (6.2).

THEOREM 6.1

Hypothesis. *The point x_0 is an ordinary point of the differential equation (6.1).*

Conclusion. *The differential equation (6.1) has two nontrivial linearly independent power series solutions of the form*

$$\sum_{n=0}^{\infty} c_n(x - x_0)^n, \tag{6.2}$$

and these power series converge in some interval $|x - x_0| < R$ (where $R > 0$) about x_0.

This theorem gives us a sufficient condition for the existence of power series solutions of the differential equation (6.1). It states that if x_0 is an ordinary point of equation (6.1), then this equation has *two* power series solutions in powers of $x - x_0$ and that these two power series solutions are *linearly independent*. Thus if x_0 is an ordinary point of (6.1), we may obtain the *general solution* of (6.1) as a linear combination of these two linearly independent power series. We shall omit the proof of this important theorem.

▶ **Example 6.3**

In Example 6.1 we noted that all points are ordinary points of the differential equation (6.4). Thus this differential equation has two linearly independent solutions of the form (6.2) about *any* point x_0.

▶ **Example 6.4**

In Example 6.2 we observed that $x = 0$ and $x = 1$ are the only singular points of the differential equation (6.5). Thus this differential equation has two linearly independent solutions of the form (6.2) about any point $x_0 \neq 0$ or 1. For example, the equation has two linearly independent solutions of the form

$$\sum_{n=0}^{\infty} c_n(x - 2)^n$$

about the ordinary point 2. However, we are *not* assured that there exists any solution of the form

$$\sum_{n=0}^{\infty} c_n x^n$$

about the singular point 0 or any solution of the form

$$\sum_{n=0}^{\infty} c_n(x - 1)^n$$

about the singular point 1.

B. The Method of Solution

Now that we are assured that under appropriate hypotheses Equation (6.1) actually does have power series solutions of the form (6.2), how do we proceed to find these solutions? In other words, how do we determine the coefficients $c_0, c_1, c_2, \ldots$ in the expression

$$\sum_{n=0}^{\infty} c_n(x - x_0)^n \tag{6.2}$$

so that this expression actually does satisfy Equation (6.1)? We shall first give a brief outline of the procedure for finding these coefficients and shall then illustrate the procedure in detail by considering specific examples.

Assuming that x_0 is an ordinary point of the differential equation (6.1), so that solutions in powers of $x - x_0$ actually do exist, we denote such a solution by

$$y = c_0 + c_1(x - x_0) + c_2(x - x_0)^2 + \cdots = \sum_{n=0}^{\infty} c_n(x - x_0)^n. \tag{6.6}$$

Since the series in (6.6) converges on an interval $|x - x_0| < R$ about x_0, it may be differentiated term by term on this interval twice in succession to obtain

$$\frac{dy}{dx} = c_1 + 2c_2(x - x_0) + 3c_3(x - x_0)^2 + \cdots = \sum_{n=1}^{\infty} nc_n(x - x_0)^{n-1} \tag{6.7}$$

and

$$\frac{d^2y}{dx^2} = 2c_2 + 6c_3(x - x_0) + 12c_4(x - x_0)^2 + \cdots = \sum_{n=2}^{\infty} n(n-1)c_n(x - x_0)^{n-2},$$

(6.8)

respectively. We now substitute the series in the right members of (6.6), (6.7), and (6.8) for y and its first two derivatives, respectively, in the differential equation (6.1). We then simplify the resulting expression so that it takes the form

$$K_0 + K_1(x - x_0) + K_2(x - x_0)^2 + \cdots = 0,$$

(6.9)

where the coefficients $K_i(i = 0, 1, 2, \ldots)$ are functions of certain coefficients c_n of the solution (6.6). In order that (6.9) be valid for all x in the interval of convergence $|x - x_0| < R$, we must set

$$K_0 = K_1 = K_2 = \cdots = 0.$$

In other words, we must equate to zero the coefficient of each power of $x - x_0$ in the left member of (6.9). This leads to a set of conditions that must be satisfied by the various coefficients c_n in the series (6.6) in order that (6.6) be a solution of the differential equation (6.1). If the c_n are chosen to satisfy the set of conditions that thus occurs, then the resulting series (6.6) is the desired solution of the differential equation (6.1). We shall illustrate this procedure in detail in the two examples which follow.

▶ **Example 6.5**

Find the power series solution of the differential equation

$$\frac{d^2y}{dx^2} + x\frac{dy}{dx} + (x^2 + 2)y = 0$$

(6.4)

in powers of x (that is, about $x_0 = 0$).

Solution. We have already observed that $x_0 = 0$ is an ordinary point of the differential equation (6.4) and that two linearly independent solutions of the desired type actually exist. Our procedure will yield both of these solutions at once.

We thus assume a solution of the form (6.6) with $x_0 = 0$. That is, we assume

$$y = \sum_{n=0}^{\infty} c_n x^n.$$

(6.10)

Differentiating term by term we obtain

$$\frac{dy}{dx} = \sum_{n=1}^{\infty} nc_n x^{n-1}$$

(6.11)

and

$$\frac{d^2y}{dx^2} = \sum_{n=2}^{\infty} n(n-1)c_n x^{n-2}.$$

(6.12)

Substituting the series (6.10), (6.11), and (6.12) into the differential equation (6.4), we obtain

$$\sum_{n=2}^{\infty} n(n-1)c_n x^{n-2} + x \sum_{n=1}^{\infty} nc_n x^{n-1} + x^2 \sum_{n=0}^{\infty} c_n x^n + 2 \sum_{n=0}^{\infty} c_n x^n = 0.$$

Since x is independent of the index of summation n, we may rewrite this as

$$\sum_{n=2}^{\infty} n(n-1)c_n x^{n-2} + \sum_{n=1}^{\infty} nc_n x^n + \sum_{n=0}^{\infty} c_n x^{n+2} + 2 \sum_{n=0}^{\infty} c_n x^n = 0. \qquad (6.13)$$

In order to put the left member of Equation (6.13) in the form (6.9), we shall rewrite the first and third summations in (6.13) so that x in each of these summations will have the exponent n. Let us consider the first summation

$$\sum_{n=2}^{\infty} n(n-1)c_n x^{n-2} \qquad (6.14)$$

in (6.13). To rewrite the summation (6.14) so that x will have the desired exponent n, we first replace the present exponent $n-2$ in (6.14) by a new variable m. That is, we let $m = n - 2$ in (6.14). Then $n = m + 2$, and since $m = 0$ for $n = 2$, the summation (6.14) takes the form

$$\sum_{m=0}^{\infty} (m+2)(m+1)c_{m+2} x^m. \qquad (6.15)$$

Now since the variable of summation is merely a "dummy" variable, we may replace m by n in (6.15) to write the first summation in (6.13) as

$$\sum_{n=0}^{\infty} (n+2)(n+1)c_{n+2} x^n. \qquad (6.16)$$

In like manner, letting $m = n + 2$, the third summation

$$\sum_{n=0}^{\infty} c_n x^{n+2} \qquad (6.17)$$

in (6.13) first takes the form

$$\sum_{m=2}^{\infty} c_{m-2} x^m. \qquad (6.18)$$

Then replacing m by n in (6.18), the third summation in (6.13) may be written as

$$\sum_{n=2}^{\infty} c_{n-2} x^n. \qquad (6.19)$$

Thus replacing (6.14) by its equivalent (6.16) and (6.17) by its equivalent (6.19), Equation (6.13) may be written

$$\sum_{n=0}^{\infty} (n+2)(n+1)c_{n+2} x^n + \sum_{n=1}^{\infty} nc_n x^n + \sum_{n=2}^{\infty} c_{n-2} x^n + 2 \sum_{n=0}^{\infty} c_n x^n = 0. \qquad (6.20)$$

Although x has the same exponent n in each summation in (6.20), the ranges of the various summations are not all the same. In the first and fourth summations n ranges from 0 to ∞, in the second n ranges from 1 to ∞, and in the third the range is from 2 to ∞. The common range is from 2 to ∞. We now write out individually the terms in each summation that do *not* belong to this common range, and we continue to employ the "sigma" notation to denote the remainder of each such summation. For example, in the first summation

$$\sum_{n=0}^{\infty} (n + 2)(n + 1)c_{n+2} x^n$$

of (6.20) we write out individually the terms corresponding to $n = 0$ and $n = 1$ and denote the remainder of this summation by

$$\sum_{n=2}^{\infty} (n + 2)(n + 1)c_{n+2} x^n.$$

We thus rewrite

$$\sum_{n=0}^{\infty} (n + 2)(n + 1)c_{n+2} x^n$$

in (6.20) as

$$2c_2 + 6c_3 x + \sum_{n=2}^{\infty} (n + 2)(n + 1)c_{n+2} x^n.$$

In like manner, we write

$$\sum_{n=1}^{\infty} nc_n x^n$$

in (6.20) as

$$c_1 x + \sum_{n=2}^{\infty} nc_n x^n$$

and

$$2 \sum_{n=0}^{\infty} c_n x^n$$

in (6.20) as

$$2c_0 + 2c_1 x + 2 \sum_{n=2}^{\infty} c_n x^n.$$

Thus Equation (6.20) is now written as

$$2c_2 + 6c_3 x + \sum_{n=2}^{\infty} (n + 2)(n + 1)c_{n+2} x^n + c_1 x + \sum_{n=2}^{\infty} nc_n x^n$$

$$+ \sum_{n=2}^{\infty} c_{n-2} x^n + 2c_0 + 2c_1 x + 2 \sum_{n=2}^{\infty} c_n x^n = 0.$$

We can now combine like powers of x and write this equation as

$$(2c_0 + 2c_2) + (3c_1 + 6c_3)x + \sum_{n=2}^{\infty} [(n + 2)(n + 1)c_{n+2} + (n + 2)c_n + c_{n-2}]x^n = 0.$$

$$(6.21)$$

Equation (6.21) is in the desired form (6.9). For (6.21) to be valid for all x in the interval of convergence $|x - x_0| < R$, the coefficient of each power of x in the left member of (6.21) must be equated to zero. This leads immediately to the conditions

$$2c_0 + 2c_2 = 0, \tag{6.22}$$

$$3c_1 + 6c_3 = 0, \tag{6.23}$$

and

$$(n + 2)(n + 1)c_{n+2} + (n + 2)c_n + c_{n-2} = 0, \qquad n \geq 2. \tag{6.24}$$

The condition (6.22) enables us to express c_2 in terms of c_0. Doing so, we find that

$$c_2 = -c_0. \tag{6.25}$$

The condition (6.23) enables us to express c_3 in terms of c_1. This leads to

$$c_3 = -\tfrac{1}{2}c_1. \tag{6.26}$$

The condition (6.24) is called a *recurrence formula*. It enables us to express each coefficient c_{n+2} for $n \geq 2$ in terms of the previous coefficients c_n and c_{n-2}, thus giving

$$c_{n+2} = -\frac{(n + 2)c_n + c_{n-2}}{(n + 1)(n + 2)}, \qquad n \geq 2. \tag{6.27}$$

For $n = 2$, formula (6.27) is

$$c_4 = -\frac{4c_2 + c_0}{12}.$$

Now using (6.25), this reduces to

$$c_4 = \tfrac{1}{4}c_0, \tag{6.28}$$

which expresses c_4 in terms of c_0. For $n = 3$, formula (6.27) is

$$c_5 = -\frac{5c_3 + c_1}{20}.$$

Now using (6.26), this reduces to

$$c_5 = \tfrac{3}{40}c_1, \tag{6.29}$$

which expresses c_5 in terms of c_1. In the same way we may express each even coefficient in terms of c_0 and each odd coefficient in terms of c_1.

Substituting the values of c_2, c_3, c_4, and c_5, given by (6.25), (6.26), (6.28), and (6.29), respectively, into the assumed solution (6.10), we have

$$y = c_0 + c_1 x - c_0 x^2 - \tfrac{1}{2}c_1 x^3 + \tfrac{1}{4}c_0 x^4 + \tfrac{3}{40}c_1 x^5 + \cdots.$$

Collecting terms in c_0 and c_1, we have finally

$$y = c_0(1 - x^2 + \tfrac{1}{4}x^4 + \cdots) + c_1(x - \tfrac{1}{2}x^3 + \tfrac{3}{40}x^5 + \cdots), \tag{6.30}$$

which gives the solution of the differential equation (6.4) in powers of x through terms in x^5. The two series in parentheses in (6.30) are the power series expansions of two linearly independent solutions of (6.4), and the constants c_0 and c_1 are arbitrary constants. Thus (6.30) represents the general solution of (6.4) in powers of x (through terms in x^5).

▶ **Example 6.6**

Find a power series solution of the initial-vaiue problem

$$(x^2 - 1)\frac{d^2y}{dx^2} + 3x\frac{dy}{dx} + xy = 0, \tag{6.31}$$

$$y(0) = 4, \tag{6.32}$$

$$y'(0) = 6. \tag{6.33}$$

Solution. We first observe that all points except $x = \pm 1$ are ordinary points for the differential equation (6.31). Thus we could assume solutions of the form (6.6) for any $x_0 \neq \pm 1$. However, since the initial values of y and its first derivative are prescribed at $x = 0$, we shall choose $x_0 = 0$ and seek solutions in powers of x. Thus we assume

$$y = \sum_{n=0}^{\infty} c_n x^n. \tag{6.34}$$

Differentiating term by term we obtain

$$\frac{dy}{dx} = \sum_{n=1}^{\infty} nc_n x^{n-1} \tag{6.35}$$

and

$$\frac{d^2y}{dx^2} = \sum_{n=2}^{\infty} n(n-1)c_n x^{n-2}. \tag{6.36}$$

Substituting the series (6.34), (6.35), and (6.36) into the differential equation (6.31), we obtain

$$\sum_{n=2}^{\infty} n(n-1)c_n x^n - \sum_{n=2}^{\infty} n(n-1)c_n x^{n-2} + 3\sum_{n=1}^{\infty} nc_n x^n + \sum_{n=0}^{\infty} c_n x^{n+1} = 0. \tag{6.37}$$

We now rewrite the second and fourth summations in (6.37) so that x in each of these summations has the exponent n. Doing this, Equation (6.37) takes the form

$$\sum_{n=2}^{\infty} n(n-1)c_n x^n - \sum_{n=0}^{\infty} (n+2)(n+1)c_{n+2} x^n + 3\sum_{n=1}^{\infty} nc_n x^n + \sum_{n=1}^{\infty} c_{n-1} x^n = 0. \tag{6.38}$$

The common range of the four summations in (6.38) is from 2 to ∞. We can write out the individual terms in each summation that do *not* belong to this common range and thus express (6.38) in the form

$$\sum_{n=2}^{\infty} n(n-1)c_n x^n - 2c_2 - 6c_3 x - \sum_{n=2}^{\infty} (n+2)(n+1)c_{n+2} x^n$$

$$+ 3c_1 x + 3\sum_{n=2}^{\infty} nc_n x^n + c_0 x + \sum_{n=2}^{\infty} c_{n-1} x^n = 0.$$

Combining like powers of x, this takes the form

$$-2c_2 + (c_0 + 3c_1 - 6c_3)x$$

$$+ \sum_{n=2}^{\infty} [-(n+2)(n+1)c_{n+2} + n(n+2)c_n + c_{n-1}]x^n = 0. \tag{6.39}$$

For (6.39) to be valid for all x in the interval of convergence $|x - x_0| < R$, the coefficient of each power of x in the left member of (6.39) must be equated to zero. In doing this, we are led to the relations

$$-2c_2 = 0, \tag{6.40}$$

$$c_0 + 3c_1 - 6c_3 = 0, \tag{6.41}$$

and

$$-(n + 2)(n + 1)c_{n+2} + n(n + 2)c_n + c_{n-1} = 0, \qquad n \geq 2. \tag{6.42}$$

From (6.40), we find that $c_2 = 0$; and from (6.41), $c_3 = \frac{1}{6}c_0 + \frac{1}{2}c_1$. The recurrence formula (6.42) gives

$$c_{n+2} = \frac{n(n + 2)c_n + c_{n-1}}{(n + 1)(n + 2)}, \qquad n \geq 2.$$

Using this, we find successively

$$c_4 = \frac{8c_2 + c_1}{12} = \frac{1}{12}c_1,$$

$$c_5 = \frac{15c_3 + c_2}{20} = \frac{1}{8}c_0 + \frac{3}{8}c_1.$$

Substituting these values of $c_2, c_3, c_4, c_5, \ldots$ into the assumed solution (6.34), we have

$$y = c_0 + c_1 x + \left(\frac{c_0}{6} + \frac{c_1}{2}\right)x^3 + \frac{c_1}{12} x^4 + \left(\frac{c_0}{8} + \frac{3c_1}{8}\right)x^5 + \cdots$$

or

$$y = c_0(1 + \tfrac{1}{6}x^3 + \tfrac{1}{8}x^5 + \cdots) + c_1(x + \tfrac{1}{2}x^3 + \tfrac{1}{12}x^4 + \tfrac{3}{8}x^5 + \cdots). \tag{6.43}$$

The solution (6.43) is the general solution of the differential equation (6.31) in powers of x (through terms in x^5).

We must now apply the initial conditions (6.32) and (6.33). Applying (6.32) to (6.43), we immediately find that

$$c_0 = 4.$$

Differentiating (6.43), we have

$$\frac{dy}{dx} = c_0(\tfrac{1}{2}x^2 + \tfrac{5}{8}x^4 + \cdots) + c_1(1 + \tfrac{3}{2}x^2 + \tfrac{1}{3}x^3 + \tfrac{15}{8}x^4 + \cdots). \tag{6.44}$$

Applying (6.33) to (6.44) we find that

$$c_1 = 6.$$

Thus the solution of the given initial-value problem in powers of x (through terms in x^5) is

$$y = 4(1 + \tfrac{1}{6}x^3 + \tfrac{1}{8}x^5 + \cdots) + 6(x + \tfrac{1}{2}x^3 + \tfrac{1}{12}x^4 + \tfrac{3}{8}x^5 + \cdots)$$

or

$$y = 4 + 6x + \tfrac{11}{3}x^3 + \tfrac{1}{2}x^4 + \tfrac{11}{4}x^5 + \cdots.$$

Remark 1. Suppose the initial values of y and its first derivative in conditions (6.32) and (6.33) of Example 6.6 are prescribed at $x = 2$, instead of $x = 0$. Then we have the initial-value problem

$$(x^2 - 1)\frac{d^2 y}{dx^2} + 3x\frac{dy}{dx} + xy = 0,$$

$$y(2) = 4, \qquad y'(2) = 6.$$

$$(6.45)$$

Since the initial values in this problem are prescribed at $x = 2$, we would seek solutions in powers of $x - 2$. That is, in this case we would seek solutions of the form

$$y = \sum_{n=0}^{\infty} c_n(x - 2)^n.$$

$$(6.46)$$

The simplest procedure for obtaining a solution of the form (6.46) is first to make the substitution $t = x - 2$. This replaces the initial-value problem (6.45) by the equivalent problem

$$(t^2 + 4t + 3)\frac{d^2 y}{dt^2} + (3t + 6)\frac{dy}{dt} + (t + 2)y = 0,$$

$$y(0) = 4, \qquad y'(0) = 6,$$

$$(6.47)$$

in which t is the independent variable and the initial values are prescribed at $t = 0$. One then seeks a solution of the problem (6.47) in powers of t,

$$y = \sum_{n=0}^{\infty} c_n t^n.$$

$$(6.48)$$

Differentiating (6.48) and substituting into the differential equation in (6.47), one determines the c_n as in Examples 6.5 and 6.6. The initial conditions in (6.47) are then applied. Replacing t by $x - 2$ in the resulting solution (6.48), one obtains the desired solution (6.46) of the original problem (6.45).

Remark 2. In Examples 6.5 and 6.6 we obtained power series solutions of the differential equations under consideration but made no attempt to discuss the convergence of these solutions. According to Theorem 6.1, if x_0 is an ordinary point of the differential equation

$$a_0(x)\frac{d^2 y}{dx^2} + a_1(x)\frac{dy}{dx} + a_2(x)y = 0,$$

$$(6.1)$$

then the power series solutions of the form (6.2) converge in some interval $|x - x_0| < R$ (where $R > 0$) about x_0. Let us again write (6.1) in the normalized form

$$\frac{d^2 y}{dx^2} + P_1(x)\frac{dy}{dx} + P_2(x)y = 0,$$

$$(6.3)$$

where

$$P_1(x) = \frac{a_1(x)}{a_0(x)} \quad \text{and} \quad P_2(x) = \frac{a_2(x)}{a_0(x)}.$$

If x_0 is an ordinary point of (6.1), the functions P_1 and P_2 have Taylor series expansions about x_0 that converge in intervals $|x - x_0| < R_1$ and $|x - x_0| < R_2$, respectively, about x_0. It can be proved that the interval of convergence $|x - x_0| < R$ of a series solution (6.2) of (6.1) is at least as great as the smaller of the intervals $|x - x_0| < R_1$ and $|x - x_0| < R_2$.

In the differential equation (6.4) of Example 6.5, $P_1(x) = x$ and $P_2(x) = x^2 + 2$. Thus in this example the Taylor series expansions for P_1 and P_2 about the ordinary point $x_0 = 0$ converge for all x. Hence the series solutions (6.30) of (6.4) also converge for all x.

In the differential equation (6.31) of Example 6.6,

$$P_1(x) = \frac{3x}{x^2 - 1} \quad \text{and} \quad P_2(x) = \frac{x}{x^2 - 1}.$$

In this example the Taylor series for P_1 and P_2 about $x_0 = 0$ both converge for $|x| < 1$. Thus the solutions (6.43) of (6.31) converge at least for $|x| < 1$.

Exercises

Find power series solutions in powers of x of each of the differential equations in Exercises 1–10.

1. $\dfrac{d^2 y}{dx^2} + x \dfrac{dy}{dx} + y = 0.$

2. $\dfrac{d^2 y}{dx^2} + 8x \dfrac{dy}{dx} - 4y = 0.$

3. $\dfrac{d^2 y}{dx^2} + x \dfrac{dy}{dx} + (2x^2 + 1)y = 0.$

4. $\dfrac{d^2 y}{dx^2} + x \dfrac{dy}{dx} + (x^2 - 4)y = 0.$

5. $\dfrac{d^2 y}{dx^2} + x \dfrac{dy}{dx} + (3x + 2)y = 0.$

6. $\dfrac{d^2 y}{dx^2} - x \dfrac{dy}{dx} + (3x - 2)y = 0.$

7. $(x^2 + 1)\dfrac{d^2 y}{dx^2} + x \dfrac{dy}{dx} + xy = 0.$

8. $(x - 1)\dfrac{d^2 y}{dx^2} - (3x - 2)\dfrac{dy}{dx} + 2xy = 0.$

9. $(x^3 - 1)\dfrac{d^2 y}{dx^2} + x^2 \dfrac{dy}{dx} + xy = 0.$

10. $(x + 3)\dfrac{d^2 y}{dx^2} + (x + 2)\dfrac{dy}{dx} + y = 0.$

Find the power series solution of each of the initial-value problems in Exercises 11–14.

11. $\dfrac{d^2 y}{dx^2} - x \dfrac{dy}{dx} - y = 0, \quad y(0) = 1, \quad y'(0) = 0.$

12. $\dfrac{d^2 y}{dx^2} + x \dfrac{dy}{dx} - 2y = 0, \quad y(0) = 0, \quad y'(0) = 1.$

13. $(x^2 + 1)\dfrac{d^2 y}{dx^2} + x \dfrac{dy}{dx} + 2xy = 0, \quad y(0) = 2, \quad y'(0) = 3.$

14. $(2x^2 - 3)\dfrac{d^2 y}{dx^2} - 2x \dfrac{dy}{dx} + y = 0, \quad y(0) = -1, \quad y'(0) = 5.$

Find power series solutions in powers of $x - 1$ of each of the differential equations in Exercises 15 and 16.

15. $x^2 \dfrac{d^2 y}{dx^2} + x \dfrac{dy}{dx} + y = 0.$

16. $x^2 \dfrac{d^2 y}{dx^2} + 3x \dfrac{dy}{dx} - y = 0.$

17. Find the power series solution in powers of $x - 1$ of the initial-value problem

$$x \frac{d^2 y}{dx^2} + \frac{dy}{dx} + 2y = 0, \qquad y(1) = 2, \qquad y'(1) = 4.$$

18. The differential equation

$$(1 - x^2) \frac{d^2 y}{dx^2} - 2x \frac{dy}{dx} + n(n + 1)y = 0,$$

where n is a constant, is called *Legendre's differential equation.*

(a) Show that $x = 0$ is an ordinary point of this differential equation, and find two linearly independent power series solutions in powers of x.

(b) Show that if n is a nonnegative integer, then one of the two solutions found in part (a) is a polynomial of degree n.

6.2 SOLUTIONS ABOUT SINGULAR POINTS; THE METHOD OF FROBENIUS

A. Regular Singular Points

We again consider the homogeneous linear differential equation

$$a_0(x) \frac{d^2 y}{dx^2} + a_1(x) \frac{dy}{dx} + a_2(x)y = 0, \tag{6.1}$$

and we assume that x_0 is a *singular* point of (6.1). Then Theorem 6.1 does *not* apply at the point x_0, and we are *not* assured of a power series solution

$$y = \sum_{n=0}^{\infty} c_n (x - x_0)^n \tag{6.2}$$

of (6.1) in powers of $x - x_0$. Indeed an equation of the form (6.1) with a singular point at x_0 does *not*, in general, have a solution of the form (6.2). Clearly we must seek a different type of solution in such a case, but what type of solution can we expect? It happens that under certain conditions we are justified in assuming a solution of the form

$$y = |x - x_0|^r \sum_{n=0}^{\infty} c_n (x - x_0)^n, \tag{6.49}$$

where r is a certain (real or complex) constant. Such a solution is clearly a power series in $x - x_0$ multiplied by a certain *power* of $|x - x_0|$. In order to state conditions under which a solution of this form is assured, we proceed to classify singular points.

We again write the differential equation (6.1) in the equivalent normalized form

$$\frac{d^2 y}{dx^2} + P_1(x)\frac{dy}{dx} + P_2(x)y = 0, \tag{6.3}$$

where

$$P_1(x) = \frac{a_1(x)}{a_0(x)} \quad \text{and} \quad P_2(x) = \frac{a_2(x)}{a_0(x)}.$$

DEFINITION

Consider the differential equation (6.1), and assume that at least one of the functions P_1 and P_2 in the equivalent normalized equation (6.3) is not analytic at x_0, so that x_0 is a singular point of (6.1). If the functions defined by the products

$$(x - x_0)P_1(x) \quad \text{and} \quad (x - x_0)^2 P_2(x) \tag{6.50}$$

are both analytic at x_0, then x_0 is called a regular singular point *of the differential equation (6.1). If either (or both) of the functions defined by the products (6.50) is not analytic at x_0, then x_0 is called an* irregular singular point *of (6.1).*

▶ **Example 6.7**

Consider the differential equation

$$2x^2 \frac{d^2 y}{dx^2} - x\frac{dy}{dx} + (x - 5)y = 0. \tag{6.51}$$

Writing this in the normalized form (6.3), we have

$$\frac{d^2 y}{dx^2} - \frac{1}{2x}\frac{dy}{dx} + \frac{x - 5}{2x^2}y = 0.$$

Here $P_1(x) = -1/2x$ and $P_2(x) = (x - 5)/2x^2$. Since both P_1 and P_2 fail to be analytic at $x = 0$, we conclude that $x = 0$ is a singular point of (6.51). We now consider the functions defined by the products

$$xP_1(x) = -\frac{1}{2} \quad \text{and} \quad x^2 P_2(x) = \frac{x - 5}{2}$$

of the form (6.50). Both of these product functions are analytic at $x = 0$, and so $x = 0$ is a *regular* singular point of the differential equation (6.51).

▶ **Example 6.8**

Consider the differential equation

$$x^2(x - 2)^2 \frac{d^2 y}{dx^2} + 2(x - 2)\frac{dy}{dx} + (x + 1)y = 0. \tag{6.52}$$

In the normalized form (6.3), this is

$$\frac{d^2y}{dx^2} + \frac{2}{x^2(x-2)}\frac{dy}{dx} + \frac{x+1}{x^2(x-2)^2}y = 0.$$

Here

$$P_1(x) = \frac{2}{x^2(x-2)} \quad \text{and} \quad P_2(x) = \frac{x+1}{x^2(x-2)^2}.$$

Clearly the singular points of the differential equation (6.52) are $x = 0$ and $x = 2$. We investigate them one at a time.

Consider $x = 0$ first, and form the functions defined by the products

$$xP_1(x) = \frac{2}{x(x-2)} \quad \text{and} \quad x^2P_2(x) = \frac{x+1}{(x-2)^2}$$

of the form (6.50). The product function defined by $x^2P_2(x)$ is analytic at $x = 0$, but that defined by $xP_1(x)$ is *not*. Thus $x = 0$ is an *irregular* singular point of (6.52).

Now consider $x = 2$. Forming the products (6.50) for this point, we have

$$(x-2)P_1(x) = \frac{2}{x^2} \quad \text{and} \quad (x-2)^2P_2(x) = \frac{x+1}{x^2}.$$

Both of the product functions thus defined are analytic at $x = 2$, and hence $x = 2$ is a *regular* singular point of (6.52).

Now that we can distinguish between regular and irregular singular points, we shall state a basic theorem concerning solutions of the form (6.49) about regular singular points. We shall later give a more complete theorem on this topic.

THEOREM 6.2

Hypothesis. *The point x_0 is a regular singular point of the differential equation (6.1).*

Conclusion. *The differential equation (6.1) has at least one nontrivial solution of the form*

$$|x - x_0|^r \sum_{n=0}^{\infty} c_n(x - x_0)^n, \tag{6.49}$$

where r is a definite (real or complex) constant which may be determined, and this solution is valid in some deleted interval $0 < |x - x_0| < R$ (where $R > 0$) about x_0.

▶ **Example 6.9**

In Example 6.7 we saw that $x = 0$ is a regular singular point of the differential equation

$$2x^2\frac{d^2y}{dx^2} - x\frac{dy}{dx} + (x-5)y = 0. \tag{6.51}$$

By Theorem 6.2 we conclude that this equation has at least one nontrivial solution of the form

$$|x|^r \sum_{n=0}^{\infty} c_n x^n,$$

valid in some deleted interval $0 < |x| < R$ about $x = 0$.

▶ **Example 6.10**

In Example 6.8 we saw that $x = 2$ is a regular singular point of the differential equation

$$x^2(x-2)^2 \frac{d^2y}{dx^2} + 2(x-2)\frac{dy}{dx} + (x+1)y = 0. \tag{6.52}$$

Thus we know that this equation has at least one nontrivial solution of the form

$$|x-2|^r \sum_{n=0}^{\infty} c_n(x-2)^n,$$

valid in some deleted interval $0 < |x - 2| < R$ about $x = 2$.

We also observed that $x = 0$ is a singular point of Equation (6.52). However, this singular point is irregular and so Theorem 6.2 does not apply to it. We are *not* assured that the differential equation (6.52) has a solution of the form

$$|x|^r \sum_{n=0}^{\infty} c_n x^n$$

in any deleted interval about $x = 0$.

B. The Method of Frobenius

Now that we are assured of at least one solution of the form (6.49) about a regular singular point x_0 of the differential equation (6.1), how do we proceed to determine the coefficients c_n and the number r in this solution? The procedure is similar to that introduced in Section 6.1 and is commonly called the *method of Frobenius*. We shall briefly outline the method and then illustrate it by applying it to the differential equation (6.51). In this outline and the illustrative example that follows we shall seek solutions valid in some interval $0 < x - x_0 < R$. Note that for all such x, $|x - x_0|$ is simply $x - x_0$. To obtain solutions valid for $-R < x - x_0 < 0$, simply replace $x - x_0$ by $-(x - x_0) > 0$ and proceed as in the outline.

Outline of the Method of Frobenius

1. Let x_0 be a regular singular point of the differential equation (6.1), seek solutions valid in some interval $0 < x - x_0 < R$, and assume a solution

$$y = (x - x_0)^r \sum_{n=0}^{\infty} c_n(x - x_0)^n$$

of the form (6.49), where $c_0 \neq 0$. We write this solution in the form

$$y = \sum_{n=0}^{\infty} c_n(x - x_0)^{n+r}, \tag{6.53}$$

where $c_0 \neq 0$.

2. Assuming term-by-term differentiation of (6.53) is valid, we obtain

$$\frac{dy}{dx} = \sum_{n=0}^{\infty} (n + r)c_n(x - x_0)^{n+r-1} \tag{6.54}$$

and

$$\frac{d^2y}{dx^2} = \sum_{n=0}^{\infty} (n + r)(n + r - 1)c_n(x - x_0)^{n+r-2}. \tag{6.55}$$

We now substitute the series (6.53), (6.54), and (6.55) for y and its first two derivatives, respectively, into the differential equation (6.1).

3. We now proceed (essentially as in Section 6.1) to simplify the resulting expression so that it takes the form

$$K_0(x - x_0)^{r+k} + K_1(x - x_0)^{r+k+1} + K_2(x - x_0)^{r+k+2} + \cdots = 0, \tag{6.56}$$

where k is a certain integer and the coefficients $K_i(i = 0, 1, 2, \ldots)$ are functions of r and certain of the coefficients c_n of the solution (6.53).

4. In order that (6.56) be valid for all x in the deleted interval $0 < x - x_0 < R$, we must set

$$K_0 = K_1 = K_2 = \cdots = 0.$$

5. Upon equating to zero the coefficient K_0 of the *lowest* power $r + k$ of $(x - x_0)$, we obtain a quadratic equation in r, called the *indicial equation* of the differential equation (6.1). The two roots of this quadratic equation are often called the *exponents* of the differential equation (6.1) and are the only possible values for the constant r in the assumed solution (6.53). Thus at this stage the "unknown" constant r is determined. We denote the roots of the indicial equation by r_1 and r_2, where $\text{Re}(r_1) \geq \text{Re}(r_2)$. Here $\text{Re}(r_j)$ denotes the real part of $r_j(j = 1, 2)$; and of course if r_j is real, then $\text{Re}(r_j)$ is simply r_j itself.

6. We now equate to zero the remaining coefficients $K_1, K_2, \ldots$ in (6.56). We are thus led to a set of conditions, involving the constant r, which must be satisfied by the various coefficients c_n in the series (6.53).

7. We now substitute the root r_1 for r into the conditions obtained in Step 6, and then choose the c_n to satisfy these conditions. If the c_n are so chosen, the resulting series (6.53) with $r = r_1$ is a solution of the desired form. Note that if r_1 and r_2 are real and unequal, then r_1 is the *larger* root.

8. If $r_2 \neq r_1$, we may repeat the procedure of Step 7 using the root r_2 instead of r_1. In this way a second solution of the desired form (6.53) may be obtained. Note that if r_1 and r_2 are real and unequal, then r_2 is the *smaller* root. However, in the case in which r_1 and r_2 are real and unequal, the second solution of the desired form (6.53) obtained in this step may not be linearly independent of the solution obtained in Step 7. Also, in the case in which r_1 and r_2 are real and equal, the solution obtained in this step is clearly identical with the one obtained in Step 7. We shall consider these "exceptional" situations after we have considered an example.

▶ **Example 6.11**

Use the method of Frobenius to find solutions of the differential equation

$$2x^2 \frac{d^2y}{dx^2} - x \frac{dy}{dx} + (x - 5)y = 0 \tag{6.51}$$

in some interval $0 < x < R$.

Solution. Since $x = 0$ is a regular singular point of the differential equation (6.51) and we seek solution for $0 < x < R$, we assume

$$y = \sum_{n=0}^{\infty} c_n x^{n+r}, \tag{6.57}$$

where $c_0 \neq 0$. Then

$$\frac{dy}{dx} = \sum_{n=0}^{\infty} (n + r)c_n x^{n+r-1} \tag{6.58}$$

and

$$\frac{d^2y}{dx^2} = \sum_{n=0}^{\infty} (n + r)(n + r - 1)c_n x^{n+r-2}. \tag{6.59}$$

Substituting the series (6.57), (6.58), and (6.59) into (6.51), we have

$$2 \sum_{n=0}^{\infty} (n + r)(n + r - 1)c_n x^{n+r} - \sum_{n=0}^{\infty} (n + r)c_n x^{n+r} + \sum_{n=0}^{\infty} c_n x^{n+r+1} - 5 \sum_{n=0}^{\infty} c_n x^{n+r} = 0.$$

Simplifying, as in the examples of Section 6.1, we may write this as

$$\sum_{n=0}^{\infty} [2(n + r)(n + r - 1) - (n + r) - 5]c_n x^{n+r} + \sum_{n=1}^{\infty} c_{n-1} x^{n+r} = 0$$

or

$$[2r(r - 1) - r - 5]c_0 x^r$$

$$+ \sum_{n=1}^{\infty} \{[2(n + r)(n + r - 1) - (n + r) - 5]c_n + c_{n-1}\} x^{n+r} = 0. \tag{6.60}$$

This is of the form (6.56), where $k = 0$.

Equating to zero the coefficient of the lowest power of x (that is, the coefficient of x^r) in (6.60), we are led to the quadratic equation

$$2r(r - 1) - r - 5 = 0$$

(since we have assumed that $c_0 \neq 0$). This is the *indicial* equation of the differential equation (6.51). We write it in the form

$$2r^2 - 3r - 5 = 0$$

and observe that its roots are

$$r_1 = \tfrac{5}{2} \quad \text{and} \quad r_2 = -1.$$

These are the so-called *exponents* of the differential equation (6.51) and are the only possible values for the previously unknown constant r in the solution (6.57). Note that they are real and unequal.

Equating to zero the coefficients of the higher powers of x in (6.60), we obtain the *recurrence formula*

$$[2(n + r)(n + r - 1) - (n + r) - 5]c_n + c_{n-1} = 0, \qquad n \geq 1. \qquad (6.61)$$

Letting $r = r_1 = \frac{5}{2}$ in (6.61), we obtain the recurrence formula

$$[2(n + \tfrac{5}{2})(n + \tfrac{3}{2}) - (n + \tfrac{5}{2}) - 5]c_n + c_{n-1} = 0, \qquad n \geq 1,$$

corresponding to the larger root $\frac{5}{2}$ of the indicial equation. This simplifies to the form

$$n(2n + 7)c_n + c_{n-1} = 0, \qquad n \geq 1,$$

or, finally,

$$c_n = -\frac{c_{n-1}}{n(2n + 7)}, \qquad n \geq 1, \qquad (6.62)$$

Using (6.62) we find that

$$c_1 = -\frac{c_0}{9}, \qquad c_2 = -\frac{c_1}{22} = \frac{c_0}{198}, \qquad c_3 = -\frac{c_2}{39} = -\frac{c_0}{7722}, \dots$$

Letting $r = \frac{5}{2}$ in (6.57), and using these values of $c_1, c_2, c_3, \dots$, we obtain the solution

$$\begin{aligned} y &= c_0(x^{5/2} - \tfrac{1}{9}x^{7/2} + \tfrac{1}{198}x^{9/2} - \tfrac{1}{7722}x^{11/2} + \cdots) \\ &= c_0 x^{5/2}(1 - \tfrac{1}{9}x + \tfrac{1}{198}x^2 - \tfrac{1}{7722}x^3 + \cdots), \end{aligned} \qquad (6.63)$$

corresponding to the larger root $r_1 = \frac{5}{2}$.

We now let $r = r_2 = -1$ in (6.61) to obtain the recurrence formula

$$[2(n - 1)(n - 2) - (n - 1) - 5]c_n + c_{n-1} = 0, \qquad n \geq 1,$$

corresponding to this smaller root of the indicial equation. This simplifies to the form

$$n(2n - 7)c_n + c_{n-1} = 0, \qquad n \geq 1,$$

or finally

$$c_n = -\frac{c_{n-1}}{n(2n - 7)}, \qquad n \geq 1.$$

Using this, we find that

$$c_1 = \tfrac{1}{5}c_0, \qquad c_2 = \tfrac{1}{6}c_1 = \tfrac{1}{30}c_0, \qquad c_3 = \tfrac{1}{3}c_2 = \tfrac{1}{90}c_0, \dots.$$

Letting $r = -1$ in (6.57), and using these values of $c_1, c_2, c_3, \dots$ we obtain the solution

$$\begin{aligned} y &= c_0(x^{-1} + \tfrac{1}{5} + \tfrac{1}{30}x + \tfrac{1}{90}x^2 - \cdots) \\ &= c_0 x^{-1}(1 + \tfrac{1}{5}x + \tfrac{1}{30}x^2 + \tfrac{1}{90}x^3 - \cdots), \end{aligned} \qquad (6.64)$$

corresponding to the smaller exponent $r_2 = -1$.

The two solutions (6.63) and (6.64) corresponding to the two roots $\frac{5}{2}$ and -1, respectively, are linearly independent. Thus the general solution of (6.51) may be written

$$y = C_1 x^{5/2}(1 - \tfrac{1}{9}x + \tfrac{1}{198}x^2 - \tfrac{1}{7722}x^3 + \cdots) + C_2 x^{-1}(1 + \tfrac{1}{5}x + \tfrac{1}{30}x^2 + \tfrac{1}{90}x^3 - \cdots),$$

where C_1 and C_2 are arbitrary constants.

Observe that in Example 6.11 *two* linearly independent solutions of the form (6.49) were obtained for $x > 0$. However, in Step 8 of the outline preceding Example 6.11, we indicated that this is not always the case. Thus we are led to ask the following questions:

1. Under what conditions are we assured that the differential equation (6.1) has *two* linearly independent solutions

$$|x - x_0|^r \sum_{n=0}^{\infty} c_n(x - x_0)^n$$

of the form (6.49) about a regular singular point x_0?

2. If the differential equation (6.1) does *not* have two linearly independent solutions of the form (6.49) about a regular singular point x_0, then what is the form of a solution that *is* linearly independent of the basic solution of the form (6.49)?

In answer to these questions we state the following theorem.

THEOREM 6.3

Hypothesis. *Let the point x_0 be a regular singular point of the differential equation (6.1). Let r_1 and r_2 [where $\mathrm{Re}(r_1) \geq \mathrm{Re}(r_2)$] be the roots of the indicial equation associated with x_0.*

Conclusion 1. *Suppose $r_1 - r_2 \neq N$, where N is a nonnegative integer (that is, $r_1 - r_2 \neq 0, 1, 2, 3, \ldots$). Then the differential equation (6.1) has two nontrivial linearly independent solutions y_1 and y_2 of the form (6.49) given respectively by*

$$y_1(x) = |x - x_0|^{r_1} \sum_{n=0}^{\infty} c_n(x - x_0)^n, \tag{6.65}$$

where $c_0 \neq 0$, and

$$y_2(x) = |x - x_0|^{r_2} \sum_{n=0}^{\infty} c_n^*(x - x_0)^n, \tag{6.66}$$

where $c_0^* \neq 0$.

Conclusion 2. *Suppose $r_1 - r_2 = N$, where N is a positive integer. Then the differential equation (6.1) has two nontrivial linearly independent solutions y_1 and y_2 given respectively by*

$$y_1(x) = |x - x_0|^{r_1} \sum_{n=0}^{\infty} c_n(x - x_0)^n, \tag{6.65}$$

where $c_0 \neq 0$, and

$$y_2(x) = |x - x_0|^{r_2} \sum_{n=0}^{\infty} c_n^*(x - x_0)^n + C y_1(x)\ln|x - x_0|, \tag{6.67}$$

where $c_0^* \neq 0$ and C is a constant which may or may not be zero.

Conclusion 3. *Suppose $r_1 - r_2 = 0$. Then the differential equation (6.1) has two nontrivial linearly independent solutions y_1 and y_2 given respectively by*

$$y_1(x) = |x - x_0|^{r_1} \sum_{n=0}^{\infty} c_n(x - x_0)^n, \qquad (6.65)$$

where $c_0 \neq 0$, and

$$y_2(x) = |x - x_0|^{r_1 + 1} \sum_{n=0}^{\infty} c_n^*(x - x_0)^n + y_1(x)\ln|x - x_0|. \qquad (6.68)$$

The solutions in Conclusions 1, 2, and 3 are valid in some deleted interval $0 < |x - x_0| < R$ about x_0.

In the illustrative examples and exercises that follow, we shall again seek solutions valid in some interval $0 < x - x_0 < R$. We shall therefore discuss the conclusions of Theorem 6.3 for such an interval. Before doing so, we again note that if $0 < x - x_0 < R$, then $|x - x_0|$ is simply $x - x_0$.

From the three conclusions of Theorem 6.3 we see that if x_0 is a regular singular point of (6.1), and $0 < x - x_0 < R$, then there is *always* a solution

$$y_1(x) = (x - x_0)^{r_1} \sum_{n=0}^{\infty} c_n(x - x_0)^n$$

of the form (6.49) for $0 < x - x_0 < R$ corresponding to the root r_1 of the indicial equation associated with x_0. Note again that the root r_1 is the *larger* root if r_1 and r_2 are real and unequal. From Conclusion 1 we see that if $0 < x - x_0 < R$ and the difference $r_1 - r_2$ between the roots of the indicial equation is *not* zero or a positive integer, then there is always a linearly independent solution

$$y_2(x) = (x - x_0)^{r_2} \sum_{n=0}^{\infty} c_n^*(x - x_0)^n$$

of the form (6.49) for $0 < x - x_0 < R$ corresponding to the root r_2. Note that the root r_2 is the *smaller* root if r_1 and r_2 are real and unequal. In particular, observe that if r_1 and r_2 are conjugate complex, then $r_1 - r_2$ is pure imaginary, and there will *always* be a linearly independent solution of the form (6.49) corresponding to r_2. However, from Conclusion 2 we see that if $0 < x - x_0 < R$ and the difference $r_1 - r_2$ is a *positive integer*, then a solution that is linearly independent of the "basic" solution of the form (6.49) for $0 < x - x_0 < R$ is of the generally more complicated form

$$y_2(x) = (x - x_0)^{r_2} \sum_{n=0}^{\infty} c_n^*(x - x_0)^n + Cy_1(x)\ln|x - x_0|$$

for $0 < x - x_0 < R$. Of course, if the constant C in this solution is zero, then it reduces to the simpler type of the form (6.49) for $0 < x - x_0 < R$. Finally, from Conclusion 3, we see that if $r_1 - r_2$ is zero, then the linearly independent solution $y_2(x)$ *always* involves the logarithmic term $y_1(x)\ln|x - x_0|$ and is *never* of the simple form (6.49) for $0 < x - x_0 < R$.

We shall now consider several examples that will (1) give further practice in the method of Frobenius, (2) illustrate the conclusions of Theorem 6.3, and (3) indicate how a linearly independent solution of the more complicated form involving the logarithmic term may be found in cases in which it exists. In each example we shall take $x_0 = 0$ and seek solutions valid in some interval $0 < x < R$. Thus note that in each example $|x - x_0| = |x| = x$.

▶   **Example 6.12**

Use the method of Frobenius to find solutions of the differential equation

$$2x^2 \frac{d^2y}{dx^2} + x \frac{dy}{dx} + (x^2 - 3)y = 0 \qquad (6.69)$$

in some interval $0 < x < R$.

Solution.   We observe at once that $x = 0$ is a regular singular point of the differential equation (6.69). Hence, since we seek solutions for $0 < x < R$, we assume a solution

$$y = \sum_{n=0}^{\infty} c_n x^{n+r}, \qquad (6.70)$$

where $c_0 \neq 0$. Differentiating (6.70), we obtain

$$\frac{dy}{dx} = \sum_{n=0}^{\infty} (n + r)c_n x^{n+r-1} \quad \text{and} \quad \frac{d^2y}{dx^2} = \sum_{n=0}^{\infty} (n + r)(n + r - 1)c_n x^{n+r-2}.$$

Upon substituting (6.70) and these derivatives into (6.69), we find

$$2 \sum_{n=0}^{\infty} (n + r)(n + r - 1)c_n x^{n+r} + \sum_{n=0}^{\infty} (n + r)c_n x^{n+r} + \sum_{n=0}^{\infty} c_n x^{n+r+2} - 3 \sum_{n=0}^{\infty} c_n x^{n+r} = 0.$$

Simplifying, as in the previous examples, we write this as

$$\sum_{n=0}^{\infty} [2(n + r)(n + r - 1) + (n + r) - 3]c_n x^{n+r} + \sum_{n=2}^{\infty} c_{n-2} x^{n+r} = 0$$

or

$$[2r(r - 1) + r - 3]c_0 x^r + [2(r + 1)r + (r + 1) - 3]c_1 x^{r+1}$$

$$+ \sum_{n=2}^{\infty} \{[2(n + r)(n + r - 1) + (n + r) - 3]c_n + c_{n-2}\} x^{n+r} = 0. \quad (6.71)$$

This is of the form (6.56), where $k = 0$.

Equating to zero the coefficient of the lowest power of x in (6.71), we obtain the indicial equation

$$2r(r - 1) + r - 3 = 0 \quad \text{or} \quad 2r^2 - r - 3 = 0.$$

The roots of this equation are

$$r_1 = \tfrac{3}{2} \quad \text{and} \quad r_2 = -1.$$

Since the difference $r_1 - r_2 = \tfrac{5}{2}$ between these roots is *not* zero or a positive integer, Conclusion 1 of Theorem 6.3 tells us that Equation (6.69) has *two* linearly independent solutions of the form (6.70), one corresponding to each of the roots r_1 and r_2.

Equating to zero the coefficients of the higher powers of x in (6.71), we obtain the condition

$$[2(r + 1)r + (r + 1) - 3]c_1 = 0 \qquad (6.72)$$

and the recurrence formula

$$[2(n + r)(n + r - 1) + (n + r) - 3]c_n + c_{n-2} = 0, \qquad n \geq 2. \qquad (6.73)$$

Letting $r = r_1 = \frac{3}{2}$ in (6.72), we obtain $7c_1 = 0$ and hence $c_1 = 0$. Letting $r = r_1 = \frac{3}{2}$ in (6.73), we obtain (after slight simplifications) the recurrence formula

$$n(2n + 5)c_n + c_{n-2} = 0, \qquad n \geq 2,$$

corresponding to the larger root $\frac{3}{2}$. Writing this in the form

$$c_n = -\frac{c_{n-2}}{n(2n + 5)}, \qquad n \geq 2,$$

we obtain

$$c_2 = -\frac{c_0}{18}, \qquad c_3 = -\frac{c_1}{33} = 0 \text{ (since } c_1 = 0), \qquad c_4 = -\frac{c_2}{52} = \frac{c_0}{936}, \ldots$$

Note that *all* odd coefficients are zero, since $c_1 = 0$. Letting $r = \frac{3}{2}$ in (6.70) and using these values of $c_1, c_2, c_3, \ldots$, we obtain the solution corresponding to the larger root $r_1 = \frac{3}{2}$. This solution is $y = y_1(x)$, where

$$y_1(x) = c_0 x^{3/2}(1 - \tfrac{1}{18}x^2 + \tfrac{1}{936}x^4 - \cdots). \tag{6.74}$$

Now let $r = r_2 = -1$ in (6.72). We obtain $-3c_1 = 0$ and hence $c_1 = 0$. Letting $r = r_2 = -1$ in (6.73), we obtain the recurrence formula

$$n(2n - 5)c_n + c_{n-2} = 0, \qquad n \geq 2,$$

corresponding to the smaller root -1. Writing this in the form

$$c_n = -\frac{c_{n-2}}{n(2n - 5)}, \qquad n \geq 2,$$

we obtain

$$c_2 = \frac{c_0}{2}, \qquad c_3 = -\frac{c_1}{3} = 0 \text{ (since } c_1 = 0), \qquad c_4 = -\frac{c_2}{12} = -\frac{c_0}{24}, \ldots$$

In this case also all odd coefficients are zero. Letting $r = -1$ in (6.70) and using these values of $c_1, c_2, c_3, \ldots$, we obtain the solution corresponding to the smaller root $r_2 = -1$. This solution is $y = y_2(x)$, where

$$y_2(x) = c_0 x^{-1}(1 + \tfrac{1}{2}x^2 - \tfrac{1}{24}x^4 + \cdots). \tag{6.75}$$

Since the solutions defined by (6.74) and (6.75) are linearly independent, the general solution of (6.69) may be written

$$y = C_1 y_1(x) + C_2 y_2(x),$$

where C_1 and C_2 are arbitrary constants and $y_1(x)$ and $y_2(x)$ are defined by (6.74) and (6.75), respectively.

▶ **Example 6.13**

Use the method of Frobenius to find solutions of the differential equation

$$x^2 \frac{d^2 y}{dx^2} - x \frac{dy}{dx} - (x^2 + \tfrac{5}{4})y = 0 \tag{6.76}$$

in some interval $0 < x < R$.

Solution. We observe that $x = 0$ is a regular singular point of this differential equation and we seek solutions for $0 < x < R$. Hence we assume a solution

$$y = \sum_{n=0}^{\infty} c_n x^{n+r}, \tag{6.77}$$

where $c_0 \neq 0$. Upon differentiating (6.77) twice and substituting into (6.76), we obtain

$$\sum_{n=0}^{\infty} (n + r)(n + r - 1)c_n x^{n+r} - \sum_{n=0}^{\infty} (n + r)c_n x^{n+r} - \sum_{n=0}^{\infty} c_n x^{n+r+2} - \frac{5}{4} \sum_{n=0}^{\infty} c_n x^{n+r} = 0.$$

Simplifying, we write this in the form

$$\sum_{n=0}^{\infty} \left[(n + r)(n + r - 1) - (n + r) - \tfrac{5}{4} \right] c_n x^{n+r} - \sum_{n=2}^{\infty} c_{n-2} x^{n+r} = 0$$

or

$$\left[r(r - 1) - r - \tfrac{5}{4} \right] c_0 x^r + \left[(r + 1)r - (r + 1) - \tfrac{5}{4} \right] c_1 x^{r+1}$$

$$+ \sum_{n=2}^{\infty} \left\{ \left[(n + r)(n + r - 1) - (n + r) - \tfrac{5}{4} \right] c_n - c_{n-2} \right\} x^{n+r} = 0. \tag{6.78}$$

Equating to zero the coefficient of the lowest power of x in (6.78), we obtain the indicial equation

$$r^2 - 2r - \tfrac{5}{4} = 0.$$

The roots of this equation are

$$r_1 = \tfrac{5}{2}, \qquad r_2 = -\tfrac{1}{2}.$$

Although these roots themselves are not integers, the difference $r_1 - r_2$ between them is the positive integer 3. By Conclusion 2 of Theorem 6.3 we know that the differential equation (6.76) has a solution of the assumed form (6.77) corresponding to the larger root $r_1 = \tfrac{5}{2}$. We proceed to obtain this solution.

Equating to zero the coefficients of the higher powers of x in (6.78), we obtain the condition

$$\left[(r + 1)r - (r + 1) - \tfrac{5}{4} \right] c_1 = 0 \tag{6.79}$$

and the recurrence formula

$$\left[(n + r)(n + r - 1) - (n + r) - \tfrac{5}{4} \right] c_n - c_{n-2} = 0, \qquad n \geq 2. \tag{6.80}$$

Letting $r = r_1 = \tfrac{5}{2}$ in (6.79), we obtain

$$4c_1 = 0 \quad \text{and hence} \quad c_1 = 0.$$

Letting $r = r_1 = \tfrac{5}{2}$ in (6.80), we obtain the recurrence formula

$$n(n + 3)c_n - c_{n-2} = 0, \qquad n \geq 2,$$

corresponding to the larger root $\tfrac{5}{2}$. Since $n \geq 2$, we may write this in the form

$$c_n = \frac{c_{n-2}}{n(n + 3)}, \qquad n \geq 2.$$

From this we obtain successively

$$c_2 = \frac{c_0}{2 \cdot 5}, \qquad c_3 = \frac{c_1}{3 \cdot 6} = 0 \text{ (since } c_1 = 0), \qquad c_4 = \frac{c_2}{4 \cdot 7} = \frac{c_0}{2 \cdot 4 \cdot 5 \cdot 7}, \cdots$$

We note that all odd coefficients are zero. The general even coefficient may be written

$$c_{2n} = \frac{c_0}{[2 \cdot 4 \cdot 6 \cdots (2n)][5 \cdot 7 \cdot 9 \cdots (2n+3)]}, \qquad n \geq 1.$$

Letting $r = \frac{5}{2}$ in (6.77) and using these values of c_{2n}, we obtain the solution corresponding to the larger root $r_1 = \frac{5}{2}$. This solution is $y = y_1(x)$, where

$$y_1(x) = c_0 x^{5/2} \left[1 + \frac{x^2}{2 \cdot 5} + \frac{x^4}{2 \cdot 4 \cdot 5 \cdot 7} + \cdots \right]$$

$$= c_0 x^{5/2} \left[1 + \sum_{n=1}^{\infty} \frac{x^{2n}}{[2 \cdot 4 \cdot 6 \cdots (2n)][5 \cdot 7 \cdot 9 \cdots (2n+3)]} \right]. \qquad (6.81)$$

We now consider the smaller root $r_2 = -\frac{1}{2}$. Theorem 6.3 does *not* assure us that the differential equation (6.76) has a linearly independent solution of the assumed form (6.77) corresponding to this smaller root. Conclusion 2 of that theorem merely tells us that there is a linearly independent solution of the form

$$\sum_{n=0}^{\infty} c_n^* x^{n+r_2} + C y_1(x) \ln x, \qquad (6.82)$$

where C may or may not be zero. Of course, if $C = 0$, then the linearly independent solution (6.82) *is* of the assumed form (6.77) and we can let $r = r_2 = -\frac{1}{2}$ in the formula (6.79) and the recurrence formula (6.80) and proceed as in the previous examples. Let us assume (hopefully, but without justification!) that this is indeed the case.

Thus we let $r = r_2 = -\frac{1}{2}$ in (6.79) and (6.80). Letting $r = -\frac{1}{2}$ in (6.79) we obtain $-2c_1 = 0$ and hence $c_1 = 0$. Letting $r = -\frac{1}{2}$ in (6.80) we obtain the recurrence formula

$$n(n-3)c_n - c_{n-2} = 0, \qquad n \geq 2, \qquad (6.83)$$

corresponding to the smaller root $-\frac{1}{2}$. For $n \neq 3$, this may be written

$$c_n = \frac{c_{n-2}}{n(n-3)}, \qquad n \geq 2, \qquad n \neq 3. \qquad (6.84)$$

For $n = 2$, formula (6.84) gives $c_2 = -c_0/2$. For $n = 3$, formula (6.84) does not apply and we must use (6.83). For $n = 3$ formula (6.83) is $0 \cdot c_3 - c_1 = 0$ or simply $0 = 0$ (since $c_1 = 0$). Hence, for $n = 3$, the recurrence formula (6.83) is automatically satisfied with *any* choice of c_3. Thus c_3 is independent of the arbitrary constant c_0; it is a second arbitrary constant! For $n > 3$, we can again use (6.84). Proceeding, we have

$$c_4 = \frac{c_2}{4} = -\frac{c_0}{2 \cdot 4}, \qquad c_5 = \frac{c_3}{2 \cdot 5},$$

$$c_6 = \frac{c_4}{6 \cdot 3} = -\frac{c_0}{2 \cdot 4 \cdot 6 \cdot 3}, \qquad c_7 = \frac{c_5}{4 \cdot 7} = \frac{c_3}{2 \cdot 4 \cdot 5 \cdot 7}, \cdots$$

We note that all even coefficients may be expressed in terms of c_0 and that all odd coefficients beyond c_3 may be expressed in terms of c_3. In fact, we may write

$$c_{2n} = -\frac{c_0}{[2 \cdot 4 \cdot 6 \cdots (2n)][3 \cdot 5 \cdot 7 \cdots (2n-3)]}, \quad n \geq 3$$

(even coefficients c_6 and beyond), and

$$c_{2n+1} = \frac{c_3}{[2 \cdot 4 \cdot 6 \cdots (2n-2)][5 \cdot 7 \cdot 9 \cdots (2n+1)]}, \quad n \geq 2$$

(odd coefficients c_5 and beyond). Letting $r = -\frac{1}{2}$ in (6.77) and using the values of c_n in terms of c_0 (for even n) and c_3 (for odd n beyond c_3), we obtain the solution corresponding to the smaller root $r_2 = -\frac{1}{2}$. This solution is $y = y_2(x)$, where

$$y_2(x) = c_0 x^{-1/2} \left[1 - \frac{x^2}{2} - \frac{x^4}{2 \cdot 4} - \frac{x^6}{2 \cdot 4 \cdot 6 \cdot 3} - \cdots \right]$$

$$+ c_3 x^{-1/2} \left[x^3 + \frac{x^5}{2 \cdot 5} + \frac{x^7}{2 \cdot 4 \cdot 5 \cdot 7} + \cdots \right]$$

$$= c_0 x^{-1/2} \left[1 - \frac{x^2}{2} - \frac{x^4}{2 \cdot 4} - \sum_{n=3}^{\infty} \frac{x^{2n}}{[2 \cdot 4 \cdot 6 \cdots (2n)][3 \cdot 5 \cdot 7 \cdots (2n-3)]} \right]$$

$$+ c_3 x^{-1/2} \left[x^3 + \sum_{n=2}^{\infty} \frac{x^{2n+1}}{[2 \cdot 4 \cdot 6 \cdots (2n-2)][5 \cdot 7 \cdot 9 \cdots (2n+1)]} \right],$$

$$\tag{6.85}$$

and c_0 and c_3 are arbitrary constants.

If we now let $c_0 = 1$ in (6.81) we obtain the particular solution $y = y_{11}(x)$, where

$$y_{11}(x) = x^{5/2} \left[1 + \sum_{n=1}^{\infty} \frac{x^{2n}}{[2 \cdot 4 \cdot 6 \cdots (2n)][5 \cdot 7 \cdot 9 \cdots (2n+3)]} \right]$$

corresponding to the larger root $\frac{5}{2}$; and if we let $c_0 = 1$ and $c_3 = 0$ in (6.85) we obtain the particular solution $y = y_{21}(x)$, where

$$y_{21}(x) = x^{-1/2} \left[1 - \frac{x^2}{2} - \frac{x^4}{2 \cdot 4} - \sum_{n=3}^{\infty} \frac{x^{2n}}{[2 \cdot 4 \cdot 6 \cdots (2n)][3 \cdot 5 \cdot 7 \cdots (2n-3)]} \right]$$

corresponding to the smaller root $-\frac{1}{2}$. These two particular solutions, which are both of the assumed form (6.77), are linearly independent. Thus the general solution of the differential equation (6.76) may be written

$$y = C_1 y_{11}(x) + C_2 y_{21}(x), \tag{6.86}$$

where C_1 and C_2 are arbitrary constants.

Now let us examine more carefully the solution y_2 defined by (6.85). The expression

$$x^{-1/2} \left[x^3 + \sum_{n=2}^{\infty} \frac{x^{2n+1}}{[2 \cdot 4 \cdot 6 \cdots (2n-2)][5 \cdot 7 \cdot 9 \cdots (2n+1)]} \right]$$

of which c_3 is the coefficient in (6.85) may be written

$$x^{5/2} \left[1 + \sum_{n=2}^{\infty} \frac{x^{2n-2}}{[2 \cdot 4 \cdot 6 \cdots (2n-2)][5 \cdot 7 \cdot 9 \cdots (2n+1)]} \right]$$

$$= x^{5/2} \left[1 + \sum_{n=1}^{\infty} \frac{x^{2n}}{[2 \cdot 4 \cdot 6 \cdots (2n)][5 \cdot 7 \cdot 9 \cdots (2n+3)]} \right]$$

and this is precisely $y_{11}(x)$. Thus we may write

$$y_2(x) = c_0 y_{21}(x) + c_3 y_{11}(x), \tag{6.87}$$

where c_0 and c_3 are arbitrary constants. Now compare (6.86) and (6.87). We see that the solution $y = y_2(x)$ by itself is actually the general solution of the differential equation (6.76), even though $y_2(x)$ was obtained using only the smaller root $-\frac{1}{2}$.

We thus observe that if the difference $r_1 - r_2$ between the roots of the indicial equation is a positive integer, it is sometimes possible to obtain the general solution using the smaller root alone, without bothering to find explicitly the solution corresponding to the larger root. Indeed, if the difference $r_1 - r_2$ is a positive integer, it is a worthwhile practice to work with the smaller root first, in the hope that this smaller root by itself may lead directly to the general solution.

▶ **Example 6.14**

Use the method of Frobenius to find solutions of the differential equation

$$x^2 \frac{d^2 y}{dx^2} + (x^2 - 3x)\frac{dy}{dx} + 3y = 0 \tag{6.88}$$

in some interval $0 < x < R$.

Solution. We observe that $x = 0$ is a regular singular point of (6.88) and we seek solutions for $0 < x < R$. Hence we assume a solution

$$y = \sum_{n=0}^{\infty} c_n x^{n+r}, \tag{6.89}$$

where $c_0 \neq 0$. Upon differentiating (6.89) twice and substituting into (6.88), we obtain

$$\sum_{n=0}^{\infty} (n+r)(n+r-1)c_n x^{n+r} + \sum_{n=0}^{\infty} (n+r)c_n x^{n+r+1}$$

$$-3\sum_{n=0}^{\infty} (n+r)c_n x^{n+r} + 3\sum_{n=0}^{\infty} c_n x^{n+r} = 0.$$

Simplifying, we write this in the form

$$\sum_{n=0}^{\infty} [(n+r)(n+r-1) - 3(n+r) + 3]c_n x^{n+r} + \sum_{n=1}^{\infty} (n+r-1)c_{n-1} x^{n+r} = 0$$

or

$$[r(r-1) - 3r + 3]c_0 x^r + \sum_{n=1}^{\infty} \{[(n+r)(n+r-1) - 3(n+r) + 3]c_n$$

$$+ (n+r-1)c_{n-1}\}x^{n+r} = 0. \tag{6.90}$$

Equating to zero the coefficient of the lowest power of x in (6.90) we obtain the indicial equation

$$r^2 - 4r + 3 = 0.$$

The roots of this equation are

$$r_1 = 3, \qquad r_2 = 1.$$

The difference $r_1 - r_2$ between these roots is the positive integer 2. We know from Theorem 6.3 that the differential equation (6.88) has a solution of the assumed form (6.89) corresponding to the larger root $r_1 = 3$. We shall find this first, even though the results of Example 6.13 suggest that we should work first with the smaller root $r_2 = 1$ in the hopes of finding the general solution directly from this smaller root.

Equating to zero the coefficients of the higher powers of x in (6.90), we obtain the recurrence formula

$$[(n + r)(n + r - 1) - 3(n + r) + 3]c_n + (n + r - 1)c_{n-1} = 0. \qquad n \geq 1. \quad (6.91)$$

Letting $r = r_1 = 3$ in (6.91), we obtain the recurrence formula

$$n(n + 2)c_n + (n + 2)c_{n-1} = 0, \qquad n \geq 1,$$

corresponding to the larger root 3. Since $n \geq 1$, we may write this in the form

$$c_n = -\frac{c_{n-1}}{n}, \qquad n \geq 1.$$

From this we find successively

$$c_1 = -c_0, \qquad c_2 = -\frac{c_1}{2} = \frac{c_0}{2!}, \qquad c_3 = -\frac{c_2}{3} = -\frac{c_0}{3!}, \ldots, c_n = \frac{(-1)^n c_0}{n!}, \ldots.$$

Letting $r = 3$ in (6.89) and using these values of c_n, we obtain the solution corresponding to the larger root $r_1 = 3$. This solution is $y = y_1(x)$, where

$$y_1(x) = c_0 x^3 \left[1 - x + \frac{x^2}{2!} - \frac{x^3}{3!} + \cdots + \frac{(-1)^n x^n}{n!} + \cdots \right].$$

We recognize the series in brackets in this solution as the Maclaurin expansion for e^{-x}. Thus we may write

$$y_1(x) = c_0 x^3 e^{-x} \qquad (6.92)$$

and express the solution corresponding to r_1 in the closed form

$$y = c_0 x^3 e^{-x},$$

where c_0 is an arbitrary constant.

We now consider the smaller root $r_2 = 1$. As in Example 6.13, we have no assurance that the differential equation has a linearly independent solution of the assumed form (6.89) corresponding to this smaller root. However, as in that example, we shall tentatively assume that such a solution actually does exist and let $r = r_2 = 1$ in (6.91) in the hopes of finding "it." Further, we are now aware that this step by itself *might* even provide us with the *general* solution.

Thus we let $r = r_2 = 1$ in (6.91) to obtain the recurrence formula

$$n(n - 2)c_n + nc_{n-1} = 0, \qquad n \geq 1, \qquad (6.93)$$

corresponding to the smaller root 1. For $n \neq 2$, this may be written

$$c_n = -\frac{c_{n-1}}{n - 2}, \qquad n \geq 1, \quad n \neq 2. \qquad (6.94)$$

For $n = 1$, formula (6.94) gives $c_1 = c_0$. For $n = 2$, formula (6.94) does not apply and we must use (6.93). For $n = 2$ formula (6.93) is $0 \cdot c_2 + 2c_1 = 0$, and hence we must have

$c_1 = 0$. But then, since $c_1 = c_0$, we must have $c_0 = 0$. However, $c_0 \neq 0$ in the assumed solution (6.89). This contradiction shows that there is no solution of the form (6.89), with $c_0 \neq 0$, corresponding to the smaller root 1.

Further, we observe that the use of (6.94) for $n \geq 3$ will only lead us to the solution y_1 already obtained. For, from the condition $0 \cdot c_2 + 2c_1 = 0$ we see that c_2 is arbitrary; and using (6.94) for $n \geq 3$, we obtain successively

$$c_3 = -c_2, \qquad c_4 = -\frac{c_3}{2} = \frac{c_2}{2!},$$

$$c_5 = -\frac{c_4}{3} = -\frac{c_2}{3!}, \ldots, c_{n+2} = \frac{(-1)^n c_2}{n!}, \ldots, n \geq 1.$$

Thus letting $r = 1$ in (6.89) and using these values of c_n, we obtain formally

$$y = c_2 x \left[x^2 - x^3 + \frac{x^4}{2!} - \frac{x^5}{3!} + \cdots + \frac{(-1)^n x^{n+2}}{n!} + \cdots \right]$$

$$= c_2 x^3 \left[1 - x + \frac{x^2}{2!} - \frac{x^3}{3!} + \cdots + \frac{(-1)^n x^n}{n!} + \cdots \right]$$

$$= c_2 x^3 e^{-x}.$$

Comparing this with (6.92), we see that it is essentially the solution $y = y_1(x)$.

We now seek a solution of (6.88) that is linearly independent of the solution y_1. From Theorem 6.3 we now know that this solution is of the form

$$\sum_{n=0}^{\infty} c_n^* x^{n+1} + C y_1(x) \ln x, \tag{6.95}$$

where $c_0^* \neq 0$ and $C \neq 0$. Various methods for obtaining such a solution are available; we shall employ the method of reduction of order (Section 4.1). We let $y = f(x)v$, where $f(x)$ is a known solution of (6.88). Choosing for f the known solution y_1 defined by (6.92), with $c_0 = 1$, we thus let

$$y = x^3 e^{-x} v. \tag{6.96}$$

From this we obtain

$$\frac{dy}{dx} = x^3 e^{-x} \frac{dv}{dx} + (3x^2 e^{-x} - x^3 e^{-x})v \tag{6.97}$$

and

$$\frac{d^2 y}{dx^2} = x^3 e^{-x} \frac{d^2 v}{dx^2} + 2(3x^2 e^{-x} - x^3 e^{-x}) \frac{dv}{dx} + (x^3 e^{-x} - 6x^2 e^{-x} + 6x e^{-x})v. \tag{6.98}$$

Substituting (6.96), (6.97), and (6.98) for y and its first two derivatives, respectively, in the differential equation (6.88), after some simplifications we obtain

$$x \frac{d^2 v}{dx^2} + (3 - x) \frac{dv}{dx} = 0. \tag{6.99}$$

Letting $w = dv/dx$, this reduces at once to the first-order differential equation

$$x \frac{dw}{dx} + (3 - x)w = 0,$$

a particular solution of which is $w = x^{-3}e^x$. Thus a particular solution of (6.99) is given by

$$v = \int x^{-3}e^x \, dx,$$

and hence $y = y_2(x)$, where

$$y_2(x) = x^3 e^{-x} \int x^{-3} e^x \, dx \qquad (6.100)$$

is a particular solution of (6.88) that is linearly independent of the solution y_1 defined by (6.92).

We now show that the solution y_2 defined by (6.100) is of the form (6.95). Introducing the Maclaurin series for e^x in (6.100) we have

$$y_2(x) = x^3 e^{-x} \int x^{-3}\left(1 + x + \frac{x^2}{2} + \frac{x^3}{6} + \frac{x^x}{24} + \cdots\right) dx$$

$$= x^3 e^{-x} \int \left(x^{-3} + x^{-2} + \frac{1}{2}x^{-1} + \frac{1}{6} + \frac{x}{24} + \cdots\right) dx.$$

Integrating term by term, we obtain

$$y_2(x) = x^3 e^{-x}\left[-\frac{1}{2x^2} - \frac{1}{x} + \frac{1}{2}\ln x + \frac{1}{6}x + \frac{1}{48}x^2 + \cdots\right].$$

Now introducing the Maclaurin series for e^{-x}, we may write

$$y_2(x) = \left(x^3 - x^4 + \frac{x^5}{2} - \frac{x^6}{6} + \cdots\right)\left(-\frac{1}{2x^2} - \frac{1}{x} + \frac{1}{6}x + \frac{1}{48}x^2 + \cdots\right)$$

$$+ \tfrac{1}{2}x^3 e^{-x} \ln x.$$

Finally, multiplying the two series involved, we have

$$y_2(x) = \left(-\tfrac{1}{2}x - \tfrac{1}{2}x^2 + \tfrac{3}{4}x^3 - \tfrac{1}{4}x^4 + \cdots\right) + \tfrac{1}{2}x^3 e^{-x} \ln x,$$

which is of the form (6.95), where $y_1(x) = x^3 e^{-x}$. The general solution of the differential equation (6.88) may thus be written

$$y = C_1 y_1(x) + C_2 y_2(x),$$

where C_1 and C_2 are arbitrary constants.

In this example it was fortunate that we were able to express the first solution y_1 in closed form, for this simplified the computations involved in finding the second solution y_2. Of course the method of reduction of order may be applied to find the second solution, even if we cannot express the first solution in closed form. In such cases the various steps of the method must be carried out in terms of the series expression for y_1. The computations that result are generally quite complicated.

Examples 6.12, 6.13, and 6.14 illustrate all of the possibilities listed in the conclusions of Theorem 6.3 except the case in which $r_1 - r_2 = 0$ (that is, the case in which the roots of the indicial equation are equal). In this case it is obvious that for $0 < x - x_0 < R$ both roots lead to the *same* solution

$$y_1 = (x - x_0)^r \sum_{n=0}^{\infty} c_n(x - x_0)^n,$$

where r is the common value of r_1 and r_2. Thus, as Conclusion 3 of Theorem 6.3 states, for $0 < x - x_0 < R$, a linearly independent solution is of the form

$$y_2 = (x - x_0)^{r+1} \sum_{n=0}^{\infty} c_n^*(x - x_0)^n + y_1(x)\ln(x - x_0).$$

Once $y_1(x)$ has been found, we may obtain y_2 by the method of reduction of order. This procedure has already been illustrated in finding the second solution of the equation in Example 6.14. A further illustration is provided in Section 6.3 by the solution of Bessel's equation of order zero.

Exercises

Locate and classify the singular points of each of the differential equations in Exercises 1–4.

1. $(x^2 - 3x)\dfrac{d^2y}{dx^2} + (x + 2)\dfrac{dy}{dx} + y = 0.$

2. $(x^3 + x^2)\dfrac{d^2y}{dx^2} + (x^2 - 2x)\dfrac{dy}{dx} + 4y = 0.$

3. $(x^4 - 2x^3 + x^2)\dfrac{d^2y}{dx^2} + 2(x - 1)\dfrac{dy}{dx} + x^2y = 0.$

4. $(x^5 + x^4 - 6x^3)\dfrac{d^2y}{dx^2} + x^2\dfrac{dy}{dx} + (x - 2)y = 0.$

Use the method of Frobenius to find solutions near $x = 0$ of each of the differential equations in Exercises 5–26.

5. $2x^2\dfrac{d^2y}{dx^2} + x\dfrac{dy}{dx} + (x^2 - 1)y = 0.$

6. $2x^2\dfrac{d^2y}{dx^2} + x\dfrac{dy}{dx} + (2x^2 - 3)y = 0.$

7. $x^2\dfrac{d^2y}{dx^2} - x\dfrac{dy}{dx} + \left(x^2 + \dfrac{8}{9}\right)y = 0.$

8. $x^2\dfrac{d^2y}{dx^2} - x\dfrac{dy}{dx} + \left(2x^2 + \dfrac{5}{9}\right)y = 0.$

9. $x^2\dfrac{d^2y}{dx^2} + x\dfrac{dy}{dx} + \left(x^2 - \dfrac{1}{9}\right)y = 0.$

10. $2x\dfrac{d^2y}{dx^2} + \dfrac{dy}{dx} + 2y = 0.$

11. $3x\dfrac{d^2y}{dx^2} - (x - 2)\dfrac{dy}{dx} - 2y = 0.$

12. $x\dfrac{d^2y}{dx^2} + 2\dfrac{dy}{dx} + xy = 0.$

13. $x^2 \dfrac{d^2 y}{dx^2} + x \dfrac{dy}{dx} + \left(x^2 - \dfrac{1}{4} \right) y = 0.$

14. $x^2 \dfrac{d^2 y}{dx^2} + (x^4 + x) \dfrac{dy}{dx} - y = 0.$

15. $x \dfrac{d^2 y}{dx^2} - (x^2 + 2) \dfrac{dy}{dx} + xy = 0.$

16. $x^2 \dfrac{d^2 y}{dx^2} + x^2 \dfrac{dy}{dx} - 2y = 0.$

17. $(2x^2 - x) \dfrac{d^2 y}{dx^2} + (2x - 2) \dfrac{dy}{dx} + (-2x^2 + 3x - 2)y = 0.$

18. $x^2 \dfrac{d^2 y}{dx^2} - x \dfrac{dy}{dx} + \dfrac{3}{4} y = 0.$

19. $x^2 \dfrac{d^2 y}{dx^2} + x \dfrac{dy}{dx} + (x - 1)y = 0.$

20. $x^2 \dfrac{d^2 y}{dx^2} + (x^3 - x) \dfrac{dy}{dx} - 3y = 0.$

21. $x^2 \dfrac{d^2 y}{dx^2} - x \dfrac{dy}{dx} + 8(x^2 - 1)y = 0.$

22. $x^2 \dfrac{d^2 y}{dx^2} + x^2 \dfrac{dy}{dx} - \dfrac{3}{4} y = 0.$

23. $x \dfrac{d^2 y}{dx^2} + \dfrac{dy}{dx} + 2y = 0.$

24. $2x \dfrac{d^2 y}{dx^2} + 6 \dfrac{dy}{dx} + y = 0.$

25. $x^2 \dfrac{d^2 y}{dx^2} - x \dfrac{dy}{dx} + (x^2 + 1)y = 0.$

26. $x^2 \dfrac{d^2 y}{dx^2} - x \dfrac{dy}{dx} + (x^2 - 3)y = 0.$

6.3 BESSEL'S EQUATION AND BESSEL FUNCTIONS

A. Bessel's Equation of Order Zero

The differential equation

$$x^2 \dfrac{d^2 y}{dx^2} + x \dfrac{dy}{dx} + (x^2 - p^2)y = 0, \qquad (6.101)$$

where p is a parameter, is called *Bessel's equation of order p*. Any solution of Bessel's equation of order p is called a *Bessel function of order p*. Bessel's equation and Bessel functions occur in connection with many problems of physics and engineering, and there is an extensive literature dealing with the theory and application of this equation and its solutions.

If $p = 0$, Equation (6.101) is equivalent to the equation

$$x \frac{d^2 y}{dx^2} + \frac{dy}{dx} + xy = 0, \tag{6.102}$$

which is called *Bessel's equation of order zero*. We shall seek solutions of this equation that are valid in an interval $0 < x < R$. We observe at once that $x = 0$ is a regular singular point of (6.102); and hence, since we seek solutions for $0 < x < R$, we assume a solution

$$y = \sum_{n=0}^{\infty} c_n x^{n+r}, \tag{6.103}$$

where $c_0 \neq 0$. Upon differentiating (6.103) twice and substituting into (6.102), we obtain

$$\sum_{n=0}^{\infty} (n+r)(n+r-1)c_n x^{n+r-1} + \sum_{n=0}^{\infty} (n+r)c_n x^{n+r-1} + \sum_{n=0}^{\infty} c_n x^{n+r+1} = 0.$$

Simplifying, we write this in the form

$$\sum_{n=0}^{\infty} (n+r)^2 c_n x^{n+r-1} + \sum_{n=2}^{\infty} c_{n-2} x^{n+r-1} = 0$$

or

$$r^2 c_0 x^{r-1} + (1+r)^2 c_1 x^r + \sum_{n=2}^{\infty} [(n+r)^2 c_n + c_{n-2}] x^{n+r-1} = 0. \tag{6.104}$$

Equating to zero the coefficient of the lowest power of x in (6.104), we obtain the indicial equation $r^2 = 0$, which has equal roots $r_1 = r_2 = 0$. Equating to zero the coefficients of the higher powers of x in (6.104) we obtain

$$(1+r)^2 c_1 = 0 \tag{6.105}$$

and the recurrence formula

$$(n+r)^2 c_n + c_{n-2} = 0, \qquad n \geq 2. \tag{6.106}$$

Letting $r = 0$ in (6.105), we find at once that $c_1 = 0$. Letting $r = 0$ in (6.106) we obtain the recurrence formula in the form

$$n^2 c_n + c_{n-2} = 0, \qquad n \geq 2,$$

or

$$c_n = -\frac{c_{n-2}}{n^2}, \qquad n \geq 2.$$

From this we obtain successively

$$c_2 = -\frac{c_0}{2^2}, \qquad c_3 = -\frac{c_1}{3^2} = 0 \text{ (since } c_1 = 0\text{)}, \qquad c_4 = -\frac{c_2}{4^2} = \frac{c_0}{2^2 \cdot 4^2}, \ldots$$

We note that all odd coefficients are zero and that the general even coefficient may be written

$$c_{2n} = \frac{(-1)^n c_0}{2^2 \cdot 4^2 \cdot 6^2 \cdots (2n)^2} = \frac{(-1)^n c_0}{(n!)^2 2^{2n}}, \qquad n \geq 1.$$

Letting $r = 0$ in (6.103) and using these values of c_{2n}, we obtain the solution $y = y_1(x)$, where

$$y_1(x) = c_0 \sum_{n=0}^{\infty} \frac{(-1)^n}{(n!)^2} \left(\frac{x}{2}\right)^{2n}$$

If we set the arbitrary constant $c_0 = 1$, we obtain an important particular solution of Equation (6.102). This particular solution defines a function denoted by J_0 and called the *Bessel function of the first kind of order zero.* That is, the function J_0 is the particular solution of Equation (6.102) defined by

$$J_0(x) = \sum_{n=0}^{\infty} \frac{(-1)^n}{(n!)^2} \left(\frac{x}{2}\right)^{2n}. \qquad (6.107)$$

Writing out the first few terms of this series solution, we see that

$$J_0(x) = 1 - \frac{1}{(1!)^2} \left(\frac{x}{2}\right)^2 + \frac{1}{(2!)^2} \left(\frac{x}{2}\right)^4 - \frac{1}{(3!)^2} \left(\frac{x}{2}\right)^6 + \cdots$$

$$= 1 - \frac{x^2}{4} + \frac{x^4}{64} - \frac{x^6}{2304} + \cdots. \qquad (6.108)$$

Since the roots of the indicial equation are equal, we know from Theorem 6.3 that a solution of Equation (6.102) which is linearly independent of J_0 must be of the form

$$y = x \sum_{n=0}^{\infty} c_n^* x^n + J_0(x) \ln x,$$

for $0 < x < R$. Also, we know that such a linearly independent solution can be found by the method of reduction of order (Section 4.1). Indeed from Theorem 4.7 we know that this linearly independent solution y_2 is given by

$$y_2(x) = J_0(x) \int \frac{e^{-\int dx/x}}{[J_0(x)]^2} \, dx$$

and hence by

$$y_2(x) = J_0(x) \int \frac{dx}{x[J_0(x)]^2}.$$

From (6.108) we find that

$$[J_0(x)]^2 = 1 - \frac{x^2}{2} + \frac{3x^4}{32} - \frac{5x^6}{576} + \cdots$$

and hence

$$\frac{1}{[J_0(x)]^2} = 1 + \frac{x^2}{2} + \frac{5x^4}{32} + \frac{23x^6}{576} + \cdots.$$

Thus

$$y_2(x) = J_0(x) \int \left(\frac{1}{x} + \frac{x}{2} + \frac{5x^3}{32} + \frac{23x^5}{576} + \cdots \right) dx$$

$$= J_0(x)\left(\ln x + \frac{x^2}{4} + \frac{5x^4}{128} + \frac{23x^6}{3456} + \cdots \right)$$

$$= J_0(x)\ln x + \left(1 - \frac{x^2}{4} + \frac{x^4}{64} - \frac{x^6}{2304} + \cdots \right)\left(\frac{x^2}{4} + \frac{5x^4}{128} + \frac{23x^6}{3456} + \cdots \right)$$

$$= J_0(x)\ln x + \frac{x^2}{4} - \frac{3x^4}{128} + \frac{11x^6}{13824} + \cdots .$$

We thus obtain the first few terms of the "second" solution y_2 by the method of reduction of order. However, our computations give no information concerning the general coefficient c_{2n}^* in the above series. Indeed, it seems unlikely that an expression for the general coefficient can be found. However, let us observe that

$$(-1)^2 \frac{1}{2^2(1!)^2}(1) = \frac{1}{2^2} = \frac{1}{4},$$

$$(-1)^3 \frac{1}{2^4(2!)^2}\left(1 + \frac{1}{2}\right) = -\frac{3}{2^4 \cdot 2^2 \cdot 2} = -\frac{3}{128},$$

$$(-1)^4 \frac{1}{2^6(3!)^2}\left(1 + \frac{1}{2} + \frac{1}{3}\right) = \frac{11}{2^6 \cdot 6^2 \cdot 6} = \frac{11}{13824}.$$

Having observed these relations, we may express the solution y_2 in the following more systematic form:

$$y_2(x) = J_0(x)\ln x + \frac{x^2}{2^2} - \frac{x^4}{2^4(2!)^2}\left(1 + \frac{1}{2}\right) + \frac{x^6}{2^6(3!)^2}\left(1 + \frac{1}{2} + \frac{1}{3}\right) + \cdots .$$

Further, we would certainly suspect that the general coefficient c_{2n}^* is given by

$$c_{2n}^* = \frac{(-1)^{n+1}}{2^{2n}(n!)^2}\left(1 + \frac{1}{2} + \frac{1}{3} + \cdots + \frac{1}{n}\right), \qquad n \geq 1.$$

It may be shown (though not without some difficulty) that this is indeed the case. This being true, we may express y_2 in the form

$$y_2(x) = J_0(x)\ln x + \sum_{n=1}^{\infty} \frac{(-1)^{n+1}x^{2n}}{2^{2n}(n!)^2}\left(1 + \frac{1}{2} + \frac{1}{3} + \cdots + \frac{1}{n}\right). \qquad (6.109)$$

Since the solution y_2 defined by (6.109) is linearly independent of J_0 we could write the general solution of the differential equation (6.102) as a general linear combination of J_0 and y_2. However, this is not usually done; instead, it has been customary to choose a certain special linear combination of J_0 and y_2 and take this special combination as the "second" solution of Equation (6.102). This special combination is defined by

$$\frac{2}{\pi}\left[y_2(x) + (\gamma - \ln 2)J_0(x)\right],$$

where γ is a number called *Euler's constant* and is defined by

$$\gamma = \lim_{n \to \infty} \left(1 + \frac{1}{2} + \frac{1}{3} + \cdots + \frac{1}{n} - \ln n \right) \approx 0.5772.$$

It is called the *Bessel function of the second kind of order zero* (Weber's form) and is commonly denoted by Y_0. Thus the second solution of (6.102) is commonly taken as the function Y_0, where

$$Y_0(x) = \frac{2}{\pi} \left[J_0(x) \ln x + \sum_{n=1}^{\infty} \frac{(-1)^{n+1} x^{2n}}{2^{2n}(n!)^2} \left(1 + \frac{1}{2} + \frac{1}{3} + \cdots + \frac{1}{n} \right) + (\gamma - \ln 2) J_0(x) \right]$$

or

$$Y_0(x) = \frac{2}{\pi} \left[\left(\ln \frac{x}{2} + \gamma \right) J_0(x) + \sum_{n=1}^{\infty} \frac{(-1)^{n+1} x^{2n}}{2^{2n}(n!)^2} \left(1 + \frac{1}{2} + \frac{1}{3} + \cdots + \frac{1}{n} \right) \right]. \quad (6.110)$$

Therefore if we choose Y_0 as the second solution of the differential equation (6.102), the general solution of (6.102) for $0 < x < R$ is given by

$$y = c_1 J_0(x) + c_2 Y_0(x), \quad (6.111)$$

where c_1 and c_2 are arbitrary constants, and J_0 and Y_0 are defined by (6.107) and (6.110), respectively.

The functions J_0 and Y_0 have been studied extensively and tabulated. Many of the interesting properties of these functions are indicated by their graphs, which are shown in Figure 6.1.

B. Bessel's Equation of Order p

We now consider Bessel's equation of order p for $x > 0$, which we have already introduced at the beginning of Section 6.3A, and seek solutions valid for $0 < x < R$. This is the equation

$$x^2 \frac{d^2 y}{dx^2} + x \frac{dy}{dx} + (x^2 - p^2)y = 0, \quad (6.101)$$

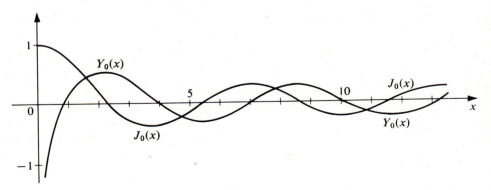

Figure 6.1

where we now assume that p is real and positive. We see at once that $x = 0$ is a regular singular point of Equation (6.101); and since we seek solutions valid for $0 < x < R$, we may assume a solution

$$y = \sum_{n=0}^{\infty} c_n x^{n+r}, \tag{6.112}$$

where $c_0 \neq 0$. Differentiating (6.112), substituting into (6.101), and simplifying as in our previous examples, we obtain

$$(r^2 - p^2)c_0 x^r + [(r+1)^2 - p^2]c_1 x^{r+1}$$

$$+ \sum_{n=2}^{\infty} \{[(n+r)^2 - p^2]c_n + c_{n-2}\}x^{n+r} = 0. \tag{6.113}$$

Equating to zero the coefficient of each power of x in (6.113), we obtain

$$r^2 - p^2 = 0, \tag{6.114}$$

$$[(r+1)^2 - p^2]c_1 = 0, \tag{6.115}$$

and

$$[(n+r)^2 - p^2]c_n + c_{n-2} = 0, \qquad n \geq 2. \tag{6.116}$$

Equation (6.114) is the indicial equation of the differential equation (6.101). Its roots are $r_1 = p > 0$ and $r_2 = -p$. If $r_1 - r_2 = 2p > 0$ is unequal to a positive integer, then from Theorem 6.3 we know that the differential equation (6.101) has *two* linearly independent solutions of the form (6.112). However, if $r_1 - r_2 = 2p$ is equal to a positive integer, we are only certain of a solution of this form corresponding to the *larger* root $r_1 = p$. We shall now proceed to obtain this one solution, the existence of which is always assured.

Letting $r = r_1 = p$ in (6.115), we obtain $(2p+1)c_1 = 0$. Thus, since $p > 0$, we must have $c_1 = 0$. Letting $r = r_1 = p$ in (6.116), we obtain the recurrence formula

$$n(n + 2p)c_n + c_{n-2} = 0, \qquad n \geq 2,$$

or

$$c_n = -\frac{c_{n-2}}{n(n+2p)}, \qquad n \geq 2, \tag{6.117}$$

corresponding to the larger root p. From this one finds that all odd coefficients are zero (since $c_1 = 0$) and that the general even coefficient is given by

$$c_{2n} = \frac{(-1)^n c_0}{[2 \cdot 4 \cdots (2n)][(2 + 2p)(4 + 2p) \cdots (2n + 2p)]}$$

$$= \frac{(-1)^n c_0}{2^{2n} n! [(1+p)(2+p) \cdots (n+p)]}, \qquad n \geq 1.$$

Hence the solution of the differential equation (6.101) corresponding to the larger root p is given by $y = y_1(x)$, where

$$y_1(x) = c_0 \sum_{n=0}^{\infty} \frac{(-1)^n x^{2n+p}}{2^{2n} n! [(1+p)(2+p) \cdots (n+p)]}. \tag{6.118}$$

If p is a positive integer, we may write this in the form

$$y_1(x) = c_0 2^p p! \sum_{n=0}^{\infty} \frac{(-1)^n}{n!(n+p)!} \left(\frac{x}{2}\right)^{2n+p}.$$ (6.119)

If p is unequal to a positive integer, we need a generalization of the factorial function in order to express $y_1(x)$ in a form analogous to that given by (6.119). Such a generalization is provided by the so-called *gamma function*, which we now introduce.

For $N > 0$ the gamma function is defined by the convergent improper integral

$$\Gamma(N) = \int_0^{\infty} e^{-x} x^{N-1}\, dx.$$ (6.120)

If N is a positive integer, it can be shown that

$$N! = \Gamma(N+1).$$ (6.121)

If N is positive but *not* an integer, we use (6.121) to *define* $N!$. The gamma function has been studied extensively. It can be shown that $\Gamma(N)$ satisfies the recurrence formula

$$\Gamma(N+1) = N\Gamma(N) \qquad (N > 0).$$ (6.122)

Values of $\Gamma(N)$ have been tabulated and are usually given for the range $1 \le N \le 2$. Using the tabulated values of $\Gamma(N)$ for $1 \le N \le 2$, one can evaluate $\Gamma(N)$ for all $N > 0$ by repeated application of formula (6.122). Suppose, for example, that we wish to evaluate $(\frac{3}{2})!$. From the definition (6.121), we have $(\frac{3}{2})! = \Gamma(\frac{5}{2})$. Then from (6.122), we find that $\Gamma(\frac{5}{2}) = \frac{3}{2}\Gamma(\frac{3}{2})$. From tables one finds that $\Gamma(\frac{3}{2}) \approx 0.8862$, and thus $(\frac{3}{2})! = \Gamma(\frac{5}{2}) \approx 1.3293$.

For $N < 0$ the integral (6.120) diverges, and thus $\Gamma(N)$ is not defined by (6.120) for negative values of N. We extend the definition of $\Gamma(N)$ to values of $N < 0$ by demanding that the recurrence formula (6.122) hold for *negative* (as well as positive) values of N. Repeated use of this formula thus defines $\Gamma(N)$ for every nonintegral negative value of N.

Thus $\Gamma(N)$ is defined for all $N \neq 0, -1, -2, -3, \ldots$. The graph of this function is shown in Figure 6.2. We now define $N!$ for all $N \neq -1, -2, -3, \ldots$ by the formula (6.121).

We now return to the solution y_1 defined by (6.118), for the case in which p is unequal to a positive integer. Applying the recurrence formula (6.122) successively with $N = n+p, n+p-1, n+p-2, \ldots, p+1$, we obtain

$$\Gamma(n+p+1) = (n+p)(n+p-1)\cdots(p+1)\Gamma(p+1).$$

Thus for p unequal to a positive integer we may write the solution defined by (6.118) in the form

$$y_1(x) = c_0 \Gamma(p+1) \sum_{n=0}^{\infty} \frac{(-1)^n x^{2n+p}}{2^{2n} n! \Gamma(n+p+1)}$$

$$= c_0 2^p \Gamma(p+1) \sum_{n=0}^{\infty} \frac{(-1)^n}{n! \Gamma(n+p+1)} \left(\frac{x}{2}\right)^{2n+p}.$$ (6.123)

Now using (6.121) with $N = p$ and $N = n+p$, we see that (6.123) takes the form (6.119). Thus the solution of the differential equation (6.101) corresponding to the larger

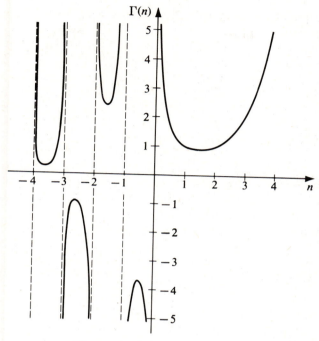

Figure 6.2. Graph of Γ (N)

root $p > 0$ is given by (6.119), where $p!$ and $(n + p)!$ are defined by $\Gamma(p + 1)$ and $\Gamma(n + p + 1)$, respectively, if p is not a positive integer.

If we set the arbitrary constant c_0 in (6.119) equal to the reciprocal of $2^p p!$, we obtain an important particular solution of (6.101). This particular solution defines a function denoted by J_p and called the *Bessel function of the first kind of order p*. That is, the function J_p is the particular solution of Equation (6.101) defined by

$$J_p(x) = \sum_{n=0}^{\infty} \frac{(-1)^n}{n!(n + p)!} \left(\frac{x}{2}\right)^{2n+p}, \tag{6.124}$$

where $(n + p)!$ is defined by $\Gamma(n + p + 1)$ if p is not a positive integer.

Throughout this discussion we have assumed that $p > 0$ in (6.101) and hence that $p > 0$ in (6.124). If $p = 0$ in (6.101), then (6.101) reduces to the Bessel equation of order zero given by (6.102) and the solution (6.124) reduces to the Bessel function of the first kind of order zero given by (6.107).

If $p = 1$ in (6.101), then Equation (6.101) becomes

$$x^2 \frac{d^2y}{dx^2} + x \frac{dy}{dx} + (x^2 - 1)y = 0, \tag{6.125}$$

which is Bessel's equation of order one. Letting $p = 1$ in (6.124) we obtain a solution of Equation (6.125) that is called the *Bessel function of the first kind of order one* and is denoted by J_1. That is, the function J_1 is defined by

$$J_1(x) = \sum_{n=0}^{\infty} \frac{(-1)^n}{n!(n + 1)!} \left(\frac{x}{2}\right)^{2n+1}. \tag{6.126}$$

The graphs of the functions J_0, defined by (6.107), and J_1, defined by (6.126), are shown in Figure 6.3.

Several interesting properties of the Bessel functions of the first kind are suggested by these graphs. For one thing, they suggest that J_0 and J_1 each have a damped oscillatory behavior and that the positive zeros of J_0 and J_1 separate each other. This is indeed the case. In fact, it may be shown that for every $p \geq 0$ the function J_p has a damped oscillatory behavior as $x \to \infty$ and the positive zeros of J_p and J_{p+1} separate each other.

We now know that for every $p \geq 0$ one solution of Bessel's equation of order p (6.101) is given by (6.124). We now consider briefly the problem of finding a linearly independent solution of (6.101). We have already found such a solution for the case in which $p = 0$; it is given by (6.110). For $p > 0$, we have observed that if $2p$ is unequal to a positive integer, then the differential equation (6.101) has a linearly independent solution of the form (6.112) corresponding to the smaller root $r_2 = -p$. We now proceed to work with this smaller root.

Letting $r = r_2 = -p$ in (6.115), we obtain

$$(-2p + 1)c_1 = 0. \tag{6.127}$$

Letting $r = r_2 = -p$ in (6.116) we obtain the recurrence formula

$$n(n - 2p)c_n + c_{n-2} = 0, \qquad n \geq 2, \tag{6.128}$$

or

$$c_n = -\frac{c_{n-2}}{n(n - 2p)}, \qquad n \geq 2, \quad n \neq 2p. \tag{6.129}$$

Using (6.127) and (6.128) or (6.129) one finds solutions $y = y_2(x)$, corresponding to the smaller root $-p$. Three distinct cases occur, leading to solutions of the following forms:

1. If $2p \neq$ a positive integer,

$$y_2(x) = c_0 x^{-p}\left(1 + \sum_{n=1}^{\infty} \alpha_{2n} x^{2n}\right), \tag{6.130}$$

where c_0 is an arbitrary constant and the $\alpha_{2n}(n = 1, 2, \ldots)$ are definite constants.

2. If $2p =$ an *odd* positive integer,

$$y_2(x) = c_0 x^{-p}\left(1 + \sum_{n=1}^{\infty} \beta_{2n} x^{2n}\right) + c_{2p} x^{p}\left(1 + \sum_{n=1}^{\infty} \gamma_{2n} x^{2n}\right), \tag{6.131}$$

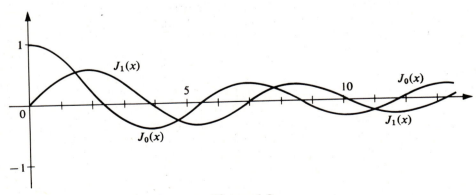

Figure 6.3

where c_0 and c_{2p} are arbitrary constants and β_{2n} and $\gamma_{2n}(n = 1, 2, \ldots)$ are definite constants.

3. If $2p = $ an *even* positive integer,

$$y_2(x) = c_{2p} x^p \left(1 + \sum_{n=1}^{\infty} \delta_{2n} x^{2n} \right), \tag{6.132}$$

where c_{2p} is an arbitrary constant and the $\delta_{2n}(n = 1, 2, \ldots)$ are definite constants.

In Case 1 the solution defined by (6.130) is linearly independent of J_p. In Case 2 the solution defined by (6.131) with $c_{2p} = 0$ is also linearly independent of J_p. However, in Case 3 the solution defined by (6.132) is merely a constant multiple of $J_p(x)$, and hence this solution is *not* linearly independent of J_p. Thus if $2p$ is unequal to an *even* positive integer, there exists a linearly independent solution of the form (6.112) corresponding to the smaller root $-p$. In other words, if p is unequal to a positive integer, the differential equation (6.101) has a solution of the form $y = y_2(x)$, where

$$y_2(x) = \sum_{n=0}^{\infty} c_{2n} x^{2n-p}, \tag{6.133}$$

and this solution y_2 is linearly independent of J_p.

It is easy to determine the coefficients c_{2n} in (6.133). We observe that the recurrence formula (6.129) corresponding to the smaller root $-p$ is obtained from the recurrence formula (6.117) corresponding to the larger root p simply by replacing p in (6.117) by $-p$. Hence a solution of the form (6.133) may be obtained from (6.124) simply by replacing p in (6.124) by $-p$. This leads at once to the solution denoted by J_{-p} and defined by

$$J_{-p}(x) = \sum_{n=0}^{\infty} \frac{(-1)^n}{n!(n-p)!} \left(\frac{x}{2} \right)^{2n-p}, \tag{6.134}$$

where $(n - p)!$ is defined by $\Gamma(n - p + 1)$.

Thus if $p > 0$ is unequal to a positive integer, two linearly independent solutions of the differential equation (6.101) are J_p defined by (6.124), and J_{-p}, defined by (6.134). Hence, if $p > 0$ is unequal to a positive integer, the general solution of Bessel's equation of order p is given by

$$y = C_1 J_p(x) + C_2 J_{-p}(x),$$

where J_p and J_{-p} are defined by (6.124) and (6.134), respectively, and C_1 and C_2 are arbitrary constants.

If p is a positive integer, the corresponding solution defined by (6.123) is *not* linearly independent of J_p, as we have already noted. Hence in this case a solution that is linearly independent of J_p must be given by $y = y_p(x)$, where

$$y_p(x) = x^{-p} \sum_{n=0}^{\infty} c_n^* x^n + C J_p(x) \ln x,$$

where $C \neq 0$. Such a linearly independent solution y_p may be found by the method of reduction of order. Then the general solution of the differential equation (6.101) may be written as a general linear combination of J_p and y_p. However, as in the case of Bessel's equation of order zero, it is customary to choose a certain special linear combination of J_p and y_p and take this special combination as the "second" solution of Equation

(6.101). This special combination is denoted by Y_p and defined by

$$Y_p(x) = \frac{2}{\pi}\left\{\left(\ln\frac{x}{2} + \gamma\right)J_p(x) - \frac{1}{2}\sum_{n=0}^{p-1}\frac{(p-n-1)!}{n!}\left(\frac{x}{2}\right)^{2n-p}\right.$$

$$\left. + \frac{1}{2}\sum_{n=0}^{\infty}(-1)^{n+1}\left(\sum_{k=1}^{n}\frac{1}{k} + \sum_{k=1}^{n+p}\frac{1}{k}\right)\left[\frac{1}{n!(n+p)!}\left(\frac{x}{2}\right)^{2n+p}\right]\right\}, \quad (6.135)$$

where γ is Euler's constant. The solution Y_p is called the *Bessel function of the second kind of order p* (Weber's form).

Thus if p is a positive integer, two linearly independent solutions of the differential equation (6.101) are J_p, defined by (6.124), and Y_p, defined by (6.135). Hence if p is a positive integer, the general solution of Bessel's equation of order p is given by

$$y = C_1 J_p(x) + C_2 Y_p(x),$$

where J_p and Y_p are defined by (6.124) and (6.135), respectively, and C_1 and C_2 are arbitrary constants.

Exercises

1. Show that $J_0(kx)$, where k is a constant, satisfies the differential equation

$$x\frac{d^2y}{dx^2} + \frac{dy}{dx} + k^2xy = 0.$$

2. Show that the transformation

$$y = \frac{u(x)}{\sqrt{x}}$$

reduces the Bessel equation of order p, Equation (6.101), to the form

$$\frac{d^2u}{dx^2} + \left[1 + \left(\frac{1}{4} - p^2\right)\frac{1}{x^2}\right]u = 0.$$

3. Use the result of Exercise 2 to obtain a solution of the Bessel equation of order $\frac{1}{2}$.

4. Using the series definition (6.124) for J_p, show that

$$\frac{d}{dx}[x^pJ_p(kx)] = kx^pJ_{p-1}(kx)$$

and

$$\frac{d}{dx}[x^{-p}J_p(kx)] = -kx^{-p}J_{p+1}(kx),$$

where k is a constant.

5. Use the results of Exercise 4 to show that

$$\frac{d}{dx}[J_p(kx)] = kJ_{p-1}(kx) - \frac{p}{x}J_p(kx),$$

$$\frac{d}{dx}[J_p(kx)] = -kJ_{p+1}(kx) + \frac{p}{x}J_p(kx).$$

Hence show that

$$\frac{d}{dx}\left[J_p(kx)\right] = \frac{k}{2}\left[J_{p-1}(kx) - J_{p+1}(kx)\right],$$

$$J_p(kx) = \frac{kx}{2p}\left[J_{p-1}(kx) + J_{p+1}(kx)\right].$$

6. Using the results of Exercise 5,

(a) express $J_1(x)$ and $\dfrac{d}{dx}\left[J_1(x)\right]$ in terms of $J_0(x)$ and $J_2(x)$;

(b) express $J_{n+1/2}(x)$ in terms of $J_{n-1/2}(x)$ and $J_{n-3/2}(x)$.

CHAPTER SEVEN

Systems of Linear Differential Equations

In the previous chapters we have been concerned with one differential equation in one unknown function. Now we shall consider systems of two differential equations in two unknown functions, and more generally, systems of n differential equations in n unknown functions. We shall restrict our attention to linear systems only, and we shall begin by considering various types of these systems. After this, we shall proceed to introduce differential operators, present an operator method of solving linear systems, and then consider some basic applications of this method. We shall then turn to a study of the fundamental theory and basic method of solution for a standard type of linear system in the special case of two equations in two unknown functions. Following this, we shall outline the most basic material about matrices and vectors. We shall then present the basic method for the standard type of linear system by means of matrices and vectors, first in the special case of two equations in two unknown functions and then in the more general case of n equations in n unknown functions. The fundamental theory for this general case will be presented in Chapter 11.

7.1 DIFFERENTIAL OPERATORS AND AN OPERATOR METHOD

A. Types of Linear Systems

We start by introducing the various types of linear systems that we shall consider. The general linear system of two first-order differential equations in two unknown functions x and y is of the form

$$a_1(t)\frac{dx}{dt} + a_2(t)\frac{dy}{dt} + a_3(t)x + a_4(t)y = F_1(t),$$

$$b_1(t)\frac{dx}{dt} + b_2(t)\frac{dy}{dt} + b_3(t)x + b_4(t)y = F_2(t). \tag{7.1}$$

We shall be concerned with systems of this type that have constant coefficients. An example of such a system is

$$2\frac{dx}{dt} + 3\frac{dy}{dt} - 2x + y = t^2, \qquad \frac{dx}{dt} - 2\frac{dy}{dt} + 3x + 4y = e^t.$$

We shall say that a *solution* of system (7.1) is an ordered pair of real functions (f, g) such that $x = f(t)$, $y = g(t)$ simultaneously satisfy both equations of the system (7.1) on some real interval $a \leq t \leq b$.

The general linear system of three first-order differential equations in three unknown functions x, y, and z is of the form

$$a_1(t)\frac{dx}{dt} + a_2(t)\frac{dy}{dt} + a_3(t)\frac{dz}{dt} + a_4(t)x + a_5(t)y + a_6(t)z = F_1(t),$$

$$b_1(t)\frac{dx}{dt} + b_2(t)\frac{dy}{dt} + b_3(t)\frac{dz}{dt} + b_4(t)x + b_5(t)y + b_6(t)z = F_2(t), \qquad (7.2)$$

$$c_1(t)\frac{dx}{dt} + c_2(t)\frac{dy}{dt} + c_3(t)\frac{dz}{dt} + c_4(t)x + c_5(t)y + c_6(t)z = F_3(t).$$

As in the case of systems of the form (7.1), so also in this case we shall be concerned with systems that have constant coefficients. An example of such a system is

$$\frac{dx}{dt} + \frac{dy}{dt} - 2\frac{dz}{dt} + 2x - 3y + z = t,$$

$$2\frac{dx}{dt} - \frac{dy}{dt} + 3\frac{dz}{dt} + x + 4y - 5z = \sin t,$$

$$\frac{dx}{dt} + 2\frac{dy}{dt} + \frac{dz}{dt} - 3x + 2y - z = \cos t.$$

We shall say that a solution of system (7.2) is an ordered triple of real functions (f, g, h) such that $x = f(t)$, $y = g(t)$, $z = h(t)$ simultaneously satisfy all three equations of the system (7.2) on some real interval $a \leq t \leq b$.

Systems of the form (7.1) and (7.2) contained only first derivatives, and we now consider the basic linear system involving higher derivatives. This is the general linear system of two second-order differential equations in two unknown fucntions x and y, and is a system of the form

$$a_1(t)\frac{d^2x}{dt^2} + a_2(t)\frac{d^2y}{dt^2} + a_3(t)\frac{dx}{dt} + a_4(t)\frac{dy}{dt} + a_5(t)x + a_6(t)y = F_1(t),$$

$$b_1(t)\frac{d^2x}{dt^2} + b_2(t)\frac{d^2y}{dt^2} + b_3(t)\frac{dx}{dt} + b_4(t)\frac{dy}{dt} + b_5(t)x + b_6(t)y = F_2(t). \qquad (7.3)$$

We shall be concerned with systems having constant coefficients in this case also, and an example is provided by

$$2\frac{d^2x}{dt^2} + 5\frac{d^2y}{dt^2} + 7\frac{dx}{dt} + 3\frac{dy}{dt} + 2y = 3t + 1,$$

$$3\frac{d^2x}{dt^2} + 2\frac{d^2y}{dt^2} - 2\frac{dy}{dt} + 4x + y = 0.$$

For given fixed positive integers m and n, we could proceed, in like manner, to exhibit other general linear systems of n mth-order differential equations in n unknown functions and give examples of each such type of system. Instead we proceed to introduce the standard type of linear system referred to in the introductory paragraph at the start of the chapter, and of which we shall make a more systematic study later. We introduce this standard type as a special case of the system (7.1) of two first-order differential equations in two unknowns functions x and y.

We consider the special type of linear system (7.1), which is of the form

$$\frac{dx}{dt} = a_{11}(t)x + a_{12}(t)y + F_1(t),$$

$$\frac{dy}{dt} = a_{21}(t)x + a_{22}(t)y + F_2(t).$$

(7.4)

This is the so-called *normal form* in the case of two linear differential equations in two unknown functions. The characteristic feature of such a system is apparent from the manner in which the derivatives appear in it. An example of such a system with variable coefficients is

$$\frac{dx}{dt} = t^2 x + (t + 1)y + t^3,$$

$$\frac{dy}{dt} = te^t x + t^3 y - e^t,$$

while one with constant coefficients is

$$\frac{dx}{dt} = 5x + 7y + t^2,$$

$$\frac{dy}{dt} = 2x - 3y + 2t.$$

The normal form in the case of a linear system of three differential equations in three unknown functions x, y, and z is

$$\frac{dx}{dt} = a_{11}(t)x + a_{12}(t)y + a_{13}(t)z + F_1(t),$$

$$\frac{dy}{dt} = a_{21}(t)x + a_{22}(t)y + a_{23}(t)z + F_2(t),$$

$$\frac{dz}{dt} = a_{31}(t)x + a_{32}(t)y + a_{33}(t)z + F_3(t).$$

An example of such a system is the constant coefficient system

$$\frac{dx}{dt} = 3x + 2y + z + t,$$

$$\frac{dy}{dt} = 2x - 4y + 5z - t^2,$$

$$\frac{dz}{dt} = 4x + y - 3z + 2t + 1.$$

The normal form in the general case of a linear system of n differential equations in n unknown functions $x_1, x_2, \ldots, x_n$ is

$$\frac{dx_1}{dt} = a_{11}(t)x_1 + a_{12}(t)x_2 + \cdots + a_{1n}(t)x_n + F_1(t),$$

$$\frac{dx_2}{dt} = a_{21}(t)x_1 + a_{22}(t)x_2 + \cdots + a_{2n}(t)x_n + F_2(t), \tag{7.5}$$

$$\vdots$$

$$\frac{dx_n}{dt} = a_{n1}(t)x_1 + a_{n2}(t)x_2 + \cdots + a_{nn}(t)x_n + F_n(t).$$

An important fundamental property of a normal linear system (7.5) is its relationship to a single nth-order linear differential equation in one unknown function. Specifically, consider the so-called normalized (meaning, the coefficient of the highest derivative is one) nth-order linear differential equation

$$\frac{d^n x}{dt^n} + a_1(t)\frac{d^{n-1}x}{dt^{n-1}} + \cdots + a_{n-1}(t)\frac{dx}{dt} + a_n(t)x = F(t) \tag{7.6}$$

in the one unknown function x. Let

$$x_1 = x, \qquad x_2 = \frac{dx}{dt},$$

$$x_3 = \frac{d^2 x}{dt^2}, \ldots, x_{n-1} = \frac{d^{n-2}x}{dt^{n-2}}, \qquad x_n = \frac{d^{n-1}x}{dt^{n-1}}. \tag{7.7}$$

From (7.7), we have

$$\frac{dx}{dt} = \frac{dx_1}{dt}, \quad \frac{d^2 x}{dt^2} = \frac{dx_2}{dt}, \ldots, \frac{d^{n-1}}{dt^{n-1}} = \frac{dx_{n-1}}{dt}, \quad \frac{d^n x}{dt^n} = \frac{dx_n}{dt}. \tag{7.8}$$

Then using both (7.7) and (7.8), the single nth-order equation (7.6) can be transformed into

$$\frac{dx_1}{dt} = x_2,$$

$$\frac{dx_2}{dt} = x_3,$$

$$\vdots \tag{7.9}$$

$$\frac{dx_{n-1}}{dt} = x_n,$$

$$\frac{dx_n}{dt} = -a_n(t)x_1 - a_{n-1}(t)x_2 - \cdots - a_1(t)x_n + F(t),$$

which is a special case of the normal linear system (7.5) of n equations in n unknown functions. Thus we see that a single nth-order linear differential equation of form (7.6) is one unknown function is indeed intimately related to a normal linear system (7.5) of n first-order differential equation in n unknown functions.

B. Differential Operators

In this section we shall present a symbolic operator method for solving linear systems with constant coefficients. This method depends upon the use of so-called *differential operators*, which we now introduce.

Let x be an n-times differentiable function of the independent variable t. We denote the operation of differentiation with respect to t by the symbol D and call D a differential operator. In terms of this differential operator the derivative dx/dt is denoted by Dx. That is,

$$Dx \equiv dx/dt.$$

In like manner, we denote the second derivative of x with respect to t by D^2x. Extending this, we denote the nth derivative of x with respect to t by D^nx. That is,

$$D^nx = \frac{d^nx}{dt^n} \qquad (n = 1, 2, \ldots).$$

Further extending this operator notation, we write

$$(D + c)x \quad \text{to denote} \quad \frac{dx}{dt} + cx$$

and

$$(aD^n + bD^m)x \quad \text{to denote} \quad a\frac{d^nx}{dt^n} + b\frac{d^mx}{dt^m},$$

where a, b, and c are constants.

In this notation the general linear differential expression with constant coefficients $a_0, a_1, \ldots, a_{n-1}, a_n$,

$$a_0\frac{d^nx}{dt^n} + a_1\frac{d^{n-1}x}{dt^{n-1}} + \cdots + a_{n-1}\frac{dx}{dt} + a_nx,$$

is written

$$(a_0D^n + a_1D^{n-1} + \cdots + a_{n-1}D + a_n)x.$$

Observe carefully that the operators $D^n, D^{n-1}, \ldots, D$ in this expression do *not* represent quantities that are to be multiplied by the function x, but rather they indicate *operations* (differentiations) that are to be carried out upon this function. The expression

$$a_0D^n + a_1D^{n-1} + \cdots + a_{n-1}D + a_n$$

by itself, where $a_0, a_1, \ldots, a_{n-1}, a_n$ are constants, is called a linear differential operator with constant coefficients.

▶ **Example 7.1**

Consider the linear differential operator

$$3D^2 + 5D - 2.$$

If x is a twice differentiable function of t, then

$$(3D^2 + 5D - 2)x \quad \text{denotes} \quad 3\frac{d^2x}{dt^2} + 5\frac{dx}{dt} - 2x.$$

For example, if $x = t^3$, we have

$$(3D^2 + 5D - 2)t^3 = 3\frac{d^2}{dt^2}(t^3) + 5\frac{d}{dt}(t^3) - 2(t^3) = 18t + 15t^2 - 2t^3.$$

We shall now discuss certain useful properties of the linear differential operator with constant coefficients. In order to facilitate our discussion, we shall let L denote this operator. That is,

$$L \equiv a_0D^n + a_1D^{n-1} + \cdots + a_{n-1}D + a_n,$$

where $a_0, a_1, \ldots, a_{n-1}, a_n$ are constants. Now suppose that f_1 and f_2 are both n-times differentiable functions of t and c_1 and c_2 are constants. Then it can be shown that

$$L[c_1 f_1 + c_2 f_2] = c_1 L[f_1] + c_2 L[f_2].$$

For example, if the operator $L \equiv 3D^2 + 5D - 2$ is applied to $3t^2 + 2\sin t$, then

$$L[3t^2 + 2\sin t] = 3L[t^2] + 2L[\sin t]$$

or

$$(3D^2 + 5D - 2)(3t^2 + 2\sin t) = 3(3D^2 + 5D - 2)t^2 + 2(3D^2 + 5D - 2)\sin t.$$

Now let

$$L_1 \equiv a_0D^m + a_1D^{m-1} + \cdots + a_{m-1}D + a_m$$

and

$$L_2 \equiv b_0D^n + b_1D^{n-1} + \cdots + b_{n-1}D + b_n$$

be two linear differential operators with constant coefficients $a_0, a_1, \ldots, a_{m-1}, a_m$, and $b_0, b_1, \ldots, b_{n-1}, b_n$, respectively. Let

$$L_1(r) \equiv a_0r^m + a_1r^{m-1} + \cdots + a_{m-1}r + a_m$$

and

$$L_2(r) \equiv b_0r^n + b_1r^{n-1} + \cdots + b_{n-1}r + b_n$$

be the two polynomials in the quantity r obtained from the operators L_1 and L_2, respectively, by formally replacing D by r, D^2 by $r^2, \ldots,$ D^k by r^k. Let us denote the product of the polynomials $L_1(r)$ and $L_2(r)$ by $L(r)$; that is,

$$L(r) = L_1(r)L_2(r).$$

Then, if f is a function possessing $n + m$ derivatives, it can be shown that

$$L_1 L_2 f = L_2 L_1 f = Lf, \tag{7.10}$$

where L is the operator obtained from the "product polynomial" $L(r)$ by formally replacing r by D, r^2 by $D^2, \ldots,$ r^{m+n} by D^{m+n}. Equation (7.10) indicates two important properties of linear differential operators with constant coefficients. First, it states the effect of first operating on f by L_2 and then operating on the resulting function by L_1 is the same as that which results from first operating on f by L_1 and then operating on this resulting function by L_2. Second, Equation (7.10) states that the effect of first operating on f by either L_1 or L_2 and then operating on the resulting function by the other is the same as that which results from operating on f by the "product operator" L. We illustrate these important properties in the following example.

▶ **Example 7.2**

Let $L_1 \equiv D^2 + 1$, $L_2 \equiv 3D + 2$, $f(t) = t^3$. Then

$$L_1 L_2 f = (D^2 + 1)(3D + 2)t^3 = (D^2 + 1)(9t^2 + 2t^3)$$
$$= 9(D^2 + 1)t^2 + 2(D^2 + 1)t^3$$
$$= 9(2 + t^2) + 2(6t + t^3) = 2t^3 + 9t^2 + 12t + 18$$

and

$$L_2 L_1 f = (3D + 2)(D^2 + 1)t^3 = (3D + 2)(6t + t^3)$$
$$= 6(3D + 2)t + (3D + 2)t^3$$
$$= 6(3 + 2t) + (9t^2 + 2t^3) = 2t^3 + 9t^2 + 12t + 18.$$

Finally, $L \equiv 3D^3 + 2D^2 + 3D + 2$ and

$$Lf = (3D^3 + 2D^2 + 3D + 2)t^3 = 3(6) + 2(6t) + 3(3t^2) + 2t^3$$
$$= 2t^3 + 9t^2 + 12t + 18.$$

Now let $L \equiv a_0 D^n + a_1 D^{n-1} + \cdots + a_{n-1}D + a_n$, where $a_0, a_1, \ldots, a_{n-1}, a_n$ are constants; and let $L(r) \equiv a_0 r^n + a_1 r^{n-1} + \cdots + a_{n-1}r + a_n$ be the polynomial in r obtained from L by formally replacing D by r, D^2 by $r^2, \ldots, D^n$ by r^n. Let $r_1, r_2, \ldots, r_n$ be the roots of the polynomial equation $L(r) = 0$. Then $L(r)$ may be written in the factored form

$$L(r) = a_0(r - r_1)(r - r_2) \cdots (r - r_n).$$

Now formally replacing r by D in the right member of this identity, we may express the operator $L \equiv a_0 D^n + a_1 D^{n-1} + \cdots + a_{n-1}D + a_n$ in the factored form

$$L = a_0(D - r_1)(D - r_2) \cdots (D - r_n).$$

We thus observe that linear differential operators with constant coefficients can be formally multiplied and factored exactly as if they were polynomials in the algebraic quantity D.

C. An Operator Method for Linear Systems with Constant Coefficients

We now proceed to explain a symbolic operator method for solving linear systems with constant coefficients. We shall outline the procedure of this method on a strictly formal basis and shall make no attempt to justify it.

We consider a linear system of the form

$$L_1 x + L_2 y = f_1(t),$$
$$L_3 x + L_4 y = f_2(t), \tag{7.11}$$

where L_1, L_2, L_3, and L_4 are linear differential operators with constant coefficients.

That is, L_1, L_2, L_3, and L_4 are operators of the forms

$$L_1 \equiv a_0 D^m + a_1 D^{m-1} + \cdots + a_{m-1} D + a_m,$$

$$L_2 \equiv b_0 D^n + b_1 D^{n-1} + \cdots + b_{n-1} D + b_n,$$

$$L_3 \equiv \alpha_0 D^p + \alpha_1 D^{p-1} + \cdots + \alpha_{p-1} D + \alpha_p,$$

$$L_4 \equiv \beta_0 D^q + \beta_1 D^{q-1} + \cdots + \beta_{q-1} D + \beta_q,$$

where the a's, b's, α's, and β's are constants.

A simple example of a system which may be expressed in the form (7.11) is provided by

$$2\frac{dx}{dt} - 2\frac{dy}{dt} - 3x = t,$$

$$2\frac{dx}{dt} + 2\frac{dy}{dt} + 3x + 8y = 2.$$

Introducing operator notation this system takes the form

$$(2D - 3)x - 2Dy = t,$$

$$(2D + 3)x + (2D + 8)y = 2.$$

This is clearly of the form (7.11), where $L_1 \equiv 2D - 3$, $L_2 \equiv -2D$, $L_3 \equiv 2D + 3$, and $L_4 \equiv 2D + 8$.

Returning now to the general system (7.11), we apply the operator L_4 to the first equation of (7.11) and the operator L_2 to the second equation of (7.11), obtaining

$$L_4 L_1 x + L_4 L_2 y = L_4 f_1,$$

$$L_2 L_3 x + L_2 L_4 y = L_2 f_2.$$

We now subtract the second of these equations from the first. Since $L_4 L_2 y = L_2 L_4 y$, we obtain

$$L_4 L_1 x - L_2 L_3 x = L_4 f_1 - L_2 f_2,$$

or

$$(L_1 L_4 - L_2 L_3)x = L_4 f_1 - L_2 f_2. \tag{7.12}$$

The expression $L_1 L_4 - L_2 L_3$ in the left member of this equation is itself a linear differential operator with constant coefficients. We assume that it is neither zero nor a nonzero constant and denote it by L_5. If we further assume that the functions f_1 and f_2 are such that the right member $L_4 f_1 - L_2 f_2$ of (7.12) exists, then this member is some function, say g_1, of t. Then Equation (7.12) may be written

$$L_5 x = g_1. \tag{7.13}$$

Equation (7.13) is a linear differential equation with constant coefficients in the single dependent variable x. We thus observe that our procedure has eliminated the other dependent variable y. We now solve the differential equation (7.13) for x using the methods developed in Chapter 4. Suppose Equation (7.13) is of order N. Then the general solution of (7.13) is of the form

$$x = c_1 u_1 + c_2 u_2 + \cdots + c_N u_N + U_1, \tag{7.14}$$

where $u_1, u_2, \ldots, u_N$ are N linearly independent solutions of the homogeneous linear equation $L_5 x = 0, c_1, c_2, \ldots, c_N$ are arbitrary constants, and U_1 is a particular solution of $L_5 x = g_1$.

We again return to the system (7.11) and this time apply the operators L_3 and L_1 to the first and second equations, respectively, of the system. We obtain

$$L_3 L_1 x + L_3 L_2 y = L_3 f_1,$$
$$L_1 L_3 x + L_1 L_4 y = L_1 f_2.$$

Subtracting the first of these from the second, we obtain

$$(L_1 L_4 - L_2 L_3) y = L_1 f_2 - L_3 f_1.$$

Assuming that f_1 and f_2 are such that the right member $L_1 f_2 - L_3 f_1$ of this equation exists, we may express it as some function, say g_2, of t. Then this equation may be written

$$L_5 y = g_2, \tag{7.15}$$

where L_5 denotes the operator $L_1 L_4 - L_2 L_3$. Equation (7.15) is a linear differential equation with constant coefficients in the single dependent variable y. This time we have eliminated the dependent variable x. Solving the differential equation (7.15) for y, we obtain its general solution in the form

$$y = k_1 u_1 + k_2 u_2 + \cdots + k_N u_N + U_2, \tag{7.16}$$

where $u_1, u_2, \ldots, u_N$ are the N linearly independent solutions of $L_5 y = 0$ (or $L_5 x = 0$) that already appear in (7.14), $k_1, k_2, \ldots, k_N$ are arbitrary constants, and U_2 is a particular solution of $L_5 y = g_2$.

We thus see that if x and y satisfy the linear system (7.11), then x satisfies the single linear differential equation (7.13) and y satisfies the single linear differential equation (7.15). Thus if x and y satisfy the system (7.11), then x is of the form (7.14) and y is of the form (7.16). However, the pairs of functions given by (7.14) and (7.16) do *not* satisfy the given system (7.11) for *all* choices of the constants $c_1, c_2, \ldots, c_N, k_1, k_2, \ldots, k_N$. That is, these pairs (7.14) and (7.16) do not simultaneously satisfy both equations of the given system (7.11) for arbitrary choices of the $2N$ constants $c_1, c_2, \ldots, c_N, k_1, k_2, \ldots, k_N$. In other words, in order for x given by (7.14) and y given by (7.16) to satisfy the given system (7.11), the $2N$ constants $c_1, c_2, \ldots, c_N, k_1, k_2, \ldots, k_N$ cannot all be independent but rather certain of them must be dependent on the others. It can be shown that the number of independent constants in the so-called general solution of the linear system (7.11) is equal to the order of the operator $L_1 L_4 - L_2 L_3$ obtained from the determinant

$$\begin{vmatrix} L_1 & L_2 \\ L_3 & L_4 \end{vmatrix}$$

of the operator "coefficients" of x and y in (7.11), provided that this determinant is not zero. We have assumed that this operator is of order N. Thus in order for the pair (7.14) and (7.16) to satisfy the system (7.11) only N of the $2N$ constants in this pair can be independent. The remaining N constants must depend upon the N that are independent. In order to determine which of these $2N$ constants may be chosen as independent and how the remaining N then relate to the N so chosen, we must substitute x as given by (7.14) and y as given by (7.16) into the system (7.11). This determines the relations that must exist among the constants $c_1, c_2, \ldots, c_N, k_1, k_2, \ldots, k_N$

in order that the pair (7.14) and (7.16) constitute the so-called general solution of (7.11). Once this has been done, appropriate substitutions based on these relations are made in (7.14) and/or (7.16) and then the resulting pair (7.14) and (7.16) contain the required number N of arbitrary constants and so does indeed constitute the so-called general solution of system (7.11).

We now illustrate the above procedure with an example.

▶ **Example 7.3**

Solve the system

$$2\frac{dx}{dt} - 2\frac{dy}{dt} - 3x = t,$$

$$2\frac{dx}{dt} + 2\frac{dy}{dt} + 3x + 8y = 2. \tag{7.17}$$

We introduce operator notation and write this system in the form

$$(2D - 3)x - 2Dy = t,$$

$$(2D + 3)x + (2D + 8)y = 2. \tag{7.18}$$

We apply the operator $(2D + 8)$ to the first equation of (7.18) and the operator $2D$ to the second equation of (7.18), obtaining

$$(2D + 8)(2D - 3)x - (2D + 8)2Dy = (2D + 8)t,$$

$$2D(2D + 3)x + 2D(2D + 8)y = (2D)2.$$

Adding these two equations, we obtain

$$[(2D + 8)(2D - 3) + 2D(2D + 3)]x = (2D + 8)t + (2D)2$$

or

$$(8D^2 + 16D - 24)x = 2 + 8t + 0$$

or, finally

$$(D^2 + 2D - 3)x = t + \tfrac{1}{4}. \tag{7.19}$$

The general solution of the differential equation (7.19) is

$$x = c_1 e^t + c_2 e^{-3t} - \tfrac{1}{3}t - \tfrac{11}{36}. \tag{7.20}$$

We now return to the system (7.18) and apply the operator $(2D + 3)$ to the first equation of (7.18) and the operator $(2D - 3)$ to the second equation of (7.18). We obtain

$$(2D + 3)(2D - 3)x - (2D + 3)2Dy = (2D + 3)t,$$

$$(2D - 3)(2D + 3)x + (2D - 3)(2D + 8)y = (2D - 3)2.$$

Subtracting the first of these equations from the second, we have

$$[(2D - 3)(2D + 8) + (2D + 3)2D]y = (2D - 3)2 - (2D + 3)t$$

or

$$(8D^2 + 16D - 24)y = 0 - 6 - 2 - 3t$$

or, finally,

$$(D^2 + 2D - 3)y = -\tfrac{3}{8}t - 1. \tag{7.21}$$

The general solution of the differential equation (7.21) is

$$y = k_1 e^t + k_2 e^{-3t} + \tfrac{1}{8}t + \tfrac{5}{12}. \tag{7.22}$$

Thus if x and y satisfy the system (7.17), then x must be of the form (7.20) and y must be of the form (7.22) for some choice of the constants c_1, c_2, k_1, k_2. The determinant of the operator "coefficients" of x and y in (7.18) is

$$\begin{vmatrix} 2D - 3 & -2D \\ 2D + 3 & 2D + 8 \end{vmatrix} = 8D^2 + 16D - 24.$$

Since this is of order 2, the number of independent constants in the general solution of the system (7.17) must also be two. Thus in order for the pair (7.20) and (7.22) to satisfy the system (7.17) only two of the four constants c_1, c_2, k_1, and k_2 can be independent. In order to determine the necessary relations that must exist among these constants, we substitute x as given by (7.20) and y as given by (7.22) into the system (7.17). Substituting into the first equation of (7.17), we have

$$\left[2c_1 e^t - 6c_2 e^{-3t} - \tfrac{2}{3}\right] - \left[2k_1 e^t - 6k_2 e^{-3t} + \tfrac{1}{4}\right] - \left[3c_1 e^t + 3c_2 e^{-3t} - t - \tfrac{11}{12}\right] = t$$

or

$$(-c_1 - 2k_1)e^t + (-9c_2 + 6k_2)e^{-3t} = 0.$$

Thus in order that the pair (7.20) and (7.22) satisfy the first equation of the system (7.17) we must have

$$\begin{aligned} -c_1 - 2k_1 &= 0, \\ -9c_2 + 6k_2 &= 0. \end{aligned} \tag{7.23}$$

Substitution of x and y into the second equation of the system (7.17) will lead to relations equivalent to (7.23). Hence in order for the pair (7.20) and (7.22) to satisfy the system (7.17), the relations (7.23) must be satisfied. Two of the four constants in (7.23) must be chosen as independent. If we choose c_1 and c_2 as independent, then we have

$$k_1 = -\tfrac{1}{2}c_1 \quad \text{and} \quad k_2 = \tfrac{3}{2}c_2.$$

Using these values for k_1 and k_2 in (7.22), the resulting pair (7.20) and (7.22) constitute the general solution of the system (7.17). That is, the general solution of (7.17) is given by

$$x = c_1 e^t + c_2 e^{-3t} - \tfrac{1}{3}t - \tfrac{11}{36},$$
$$y = -\tfrac{1}{2}c_1 e^t + \tfrac{3}{2}c_2 e^{-3t} + \tfrac{1}{8}t + \tfrac{5}{12},$$

where c_1 and c_2 are arbitrary constants. If we had chosen k_1 and k_2 as the independent constants in (7.23), then the general solution of the system (7.17) would have been written

$$x = -2k_1 e^t + \tfrac{2}{3}k_2 e^{-3t} - \tfrac{1}{3}t - \tfrac{11}{36},$$
$$y = k_1 e^t + k_2 e^{-3t} + \tfrac{1}{8}t + \tfrac{5}{12}.$$

An Alternative Procedure. Here we present an alternative procedure for solving a linear system of the form

$$L_1 x + L_2 y = f_1(t),$$
$$L_3 x + L_4 y = f_2(t), \tag{7.11}$$

where L_1, L_2, L_3, and L_4 are linear differential operators with constant coefficients. This alternative procedure begins in exactly the same way as the procedure already described. That is, we first apply the operator L_4 to the first equation of (7.11) and the operator L_2 to the second equation of (7.11), obtaining

$$L_4 L_1 x + L_4 L_2 y = L_4 f_1,$$
$$L_2 L_3 x + L_2 L_4 y = L_2 f_2.$$

We next subtract the second from the first, obtaining

$$(L_1 L_4 - L_2 L_3)x = L_4 f_1 - L_2 f_2, \tag{7.12}$$

which, under the same assumptions as we previously made at this point, may be written

$$L_5 x = g_1. \tag{7.13}$$

Then we solve this single linear differential equation with constant coefficients in the single dependent variable x. Assuming its order is N, we obtain its general solution in the form

$$x = c_1 u_1 + c_2 u_2 + \cdots + c_N u_N + U_1, \tag{7.14}$$

where $u_1, u_2, \ldots, u_N$ are N linearly independent solutions of the homogeneous linear equation $L_5 x = 0, c_1, c_2, \ldots, c_N$ are N arbitrary constants, and U_1 is a particular solution of $L_5 x = g_1$.

Up to this point, we have indeed proceeded just exactly as before. However, we now return to system (7.11) and attempt to eliminate from it *all* terms that involve the derivatives of the *other* dependent variable y. In other words, we attempt to obtain from system (7.11) a relation R that involves the still unknown y but *none of the derivatives of* y. This relation R will involve x and/or certain of the derivatives of x; but x is given by (7.14) and its derivatives can readily be found from (7.14). Finding these derivatives of x and substituting them and the known x itself into the relation R, we see that the result is merely a single linear *algebraic* equation in the one unknown y. Solving it, we thus determine y without the need to find (7.15) and (7.16) or to relate the arbitrary constants.

As we shall see, this alternative procedure always applies in an easy straightforward manner if the operators L_1, L_2, L_3, and L_4 are all of the first order. However, for systems involving one or more higher-order operators, it is generally difficult to eliminate *all* the derivatives of y.

We now give an explicit presentation of the procedure for finding y when L_1, L_2, L_3, and L_4 are all first-order operators.

Specifically, suppose

$$L_1 \equiv a_0 D + a_1,$$
$$L_2 \equiv b_0 D + b_1,$$
$$L_3 \equiv \alpha_0 D + \alpha_1,$$
$$L_4 \equiv \beta_0 D + \beta_1.$$

Then (7.11) is

$$(a_0D + a_1)x + (b_0D + b_1)y = f_1(t),$$
$$(\alpha_0 D + \alpha_1)x + (\beta_0 D + \beta_1)y = f_2(t). \tag{7.24}$$

Multiplying the first equation of (7.24) by β_0 and the second by $-b_0$ and adding, we obtain

$$[(a_0\beta_0 - b_0\alpha_0)D + (a_1\beta_0 - b_0\alpha_1)]x + (b_1\beta_0 - b_0\beta_1)y = \beta_0 f_1(t) - b_0 f_2(t).$$

Note that this involves y but *none of the derivatives of* y. From this, we at once obtain

$$y = \frac{(b_0\alpha_0 - a_0\beta_0)Dx + (b_0\alpha_1 - a_1\beta_0)x + \beta_0 f_1(t) - b_0 f_2(t)}{b_1\beta_0 - b_0\beta_1}, \tag{7.25}$$

assuming $b_1\beta_0 - b_0\beta_1 \neq 0$. Now x is given by (7.14) and Dx may be found from (7.14) by straightforward differentiation. Then substituting these known expressions for x and Dx into (7.25), we at once obtain y without the need of obtaining (7.15) and (7.16) and hence without having to determine any relations between constants c_i and $k_i(i = 1, 2, \ldots, N)$, as in the original procedure.

We illustrate the alternative procedure by applying it to the system of Example 7.3.

▶ **Example 7.4**

Solve the system

$$2\frac{dx}{dt} - 2\frac{dy}{dt} - 3x = t,$$
$$2\frac{dx}{dt} + 2\frac{dy}{dt} + 3x + 8y = 2. \tag{7.17}$$

of Example 7.3 by the alternative procedure that we have just described.

Following this alternative procedure, we introduce operator notation and write the system (7.17) in the form

$$(2D - 3)x - 2Dy = t,$$
$$(2D + 3)x + (2D + 8)y = 2. \tag{7.18}$$

Now we eliminate y, obtain the differential equation

$$(D^2 + 2D - 3)x = t + \tfrac{1}{4} \tag{7.19}$$

for x, and find its general solution

$$x = c_1 e^t + c_2 e^{-3t} - \tfrac{1}{3}t - \tfrac{11}{36}, \tag{7.20}$$

exactly as in Example 7.3.

We now proceed using the alternative method. We first obtain from (7.18) a relation that involves the unknown y but *not* the derivative Dy. The system (7.18) of this example is so very simple that we do so by merely adding the equations (7.18). Doing so, we at once obtain

$$4Dx + 8y = t + 2,$$

which does indeed involve y but *not* the derivative Dy, as desired. From this, we at once find

$$y = \tfrac{1}{8}(t + 2 - 4Dx). \tag{7.26}$$

From (7.20), we find

$$Dx = c_1 e^t - 3c_2 e^{-3t} - \tfrac{1}{3}.$$

Substituting into (7.26), we get

$$y = \tfrac{1}{8}(t + 2 - 4c_1 e^t + 12c_2 e^{-3t} + \tfrac{4}{3})$$
$$= -\tfrac{1}{2}c_1 e^t + \tfrac{3}{2}c_2 e^{-3t} + \tfrac{1}{8}t + \tfrac{5}{12}.$$

Thus the general solution of the system may be written

$$x = c_1 e^t + c_2 e^{-3t} - \tfrac{1}{3}t - \tfrac{11}{36},$$
$$y = -\tfrac{1}{2}c_1 e^t + \tfrac{3}{2}c_2 e^{-3t} + \tfrac{1}{8}t + \tfrac{5}{12},$$

where c_1 and c_2 are arbitrary constants.

Exercises

Use the operator method described in this section to find the general solution of each of the following linear systems.

1. $\dfrac{dx}{dt} + \dfrac{dy}{dt} - 2x - 4y = e^t,$

    $\dfrac{dx}{dt} + \dfrac{dy}{dt} - y = e^{4t}.$

2. $\dfrac{dx}{dt} + \dfrac{dy}{dt} - x = -2t,$

    $\dfrac{dx}{dt} + \dfrac{dy}{dt} - 3x - y = t^2.$

3. $\dfrac{dx}{dt} + \dfrac{dy}{dt} - x - 3y = e^t,$

    $\dfrac{dx}{dt} + \dfrac{dy}{dt} + x = e^{3t}.$

4. $\dfrac{dx}{dt} + \dfrac{dy}{dt} - x - 2y = 2e^t,$

    $\dfrac{dx}{dt} + \dfrac{dy}{dt} - 3x - 4y = e^{2t}.$

5. $2\dfrac{dx}{dt} + \dfrac{dy}{dt} - x - y = e^{-t},$

    $\dfrac{dx}{dt} + \dfrac{dy}{dt} + 2x + y = e^t.$

6. $2\dfrac{dx}{dt} + \dfrac{dy}{dt} - 3x - y = t,$

    $\dfrac{dx}{dt} + \dfrac{dy}{dt} - 4x - y = e^t.$

7. $\dfrac{dx}{dt} + \dfrac{dy}{dt} - x - 6y = e^{3t},$

    $\dfrac{dx}{dt} + 2\dfrac{dy}{dt} - 2x - 6y = t.$

8. $\dfrac{dx}{dt} + \dfrac{dy}{dt} - x - 3y = 3t,$

    $\dfrac{dx}{dt} + 2\dfrac{dy}{dt} - 2x - 3y = 1.$

9. $\dfrac{dx}{dt} + \dfrac{dy}{dt} + 2y = \sin t,$

    $\dfrac{dx}{dt} + \dfrac{dy}{dt} - x - y = 0.$

10. $\dfrac{dx}{dt} - \dfrac{dy}{dt} - 2x + 4y = t,$

    $\dfrac{dx}{dt} + \dfrac{dy}{dt} - x - y = 1.$

11. $2\dfrac{dx}{dt} + \dfrac{dy}{dt} + x + 5y = 4t,$

$\dfrac{dx}{dt} + \dfrac{dy}{dt} + 2x + 2y = 2.$

12. $\dfrac{dx}{dt} + \dfrac{dy}{dt} - x + 5y = t^2,$

$\dfrac{dx}{dt} + 2\dfrac{dy}{dt} - 2x + 4y = 2t + 1.$

13. $2\dfrac{dx}{dt} + \dfrac{dy}{dt} + x + y = t^2 + 4t,$

$\dfrac{dx}{dt} + \dfrac{dy}{dt} + 2x + 2y = 2t^2 - 2t.$

14. $3\dfrac{dx}{dt} + 2\dfrac{dy}{dt} - x + y = t - 1,$

$\dfrac{dx}{dt} + \dfrac{dy}{dt} - x = t + 2.$

15. $2\dfrac{dx}{dt} + 4\dfrac{dy}{dt} + x - y = 3e^t,$

$\dfrac{dx}{dt} + \dfrac{dy}{dt} + 2x + 2y = e^t.$

16. $2\dfrac{dx}{dt} + \dfrac{dy}{dt} - x - y = -2t,$

$\dfrac{dx}{dt} + \dfrac{dy}{dt} + x - y = t^2.$

17. $2\dfrac{dx}{dt} + \dfrac{dy}{dt} - x - y = 1,$

$\dfrac{dx}{dt} + \dfrac{dy}{dt} + 2x - y = t.$

18. $\dfrac{d^2x}{dt^2} + \dfrac{dy}{dt} = e^{2t},$

$\dfrac{dx}{dt} + \dfrac{dy}{dt} - x - y = 0.$

19. $\dfrac{d^2x}{dt^2} + \dfrac{dy}{dt} - x + y = 1,$

$\dfrac{d^2y}{dt^2} + \dfrac{dx}{dt} - x + y = 0.$

20. $\dfrac{d^2x}{dt^2} - \dfrac{dy}{dt} = e^t,$

$\dfrac{dx}{dt} + \dfrac{dy}{dt} - 4x - y = 2e^t.$

21. $\dfrac{d^2x}{dt^2} - \dfrac{dy}{dt} = t + 1,$

$\dfrac{dx}{dt} + \dfrac{dy}{dt} - 3x + y = 2t - 1.$

22. $\dfrac{d^2x}{dt^2} + 4\dfrac{dy}{dt} + x - 4y = 0,$

$\dfrac{dx}{dt} + \dfrac{dy}{dt} - x + 9y = e^{2t}.$

In each of Exercises 23–26, transform the single linear differential equation of the form (7.6) into a system of first-order differential equations of the form (7.9).

23. $\dfrac{d^2x}{dt^2} - 3\dfrac{dx}{dt} + 2x = t^2.$

24. $\dfrac{d^3x}{dt^3} + 2\dfrac{d^2x}{dt^2} - \dfrac{dx}{dt} - 2x = e^{3t}.$

25. $\dfrac{d^3x}{dt^3} + t\dfrac{d^2x}{dt^2} + 2t^3\dfrac{dx}{dt} - 5t^4 = 0.$

26. $\dfrac{d^4x}{dt^4} - t^2\dfrac{d^2x}{dt^2} + 2tx = \cos t.$

7.2 APPLICATIONS

A. Applications to Mechanics

Systems of linear differential equations originate in the mathematical formulation of numerous problems in mechanics. We consider one such problem in the following example. Another mechanics problem leading to a linear system in given in the exercises at the end of this section.

▶ **Example 7.5**

On a smooth horizontal plane BC (for example, a smooth table top) an object A_1 is connected to a fixed point P by a massless spring S_1 of natural length L_1. An object A_2 is then connected to A_1 by a massless spring S_2 of natural length L_2 in such a way that the fixed point P and the centers of gravity A_1 and A_2 all lie in a straight line (see Figure 7.1).

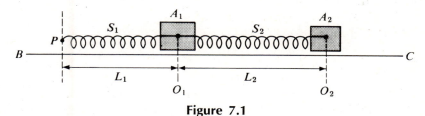

Figure 7.1

The object A_1 is then displaced a distance a_1 to the right or left of its equilibrium position O_1, the object A_2 is displaced a distance a_2 to the right or left of its equilibrium position O_2, and at time $t = 0$ the two objects are released (see Figure 7.2). What are the positions of the two objects at any time $t > 0$?

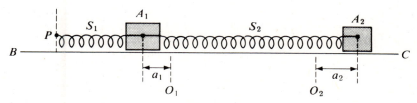

Figure 7.2

Formulation. We assume first that the plane BC is so smooth that frictional forces may be neglected. We also assume that no external forces act upon the system. Suppose object A_1 has mass m_1 and object A_2 has mass m_2. Further suppose spring S_1 has spring constant k_1 and spring S_2 has spring constant k_2.

Let x_1 denote the displacement of A_1 from its equilibrium position O_1 at time $t \geq 0$ and assume that x_1 is positive when A_1 is to the right of O_1. In like manner, let x_2 denote the displacement of A_2 from its equilibrium position O_2 at time $t \geq 0$ and assume that x_2 is positive when A_2 is to the right of O_2 (see Figure 7.3).

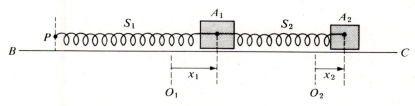

Figure 7.3

Consider the forces acting on A_1 at time $t > 0$. There are two such forces, F_1 and F_2, where F_1 is exerted by spring S_1 and F_2 is exerted by spring S_2. By Hooke's law (Section 5.1) the force F_1 is of magnitude $k_1|x_1|$. Since this force is exerted toward the left when A_1 is to the right of O_1 and toward the right when A_1 is to the left of O_1, we have $F_1 = -k_1x_1$. Again using Hooke's law, the force F_2 is of magnitude k_2s, where s is the elongation of S_2 at time t. Since $s = |x_2 - x_1|$, we see that the magnitude of F_2 is $k_2|x_2 - x_1|$. Further, since this force is exerted toward the left when $x_2 - x_1 < 0$ and toward the right when $x_2 - x_1 > 0$, we see that $F_2 = k_2(x_2 - x_1)$.

Now applying Newton's second law (Section 3.2) to the object A_1, we obtain the differential equation

$$m_1 \frac{d^2x_1}{dt^2} = -k_1x_1 + k_2(x_2 - x_1). \tag{7.27}$$

We now turn to the object A_2 and consider the forces that act upon it at time $t > 0$. There is one such force, F_3, and this is exerted by spring S_2. Applying Hooke's law once again, we observe that this force is also of magnitude $k_2s = k_2|x_2 - x_1|$. Since F_3 is exerted toward the left when $x_2 - x_1 > 0$ and toward the right when $x_2 - x_1 < 0$, we see that $F_3 = -k(x_2 - x_1)$. Applying Newton's second law to the object A_2, we obtain the differential equation

$$m_2 \frac{d^2x_2}{dt^2} = -k_2(x_2 - x_1). \tag{7.28}$$

In addition to the differential equations (7.27) and (7.28), we see from the statement of the problem that the initial conditions are given by

$$x_1(0) = a_1, \qquad x_1'(0) = 0, \qquad x_2(0) = a_2, \qquad x_2'(0) = 0. \tag{7.29}$$

The mathematical formulation of the problem thus consists of the differential equations (7.27) and (7.28) and the initial conditions (7.29). Writing the differential equations in the form

$$m_1 \frac{d^2x_1}{dt^2} + (k_1 + k_2)x_1 - k_2x_2 = 0,$$

$$m_2 \frac{d^2x_2}{dt^2} - k_2x_1 + k_2x_2 = 0, \tag{7.30}$$

we see that they form a system of homogeneous linear differential equations with constant coefficients.

Solution of a Specific Case. Rather than solve the general problem consisting of the system (7.30) and conditions (7.29), we shall carry through the solution in a particular case that was chosen to facilitate the work. Suppose the two objects A_1 and A_2 are each of unit mass, so that $m_1 = m_2 = 1$. Further, suppose that the springs S_1 and S_2 have spring constants $k_1 = 3$ and $k_2 = 2$, respectively. Also, we shall take $a_1 = -1$ and $a_2 = 2$. Then the system (7.30) reduces to

$$\frac{d^2x_1}{dt^2} + 5x_1 - 2x_2 = 0,$$

$$\frac{d^2x_2}{dt^2} - 2x_1 + 2x_2 = 0, \tag{7.31}$$

and the initial conditions (7.29) become

$$x_1(0) = -1, \qquad x_1'(0) = 0, \qquad x_2(0) = 2, \qquad x_2'(0) = 0. \tag{7.32}$$

Writing the system (7.31) in operator notation, we have

$$(D^2 + 5)x_1 - 2x_2 = 0,$$
$$-2x_1 + (D^2 + 2)x_2 = 0. \tag{7.33}$$

We apply the operator $(D^2 + 2)$ to the first equation of (7.33), multiply the second equation of (7.33) by 2, and add the two equations to obtain

$$[(D^2 + 2)(D^2 + 5) - 4]x_1 = 0$$

or

$$(D^4 + 7D^2 + 6)x_1 = 0. \tag{7.34}$$

The auxiliary equation corresponding to the fourth-order differential equation (7.34) is

$$m^4 + 7m^2 + 6 = 0 \quad \text{or} \quad (m^2 + 6)(m^2 + 1) = 0.$$

Thus the general solution of the differential equation (7.34) is

$$x_1 = c_1 \sin t + c_2 \cos t + c_3 \sin \sqrt{6}t + c_4 \cos \sqrt{6}t. \tag{7.35}$$

We now multiply the first equation of (7.33) by 2, apply the operator $(D^2 + 5)$ to the second equation of (7.33), and add to obtain the differential equation

$$(D^4 + 7D^2 + 6)x_2 = 0 \tag{7.36}$$

for x_2. The general solution of (7.36) is clearly

$$x_2 = k_1 \sin t + k_2 \cos t + k_3 \sin \sqrt{6}t + k_4 \cos \sqrt{6}t. \tag{7.37}$$

The determinant of the operator "coefficients" in the system (7.33) is

$$\begin{vmatrix} D^2 + 5 & -2 \\ -2 & D^2 + 2 \end{vmatrix} = D^4 + 7D^2 + 6.$$

Since this is a fourth-order operator, the general solution of (7.31) must contain four independent constants. We must substitute x_1 given by (7.35) and x_2 given by (7.37) into the equations of the system (7.31) to determine the relations that must exist among the constants $c_1, c_2, c_3, c_4, k_1, k_2, k_3$, and k_4 in order that the pair (7.35) and (7.37) represent the general solution of (7.31). Substituting, we find that

$$k_1 = 2c_1, \qquad k_2 = 2c_2, \qquad k_3 = -\tfrac{1}{2}c_3, \qquad k_4 = -\tfrac{1}{2}c_4.$$

Thus the general solution of the system (7.31) is given by

$$x_1 = c_1 \sin t + c_2 \cos t + c_3 \sin \sqrt{6}t + c_4 \cos \sqrt{6}t,$$
$$x_2 = 2c_1 \sin t + 2c_2 \cos t - \tfrac{1}{2}c_3 \sin \sqrt{6}t - \tfrac{1}{2}c_4 \cos \sqrt{6}t. \tag{7.38}$$

We now apply the initial conditions (7.32). Applying the conditions $x_1 = -1$, $dx_1/dt = 0$ at $t = 0$ to the first of the pair (7.38), we find

$$-1 = c_2 + c_4,$$
$$0 = c_1 + \sqrt{6}c_3. \tag{7.39}$$

Applying the conditions $x_2 = 2$, $dx_2/dt = 0$ at $t = 0$ to the second of the pair (7.38), we obtain

$$2 = 2c_2 - \tfrac{1}{2}c_4,$$

$$0 = 2c_1 - \frac{\sqrt{6}}{2}c_3.$$

(7.40)

From Equations (7.39) and (7.40), we find that

$$\dot{c}_1 = 0, \qquad c_2 = \tfrac{3}{5}, \qquad c_3 = 0, \qquad c_4 = -\tfrac{8}{5}.$$

Thus the particular solution of the specific problem consisting of the system (7.31) and the conditions (7.32) is

$$x_1 = \tfrac{3}{5}\cos t - \tfrac{8}{5}\cos\sqrt{6}t,$$

$$x_2 = \tfrac{6}{5}\cos t + \tfrac{4}{5}\cos\sqrt{6}t.$$

B. Applications to Electric Circuits

In Section 5.6 we considered the application of differential equations to electric circuits consisting of a single closed path. A closed path in an electrical network is called a *loop*. We shall now consider electrical networks that consist of several loops. For example, consider the network shown in Figure 7.4.

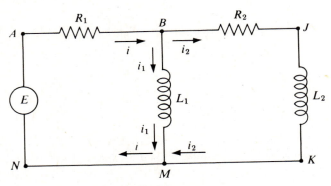

Figure 7.4

This network consists of the three loops $ABMNA$, $BJKMB$, and $ABJKMNA$. Points such as B and M at which two or more circuits join are called *junction points* or *branch points*. The direction of current flow has been arbitrarily assigned and indicated by arrows.

In order to solve problems involving multiple loop networks we shall need two fundamental laws of circuit theory. One of these is Kirchhoff's voltage law, which we have already stated and applied in Section 5.6. The other basic law that we shall employ is the following:

Kirchhoff's Current Law. In an electrical network the total current flowing into a junction point is equal to the total current flowing away from the junction point.

As an application of these laws we consider the following problem dealing with the circuit of Figure 7.4.

▶ **Example 7.6**

Determine the currents in the electrical network of Figure 7.4, if E is an electromotive force of 30 V, R_1 is a resistor of 10 Ω, R_2 is a resistor of 20 Ω, L_1 is an inductor of 0.02 H, L_2 is an inductor of 0.04 H, and the currents are initially zero.

Formulation. The current flowing in the branch $MNAB$ is denoted by i, that flowing on the branch BM by i_1, and that flowing on the branch $BJKM$ by i_2.

We now apply Kirchhoff's voltage law (Section 5.6) to each of the three loops $ABMNA$, $BJKMB$, and $ABJKMNA$.

For the loop $ABMNA$ the voltage drops are as follows:

1. Across the resistor R_1: $10i$.

2. Across the inductor L_1: $0.02 \dfrac{di_1}{dt}$.

Thus applying the voltage law to the loop $ABMNA$, we have the equation

$$0.02 \frac{di_1}{dt} + 10i = 30. \tag{7.41}$$

For the loop $BJKMB$, the voltage drops are as follows:

1. Across the resistor R_2: $20i_2$.

2. Across the inductor L_2: $0.04 \dfrac{di_2}{dt}$.

3. Across the inductor L_1: $-0.02 \dfrac{di_1}{dt}$.

The minus sign enters into 3 since we traverse the branch MB in the direction opposite to that of the current i_1 as we complete the loop $BJKMB$. Since the loop $BJKMB$ contains no electromotive force, upon applying the voltage law to this loop we obtain the equation

$$-0.02 \frac{di_1}{dt} + 0.04 \frac{di_2}{dt} + 20i_2 = 0. \tag{7.42}$$

For the loop $ABJKMNA$, the voltage drops are as follows:

1. Across the resistor R_1: $10i$.

2. Across the resistor R_2: $20i_2$.

3. Across the inductor L_2: $0.04 \dfrac{di_2}{dt}$.

Applying the voltage law to this loop, we obtain the equation

$$10i + 0.04\frac{di_2}{dt} + 20i_2 = 30. \tag{7.43}$$

We observe that the three equations (7.41), (7.42), and (7.43) are not all independent. For example, we note that (7.42) may be obtained by subtracting (7.41) from (7.43). Thus we need to retain only the two equations (7.41) and (7.43).

We now apply Kirchhoff's current law to the junction point B. From this we see at once that

$$i = i_1 + i_2. \tag{7.44}$$

In accordance with this we replace i by $i_1 + i_2$ in (7.41) and (7.43) and thus obtain the linear system

$$0.02\frac{di_1}{dt} + 10i_1 + 10i_2 = 30,$$

$$10i_1 + 0.04\frac{di_2}{dt} + 30i_2 = 30. \tag{7.45}$$

Since the currents are initially zero, we have the initial conditions

$$i_1(0) = 0 \quad \text{and} \quad i_2(0) = 0. \tag{7.46}$$

Solution. We introduce operator notation and write the system (7.45) in the form

$$(0.02D + 10)i_1 + 10i_2 = 30,$$

$$10i_1 + (0.04D + 30)i_2 = 30. \tag{7.47}$$

We apply the operator $(0.04D + 30)$ to the first equation of (7.47), multiply the second by 10, and subtract to obtain

$$[(0.04D + 30)(0.02D + 10) - 100]i_1 = (0.04D + 30)30 - 300$$

or

$$(0.0008D^2 + D + 200)i_1 = 600$$

or, finally,

$$(D^2 + 1250D + 250,000)i_1 = 750,000. \tag{7.48}$$

We now solve the differential equation (7.48) for i_1. The auxiliary equation is

$$m^2 + 1250m + 250,000 = 0$$

or

$$(m + 250)(m + 1000) = 0.$$

Thus the complementary function of Equation (7.48) is

$$i_{1,c} = c_1 e^{-250t} + c_2 e^{-1000t},$$

and a particular integral is obviously $i_{1,p} = 3$. Hence the general solution of the differential equation (7.48) is

$$i_1 = c_1 e^{-250t} + c_2 e^{-1000t} + 3. \tag{7.49}$$

Now returning to the system (7.47), we multiply the first equation of the system by 10, apply the operator $(0.02 + 10)$ to the second equation, and subtract the first from the second. After simplifications we obtain the differential equation

$$(D^2 + 1250D + 250{,}000)i_2 = 0$$

for i_2. The general solution of this differential equation is clearly

$$i_2 = k_1 e^{-250t} + k_2 e^{-1000t}. \tag{7.50}$$

Since the determinant of the operator "coefficients" in the system (7.47) is a second-order operator, the general solution of the system (7.45) must contain two independent constants. We must substitute i_1 given by (7.49) and i_2 given by (7.50) into the equations of the system (7.45) to determine the relations that must exist among the constants c_1, c_2, k_1, k_2 in order that the pair (7.49) and (7.50) represent the general solution of (7.45). Substituting, we find that

$$k_1 = -\tfrac{1}{2}c_1, \qquad k_2 = c_2. \tag{7.51}$$

Thus the general solution of the system (7.45) is given by

$$\begin{aligned} i_1 &= c_1 e^{-250t} + c_2 e^{-1000t} + 3, \\ i_2 &= -\tfrac{1}{2}c_1 e^{-250t} + c_2 e^{-1000t}. \end{aligned} \tag{7.52}$$

Now applying the initial conditions (7.46), we find that $c_1 + c_2 + 3 = 0$ and $-\tfrac{1}{2}c_1 + c_2 = 0$ and hence $c_1 = -2$ and $c_2 = -1$. Thus the solution of the linear system (7.45) that satisfies the conditions (7.46) is

$$\begin{aligned} i_1 &= -2e^{-250t} - e^{-1000t} + 3, \\ i_2 &= e^{-250t} - e^{-1000t}. \end{aligned}$$

Finally, using (7.44) we find that

$$i = -e^{-250t} - 2e^{-1000t} + 3.$$

We observe that the current i_2 rapidly approaches zero. On the other hand, the currents, i_1 and $i = i_1 + i_2$ rapidly approach the value 3.

C. Applications to Mixture Problems

In Section 3.3C, we considered mixture involving the amount of a substance S in a mixture in a single container, into and from which there flowed mixtures containing S. Here we extend the problem to situations involving *two* containers. That is, we consider a substance S in mixtures in two interconnected containers, into and from which there flow mixtures containing S. The mixture in each container is kept uniform by stirring; and we seek to determine the amount of substance S present in each container at time t.

Let x denote the amount of S present in tank X at time t, and let y denote the amount of S present in tank Y at time t. We then apply the basic equation (3.55) of Section 3.3C in the case of *each* of the unknowns x and y. Now, however, the "IN" and "OUT" terms in each of the resulting equations also depends on just how the two containers are interconnected. We illustrate with the following example.

▶ **Example 7.7**

Two tanks X and Y are interconnected (see Figure 7.5). Tank X initially contains 100 liters of brine in which there is dissolved 5 kg of salt, and tank Y initially contains 100 liters of brine in which there is dissolved 2 kg of salt. Starting at time $t = 0$, (1) pure water flows into tank X at the rate of 6 liters/min, (2) brine flows from tank X into tank Y at the rate of 8 liters/min, (3) brine is pumped from tank Y back into tank X at the rate of 2 liters/min, and (4) brine flows out of tank Y and away from the system at the rate of 6 liters/min. The mixture in each tank is kept uniform by stirring. How much salt is in each tank at any time $t > 0$?

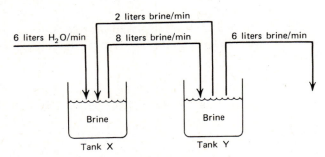

Figure 7.5

Formulation. Let $x =$ the amount of salt in tank X at time t, and let $y =$ the amount of salt in tank Y at time t, each measured in kilograms. Each of these tanks initially contains 100 liters of fluid, and fluid flows both in and out of each tank at the same rate, 8 liters/min, so each tank always contains 100 liters of fluid. Thus the concentration of salt at time t in tank X is $x/100$ (kg/liter) and that in tank Y is $y/100$ (kg/liters).

The only salt entering tank X is in the brine that is pumped from tank Y back into tank X. Since this enters at the rate of 2 liters/min and contains $y/100$ kg/liter, the rate at which salt enters tank X is $2y/100$. Similarly, the only salt leaving tank X is in the brine that flows from tank X into tank Y. Since this leaves at the rate of 8 liters/min and contains $x/100$ kg/liter, the rate at which salt leaves tank X is $8x/100$. Thus we obtain the differential equation

$$\frac{dx}{dt} = \frac{2y}{100} - \frac{8x}{100} \tag{7.53}$$

for the amount of salt in tank X at time t. In a similar way, we obtain the differential equation

$$\frac{dy}{dt} = \frac{8x}{100} - \frac{8y}{100} \tag{7.54}$$

for the amount of salt in tank Y at time t. Since initially there was 5 kg of salt in tank X and 2 kg in tank Y, we have the initial conditions

$$x(0) = 5, \qquad y(0) = 2 \tag{7.55}$$

Thus we have the linear system consisting of differential equations (7.53) and (7.54) and initial conditions (7.55).

Solution. We introduce operator notation and write the differential equations (7.53) and (7.54) in the forms

$$\left(D + \frac{8}{100}\right)x - \frac{2}{100}y = 0,$$

$$-\frac{8}{100}x + \left(D + \frac{8}{100}\right)y = 0. \tag{7.56}$$

We apply the operator $\left(D + \frac{8}{100}\right)$ to the first equation of (7.56), multiply the second equation by $\frac{2}{100}$, and add to obtain

$$\left[\left(D + \frac{8}{100}\right)\left(D + \frac{8}{100}\right) - \frac{16}{(100)^2}\right]x = 0,$$

which quickly reduces to

$$\left[D^2 + \frac{16}{100}D + \frac{48}{(100)^2}\right]x = 0. \tag{7.57}$$

We now solve the homogeneous differential equation (7.57) for x. The auxiliary equation is

$$m^2 + \frac{16}{100}m + \frac{48}{(100)^2} = 0,$$

or

$$\left(m + \frac{4}{100}\right)\left(m + \frac{12}{100}\right) = 0,$$

with real distinct roots $(-1)/25$ and $(-3)/25$. Thus the general solution of equation (7.57) is

$$x = c_1 e^{-(1/25)t} + c_2 e^{-(3/25)t}. \tag{7.58}$$

Now applying the so-called alternative procedure of Section 7.1C, we obtain from system (7.56) a relation that involves the unknown y but *not* the derivative Dy. The system (7.56) is so especially simple that the first equation of this system is itself such a relation. Solving this for y, we at once obtain

$$y = 50Dx + 4x. \tag{7.59}$$

From (7.58), we find

$$Dx = -\frac{c_1}{25}e^{-(1/25)t} - \frac{3c_2}{25}e^{-(3/25)t}.$$

Substituting into (7.59), we get

$$y = 2c_1 e^{-(1/25)t} - 2c_2 e^{-(3/25)t}.$$

Thus the general solution of the system (7.56) is

$$x = c_1 e^{-(1/25)t} + c_2 e^{-(3/25)t},$$
$$y = 2c_1 e^{-(1/25)t} - 2c_2 e^{-(3/25)t}. \tag{7.60}$$

We now apply the initial conditions (7.55). We at once obtain

$$c_1 + c_2 = 5,$$
$$2c_1 - 2c_2 = 2,$$

from which we find

$$c_1 = 3, \qquad c_2 = 2.$$

Thus the solution of the linear system (7.56) that satisfies the initial conditions (7.55) is

$$x = 3e^{-(1/25)t} + 2e^{-(3/25)t},$$
$$y = 6e^{-(1/25)t} - 4e^{-(3/25)t}.$$

These expressions give the amount of salt x in tank X and the amount y in tank Y, respectively, each measured in kilograms, at any time t (min) > 0. Thus, for example, after 25 min, we find

$$x = 3e^{-1} + 2e^{-3} \approx 1.203 \text{ (kg)},$$
$$y = 6e^{-1} - 4e^{-3} \approx 2.008 \text{ (kg)}.$$

Note that as $t \to \infty$, both x and $y \to 0$. Thus is in accordance with the fact that no salt at all (but only pure water) flows into the system from outside.

Exercises

1. Solve the problem of Example 7.5 for the case in which the object A_1 has mass $m_1 = 2$, the object A_2 has mass $m_2 = 1$, the spring S_1 has spring constant $k_1 = 4$, the spring S_2 has spring constant $k_2 = 2$, and the initial conditions are $x_1(0) = 1$, $x_1'(0) = 0$, $x_2(0) = 5$, and $x_2'(0) = 0$.

2. A projectile of mass m is fired into the air from a gun that is inclined at an angle θ with the horizontal, and suppose the initial velocity of the projectile is v_0 feet per second. Neglect all forces except that of gravity and the air resistance, and assume that this latter force (in pounds) is numerically equal to k times the velocity (in ft/sec).

   (a) Taking the origin at the position of the gun, with the x axis horizontal and the y axis vertical, show that the differential equations of the resulting motion are

   $$m\frac{d^2x}{dt^2} + k\frac{dx}{dt} = 0,$$

   $$m\frac{d^2y}{dt^2} + k\frac{dy}{dt} + mg = 0.$$

   (b) Find the solution of the system of differential equations of part (a).

3. Determine the currents in the electrical network of Figure 7.6 if E is an electromotive force of 100 V, R_1 is a resistor of 20 Ω, R_2 is a resistor of 40 Ω, L_1 is an inductor of 0.01 H, L_2 is an inductor of 0.02 H, and the currents are initially zero.

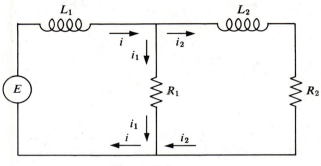

Figure 7.6

4. Set up differential equations for the currents in each of the electrical networks shown in Figure 7.7. Do not solve the equations.
   (a) For the network in Figure 7.7a assume that E is an electromotive force of 15 V, R is a resistor of 20 Ω, L is an inductor of 0.02 H, and C is a capacitor of 10^{-4} farads.
   (b) For the network in Figure 7.7b assume that E is an electromotive force of $100 \sin 130t$ V, R_1 is a resistor of 20 Ω, R_2 is a resistor of 30 Ω, and L is an inductor of 0.05 H.
   (c) For the network of Figure 7.7c assume that E is an electromotive force of 100 V, R_1 is a resistor of 20 Ω, R_2 is a resistor of 10 Ω, C_1 is a capacitor of 10^{-4} farads, and C_2 is a capacitor of 2×10^{-4} farads.

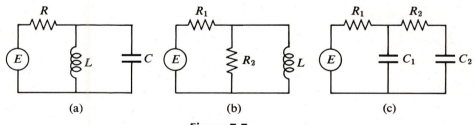

| (a) | (b) | (c) |

Figure 7.7

5. Two tanks are interconnected. Tank X initially contains 90 liters of brine in which there is dissolved 3 kg of salt, and tank Y initially contains 90 liters of brine in which there is dissolved 2 kg of salt. Starting at time $t = 0$, (1) pure water flows into tank X at the rate of 4.5 liters/min, (2) brine flows from tank X into tank Y at the rate of 6 liters/min, (3) brine is pumped from tank Y back into tank X at the rate of 1.5 liters/min, and (4) brine flows out of tank Y and away from the system at the rate of 4.5 liters/min. The mixture in each tank is kept uniform by stirring. How much salt is in each tank at any time $t > 0$?

6. Two tanks X and Y are interconnected. Tank X initially contains 30 liters of brine in which there is dissolved 30 kg of salt, and tank Y initially contains 30 liters of pure water. Starting at time $t = 0$, (1) brine containing 1 kg of salt per liter flows into tank X at the rate of 2 liters/min and pure water also flows into tank X at the

rate of 1 liter/min, (2) brine flows from tank X into tank Y at the rate of 4 liters/min, (3) brine is pumped from tank Y back into tank X at the rate of 1 liter/min, and (4) brine flows out of tank Y and away from the system at the rate of 3 liters/min. The mixture in each tank is kept uniform by stirring. How much salt is in each tank at any time $t > 0$?

7.3 BASIC THEORY OF LINEAR SYSTEMS IN NORMAL FORM: TWO EQUATIONS IN TWO UNKNOWN FUNCTIONS

A. Introduction

We shall begin by considering a basic type of system of two linear differential equations in two unknown functions. This system is of the form

$$\frac{dx}{dt} = a_{11}(t)x + a_{12}(t)y + F_1(t),$$

$$\frac{dy}{dt} = a_{21}(t)x + a_{22}(t)y + F_2(t). \tag{7.61}$$

We shall assume that the functions a_{11}, a_{12}, F_1, a_{21}, a_{22}, and F_2 are all continuous on a real interval $a \leq t \leq b$. If $F_1(t)$ and $F_2(t)$ are zero for all t, then the system (7.61) is called *homogeneous*; otherwise, the system is said to be *nonhomogeneous*.

▶ **Example 7.8**

The system

$$\frac{dx}{dt} = 2x - y,$$

$$\frac{dy}{dt} = 3x + 6y, \tag{7.62}$$

is homogeneous; the system

$$\frac{dx}{dt} = 2x - y - 5t,$$

$$\frac{dy}{dt} = 3x + 6y - 4, \tag{7.63}$$

is nonhomogeneous.

DEFINITION

By a solution of the system (7.61) we shall mean an ordered pair of real functions

$$(f, g), \tag{7.64}$$

each having a continuous derivative on the real interval $a \leq t \leq b$, such that

$$\frac{df(t)}{dt} = a_{11}(t)f(t) + a_{12}(t)g(t) + F_1(t),$$

$$\frac{dg(t)}{dt} = a_{21}(t)f(t) + a_{22}(t)g(t) + F_2(t),$$

for all t such that $a \leq t \leq b$. In other words,

$$x = f(t),$$
$$y = g(t), \tag{7.65}$$

simultaneously satisfy both equations of the system (7.61) identically for $a \leq t \leq b$.

Notation.   We shall use the notation

$$x = f(t),$$
$$y = g(t), \tag{7.65}$$

to denote a solution of the system (7.61) and shall speak of "the solution

$$x = f(t),$$
$$y = g(t)."$$

Whenever we do this, we must remember that the solution thus referred to is really the ordered pair of functions (f, g) such that (7.65) simultaneously satisfy both equations of the system (7.61) identically on $a \leq t \leq b$.

▶ **Example 7.9**

The ordered pair of functions defined for all t by $(e^{5t}, -3e^{5t})$, which we denote by

$$x = e^{5t},$$
$$y = -3e^{5t}, \tag{7.66}$$

is a solution of the system (7.62). That is,

$$x = e^{5t},$$
$$y = -3e^{5t}, \tag{7.66}$$

simultaneously satisfy both equations of the system (7.62). Let us verify this by directly substituting (7.66) into (7.62). We have

$$\frac{d}{dt}(e^{5t}) = 2(e^{5t}) - (-3e^{5t}),$$

$$\frac{d}{dt}(-3e^{5t}) = 3(e^{5t}) + 6(-3e^{5t}),$$

or

$$5e^{5t} = 2e^{5t} + 3e^{5t},$$
$$-15e^{5t} = 3e^{5t} - 18e^{5t}.$$

Hence (7.66) is indeed a solution of the system (7.62). The reader should verify that the ordered pair of functions defined for all t by $(e^{3t}, -e^{3t})$, which we denote by

$$x = e^{3t},$$

$$y = -e^{3t},$$

is also a solution of the system (7.62).

We shall now proceed to survey the basic theory of linear systems. We shall observe a close analogy between this theory and that introduced in Section 4.1 for the single linear equation of higher order. Theorem 7.1 is the basic existence theorem dealing with the system (7.61).

THEOREM 7.1

Hypothesis. *Let the functions $a_{11}, a_{12}, F_1, a_{21}, a_{22}$, and F_2 in the system (7.61) all be continuous on the interval $a \leq t \leq b$. Let t_0 be any point of the interval $a \leq t \leq b$; and let c_1 and c_2 be two arbitrary constants.*

Conclusion. *There exists a unique solution*

$$x = f(t),$$

$$y = g(t),$$

of the system (7.61) such that

$$f(t_0) = c_1 \quad and \quad g(t_0) = c_2,$$

and this solution is defined on the entire interval $a \leq t \leq b$.

▶ **Example 7.10**

Let us consider the system (7.63). The continuity requirements of the hypothesis of Theorem 7.1 are satisfied on every closed interval $a \leq t \leq b$. Hence, given any point t_0 and any two constants c_1 and c_2, there exists a unique solution $x = f(t), y = g(t)$ of the system (7.63) that satisfies the conditions $f(t_0) = c_1, g(t_0) = c_2$. For example, there exists one and only one solution $x = f(t), y = g(t)$ such that $f(2) = 5, g(2) = -7$.

B. Homogeneous Linear Systems

We shall now assume that $F_1(t)$ and $F_2(t)$ in the system (7.61) are both zero for all t and consider the basic theory of the resulting *homogeneous* linear system

$$\frac{dx}{dt} = a_{11}(t)x + a_{12}(t)y,$$

$$\frac{dy}{dt} = a_{21}(t)x + a_{22}(t)y.$$

(7.67)

We shall see that this theory is analogous to that of the single nth-order homogeneous linear differential equation presented in Section 4.1B. Our first result concerning the system (7.67) is the following.

THEOREM 7.2

Hypothesis. *Let*

$$x = f_1(t), \qquad\qquad x = f_2(t),$$
$$\textit{and} \qquad\qquad\qquad\qquad (7.68)$$
$$y = g_1(t), \qquad\qquad y = g_2(t),$$

be two solutions of the homogeneous linear system (7.67). Let c_1 and c_2 be two arbitrary constants.

Conclusion. *Then*

$$x = c_1 f_1(t) + c_2 f_2(t),$$
$$y = c_1 g_1(t) + c_2 g_2(t), \qquad\qquad (7.69)$$

is also a solution of the system (7.67).

DEFINITION

The solution (7.69) is called a linear combination *of the solutions (7.68). This definition enables us to express Theorem 7.2 in the following alternative form.*

THEOREM 7.2 RESTATED

Any linear combination of two solutions of the homogeneous linear system (7.67) is itself a solution of the system (7.67).

▶ **Example 7.11**

We have already observed that

$$x = e^{5t}, \qquad\qquad x = e^{3t},$$
$$\text{and} \qquad\qquad\qquad\qquad$$
$$y = -3e^{5t}, \qquad\qquad y = -e^{3t},$$

are solutions of the homogeneous linear system (7.62). Theorem 7.2 tells us that

$$x = c_1 e^{5t} + c_2 e^{3t},$$
$$y = -3c_1 e^{5t} - c_2 e^{3t},$$

where c_1 and c_2 are arbitrary constants, is also a solution of the system (7.62). For example, if $c_1 = 4$ and $c_2 = -2$, we have the solution

$$x = 4e^{5t} - 2e^{3t},$$
$$y = -12e^{5t} + 2e^{3t}.$$

DEFINITION

Let

$$x = f_1(t), \qquad\qquad x = f_2(t),$$
$$\text{and}$$
$$y = g_1(t), \qquad\qquad y = g_2(t),$$

be two solutions of the homogeneous linear system (7.67). *These two solutions are* linearly dependent *on the interval* $a \leq t \leq b$ *if there exist constants* c_1 *and* c_2, *not both zero, such that*

$$c_1 f_1(t) + c_2 f_2(t) = 0,$$
$$c_1 g_1(t) + c_2 g_2(t) = 0,$$

$$(7.70)$$

for all t such that $a \leq t \leq b$.

DEFINITION

Let

$$x = f_1(t), \qquad\qquad x = f_2(t),$$
$$\text{and}$$
$$y = g_1(t), \qquad\qquad y = g_2(t),$$

be two solutions of the homogeneous linear system (7.67). *These two solutions are* linearly independent *on* $a \leq t \leq b$ *if they are not linearly dependent on* $a \leq t \leq b$. *That is, the solutions* $x = f_1(t)$, $y = g_1(t)$ *and* $x = f_2(t)$, $y = g_2(t)$ *are linearly independent on* $a \leq t \leq b$ *if*

$$c_1 f_1(t) + c_2 f_2(t) = 0,$$
$$c_1 g_1(t) + c_2 g_2(t) = 0,$$

$$(7.71)$$

for all t such that $a \leq t \leq b$ *implies that*

$$c_1 = c_2 = 0.$$

▶ **Example 7.12**

The solutions

$$x = e^{5t}, \qquad\qquad x = 2e^{5t},$$
$$\text{and}$$
$$y = -3e^{5t}, \qquad\qquad y = -6e^{5t},$$

of the system (7.62) are linearly dependent on every interval $a \leq t \leq b$. For in this case the conditions (7.70) become

$$c_1 e^{5t} + 2c_2 e^{5t} = 0,$$
$$-3c_1 e^{5t} - 6c_2 e^{5t} = 0, \qquad (7.72)$$

and clearly there exist constants c_1 and c_2, not both zero, such that the conditions (7.72) hold on $a \le t \le b$. For example, let $c_1 = 2$ and $c_2 = -1$.

On the other hand, the solutions

$$x = e^{5t}, \qquad\qquad x = e^{3t},$$
$$\text{and}$$
$$y = -3e^{5t}, \qquad\qquad y = -e^{3t},$$

of system (7.62) are linearly independent on $a \le t \le b$. For in this case the conditions (7.71) are

$$c_1 e^{5t} + c_2 e^{3t} = 0,$$
$$-3c_1 e^{5t} - c_2 e^{3t} = 0.$$

If these conditions hold for all t such that $a \le t \le b$, then we must have $c_1 = c_2 = 0$.

We now state the following basic theorem concerning sets of linearly independent solutions of the homogeneous linear system (7.67).

THEOREM 7.3

There exist sets of two linearly independent solutions of the homogeneous linear system (7.67). Every solution of the system (7.67) can be written as a linear combination of any two linearly independent solutions of (7.67).

▶ **Example 7.13**

We have seen that

$$x = e^{5t}, \qquad\qquad x = e^{3t},$$
$$\text{and}$$
$$y = -3e^{5t}, \qquad\qquad y = -e^{3t},$$

constitute a pair of linearly independent solutions of the system (7.62). This illustrates the first part of Theorem 7.3. The second part of the theorem tells us that every solution of the system (7.62) can be written in the form

$$x = c_1 e^{5t} + c_2 e^{3t},$$
$$y = -3c_1 e^{5t} - c_2 e^{3t},$$

where c_1 and c_2 are suitably chosen constants.

Recall that in Section 4.1 in connection with the single nth-order homogeneous linear differential equation, we defined the general solution of such an equation to be a linear combination of n linearly independent solutions. As a result of Theorems 7.2 and 7.3 we now give an analogous definition of general solution for the homogeneous linear system (7.67).

DEFINITION

Let

$$x = f_1(t), \qquad\qquad x = f_2(t),$$
$$\text{and}$$
$$y = g_1(t), \qquad\qquad y = g_2(t),$$

be two linearly independent solutions of the homogeneous linear system (7.67). *Let* c_1 *and* c_2 *be two arbitrary constants. Then the solution*

$$x = c_1 f_1(t) + c_2 f_2(t),$$
$$y = c_1 g_1(t) + c_2 g_2(t),$$

is called a general solution *of the system* (7.67).

▶ **Example 7.14**

Since

$$x = e^{5t}, \qquad\qquad x = e^{3t},$$
$$\text{and}$$
$$y = -3e^{5t}, \qquad\qquad y = -e^{3t},$$

are linearly independent solutions of the system (7.62), we may write the general solution of (7.62) in the form

$$x = c_1 e^{5t} + c_2 e^{3t},$$
$$y = -3c_1 e^{5t} - c_2 e^{3t},$$

where c_1 and c_2 are arbitrary constants.

In order to state a useful criterion for the linear independence of two solutions of system (7.67), we introduce the Wronskian of two solutions. Note that this is similar to, but different from, the now familiar Wronskian of two solutions of a single second-order linear equation, introduced in Section 4.1.

DEFINITION

Let

$$x = f_1(t), \qquad\qquad x = f_2(t),$$
$$\text{and}$$
$$y = g_1(t), \qquad\qquad y = g_2(t),$$

be two solutions of the homogeneous linear system (7.67). *The determinant*

$$\begin{vmatrix} f_1(t) & f_2(t) \\ g_1(t) & g_2(t) \end{vmatrix} \tag{7.73}$$

is called the Wronskian *of these two solutions. We denote it by* $W(t)$.

We may now state the following useful criterion for the linear independence of two solutions of system (7.67).

THEOREM 7.4

Two solutions

$$x = f_1(t), \qquad\qquad x = f_2(t),$$
$$\text{and}$$
$$y = g_1(t), \qquad\qquad y = g_2(t),$$

of the homogeneous linear system (7.67) are linearly independent on an interval $a \le t \le b$ if and only if their Wronskian determinant

$$W(t) = \begin{vmatrix} f_1(t) & f_2(t) \\ g_1(t) & g_2(t) \end{vmatrix} \qquad\qquad (7.73)$$

is different from zero for all t such that $a \le t \le b$.

Concerning the values of $W(t)$, we also state the following result.

THEOREM 7.5

Let $W(t)$ be the Wronskian of two solutions of homogeneous linear system (7.67) on an interval $a \le t \le b$. Then either $W(t) = 0$ for all $t \in [a, b]$ or $W(t) = 0$ for no $t \in [a, b]$.

► **Example 7.15**

Let us employ Theorem 7.4 to verify the linear independence of the solutions

$$x = e^{5t}, \qquad\qquad x = e^{3t},$$
$$\text{and}$$
$$y = -3e^{5t}, \qquad\qquad y = -e^{3t},$$

of the system (7.62). We have

$$W(t) = \begin{vmatrix} e^{5t} & e^{3t} \\ -3e^{5t} & -e^{3t} \end{vmatrix} = 2e^{8t} \neq 0$$

on every closed interval $a \le t \le b$. Thus by Theorem 7.4 the two solutions are indeed linearly independent on $a \le t \le b$.

C. Nonhomogeneous Linear Systems

Let us now return briefly to the nonhomogeneous system (7.61). A theorem and a definition, illustrated by a simple example, will suffice for our purposes here.

THEOREM 7.6

Hypothesis. *Let*

$$x = f_0(t),$$
$$y = g_0(t),$$

be any solution of the nonhomogeneous system (7.61), and let

$$x = f(t),$$
$$y = g(t),$$

be any solution of the corresponding homogeneous system (7.67).

Conclusion.   *Then*

$$x = f(t) + f_0(t),$$
$$y = g(t) + g_0(t),$$

is also a solution of the nonhomogeneous system (7.61).

DEFINITION

Let

$$x = f_0(t),$$
$$y = g_0(t),$$

be any solution of the nonhomogeneous system (7.61), and let

$$x = f_1(t), \qquad\qquad x = f_2(t),$$
$$\text{and}$$
$$y = g_1(t) \qquad\qquad y = g_2(t),$$

be two linearly independent solutions of the corresponding homogeneous system (7.67). Let c_1 and c_2 be two arbitrary constants. Then the solution

$$x = c_1 f_1(t) + c_2 f_2(t) + f_0(t),$$
$$y = c_1 g_1(t) + c_2 g_2(t) + g_0(t),$$

will be called a general solution *of the nonhomogeneous system (7.61).*

► **Example 7.16**

The student may verify that

$$x = 2t + 1,$$
$$y = -t,$$

is a solution of the nonhomogeneous system (7.63). The corresponding homogeneous system is the system (7.62), and we have already seen that

$$x = e^{5t}, \qquad\qquad x = e^{3t},$$
$$\text{and}$$
$$y = -3e^{5t}, \qquad\qquad y = -e^{3t},$$

are linearly independent solutions of this homogeneous system. Theorem 7.6 tells us, for example, that

$$x = e^{5t} + 2t + 1,$$
$$y = -3e^{5t} - t,$$

is a solution of the nonhomogeneous system (7.63). From the preceding definition we see that the general solution of (7.63) may be written in the form

$$x = c_1 e^{5t} + c_2 e^{3t} + 2t + 1,$$
$$y = -3c_1 e^{5t} - c_2 e^{3t} - t,$$

where c_1 and c_2 are arbitrary constants.

Exercises

1. Consider the linear system

$$\frac{dx}{dt} = 3x + 4y,$$

$$\frac{dy}{dt} = 2x + y.$$

(a) Show that

$$x = 2e^{5t}, \qquad\qquad x = e^{-t},$$
$$\text{and}$$
$$y = e^{5t}, \qquad\qquad y = -e^{-t},$$

are solutions of this system.

(b) Show that the two solutions of part (a) are linearly independent on every interval $a \le t \le b$, and write the general solution of the system.

(c) Find the solution

$$x = f(t),$$
$$y = g(t),$$

of the system which is such that $f(0) = 1$ and $g(0) = 2$. Why is this solution unique? Over what interval is it defined?

2. Consider the linear system

$$\frac{dx}{dt} = 5x + 3y,$$

$$\frac{dy}{dt} = 4x + y.$$

(a) Show that

$$x = 3e^{7t}, \qquad\qquad x = e^{-t},$$
$$\text{and}$$
$$y = 2e^{7t}, \qquad\qquad y = -2e^{-t},$$

are solutions of this system.

(b) Show that the two solutions of part (a) are linearly independent on every interval $a \le t \le b$, and write the general solution of the system.

(c) Find the solution

$$x = f(t),$$

$$y = g(t),$$

of the system which is such that $f(0) = 0$ and $g(0) = 8$.

3. (a) Show that

$$x = 2e^{2t}, \qquad\qquad x = e^{7t},$$

$$\text{and}$$

$$y = -3e^{2t}, \qquad\qquad y = e^{7t},$$

are solutions of the homogeneous linear system

$$\frac{dx}{dt} = 5x + 2y,$$

$$\frac{dy}{dt} = 3x + 4y.$$

(b) Show that the two solutions defined in part (a) are linearly independent on every interval $a \le t \le b$, and write the general solution of the homogeneous system of part (a).

(c) Show that

$$x = t + 1,$$

$$y = -5t - 2,$$

is a particular solution of the nonhomogeneous linear system

$$\frac{dx}{dt} = 5x + 2y + 5t,$$

$$\frac{dy}{dt} = 3x + 4y + 17t,$$

and write the general solution of this system.

4. Let

$$x = f_1(t), \qquad\qquad x = f_2(t),$$

$$\text{and}$$

$$y = g_1(t), \qquad\qquad y = g_2(t),$$

be two linearly independent solutions of the homogeneous linear system (7.67), and let $W(t)$ be their Wronskian determinant (defined by (7.73)). Show that W satisfies the first-order differential equation

$$\frac{dW}{dt} = [a_{11}(t) + a_{22}(t)]W.$$

5. Prove Theorem 7.2.

6. Prove Theorem 7.6.

7.4 HOMOGENEOUS LINEAR SYSTEMS WITH CONSTANT COEFFICIENTS: TWO EQUATIONS IN TWO UNKNOWN FUNCTIONS

A. Introduction

In this section we shall be concerned with the homogeneous linear system

$$
\frac{dx}{dt} = a_1 x + b_1 y,
$$
$$
\frac{dy}{dt} = a_2 x + b_2 y,
\tag{7.74}
$$

where the coefficients a_1, b_1, a_2, and b_2 are real constants. We seek solutions of this system; but how shall we proceed? Recall that in Section 4.2 we sought and found exponential solutions of the single nth-order linear equation with constant coefficients. Remembering the analogy that exists between linear systems and single higher-order linear equations, we might now attempt to find exponential solutions of the system (7.74). Let us therefore attempt to determine a solution of the form

$$
x = Ae^{\lambda t},
$$
$$
y = Be^{\lambda t},
\tag{7.75}
$$

where A, B, and λ are constants. If we substitute (7.75) into (7.74), we obtain

$$
A\lambda e^{\lambda t} = a_1 A e^{\lambda t} + b_1 B e^{\lambda t},
$$
$$
B\lambda e^{\lambda t} = a_2 A e^{\lambda t} + b_2 B e^{\lambda t}.
$$

These equations lead at once to the system

$$
(a_1 - \lambda)A + b_1 B = 0,
$$
$$
a_2 A + (b_2 - \lambda)B = 0,
\tag{7.76}
$$

in the unknowns A and B. This system obviously has the trivial solution $A = B = 0$. But this would only lead to the trivial solution $x = 0$, $y = 0$ of the system (7.74). Thus we seek nontrivial solutions of the system (7.76). A necessary and sufficient condition (see Section 7.5c, Theorem A) that this system have a nontrivial solution is that the determinant

$$
\begin{vmatrix} a_1 - \lambda & b_1 \\ a_2 & b_2 - \lambda \end{vmatrix} = 0.
\tag{7.77}
$$

Expanding this determinant we are led at once to the quadratic equation

$$
\lambda^2 - (a_1 + b_2)\lambda + (a_1 b_2 - a_2 b_1) = 0
\tag{7.78}
$$

in the unknown λ. This equation is called the *characteristic equation* associated with the system (7.74). Its roots λ_1 and λ_2 are called the *characteristic roots*. If the pair (7.75) is to be a solution of the system (7.74), then λ in (7.75) must be one of these roots. Suppose $\lambda = \lambda_1$. Then substituting $\lambda = \lambda_1$ into the algebraic system (7.76), we may obtain a nontrivial solution A_1, B_1, of this algebraic system. With these values A_1, B_1 we obtain the nontrivial solution

$$x = A_1 e^{\lambda_1 t},$$

$$y = B_1 e^{\lambda_1 t},$$

of the given system (7.74).

Three cases must now be considered:

1. The roots λ_1 and λ_2 are real and distinct.
2. The roots λ_1 and λ_2 are conjugate complex.
3. The roots λ_1 and λ_2 are real and equal.

B. Case 1. The Roots of the Characteristic Equations (7.78) are Real and Distinct

If the roots λ_1 and λ_2 of the characteristic equation (7.78) are real and distinct, it appears that we should expect two distinct solutions of the form (7.75), one corresponding to each of the two distinct roots. This is indeed the case. Furthermore, these two distinct solutions are linearly independent. We summarize this case in the following theorem.

THEOREM 7.7

Hypothesis. *The roots λ_1 and λ_2, of the characteristic equation (7.78) associated with the system (7.74) are real and distinct.*

Conclusion. *The system (7.74) has two nontrivial linearly independent solutions of the form*

$$x = A_1 e^{\lambda_1 t}, \qquad\qquad x = A_2 e^{\lambda_2 t},$$
$$and$$
$$y = B_1 e^{\lambda_1 t}, \qquad\qquad y = B_2 e^{\lambda_2 t},$$

where $A_1, B_1, A_2,$ and B_2 are definite constants. The general solution of the system (7.74) may thus be written

$$x = c_1 A_1 e^{\lambda_1 t} + c_2 A_2 e^{\lambda_2 t},$$

$$y = c_1 B_1 e^{\lambda_1 t} + c_2 B_2 e^{\lambda_2 t},$$

where c_1 and c_2 are arbitrary constants.

▶ **Example 7.17**

$$\frac{dx}{dt} = 6x - 3y,$$

$$\frac{dy}{dt} = 2x + y. \tag{7.79}$$

We assume a solution of the form (7.75):

$$x = Ae^{\lambda t},$$
$$y = Be^{\lambda t}. \tag{7.80}$$

Substituting (7.80) into (7.79) we obtain

$$A\lambda e^{\lambda t} = 6Ae^{\lambda t} - 3Be^{\lambda t},$$
$$B\lambda e^{\lambda t} = 2Ae^{\lambda t} + Be^{\lambda t},$$

and this leads at once to the algebraic system

$$(6 - \lambda)A - 3B = 0,$$
$$2A + (1 - \lambda)B = 0, \tag{7.81}$$

in the unknown λ. For nontrivial solutions of this system we must have

$$\begin{vmatrix} 6 - \lambda & -3 \\ 2 & 1 - \lambda \end{vmatrix} = 0.$$

Expanding this we obtain the characteristic equation

$$\lambda^2 - 7\lambda + 12 = 0.$$

Solving this, we find the roots $\lambda_1 = 3$, $\lambda_2 = 4$.
  Setting $\lambda = \lambda_1 = 3$ in (7.81), we obtain

$$3A - 3B = 0,$$
$$2A - 2B = 0.$$

A simple nontrivial solution of this system is obviously $A = B = 1$. With these values of A, B, and λ we find the nontrivial solution

$$x = e^{3t},$$
$$y = e^{3t}. \tag{7.82}$$

We note that a different solution for A and B here (for instance, $A = B = 2$) would only lead to a solution which is linearly dependent of (and generally less simple than) solution (7.82).
  Now setting $\lambda = \lambda_2 = 4$ in (7.81), we find

$$2A - 3B = 0,$$
$$2A - 3B = 0.$$

A simple nontrivial solution of this system is $A = 3$, $B = 2$. Using these values of A, B, and λ we find the nontrivial solution

$$x = 3e^{4t},$$
$$y = 2e^{4t}. \tag{7.83}$$

By Theorem 7.7 the solutions (7.82) and (7.83) are linearly independent (one may check this using Theorem 7.4) and the general solution of the system (7.79) may be written

$$x = c_1 e^{3t} + 3c_2 e^{4t},$$
$$y = c_1 e^{3t} + 2c_2 e^{4t},$$

where c_1 and c_2 are arbitrary constants.

C. Case 2. The Roots of the Characteristic Equation (7.78) are Conjugate Complex

If the roots λ_1 and λ_2 of the characteristic equation (7.78) are the conjugate complex numbers $a + bi$ and $a - bi$, then we still obtain two distinct solutions

$$x = A_1^* e^{(a+bi)t}, \qquad x = A_2^* e^{(a-bi)t},$$
$$\text{and} \qquad\qquad (7.84)$$
$$y = B_1^* e^{(a+bi)t}, \qquad y = B_2^* e^{(a-bi)t},$$

of the form (7.75), one corresponding to each of the complex roots. However, the solutions (7.84) are *complex* solutions. In order to obtain *real* solutions in this case we consider the first of the two solutions (7.84) and proceed as follows: We first express the complex constants A_1^* and B_1^* in this solution in the forms $A_1^* = A_1 + iA_2$ and $B_1^* = B_1 + iB_2$, where A_1, A_2, B_1, and B_2 are real. We then apply Euler's formula $e^{i\theta} = \cos\theta + i\sin\theta$ and express the first solution (7.84) in the form

$$x = (A_1 + iA_2)e^{at}(\cos bt + i\sin bt),$$
$$y = (B_1 + iB_2)e^{at}(\cos bt + i\sin bt).$$

Rewriting this, we have

$$x = e^{at}[(A_1 \cos bt - A_2 \sin bt) + i(A_2 \cos bt + A_1 \sin bt)],$$
$$y = e^{at}[(B_1 \cos bt - B_2 \sin bt) + i(B_2 \cos bt + B_1 \sin bt)]. \qquad (7.85)$$

It can be shown that a pair $[f_1(t) + if_2(t), g_1(t) + ig_2(t)]$ of complex functions is a solution of the system (7.74) if and only if both the pair $[f_1(t), g_1(t)]$ consisting of their real parts and the pair $[f_2(t), g_2(t)]$ consisting of their imaginary parts are solutions of (7.74). Thus both the real part

$$x = e^{at}(A_1 \cos bt - A_2 \sin bt),$$
$$y = e^{at}(B_1 \cos bt - B_2 \sin bt), \qquad (7.86)$$

and the imaginary part

$$x = e^{at}(A_2 \cos bt + A_1 \sin bt),$$
$$y = e^{at}(B_2 \cos bt + B_1 \sin bt), \qquad (7.87)$$

of the solution (7.85) of the system (7.74) are also solutions of (7.74). Furthermore, the solutions (7.86) and (7.87) are linearly independent. We verify this by evaluating the Wronskian determinant (7.73) for these solutions. We find

$$W(t) = \begin{vmatrix} e^{at}(A_1 \cos bt - A_2 \sin bt) & e^{at}(A_2 \cos bt + A_1 \sin bt) \\ e^{at}(B_1 \cos bt - B_2 \sin bt) & e^{at}(B_2 \cos bt + B_1 \sin bt) \end{vmatrix}$$
$$= e^{2at}(A_1 B_2 - A_2 B_1). \qquad (7.88)$$

Now, the constant B_1^* is a *nonreal* multiple of the constant A_1^*. If we assume that $A_1 B_2 - A_2 B_1 = 0$, then it follows that B_1^* is a *real* multiple of A_1^*, which contradicts the result stated in the previous sentence. Thus $A_1 B_2 - A_2 B_1 \neq 0$ and so the Wronskian determinant $W(t)$ in (7.88) is unequal to zero. Thus by Theorem 7.4 the solutions (7.86) and (7.87) are indeed linearly independent. Hence a linear combination of these two real solutions provides the general solution of the system (7.74) in this case. There is no need to consider the second of the two solutions (7.84). We summarize the above results in the following theorem:

THEOREM 7.8

Hypothesis. *The roots λ_1 and λ_2 of the characteristic equation (7.78) associated with the system (7.74) are the conjugate complex numbers $a \pm bi$.*

Conclusion. *The system (7.74) has two real linearly independent solutions of the form*

$$x = e^{at}(A_1 \cos bt - A_2 \sin bt), \qquad x = e^{at}(A_2 \cos bt + A_1 \sin bt),$$
$$and$$
$$y = e^{at}(B_1 \cos bt - B_2 \sin bt), \qquad y = e^{at}(B_2 \cos bt + B_1 \sin bt),$$

where $A_1, A_2, B_1,$ and B_2 are definite real constants. The general solution of the system (7.74) may thus be written

$$x = e^{at}[c_1(A_1 \cos bt - A_2 \sin bt) + c_2(A_2 \cos bt + A_1 \sin bt)],$$
$$y = e^{at}[c_1(B_1 \cos bt - B_2 \sin bt) + c_2(B_2 \cos bt + B_1 \sin bt)],$$

where c_1 and c_2 are arbitrary constants.

▶ **Example 7.18**

$$\frac{dx}{dt} = 3x + 2y,$$

$$\frac{dy}{dt} = -5x + y. \tag{7.89}$$

We assume a solution of the form (7.75):

$$x = Ae^{\lambda t},$$

$$y = Be^{\lambda t}. \tag{7.90}$$

Substituting (7.90) into (7.89) we obtain

$$A\lambda e^{\lambda t} = 3Ae^{\lambda t} + 2Be^{\lambda t},$$

$$B\lambda e^{\lambda t} = -5Ae^{\lambda t} + Be^{\lambda t},$$

and this leads at once to the algebraic system

$$(3 - \lambda)A + 2B = 0,$$

$$-5A + (1 - \lambda)B = 0, \tag{7.91}$$

in the unknown λ. For nontrivial solutions of this system we must have

$$\begin{vmatrix} 3 - \lambda & 2 \\ -5 & 1 - \lambda \end{vmatrix} = 0.$$

Expanding this, we obtain the characteristic equation

$$\lambda^2 - 4\lambda + 13 = 0.$$

The roots of this equation are the conjugate complex numbers $2 \pm 3i$.

Setting $\lambda = 2 + 3i$ in (7.91), we obtain

$$(1 - 3i)A + 2B = 0,$$

$$-5A + (-1 - 3i)B = 0.$$

A simple nontrivial solution of this system is $A = 2, B = -1 + 3i$. Using these values we obtain the complex solution

$$x = 2e^{(2 + 3i)t},$$

$$y = (-1 + 3i)e^{(2 + 3i)t},$$

of the given system (7.89). Using Euler's formula this takes the form

$$x = e^{2t}[(2 \cos 3t) + i(2 \sin 3t)],$$

$$y = e^{2t}[(-\cos 3t - 3 \sin 3t) + i(3 \cos 3t - \sin 3t)].$$

Since both the real and imaginary parts of this solution of system (7.89) are themselves solutions of (7.89), we thus obtain the two real solutions

$$x = 2e^{2t} \cos 3t,$$
$$y = -e^{2t}(\cos 3t + 3 \sin 3t), \tag{7.92}$$

and

$$x = 2e^{2t} \sin 3t,$$
$$y = e^{2t}(3 \cos 3t - \sin 3t). \tag{7.93}$$

Finally, since the two solutions (7.92) and (7.93) are linearly independent we may write the general solution of the system (7.89) in the form

$$x = 2e^{2t}(c_1 \cos 3t + c_2 \sin 3t),$$

$$y = e^{2t}[c_1(-\cos 3t - 3 \sin 3t) + c_2(3 \cos 3t - \sin 3t)],$$

where c_1 and c_2 are arbitrary constants.

D. Case 3. The Roots of the Characteristic Equation (7.78) are Real and Equal

If the two roots of the characteristic equation (7.78) are real and equal, it would appear that we could find only one solution of the form (7.75). Except in the special subcase in which $a_1 = b_2 \neq 0, a_2 = b_1 = 0$ (see Exercise 33 at the end of this section) this is indeed true. In general, how shall we then proceed to find a second, linearly independent solution? Recall the analogous situation in which the auxiliary equation corresponding to a single nth-order linear equation has a double root. This would lead us to expect a second solution of the form

$$x = Ate^{\lambda t},$$

$$y = Bte^{\lambda t}.$$

However, the situation here is not quite so simple (see Exercise 34 at the end of this section). We must actually seek a second solution of the form

$$x = (A_1 t + A_2)e^{\lambda t},$$

$$y = (B_1 t + B_2)e^{\lambda t}. \tag{7.94}$$

We shall illustrate this in Example 7.19. We first summarize Case 3 in the following theorem.

THEOREM 7.9

Hypothesis. *The roots λ_1 and λ_2 of the characteristic equation (7.78) associated with the system (7.74) are real and equal. Let λ denote their common value. Further assume that system (7.74) is not such that $a_1 = b_2 \neq 0$, $a_2 = b_1 = 0$.*

Conclusion. *The system (7.74) has two linearly independent solutions of the form*

$$x = Ae^{\lambda t}, \qquad\qquad x = (A_1 t + A_2)e^{\lambda t},$$
$$\qquad\qquad and$$
$$y = Be^{\lambda t}, \qquad\qquad y = (B_1 t + B_2)e^{\lambda t},$$

where A, B, A_1, A_2, B_1, and B_2 are definite constants, A_1 and B_1 are not both zero, and $B_1/A_1 = B/A$. The general solution of the system (7.74) may thus be written

$$x = c_1 Ae^{\lambda t} + c_2(A_1 t + A_2)e^{\lambda t},$$

$$y = c_1 Be^{\lambda t} + c_2(B_1 t + B_2)e^{\lambda t},$$

where c_1 and c_2 are arbitrary constants.

► Example 7.19

$$\frac{dx}{dt} = 4x - y,$$

$$\frac{dy}{dt} = x + 2y. \tag{7.95}$$

We assume a solution of the form (7.75):

$$x = Ae^{\lambda t},$$

$$y = Be^{\lambda t}. \tag{7.96}$$

Substituting (7.96) into (7.95) we obtain

$$A\lambda e^{\lambda t} = 4Ae^{\lambda t} - Be^{\lambda t},$$

$$B\lambda e^{\lambda t} = Ae^{\lambda t} + 2Be^{\lambda t},$$

and this leads at once to the algebraic system

$$(4 - \lambda)A - B = 0,$$
$$A + (2 - \lambda)B = 0,$$

(7.97)

in the unknown λ. For nontrivial solutions of this system we must have

$$\begin{vmatrix} 4 - \lambda & -1 \\ 1 & 2 - \lambda \end{vmatrix} = 0.$$

Expanding this we obtain the characteristic equation

$$\lambda^2 - 6\lambda + 9 = 0$$

(7.98)

or

$$(\lambda - 3)^2 = 0.$$

Thus the characteristic equation (7.98) has the real and equal roots 3, 3.
   Setting $\lambda = 3$ in (7.97), we obtain

$$A - B = 0,$$
$$A - B = 0.$$

A simple nontrivial solution of this system being $A = B = 1$, we obtain the nontrivial solution

$$x = e^{3t},$$
$$y = e^{3t},$$

(7.99)

of the given system (7.95).
   Since the roots of the characteristic equation are both equal to 3, we must seek a second solution of the form (7.94), with $\lambda = 3$. That is, we must determine $A_1, A_2, B_1,$ and B_2 (with A_1 and B_1 not both zero) such that

$$x = (A_1 t + A_2)e^{3t},$$
$$y = (B_1 t + B_2)e^{3t},$$

(7.100)

is a solution of the system (7.95). Substituting (7.100) into (7.95), we obtain

$$(3A_1 t + 3A_2 + A_1)e^{3t} = 4(A_1 t + A_2)e^{3t} - (B_1 t + B_2)e^{3t},$$
$$(3B_1 t + 3B_2 + B_1)e^{3t} = (A_1 t + A_2)e^{3t} + 2(B_1 t + B_2)e^{3t}.$$

These equations reduce at once to

$$(A_1 - B_1)t + (A_2 - A_1 - B_2) = 0,$$
$$(A_1 - B_1)t + (A_2 - B_1 - B_2) = 0.$$

In order for these equations to be identities, we must have

$$A_1 - B_1 = 0, \qquad A_2 - A_1 - B_2 = 0,$$
$$A_1 - B_1 = 0, \qquad A_2 - B_1 - B_2 = 0.$$

(7.101)

Thus in order for (7.100) to be a solution of the system (7.95), the constants $A_1, A_2,$

B_1, and B_2 must be chosen to satisfy the equations (7.101). From the equations $A_1 - B_1 = 0$, we see that $A_1 = B_1$. The other two equations of (7.101) show that A_2 and B_2 must satisfy

$$A_2 - B_2 = A_1 = B_1. \tag{7.102}$$

We may choose any convenient nonzero values for A_1 and B_1. We choose $A_1 = B_1 = 1$. Then (7.102) reduces to $A_2 - B_2 = 1$, and we can choose any convenient values for A_2 and B_2 that will satisfy this equation. We choose $A_2 = 1, B_2 = 0$. We are thus led to the solution

$$x = (t + 1)e^{3t},$$
$$y = te^{3t}. \tag{7.103}$$

By Theorem 7.9 the solutions (7.99) and (7.103) are linearly independent. We may thus write the general solution of the system (7.95) in the form

$$x = c_1e^{3t} + c_2(t + 1)e^{3t},$$
$$y = c_1e^{3t} + c_2te^{3t},$$

where c_1 and c_2 are arbitrary constants.

We note that a different choice of nonzero values for A_1 and B_1 in (7.102) and/or a different choice for A_2 and B_2 in the resulting equation for A_2 and B_2 will lead to a second solution which is different from solution (7.103). But this different second solution will also be linearly independent of the basic solution (7.99), and hence could serve along with (7.99) as one of the two constituent parts of a general solution. For example, if we choose $A_1 = B_1 = 2$ in (7.102), then (7.102) reduces to $A_2 - B_2 = 2$; and if we then choose $A_2 = 3, B_2 = 1$, we are led to the different second solution

$$x = (2t + 3)e^{3t},$$
$$y = (2t + 1)e^{3t}.$$

This is linearly independent of solution (7.99) and could serve along with (7.99) as one of the two constituent parts of a general solution.

Exercises

Find the general solution of each of the linear systems in Exercises 1–22.

1. $\dfrac{dx}{dt} = 5x - 2y,$

   $\dfrac{dy}{dt} = 4x - y.$

2. $\dfrac{dx}{dt} = 5x - y,$

   $\dfrac{dy}{dt} = 3x + y.$

3. $\dfrac{dx}{dt} = x + 2y,$

   $\dfrac{dy}{dt} = 3x + 2y.$

4. $\dfrac{dx}{dt} = 2x + 3y,$

   $\dfrac{dy}{dt} = -x - 2y.$

5. $\dfrac{dx}{dt} = 3x + y,$

$\dfrac{dy}{dt} = 4x + 3y.$

6. $\dfrac{dx}{dt} = 6x - y,$

$\dfrac{dy}{dt} = 3x + 2y.$

7. $\dfrac{dx}{dt} = 3x - 4y,$

$\dfrac{dy}{dt} = 2x - 3y.$

8. $\dfrac{dx}{dt} = 2x - y,$

$\dfrac{dy}{dt} = 9x + 2y.$

9. $\dfrac{dx}{dt} = x + 3y,$

$\dfrac{dy}{dt} = 3x + y.$

10. $\dfrac{dx}{dt} = 3x + 2y,$

$\dfrac{dy}{dt} = 6x - y.$

11. $\dfrac{dx}{dt} = x - 4y,$

$\dfrac{dy}{dt} = x + y.$

12. $\dfrac{dx}{dt} = 2x - 3y,$

$\dfrac{dy}{dt} = 3x + 2y.$

13. $\dfrac{dx}{dt} = x - 3y,$

$\dfrac{dy}{dt} = 3x + y.$

14. $\dfrac{dx}{dt} = 5x - 4y,$

$\dfrac{dy}{dt} = 2x + y.$

15. $\dfrac{dx}{dt} = 4x - 2y,$

$\dfrac{dy}{dt} = 5x + 2y.$

16. $\dfrac{dx}{dt} = x - 5y,$

$\dfrac{dy}{dt} = 2x - y.$

17. $\dfrac{dx}{dt} = 3x - 2y,$

$\dfrac{dy}{dt} = 2x + 3y.$

18. $\dfrac{dx}{dt} = 6x - 5y,$

$\dfrac{dy}{dt} = x + 2y.$

19. $\dfrac{dx}{dt} = 3x - y,$

$\dfrac{dy}{dt} = 4x - y.$

20. $\dfrac{dx}{dt} = 7x + 4y,$

$\dfrac{dy}{dt} = -x + 3y.$

21. $\dfrac{dx}{dt} = 5x + 4y,$

$\dfrac{dy}{dt} = -x + y.$

22. $\dfrac{dx}{dt} = x - 2y,$

$\dfrac{dy}{dt} = 2x - 3y.$

In each of Exercises 23–28, find the particular solution of the linear system that satisfies the stated initial conditions.

23. $\dfrac{dx}{dt} = -2x + 7y,$

$\dfrac{dy}{dt} = 3x + 2y,$

$x(0) = 9, \; y(0) = -1.$

24. $\dfrac{dx}{dt} = -2x + y,$

$\dfrac{dy}{dt} = 7x + 4y,$

$x(0) = 6, \; y(0) = 2.$

25. $\dfrac{dx}{dt} = 2x - 8y,$

$\dfrac{dy}{dt} = x + 6y.$

$x(0) = 4, \; y(0) = 1.$

26. $\dfrac{dx}{dt} = 3x + 5y,$

$\dfrac{dy}{dt} = -2x + 5y,$

$x(0) = 5, \; y(0) = -1.$

27. $\dfrac{dx}{dt} = 6x - 4y,$

$\dfrac{dy}{dt} = x + 2y,$

$x(0) = 2, \; y(0) = 3.$

28. $\dfrac{dx}{dt} = 7x - y,$

$\dfrac{dy}{dt} = 4x + 3y,$

$x(0) = 1, \; y(0) = 3.$

29. Consider the linear system

$$t\frac{dx}{dt} = a_1 x + b_1 y,$$

$$t\frac{dy}{dt} = a_2 x + b_2 y,$$

where $a_1, b_1, a_2,$ and b_2 are real constants. Show that the transformation $t = e^w$ transforms this system into a linear system with constant coefficients.

30. Use the result of Exercise 29 to solve the system

$$t\frac{dx}{dt} = x + y,$$

$$t\frac{dy}{dt} = -3x + 5y.$$

31. Use the result of Exercise 29 to solve the system

$$t\frac{dx}{dt} = 2x + 3y,$$

$$t\frac{dy}{dt} = 2x + y.$$

32. Consider the linear system

$$\frac{dx}{dt} = a_1 x + b_1 y,$$

$$\frac{dy}{dt} = a_2 x + b_2 y,$$

where a_1, b_1, a_2, and b_2 are real constants. Show that the condition $a_2 b_1 > 0$ is sufficient, but not necessary, for the system to have two real linearly independent solutions of the form

$$x = Ae^{\lambda t}, \qquad y = Be^{\lambda t}.$$

33. Suppose that the roots of the characteristic equation (7.78) of the system (7.74) are real and equal; and let λ denote their common value. Also assume that the system (7.74) *is* such that $a_1 = b_2 \neq 0$ and $a_2 = b_1 = 0$. Show that in this special subcase there exist two linearly independent solutions of the form (7.75).

34. Suppose that the roots of the characteristic equation (7.78) of the system (7.74) are real and equal; and let λ denote their common value. Also assume that the system (7.74) is *not* such that $a_1 = b_2 \neq 0$ and $a_2 = b_1 = 0$. Then show that there exists *no* nontrivial solution of the form

$$x = Ate^{\lambda t}, \qquad y = Bte^{\lambda t},$$

which is linearly independent of the "basic" solution of the form (7.75).

35. Referring to the conclusion of Theorem 7.9, show that $B_1/A_1 = B/A$ in the case under consideration.

7.5 MATRICES AND VECTORS

A. The Most Basic Concepts

The study of matrices and vectors is a large and important subject, and an entire chapter the size of this one would be needed to present all of the most fundamental concepts and results. Therefore, after defining a matrix, we shall introduce only those very special concepts and results that we shall need and use in this book. For the reader familiar with matrices and vectors, this section will be a very simple review of a few select topics. On the other hand, for the reader unfamiliar with matrices and vectors, the detailed treatment here will provide just what is needed for an understanding of the material presented in this book.

DEFINITIONS

A matrix is defined to be a rectangular array

$$\mathbf{A} = \begin{pmatrix} a_{11} & a_{12} & \cdots & a_{1n} \\ a_{21} & a_{22} & \cdots & a_{2n} \\ \vdots & \vdots & & \vdots \\ a_{m1} & a_{m2} & \cdots & a_{mn} \end{pmatrix}$$

of elements a_{ij} $(i = 1, 2, \ldots, m; j = 1, 2, \ldots, n)$, *arranged in m (horizontal) rows and n (vertical) columns. The matrix is denoted by the boldface letter* **A**, *as indicated; and the element in its ith row and jth column, by* a_{ij}, *as suggested. We write* $\mathbf{A} = (a_{ij})$, *and call* **A** *an m × n matrix.*

We shall be concerned with the two following special sizes of matrices.

1. A *square matrix* is a matrix for which the number of rows is the same as the number of columns. If the common number of rows and columns is *n*, we call the matrix an *n × n square matrix*. We write

$$\mathbf{A} = \begin{pmatrix} a_{11} & a_{12} & \cdots & a_{1n} \\ a_{21} & a_{22} & \cdots & a_{2n} \\ \vdots & \vdots & & \vdots \\ a_{n1} & a_{n2} & \cdots & a_{nn} \end{pmatrix}.$$

2. A *vector* (or *column vector*) is a matrix having just one column. If the vector has *n* rows (and, of course, one column), we call it an *n × 1 vector* (or *n × 1 column vector*). We write

$$\mathbf{x} = \begin{pmatrix} x_1 \\ x_2 \\ \vdots \\ x_n \end{pmatrix}.$$

The elements of a matrix (and hence, in particular, of a vector) may be real numbers, real functions, real function values, or simply "variables." We usually denote square matrices by boldface Roman or Greek capital letters and vectors by boldface Roman or Greek lowercase letters.

Let us give a few specific examples.

The matrix

$$\mathbf{A} = \begin{pmatrix} 1 & 3 & 6 & 1 \\ -2 & 0 & 4 & -5 \\ 7 & 5 & -3 & 2 \\ 4 & -1 & 3 & -6 \end{pmatrix}$$

is a 4 × 4 square matrix of real numbers; whereas $\boldsymbol{\Phi}$ defined by

$$\boldsymbol{\Phi}(t) = \begin{pmatrix} t^2 & t+1 & 5 \\ t & t^2 & 3t \\ 1 & 0 & 2t-1 \end{pmatrix}$$

is a 3 × 3 square matrix of real functions defined for all real *t*.

The vector

$$\mathbf{c} = \begin{pmatrix} 3 \\ 1 \\ 2 \\ 5 \\ -2 \end{pmatrix}$$

is a 5 × 1 column vector of real numbers; the vector $\boldsymbol{\phi}$ defined by

$$\boldsymbol{\phi}(t) = \begin{pmatrix} e^t \\ te^t \\ 2e^t + 1 \end{pmatrix}$$

is a 3 × 1 column vector of real functions defined for all real t; and the vector

$$\mathbf{x} = \begin{pmatrix} x_1 \\ x_2 \\ x_3 \\ x_4 \end{pmatrix}$$

is a 4 × 1 column vector in the four variables x_1, x_2, x_3, x_4.

The elements of a vector are usually called its *components*. Given an $n \times 1$ column vector, the element in the ith row, for each $i = 1, 2, \ldots, n$, is then called its *ith component*.

For example, the third component of the column vector $\mathbf{c}$ illustrated above is 2, and its fourth component is 5.

For any given positive integer n, the $n \times 1$ column vector with all components equal to zero is called the *zero vector* and is denoted by $\mathbf{0}$. Thus if $n = 4$, we have the 4 × 1 zero vector

$$\mathbf{0} = \begin{pmatrix} 0 \\ 0 \\ 0 \\ 0 \end{pmatrix}.$$

DEFINITION

We say that two $m \times n$ matrices $\mathbf{A} = (a_{ij})$ and $\mathbf{B} = (b_{ij})$ are equal if and only if each element of one is equal to the corresponding element of the other. That is, $\mathbf{A}$ and $\mathbf{B}$ are equal if and only if $a_{ij} = b_{ij}$ for $i = 1, 2, \ldots, m$ and $j = 1, 2, \ldots, n$. If $\mathbf{A}$ and $\mathbf{B}$ are equal, we write $\mathbf{A} = \mathbf{B}$.

Thus, for example, the two 3 × 3 square matrices

$$\mathbf{A} = \begin{pmatrix} a_{11} & a_{12} & a_{13} \\ a_{21} & a_{22} & a_{23} \\ a_{31} & a_{32} & a_{33} \end{pmatrix} \quad \text{and} \quad \mathbf{B} = \begin{pmatrix} 6 & 5 & 7 \\ -1 & 0 & 8 \\ 0 & -2 & -4 \end{pmatrix}$$

are equal if and only if

$$a_{11} = 6, \quad a_{12} = 5, \quad a_{13} = 7, \quad a_{21} = -1, \quad a_{22} = 0,$$
$$a_{23} = 8, \quad a_{31} = 0, \quad a_{32} = -2, \quad \text{and} \quad a_{33} = -4.$$

We then write $\mathbf{A} = \mathbf{B}$.

Likewise, the vectors

$$\mathbf{x} = \begin{pmatrix} x_1 \\ x_2 \\ x_3 \\ x_4 \end{pmatrix} \quad \text{and} \quad \mathbf{c} = \begin{pmatrix} -3 \\ 7 \\ 2 \\ -6 \end{pmatrix}$$

are equal if and only if

$$x_1 = -3, \qquad x_2 = 7, \qquad x_3 = 2, \qquad x_4 = -6.$$

We then write $\mathbf{x} = \mathbf{c}$.

DEFINITION Addition of Matrices

The sum *of two* $m \times n$ *matrices* $\mathbf{A} = (a_{ij})$ *and* $\mathbf{B} = (b_{ij})$ *is defined to be the* $m \times n$ *matrix* $\mathbf{C} = (c_{ij})$, *where* $c_{ij} = a_{ij} + b_{ij}$, *for* $i = 1, 2, \ldots, m$ *and* $j = 1, 2, \ldots, n$. *We write* $\mathbf{C} = \mathbf{A} + \mathbf{B}$.

We may describe the addition of matrices by saying that the sum of two $m \times n$ matrices is the $m \times n$ matrix obtained by adding element-by-element.

Thus, for example, the sum of the two 3×3 square matrices

$$\mathbf{A} = \begin{pmatrix} 1 & 4 & -5 \\ 6 & 2 & 0 \\ 9 & 8 & 3 \end{pmatrix} \quad \text{and} \quad \mathbf{B} = \begin{pmatrix} -5 & 4 & 0 \\ 1 & -1 & 3 \\ 6 & 2 & 7 \end{pmatrix}$$

is the 3×3 square matrix

$$\mathbf{C} = \begin{pmatrix} 1-5 & 4+4 & -5+0 \\ 6+1 & 2-1 & 0+3 \\ 9+6 & 8+2 & 3+7 \end{pmatrix} = \begin{pmatrix} -4 & 8 & -5 \\ 7 & 1 & 3 \\ 15 & 10 & 10 \end{pmatrix}.$$

We write $\mathbf{C} = \mathbf{A} + \mathbf{B}$.

Likewise, the sum of the vectors

$$\mathbf{x} = \begin{pmatrix} x_1 \\ x_2 \\ x_3 \\ x_4 \end{pmatrix} \quad \text{and} \quad \mathbf{y} = \begin{pmatrix} y_1 \\ y_2 \\ y_3 \\ y_4 \end{pmatrix} \quad \text{is} \quad \mathbf{x} + \mathbf{y} = \begin{pmatrix} x_1 + y_1 \\ x_2 + y_2 \\ x_3 + y_3 \\ x_4 + y_4 \end{pmatrix}.$$

DEFINITION Multiplication by a Number

The product *of the* $m \times n$ *matrix* $\mathbf{A} = (a_{ij})$ *and the number* c *is defined to be the* $m \times n$ *matrix* $\mathbf{B} = (b_{ij})$, *where* $b_{ij} = ca_{ij}$ *for* $i = 1, 2, \ldots, m$ *and* $j = 1, 2, \ldots, n$. *We write* $\mathbf{B} = c\mathbf{A}$.

We may describe the multiplication of an $m \times n$ matrix $\mathbf{A}$ by a number c by saying that the product so formed is the $m \times n$ matrix that results from multiplying each individual element of $\mathbf{A}$ by the number c.

Thus, for example, the product of the 3×3 square matrix

$$\mathbf{A} = \begin{pmatrix} 2 & -3 & 6 \\ -7 & 4 & -1 \\ 0 & 5 & 2 \end{pmatrix}$$

by the number 3 is the 3×3 square matrix

$$\mathbf{B} = \begin{pmatrix} 3 \cdot 2 & 3 \cdot (-3) & 3 \cdot 6 \\ 3 \cdot (-7) & 3 \cdot 4 & 3 \cdot (-1) \\ 3 \cdot 0 & 3 \cdot 5 & 3 \cdot 2 \end{pmatrix} = \begin{pmatrix} 6 & -9 & 18 \\ -21 & 12 & -3 \\ 0 & 15 & 6 \end{pmatrix}.$$

We write $\mathbf{B} = 3\mathbf{A}$.

Likewise, the product of the vector

$$\mathbf{x} = \begin{pmatrix} x_1 \\ x_2 \\ x_3 \\ x_4 \end{pmatrix}$$

by the number 5 is the vector

$$\mathbf{y} = \begin{pmatrix} 5x_1 \\ 5x_2 \\ 5x_3 \\ 5x_4 \end{pmatrix}.$$

We write $\mathbf{y} = 5\mathbf{x}$.

DEFINITION

Let $\mathbf{x}_1, \mathbf{x}_2, \ldots, \mathbf{x}_m$ be m $n \times 1$ vectors, and let $c_1, c_2, \ldots, c_m$ be m numbers. Then an element of the form

$$c_1 \mathbf{x}_1 + c_2 \mathbf{x}_2 + \cdots + c_m \mathbf{x}_m$$

is an $n \times 1$ vector called a linear combination *of the vectors $\mathbf{x}_1, \mathbf{x}_2, \ldots, \mathbf{x}_m$.*

For example, consider the four 3×1 vectors

$$\mathbf{x}_1 = \begin{pmatrix} 2 \\ -1 \\ 3 \end{pmatrix}, \qquad \mathbf{x}_2 = \begin{pmatrix} 3 \\ 2 \\ 1 \end{pmatrix}, \qquad \mathbf{x}_3 = \begin{pmatrix} 1 \\ -3 \\ -2 \end{pmatrix}, \qquad \mathbf{x}_4 = \begin{pmatrix} 4 \\ 5 \\ 0 \end{pmatrix}$$

and the four real numbers 2, 4, 5, and -3. Then

$$2\mathbf{x}_1 + 4\mathbf{x}_2 + 5\mathbf{x}_3 - 3\mathbf{x}_4 = 2\begin{pmatrix} 2 \\ -1 \\ 3 \end{pmatrix} + 4\begin{pmatrix} 3 \\ 2 \\ 1 \end{pmatrix} + 5\begin{pmatrix} 1 \\ -3 \\ -2 \end{pmatrix} - 3\begin{pmatrix} 4 \\ 5 \\ 0 \end{pmatrix}$$

$$= \begin{pmatrix} 4 \\ -2 \\ 6 \end{pmatrix} + \begin{pmatrix} 12 \\ 8 \\ 4 \end{pmatrix} + \begin{pmatrix} 5 \\ -15 \\ -10 \end{pmatrix} + \begin{pmatrix} -12 \\ -15 \\ 0 \end{pmatrix}$$

$$= \begin{pmatrix} 4 + 12 + 5 - 12 \\ -2 + 8 - 15 - 15 \\ 6 + 4 - 10 + 0 \end{pmatrix} = \begin{pmatrix} 9 \\ -24 \\ 0 \end{pmatrix}$$

is a linear combination of $\mathbf{x}_1, \mathbf{x}_2, \mathbf{x}_3$, and $\mathbf{x}_4$.

DEFINITION

Let

$$\mathbf{A} = \begin{pmatrix} a_{11} & a_{12} & \cdots & a_{1n} \\ a_{21} & a_{22} & \cdots & a_{2n} \\ \vdots & \vdots & & \vdots \\ a_{n1} & a_{n2} & \cdots & a_{nn} \end{pmatrix} \quad and \quad \mathbf{x} = \begin{pmatrix} x_1 \\ x_2 \\ \vdots \\ x_n \end{pmatrix}$$

be an $n \times n$ square matrix and an $n \times 1$ vector, respectively. Then the product $\mathbf{Ax}$ of the $n \times n$ matrix $\mathbf{A}$ by the $n \times 1$ vector $\mathbf{x}$ is defined to be the $n \times 1$ vector

$$\mathbf{Ax} = \begin{pmatrix} a_{11}x_1 + a_{12}x_2 + \cdots + a_{1n}x_n \\ a_{21}x_1 + a_{22}x_2 + \cdots + a_{2n}x_n \\ \vdots \\ a_{n1}x_1 + a_{n2}x_2 + \cdots + a_{nn}x_n \end{pmatrix}.$$

Note that $\mathbf{Ax}$ is a vector. If we denote it by $\mathbf{y}$ and write

$$\mathbf{y} = \mathbf{Ax},$$

where

$$\mathbf{y} = \begin{pmatrix} y_1 \\ y_2 \\ \vdots \\ y_n \end{pmatrix},$$

then we have

$$y_1 = a_{11}x_1 + a_{12}x_2 + \cdots + a_{1n}x_n,$$

$$y_2 = a_{21}x_1 + a_{22}x_2 + \cdots + a_{2n}x_n,$$

$$\vdots$$

$$y_n = a_{n1}x_1 + a_{n2}x_2 + \cdots + a_{nn}x_n;$$

that is, in general, for each $i = 1, 2, \ldots, n$,

$$y_i = a_{i1}x_1 + a_{i2}x_2 + \cdots + a_{in}x_n = \sum_{j=1}^{n} a_{ij}x_j.$$

For example, if

$$\mathbf{A} = \begin{pmatrix} 2 & -4 & 7 \\ 5 & 3 & -8 \\ -3 & 6 & 1 \end{pmatrix} \quad and \quad \mathbf{x} = \begin{pmatrix} x_1 \\ x_2 \\ x_3 \end{pmatrix},$$

then

$$\mathbf{y} = \mathbf{Ax} = \begin{pmatrix} 2x_1 - 4x_2 + 7x_3 \\ 5x_1 + 3x_2 - 8x_3 \\ -3x_1 + 6x_2 + x_3 \end{pmatrix}.$$

Before introducing the next concept, we state and illustrate two useful results, leaving their proofs to the reader (see Exercise 7 at the end of this section).

Result A. *If **A** is an n × n square matrix and **x** and **y** are n × 1 column vectors, then*

$$\mathbf{A}(\mathbf{x} + \mathbf{y}) = \mathbf{A}\mathbf{x} + \mathbf{A}\mathbf{y}.$$

Result B. *If **A** is an n × n square matrix, **x** is an n × 1 column vector, and c is a number, then*

$$\mathbf{A}(c\mathbf{x}) = c(\mathbf{A}\mathbf{x}).$$

▶ **Example 7.20**

We illustrate Results A and B using the matrix

$$\mathbf{A} = \begin{pmatrix} 2 & 1 & 3 \\ -1 & 4 & 1 \\ 3 & -2 & 5 \end{pmatrix},$$

the vectors

$$\mathbf{x} = \begin{pmatrix} x_1 \\ x_2 \\ x_3 \end{pmatrix} \quad \text{and} \quad \mathbf{y} = \begin{pmatrix} y_1 \\ y_2 \\ y_3 \end{pmatrix},$$

and the number $c = 3$.

Illustrating Result A, we have

$$\mathbf{A}(\mathbf{x} + \mathbf{y}) = \begin{pmatrix} 2 & 1 & 3 \\ -1 & 4 & 1 \\ 3 & -2 & 5 \end{pmatrix} \begin{pmatrix} x_1 + y_1 \\ x_2 + y_2 \\ x_3 + y_3 \end{pmatrix}$$

$$= \begin{pmatrix} 2(x_1 + y_1) + (x_2 + y_2) + 3(x_3 + y_3) \\ -(x_1 + y_1) + 4(x_2 + y_2) + (x_3 + y_3) \\ 3(x_1 + y_1) - 2(x_2 + y_2) + 5(x_3 + y_3) \end{pmatrix}$$

$$= \begin{pmatrix} 2x_1 + x_2 + 3x_3 \\ -x_1 + 4x_2 + x_3 \\ 3x_1 - 2x_2 + 5x_3 \end{pmatrix} + \begin{pmatrix} 2y_1 + y_2 + 3y_3 \\ -y_1 + 4y_2 + y_3 \\ 3y_1 - 2y_2 + 5y_3 \end{pmatrix}$$

$$= \begin{pmatrix} 2 & 1 & 3 \\ -1 & 4 & 1 \\ 3 & -2 & 5 \end{pmatrix} \begin{pmatrix} x_1 \\ x_2 \\ x_3 \end{pmatrix} + \begin{pmatrix} 2 & 1 & 3 \\ -1 & 4 & 1 \\ 3 & -2 & 5 \end{pmatrix} \begin{pmatrix} y_1 \\ y_2 \\ y_3 \end{pmatrix} = \mathbf{A}\mathbf{x} + \mathbf{A}\mathbf{y}.$$

Illustrating Result B, we have

$$\mathbf{A}(c\mathbf{x}) = \begin{pmatrix} 2 & 1 & 3 \\ -1 & 4 & 1 \\ 3 & -2 & 5 \end{pmatrix} \begin{pmatrix} 3x_1 \\ 3x_2 \\ 3x_3 \end{pmatrix} = \begin{pmatrix} 6x_1 + 3x_2 + 9x_3 \\ -3x_1 + 12x_2 + 3x_3 \\ 9x_1 - 6x_2 + 15x_3 \end{pmatrix}$$

$$= 3 \begin{pmatrix} 2x_1 + x_2 + 3x_3 \\ -x_1 + 4x_2 + x_3 \\ 3x_1 - 2x_2 + 5x_3 \end{pmatrix} = c(\mathbf{A}\mathbf{x}).$$

We have seen examples of vectors whose components are numbers and also of vectors whose components are real functions defined on an interval $[a, b]$. We now distinguish between these two types by means of the following definitions:

DEFINITIONS

1. A vector all of whose components are numbers is called a constant vector.
2. A vector all of whose components are real functions defined on an interval $[a, b]$ is called a vector function.

DEFINITIONS

Let ϕ be the $n \times 1$ vector function defined by

$$\phi(t) = \begin{pmatrix} \phi_1(t) \\ \phi_2(t) \\ \vdots \\ \phi_n(t) \end{pmatrix}$$

for all t on a real interval $[a, b]$.

(1) *Suppose the components $\phi_1(t), \phi_2(t), \ldots, \phi_n(t)$ are differentiable on $[a, b]$. Then the derivative of ϕ is the vector function defined by*

$$\frac{d\phi(t)}{dt} = \begin{pmatrix} \dfrac{d\phi_1(t)}{dt} \\[2ex] \dfrac{d\phi_2(t)}{dt} \\[2ex] \vdots \\[2ex] \dfrac{d\phi_n(t)}{dt} \end{pmatrix}$$

for all $t \in [a, b]$.

(2) *Suppose the components $\phi_1(t), \phi_2(t), \ldots, \phi_n(t)$ are integrable on $[a, b]$; and let t_0 and $t \in [a, b]$, where $t_0 < t$. Then the integral of ϕ from t_0 to t is the vector function defined by*

$$\int_{t_0}^{t} \phi(u)\, du = \begin{pmatrix} \displaystyle\int_{t_0}^{t} \phi_1(u)\, du \\[2ex] \displaystyle\int_{t_0}^{t} \phi_2(u)\, du \\[2ex] \vdots \\[2ex] \displaystyle\int_{t_0}^{t} \phi_n(u)\, du \end{pmatrix}.$$

Thus the derivative of a given vector function all of whose components are differentiable is the vector function obtained from the given vector function by differentiating each component of the given vector function. Likewise, the integral from t_0 to t of a given vector function, all of whose components are integrable on the given interval, is the vector function obtained from the given vector function by integrating each component of the given vector function from t_0 to t.

▶ **Example 7.21**

The derivative of the vector function ϕ defined for all t by

$$\phi(t) = \begin{pmatrix} 4t^3 \\ 2t^2 + 3t \\ 2e^{3t} \end{pmatrix}$$

is the vector function defined for all t by

$$\frac{d\phi(t)}{dt} = \begin{pmatrix} 12t^2 \\ 4t + 3 \\ 6e^{3t} \end{pmatrix}.$$

Also, the integral of ϕ from 0 to t is the vector function defined for all t by

$$\int_0^t \phi(u)\,du = \begin{pmatrix} \int_0^t 4u^3\,du \\ \int_0^t (2u^2 + 3u)\,du \\ \int_0^t 2e^{3u}\,du \end{pmatrix} = \begin{pmatrix} t^4 \\ \frac{2}{3}t^3 + \frac{3}{2}t^2 \\ \frac{2}{3}(e^{3t} - 1) \end{pmatrix}.$$

Exercises

1. In each case, find the sum $A + B$ of the given matrices.

(a) $A = \begin{pmatrix} 2 & 3 \\ 5 & 4 \end{pmatrix}$ and $B = \begin{pmatrix} 6 & -1 \\ -7 & 2 \end{pmatrix}.$

(b) $A = \begin{pmatrix} 2 & 1 & 3 \\ -1 & 0 & 5 \\ -4 & 3 & -2 \end{pmatrix}$ and $B = \begin{pmatrix} 7 & -1 & 6 \\ 2 & 4 & -3 \\ 5 & -5 & 1 \end{pmatrix}.$

(c) $A = \begin{pmatrix} -5 & 0 & 4 \\ -2 & -1 & -3 \\ 6 & 2 & 5 \end{pmatrix}$ and $B = \begin{pmatrix} 7 & -2 & -3 \\ 6 & -3 & 1 \\ -2 & 1 & -3 \end{pmatrix}.$

2. In each case, find the product cA of the given matrix A and the number c.

(a) $A = \begin{pmatrix} 1 & 2 \\ 7 & -3 \end{pmatrix}$, $c = 3.$

(b) $\mathbf{A} = \begin{pmatrix} 1 & -3 & 5 \\ 6 & -2 & 0 \\ -3 & 1 & 2 \end{pmatrix}$, $c = -4$.

(c) $\mathbf{A} = \begin{pmatrix} 5 & -1 & 2 \\ 4 & -3 & -2 \\ 0 & 3 & -6 \end{pmatrix}$, $c = -3$.

3.   In each case, find the indicated linear combination of the given vectors

$$\mathbf{x}_1 = \begin{pmatrix} 3 \\ 2 \\ -1 \\ 4 \end{pmatrix}, \quad \mathbf{x}_2 = \begin{pmatrix} -1 \\ 3 \\ 5 \\ -2 \end{pmatrix}, \quad \mathbf{x}_3 = \begin{pmatrix} 2 \\ 4 \\ 0 \\ 6 \end{pmatrix}, \quad \text{and} \quad \mathbf{x}_4 = \begin{pmatrix} -1 \\ 2 \\ -3 \\ 5 \end{pmatrix}.$$

(a) $2\mathbf{x}_1 + 3\mathbf{x}_2 - \mathbf{x}_3$.

(b) $3\mathbf{x}_1 - 2\mathbf{x}_2 + 4\mathbf{x}_4$.

(c) $-\mathbf{x}_1 + 5\mathbf{x}_2 - 2\mathbf{x}_3 + 3\mathbf{x}_4$.

4.   In each case, find the product $\mathbf{A}\mathbf{x}$ of the given matrix $\mathbf{A}$ by the given vector $\mathbf{x}$.

(a) $\mathbf{A} = \begin{pmatrix} 2 & 1 & -4 \\ 5 & -2 & 3 \\ 1 & -3 & 2 \end{pmatrix}$, $\mathbf{x} = \begin{pmatrix} x_1 \\ x_2 \\ x_3 \end{pmatrix}$.

(b) $\mathbf{A} = \begin{pmatrix} -3 & -5 & 7 \\ 0 & 4 & 1 \\ -2 & 1 & 3 \end{pmatrix}$, $\mathbf{x} = \begin{pmatrix} 2 \\ 3 \\ -2 \end{pmatrix}$.

(c) $\mathbf{A} = \begin{pmatrix} 1 & 0 & -3 \\ 2 & -5 & 4 \\ -3 & 1 & 2 \end{pmatrix}$, $\mathbf{x} = \begin{pmatrix} x_1 + x_2 \\ x_1 + 2x_2 \\ x_2 - x_3 \end{pmatrix}$.

5.   Illustrate Results A and B of this subsection using the matrix

$$\mathbf{A} = \begin{pmatrix} 3 & -1 & 2 \\ 5 & 4 & -3 \\ -5 & 1 & 2 \end{pmatrix},$$

the vectors

$$\mathbf{x} = \begin{pmatrix} x_1 \\ x_2 \\ x_3 \end{pmatrix} \quad \text{and} \quad \mathbf{y} = \begin{pmatrix} y_1 \\ y_2 \\ y_3 \end{pmatrix},$$

and the number $c = 4$.

6.   In each case, find (i) the derivative and (ii) the integral from 0 to t of the vector function $\boldsymbol{\phi}$ that is defined.

(a) $\boldsymbol{\phi}(t) = \begin{pmatrix} 5t^2 \\ -6t^3 + t^2 \\ 2t^2 - 5t \end{pmatrix}$.

(b) $\phi(t) = \begin{pmatrix} e^{3t} \\ (2t + 3)e^{3t} \\ t^2 e^{3t} \end{pmatrix}.$

(c) $\phi(t) = \begin{pmatrix} \sin 3t \\ \cos 3t \\ t \sin 3t \\ t \cos 3t \end{pmatrix}.$

7.   Prove Results A and B of the text (page 318).

B. Matrix Multiplication and Inversion

DEFINITION   Multiplication of Matrices

Let $\mathbf{A} = (a_{ij})$ *be an* $m \times n$ *matrix and* $\mathbf{B} = (b_{ij})$ *be an* $n \times p$ *matrix. The product* $\mathbf{AB}$ *is defined to be the* $m \times p$ *matrix* $\mathbf{C} = (c_{ij})$, *where*

$$c_{ij} = a_{i1}b_{1j} + a_{i2}b_{2j} + \cdots + a_{in}b_{nj} = \sum_{k=1}^{n} a_{ik}b_{kj} \tag{7.104}$$

for all $i = 1, 2, \ldots, m$ *and all* $j = 1, 2, \ldots, p$. *We write* $\mathbf{C} = \mathbf{AB}$.

Concerning this definition, we first note that $\mathbf{A}$ has m rows and n columns, $\mathbf{B}$ has n rows and p columns, and the product $\mathbf{C} = \mathbf{AB}$ has m rows (the same number as $\mathbf{A}$ has) and p columns (the same number as $\mathbf{B}$ has). Next observe that the number of columns in $\mathbf{A}$ is equal to the number of rows in $\mathbf{B}$. Matrix multiplication is defined only when this is the case, that is, only when the number of columns in the first matrix is the same as the number of rows in the second. Then the element c_{ij} in the ith row and jth column of the product matrix $\mathbf{C} = \mathbf{AB}$ is the sum of the n products that are obtained by multiplying each element in the ith row of A by the corresponding element in the jth column of B.

▶ **Example 7.22**

Let

$$\mathbf{A} = \begin{pmatrix} 2 & 1 \\ -1 & 3 \\ 4 & 5 \end{pmatrix} \quad \text{and} \quad \mathbf{B} = \begin{pmatrix} 4 & 2 & -2 \\ -1 & 3 & 0 \end{pmatrix}.$$

Matrix $\mathbf{A}$ is a 3×2 matrix (3 rows, 2 columns), and matrix $\mathbf{B}$ is a 2×3 matrix (2 rows, 3 columns). Thus the number of columns in $\mathbf{A}$ is equal to the number of rows in $\mathbf{B}$, and so $\mathbf{AB}$ is defined. Also, the number of columns in $\mathbf{B}$ is equal to the number of rows in $\mathbf{A}$, so $\mathbf{BA}$ is defined. Thus, for these two matrices $\mathbf{A}$ and $\mathbf{B}$, both of the products $\mathbf{AB}$ and $\mathbf{BA}$ are defined.

We first find the 3×3 matrix $\mathbf{AB}$:

$$\mathbf{AB} = \begin{pmatrix} 2 & 1 \\ -1 & 3 \\ 4 & 5 \end{pmatrix} \begin{pmatrix} 4 & 2 & -2 \\ -1 & 3 & 0 \end{pmatrix}$$

$$= \begin{pmatrix} (2)(4) + (1)(-1) & (2)(2) + (1)(3) & (2)(-2) + (1)(0) \\ (-1)(4) + (3)(-1) & (-1)(2) + (3)(3) & (-1)(-2) + (3)(0) \\ (4)(4) + (5)(-1) & (4)(2) + (5)(3) & (4)(-2) + (5)(0) \end{pmatrix}$$

$$= \begin{pmatrix} 7 & 7 & -4 \\ -7 & 7 & 2 \\ 11 & 23 & -8 \end{pmatrix}.$$

We now find the 2×2 matrix **BA**:

$$\mathbf{BA} = \begin{pmatrix} 4 & 2 & -2 \\ -1 & 3 & 0 \end{pmatrix} \begin{pmatrix} 2 & 1 \\ -1 & 3 \\ 4 & 5 \end{pmatrix}$$

$$= \begin{pmatrix} (4)(2) + (2)(-1) + (-2)(4) & (4)(1) + (2)(3) + (-2)(5) \\ (-1)(2) + (3)(-1) + (0)(4) & (-1)(1) + (3)(3) + (0)(5) \end{pmatrix} = \begin{pmatrix} -2 & 0 \\ -5 & 8 \end{pmatrix}$$

For the matrices **A** and **B** of Example 7.22, we note that although the products **AB** and **BA** are both defined, they are not of the same size and hence are necessarily not equal. Now suppose two matrices **A** and **B** are such that **AB** and **BA** are both defined and of the same size. (Note that this will always be the case when **A** and **B** themselves are both $n \times n$ square matrices of the same size.) We are led to inquire whether or not **AB** and **BA** are necessarily equal in such a case. The following example shows that they are not:

▶ **Example 7.23**

Let

$$\mathbf{A} = \begin{pmatrix} 2 & 1 & -1 \\ 4 & 3 & -2 \\ -6 & 2 & 5 \end{pmatrix} \quad \text{and} \quad \mathbf{B} = \begin{pmatrix} 1 & 3 & -2 \\ 0 & -1 & 3 \\ 2 & 1 & -1 \end{pmatrix}.$$

We observe that **A**, **B**, **AB**, and **BA** are all 3×3 matrices. We find

$$\mathbf{AB} = \begin{pmatrix} 2 & 1 & -1 \\ 4 & 3 & -2 \\ -6 & 2 & 5 \end{pmatrix} \begin{pmatrix} 1 & 3 & -2 \\ 4 & -1 & 3 \\ 2 & 1 & -1 \end{pmatrix}$$

$$= \begin{pmatrix} (2)(1) + (1)(4) + (-1)(2) & (2)(3) + (1)(-1) + (-1)(1) \\ (4)(1) + (3)(4) + (-2)(2) & (4)(3) + (3)(-1) + (-2)(1) \\ (-6)(1) + (2)(4) + (5)(2) & (-6)(3) + (2)(-1) + (5)(1) \end{pmatrix}$$

$$\begin{pmatrix} (2)(-2) + (1)(3) + (-1)(-1) \\ (4)(-2) + (3)(3) + (-2)(-1) \\ (-6)(-2) + (2)(3) + (5)(-1) \end{pmatrix}$$

$$= \begin{pmatrix} 4 & 4 & 0 \\ 12 & 7 & 3 \\ 12 & -15 & 13 \end{pmatrix}$$

and

$$\mathbf{BA} = \begin{pmatrix} 1 & 3 & -2 \\ 4 & -1 & 3 \\ 2 & 1 & -1 \end{pmatrix} \begin{pmatrix} 2 & 1 & -1 \\ 4 & 3 & -2 \\ -6 & 2 & 5 \end{pmatrix}$$

$$= \begin{pmatrix} (1)(2) + (3)(4) + (-2)(-6) & (1)(1) + (3)(3) + (-2)(2) \\ (4)(2) + (-1)(4) + (3)(-6) & (4)(1) + (-1)(3) + (3)(2) \\ (2)(2) + (1)(4) + (-1)(-6) & (2)(1) + (1)(3) + (-1)(2) \end{pmatrix}$$

$$\begin{pmatrix} (1)(-1) + (3)(-2) + (-2)(5) \\ (4)(-1) + (-1)(-2) + (3)(5) \\ (2)(-1) + (1)(-2) + (-1)(5) \end{pmatrix}$$

$$= \begin{pmatrix} 26 & 6 & -17 \\ -14 & 7 & 13 \\ 14 & 3 & -9 \end{pmatrix}.$$

Thus, even though $\mathbf{AB}$ and $\mathbf{BA}$ are both of the same size, they are *not* equal. We write $\mathbf{AB} \neq \mathbf{BA}$ in this case.

The preceding example clearly illustrates the fact that *matrix multiplication is not commutative*. That is, we do *not* have $\mathbf{AB} = \mathbf{BA}$ in general, even when $\mathbf{A}$ and $\mathbf{B}$ and the two products $\mathbf{AB}$ and $\mathbf{BA}$ are all $n \times n$ square matrices of the same size. However, we point out that matrix multiplication is *associative*,

$$(\mathbf{AB})\mathbf{C} = \mathbf{A}(\mathbf{BC}),$$

and *distributive*,

$$\mathbf{A}(\mathbf{B} + \mathbf{C}) = \mathbf{AB} + \mathbf{AC},$$

for any three matrices $\mathbf{A}$, $\mathbf{B}$, $\mathbf{C}$ for which the operations involved are defined.

Now consider an $n \times n$ square matrix $\mathbf{A} = (a_{ij})$. The *principal diagonal* of $\mathbf{A}$ is the diagonal of elements from the upper left corner to the lower right corner of $\mathbf{A}$; and the *diagonal elements* of $\mathbf{A}$ are the set of elements that lie along this principal diagonal, that is, the elements $a_{11}, a_{22}, \ldots, a_{nn}$. Now, for any given positive integer n, the $n \times n$ square matrix in which all the diagonal elements are one and all the other elements are zero is called the *identity matrix* and is denoted by $\mathbf{I}$. That is, $\mathbf{I} = (a_{ij})$, where $a_{ij} = 1$ for all $i = j$ and $a_{ij} = 0$ for $i \neq j$ ($i = 1, 2, \ldots, n; j = 1, 2, \ldots, n$). Thus if $n = 4$, the 4×4 identity matrix is

$$\mathbf{I} = \begin{pmatrix} 1 & 0 & 0 & 0 \\ 0 & 1 & 0 & 0 \\ 0 & 0 & 1 & 0 \\ 0 & 0 & 0 & 1 \end{pmatrix}.$$

If $\mathbf{A}$ is an arbitrary $n \times n$ square matrix and $\mathbf{I}$ is the $n \times n$ identity matrix, then it follows at once from the definition of matrix multiplication that

$$\mathbf{AI} = \mathbf{IA} = \mathbf{A}.$$

We now consider the following problem involving matrix multiplication: Given an $n \times n$ square matrix $\mathbf{A}$, we seek another $n \times n$ square matrix $\mathbf{B}$ such that $\mathbf{AB} = \mathbf{I}$, where $\mathbf{I}$ is the identity matrix. In order to discuss the existence of such a matrix $\mathbf{B}$, we need the following definition.

DEFINITION

Let $\mathbf{A}$ be an $n \times n$ square matrix. The matrix $\mathbf{A}$ is called nonsingular *if and only if its determinant is unequal to zero:* $|\mathbf{A}| \neq 0$. *Otherwise, the matrix $\mathbf{A}$ is called* singular.

▶ **Example 7.24**

The matrix

$$\mathbf{A} = \begin{pmatrix} 2 & 1 & -1 \\ 4 & 3 & -2 \\ -6 & 2 & 5 \end{pmatrix}$$

of Example 7.23 is nonsingular; for we find

$$|\mathbf{A}| = \begin{vmatrix} 2 & 1 & -1 \\ 4 & 3 & -2 \\ -6 & 2 & 5 \end{vmatrix} = 2\begin{vmatrix} 3 & -2 \\ 2 & 5 \end{vmatrix} - \begin{vmatrix} 4 & -2 \\ -6 & 5 \end{vmatrix} - \begin{vmatrix} 4 & 3 \\ -6 & 2 \end{vmatrix} = 4 \neq 0.$$

We are now in a position to state the following basic result concerning the existence of a matrix $\mathbf{B}$ such that $\mathbf{AB} = \mathbf{I}$.

Result C. *Let $\mathbf{A}$ be an $n \times n$ square matrix, and let $\mathbf{I}$ be the $n \times n$ identity matrix. Then there exists a unique $n \times n$ square matrix $\mathbf{B}$ such that $\mathbf{AB} = \mathbf{BA} = \mathbf{I}$ if and only if $\mathbf{A}$ is nonsingular, that is, if and only if $|\mathbf{A}| \neq 0$.*

DEFINITION

Let $\mathbf{A}$ be an $n \times n$ nonsingular matrix, and let $\mathbf{I}$ be the $n \times n$ identity matrix. The unique $n \times n$ matrix $\mathbf{B}$ such that $\mathbf{AB} = \mathbf{BA} = \mathbf{I}$ is called the inverse *of $\mathbf{A}$. We denote the unique inverse of $\mathbf{A}$ by $\mathbf{A}^{-1}$ and thus write the defining relation between a given matrix $\mathbf{A}$ and its inverse $\mathbf{A}^{-1}$ in the form*

$$\mathbf{AA}^{-1} = \mathbf{A}^{-1}\mathbf{A} = \mathbf{I}. \tag{7.105}$$

Now suppose that $\mathbf{A}$ is a nonsingular $n \times n$ matrix, so that the existence and uniqueness of its inverse $\mathbf{A}^{-1}$ is assured. We now consider the question of finding $\mathbf{A}^{-1}$. Several distinct methods are known. We shall introduce, illustrate, and use a method that involves the use of determinants. Although this method is not very efficient except when $n = 2$ or $n = 3$, it will be sufficiently useful for our purposes in this text. In order to describe the procedure, several preliminary concepts will be introduced and illustrated.

DEFINITIONS

Let **A** *be an* $n \times n$ *matrix, and let* a_{ij} *be the element in the ith row and jth column of* **A**.

(1) *The* minor *of* a_{ij} *is defined to be the determinant of the* $(n-1) \times (n-1)$ *matrix obtained from* **A** *by deleting the ith row and jth column of* **A**. *We denote it by* M_{ij}.

(2) *The* cofactor *of* a_{ij} *is defined to be the minor* M_{ij} *of* a_{ij} *multiplied by the number* $(-1)^{i+j}$. *We denote it by* C_{ij}. *Thus we have* $C_{ij} = (-1)^{i+j}M_{ij}$.

(3) *The* matrix of cofactors *of the elements of* **A** *is defined to be the matrix obtained from* **A** *by replacing each element* a_{ij} *of* **A** *by its cofactor* C_{ij} $(i = 1, 2, \ldots, n; j = 1, 2, \ldots, n)$. *We denote it by* cof **A**.

Thus the minor M_{ij} of a_{ij} is formally obtained simply by crossing out the row and column in which a_{ij} appears and finding the determinant of the resulting matrix that remains. Then the cofactor C_{ij} of a_{ij} is obtained simply by multiplying the minor M_{ij} by $+1$ or -1, depending upon whether the sum $i + j$ of the row and column of a_{ij} is respectively even or odd.

▶ **Example 7.25**

Consider the 3×3 matrix

$$A = \begin{pmatrix} 2 & 1 & -1 \\ 4 & 3 & -2 \\ -6 & 2 & 5 \end{pmatrix}.$$

We find the cofactor of the element $a_{23} = -2$ in the second row and third column of A. We first find the minor M_{23} of a_{23}. By definition, this is the determinant of the 2×2 matrix obtained from **A** by deleting the second row and third column of **A**. Clearly this is

$$M_{23} = \begin{vmatrix} 2 & 1 \\ -6 & 2 \end{vmatrix} = 10.$$

Now, by definition, the cofactor C_{23} of a_{23} is given by

$$C_{23} = (-1)^{2+3}M_{23} = -10.$$

In like manner, we find the cofactors of the other elements of **A**. Then, replacing each element a_{ij} of **A** by its cofactor C_{ij}, we obtain the matrix of cofactors of **A**,

$$\text{cof } A = \begin{pmatrix} \begin{vmatrix} 3 & -2 \\ 2 & 5 \end{vmatrix} & -\begin{vmatrix} 4 & -2 \\ -6 & 5 \end{vmatrix} & \begin{vmatrix} 4 & 3 \\ -6 & 2 \end{vmatrix} \\ -\begin{vmatrix} 1 & -1 \\ 2 & 5 \end{vmatrix} & \begin{vmatrix} 2 & -1 \\ -6 & 5 \end{vmatrix} & -\begin{vmatrix} 2 & 1 \\ -6 & 2 \end{vmatrix} \\ \begin{vmatrix} 1 & -1 \\ 3 & -2 \end{vmatrix} & -\begin{vmatrix} 2 & -1 \\ 4 & -2 \end{vmatrix} & \begin{vmatrix} 2 & 1 \\ 4 & 3 \end{vmatrix} \end{pmatrix}$$

$$= \begin{pmatrix} 19 & -8 & 26 \\ -7 & 4 & -10 \\ 1 & 0 & 2 \end{pmatrix}.$$

DEFINITION

Let $\mathbf{A} = (a_{ij})$ *be an* $m \times n$ *matrix. The transpose of* $\mathbf{A}$ *is the* $n \times m$ *matrix* $\mathbf{B} = (b_{ij})$*, where* $b_{ij} = a_{ji}$, $i = 1, 2, \ldots, n$, $j = 1, 2, \ldots, m$*. That is, the transpose of an* $m \times n$ *matrix* $\mathbf{A}$ *is the* $n \times m$ *matrix obtained from* $\mathbf{A}$ *by interchanging the rows and columns of* $\mathbf{A}$*. We denote the transpose of* $\mathbf{A}$ *by* $\mathbf{A}^T$*.*

For example, the transpose of the 3×3 matrix

$$\mathbf{A} = \begin{pmatrix} 2 & 1 & -1 \\ 4 & 3 & -2 \\ -6 & 2 & 5 \end{pmatrix}$$

of Example 7.25 is the 3×3 matrix

$$\mathbf{A}^T = \begin{pmatrix} 2 & 4 & -6 \\ 1 & 3 & 2 \\ -1 & -2 & 5 \end{pmatrix}.$$

Note, in particular, that the transpose of the $n \times 1$ *column* vector

$$\mathbf{x} = \begin{pmatrix} x_1 \\ x_2 \\ \vdots \\ x_n \end{pmatrix}$$

is the $1 \times n$ *row* vector

$$\mathbf{x}^T = (x_1, x_2, \ldots, x_n).$$

DEFINITION

Let $\mathbf{A}$ *be an* $n \times n$ *square matrix, and let cof* $\mathbf{A}$ *be the matrix of cofactors of* $\mathbf{A}$*. The adjoint of* $\mathbf{A}$ *is defined to be the transpose of the matrix of cofactors of* $\mathbf{A}$*. We denote the adjoint of* $\mathbf{A}$ *by adj* $\mathbf{A}$*, and thus write*

$$\text{adj } \mathbf{A} = (\text{cof } \mathbf{A})^T.$$

▶ **Example 7.26**

In Example 7.25 we found that the matrix of cofactors of the matrix

$$\mathbf{A} = \begin{pmatrix} 2 & 1 & -1 \\ 4 & 3 & -2 \\ -6 & 2 & 5 \end{pmatrix}$$

is the matrix

$$\text{cof } \mathbf{A} = \begin{pmatrix} 19 & -8 & 26 \\ -7 & 4 & -10 \\ 1 & 0 & 2 \end{pmatrix}.$$

Thus the adjoint of A, being the transpose of cof A, is given by

$$\text{adj } A = (\text{cof } A)^T = \begin{pmatrix} 19 & -7 & 1 \\ -8 & 4 & 0 \\ 26 & -10 & 2 \end{pmatrix}.$$

We are now in a position to state the following important result giving a formula for the inverse of a nonsingular matrix.

Result D. *Let A be an $n \times n$ nonsingular matrix. Then the unique inverse A^{-1} of A is given by the formula*

$$A^{-1} = \frac{1}{|A|} (\text{adj } A), \tag{7.106}$$

where $|A|$ is the determinant of A and adj A is the adjoint matrix of A.

Looking back over the preceding definitions, we observe that to find the inverse A^{-1} of a given nonsingular matrix A, we proceed as follows:

(1) Replace each element a_{ij} of A by its cofactor C_{ij} to find the matrix of cofactors c of A of A;

(2) Take the transpose (interchange the rows and columns) of the matrix of A found in step (1) to find the adjoint matrix adj A of A; and

(3) Divide each element of the matrix adj A found in step (2) by the determinant $|A|$ of A. This gives the inverse matrix A^{-1}.

▶ **Example 7.27**

Consider again the 3×3 matrix

$$A = \begin{pmatrix} 2 & 1 & -1 \\ 4 & 3 & -2 \\ -6 & 2 & 5 \end{pmatrix},$$

already considered in Examples 7.24, 7.25, and 7.26. In Example 7.24 we found that

$$|A| = 4$$

and thus noted that since $|A| \neq 0$, A is nonsingular. Thus by Result C and the definition immediately following, we know that A has a unique inverse A^{-1}. Now by Result D we know that A^{-1} is given by

$$A^{-1} = \frac{1}{|A|} (\text{adj } A) = \tfrac{1}{4}(\text{adj } A).$$

In Example 7.25 we found

$$\text{cof } A = \begin{pmatrix} 19 & -8 & 26 \\ -7 & 4 & -10 \\ 1 & 0 & 2 \end{pmatrix};$$

and then, in Example 7.26, we obtained

$$\text{adj } \mathbf{A} = (\text{cof } \mathbf{A})^T = \begin{pmatrix} 19 & -7 & 1 \\ -8 & 4 & 0 \\ 26 & -10 & 2 \end{pmatrix}.$$

Thus we find

$$\mathbf{A}^{-1} = \tfrac{1}{4}(\text{adj } \mathbf{A}) = \frac{1}{4}\begin{pmatrix} 19 & -7 & 1 \\ -8 & 4 & 0 \\ 26 & -10 & 2 \end{pmatrix} = \begin{pmatrix} \dfrac{19}{4} & -\dfrac{7}{4} & \dfrac{1}{4} \\ -2 & 1 & 0 \\ \dfrac{13}{2} & -\dfrac{5}{2} & \dfrac{1}{2} \end{pmatrix}.$$

The reader should now calculate the products $\mathbf{AA}^{-1}$ and $\mathbf{A}^{-1}\mathbf{A}$ and observe that both of these products are indeed the identity matrix $\mathbf{I}$.

Just as in the special case of vectors, there are matrices whose elements are numbers and matrices whose elements are real functions defined on an interval $[a, b]$. We distinguish between these two types by means of the following definitions:

DEFINITIONS

   (1) A matrix all of whose components are numbers is called a constant matrix.
   (2) A matrix all of whose components are real functions defined on an interval $[a, b]$ is called a matrix function.

DEFINITION

Let A defined by

$$\mathbf{A}(t) = \begin{pmatrix} a_{11}(t) & a_{12}(t) & \cdots & a_{1n}(t) \\ a_{21}(t) & a_{22}(t) & \cdots & a_{2n}(t) \\ \vdots & \vdots & & \vdots \\ a_{n1}(t) & a_{n2}(t) & \cdots & a_{nn}(t) \end{pmatrix}$$

be an $n \times n$ matrix function whose elements $a_{ij}(t)$ ($i = 1, 2, \ldots, n$; $j = 1, 2, \ldots, n$) are differentiable on an interval $[a, b]$. Then $\mathbf{A}$ is said to be differentiable *on $[a, b]$, and the* derivative *of $\mathbf{A}$ is the matrix function defined by*

$$\frac{d\mathbf{A}(t)}{dt} = \begin{pmatrix} a'_{11}(t) & a'_{12}(t) & a'_{1n}(t) \\ a'_{21}(t) & a'_{22}(t) & a'_{2n}(t) \\ \vdots & \vdots & \vdots \\ a'_{n1}(t) & a'_{n2}(t) & a'_{nn}(t) \end{pmatrix}$$

for all $t \in [a, b]$.

Thus the derivative of a given matrix function all of whose elements are differentiable is the matrix function obtained from the given matrix function by differentiating each element of the given matrix function.

▶ **Example 7.28**

The derivative of the matrix function **A** defined for all t by

$$\mathbf{A}(t) = \begin{pmatrix} 3t^2 & e^{4t} \\ 5t^3 & 3e^{2t} \end{pmatrix}$$

is the matrix function defined for all t by

$$\frac{d\mathbf{A}(t)}{dt} = \begin{pmatrix} 6t & 4e^{4t} \\ 15t^2 & 6e^{2t} \end{pmatrix}.$$

We close this section by stating the following useful result on the differentiation of a product of differentiable matrices. Note that the order of the factors in each product cannot be reversed, since matrix multiplication is not commutative.

Result E.   Let **A** and **B** be differentiable matrices on $[a, b]$. Then the product **AB** is differentiable on $[a, b]$, and

$$\frac{d}{dt}[\mathbf{A}(t)\mathbf{B}(t)] = \frac{d\mathbf{A}(t)}{dt}\mathbf{B}(t) + \mathbf{A}(t)\frac{d\mathbf{B}(t)}{dt}$$

for all $t \in [a, b]$.

Exercises

In each of Exercises 1–10, given the matrices **A** and **B**, find the product **AB**. Also, find the product **BA** in each case in which it is defined.

1. $\mathbf{A} = \begin{pmatrix} 3 & 5 \\ 1 & 7 \end{pmatrix}$, $\mathbf{B} = \begin{pmatrix} 4 & 6 \\ 2 & 1 \end{pmatrix}$.

2. $\mathbf{A} = \begin{pmatrix} 5 & -2 \\ 4 & 3 \end{pmatrix}$, $\mathbf{B} = \begin{pmatrix} -1 & 8 \\ 2 & -5 \end{pmatrix}$.

3. $\mathbf{A} = \begin{pmatrix} 3 & 2 \\ -1 & 4 \\ 5 & -2 \end{pmatrix}$, $\mathbf{B} = \begin{pmatrix} 3 & 0 & 2 \\ -1 & 2 & -3 \end{pmatrix}$.

4. $\mathbf{A} = \begin{pmatrix} 6 & 1 \\ 5 & -2 \end{pmatrix}$, $\mathbf{B} = \begin{pmatrix} 1 & 3 & 2 & 4 \\ 0 & 2 & -1 & -3 \end{pmatrix}$.

5. $\mathbf{A} = \begin{pmatrix} 1 & 2 & -3 \\ 3 & -1 & 0 \\ 0 & 2 & 4 \end{pmatrix}$, $\mathbf{B} = \begin{pmatrix} 4 & 1 \\ 3 & 2 \\ 1 & 0 \end{pmatrix}$.

6. $\mathbf{A} = \begin{pmatrix} 3 & 2 & 1 \\ 0 & 1 & 2 \\ 5 & 4 & 3 \end{pmatrix}$, $\mathbf{B} = \begin{pmatrix} 2 & -1 & -3 \\ -6 & 0 & 1 \\ 1 & -3 & 4 \end{pmatrix}$.

7. $\mathbf{A} = \begin{pmatrix} -2 & 4 & 6 \\ 1 & 3 & 5 \\ 0 & 2 & 0 \end{pmatrix}$, $\mathbf{B} = \begin{pmatrix} 0 & 1 & 2 \\ 3 & -2 & -1 \\ 5 & 4 & 2 \end{pmatrix}$.

8. $\mathbf{A} = \begin{pmatrix} 4 & 2 & -1 \\ 1 & 1 & 2 \\ 3 & 2 & -2 \end{pmatrix}$, $\mathbf{B} = \begin{pmatrix} 1 & 1 & 2 \\ 2 & 3 & 6 \\ 1 & -4 & -1 \end{pmatrix}$.

9. $\mathbf{A} = \begin{pmatrix} 2 & 1 & 0 & 1 \\ 0 & 2 & 3 & -1 \\ 1 & -2 & 1 & 0 \end{pmatrix}$, $\mathbf{B} = \begin{pmatrix} 1 & 2 \\ 0 & 3 \\ -1 & 0 \\ 1 & -2 \end{pmatrix}$.

10. $\mathbf{A} = \begin{pmatrix} 1 & 2 & 3 \\ 4 & -1 & -2 \\ -1 & 0 & 5 \end{pmatrix}$, $\mathbf{B} = \begin{pmatrix} 6 & 4 & 2 \\ 3 & -1 & -3 \\ 0 & 2 & -4 \end{pmatrix}$.

11. Given a square matrix $\mathbf{A}$, we define $\mathbf{A}^2 = \mathbf{AA}$, $\mathbf{A}^3 = \mathbf{AAA}$, and so forth. Using this definition, find $\mathbf{A}^2$ and $\mathbf{A}^3$ if

$$\mathbf{A} = \begin{pmatrix} 1 & 2 & -1 \\ -2 & 1 & 3 \\ 2 & 0 & 1 \end{pmatrix}.$$

12. Given the matrix

$$\mathbf{A} = \begin{pmatrix} 2 & 3 & 3 \\ 1 & -2 & 1 \\ -3 & -1 & 0 \end{pmatrix}.$$

find $\mathbf{A}^2 + 3\mathbf{A} + 2\mathbf{I}$, where $\mathbf{I}$ is the 3×3 identity matrix.

Find the inverse of the given matrix $\mathbf{A}$ in each of Exercises 13–24.

13. $\mathbf{A} = \begin{pmatrix} 1 & 3 \\ 2 & 5 \end{pmatrix}$.

14. $\mathbf{A} = \begin{pmatrix} -1 & 5 \\ -2 & 8 \end{pmatrix}$.

15. $\mathbf{A} = \begin{pmatrix} 4 & -2 \\ -6 & -1 \end{pmatrix}$.

16. $\mathbf{A} = \begin{pmatrix} 3 & -6 \\ -2 & 5 \end{pmatrix}$.

17. $\mathbf{A} = \begin{pmatrix} 4 & 3 & 1 \\ 3 & 3 & 2 \\ -1 & 1 & 1 \end{pmatrix}$.

18. $\mathbf{A} = \begin{pmatrix} 2 & -1 & 1 \\ 1 & 1 & 2 \\ -1 & 3 & 4 \end{pmatrix}$.

19. $\mathbf{A} = \begin{pmatrix} 3 & 4 & 7 \\ 1 & 1 & 2 \\ 2 & 5 & 4 \end{pmatrix}$.

20. $\mathbf{A} = \begin{pmatrix} 1 & 2 & -1 \\ 4 & 3 & 1 \\ 2 & 0 & 1 \end{pmatrix}$.

21. $\mathbf{A} = \begin{pmatrix} 3 & 2 & 4 \\ 1 & 4 & 2 \\ 2 & 2 & 3 \end{pmatrix}.$

22. $\mathbf{A} = \begin{pmatrix} 1 & 2 & 0 \\ 3 & 7 & 1 \\ 1 & 3 & 3 \end{pmatrix}.$

23. $\mathbf{A} = \begin{pmatrix} 1 & 0 & -1 \\ 4 & 2 & 1 \\ 2 & 1 & 3 \end{pmatrix}.$

24. $\mathbf{A} = \begin{pmatrix} -5 & -12 & 6 \\ 1 & 5 & -1 \\ -7 & -10 & 8 \end{pmatrix}.$

25. Verify Result E for matrices **A** and **B** defined on an arbitrary real interval $[a, b]$ as follows:

$$\mathbf{A}(t) = \begin{pmatrix} t^2 & t & 1 \\ 2t & 3t & t \\ t & t^2 & t^3 \end{pmatrix}, \quad \mathbf{B}(t) = \begin{pmatrix} t^3 & t^2 & t \\ 3t^2 & 2t & 1 \\ 6t & 2 & 0 \end{pmatrix}.$$

C. Linear Independence and Dependence

Before proceeding, we state without proof the following two theorems from algebra.

THEOREM A

A system of n homogeneous linear algebraic equations in n unknowns has a nontrivial solution if and only if the determinant of coefficients of the system is equal to zero.

THEOREM B

A system of n linear algebraic equations in n unknowns has a unique solution if and only if the determinant of coefficients of the system is unequal to zero.

DEFINITION

A set of m constant vectors $\mathbf{v}_1, \mathbf{v}_2, \ldots, \mathbf{v}_m$ *is linearly dependent if there exists a set of m numbers* $c_1, c_2, \ldots, c_m$, *not all of which are zero, such that*

$$c_1 \mathbf{v}_1 + c_2 \mathbf{v}_2 + \cdots + c_m \mathbf{v}_m = \mathbf{0}.$$

▶ **Example 7.29**

The set of three constant vectors

$$\mathbf{v}_1 = \begin{pmatrix} 2 \\ 3 \\ 1 \end{pmatrix}, \quad \mathbf{v}_2 = \begin{pmatrix} 1 \\ -1 \\ 2 \end{pmatrix}, \quad \mathbf{v}_3 = \begin{pmatrix} 7 \\ 3 \\ 8 \end{pmatrix}$$

is linearly dependent, since there exists the set of three numbers 2, 3, and -1, none of which are zero, such that

$$2\mathbf{v}_1 + 3\mathbf{v}_2 + (-1)\mathbf{v}_3 = \mathbf{0}.$$

DEFINITION

A set of m constant vectors is linearly independent *if and only if the set is not linearly dependent. That is, a set of m constant vectors* $\mathbf{v}_1, \mathbf{v}_2, \ldots, \mathbf{v}_m$ *is linearly independent if the relation*

$$c_1\mathbf{v}_1 + c_2\mathbf{v}_2 + \cdots + c_m\mathbf{v}_m = \mathbf{0}$$

implies that

$$c_1 = c_2 = \cdots = c_m = 0.$$

▶ **Example 7.30**

The set of three constant vectors

$$\mathbf{v}_1 = \begin{pmatrix} 1 \\ 1 \\ 1 \end{pmatrix}, \qquad \mathbf{v}_2 = \begin{pmatrix} -1 \\ 2 \\ 0 \end{pmatrix}, \quad \text{and} \quad \mathbf{v}_3 = \begin{pmatrix} 0 \\ 2 \\ 1 \end{pmatrix}$$

is linearly independent. For suppose we have

$$c_1\mathbf{v}_1 + c_2\mathbf{v}_2 + c_3\mathbf{v}_3 = \mathbf{0}; \tag{7.107}$$

that is,

$$c_1\begin{pmatrix} 1 \\ 1 \\ 1 \end{pmatrix} + c_2\begin{pmatrix} -1 \\ 2 \\ 0 \end{pmatrix} + c_3\begin{pmatrix} 0 \\ 2 \\ 1 \end{pmatrix} = \begin{pmatrix} 0 \\ 0 \\ 0 \end{pmatrix}.$$

This is equivalent to the system

$$c_1 - c_2 = 0,$$
$$c_1 + 2c_2 + 2c_3 = 0, \tag{7.108}$$
$$c_1 + c_3 = 0,$$

of three homogeneous linear algebraic equations in the three unknowns c_1, c_2, c_3. The determinant of coefficients of this system is

$$\begin{vmatrix} 1 & -1 & 0 \\ 1 & 2 & 2 \\ 1 & 0 & 1 \end{vmatrix} = 1 \neq 0.$$

Thus by Theorem A, with $n = 3$, the system (7.108) has only the trivial solution $c_1 = c_2 = c_3 = 0$. Thus for the three given constant vectors, the relation (7.107) implies $c_1 = c_2 = c_3 = 0$; and so these three vectors are indeed linearly independent.

DEFINITION

The set of m vector functions $\phi_1, \phi_2, \ldots, \phi_m$ *is* linearly dependent *on an interval* $a \leq t \leq b$ *if there exists a set of m numbers* $c_1, c_2, \ldots, c_m$, *not all zero, such that*

$$c_1 \phi_1(t) + c_2 \phi_2(t) + \cdots + c_m \phi_m(t) = 0$$

for all $t \in [a, b]$.

▶ **Example 7.31**

Consider the set of three vector functions ϕ_1, ϕ_2, and ϕ_3, defined for all t by

$$\phi_1(t) = \begin{pmatrix} e^{2t} \\ 2e^{2t} \\ 5e^{2t} \end{pmatrix}, \quad \phi_2(t) = \begin{pmatrix} e^{2t} \\ 4e^{2t} \\ 11e^{2t} \end{pmatrix}, \quad \text{and} \quad \phi_3(t) = \begin{pmatrix} e^{2t} \\ e^{2t} \\ 2e^{2t} \end{pmatrix},$$

respectively. This set of vector functions is linearly dependent on any interval $a \leq t \leq b$. To see this, note that

$$3 \begin{pmatrix} e^{2t} \\ 2e^{2t} \\ 5e^{2t} \end{pmatrix} + (-1) \begin{pmatrix} e^{2t} \\ 4e^{2t} \\ 11e^{2t} \end{pmatrix} + (-2) \begin{pmatrix} e^{2t} \\ e^{2t} \\ 2e^{2t} \end{pmatrix} = \begin{pmatrix} 0 \\ 0 \\ 0 \end{pmatrix};$$

and hence there exists the set of three numbers 3, -1, and -2, none of which are zero, such that

$$3\phi_1(t) + (-1)\phi_2(t) + (-2)\phi_3(t) = 0$$

for all $t \in [a, b]$.

DEFINITION

A set of m vector functions is linearly independent *on an interval if and only if the set is not linearly dependent on that interval. That is, a set of m vector functions* $\phi_1, \phi_2, \ldots, \phi_m$ *is* linearly independent *on an interval* $a \leq t \leq b$ *if the relation*

$$c_1 \phi_1(t) + c_2 \phi_2(t) + \cdots + c_m \phi_m(t) = 0$$

for all $t \in [a, b]$ *implies that*

$$c_1 = c_2 = \cdots = c_m = 0.$$

▶ **Example 7.32**

Consider the set of two vector functions ϕ_1 and ϕ_2, defined for all t by

$$\phi_1(t) = \begin{pmatrix} e^t \\ e^t \end{pmatrix} \quad \text{and} \quad \phi_2(t) = \begin{pmatrix} e^{2t} \\ 2e^{2t} \end{pmatrix},$$

respectively. We shall show that ϕ_1 and ϕ_2 are linearly independent on any interval $a \leq t \leq b$. To do this, we assume the contrary; that is, we assume that ϕ_1 and ϕ_2 are linear *dependent* on $[a, b]$. Then there exist numbers c_1 and c_2, not both zero, such that

$$c_1 \boldsymbol{\phi}_1(t) + c_2 \boldsymbol{\phi}_2(t) = \mathbf{0},$$

for all $t \in [a, b]$. Then

$$c_1 e^t + c_2 e^{2t} = 0,$$
$$c_1 e^t + 2c_2 e^{2t} = 0;$$

and multiplying each equation through by e^{-t}, we have

$$c_1 + c_2 e^t = 0,$$
$$c_1 + 2c_2 e^t = 0,$$

for all $t \in [a, b]$. This implies that $c_1 + c_2 e^t = c_1 + 2c_2 e^t$ and hence $1 = 2$, which is an obvious contradiction. Thus the assumption that $\boldsymbol{\phi}_1$ and $\boldsymbol{\phi}_2$ are linearly dependent on $[a, b]$ is false, and so these two vector functions are linearly independent on that interval.

Note. If a set of m vector functions $\boldsymbol{\phi}_1, \boldsymbol{\phi}_2, \ldots, \boldsymbol{\phi}_m$ is linearly dependent on an interval $a \leq t \leq b$, then it readily follows that for each fixed $t_0 \in [a, b]$, the corresponding set of m constant vectors $\boldsymbol{\phi}_1(t_0), \boldsymbol{\phi}_2(t_0), \ldots, \boldsymbol{\phi}_m(t_0)$ is linearly dependent. However, the analogous statement for a set of m linearly independent vector functions is not valid. That is, if a set of m vector functions $\boldsymbol{\phi}_1, \boldsymbol{\phi}_2, \ldots, \boldsymbol{\phi}_m$ is linearly independent on an interval $a \leq t \leq b$, then it is *not* necessarily true that for each fixed $t_0 \in [a, b]$, the corresponding set of m constant vectors $\boldsymbol{\phi}_1(t_0), \boldsymbol{\phi}_2(t_0), \ldots, \boldsymbol{\phi}_m(t_0)$ is linearly independent. Indeed, the corresponding set of constant vectors $\boldsymbol{\phi}_1(t_0), \boldsymbol{\phi}_2(t_0), \ldots, \boldsymbol{\phi}_m(t_0)$ may be linearly *dependent* for *each* $t_0 \in [a, b]$. See Exercise 7 at the end of this section.

Exercises

1. In each case, show that the given set of constant vectors is linearly dependent.

   (a) $\quad \mathbf{v}_1 = \begin{pmatrix} 3 \\ -1 \\ 2 \end{pmatrix}, \quad \mathbf{v}_2 = \begin{pmatrix} 13 \\ 5 \\ -4 \end{pmatrix}, \quad \mathbf{v}_3 = \begin{pmatrix} 2 \\ 4 \\ -5 \end{pmatrix}.$

   (b) $\quad \mathbf{v}_1 = \begin{pmatrix} 2 \\ 4 \\ 5 \end{pmatrix}, \quad \mathbf{v}_2 = \begin{pmatrix} 2 \\ 6 \\ 8 \end{pmatrix}, \quad \mathbf{v}_3 = \begin{pmatrix} 1 \\ -1 \\ -2 \end{pmatrix}.$

   (c) $\quad \mathbf{v}_1 = \begin{pmatrix} 1 \\ 2 \\ -3 \end{pmatrix}, \quad \mathbf{v}_2 = \begin{pmatrix} 2 \\ 1 \\ -3 \end{pmatrix}, \quad \mathbf{v}_3 = \begin{pmatrix} 7 \\ -1 \\ -6 \end{pmatrix}.$

2. In each case, show that the given set of constant vectors is linearly independent.

   (a) $\quad \mathbf{v}_1 = \begin{pmatrix} 2 \\ 1 \\ 0 \end{pmatrix}, \quad \mathbf{v}_2 = \begin{pmatrix} 1 \\ 0 \\ 3 \end{pmatrix}, \quad \mathbf{v}_3 = \begin{pmatrix} 0 \\ -1 \\ 1 \end{pmatrix}.$

(b) $\quad \mathbf{v}_1 = \begin{pmatrix} 1 \\ 2 \\ 0 \end{pmatrix}, \quad \mathbf{v}_2 = \begin{pmatrix} -2 \\ 0 \\ 1 \end{pmatrix}, \quad \mathbf{v}_3 = \begin{pmatrix} 1 \\ -1 \\ 2 \end{pmatrix}.$

(c) $\quad \mathbf{v}_1 = \begin{pmatrix} 1 \\ 2 \\ -1 \end{pmatrix}, \quad \mathbf{v}_2 = \begin{pmatrix} 1 \\ 1 \\ 1 \end{pmatrix}, \quad \mathbf{v}_3 = \begin{pmatrix} 1 \\ -1 \\ 3 \end{pmatrix}.$

3. In each case, determine the value of k so that the given set of constant vectors is linearly dependent.

(a) $\quad \mathbf{v}_1 = \begin{pmatrix} k \\ 2 \\ -1 \end{pmatrix}, \quad \mathbf{v}_2 = \begin{pmatrix} 1 \\ -1 \\ 2 \end{pmatrix}, \quad \mathbf{v}_3 = \begin{pmatrix} 3 \\ 7 \\ -8 \end{pmatrix}.$

(b) $\quad \mathbf{v}_1 = \begin{pmatrix} k \\ 3 \\ -1 \end{pmatrix}, \quad \mathbf{v}_2 = \begin{pmatrix} 1 \\ -3 \\ 2 \end{pmatrix}, \quad \mathbf{v}_3 = \begin{pmatrix} 8 \\ 3 \\ 1 \end{pmatrix}.$

4. In each case, show that the set of vector functions $\boldsymbol{\phi}_1, \boldsymbol{\phi}_2, \boldsymbol{\phi}_3$, defined for all t as indicated, is linearly dependent on any interval $a \leq t \leq b$.

(a) $\quad \boldsymbol{\phi}_1(t) = \begin{pmatrix} 2e^{3t} \\ 3e^{3t} \\ -e^{3t} \end{pmatrix}, \quad \boldsymbol{\phi}_2(t) = \begin{pmatrix} 4e^{3t} \\ -5e^{3t} \\ 5e^{3t} \end{pmatrix}, \quad \boldsymbol{\phi}_3(t) = \begin{pmatrix} 5e^{3t} \\ 2e^{3t} \\ e^{3t} \end{pmatrix}.$

(b) $\quad \boldsymbol{\phi}_1(t) = \begin{pmatrix} \sin t + \cos t \\ 2 \sin t \\ -\cos t \end{pmatrix}, \quad \boldsymbol{\phi}_2(t) = \begin{pmatrix} 2 \sin t \\ 4 \sin t - \cos t \\ -\sin t \end{pmatrix},$

$\quad \boldsymbol{\phi}_3(t) = \begin{pmatrix} 4 \cos t \\ 2 \cos t \\ 2 \sin t - 4 \cos t \end{pmatrix}.$

5. In each case, show that the set of vector functions $\boldsymbol{\phi}_1$ and $\boldsymbol{\phi}_2$ defined for all t as indicated, is linearly independent on any interval $a \leq t \leq b$.

(a) $\quad \boldsymbol{\phi}_1(t) = \begin{pmatrix} e^t \\ 2e^t \end{pmatrix} \quad$ and $\quad \boldsymbol{\phi}_2(t) = \begin{pmatrix} e^{3t} \\ 4e^{3t} \end{pmatrix}.$

(b) $\quad \boldsymbol{\phi}_1(t) = \begin{pmatrix} 2e^{2t} \\ -e^{2t} \end{pmatrix} \quad$ and $\quad \boldsymbol{\phi}_2(t) = \begin{pmatrix} e^{-t} \\ 3e^{-t} \end{pmatrix}.$

6. Show that the set of two vector functions $\boldsymbol{\phi}_1$ and $\boldsymbol{\phi}_2$ defined for all t by

$$\boldsymbol{\phi}_1(t) = \begin{pmatrix} t \\ 0 \end{pmatrix} \quad \text{and} \quad \boldsymbol{\phi}_2(t) = \begin{pmatrix} t^2 \\ 0 \end{pmatrix},$$

respectively, is linearly independent on any interval $a \leq t \leq b$.

7. Consider the vector functions $\boldsymbol{\phi}_1$ and $\boldsymbol{\phi}_2$ defined by

$$\boldsymbol{\phi}_1(t) = \begin{pmatrix} t \\ 1 \end{pmatrix} \quad \text{and} \quad \boldsymbol{\phi}_2(t) = \begin{pmatrix} te^t \\ e^t \end{pmatrix},$$

respectively. Show that the constant vectors $\boldsymbol{\phi}_1(t_0)$ and $\boldsymbol{\phi}_2(t_0)$ are linearly dependent for each t_0 in the interval $0 \leq t \leq 1$, but that the vector functions $\boldsymbol{\phi}_1$ and $\boldsymbol{\phi}_2$ are linearly independent on $0 \leq t \leq 1$.

8. Let

$$\boldsymbol{\alpha}^{(i)} = \begin{pmatrix} \alpha_{1i} \\ \alpha_{2i} \\ \vdots \\ \alpha_{ni} \end{pmatrix} \qquad (i = 1, 2, \ldots, n),$$

be a set of n linearly independent vectors. Show that

$$\begin{vmatrix} \alpha_{11} & \alpha_{12} & \cdots & \alpha_{1n} \\ \alpha_{12} & \alpha_{22} & \cdots & \alpha_{2n} \\ \vdots & \vdots & & \vdots \\ \alpha_{n1} & \alpha_{n2} & \cdots & \alpha_{nn} \end{vmatrix} \neq 0.$$

D. Characteristic Values and Characteristic Vectors

Let $\mathbf{A}$ be a given $n \times n$ square matrix of real numbers, and let S denote the set of all $n \times 1$ column vectors of numbers. Now consider the equation

$$\mathbf{Ax} = \lambda \mathbf{x} \tag{7.109}$$

in the unknown vector $\mathbf{x} \in S$, where λ is a number. Clearly the zero vector $\mathbf{0}$ is a solution of this equation for every number λ. We investigate the possibility of finding nonzero vectors $\mathbf{x} \in S$ which are solutions of (7.109) for some choice of the number λ. In other words, we seek numbers λ corresponding to which there exist nonzero vectors $\mathbf{x}$ that satisfy (7.109). These desired values of λ and the corresponding desired nonzero vectors are designated in the following definitions.

DEFINITIONS

A characteristic value (or eigenvalue) of the matrix $\mathbf{A}$ is a number λ for which the equation $\mathbf{Ax} = \lambda \mathbf{x}$ has a nonzero vector solution $\mathbf{x}$.

A characteristic vector (or eigenvector) of $\mathbf{A}$ is a nonzero vector $\mathbf{x}$ such that $\mathbf{Ax} = \lambda \mathbf{x}$ for some number λ.

▶ **Example 7.33**

Consider the 2×2 square matrix

$$\mathbf{A} = \begin{pmatrix} 6 & -3 \\ 2 & 1 \end{pmatrix}.$$

Observe that

$$\begin{pmatrix} 6 & -3 \\ 2 & 1 \end{pmatrix} \begin{pmatrix} 3 \\ 2 \end{pmatrix} = \begin{pmatrix} 12 \\ 8 \end{pmatrix} \quad \text{and} \quad 4 \begin{pmatrix} 3 \\ 2 \end{pmatrix} = \begin{pmatrix} 12 \\ 8 \end{pmatrix},$$

and so

$$\begin{pmatrix} 6 & -3 \\ 2 & 1 \end{pmatrix} \begin{pmatrix} 3 \\ 2 \end{pmatrix} = 4 \begin{pmatrix} 3 \\ 2 \end{pmatrix}.$$

This is of the form $\mathbf{Ax} = \lambda \mathbf{x}$, where $\mathbf{A}$ is the given 2×2 matrix, $\lambda = 4$, and $\mathbf{x} = \begin{pmatrix} 3 \\ 2 \end{pmatrix}$. Thus $\lambda = 4$ is a characteristic value of the given matrix $\mathbf{A}$ and $\mathbf{x} = \begin{pmatrix} 3 \\ 2 \end{pmatrix}$ is a corresponding characteristic vector of $\mathbf{A}$.

On the other hand, we shall now show that $\lambda = 2$ is *not* a characteristic value of this matrix $\mathbf{A}$. For, if it were, then there would exist a *nonzero* vector $\mathbf{x} = \begin{pmatrix} x_1 \\ x_2 \end{pmatrix}$ such that

$$\begin{pmatrix} 6 & -3 \\ 2 & 1 \end{pmatrix} \begin{pmatrix} x_1 \\ x_2 \end{pmatrix} = 2 \begin{pmatrix} x_1 \\ x_2 \end{pmatrix}.$$

Performing the indicated multiplications on each side of this equation and then equating the corresponding components, we would have

$$6x_1 - 3x_2 = 2x_1,$$
$$2x_1 + x_2 = 2x_2,$$

or simply

$$4x_1 - 3x_2 = 0,$$
$$2x_1 - x_2 = 0.$$

Since the determinant of coefficients of this homogeneous linear algebraic system is unequal to zero, by Theorem A the only solution of the system is the trivial solution $x_1 = x_2 = 0$. That is, we must have

$$\mathbf{x} = \mathbf{0},$$

which is a contradiction.

We proceed to solve the problem of determining the characteristic values and vectors of an $n \times n$ square matrix $\mathbf{A}$. Suppose

$$\mathbf{A} = \begin{pmatrix} a_{11} & a_{12} & \cdots & a_{1n} \\ a_{21} & a_{22} & \cdots & a_{2n} \\ \vdots & \vdots & & \vdots \\ a_{n1} & a_{n2} & \cdots & a_{nn} \end{pmatrix}$$

is the given $n \times n$ square matrix of real numbers, and let

$$\mathbf{x} = \begin{pmatrix} x_1 \\ x_2 \\ \vdots \\ x_n \end{pmatrix}.$$

Then Equation (7.109) may be written

$$\begin{pmatrix} a_{11} & a_{12} & \cdots & a_{1n} \\ a_{21} & a_{22} & \cdots & a_{2n} \\ \vdots & \vdots & & \vdots \\ a_{n1} & a_{n2} & \cdots & a_{nn} \end{pmatrix} \begin{pmatrix} x_1 \\ x_2 \\ \vdots \\ x_n \end{pmatrix} = \lambda \begin{pmatrix} x_1 \\ x_1 \\ \vdots \\ x_n \end{pmatrix},$$

and hence, multiplying the indicated entities,

$$\begin{pmatrix} a_{11}x_1 + a_{12}x_2 + \cdots + a_{1n}x_n \\ a_{21}x_1 + a_{22}x_2 + \cdots + a_{2n}x_n \\ \vdots & \vdots & \vdots \\ a_{n1}x_1 + a_{n2}x_2 + \cdots + a_{nn}x_n \end{pmatrix} = \begin{pmatrix} \lambda x_1 \\ \lambda x_2 \\ \vdots \\ \lambda x_n \end{pmatrix}.$$

Equating corresponding components of these two equal vectors, we have

$$a_{11}x_1 + a_{12}x_2 + \cdots + a_{1n}x_n = \lambda x_1,$$

$$a_{21}x_1 + a_{22}x_2 + \cdots + a_{2n}x_n = \lambda x_2,$$

$$\vdots$$

$$a_{n1}x_1 + a_{n2}x_2 + \cdots + a_{nn}x_n = \lambda x_n;$$

and rewriting this, we obtain

$$(a_{11} - \lambda)x_1 + a_{12}x_2 + \cdots + a_{1n}x_n = 0,$$

$$a_{21}x_1 + (a_{22} - \lambda)x_2 + \cdots + a_{2n}x_n = 0,$$

$$\vdots$$

$$a_{n1}x_1 + a_{n2}x_2 + \cdots + (a_{nn} - \lambda)x_n = 0.$$

(7.110)

Thus we see that (7.109) holds if and only if (7.110) does. Now we are seeking nonzero vectors **x** that satisfy (7.109). Thus a nonzero vector **x** satisfies (7.109) if and only if its set of components $x_1, x_2, \ldots, x_n$ is a nontrivial solution of (7.110). By Theorem A of Section 7.5C, the system (7.110) has nontrivial solutions if and only if its determinant of coefficients is equal to zero, that is, if and only if

$$\begin{vmatrix} a_{11} - \lambda & a_{12} & \cdots & a_{1n} \\ a_{21} & a_{22} - \lambda & \cdots & a_{2n} \\ \vdots & \vdots & & \vdots \\ a_{n1} & a_{n2} & \cdots & a_{nn} - \lambda \end{vmatrix} = 0.$$

(7.111)

It is easy to see that (7.111) is a polynomial equation of the nth degree in the unknown λ. In matrix notation it is written

$$|A - \lambda \mathbf{I}| = 0,$$

where **I** is the $n \times n$ identity matrix (see Section 7.5B). Thus Equation (7.109) has a nonzero vector solution **x** for a certain value of λ if and only if λ satisfies the nth-degree polynomial equation (7.111). That is, the number λ is a characteristic value of the matrix **A** if and only if it satisfies this polynomial equation. We now designate this equation and also state the alternative definition of characteristic value that we have thus obtained.

DEFINITION

Let $\mathbf{A} = (a_{ij})$ *be an* $n \times n$ *square matrix of real numbers. The* characteristic equation *of* **A** *is the nth-degree polynomial equation*

$$\begin{vmatrix} a_{11} - \lambda & a_{12} & \cdots & a_{1n} \\ a_{21} & a_{22} - \lambda & \cdots & a_{2n} \\ \vdots & \vdots & & \vdots \\ a_{n1} & a_{n2} & \cdots & a_{nn} - \lambda \end{vmatrix} = 0 \tag{7.111}$$

in the unknown λ; *and the* characteristic values *of* **A** *are the roots of this equation.*

Since the characteristic equation (7.111) of **A** is a polynomial equation of the *n*th degree, it has *n* roots. These roots may be real or complex, but of course they may or may not all be distinct. If a certain repeated root occurs *m* times, where $1 < m \leq n$, then we say that that root has *multiplicity m*. If we count each nonrepeated root once and each repeated root according to its multiplicity, then we can say that the $n \times n$ matrix **A** has precisely *n* characteristic values, say $\lambda_1, \lambda_2, \ldots, \lambda_n$.

Corresponding to each characteristic value λ_k of **A** there is a characteristic vector $\mathbf{x}_k$ $(k = 1, 2, \ldots, n)$. Further, if $\mathbf{x}_k$ is a characteristic vector of **A** corresponding to characteristic value λ_k, then so is $c\mathbf{x}_k$ for any nonzero number *c*. We shall be concerned with the linear independence of the various characteristic vectors of **A**. Concerning this, we state the following two results without proof.

Result F. *Suppose each of the n characteristic values* $\lambda_1, \lambda_2, \ldots, \lambda_n$ *of the* $n \times n$ *square matrix* **A** *is distinct (that is, nonrepeated); and let* $\mathbf{x}_1, \mathbf{x}_2, \ldots, \mathbf{x}_n$ *be a set of n respective corresponding characteristic vectors of* **A**. *Then the set of these n characteristic vectors is linearly independent.*

Result G. *Suppose the* $n \times n$ *square matrix* **A** *has a characteristic value of multiplicity m, where* $1 < m \leq n$. *Then this repeated characteristic value having multiplicity m has p linearly independent characteristic vectors corresponding to it, where* $1 \leq p \leq m$.

Now suppose **A** has at least one characteristic value of multiplicity *m*, where $1 < m \leq n$; and further suppose that for this repeated characteristic value, the number *p* of Result G is strictly less than *m*; that is, *p* is such that $1 \leq p < m$. Then corresponding to this characteristic value of multiplicity *m*, there are less than *m* linearly independent characteristic vectors. It follows at once that the matrix **A** must then have *less than n* linearly independent characteristic vectors. Thus we are led to the following result:

Result H. *If the* $n \times n$ *matrix* **A** *has one or more repeated characteristic values, then there may exist less than n linearly independent characteristic vectors of* **A**.

Before giving an example of finding the characteristic values and corresponding characteristic vectors of a matrix, we introduce a very special class of matrices whose characteristic values and vectors have some interesting special properties. This is the class of so-called real *symmetric* matrices, which we shall now define.

DEFINITION

A square matrix **A** *of real numbers is called a real symmetric matrix of* $\mathbf{A}^T = \mathbf{A}$.

For example, the 3×3 square matrix

$$\mathbf{A} = \begin{pmatrix} 2 & -1 & 4 \\ -1 & 0 & 3 \\ 4 & 3 & 1 \end{pmatrix}$$

is a real symmetric matrix, since $\mathbf{A}^T = \mathbf{A}$.

Concerning real symmetric matrices, we state without proof the following interesting results:

Result I. *All of the characteristic values of a real symmetric matrix are real numbers.*

Result J. *If* **A** *is an* $n \times n$ *real symmetric square matrix, then there exist n linearly independent characteristic vectors of* **A***, whether the n characteristic values of* **A** *are all distinct or whether one or more of these characteristic values is repeated.*

▶ **Example 7.34**

Find the characteristic values and characteristic vectors of the matrix

$$\mathbf{A} = \begin{pmatrix} 1 & 2 \\ 3 & 2 \end{pmatrix}$$

Solution. The characteristic equation of **A** is

$$\begin{vmatrix} 1 - \lambda & 2 \\ 3 & 2 - \lambda \end{vmatrix} = 0.$$

Evaluating the determinant in the left member, we find that this equation may be written in the form

$$\lambda^2 - 3\lambda - 4 = 0$$

or

$$(\lambda - 4)(\lambda + 1) = 0.$$

Thus the characteristic values of **A** are

$$\lambda = 4 \quad \text{and} \quad \lambda = -1.$$

The characteristic vectors **x** corresponding to a characteristic value λ are the nonzero vectors

$$\mathbf{x} = \begin{pmatrix} x_1 \\ x_2 \end{pmatrix} \quad \text{such that} \quad \mathbf{Ax} = \lambda\mathbf{x}.$$

With $\lambda = 4$, this is

$$\begin{pmatrix} 1 & 2 \\ 3 & 2 \end{pmatrix} \begin{pmatrix} x_1 \\ x_2 \end{pmatrix} = 4\begin{pmatrix} x_1 \\ x_2 \end{pmatrix}.$$

Performing the indicated multiplications and equating corresponding components, we see that x_1 and x_2 must satisfy

$$x_1 + 2x_2 = 4x_1,$$
$$3x_1 + 2x_2 = 4x_2;$$

that is,

$$3x_1 = 2x_2,$$
$$3x_1 = 2x_2.$$

We see at once that $x_1 = 2k$, $x_2 = 3k$ is a solution for every real k. Hence the characteristic vectors corresponding to the characteristic value $\lambda = 4$ are the vectors

$$\mathbf{x} = \begin{pmatrix} 2k \\ 3k \end{pmatrix},$$

where k is an arbitrary nonzero number. In particular, letting $k = 1$, we obtain the particular characteristic vector

$$\begin{pmatrix} 2 \\ 3 \end{pmatrix}$$

corresponding to the characteristic value $\lambda = 4$.

Proceeding in like manner, we can find the characteristic vectors corresponding to $\lambda = -1$. With $\lambda = -1$ and $\mathbf{x} = \begin{pmatrix} x_1 \\ x_2 \end{pmatrix}$, the basic equation $\mathbf{Ax} = \lambda\mathbf{x}$ is

$$\begin{pmatrix} 1 & 2 \\ 3 & 2 \end{pmatrix} \begin{pmatrix} x_1 \\ x_2 \end{pmatrix} = -1\begin{pmatrix} x_1 \\ x_2 \end{pmatrix}.$$

We leave it to the reader to show that we must have $x_2 = -x_1$. Then $x_1 = k$, $x_2 = -k$ is a solution for every real k. Hence the characteristic vectors corresponding to the characteristic value $\lambda = -1$ are the vectors

$$\mathbf{x} = \begin{pmatrix} k \\ -k \end{pmatrix},$$

where k is an arbitrary nonzero number. In particular, letting $k = 1$, we obtain the particular characteristic vector

$$\begin{pmatrix} 1 \\ -1 \end{pmatrix}$$

corresponding to the characteristic value $\lambda = -1$.

▶ **Example 7.35**

Find the characteristic values and characteristic vectors of the matrix

$$\mathbf{A} = \begin{pmatrix} 7 & -1 & 6 \\ -10 & 4 & -12 \\ -2 & 1 & -1 \end{pmatrix}.$$

Solution. The characteristic equation of **A** is

$$\begin{vmatrix} 7-\lambda & -1 & 6 \\ -10 & 4-\lambda & -12 \\ -2 & 1 & -1-\lambda \end{vmatrix} = 0.$$

Evaluating the determinant in the left member, we find that this equation may be written in the form

$$\lambda^3 - 10\lambda^2 + 31\lambda - 30 = 0.$$

or

$$(\lambda - 2)(\lambda - 3)(\lambda - 5) = 0.$$

Thus the characteristic values of **A** are

$$\lambda = 2, \qquad \lambda = 3, \quad \text{and} \quad \lambda = 5.$$

The characteristic vectors **x** corresponding to a characterictic value λ are the nonzero vectors

$$\mathbf{x} = \begin{pmatrix} x_1 \\ x_2 \\ x_3 \end{pmatrix} \quad \text{such that} \quad \mathbf{Ax} = \lambda\mathbf{x}.$$

With $\lambda = 2$, this is

$$\begin{pmatrix} 7 & -1 & 6 \\ -10 & 4 & -12 \\ -2 & 1 & -1 \end{pmatrix} \begin{pmatrix} x_1 \\ x_2 \\ x_3 \end{pmatrix} = 2 \begin{pmatrix} x_1 \\ x_2 \\ x_3 \end{pmatrix}.$$

Performing the indicated multiplications and equating corresponding components, we see that x_1, x_2, x_3 must be a nontrivial solution of the system

$$7x_1 - x_2 + 6x_3 = 2x_1,$$
$$-10x_1 + 4x_2 - 12x_3 = 2x_2,$$
$$-2x_1 + x_2 - x_3 = 2x_3;$$

that is,

$$5x_1 - x_2 + 6x_3 = 0,$$
$$-10x_1 + 2x_2 - 12x_3 = 0,$$
$$-2x_1 + x_2 - 3x_3 = 0.$$

Note that the second of these three equations is merely a constant multiple of the first.

Thus we seek nonzero numbers x_1, x_2, x_3 that satisfy the first and third of these equations. Writing these two as equations in the unknowns x_2 and x_3, we have

$$-x_2 + 6x_3 = -5x_1,$$

$$x_2 - 3x_3 = 2x_1.$$

Solving for x_2 and x_3, we find

$$x_2 = -x_1 \quad \text{and} \quad x_3 = -x_1.$$

We see at once that $x_1 = k$, $x_2 = -k$, $x_3 = -k$ is a solution of this for every real k. Hence the characteristic vectors corresponding to the characteristic value $\lambda = 2$ are the vectors

$$\mathbf{x} = \begin{pmatrix} k \\ -k \\ -k \end{pmatrix},$$

where k is an arbitrary nonzero number. In particular, letting $k = 1$, we obtain the particular characteristic vector

$$\begin{pmatrix} 1 \\ -1 \\ -1 \end{pmatrix}$$

corresponding to the characteristic value $\lambda = 2$.

Proceeding in like manner, one can find the characteristic vectors corresponding to $\lambda = 3$ and those corresponding to $\lambda = 5$. We give only a few highlights of these computations and leave the details to the reader. We find that the components x_1, x_2, x_3 of the characteristic vectors corresponding to $\lambda = 3$ must be a nontrivial solution of the system

$$4x_1 - x_2 + 6x_3 = 0,$$

$$-10x_1 + x_2 - 12x_3 = 0,$$

$$-2x_1 + x_2 - 4x_3 = 0.$$

From these we find that

$$x_2 = -2x_1 \quad \text{and} \quad x_3 = -x_1,$$

and hence $x_1 = k$, $x_2 = -2k$, $x_3 = -k$ is a solution for every real k. Hence the characteristic vectors corresponding to the characteristic value $\lambda = 3$ are the vectors

$$\mathbf{x} = \begin{pmatrix} k \\ -2k \\ -k \end{pmatrix},$$

where k is an arbitrary nonzero number. In particular, letting $k = 1$, we obtain the particular characteristic vector

$$\begin{pmatrix} 1 \\ -2 \\ -1 \end{pmatrix}$$

corresponding to the characteristic value $\lambda = 3$.

Finally, we proceed to find the characteristic vectors corresponding to $\lambda = 5$. We find that the components x_1, x_2, x_3 of these vectors must be a nontrivial solution of the system

$$2x_1 - x_2 + 6x_3 = 0,$$
$$-10x_1 - x_2 - 12x_3 = 0,$$
$$-2x_1 + x_2 - 6x_3 = 0.$$

From these we find that

$$x_2 = -2x_1 \quad \text{and} \quad 3x_3 = -2x_1.$$

We find that $x_1 = 3k$, $x_2 = -6k$, $x_3 = -2k$ satisfies this for every real k. Hence the characteristic vectors corresponding to the characteristic value $\lambda = 5$ are the vectors

$$\mathbf{x} = \begin{pmatrix} 3k \\ -6k \\ -2k \end{pmatrix},$$

where k is an arbitrary nonzero number. In particular, letting $k = 1$, we obtain the particular characteristic vector

$$\begin{pmatrix} 3 \\ -6 \\ -2 \end{pmatrix}$$

corresponding to the characteristic value $\lambda = 5$.

Exercises

In each of Exercises 1–14 find all the characteristic values and vectors of the matrix.

1. $\begin{pmatrix} 1 & 2 \\ 3 & 2 \end{pmatrix}$.

2. $\begin{pmatrix} 3 & 2 \\ 6 & -1 \end{pmatrix}$.

3. $\begin{pmatrix} 3 & 1 \\ 12 & 2 \end{pmatrix}$.

4. $\begin{pmatrix} -2 & 7 \\ 3 & 2 \end{pmatrix}$.

5. $\begin{pmatrix} 3 & 4 \\ 5 & 2 \end{pmatrix}$.

6. $\begin{pmatrix} 3 & -5 \\ -4 & 2 \end{pmatrix}$.

7. $\begin{pmatrix} 1 & 1 & -1 \\ 2 & 3 & -4 \\ 4 & 1 & -4 \end{pmatrix}$.

8. $\begin{pmatrix} 1 & -1 & -1 \\ 1 & 3 & 1 \\ -3 & -6 & 6 \end{pmatrix}$.

9. $\begin{pmatrix} 1 & -1 & -1 \\ 1 & 3 & 1 \\ -3 & 1 & -1 \end{pmatrix}$.

10. $\begin{pmatrix} 1 & 1 & 0 \\ 1 & 0 & 1 \\ 0 & 1 & 1 \end{pmatrix}$.

11. $\begin{pmatrix} 1 & 3 & -6 \\ 0 & 2 & 2 \\ 0 & -1 & 5 \end{pmatrix}$.

12. $\begin{pmatrix} -5 & -12 & 6 \\ 1 & 5 & -1 \\ -7 & -10 & 8 \end{pmatrix}$.

13. $\begin{pmatrix} -2 & 5 & 5 \\ -1 & 4 & 5 \\ 3 & -3 & 2 \end{pmatrix}$.

14. $\begin{pmatrix} -2 & 6 & -18 \\ 12 & -23 & 66 \\ 5 & -10 & 29 \end{pmatrix}$.

7.6 THE MATRIX METHOD FOR HOMOGENEOUS LINEAR SYSTEMS WITH CONSTANT COEFFICIENTS: TWO EQUATIONS IN TWO UNKNOWN FUNCTIONS

A. Introduction

We now return to the homogeneous linear systems of Section 7.4 and obtain solutions using matrix methods. In anticipation of more general systems, we change notation and consider the homogeneous linear system in the form

$$\frac{dx_1}{dt} = a_{11}x_1 + a_{12}x_2,$$

$$\frac{dx_2}{dt} = a_{21}x_1 + a_{22}x_2, \tag{7.112}$$

where the coefficients a_{11}, a_{12}, a_{21}, a_{22} are real constants.

We shall now proceed to express this system in a compact manner using vectors and matrices. We introduce the 2×2 constant matrix of real numbers,

$$\mathbf{A} = \begin{pmatrix} a_{11} & a_{12} \\ a_{21} & a_{22} \end{pmatrix}, \tag{7.113}$$

and the vector

$$\mathbf{x} = \begin{pmatrix} x_1 \\ x_2 \end{pmatrix}. \tag{7.114}$$

Then by definition of the derivative of a vector,

$$\frac{dx}{dt} = \begin{pmatrix} \dfrac{dx_1}{dt} \\ \dfrac{dx_2}{dt} \end{pmatrix};$$

and by multiplication of a matrix by a vector, we have

$$\mathbf{Ax} = \begin{pmatrix} a_{11} & a_{12} \\ a_{21} & a_{22} \end{pmatrix} \begin{pmatrix} x_1 \\ x_2 \end{pmatrix} = \begin{pmatrix} a_{11}x_1 + a_{12}x_2 \\ a_{21}x_1 + a_{22}x_2 \end{pmatrix}.$$

Comparing the components of dx/dt with the left members of (7.112) and the components of $\mathbf{Ax}$ with the right members of (7.112), we see that system (7.112) can be expressed as the homogeneous linear *vector* differential equation

$$\frac{dx}{dt} = \mathbf{Ax}. \tag{7.115}$$

The real constant matrix $\mathbf{A}$ that appears in (7.115) and is defined by (7.113) is called the *coefficient matrix* of (7.115).

We seek solutions of the system (7.112), that is, of the corresponding vector differential equation (7.115). We proceed as in Section 7.4A, but now employing vector and matrix notation. We seek nontrivial solutions of the form

$$x_1 = \alpha_1 e^{\lambda t},$$
$$x_2 = \alpha_2 e^{\lambda t}, \tag{7.116}$$

where α_1, α_2, and λ are constants. Letting

$$\boldsymbol{\alpha} = \begin{pmatrix} \alpha_1 \\ \alpha_2 \end{pmatrix} \quad \text{and} \quad \mathbf{x} = \begin{pmatrix} x_1 \\ x_2 \end{pmatrix},$$

we see that the vector form of the desired solution (7.116) is $\mathbf{x} = \boldsymbol{\alpha} e^{\lambda t}$. Thus we seek solutions of the vector differential equation (7.115) of the form

$$\mathbf{x} = \boldsymbol{\alpha} e^{\lambda t}, \tag{7.117}$$

where $\boldsymbol{\alpha}$ is a constant vector and λ is a number.

Now substituting (7.117) into (7.115) we obtain

$$\lambda \boldsymbol{\alpha} e^{\lambda t} = \mathbf{A} \boldsymbol{\alpha} e^{\lambda t}$$

which reduces at once to

$$\mathbf{A} \boldsymbol{\alpha} = \lambda \boldsymbol{\alpha} \tag{7.118}$$

and hence to

$$(\mathbf{A} - \lambda \mathbf{I}) \boldsymbol{\alpha} = \mathbf{0},$$

where $\mathbf{I}$ is the 2×2 identity matrix. Written out in terms of components, this is the system of two homogeneous linear algebraic equations

$$(a_{11} - \lambda)\alpha_1 + a_{12}\alpha_2 = 0,$$
$$a_{21}\alpha_1 + (a_{22} - \lambda)\alpha_2 = 0, \tag{7.119}$$

in the two unknowns α_1 and α_2. By Theorem A of Section 7.5C, this system has a nontrivial solution if and only if

$$\begin{vmatrix} a_{11} - \lambda & a_{12} \\ a_{21} & a_{22} - \lambda \end{vmatrix} = 0, \tag{7.120}$$

that is, in matrix notation,

$$|\mathbf{A} - \lambda \mathbf{I}| = 0. \tag{7.121}$$

Looking back at Section 7.5D, we recognize (7.120) as the *characteristic equation* of the coefficient matrix $\mathbf{A} = (a_{ij})$ of the vector differential equation (7.115). Expanding the determinant in (7.120), we express the characteristic equation (7.120) as the quadratic equation

$$\lambda^2 - (a_{11} + a_{22})\lambda + (a_{11}a_{22} - a_{12}a_{21}) = 0 \tag{7.122}$$

in the unknown λ. We recall that the roots λ_1 and λ_2 of this equation are the *characteristic values* of $\mathbf{A}$. Substituting each characteristic value λ_i, $(i = 1, 2)$, into system (7.119), or the equivalent vector equation (7.118), we obtain the corresponding nontrivial solution

$$\boldsymbol{\alpha}^{(i)} = \begin{pmatrix} \alpha_{1i} \\ \alpha_{2i} \end{pmatrix},$$

($i = 1, 2$), of (7.119). We recognize the vector $\boldsymbol{\alpha}^{(i)}$ as the *characteristic vector* corresponding to the characteristic value λ_i, ($i = 1, 2$).

We thus see that if the vector differential equation

$$\frac{dx}{dt} = \mathbf{A}\mathbf{x} \qquad (7.115)$$

has a solution of the form

$$\mathbf{x} = \boldsymbol{\alpha} e^{\lambda t}, \qquad (7.117)$$

then the number λ must be a characteristic value λ_i of the coefficient matrix $\mathbf{A}$ and the vector $\boldsymbol{\alpha}$ must be a characteristic vector $\boldsymbol{\alpha}^{(i)}$ corresponding to this characteristic value λ_i.

B. Case of Two Distinct Characteristic Values

Suppose that the two characteristic values λ_1 and λ_2 of the coefficient matrix $\mathbf{A}$ of the vector differential equation (7.115) are *distinct*, and let $\boldsymbol{\alpha}^{(1)}$ and $\boldsymbol{\alpha}^{(2)}$ be a pair of respective corresponding characteristic vectors of $\mathbf{A}$. Then the two distinct vector functions $\mathbf{x}^{(1)}$ and $\mathbf{x}^{(2)}$ defined, respectively, by

$$\mathbf{x}^{(1)}(t) = \boldsymbol{\alpha}^{(1)} e^{\lambda_1 t}, \qquad \mathbf{x}^{(2)}(t) = \boldsymbol{\alpha}^{(2)} e^{\lambda_2 t} \qquad (7.123)$$

are solutions of the vector differential equation (7.115) on every real interval $[a, b]$. We show this for the solution $\mathbf{x}^{(1)}$ as follows: From (7.118), we have

$$\lambda_1 \boldsymbol{\alpha}^{(1)} = \mathbf{A}\boldsymbol{\alpha}^{(1)};$$

and using this and the definition (7.123) of $\mathbf{x}^{(1)}(t)$, we obtain

$$\frac{d\mathbf{x}^{(1)}(t)}{dt} = \lambda_1 \boldsymbol{\alpha}^{(1)} e^{\lambda_1 t} = \mathbf{A}\boldsymbol{\alpha}^{(1)} e^{\lambda_1 t} = \mathbf{A}\mathbf{x}^{(1)}(t),$$

which states that $\mathbf{x}^{(1)}(t)$ satisfies the vector differential equation (7.115) on $[a, b]$. Similarly, one shows that $\mathbf{x}^{(2)}$ is a solution of (7.115).

The Wronskian of solutions $\mathbf{x}^{(1)}$ and $\mathbf{x}^{(2)}$ is

$$W(\mathbf{x}^{(1)}, \mathbf{x}^{(2)})(t) = \begin{vmatrix} \alpha_{11} e^{\lambda_1 t} & \alpha_{12} e^{\lambda_2 t} \\ \alpha_{21} e^{\lambda_1 t} & \alpha_{22} e^{\lambda_2 t} \end{vmatrix} = e^{(\lambda_1 + \lambda_2)t} \begin{vmatrix} \alpha_{11} & \alpha_{12} \\ \alpha_{21} & \alpha_{22} \end{vmatrix}.$$

By Result F of Section 7.5D, the characteristic vectors $\boldsymbol{\alpha}^{(1)}$ and $\boldsymbol{\alpha}^{(2)}$ are linearly independent. Therefore, using Exercise 8 at the end of Section 7.5C, we have

$$\begin{vmatrix} \alpha_{11} & \alpha_{12} \\ \alpha_{21} & \alpha_{22} \end{vmatrix} \neq 0.$$

Then since $e^{(\lambda_1 + \lambda_2)t} \neq 0$ for all t, we have $W(\mathbf{x}^{(1)}, \mathbf{x}^{(2)})(t) \neq 0$ for all t on $[a, b]$. Thus by Theorem 7.4 the solutions $\mathbf{x}^{(1)}$ and $\mathbf{x}^{(2)}$ of (7.115) defined by (7.123) are linearly independent on $[a, b]$; and so a general solution is given by $c_1\mathbf{x}^{(1)} + c_2\mathbf{x}^{(2)}$ where c_1 and c_2 are arbitrary constants. We summarize the results obtained in the following theorem.

THEOREM 7.10

Consider the homogeneous linear system

$$\frac{dx_1}{dt} = a_{11}x_1 + a_{12}x_2,$$

$$\frac{dx_2}{dt} = a_{21}x_1 + a_{22}x_2,$$

(7.112)

that is, the vector differential equation

$$\frac{d\mathbf{x}}{dt} = \mathbf{A}\mathbf{x},$$

(7.115)

where

$$\mathbf{A} = \begin{pmatrix} a_{11} & a_{12} \\ a_{21} & a_{22} \end{pmatrix}, \quad \mathbf{x} = \begin{pmatrix} x_1 \\ x_2 \end{pmatrix},$$

and $a_{11}, a_{12}, a_{21}, a_{22}$ are real constants.

Suppose the two characteristic values λ_1 and λ_2 of $\mathbf{A}$ are distinct; and let $\boldsymbol{\alpha}^{(1)}$ and $\boldsymbol{\alpha}^{(2)}$ be a pair of respective corresponding characteristic vectors of $\mathbf{A}$.

Then on every real interval, the vector functions defined by $\boldsymbol{\alpha}^{(1)}e^{\lambda_1 t}$ and $\boldsymbol{\alpha}^{(2)}e^{\lambda_2 t}$ form a linearly independent set of solutions of (7.115); and

$$\mathbf{x} = c_1\boldsymbol{\alpha}^{(1)}e^{\lambda_1 t} + c_2\boldsymbol{\alpha}^{(2)}e^{\lambda_2 t},$$

where c_1 and c_2 are arbitrary constants, is a general solution of (7.115) on $[a, b]$.

▶ **Example 7.36**

Consider the homogeneous linear system

$$\frac{dx_1}{dt} = 6x_1 - 3x_2,$$

$$\frac{dx_2}{dt} = 2x_1 + x_2,$$

(7.124)

that is, the vector differential equation

$$\frac{d\mathbf{x}}{dt} = \begin{pmatrix} 6 & -3 \\ 2 & 1 \end{pmatrix}\mathbf{x}, \quad \text{where} \quad \mathbf{x} = \begin{pmatrix} x_1 \\ x_2 \end{pmatrix}.$$

(7.125)

The characteristic equation of the coefficient matrix $\mathbf{A} = \begin{pmatrix} 6 & -3 \\ 2 & 1 \end{pmatrix}$ is

$$|\mathbf{A} - \lambda\mathbf{I}| = \begin{vmatrix} 6 - \lambda & -3 \\ 2 & 1 - \lambda \end{vmatrix} = 0.$$

Expanding the determinant and simplifying, this takes the form $\lambda^2 - 7\lambda + 12 = 0$ with roots $\lambda_1 = 3$, $\lambda_2 = 4$. These are the characteristic values of **A.** They are distinct (and real), and so Theorem 7.10 applies. We thus proceed to find respective corresponding characteristic vectors $\boldsymbol{\alpha}^{(1)}$ and $\boldsymbol{\alpha}^{(2)}$. We use (7.118) to do this.

With $\lambda = \lambda_1 = 3$ and $\boldsymbol{\alpha} = \boldsymbol{\alpha}^{(1)} = \binom{\alpha_1}{\alpha_2}$, (7.118) becomes

$$\begin{pmatrix} 6 & -3 \\ 2 & 1 \end{pmatrix}\begin{pmatrix} \alpha_1 \\ \alpha_2 \end{pmatrix} = 3\begin{pmatrix} \alpha_1 \\ \alpha_2 \end{pmatrix},$$

from which we at once find that α_1 and α_2 must satisfy

$$6\alpha_1 - 3\alpha_2 = 3\alpha_1, \qquad\qquad \alpha_1 = \alpha_2,$$
$$\text{or}$$
$$2\alpha_1 + \alpha_2 = 3\alpha_2, \qquad\qquad \alpha_1 = \alpha_2.$$

A simple nontrivial solution of this system is obviously $\alpha_1 = \alpha_2 = 1$, and thus a characteristic vector corresponding to $\lambda_1 = 3$ is

$$\boldsymbol{\alpha}^{(1)} = \begin{pmatrix} 1 \\ 1 \end{pmatrix}.$$

Then by Theorem 7.10,

$$\mathbf{x} = \begin{pmatrix} 1 \\ 1 \end{pmatrix}e^{3t}, \quad \text{that is,} \quad \mathbf{x} = \begin{pmatrix} e^{3t} \\ e^{3t} \end{pmatrix}, \tag{7.126}$$

is a solution of (7.125).

For $\lambda = \lambda_2 = 4$ and $\boldsymbol{\alpha} = \boldsymbol{\alpha}^{(2)} = \binom{\alpha_1}{\alpha_2}$, (7.118) becomes

$$\begin{pmatrix} 6 & -3 \\ 2 & 1 \end{pmatrix}\begin{pmatrix} \alpha_1 \\ \alpha_2 \end{pmatrix} = 4\begin{pmatrix} \alpha_1 \\ \alpha_2 \end{pmatrix},$$

from which we at once find that α_1 and α_2 must satisfy

$$6\alpha_1 - 3\alpha_2 = 4\alpha_1, \qquad\qquad 2\alpha_1 = 3\alpha_2,$$
$$\text{or}$$
$$2\alpha_1 + \alpha_2 = 4\alpha_2, \qquad\qquad 2\alpha_1 = 3\alpha_2.$$

A simple nontrivial solution of this system is obviously $\alpha_1 = 3$, $\alpha_2 = 2$, and thus a characteristic vector corresponding to $\lambda_2 = 4$ is

$$\boldsymbol{\alpha}^{(2)} = \begin{pmatrix} 3 \\ 2 \end{pmatrix}.$$

Then by Theorem 7.10,

$$\mathbf{x} = \begin{pmatrix} 3 \\ 2 \end{pmatrix}e^{4t}, \quad \text{that is,} \quad \mathbf{x} = \begin{pmatrix} 3e^{4t} \\ 2e^{4t} \end{pmatrix}, \tag{7.127}$$

is a solution of (7.125).

Also by Theorem 7.10 the solutions (7.126) and (7.127) of (7.125) are linearly independent, and a general solution is

$$\mathbf{x} = c_1 \begin{pmatrix} e^{3t} \\ e^{3t} \end{pmatrix} + c_2 \begin{pmatrix} 3e^{4t} \\ 2e^{4t} \end{pmatrix},$$

where c_1 and c_2 are arbitrary constants. That is, in scalar language, a general solution of the homogeneous linear system (7.124) is

$$x_1 = c_1 e^{3t} + 3c_2 e^{4t}, \qquad x_2 = c_1 e^{3t} + 2c_2 e^{4t},$$

where c_1 and c_2 are arbitrary constants.

The system (7.124) of this example is, aside from notation, precisely the system (7.79) of Example 7.17. A comparison of these two illustrations provides useful insight.

Let us return to the homogeneous linear system (7.112) and further consider the result stated in Theorem 7.10. In that theorem we stated that if the two characteristic values λ_1 and λ_2 of **A** are distinct and if $\boldsymbol{\alpha}^{(1)}$ and $\boldsymbol{\alpha}^{(2)}$ are two respective corresponding characteristic vectors of **A**, then the two functions defined by

$$\boldsymbol{\alpha}^{(1)} e^{\lambda_1 t} \quad \text{and} \quad \boldsymbol{\alpha}^{(2)} e^{\lambda_2 t}$$

constitute a pair of linearly independent solutions of (7.112). Note that although we assume that λ_1 and λ_2 are *distinct*, we do *not* require that they be *real*. Thus the case of distinct *complex* characteristic values is also included here. Since the coefficients in system (7.112) are real, if complex characteristic values exist, then they must be a conjugate-complex pair.

Suppose $\lambda_1 = a + bi$ and $\lambda_2 = a - bi$ are a pair of conjugate-complex characteristic values of the coefficient matrix **A** of system (7.112). Then the corresponding linearly independent solutions

$$\boldsymbol{\alpha}^{(1)} e^{(a + bi)t} \quad \text{and} \quad \boldsymbol{\alpha}^{(2)} e^{(a - bi)t}$$

are *complex* solutions. In such a case, we may use Euler's formula $e^{i\theta} = \cos\theta + i\sin\theta$ to replace this linearly independent complex pair of solutions by a related linearly independent *real* pair of solutions. This is accomplished exactly as explained in Section 7.4C and is illustrated in Example 7.18.

C. Case of a Double Characteristic Value

Suppose that the two characteristic values λ_1 and λ_2 of the coefficient matrix **A** of the vector differential equation (7.115) are real and equal. Let λ denote this common characteristic value, and let $\boldsymbol{\alpha}$ be a corresponding characteristic vector. Then just as in the previous subsection, it is readily shown that the vector function defined by $\boldsymbol{\alpha} e^{\lambda t}$ is a solution of (7.115) on every real interval $[a, b]$. But now, except in the special subcase in which $a_{11} = a_{22} \neq 0$, $a_{12} = a_{21} = 0$, there is only one solution of this form. Looking back at the results of Section 7.4D and expressing them in vector notation, we would now seek a second solution of the form

$$\mathbf{x} = (\boldsymbol{\gamma} t + \boldsymbol{\beta}) e^{\lambda t}, \tag{7.128}$$

where $\boldsymbol{\gamma}$ and $\boldsymbol{\beta}$ are to be determined so that this is indeed a solution.

We thus assume a solution of differential equation (7.115) of this form (7.128), differentiate, and substitute into (7.115). We at once obtain

$$(\boldsymbol{\gamma} t + \boldsymbol{\beta})\lambda e^{\lambda t} + \boldsymbol{\gamma} e^{\lambda t} = A(\boldsymbol{\gamma} t + \boldsymbol{\beta}) e^{\lambda t}.$$

Then dividing through by $e^{\lambda t} \neq 0$ and collecting terms in powers of t, we readily find

$$(\lambda\boldsymbol{\gamma} - A\boldsymbol{\gamma})t + (\lambda\boldsymbol{\beta} + \boldsymbol{\gamma} - A\boldsymbol{\beta}) = \mathbf{0}.$$

This holds for all $t \in [a, b]$ if and only if

$$\lambda \gamma - \mathbf{A}\gamma = \mathbf{0},$$

$$\lambda \beta + \gamma - \mathbf{A}\beta = \mathbf{0}. \tag{7.129}$$

The first of these gives $\mathbf{A}\gamma = \lambda \gamma$ or $(\mathbf{A} - \lambda \mathbf{I})\gamma = \mathbf{0}$. Thus we see that γ is, in fact, a characteristic vector α of $\mathbf{A}$ corresponding to characteristic value λ. The second of (7.129) with $\gamma = \alpha$ gives $\mathbf{A}\beta - \lambda\beta = \alpha$, from which we have

$$(\mathbf{A} - \lambda\mathbf{I})\beta = \alpha \tag{7.130}$$

as the equation for the determination of β in (7.128). Thus in the assumed solution (7.128), the vector γ is, in fact, a characteristic vector of $\mathbf{A}$ corresponding to characteristic value λ, and the vector β is determined from (7.130). Direct substitution of (7.128) with these choices of γ and β verifies that it is indeed a solution. Moreover, it can be shown that the two solutions thus obtained,

$$\alpha e^{\lambda t} \quad \text{and} \quad (\alpha t + \beta)e^{\lambda t},$$

are linearly independent; therefore, a linear combination of them constitutes a general solution of (7.115). We summarize these results in the following theorem.

THEOREM 7.11

Consider the homogeneous linear system

$$\frac{dx_1}{dt} = a_{11}x_1 + a_{12}x_2,$$

$$\frac{dx_2}{dt} = a_{21}x_1 + a_{22}x_2, \tag{7.112}$$

that is, the vector differential equation

$$\frac{d\mathbf{x}}{dt} = \mathbf{A}\mathbf{x}, \tag{7.115}$$

where

$$\mathbf{A} = \begin{pmatrix} a_{11} & a_{12} \\ a_{21} & a_{22} \end{pmatrix}, \quad \mathbf{x} = \begin{pmatrix} x_1 \\ x_2 \end{pmatrix},$$

and $a_{11}, a_{12}, a_{21}, a_{22}$ are real constants which are not such that $a_{11} = a_{22} \neq 0$, $a_{12} = a_{21} = 0$. Suppose the two characteristic values λ_1 and λ_2 of $\mathbf{A}$ are real and equal; and let λ denote their common value. Let α be a corresponding characteristic vector of $\mathbf{A}$ and let β be a vector satisfying the equation

$$(\mathbf{A} - \lambda\mathbf{I})\beta = \alpha. \tag{7.130}$$

Then on every real interval, the vector functions defined by

$$\alpha e^{\lambda t} \quad \text{and} \quad (\alpha t + \beta)e^{\lambda t} \tag{7.131}$$

form a linearly independent set of solutions of (7.115); and

$$\mathbf{x} = c_1\alpha e^{\lambda t} + c_2(\alpha t + \beta)e^{\lambda t},$$

where c_1 and c_2 are arbitrary constants, is a general solution of (7.115) on $[a, b]$.

▶ **Example 7.37**

Consider the homogeneous linear system

$$\frac{dx_1}{dt} = 4x_1 - x_2$$

$$\frac{dx_2}{dt} = x_1 + 2x_2, \tag{7.132}$$

that is, the vector differential equation

$$\frac{d\mathbf{x}}{dt} = \begin{pmatrix} 4 & -1 \\ 1 & 2 \end{pmatrix}\mathbf{x}, \quad \text{where} \quad \mathbf{x} = \begin{pmatrix} x_1 \\ x_2 \end{pmatrix}. \tag{7.133}$$

The characteristic equation of the coefficient matrix

$$\mathbf{A} = \begin{pmatrix} 4 & -1 \\ 1 & 2 \end{pmatrix} \quad \text{is} \quad |\mathbf{A} - \lambda\mathbf{I}| = \begin{vmatrix} 4-\lambda & -1 \\ 1 & 2-\lambda \end{vmatrix} = 0.$$

Expanding the determinant and simplifying, this takes the form $\lambda^2 - 6\lambda + 9 = 0$, with the double root $\lambda = 3$. That is, the characteristic values of A are real and equal and so Theorem 7.11 applies.

Proceeding to apply it, we first find a characteristic vector $\boldsymbol{\alpha}$ corresponding to the characteristic value $\lambda = 3$. With $\lambda = 3$ and $\boldsymbol{\alpha} = \begin{pmatrix} \alpha_1 \\ \alpha_2 \end{pmatrix}$, (7.118) becomes

$$\begin{pmatrix} 4 & -1 \\ 1 & 2 \end{pmatrix}\begin{pmatrix} \alpha_1 \\ \alpha_2 \end{pmatrix} = 3\begin{pmatrix} \alpha_1 \\ \alpha_2 \end{pmatrix},$$

from which we at once find that α_1 and α_2 must satisfy

$$4\alpha_1 - \alpha_2 = 3\alpha_1 \qquad\qquad \alpha_1 = \alpha_1 = \alpha_2,$$

$$\text{or}$$

$$\alpha_1 + 2\alpha_2 = 3\alpha_2, \qquad\qquad \alpha_1 = \alpha_1 = \alpha_2.$$

A simple nontrivial solution of this system is obviously $\alpha_1 = \alpha_2 = 1$, and thus a characteristic vector corresponding to $\lambda = 3$ is

$$\alpha = \begin{pmatrix} 1 \\ 1 \end{pmatrix}.$$

Then by Theorem 7.11,

$$\mathbf{x} = \begin{pmatrix} 1 \\ 1 \end{pmatrix}e^{3t}, \quad \text{that is,} \quad \mathbf{x} = \begin{pmatrix} e^{3t} \\ e^{3t} \end{pmatrix}, \tag{7.134}$$

is a solution of (7.133).

By Theorem 7.11, a linearly independent solution is of the form $(\boldsymbol{\alpha}t + \boldsymbol{\beta})e^{\lambda t}$, where $\boldsymbol{\alpha} = \begin{pmatrix} 1 \\ 1 \end{pmatrix}$, $\lambda = 3$, and $\boldsymbol{\beta}$ satisfies $(\mathbf{A} - \lambda\mathbf{I})\boldsymbol{\beta} = \boldsymbol{\alpha}$. Thus $\boldsymbol{\beta} = \begin{pmatrix} \beta_1 \\ \beta_2 \end{pmatrix}$ satisfies

$$\left[\begin{pmatrix} 4 & -1 \\ 1 & 2 \end{pmatrix} - 3\begin{pmatrix} 1 & 0 \\ 0 & 1 \end{pmatrix}\right]\begin{pmatrix} \beta_1 \\ \beta_2 \end{pmatrix} = \begin{pmatrix} 1 \\ 1 \end{pmatrix},$$

which quickly reduces to

$$\begin{pmatrix} 1 & -1 \\ 1 & -1 \end{pmatrix} \begin{pmatrix} \beta_1 \\ \beta_2 \end{pmatrix} = \begin{pmatrix} 1 \\ 1 \end{pmatrix}.$$

From this we at once find that β_1 and β_2 must satisfy

$$\beta_1 - \beta_2 = 1,$$
$$\beta_1 - \beta_2 = 1.$$

A simple nontrivial solution of this system is $\beta_1 = 1, \beta_2 = 0$. Thus we find the desired vector

$$\boldsymbol{\beta} = \begin{pmatrix} 1 \\ 0 \end{pmatrix}.$$

Then by Theorem 7.11,

$$\mathbf{x} = \left[\begin{pmatrix} 1 \\ 1 \end{pmatrix} t + \begin{pmatrix} 1 \\ 0 \end{pmatrix} \right] e^{3t}, \quad \text{that is,} \quad \mathbf{x} = \begin{pmatrix} (t+1)e^{3t} \\ te^{3t} \end{pmatrix}, \qquad (7.135)$$

is a solution of (7.133).

Finally, by the same theorem, solutions (7.134) and (7.135) are linearly independent, and a general solution is

$$\mathbf{x} = c_1 \begin{pmatrix} e^{3t} \\ e^{3t} \end{pmatrix} + c_2 \begin{pmatrix} (t+1)e^{3t} \\ te^{3t} \end{pmatrix},$$

where c_1 and c_2 are arbitrary constants. That is, in scalar language, a general solution of the homogeneous linear system (7.132) is

$$x_1 = c_1 e^{3t} + c_2(t+1)e^{3t},$$
$$x_2 = c_1 e^{3t} + c_2 te^{3t},$$

where c_1 and c_2 are arbitrary constants.

Exercises

Find the general solution of each of the homogeneous linear systems in Exercises 1–18, using the vector–matrix methods of this section, where in each exercise

$$\mathbf{x} = \begin{pmatrix} x_1 \\ x_2 \end{pmatrix}.$$

1. $\dfrac{d\mathbf{x}}{dt} = \begin{pmatrix} 5 & -2 \\ 4 & -1 \end{pmatrix} \mathbf{x}$

2. $\dfrac{d\mathbf{x}}{dt} = \begin{pmatrix} 5 & -1 \\ 3 & 1 \end{pmatrix} \mathbf{x}$

3. $\dfrac{d\mathbf{x}}{dt} = \begin{pmatrix} 1 & 2 \\ 3 & 2 \end{pmatrix} \mathbf{x}$

4. $\dfrac{d\mathbf{x}}{dt} = \begin{pmatrix} 2 & 3 \\ -1 & -2 \end{pmatrix} \mathbf{x}$

5. $\dfrac{d\mathbf{x}}{dt} = \begin{pmatrix} 3 & 1 \\ 4 & 3 \end{pmatrix} \mathbf{x}$

6. $\dfrac{d\mathbf{x}}{dt} = \begin{pmatrix} 6 & -1 \\ 3 & 2 \end{pmatrix} \mathbf{x}$

7. $\dfrac{d\mathbf{x}}{dt} = \begin{pmatrix} 1 & 3 \\ 3 & 1 \end{pmatrix} \mathbf{x}$

8. $\dfrac{d\mathbf{x}}{dt} = \begin{pmatrix} 3 & 2 \\ 6 & -1 \end{pmatrix} \mathbf{x}$

9. $\dfrac{d\mathbf{x}}{dt} = \begin{pmatrix} 1 & -4 \\ 1 & 1 \end{pmatrix} \mathbf{x}$

10. $\dfrac{d\mathbf{x}}{dt} = \begin{pmatrix} 2 & -3 \\ 3 & 2 \end{pmatrix} \mathbf{x}$

11. $\dfrac{d\mathbf{x}}{dt} = \begin{pmatrix} 1 & -3 \\ 3 & 1 \end{pmatrix} \mathbf{x}$

12. $\dfrac{d\mathbf{x}}{dt} = \begin{pmatrix} 5 & -4 \\ 2 & 1 \end{pmatrix} \mathbf{x}$

13. $\dfrac{d\mathbf{x}}{dt} = \begin{pmatrix} 3 & -1 \\ 4 & -1 \end{pmatrix} \mathbf{x}$

14. $\dfrac{d\mathbf{x}}{dt} = \begin{pmatrix} 7 & 4 \\ -1 & 3 \end{pmatrix} \mathbf{x}$

15. $\dfrac{d\mathbf{x}}{dt} = \begin{pmatrix} 5 & 4 \\ -1 & 1 \end{pmatrix} x$

16. $\dfrac{d\mathbf{x}}{dt} = \begin{pmatrix} 1 & -2 \\ 2 & -3 \end{pmatrix} \mathbf{x}$

17. $\dfrac{d\mathbf{x}}{dt} = \begin{pmatrix} 6 & -4 \\ 1 & 2 \end{pmatrix} \mathbf{x}$

18. $\begin{pmatrix} 7 & -1 \\ 4 & 3 \end{pmatrix} \mathbf{x}$

7.7 THE MATRIX METHOD FOR HOMOGENEOUS LINEAR SYSTEMS WITH CONSTANT COEFFICIENTS: *n* EQUATIONS IN *n* UNKNOWN FUNCTIONS

A. Introduction

In this section we extend the methods of the previous section to a homogeneous linear system of n first-order differential equations in n unknown functions and having real constant coefficients. More specifically we consider a homogeneous linear system of the form

$$\frac{dx_1}{dt} = a_{11}x_1 + a_{12}x_2 + \cdots + a_{1n}x_n,$$

$$\frac{dx_2}{dt} = a_{21}x_1 + a_{22}x_2 + \cdots + a_{2n}x_n, \tag{7.136}$$

$$\vdots$$

$$\frac{dx_n}{dt} = a_{n1}x_1 + a_{n2}x_2 + \cdots + a_{nn}x_n,$$

where the coefficients a_{ij}, $(i = 1, 2, \ldots, n; j = 1, 2, \ldots, n)$, are real constants.

We proceed to express this system in vector–matrix notation. We introduce the $n \times n$ constant matrix of real numbers

$$\mathbf{A} = \begin{pmatrix} a_{11} & a_{12} & \cdots & a_{1n} \\ a_{21} & a_{22} & \cdots & a_{2n} \\ \vdots & \vdots & & \vdots \\ a_{n1} & a_{n2} & \cdots & a_{nn} \end{pmatrix} \tag{7.137}$$

and the vector

$$\mathbf{x} = \begin{pmatrix} x_1 \\ x_2 \\ \vdots \\ x_n \end{pmatrix}. \tag{7.138}$$

Then by definition of the derivative of a vector,

$$\frac{d\mathbf{x}}{dt} = \begin{pmatrix} \dfrac{dx_1}{dt} \\[2mm] \dfrac{dx_2}{dt} \\[2mm] \vdots \\[2mm] \dfrac{dx_n}{dt} \end{pmatrix};$$

and by multiplication of a matrix by a vector, we have

$$\mathbf{Ax} = \begin{pmatrix} a_{11} & a_{12} \cdots a_{1n} \\ a_{21} & a_{22} \cdots a_{2n} \\ \vdots & \vdots \quad \vdots \\ a_{n1} & a_{n2} \cdots a_{nn} \end{pmatrix} \begin{pmatrix} x_1 \\ x_2 \\ \vdots \\ x_n \end{pmatrix} = \begin{pmatrix} a_{11}x_1 + a_{12}x_2 + \cdots + a_{1n}x_n \\ a_{21}x_1 + a_{22}x_2 + \cdots + a_{2n}x_n \\ \vdots \\ a_{n1}x_1 + a_{n2}x_2 + \cdots + a_{nn}x_n \end{pmatrix}.$$

Comparing the components of $d\mathbf{x}/dt$ with the left members of (7.136) and the components of $\mathbf{Ax}$ with the right members of (7.136), we see that system (7.136) can be expressed as the homogeneous linear *vector* differential equation

$$\frac{d\mathbf{x}}{dt} = \mathbf{Ax}. \tag{7.139}$$

The real constant matrix $\mathbf{A}$ that appears in (7.139) and is defined by (7.137) is called the *coefficient matrix* of (7.139).

DEFINITION

By a solution of the system (7.136), that is, of the vector differential equation (7.139), we mean an n × 1 column-vector function

$$\boldsymbol{\phi} = \begin{pmatrix} \phi_1 \\ \phi_2 \\ \vdots \\ \phi_n \end{pmatrix},$$

whose components $\phi_1, \phi_2, \ldots, \phi_n$ each have a continuous derivative on the real interval $a \le t \le b$, and which is such that

$$\frac{d\phi_1(t)}{dt} = a_{11}\phi_1(t) + a_{12}\phi_2(t) + \cdots + a_{1n}\phi_n(t),$$

$$\frac{d\phi_2(t)}{dt} = a_{21}\phi_1(t) + a_{22}\phi_2(t) + \cdots + a_{2n}\phi_n(t),$$

$$\vdots$$

$$\frac{d\phi_n(t)}{dt} = a_{n1}\phi_1(t) + a_{n2}\phi_2(t) + \cdots + a_{nn}\phi_n(t),$$

for all t such that $a \leq t \leq b$. In other words, the components $\phi_1, \phi_2, \ldots, \phi_n$ of ϕ are such that

$$x_1 = \phi_1(t)$$
$$x_2 = \phi_2(t)$$
$$\vdots$$
$$x_n = \phi_n(t)$$

simultaneously satisfy all n equations of the system (7.136) identically on $a \leq t \leq b$.

We proceed to introduce the concept of a general solution of system (7.136). In the process of doing so, we state several pertinent theorems. The proofs of these results will be found in Section 11.2. Our first basic result is the following.

THEOREM 7.12

Any linear combination of n solutions of the homogeneous linear system (7.136) is itself a solution of the system (7.136).

Before going on to our next result, the reader should return to Section 7.5C and review the concepts of linear dependence and linear independence of vector functions.
We now state the following basic theorem concerning sets of linearly independent solutions of the homogeneous linear system (7.136).

THEOREM 7.13

There exist sets of n linearly independent solutions of the homogeneous linear system (7.136). Every solution of the system (7.136) can be written as a linear combination of any n linearly independent solutions of (7.136).

As a result of Theorems 7.12 and 7.13, we now give the following definition of a general solution for the homogeneous linear system (7.136) of n equations in n unknown functions.

DEFINITION

Let

$$\phi_1 = \begin{pmatrix} \phi_{11} \\ \phi_{21} \\ \vdots \\ \phi_{n1} \end{pmatrix}, \phi_2 = \begin{pmatrix} \phi_{12} \\ \phi_{22} \\ \vdots \\ \phi_{n2} \end{pmatrix}, \ldots, \phi_n = \begin{pmatrix} \phi_{1n} \\ \phi_{2n} \\ \vdots \\ \phi_{nn} \end{pmatrix}$$

be n linearly independent solutions of the homogeneous linear system (7.136). Let $c_1, c_2, \ldots, c_n$ be n arbitrary constants. Then the solution

$$\mathbf{x} = c_1\phi_1(t) + c_2\phi_2(t) + \cdots + c_n\phi_n(t),$$

that is,

$$x_1 = c_1\phi_{11}(t) + c_2\phi_{12}(t) + \cdots + c_n\phi_{1n}(t),$$
$$x_2 = c_1\phi_{21}(t) + c_2\phi_{22}(t) + \cdots + c_n\phi_{2n}(t),$$
$$\vdots$$
$$x_n = c_1\phi_{n1}(t) + c_2\phi_{n2}(t) + \cdots + c_n\phi_{nn}(t),$$

is called a general solution *of the system (7.136).*

In order to state a useful criterion for the linear independence of *n* solutions of the system (7.136), we introduce the following concept.

DEFINITION

Consider the n vector functions $\phi_1, \phi_2, \ldots, \phi_n$ *defined, respectively, by*

$$\phi_1(t) = \begin{pmatrix} \phi_{11}(t) \\ \phi_{21}(t) \\ \vdots \\ \phi_{n1}(t) \end{pmatrix}, \phi_2(t) = \begin{pmatrix} \phi_{12}(t) \\ \phi_{22}(t) \\ \vdots \\ \phi_{n2}(t) \end{pmatrix}, \ldots, \phi_n(t) = \begin{pmatrix} \phi_{1n}(t) \\ \phi_{2n}(t) \\ \vdots \\ \phi_{nn}(t) \end{pmatrix}. \tag{7.140}$$

The n × n determinant

$$\begin{vmatrix} \phi_{11} & \phi_{12} & \cdots & \phi_{1n} \\ \phi_{21} & \phi_{22} & \cdots & \phi_{2n} \\ \vdots & \vdots & \bullet & \vdots \\ \phi_{n1} & \phi_{n2} & \cdots & \phi_{nn} \end{vmatrix} \tag{7.141}$$

is called the Wronskian *of the n vector functions* $\phi_1, \phi_2, \ldots, \phi_n$ *defined by (7.140). We will denote it by* $W(\phi_1, \phi_2, \ldots, \phi_n)$ *and its value at t by* $W(\phi_1, \phi_2, \ldots, \phi_n)(t)$.

We may now state the following useful criterion for the linear independence of *n* solutions of the homogeneous linear system (7.136).

THEOREM 7.14

n solutions $\phi_1, \phi_2, \ldots, \phi_n$ *of the homogeneous linear system (7.136) are linearly independent on an interval* $a \le t \le b$ *if and only if*

$$W(\phi_1, \phi_2, \ldots, \phi_n)(t) \ne 0$$

for all $t \in [a, b]$.

Concerning the values of $W(\phi_1, \phi_2, \ldots, \phi_n)$, we also state the following result.

THEOREM 7.15

Let $\phi_1, \phi_2, \ldots, \phi_n$ *be n solutions of the homogeneous linear system (7.136) on an interval* $a \le t \le b$. *Then either* $W(\phi_1, \phi_2, \ldots, \phi_n)(t) = 0$ *for all* $t \in [a, b]$ *or* $W(\phi_1, \phi_2, \ldots, \phi_n)(t) = 0$ *for no* $t \in [a, b]$.

Having introduced these basic concepts and results, we now seek solutions of the system (7.136). We shall proceed by analogy with the presentation in Section 7.6A. Doing this, we seek nontrivial solutions of system (7.136) of the form

$$x_1 = \alpha_1 e^{\lambda t},$$
$$x_2 = \alpha_2 e^{\lambda t},$$
$$\vdots$$
$$x_n = \alpha_n e^{\lambda t}, \tag{7.142}$$

where $\alpha_1, \alpha_2, \ldots, \alpha_n$, and λ are constants. Letting

$$\boldsymbol{\alpha} = \begin{pmatrix} \alpha_1 \\ \alpha_2 \\ \vdots \\ \alpha_n \end{pmatrix} \quad \text{and} \quad \mathbf{x} = \begin{pmatrix} x_1 \\ x_2 \\ \vdots \\ x_n \end{pmatrix}, \tag{7.143}$$

we see that the vector form of the desired solution (7.142) is

$$\mathbf{x} = \boldsymbol{\alpha} e^{\lambda t}.$$

Thus we seek solutions of the vector differential equation (7.139) which are of the form

$$\mathbf{x} = \boldsymbol{\alpha} e^{\lambda t}, \tag{7.144}$$

where $\boldsymbol{\alpha}$ is a constant vector and λ is a number.

Now substituting (7.144) into (7.139), we obtain

$$\lambda \boldsymbol{\alpha} e^{\lambda t} = \mathbf{A} \boldsymbol{\alpha} e^{\lambda t}$$

which reduces at once to

$$\mathbf{A} \boldsymbol{\alpha} = \lambda \boldsymbol{\alpha} \tag{7.145}$$

and hence to

$$(\mathbf{A} - \lambda \mathbf{I}) \boldsymbol{\alpha} = \mathbf{0},$$

where $\mathbf{I}$ is the $n \times n$ identity matrix. Written out in terms of components, this is the system of n homogeneous linear algebraic equations

$$(a_{11} - \lambda)\alpha_1 + a_{12}\alpha_2 + \cdots + a_{1n}\alpha_n = 0,$$
$$a_{21}\alpha_1 + (a_{22} - \lambda)\alpha_2 + \cdots + a_{2n}\alpha_n = 0, \tag{7.146}$$
$$\cdots$$
$$a_{n1}\alpha_1 + a_{n2}\alpha_2 + \cdots + (a_{nn} - \lambda)\alpha_n = 0,$$

in the n unknowns $\alpha_1, \alpha_2, \ldots, \alpha_n$. By Theorem A of Section 7.5C, this system has a nontrivial solution if and only if

$$\begin{vmatrix} a_{11} - \lambda & a_{12} & \cdots & a_{1n} \\ a_{21} & a_{22} - \lambda & \cdots & a_{2n} \\ \vdots & \vdots & & \vdots \\ a_{n1} & a_{n2} & \cdots & a_{nn} - \lambda \end{vmatrix} = 0; \tag{7.147}$$

that is, in matrix notation,

$$|\mathbf{A} - \lambda \mathbf{I}| = 0.$$

Looking back at Section 7.5D, we recognize Equation (7.147) as the *characteristic equation* of the coefficient matrix $\mathbf{A} = (a_{ij})$ of the vector differential equation (7.139). We know that this is an *n*th-degree polynomial equation in λ, and we recall that its roots $\lambda_1, \lambda_2, \ldots, \lambda_n$ are the *characteristic values* of $\mathbf{A}$. Substituting each characteristic value $\lambda_i, (i = 1, 2, \ldots, n)$, into system (7.146), we obtain the corresponding nontrivial solution

$$\alpha_1 = \alpha_{1i}, \alpha_2 = \alpha_{2i}, \ldots, \alpha_n = \alpha_{ni},$$

$(i = 1, 2, \ldots, n)$, of system (7.146). Since (7.146) is merely the component form of (7.145), we recognize that the vector defined by

$$\boldsymbol{\alpha}^{(i)} = \begin{pmatrix} \alpha_{1i} \\ \alpha_{2i} \\ \vdots \\ \alpha_{ni} \end{pmatrix}, \qquad (i = 1, 2, \ldots, n), \tag{7.148}$$

is a *characteristic vector* corresponding to the characteristic value $\lambda_i (i = 1, 2, \ldots, n)$.

We thus see that if the vector differential equation

$$\frac{d\mathbf{x}}{dt} = \mathbf{A}\mathbf{x} \tag{7.139}$$

has a solution of the form

$$\mathbf{x} = \boldsymbol{\alpha} e^{\lambda t} \tag{7.144}$$

then the number λ must be a characteristic value λ_i of the coefficient matrix $\mathbf{A}$ and the vector $\boldsymbol{\alpha}$ must be a characteristic vector $\boldsymbol{\alpha}^{(i)}$ corresponding to this characteristic value λ_i.

B. Case of *n* Distinct Characteristic Values

Suppose that each of the *n* characteristic values $\lambda_1, \lambda_2, \ldots, \lambda_n$ of the $n \times n$ square coefficient matrix $\mathbf{A}$ of the vector differential equation is *distinct* (that is, nonrepeated); and let $\boldsymbol{\alpha}^{(1)}, \boldsymbol{\alpha}^{(2)}, \ldots, \boldsymbol{\alpha}^{(n)}$ be a set of *n* respective corresponding characteristic vectors of $\mathbf{A}$. Then the *n* distinct vector functions $\mathbf{x}_1, \mathbf{x}_2, \ldots, \mathbf{x}_n$ defined, respectively, by

$$\mathbf{x}^{(1)}(t) = \boldsymbol{\alpha}^{(1)} e^{\lambda_1 t}, \mathbf{x}^{(2)}(t) = \boldsymbol{\alpha}^{(2)} e^{\lambda_2 t}, \ldots, \mathbf{x}^{(n)}(t) = \boldsymbol{\alpha}^{(n)} e^{\lambda_n t} \tag{7.149}$$

are solutions of the vector differential equation (7.139) on every real interval $[a, b]$. This is readily seen as follows: From (7.145), for each $i = 1, 2, \ldots, n$, we have

$$\lambda_i \boldsymbol{\alpha}^{(i)} = \mathbf{A}\boldsymbol{\alpha}^{(i)};$$

and using this and the definition (7.149) of $\mathbf{x}^{(i)}(t)$, we obtain

$$\frac{d\mathbf{x}^{(i)}(t)}{dt} = \lambda_i \boldsymbol{\alpha}^{(i)} e^{\lambda_i t} = \mathbf{A}\boldsymbol{\alpha}^{(i)} e^{\lambda_i t} = \mathbf{A}\mathbf{x}^{(i)}(t),$$

which states that $\mathbf{x}^{(i)}(t)$ satisfies the vector differential equation

$$\frac{d\mathbf{x}}{dt} = \mathbf{A}\mathbf{x}, \tag{7.139}$$

on $[a, b]$.

Now consider the Wronskian of the n solutions $\mathbf{x}^{(1)}, \mathbf{x}^{(2)}, \ldots, \mathbf{x}^{(n)}$, defined by (7.149). We find

$$W\left(\mathbf{x}^{(1)}, \mathbf{x}^{(2)}, \ldots, \mathbf{x}^{(n)}\right)(t) = \begin{vmatrix} \alpha_{11}e^{\lambda_1 t} & \alpha_{12}e^{\lambda_2 t} \cdots \alpha_{1n}e^{\lambda_n t} \\ \alpha_{21}e^{\lambda_1 t} & \alpha_{22}e^{\lambda_2 t} \cdots \alpha_{2n}e^{\lambda_n t} \\ \vdots & \vdots \qquad \qquad \vdots \\ \alpha_{n1}e^{\lambda_1 t} & \alpha_{n2}e^{\lambda_2 t} \cdots \alpha_{nn}e^{\lambda_n t} \end{vmatrix}$$

$$= e^{(\lambda_1 + \lambda_2 + \cdots + \lambda_n)t} \begin{vmatrix} \alpha_{11} & \alpha_{12} & \cdots & \alpha_{1n} \\ \alpha_{21} & \alpha_{22} & \cdots & \alpha_{2n} \\ \vdots & \vdots & & \vdots \\ \alpha_{n1} & \alpha_{n2} & \cdots & \alpha_{nn} \end{vmatrix}.$$

By Result F of Section 7.5D, the n characteristic vectors $\boldsymbol{\alpha}^{(1)}, \boldsymbol{\alpha}^{(2)}, \ldots, \boldsymbol{\alpha}^{(n)}$ are linearly independent. Therefore, using Exercise 8 at the end of Section 7.5C, we have

$$\begin{vmatrix} \alpha_{11} & \alpha_{12} & \cdots & \alpha_{1n} \\ \alpha_{21} & \alpha_{22} & \cdots & \alpha_{2n} \\ \vdots & \vdots & & \vdots \\ \alpha_{n1} & \alpha_{n2} & \cdots & \alpha_{nn} \end{vmatrix} \neq 0.$$

Further, it is clear that

$$e^{(\lambda_1 + \lambda_2 + \cdots + \lambda_n)t} \neq 0$$

for all t. Thus $W\left(\mathbf{x}^{(1)}, \mathbf{x}^{(2)}, \ldots, \mathbf{x}^{(n)}\right)(t) \neq 0$ for all t on $[a, b]$. Hence by Theorem 7.14, the solutions $\mathbf{x}^{(1)}, \mathbf{x}^{(2)}, \ldots, \mathbf{x}^{(n)}$, of vector differential equation (7.139) defined by (7.149), are linearly independent on $[a, b]$. Thus a general solution of (7.139) is given by

$$c_1 \mathbf{x}^{(1)} + c_2 \mathbf{x}^{(2)} + \cdots + c_n \mathbf{x}^{(n)}$$

where $c_1, c_2 \ldots, c_n$ are n arbitrary numbers. We summarize the results obtained in the following theorem:

THEOREM 7.16

Consider the homogeneous linear system

$$\frac{dx_1}{dt} = a_{11}x_1 + a_{12}x_2 + \cdots + a_{1n}x_n,$$

$$\frac{dx_2}{dt} = a_{21}x_1 + a_{22}x_2 + \cdots + a_{2n}x_n, \qquad (7.136)$$

$$\vdots$$

$$\frac{dx_n}{dt} = a_{n1}x_1 + a_{n2}x_2 + \cdots + a_{nn}x_n,$$

that is, the vector differential equation

$$\frac{d\mathbf{x}}{dt} = \mathbf{A}\mathbf{x}, \qquad (7.139)$$

where

$$\mathbf{A} = \begin{pmatrix} a_{11} & a_{12} & \cdots & a_{1n} \\ a_{21} & a_{22} & \cdots & a_{2n} \\ \vdots & \vdots & & \vdots \\ a_{n1} & a_{n2} & \cdots & a_{nn} \end{pmatrix} \qquad \mathbf{x} = \begin{pmatrix} x_1 \\ x_2 \\ \vdots \\ x_n \end{pmatrix},$$

and the a_{ij}, $(i = 1, 2, \ldots, n; j = 1, 2, \ldots, n)$, are real constants.

Suppose each of the n characteristic values $\lambda_1, \lambda_2, \ldots, \lambda_n$ of $\mathbf{A}$ is distinct; and let $\boldsymbol{\alpha}^{(1)}, \boldsymbol{\alpha}^{(2)}, \ldots, \boldsymbol{\alpha}^{(n)}$ be a set of n respective corresponding characteristic vectors of $\mathbf{A}$.

Then on every real interval, the n vector functions defined by

$$\boldsymbol{\alpha}^{(1)} e^{\lambda_1 t}, \boldsymbol{\alpha}^{(2)} e^{\lambda_2 t}, \ldots, \boldsymbol{\alpha}^{(n)} e^{\lambda_n t}$$

form a linearly independent set of solutions of (7.136), that is, (7.139); and

$$\mathbf{x} = c_1 \boldsymbol{\alpha}^{(1)} e^{\lambda_1 t} + c_2 \boldsymbol{\alpha}^{(2)} e^{\lambda_2 t} + \cdots + c_n \boldsymbol{\alpha}^{(n)} e^{\lambda_n t},$$

where $c_1, c_2, \ldots, c_n$ are n arbitrary constants, is a general solution of (7.136).

▶ **Example 7.38**

Consider the homogeneous linear system

$$\frac{dx_1}{dt} = 7x_1 - x_2 + 6x_3,$$

$$\frac{dx_2}{dt} = -10x_1 + 4x_2 - 12x_3, \tag{7.150}$$

$$\frac{dx_3}{dt} = -2x_1 + x_2 - x_3,$$

that is, the vector differential equation

$$\frac{d\mathbf{x}}{dt} = \begin{pmatrix} 7 & -1 & 6 \\ -10 & 4 & -12 \\ -2 & 1 & -1 \end{pmatrix} \mathbf{x}, \quad \text{where} \quad \mathbf{x} = \begin{pmatrix} x_1 \\ x_2 \\ x_3 \end{pmatrix}. \tag{7.151}$$

Assuming a solution of (7.151) of the form

$$\mathbf{x} = \boldsymbol{\alpha} e^{\lambda t},$$

that is,

$$x_1 = \alpha_1 e^{\lambda t}, \quad x_2 = \alpha_2 e^{\lambda t}, \quad x_3 = \alpha_3 e^{\lambda t},$$

we know that λ must be a solution of the characteristic equation of the coefficient matrix

$$\mathbf{A} = \begin{pmatrix} 7 & -1 & 6 \\ -10 & 4 & -12 \\ -2 & 1 & -1 \end{pmatrix}.$$

This characteristic equation is

$$|\mathbf{A} - \lambda\mathbf{I}| = \begin{vmatrix} 7 - \lambda & -1 & 6 \\ -10 & 4 - \lambda & -12 \\ -2 & 1 & -1 - \lambda \end{vmatrix} = 0.$$

Expanding the determinant and simplifying, we see that the characteristic equation is

$$\lambda^3 - 10\lambda^2 + 31\lambda - 30 = 0,$$

the factored form of which is

$$(\lambda - 2)(\lambda - 3)(\lambda - 5) = 0.$$

We thus see that the characteristic values of $\mathbf{A}$ are

$$\lambda_1 = 2, \quad \lambda_2 = 3, \quad \text{and} \quad \lambda_3 = 5.$$

These are distinct (and real), and so Theorem 7.16 applies. We thus proceed to find characteristic vectors $\boldsymbol{\alpha}^{(1)}, \boldsymbol{\alpha}^{(2)}, \boldsymbol{\alpha}^{(3)}$ corresponding respectively to $\lambda_1, \lambda_2, \lambda_3$. We use the defining equation

$$\mathbf{A}\boldsymbol{\alpha} = \lambda\boldsymbol{\alpha} \tag{7.145}$$

to do this.

For $\lambda = \lambda_1 = 2$ and

$$\boldsymbol{\alpha} = \boldsymbol{\alpha}^{(1)} = \begin{pmatrix} \alpha_1 \\ \alpha_2 \\ \alpha_3 \end{pmatrix},$$

defining equation (7.145) becomes

$$\begin{pmatrix} 7 & -1 & 6 \\ -10 & 4 & -12 \\ -2 & 1 & -1 \end{pmatrix} \begin{pmatrix} \alpha_1 \\ \alpha_2 \\ \alpha_3 \end{pmatrix} = 2 \begin{pmatrix} \alpha_1 \\ \alpha_2 \\ \alpha_3 \end{pmatrix}.$$

Performing the indicated multiplications and equating corresponding components of this, we find

$$7\alpha_1 - \alpha_2 + 6\alpha_3 = 2\alpha_1,$$
$$-10\alpha_1 + 4\alpha_2 - 12\alpha_3 = 2\alpha_2,$$
$$-2\alpha_1 + \alpha_2 - \alpha_3 = 2\alpha_3.$$

Simplifying, we find that $\alpha_1, \alpha_2, \alpha_3$ must satisfy

$$5\alpha_1 - \alpha_2 + 6\alpha_3 = 0,$$
$$-10\alpha_1 + 2\alpha_2 - 12\alpha_3 = 0, \tag{7.152}$$
$$-2\alpha_1 + \alpha_2 - 3\alpha_3 = 0,$$

The second of these three equations is merely a constant multiple of the first. Thus we seek nonzero numbers $\alpha_1, \alpha_2, \alpha_3$ that satisfy the first and third of these equations. Writing these two as equations in the unknowns α_2 and α_3, we have

$$-\alpha_2 + 6\alpha_3 = -5\alpha_1,$$
$$\alpha_2 - 3\alpha_3 = 2\alpha_1.$$

Solving for α_2 and α_3, we find

$$\alpha_2 = -\alpha_1 \quad \text{and} \quad \alpha_3 = -\alpha_1.$$

A simple nontrivial solution of this is $\alpha_1 = 1, \alpha_2 = -1, \alpha_3 = -1$. That is, $\alpha_1 = 1, \alpha_2 = -1, \alpha_3 = -1$ is a simple nontrivial solution of the system (7.152). Thus a characteristic vector corresponding to $\lambda_1 = 2$ is

$$\boldsymbol{\alpha}^{(1)} = \begin{pmatrix} 1 \\ -1 \\ -1 \end{pmatrix}.$$

Then by Theorem 7.16,

$$\mathbf{x} = \begin{pmatrix} 1 \\ -1 \\ -1 \end{pmatrix} e^{2t}, \quad \text{that is,} \quad \begin{pmatrix} e^{2t} \\ -e^{2t} \\ -e^{2t} \end{pmatrix}, \tag{7.153}$$

is a solution of (7.151).

For $\lambda = \lambda_2 = 3$ and

$$\boldsymbol{\alpha} = \boldsymbol{\alpha}^{(2)} = \begin{pmatrix} \alpha_1 \\ \alpha_2 \\ \alpha_3 \end{pmatrix},$$

(7.145) becomes

$$\begin{pmatrix} 7 & -1 & 6 \\ -10 & 4 & -12 \\ -2 & 1 & -1 \end{pmatrix} \begin{pmatrix} \alpha_1 \\ \alpha_2 \\ \alpha_3 \end{pmatrix} = 3 \begin{pmatrix} \alpha_1 \\ \alpha_2 \\ \alpha_3 \end{pmatrix}.$$

Performing the indicated multiplications and equating corresponding components of this and then simplifying, we find that $\alpha_1, \alpha_2, \alpha_3$ must satisfy

$$4\alpha_1 - \alpha_2 + 6\alpha_3 = 0,$$
$$-10\alpha_1 + \alpha_2 - 12\alpha_3 = 0,$$
$$-2\alpha_1 + \alpha_2 - 4\alpha_3 = 0.$$

From these we find that

$$\alpha_2 = -2\alpha_1 \quad \text{and} \quad \alpha_3 = -\alpha_1.$$

A simple nontrivial solution of this is $\alpha_1 = 1, \alpha_2 = -2, \alpha_3 = -1$. Thus a characteristic vector corresponding to $\lambda_2 = 3$ is

$$\boldsymbol{\alpha}^{(2)} = \begin{pmatrix} 1 \\ -2 \\ -1 \end{pmatrix}.$$

Then by Theorem 7.16,

$$\mathbf{x} = \begin{pmatrix} 1 \\ -2 \\ -1 \end{pmatrix} e^{3t}, \quad \text{that is,} \quad \begin{pmatrix} e^{3t} \\ -2e^{3t} \\ -e^{3t} \end{pmatrix}, \tag{7.154}$$

is a solution of (7.151).

For $\lambda = \lambda_3 = 5$ and

$$\alpha = \alpha^{(3)} = \begin{pmatrix} \alpha_1 \\ \alpha_2 \\ \alpha_3 \end{pmatrix},$$

(7.145) becomes

$$\begin{pmatrix} 7 & -1 & 6 \\ -10 & 4 & -12 \\ -2 & 1 & -1 \end{pmatrix} \begin{pmatrix} \alpha_1 \\ \alpha_2 \\ \alpha_3 \end{pmatrix} = 5 \begin{pmatrix} \alpha_1 \\ \alpha_2 \\ \alpha_3 \end{pmatrix}.$$

Performing the indicated multiplications and equating corresponding components of this, and then simplifying, we find that $\alpha_1, \alpha_2, \alpha_3$ must satisfy

$$2\alpha_1 - \alpha_2 + 6\alpha_3 = 0,$$

$$-10\alpha_1 - \alpha_2 - 12\alpha_3 = 0,$$

$$-2\alpha_1 + \alpha_2 - 6\alpha_3 = 0.$$

From these we find that

$$\alpha_2 = -2\alpha_1 \quad \text{and} \quad 3\alpha_3 = -2\alpha_1.$$

A simple nontrivial solution of this is $\alpha_1 = 3, \alpha_2 = -6, \alpha_3 = -2$. Thus a characteristic vector corresponding to $\lambda_3 = 5$ is

$$\alpha^{(3)} = \begin{pmatrix} 3 \\ -6 \\ -2 \end{pmatrix}.$$

Then by Theorem 7.16,

$$\mathbf{x} = \begin{pmatrix} 3 \\ -6 \\ -2 \end{pmatrix} e^{5t}, \quad \text{that is,} \quad \begin{pmatrix} 3e^{5t} \\ -6e^{5t} \\ -2e^{5t} \end{pmatrix} \tag{7.155}$$

is a solution of (7.151).

Also by Theorem 7.16, the solutions (7.153), (7.154), and (7.155) are linearly independent, and a general solution is

$$\mathbf{x} = c_1 \begin{pmatrix} e^{2t} \\ -e^{2t} \\ -e^{2t} \end{pmatrix} + c_2 \begin{pmatrix} e^{3t} \\ -2e^{3t} \\ -e^{3t} \end{pmatrix} + c_3 \begin{pmatrix} 3e^{5t} \\ -6e^{5t} \\ -2e^{5t} \end{pmatrix},$$

where c_1, c_2, and c_3 are arbitrary constants. That is, in scalar language, a general solution of the homogeneous linear system (7.150) is

$$x_1 = c_1 e^{2t} + c_2 e^{3t} + 3c_3 e^{5t},$$

$$x_2 = -c_1 e^{2t} - 2c_2 e^{3t} - 6c_3 e^{5t},$$

$$x_3 = -c_1 e^{2t} - c_2 e^{3t} - 2c_3 e^{5t},$$

where c_1, c_2, and c_3 are arbitrary constants.

We return to the homogeneous linear system (7.136), that is, the vector differential equation

$$\frac{d\mathbf{x}}{dt} = \mathbf{A}\mathbf{x}, \tag{7.139}$$

where $\mathbf{A}$ is an $n \times n$ real constant matrix, and reconsider the result stated in Theorem 7.16. In that theorem we stated that if each of the n characteristic values $\lambda_1, \lambda_2, \ldots, \lambda_n$ of $\mathbf{A}$ is *distinct* and if $\boldsymbol{\alpha}^{(1)}, \boldsymbol{\alpha}^{(2)}, \ldots, \boldsymbol{\alpha}^{(n)}$ is a set of n respective corresponding characteristic vectors of $\mathbf{A}$, then the n functions defined by

$$\boldsymbol{\alpha}^{(1)}e^{\lambda_1 t}, \boldsymbol{\alpha}^{(2)}e^{\lambda_2 t}, \ldots, \boldsymbol{\alpha}^{(n)}e^{\lambda_n t}$$

form a fundamental set of solutions of (7.139). Note that although we assume that $\lambda_1, \lambda_2, \ldots, \lambda_n$ are *distinct*, we do *not* require that they be *real*. Thus distinct *complex* characteristic values may be present. However, since $\mathbf{A}$ is a real matrix, any complex characteristic values must occur in conjugate pairs. Suppose $\lambda_1 = a + bi$ and $\lambda_2 = a - bi$ form such a pair. Then the corresponding solutions are

$$\boldsymbol{\alpha}^{(1)}e^{(a+bi)t} \quad \text{and} \quad \boldsymbol{\alpha}^{(2)}e^{(a-bi)t},$$

and these solutions are *complex* solutions. Thus if one or more distinct conjugate-complex pairs of characteristic values occur, the fundamental set defined by $\boldsymbol{\alpha}^{(i)}e^{\lambda_i t}$, $i = 1, 2, \ldots, n$, contains *complex* functions. However, in such a case, this fundamental set may be replaced by another fundamental set, all of whose members are *real* functions. This is accomplished exactly as explained in Section 7.4C and illustrated in Example 7.18.

C. Case of Repeated Characteristic Values

We again consider the vector differential equation

$$\frac{d\mathbf{x}}{dt} = \mathbf{A}\mathbf{x}, \tag{7.139}$$

where $\mathbf{A}$ is an $n \times n$ real constant matrix; but here we give an introduction to the case in which $\mathbf{A}$ has a repeated characteristic value. To be definite, we suppose that $\mathbf{A}$ has a real characteristic value λ_1 of multiplicity m, where $1 < m \leq n$, and that all the other characteristic values $\lambda_{m+1}, \lambda_{m+2}, \ldots, \lambda_n$ (if there are any) are distinct. By result G of Section 7.5D, we know that the repeated characteristic value λ_1 of multiplicity m has p linearly independent characteristic vectors, where $1 \leq p \leq m$. Now consider two subcases: (1) $p = m$; and (2) $p < m$.

In Subcase (1), $p = m$, there are m linearly independent characteristic vectors $\boldsymbol{\alpha}^{(1)}, \boldsymbol{\alpha}^{(2)}, \ldots, \boldsymbol{\alpha}^{(m)}$ corresponding to the characteristic value λ_1 of multiplicity m. Then the n functions defined by

$$\boldsymbol{\alpha}^{(1)}e^{\lambda_1 t}, \boldsymbol{\alpha}^{(2)}e^{\lambda_1 t}, \ldots, \boldsymbol{\alpha}^{(m)}e^{\lambda_1 t}, \boldsymbol{\alpha}^{(m+1)}e^{\lambda_{m+1} t}, \ldots, \boldsymbol{\alpha}^{(n)}e^{\lambda_n t}$$

form a linearly independent set of n solutions of differential equation (7.139); and a general solution of (7.139) is a linear combination of these n solutions having n arbitrary numbers as the "constants of combination."

▶ **Example 7.39**

Consider the homogeneous linear system

$$\frac{dx_1}{dt} = 3x_1 + x_2 - x_3,$$

$$\frac{dx_2}{dt} = x_1 + 3x_2 - x_3, \tag{7.156}$$

$$\frac{dx_3}{dt} = 3x_1 + 3x_2 - x_3,$$

or in matrix form,

$$\frac{d\mathbf{x}}{dt} = \begin{pmatrix} 3 & 1 & -1 \\ 1 & 3 & -1 \\ 3 & 3 & -1 \end{pmatrix} \mathbf{x}, \quad \text{where} \quad \mathbf{x} = \begin{pmatrix} x_1 \\ x_2 \\ x_3 \end{pmatrix}. \tag{7.157}$$

Assuming a solution of the form

$$\mathbf{x} = \boldsymbol{\alpha} e^{\lambda t},$$

that is,

$$x_1 = \alpha_1 e^{\lambda t}, \quad x_2 = \alpha_2 e^{\lambda t}, \quad x_3 = \alpha_3 e^{\lambda t},$$

we know that λ must be a solution of the characteristic equation of the coefficient matrix

$$\mathbf{A} = \begin{pmatrix} 3 & 1 & -1 \\ 1 & 3 & -1 \\ 3 & 3 & -1 \end{pmatrix}.$$

This characteristic equation is

$$|\mathbf{A} - \lambda \mathbf{I}| = \begin{vmatrix} 3 - \lambda & 1 & -1 \\ 1 & 3 - \lambda & -1 \\ 3 & 3 & -1 - \lambda \end{vmatrix} = 0.$$

Expanding the determinant and simplifying, we see that the characteristic equation is

$$\lambda^3 - 5\lambda^2 + 8\lambda - 4 = 0,$$

the factored form of which is

$$(\lambda - 1)(\lambda - 2)(\lambda - 2) = 0.$$

We thus see that the characteristic values of **A** are

$$\lambda_1 = 1, \quad \lambda_2 = 2 \quad \text{and} \quad \lambda_3 = 2.$$

Note that whereas the number 1 is a *distinct* characteristic value of **A**, the number 2 is a *repeated* characteristic value. We again use

$$\mathbf{A}\boldsymbol{\alpha} = \lambda\boldsymbol{\alpha} \tag{7.145}$$

to find characteristic vectors corresponding to these characteristic values.

For $\lambda = 1$, and

$$\boldsymbol{\alpha} = \begin{pmatrix} \alpha_1 \\ \alpha_2 \\ \alpha_3 \end{pmatrix},$$

(7.145) becomes

$$\begin{pmatrix} 3 & 1 & -1 \\ 1 & 3 & -1 \\ 3 & 3 & -1 \end{pmatrix} \begin{pmatrix} \alpha_1 \\ \alpha_2 \\ \alpha_3 \end{pmatrix} = 1 \begin{pmatrix} \alpha_1 \\ \alpha_2 \\ \alpha_3 \end{pmatrix}.$$

Performing the indicated multiplications and equating corresponding components of this and then simplifying, we find that α_1, α_2, α_3 must be a nontrivial solution of the system

$$2\alpha_1 + \alpha_2 - \alpha_3 = 0,$$
$$\alpha_1 + 2\alpha_2 - \alpha_3 = 0,$$
$$3\alpha_1 + 3\alpha_2 - 2\alpha_3 = 0.$$

One readily sees that such a solution is given by

$$\alpha_1 = 1, \quad \alpha_2 = 1, \quad \alpha_3 = 3.$$

Thus a characteristic vector corresponding to $\lambda_1 = 1$ is

$$\boldsymbol{\alpha}^{(1)} = \begin{pmatrix} 1 \\ 1 \\ 3 \end{pmatrix}.$$

Then

$$\mathbf{x} = \begin{pmatrix} 1 \\ 1 \\ 3 \end{pmatrix} e^t, \quad \text{that is,} \quad \begin{pmatrix} e^t \\ e^t \\ 3e^t \end{pmatrix}, \tag{7.158}$$

is a solution of (7.157).

We now turn to the repeated characteristic value $\lambda_2 = \lambda_3 = 2$. In terms of the discussion just preceding this example, this characteristic value 2 has multiplicity $m = 2 < 3 = n$, where $n = 3$ is the common number of rows and columns of the coefficient matrix $\mathbf{A}$. For $\lambda = 2$ and

$$\boldsymbol{\alpha} = \begin{pmatrix} \alpha_1 \\ \alpha_2 \\ \alpha_3 \end{pmatrix}, \tag{7.159}$$

(7.145) becomes

$$\begin{pmatrix} 3 & 1 & -1 \\ 1 & 3 & -1 \\ 3 & 3 & -1 \end{pmatrix} \begin{pmatrix} \alpha_1 \\ \alpha_2 \\ \alpha_3 \end{pmatrix} = 2 \begin{pmatrix} \alpha_1 \\ \alpha_2 \\ \alpha_3 \end{pmatrix}.$$

Equating corresponding coefficients of this and simplifying, we find that $\alpha_1, \alpha_2, \alpha_3$ must be a nontrivial solution of the system

$$\alpha_1 + \alpha_2 - \alpha_3 = 0,$$
$$\alpha_1 + \alpha_2 - \alpha_3 = 0,$$
$$3\alpha_1 + 3\alpha_2 - 3\alpha_3 = 0.$$

Note that each of these three relations is equivalent to each of the other two, and so the only relationship among $\alpha_1, \alpha_2, \alpha_3$ is that given most simply by

$$\alpha_1 + \alpha_2 - \alpha_3 = 0. \tag{7.160}$$

Observe that

$$\alpha_1 = 1, \quad \alpha_2 = -1, \quad \alpha_3 = 0$$

and

$$\alpha_1 = 1, \quad \alpha_2 = 0, \quad \alpha_3 = 1$$

are two distinct solutions of this relation (7.160). The corresponding vectors of the form (7.159) are thus

$$\boldsymbol{\alpha}^{(2)} = \begin{pmatrix} 1 \\ -1 \\ 0 \end{pmatrix} \quad \text{and} \mid \boldsymbol{\alpha}^{(3)} = \begin{pmatrix} 1 \\ 0 \\ 1 \end{pmatrix},$$

respectively. Since each satisfies (7.145) with $\lambda = 2$, each is a characteristic vector corresponding to the double root $\lambda_2 = \lambda_3 = 2$. Furthermore, using the definition of linear independence of a set of constant vectors, one sees that these vectors $\boldsymbol{\alpha}^{(2)}$ and $\boldsymbol{\alpha}^{(3)}$ are linearly independent. Thus the characteristic value $\lambda = 2$ of multiplicity $m = 2$ has the $p = 2$ linearly independent characteristic vectors

$$\boldsymbol{\alpha}^{(2)} = \begin{pmatrix} 1 \\ -1 \\ 0 \end{pmatrix} \quad \text{and} \quad \boldsymbol{\alpha}^{(3)} = \begin{pmatrix} 1 \\ 0 \\ 1 \end{pmatrix}$$

corresponding to it. Hence this is an illustration of Subcase (1) of the discussion preceding this example. Thus, corresponding to the twofold characteristic value $\lambda = 2$, there are two linearly independent solutions of system (7.157) of the form $\boldsymbol{\alpha}e^{\lambda t}$. These are

$$\boldsymbol{\alpha}^{(2)}e^{2t} \quad \text{and} \quad \boldsymbol{\alpha}^{(3)}e^{2t},$$

that is,

$$\begin{pmatrix} 1 \\ -1 \\ 0 \end{pmatrix} e^{2t} \quad \text{and} \quad \begin{pmatrix} 1 \\ 0 \\ 1 \end{pmatrix} e^{2t},$$

or

$$\begin{pmatrix} e^{2t} \\ -e^{2t} \\ 0 \end{pmatrix} \quad \text{and} \quad \begin{pmatrix} e^{2t} \\ 0 \\ e^{2t} \end{pmatrix}, \tag{7.161}$$

respectively.

The three solutions

$$\begin{pmatrix} e^t \\ e^t \\ 3e^t \end{pmatrix}, \quad \begin{pmatrix} e^{2t} \\ -e^{2t} \\ 0 \end{pmatrix}, \quad \text{and} \quad \begin{pmatrix} e^{2t} \\ 0 \\ e^{2t} \end{pmatrix}$$

given by (7.158) and (7.161) are linearly independent, and a general solution is

$$\mathbf{x} = c_1 \begin{pmatrix} e^t \\ e^t \\ 3e^t \end{pmatrix} + c_2 \begin{pmatrix} e^{2t} \\ -e^{2t} \\ 0 \end{pmatrix} + c_3 \begin{pmatrix} e^{2t} \\ 0 \\ e^{2t} \end{pmatrix},$$

where c_1, c_2, and c_3 are arbitrary constants. That is, in scalar language, a general solution of the homogeneous linear system (7.156) is

$$x_1 = c_1 e^t + (c_2 + c_3)e^{2t},$$

$$x_2 = c_1 e^t - c_2 e^{2t},$$

$$x_3 = 3c_1 e^t + c_3 e^{2t}.$$

where c_1, c_2, and c_3 are arbitrary numbers.

One type of vector differential equation (7.139) which *always* leads to Subcase (1), $p = m$, in the case of a repeated characteristic value λ_1 is that in which the $n \times n$ coefficient matrix $\mathbf{A}$ of (7.139) is a real symmetric matrix. For then, by Result J of Section 7.5D, there always exist n linearly independent characteristic vectors of $\mathbf{A}$, regardless of whether the n characteristic values of $\mathbf{A}$ are all distinct or not.

We now turn to a consideration of Subcase (2), $p < m$. In this case, there are less than m linearly independent characteristic vectors $\boldsymbol{\alpha}^{(1)}$ corresponding to the characteristic value λ_1 of multiplicity m. Hence there are less than m linearly independent solutions of system (7.136) of the form $\boldsymbol{\alpha}^{(1)}e^{\lambda_1 t}$ corresponding to λ_1. Thus there is *not* a full set of n linearly independent solutions of (7.136) of the basic exponential form $\boldsymbol{\alpha}^{(k)}e^{\lambda_k t}$, where λ_k is a characteristic value of $\mathbf{A}$ and $\boldsymbol{\alpha}^{(k)}$ is a characteristic vector corresponding to λ_k. Clearly we must seek linearly independent solutions of another form.

To discover what other forms of solution to seek, we first look back at the analogous situation in Section 7.6C. The results there suggest the following:

Let λ be a characteristic value of multiplicity $m = 2$. Suppose $p = 1 < m$, so that there is only one type of characteristic vector $\boldsymbol{\alpha}$ and hence only one type of solution of the basic exponential form $\boldsymbol{\alpha}e^{\lambda t}$ corresponding to λ. Then a linearly independent solution is of the form

$$(\boldsymbol{\alpha}t + \boldsymbol{\beta})e^{\lambda t},$$

where $\boldsymbol{\alpha}$ is a characteristic vector corresponding to λ, that is, $\boldsymbol{\alpha}$ satisfies

$$(\mathbf{A} - \lambda\mathbf{I})\boldsymbol{\alpha} = \mathbf{0};$$

and $\boldsymbol{\beta}$ is a vector which satisfies the equation

$$(\mathbf{A} - \lambda\mathbf{I})\boldsymbol{\beta} = \boldsymbol{\alpha}.$$

Now let λ be a characteristic value of multiplicity $m = 3$ and suppose $p < m$. Here there are two possibilities: $p = 1$ and $p = 2$.

If $p = 1$, there is only one type of characteristic vector $\boldsymbol{\alpha}$ and hence only one type of solution of the form

$$\boldsymbol{\alpha}e^{\lambda t} \tag{7.162}$$

corresponding to λ. Then a second solution corresponding to λ is of the form

$$(\boldsymbol{\alpha}t + \boldsymbol{\beta})e^{\lambda t}, \tag{7.163}$$

where $\boldsymbol{\alpha}$ is a characteristic value corresponding to λ, that is, $\boldsymbol{\alpha}$ satisfies

$$(\mathbf{A} - \lambda\mathbf{I})\boldsymbol{\alpha} = \mathbf{0}; \tag{7.164}$$

and $\boldsymbol{\beta}$ is a vector which satisfies the equation

$$(\mathbf{A} - \lambda\mathbf{I})\boldsymbol{\beta} = \boldsymbol{\alpha}. \tag{7.165}$$

In this case, a third solution corresponding to λ is of the form

$$\left(\boldsymbol{\alpha}\frac{t^2}{2!} + \boldsymbol{\beta}t + \boldsymbol{\gamma}\right)e^{\lambda t}, \tag{7.166}$$

where $\boldsymbol{\alpha}$ satisfies (7.164), $\boldsymbol{\beta}$ satisfies (7.165), and $\boldsymbol{\gamma}$ satisfies

$$(\mathbf{A} - \lambda\mathbf{I})\boldsymbol{\gamma} = \boldsymbol{\beta}. \tag{7.167}$$

The three solutions (7.162), (7.163), and (7.166) so found are linearly independent.

If $p = 2$, there are two linearly independent characteristic vectors $\boldsymbol{\alpha}^{(1)}$ and $\boldsymbol{\alpha}^{(2)}$ corresponding to λ and hence there are two linearly independent solutions of the form

$$\boldsymbol{\alpha}^{(1)}e^{\lambda t} \quad \text{and} \quad \boldsymbol{\alpha}^{(2)}e^{\lambda t}. \tag{7.168}$$

Then a third solution corresponding to λ is of the form

$$(\boldsymbol{\alpha}t + \boldsymbol{\beta})e^{\lambda t}, \tag{7.169}$$

where $\boldsymbol{\alpha}$ satisfies

$$(\mathbf{A} - \lambda\mathbf{I})\boldsymbol{\alpha} = \mathbf{0}, \tag{7.170}$$

and $\boldsymbol{\beta}$ satisfies

$$(\mathbf{A} - \lambda\mathbf{I})\boldsymbol{\beta} = \boldsymbol{\alpha}. \tag{7.171}$$

Now we must be careful here. Let us explain: Since $\boldsymbol{\alpha}^{(1)}$ and $\boldsymbol{\alpha}^{(2)}$ are both characteristic vectors corresponding to λ, both $\boldsymbol{\alpha} = \boldsymbol{\alpha}^{(1)}$ and $\boldsymbol{\alpha} = \boldsymbol{\alpha}^{(2)}$ satisfy (7.170). However, in general, neither of these values of $\boldsymbol{\alpha}$ will be such that the resulting equation (7.171) in $\boldsymbol{\beta}$ will have a nontrivial solution for $\boldsymbol{\beta}$. Thus, instead of using the simple solutions $\boldsymbol{\alpha}^{(1)}$ or $\boldsymbol{\alpha}^{(2)}$ of (7.170), a more general solution of that equation is needed. Such a solution is provided by

$$\boldsymbol{\alpha} = k_1\boldsymbol{\alpha}^{(1)} + k_2\boldsymbol{\alpha}^{(2)}, \tag{7.172}$$

where k_1 and k_2 are suitable constants. We now substitute (7.172) for $\boldsymbol{\alpha}$ in (7.171) and determine k_1 and k_2 so that the resulting equation in $\boldsymbol{\beta}$ *will have* a nontrivial solution for $\boldsymbol{\beta}$. With these values chosen for k_1 and k_2, we thus have the required $\boldsymbol{\alpha}$ and now find the desired nontrivial $\boldsymbol{\beta}$. The three resulting solutions (7.168) and (7.169) thus determined are linearly independent. We illustrate this situation in the following example.

▶ **Example 7.40**

Consider the homogeneous linear system

$$\frac{dx_1}{dt} = 4x_1 + 3x_2 + x_3,$$

$$\frac{dx_2}{dt} = -4x_1 - 4x_2 - 2x_3, \tag{7.173}$$

$$\frac{dx_3}{dt} = 8x_1 + 12x_2 + 6x_3,$$

or in matrix form,

$$\frac{dx}{dt} = \begin{pmatrix} 4 & 3 & 1 \\ -4 & -4 & -2 \\ 8 & 12 & 6 \end{pmatrix} \mathbf{x}, \quad \text{where} \quad \mathbf{x} = \begin{pmatrix} x_1 \\ x_2 \\ x_3 \end{pmatrix}. \tag{7.174}$$

Assuming a solution of the form

$$\mathbf{x} = \boldsymbol{\alpha} e^{\lambda t},$$

that is,

$$x_1 = \alpha_1 e^{\lambda t}, \quad x_2 = \alpha_2 e^{\lambda t}, \quad x_3 = \alpha_3 e^{\lambda t},$$

we know that λ must be a solution of the characteristic equation of the coefficient matrix

$$\mathbf{A} = \begin{pmatrix} 4 & 3 & 1 \\ -4 & -4 & -2 \\ 8 & 12 & 6 \end{pmatrix}.$$

This characteristic equation is

$$|\mathbf{A} - \lambda\mathbf{I}| = \begin{vmatrix} 4 - \lambda & 3 & 1 \\ -4 & -4 - \lambda & -2 \\ 8 & 12 & 6 - \lambda \end{vmatrix} = 0.$$

Expanding the determinant and simplifying, we see that the characteristic equation is

$$\lambda^3 - 6\lambda^2 + 12\lambda - 8 = 0,$$

the factored form of which is $(\lambda - 2)^3 = 0$. We thus see that the characteristic values of **A** are

$$\lambda_1 = \lambda_2 = \lambda_3 = 2.$$

That is, the number 2 is a *triple* characteristic value of **A**. We again use

$$\mathbf{A}\boldsymbol{\alpha} = \lambda\boldsymbol{\alpha} \tag{7.145}$$

to find the corresponding characteristic vector(s) $\boldsymbol{\alpha}$.

With $\lambda = 2$ and

$$\boldsymbol{\alpha} = \begin{pmatrix} \alpha_1 \\ \alpha_2 \\ \alpha_3 \end{pmatrix}, \tag{7.175}$$

(7.145) becomes

$$\begin{pmatrix} 4 & 3 & 1 \\ -4 & -4 & -2 \\ 8 & 12 & 6 \end{pmatrix} \begin{pmatrix} \alpha_1 \\ \alpha_2 \\ \alpha_3 \end{pmatrix} = 2 \begin{pmatrix} \alpha_1 \\ \alpha_2 \\ \alpha_3 \end{pmatrix}.$$

Performing the indicated multiplications and equating corresponding components of this and then simplifying, we find that $\alpha_1, \alpha_2, \alpha_3$ must be a nontrivial solution of the system

$$2\alpha_1 + 3\alpha_2 + \alpha_3 = 0,$$
$$-4\alpha_1 - 6\alpha_2 - 2\alpha_3 = 0,$$
$$8\alpha_1 + 12\alpha_2 + 4\alpha_3 = 0.$$

Each of these three relationships is equivalent to each of the other two, and so the only relationship among $\alpha_1, \alpha_2, \alpha_3$ is that given most simply by

$$2\alpha_1 + 3\alpha_2 + \alpha_3 = 0. \tag{7.176}$$

Observe that

$$\alpha_1 = 1, \quad \alpha_2 = 0, \quad \alpha_3 = -2$$

and

$$\alpha_1 = 0, \quad \alpha_2 = 1, \quad \alpha_3 = -3$$

are two distinct solutions of relation (7.176). The corresponding vectors of the form (7.175) are thus

$$\boldsymbol{\alpha}^{(1)} = \begin{pmatrix} 1 \\ 0 \\ -2 \end{pmatrix} \quad \text{and} \quad \boldsymbol{\alpha}^{(2)} = \begin{pmatrix} 0 \\ 1 \\ -3 \end{pmatrix},$$

respectively. Since each satisfies (7.145) with $\lambda = 2$, each is a characteristic vector corresponding to the triple characteristic value 2. Furthermore, it is easy to see that the two characteristic vectors $\boldsymbol{\alpha}^{(1)}$ and $\boldsymbol{\alpha}^{(2)}$ are linearly independent, whereas every set of three characteristic vectors corresponding to characteristic value 2 are linearly dependent. Thus the characteristic value $\lambda = 2$ of multiplicity $m = 3$ has the $p = 2$ linearly independent characteristic vectors

$$\boldsymbol{\alpha}^{(1)} = \begin{pmatrix} 1 \\ 0 \\ -2 \end{pmatrix} \quad \text{and} \quad \boldsymbol{\alpha}^{(2)} = \begin{pmatrix} 0 \\ 1 \\ -3 \end{pmatrix} \tag{7.177}$$

corresponding to it. Hence this is an illustration of the situation described in the paragraph immediately preceding this example. Thus corresponding to the triple characteristic value $\lambda = 2$ there are two linearly independent solutions of system (7.173) of the form $\boldsymbol{\alpha} e^{\lambda t}$. These are $\boldsymbol{\alpha}^{(1)} e^{2t}$ and $\boldsymbol{\alpha}^{(2)} e^{2t}$, that is,

$$\begin{pmatrix} 1 \\ 0 \\ -2 \end{pmatrix} e^{2t} \quad \text{and} \quad \begin{pmatrix} 0 \\ 1 \\ -3 \end{pmatrix} e^{2t},$$

or

$$\begin{pmatrix} e^{2t} \\ 0 \\ -2e^{2t} \end{pmatrix} \quad \text{and} \quad \begin{pmatrix} 0 \\ e^{2t} \\ -3e^{2t} \end{pmatrix}, \tag{7.178}$$

respectively.

A third solution corresponding to $\lambda = 2$ is of the form

$$(\alpha t + \beta)e^{2t}, \tag{7.179}$$

where α satisfies

$$(\mathbf{A} - 2\mathbf{I})\alpha = \mathbf{0} \tag{7.180}$$

and β satisfies

$$(\mathbf{A} - 2\mathbf{I})\beta = \alpha. \tag{7.181}$$

Since both $\alpha^{(1)}$ and $\alpha^{(2)}$ given by (7.177) are characteristic vectors of $\mathbf{A}$ corresponding to $\lambda = 2$, they both satisfy (7.180). But, as noted in the paragraph immediately preceding this example, we need to use the more general solution

$$\alpha = k_1\alpha^{(1)} + k_2\alpha^{(2)}$$

of (7.180) in order to obtain a nontrivial solution for β in (7.181).

Thus we let

$$\alpha = k_1\alpha^{(1)} + k_2\alpha^{(2)} = \begin{pmatrix} k_1 \\ k_2 \\ -2k_1 - 3k_2 \end{pmatrix} \quad \text{and} \quad \beta = \begin{pmatrix} \beta_1 \\ \beta_2 \\ \beta_3 \end{pmatrix},$$

and then (7.181) becomes

$$\begin{pmatrix} 2 & 3 & 1 \\ -4 & -6 & -2 \\ 8 & 12 & 4 \end{pmatrix}\begin{pmatrix} \beta_1 \\ \beta_2 \\ \beta_3 \end{pmatrix} = \begin{pmatrix} k_1 \\ k_2 \\ -2k_1 - 3k_2 \end{pmatrix}.$$

Performing the indicated multiplications and equating corresponding components of this, we obtain

$$\begin{aligned} 2\beta_1 + 3\beta_2 + \beta_3 &= k_1, \\ -4\beta_1 - 6\beta_2 - 2\beta_3 &= k_2, \\ 8\beta_1 + 12\beta_2 + 4\beta_3 &= -2k_1 - 3k_2. \end{aligned} \tag{7.182}$$

Observe that the left members of these three relations are all proportional to one another. Using any two of the relations, we find that $k_2 = -2k_1$. A simple nontrivial solution of this last relation is $k_1 = 1, k_2 = -2$. With this choice of k_1 and k_2, we find

$$\alpha = \begin{pmatrix} 1 \\ -2 \\ 4 \end{pmatrix}; \tag{7.183}$$

and the relations (7.182) become

$$
\begin{aligned}
2\beta_1 + 3\beta_2 + \beta_3 &= 1, \\
-4\beta_1 - 6\beta_2 - 2\beta_3 &= -2, \\
8\beta_1 + 12\beta_2 + 4\beta_3 &= 4.
\end{aligned}
$$

Each of these is equivalent to

$$
2\beta_1 + 3\beta_2 + \beta_3 = 1.
$$

A nontrivial solution of this is

$$
\beta_1 = \beta_2 = 0, \quad \beta_3 = 1;
$$

and thus we obtain

$$
\boldsymbol{\beta} = \begin{pmatrix} 0 \\ 0 \\ 1 \end{pmatrix}. \tag{7.184}
$$

Therefore, with $\boldsymbol{\alpha}$ given by (7.183) and $\boldsymbol{\beta}$ given by (7.184), the third solution (7.179) is

$$
\left[\begin{pmatrix} 1 \\ -2 \\ 4 \end{pmatrix} t + \begin{pmatrix} 0 \\ 0 \\ 1 \end{pmatrix} \right] e^{2t},
$$

that is,

$$
\begin{pmatrix} te^{2t} \\ -2te^{2t} \\ (4t+1)e^{2t} \end{pmatrix} \tag{7.185}
$$

The three solutions defined by (7.178) and (7.185) are linearly independent, and a general solution is the linear combination

$$
c_1 \begin{pmatrix} e^{2t} \\ 0 \\ -2e^{2t} \end{pmatrix} + c_2 \begin{pmatrix} 0 \\ e^{2t} \\ -3e^{2t} \end{pmatrix} + c_3 \begin{pmatrix} te^{2t} \\ -2te^{2t} \\ (4t+1)e^{2t} \end{pmatrix}
$$

of these three, where c_1, c_2, c_3 are arbitrary constants. That is, in component form, a general solution of system (7.173) is

$$
\begin{aligned}
x_1 &= c_1 e^{2t} + c_3 t e^{2t}, \\
x_2 &= c_2 e^{2t} - 2c_3 t e^{2t}, \\
x_3 &= -2c_1 e^{2t} - 3c_2 e^{2t} + c_3(4t+1)e^{2t},
\end{aligned}
$$

where c_1, c_2, c_3 are arbitrary constants.

Exercises

Find the general solution of each of the homogeneous linear systems in Exercises 1–24, where in each exercise

$$
\mathbf{x} = \begin{pmatrix} x_1 \\ x_2 \\ x_3 \end{pmatrix}.
$$

1. $\dfrac{d\mathbf{x}}{dt} = \begin{pmatrix} 1 & 1 & -1 \\ 2 & 3 & -4 \\ 4 & 1 & -4 \end{pmatrix} \mathbf{x}.$

2. $\dfrac{d\mathbf{x}}{dt} = \begin{pmatrix} 1 & -1 & -1 \\ 1 & 3 & 1 \\ -3 & -6 & 6 \end{pmatrix} \mathbf{x}.$

3. $\dfrac{d\mathbf{x}}{dt} = \begin{pmatrix} 1 & -1 & -1 \\ 1 & 3 & 1 \\ -3 & 1 & -1 \end{pmatrix} \mathbf{x}.$

4. $\dfrac{d\mathbf{x}}{dt} = \begin{pmatrix} 1 & 3 & -6 \\ 0 & 2 & 2 \\ 0 & -1 & 5 \end{pmatrix} \mathbf{x}.$

5. $\dfrac{d\mathbf{x}}{dt} = \begin{pmatrix} 1 & 2 & 2 \\ 2 & 0 & 3 \\ 2 & 3 & 0 \end{pmatrix} \mathbf{x}.$

6. $\dfrac{d\mathbf{x}}{dt} = \begin{pmatrix} 1 & 1 & 0 \\ 1 & 0 & 1 \\ 0 & 1 & 1 \end{pmatrix} \mathbf{x}.$

7. $\dfrac{d\mathbf{x}}{dt} = \begin{pmatrix} 1 & -2 & 0 \\ -2 & 3 & 0 \\ 0 & 0 & 2 \end{pmatrix} \mathbf{x}.$

8. $\dfrac{d\mathbf{x}}{dt} = \begin{pmatrix} -1 & 3 & -3 \\ -3 & 5 & -3 \\ 3 & 3 & -7 \end{pmatrix} \mathbf{x}.$

9. $\dfrac{d\mathbf{x}}{dt} = \begin{pmatrix} 7 & 0 & 4 \\ 8 & 3 & 8 \\ -8 & 0 & -5 \end{pmatrix} \mathbf{x}.$

10. $\dfrac{d\mathbf{x}}{dt} = \begin{pmatrix} \frac{7}{5} & \frac{2}{5} & -\frac{1}{5} \\ \frac{4}{5} & \frac{9}{5} & -\frac{2}{5} \\ \frac{2}{5} & \frac{2}{5} & \frac{4}{5} \end{pmatrix} \mathbf{x}.$

11. $\dfrac{d\mathbf{x}}{dt} = \begin{pmatrix} 1 & -3 & 9 \\ 0 & -5 & 18 \\ 0 & -3 & 10 \end{pmatrix} \mathbf{x}.$

12. $\dfrac{d\mathbf{x}}{dt} = \begin{pmatrix} 3 & \frac{2}{7} & -\frac{4}{7} \\ 0 & \frac{19}{7} & \frac{4}{7} \\ 0 & \frac{6}{7} & \frac{9}{7} \end{pmatrix} \mathbf{x}.$

13. $\dfrac{d\mathbf{x}}{dt} = \begin{pmatrix} 11 & 6 & 18 \\ 9 & 8 & 18 \\ -9 & -6 & -16 \end{pmatrix} \mathbf{x}.$

14. $\dfrac{d\mathbf{x}}{dt} = \begin{pmatrix} 1 & 9 & 9 \\ 0 & 19 & 18 \\ 0 & 9 & 10 \end{pmatrix} \mathbf{x}.$

15. $\dfrac{d\mathbf{x}}{dt} = \begin{pmatrix} -5 & -12 & 6 \\ 1 & 5 & -1 \\ -7 & -10 & 8 \end{pmatrix} \mathbf{x}.$

16. $\dfrac{d\mathbf{x}}{dt} = \begin{pmatrix} -2 & 5 & 5 \\ -1 & 4 & 5 \\ 3 & -3 & 2 \end{pmatrix} \mathbf{x}.$

17. $\dfrac{d\mathbf{x}}{dt} = \begin{pmatrix} -5 & -3 & -3 \\ 8 & 5 & 7 \\ -2 & -1 & -3 \end{pmatrix} \mathbf{x}.$

18. $\dfrac{d\mathbf{x}}{dt} = \begin{pmatrix} 4 & 2 & 1 \\ -4 & -3 & -4 \\ 1 & 1 & 4 \end{pmatrix} \mathbf{x}.$

19. $\dfrac{d\mathbf{x}}{dt} = \begin{pmatrix} 7 & 4 & 4 \\ -6 & -4 & -7 \\ -2 & -1 & 2 \end{pmatrix} \mathbf{x}.$

20. $\dfrac{d\mathbf{x}}{dt} = \begin{pmatrix} 1 & 4 & 5 \\ 2 & 7 & 9 \\ -4 & -4 & -7 \end{pmatrix} \mathbf{x}.$

21. $\dfrac{d\mathbf{x}}{dt} = \begin{pmatrix} 4 & 3 & 1 \\ -4 & -4 & -2 \\ 8 & 12 & 6 \end{pmatrix} \mathbf{x}.$

22. $\dfrac{d\mathbf{x}}{dt} = \begin{pmatrix} 4 & -1 & -1 \\ 2 & 1 & -1 \\ 2 & -1 & 1 \end{pmatrix} \mathbf{x}.$

23. $\dfrac{d\mathbf{x}}{dt} = \begin{pmatrix} 8 & 12 & -2 \\ -3 & -4 & 1 \\ -1 & -2 & 2 \end{pmatrix} \mathbf{x}.$

24. $\dfrac{d\mathbf{x}}{dt} = \begin{pmatrix} 4 & 6 & -1 \\ -1 & -2 & 1 \\ -2 & -8 & 4 \end{pmatrix} \mathbf{x}.$

CHAPTER EIGHT

Approximate Methods of Solving First-Order Equations

In Chapter 2 we considered certain special types of first-order differential equations having closed-form solutions that can be obtained exactly. For a first-order differential equation that is not of one or another of these special types, it usually is not apparent how one should proceed in an attempt to obtain a solution exactly. Indeed, in most such cases the discovery of an exact closed-form solution in terms of elementary functions would be an unexpected luxury! Therefore, one considers the possibilities of obtaining approximate solutions of first-order differential equations. In this chapter we shall introduce several approximate methods. In the study of each method in this chapter our primary concern will be to obtain familiarity with the procedure itself and to develop skill in applying it. In general we shall not be concerned here with theoretical justifications and extended discussions of accuracy and error. We shall leave such matters, important as they are, to more specialized and advanced treatises and instead shall concentrate on the formal details of the various procedures.

8.1 GRAPHICAL METHODS

A. Line Elements and Direction Fields

In Chapter 1 we considered briefly the geometric significance of the first-order differential equation

$$\frac{dy}{dx} = f(x, y), \tag{8.1}$$

where f is a real function of x and y. The explicit solutions of (8.1) are certain real functions, and the graphs of these solution functions are curves in the xy plane called the *integral curves* of (8.1). At each point (x, y) at which $f(x, y)$ is defined, the differential equation (8.1) defines the slope $f(x, y)$ at the point (x, y) of the integral curve of (8.1) that passes through this point. Thus we may construct the tangent to an integral curve of (8.1) at a given point (x, y) without actually knowing the solution function of which this integral curve is the graph.

We proceed to do this. Through the point (x, y) we draw a short segment of the tangent to the integral curve of (8.1) that passes through this point. That is, through (x, y) we construct a short segment the slope of which is $f(x, y)$, as given by the differential equation (8.1). Such a segment is called a *line element* of the differential equation (8.1).

For example, let us consider the differential equation

$$\frac{dy}{dx} = 2x + y. \tag{8.2}$$

Here $f(x, y) = 2x + y$, and the slope of the integral curve of (8.2) that passes through the point $(1, 2)$ has at this point the value

$$f(1, 2) = 4.$$

Thus through the point $(1, 2)$ we construct a short segment of slope 4 or, in other words, of angle of inclination approximately $76°$ (see Figure 8.1). This short segment is the line element of the differential equation (8.2) at the point $(1, 2)$. It is tangent to the integral curve of (8.2) which passes through this point.

Let us now return to the general equation (8.1). A line element of (8.1) can be constructed at every point (x, y) at which $f(x, y)$ in (8.1) is defined. Doing so for a selection of different points (x, y) leads to a configuration of selected line elements that indicates the directions of the integral curves at the various selected points. We shall refer to such a configuration as a *line element configuration*.

For each point (x, y) at which $f(x, y)$ is defined, the differential equation (8.1) thus defines a line segment with slope $f(x, y)$, or, in other words, a direction. Each such point, taken together with the corresponding direction so defined, constitutes the so-called *direction field* of the differential equation (8.1). We say that the differential equation (8.1) defines this direction field, and this direction field is represented graphically by a line element configuration. Clearly a more thorough and carefully constructed line element configuration gives a more accurate graphical representation of the direction field.

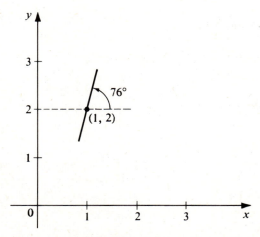

Figure 8.1   Line element of differential equation (8.2) at point (1, 2).

For a given differential equation of the form (8.1), let us assume that a "thorough and carefully constructed" line element configuration has been drawn. That is, we assume that line elements have been carefully constructed at a relatively large number of carefully chosen points. Then this resulting line element configuration will indicate the presence of a family of curves tangent to the various line elements constructed at the different points. This indicated family of curves is approximately the family of integral curves of the given differential equation. Actual smooth curves drawn tangent to the line elements as the configuration indicates will thus provide approximate graphs of the true integral curves.

Thus the construction of the line element configuration provides a procedure for approximately obtaining the solution of the differential equation in graphical form. We now summarize this basic graphical procedure and illustrate it with a simple example.

Summary of Basic Graphical Procedure

1. Carefully construct a line element configuration, proceeding until the family of "approximate integral curves" begins to appear.
2. Draw smooth curves as indicated by the configuration constructed in Step 1.

▶ **Example 8.1**

Construct a line element configuration for the differential equation

$$\frac{dy}{dx} = 2x + y,$$ (8.2)

and use this configuration to sketch the approximate integral curves.

Solution. The slope of the exact integral curve of (8.2) at any point (x, y) is given by

$$f(x, y) = 2x + y.$$

We evaluate this slope at a number of selected points and so determine the approximate inclination of the corresponding line element at each point selected. We then construct the line elements so determined. From the resulting configuration we sketch several of the approximate integral curves. A few typical inclinations are listed in Table 8.1 and the completed configuration with the approximate integral curves appears in Figure 8.2.

TABLE 8.1

x	y	dy/dx (Slope)	Approximate inclination of line element
$\frac{1}{2}$	$-\frac{1}{2}$	$\frac{1}{2}$	$27°$
$\frac{1}{2}$	0	1	$45°$
$\frac{1}{2}$	$\frac{1}{2}$	$\frac{3}{2}$	$56°$
$\frac{1}{2}$	1	2	$63°$
1	$-\frac{1}{2}$	$\frac{3}{2}$	$56°$
1	0	2	$63°$
1	$\frac{1}{2}$	$\frac{5}{2}$	$68°$
1	1	3	$72°$

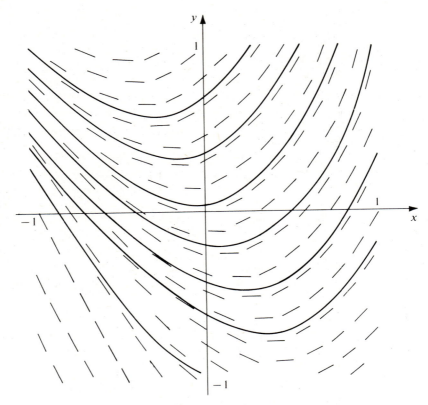

Figure 8.2

Comments. The basic graphical procedure outlined here is very general since it can be applied to any first-order differential equation of the form (8.1). However, the method has several obvious disadvantages. For one thing, although it provides the approximate graphs of the integral curves, it does not furnish analytic expressions for the solutions, either exactly or approximately. Furthermore, it is extremely tedious and time-consuming. Finally, the graphs obtained are only approximations to the graphs of the exact integral curves, and the accuracy of these approximate graphs is uncertain. Of course, apparently better approximations can be obtained by constructing more complete and careful line element configurations, but this in turn increases the time and labor involved. We shall now consider a procedure by which the process may be speeded up considerably. This is the so-called *method of isoclines.*

B. The Method of Isoclines

DEFINITION

Consider the differential equation

$$\frac{dy}{dx} = f(x, y). \tag{8.1}$$

A curve along which the slope $f(x, y)$ has a constant value c is called an isocline *of the* *differential equation (8.1). That is, the isoclines of (8.1) are the curves $f(x, y) = c$, for different values of the parameter c.*

For example, the isoclines of the differential equation

$$\frac{dy}{dx} = 2x + y \qquad (8.2)$$

are the straight lines $2x + y = c$. These are of course the straight lines $y = -2x + c$ of slope -2 and y-intercept c.

Caution. Note carefully that the isoclines of the differential equation (8.1) are *not* in general integral curves of (8.1). An isocline is merely a curve along which all of the line elements have a single, fixed inclination. This is precisely why isoclines are useful. Since the line elements along a given isocline all have the same inclination, a great number of line elements can be constructed with ease and speed, once the given isocline is drawn and *one* line element has been constructed upon it. This is exactly the procedure that we shall now outline.

Method of Isoclines Procedure

1. From the differential equation

$$\frac{dy}{dx} = f(x, y) \qquad (8.1)$$

determine the family of isoclines

$$f(x, y) = c, \qquad (8.3)$$

and carefully construct several members of this family.

2. Consider a particular isocline $f(x, y) = c_0$ of the family (8.3). At all points (x, y) on this isocline the line elements have the same slope c_0 and hence the same inclination $\alpha_0 = \arctan c_0, 0° \le \alpha_0 < 180°$. At a series of points along this isocline construct line elements having this inclination α_0.

3. Repeat Step 2 for each of the isoclines of the family (8.3) constructed in Step 1. In this way the line element configuration begins to take shape.

4. Finally, draw the smooth curves (the approximate integral curves) indicated by the line element configuration obtained in Step 3.

▶ **Example 8.2**

Employ the method of isoclines to sketch the approximate integral curves of

$$\frac{dy}{dx} = 2x + y. \qquad (8.2)$$

Solution. We have already noted that the isoclines of the differential equation (8.2) are the straight lines $2x + y = c$ or

$$y = -2x + c. \qquad (8.4)$$

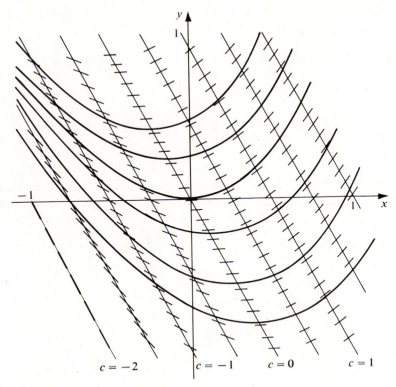

Figure 8.3

In Figure 8.3 we construct these lines for $c = -2, -\frac{3}{2}, -1, -\frac{1}{2}, 0, \frac{1}{2}, 1, \frac{3}{2}$, and 2. On each of these we then construct a number of line elements having the appropriate inclination $\alpha = \arctan c$, $0° \le \alpha < 180°$. For example, for $c = 1$, the corresponding isocline is $y = -2x + 1$, and on this line we construct line elements of inclination $\arctan 1 = 45°$. In the figure the isoclines are drawn with dashes and several of the approximate integral curves are shown (drawn solidly).

▶ **Example 8.3**

Employ the method of isoclines to sketch the approximate integral curves of

$$\frac{dy}{dx} = x^2 + y^2. \tag{8.5}$$

Solution. The isoclines of the differential equation (8.5) are the concentric circles $x^2 + y^2 = c$, $c > 0$. In Figure 8.4 the circles for which $c = \frac{1}{16}, \frac{1}{4}, \frac{9}{16}, 1, \frac{25}{16}, \frac{9}{4}, \frac{49}{16}$, and 4 have been drawn with dashes, and several line elements having the appropriate inclination have been drawn along each. For example, for $c = 4$, the corresponding isocline is the circle $x^2 + y^2 = 4$ of radius 2, and along this circle the line elements have inclination $\arctan 4 \approx 76°$. Several approximate integral curves are shown (drawn solidly).

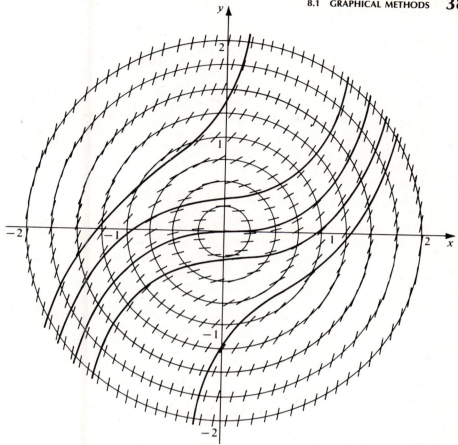

Figure 8.4

Exercises

Employ the method of isoclines to sketch the approximate integral curves of each of the differential equations in Exercises 1–12.

1. $\dfrac{dy}{dx} = 3x - y.$

2. $\dfrac{dy}{dx} = \dfrac{y}{x}.$

3. $\dfrac{dy}{dx} = \dfrac{y}{x^2}.$

4. $\dfrac{dy}{dx} = x^2 + 2y^2.$

5. $\dfrac{dy}{dx} = \dfrac{3x - y}{x + y}.$

6. $\dfrac{dy}{dx} = \sin x - y.$

7. $\dfrac{dy}{dx} = y^3 - x^2.$

8. $\dfrac{dy}{dx} = \dfrac{3x + 2y + x^2}{x + 2y}.$

9. $\dfrac{dy}{dx} = \dfrac{3x + y + x^3}{5x - y}.$

10. $\dfrac{dy}{dx} = \dfrac{\sin x + y}{x - y}.$

11. $\dfrac{dy}{dx} = \dfrac{(1 - x^2)y - x}{y}.$

12. $\dfrac{dy}{dx} = \dfrac{10(1 - x^2)y - x}{y}.$

8.2 POWER SERIES METHODS

A. Introduction

Although the graphical methods of the preceding section are very general, they suffer from several serious disadvantages. Not only are they tedious and subject to possible errors of construction, but they merely provide us with the approximate *graphs* of the solutions and do not furnish any analytic *expressions* for these solutions. Although we have now passed the naive state of searching for a closed-form solution in terms of elementary functions, we might still hope for solutions that can be represented as some type of infinite series. In particular, we shall seek solutions that are representable as power series.

We point out that not all first-order differential equations possess solutions that can be represented as power series, and it is beyond the scope of this book to consider conditions under which a first-order differential equation does possess such solutions. In order to explain the power series methods we shall *assume* that power series solutions actually do exist, realizing that this is an assumption that is not always justified.

Specifically, we consider the initial-value problem consisting of the differential equation

$$\frac{dy}{dx} = f(x, y) \tag{8.1}$$

and the initial condition

$$y(x_0) = y_0 \tag{8.6}$$

and *assume* that the differential equation (8.1) possesses a solution that is representable as a power series in powers of $(x - x_0)$. That is, we assume that the differential equation (8.1) has a solution of the form

$$y = c_0 + c_1(x - x_0) + c_2(x - x_0)^2 + \cdots = \sum_{n=0}^{\infty} c_n(x - x_0)^n \tag{8.7}$$

that is valid in some interval about the point x_0. We now consider methods of determining the coefficients $c_0, c_1, c_2, \ldots$ in (8.7) so that the series (8.7) actually does satisfy the differential equation (8.1).

B. The Taylor Series Method

We thus assume that the initial-value problem consisting of the differential equation (8.1) and the initial condition (8.6) has a solution of the form (8.7) that is valid in some interval about x_0. Then by Taylor's theorem, for each x in this interval the value $y(x)$ of this solution is given by

$$y(x) = y(x_0) + y'(x_0)(x - x_0) + \frac{y''(x_0)}{2!}(x - x_0)^2 + \cdots$$

$$= \sum_{n=0}^{\infty} \frac{y^{(n)}(x_0)}{n!}(x - x_0)^n. \tag{8.8}$$

From the initial condition (8.6), we have

$$y(x_0) = y_0,$$

and from the differential equation (8.1) itself,

$$y'(x_0) = f(x_0, y_0).$$

Substituting these values of $y(x_0)$ and $y'(x_0)$ into the series in (8.8), we obtain the first two coefficients of the desired series solution (8.7). Now differentiating the differential equation (8.1), we obtain

$$\frac{d^2 y}{dx^2} = \frac{d}{dx}[f(x, y)] = f_x(x, y) + f_y(x, y)\frac{dy}{dx}$$

$$= f_x(x, y) + f_y(x, y)f(x, y), \tag{8.9}$$

where we use subscripts to denote partial differentiations. From this we obtain

$$y''(x_0) = f_x(x_0, y_0) + f_y(x_0, y_0)f(x_0, y_0).$$

Substituting this value of $y''(x_0)$ into (8.8), we obtain the third coefficient in the series solution (8.7). Proceeding in like manner, we differentiate (8.9) successively to obtain

$$\frac{d^3 y}{dx^3}, \frac{d^4 y}{dx^4}, \ldots, \frac{d^n y}{dx^n}, \ldots$$

From these we obtain the values

$$y'''(x_0), y^{(iv)}(x_0), \ldots, y^{(n)}(x_0), \ldots.$$

Substituting these values into (8.8), we obtain the fourth and following coefficients in the series solution (8.7). Thus the coefficients in the series solution (8.7) are successively determined.

▶ **Example 8.4**

Use the Taylor series method to obtain a power series solution of the initial-value problem

$$\frac{dy}{dx} = x^2 + y^2, \tag{8.10}$$

$$y(0) = 1, \tag{8.11}$$

in powers of x.

Solution. Since we seek a solution in powers of x, we set $x_0 = 0$ in (8.7) and thus assume a solution of the form

$$y = c_0 + c_1 x + c_2 x^2 + \cdots = \sum_{n=0}^{\infty} c_n x^n.$$

By Taylor's theorem, we know that for each x in the interval where this solution is valid

$$y(x) = y(0) + y'(0)x + \frac{y''(0)}{2!}x^2 + \cdots = \sum_{n=0}^{\infty} \frac{y^{(n)}(0)}{n!} x^n. \tag{8.12}$$

The initial condition (8.11) states that

$$y(0) = 1, \tag{8.13}$$

and from the differential equation (8.10) we see that

$$y'(0) = 0^2 + 1^2 = 1. \tag{8.14}$$

Differentiating (8.10) successively, we obtain

$$\frac{d^2y}{dx^2} = 2x + 2y \frac{dy}{dx}, \tag{8.15}$$

$$\frac{d^3y}{dx^3} = 2 + 2y \frac{d^2y}{dx^2} + 2\left(\frac{dy}{dx}\right)^2, \tag{8.16}$$

$$\frac{d^4y}{dx^4} = 2y \frac{d^3y}{dx^3} + 6 \frac{dy}{dx} \frac{d^2y}{dx^2}. \tag{8.17}$$

Substituting $x = 0$, $y = 1$, $\dfrac{dy}{dx} = 1$, into (8.15), we obtain

$$y''(0) = 2(0) + 2(1)(1) = 2. \tag{8.18}$$

Substituting $y = 1$, $\dfrac{dy}{dx} = 1$, $\dfrac{d^2y}{dx^2} = 2$ into (8.16), we obtain

$$y'''(0) = 2 + 2(1)(2) + 2(1)^2 = 8. \tag{8.19}$$

Finally, substituting $y = 1$, $\dfrac{dy}{dx} = 1$, $\dfrac{d^2y}{dx^2} = 2$, $\dfrac{d^3y}{dx^3} = 8$ into (8.17), we find that

$$y^{(iv)}(0) = (2)(1)(8) + (6)(1)(2) = 28. \tag{8.20}$$

By successive differentiation of (8.17), we could proceed to determine

$$\frac{d^5y}{dx^5}, \frac{d^6y}{dx^6}, \dots,$$

and hence obtain

$$y^{(v)}(0), y^{(vi)}(0), \dots.$$

Now substituting the values given by (8.13), (8.14), (8.18), (8.19), and (8.20) into (8.12), we obtain the first five coefficients of the desired series solution. We thus find the solution

$$y = 1 + x + \frac{2}{2!} x^2 + \frac{8}{3!} x^3 + \frac{28}{4!} x^4 + \cdots$$

$$= 1 + x + x^2 + \frac{4}{3} x^3 + \frac{7}{6} x^4 + \cdots. \tag{8.21}$$

C. The Method of Undetermined Coefficients

We now consider an alternative method for obtaining the coefficients $c_0, c_1, c_2, \dots$ in the assumed series solution

$$y = c_0 + c_1(x - x_0) + c_2(x - x_0)^2 + \cdots = \sum_{n=0}^{\infty} c_n(x - x_0)^n \tag{8.7}$$

of the problem consisting of the differential equation (8.1) with initial condition (8.6). We shall refer to this alternative method as the method of undetermined coefficients. In order to apply it we assume that $f(x, y)$ in the differential equation (8.1) is representable in the form

$$
\begin{aligned}
f(x, y) = a_{00} &+ a_{10}(x - x_0) + a_{01}(y - y_0) + a_{20}(x - x_0)^2 \\
&+ a_{11}(x - x_0)(y - y_0) + a_{02}(y - y_0)^2 + \cdots.
\end{aligned} \tag{8.22}
$$

The coefficients a_{ij} in (8.22) may be found by Taylor's formula for functions of two variables, although in many simple cases the use of this formula is unnecessary. Using the representation (8.22) for $f(x, y)$, the differential equation (8.1) takes the form

$$
\begin{aligned}
\frac{dy}{dx} = a_{00} &+ a_{10}(x - x_0) + a_{01}(y - y_0) + a_{20}(x - x_0)^2 \\
&+ a_{11}(x - x_0)(y - y_0) + a_{02}(y - y_0)^2 + \cdots.
\end{aligned} \tag{8.23}
$$

Now assuming that the series (8.7) converges in some interval $|x - x_0| < r(r > 0)$ about x_0, we may differentiate (8.7) term by term and the resulting series will also converge on $|x - x_0| < r$ and represent $y'(x)$ there. Doing this we thus obtain

$$
\frac{dy}{dx} = c_1 + 2c_2(x - x_0) + 3c_3(x - x_0)^2 + \cdots. \tag{8.24}
$$

We note that in order for the series (8.7) to satisfy the initial condition (8.6) that $y = y_0$ at $x = x_0$, we must have $c_0 = y_0$ and hence

$$
y - y_0 = c_1(x - x_0) + c_2(x - x_0)^2 + \cdots. \tag{8.25}
$$

Now substituting (8.7) and (8.24) into the differential equation (8.23), and making use of (8.25), we find that

$$
\begin{aligned}
c_1 &+ 2c_2(x - x_0) + 3c_3(x - x_0)^2 + \cdots \\
&= a_{00} + a_{10}(x - x_0) + a_{01}[c_1(x - x_0) + c_2(x - x_0)^2 + \cdots] \\
&\quad + a_{20}(x - x_0)^2 + a_{11}(x - x_0)[c_1(x - x_0) + c_2(x - x_0)^2 + \cdots] \\
&\quad + a_{02}[c_1(x - x_0) + c_2(x - x_0)^2 + \cdots]^2 + \cdots.
\end{aligned} \tag{8.26}
$$

Performing the multiplications indicated in the right member of (8.26) and then combining like powers of $(x - x_0)$, we see that (8.26) takes the form

$$
\begin{aligned}
c_1 &+ 2c_2(x - x_0) + 3c_3(x - x_0)^2 + \cdots \\
&= a_{00} + (a_{10} + a_{01}c_1)(x - x_0) \\
&\quad + (a_{01}c_2 + a_{20} + a_{11}c_1 + a_{02}c_1^2)(x - x_0)^2 + \cdots.
\end{aligned} \tag{8.27}
$$

In order that (8.27) be satisfied for all values of x in the interval $|x - x_0| < r$, the coefficients of like powers of $(x - x_0)$ on both sides of (8.27) must be equal. Equating these coefficients, we obtain

$$
\begin{aligned}
c_1 &= a_{00}, \\
2c_2 &= a_{10} + a_{01}c_1, \\
3c_3 &= a_{01}c_2 + a_{20} + a_{11}c_1 + a_{02}c_1^2,
\end{aligned} \tag{8.28}
$$

$$\vdots$$

From the conditions (8.28) we determine successively the coefficients $c_1, c_2, c_3, \ldots$ of the series solution (8.7). From the first of conditions (8.28) we first obtain c_1 as the known coefficient a_{00}. Then from the second of conditions (8.28) we obtain c_2 in terms of the known coefficients a_{10} and a_{01} and the coefficient c_1 just determined. Thus we obtain $c_2 = \frac{1}{2}(a_{10} + a_{01}a_{00})$. In like manner, we proceed to determine $c_3, c_4, \ldots$. We observe that in general each coefficient c_n is thus given in terms of the known coefficients a_{ij} in the expansion (8.22) and the previously determined coefficients $c_1, c_2, \ldots, c_{n-1}$.

Finally, we substitute the coefficients $c_0, c_1, c_2, \ldots$ so determined into the series (8.7) and thereby obtain the desired solution.

▶ **Example 8.5**

Use the method of undetermined coefficients to obtain a power series solution of the initial-value problem

$$\frac{dy}{dx} = x^2 + y^2, \tag{8.10}$$

$$y(0) = 1, \tag{8.11}$$

in powers of x.

Solution. Since $x_0 = 0$, the assumed solution (8.7) is of the form

$$y = c_0 + c_1 x + c_2 x^2 + c_3 x^3 + \cdots. \tag{8.29}$$

In order to satisfy the initial condition (8.11), we must have $c_0 = 1$ and hence the series (8.29) takes the form

$$y = 1 + c_1 x + c_2 x^2 + c_3 x^3 + \cdots. \tag{8.30}$$

Differentiating (8.30), we obtain

$$\frac{dy}{dx} = c_1 + 2c_2 x + 3c_3 x^2 + 4c_4 x^3 + \cdots. \tag{8.31}$$

For the differential equation (8.10) we have $f(x, y) = x^2 + y^2$. Since $x_0 = 0$ and $y_0 = 1$, we must expand $x^2 + y^2$ in the form

$$\sum_{i,j=0}^{\infty} a_{ij} x^i (y - 1)^j.$$

Since

$$y^2 = (y - 1)^2 + 2(y - 1) + 1,$$

the desired expansion is given by

$$x^2 + y^2 = 1 + 2(y - 1) + x^2 + (y - 1)^2.$$

Thus the differential equation (8.10) takes the form

$$\frac{dy}{dx} = 1 + 2(y - 1) + x^2 + (y - 1)^2. \tag{8.32}$$

Now substituting (8.30) and (8.31) into the differential equation (8.32), we obtain

$$c_1 + 2c_2 x + 3c_3 x^2 + 4c_4 x^3 + \cdots$$
$$= 1 + 2(c_1 x + c_2 x^2 + c_3 x^3 + \cdots) + x^2 + (c_1 x + c_2 x^2 + \cdots)^2. \quad (8.33)$$

Performing the indicated multiplications and collecting like powers of x in the right member of (8.33), we see that it takes the form

$$c_1 + 2c_2 x + 3c_3 x^2 + 4c_4 x^3 + \cdots$$
$$= 1 + 2c_1 x + (2c_2 + 1 + c_1^2)x^2 + (2c_3 + 2c_1 c_2)x^3 + \cdots. \quad (8.34)$$

Equating the coefficients of the like powers of x in both members of (8.34), we obtain the conditions

$$
\begin{aligned}
c_1 &= 1, \\
2c_2 &= 2c_1, \\
3c_3 &= 2c_2 + 1 + c_1^2, \\
4c_4 &= 2c_3 + 2c_1 c_2, \\
&\;\;\vdots
\end{aligned}
\qquad (8.35)
$$

From the conditions (8.35), we obtain successively

$$
\begin{aligned}
c_1 &= 1, \\
c_2 &= c_1 = 1, \\
c_3 &= \tfrac{1}{3}(2c_2 + 1 + c_1^2) = \tfrac{4}{3}, \\
c_4 &= \tfrac{1}{4}(2c_3 + 2c_1 c_2) = \tfrac{1}{4}\left(\tfrac{14}{3}\right) = \tfrac{7}{6}, \\
&\;\;\vdots
\end{aligned}
\qquad (8.36)
$$

Substituting the coefficients so determined in (8.36) into the series (8.30), we obtain the first five terms of the desired series solution. We thus find

$$y = 1 + x + x^2 + \tfrac{4}{3}x^3 + \tfrac{7}{6}x^4 + \cdots.$$

We note that this is of course the same series previously obtained by the Taylor series method and already given by (8.21).

Remark. We have made but little mention of the interval of convergence of the series involved in our discussion. We have merely assumed that a power series solution exists and converges on some interval $|x - x_0| < r (r > 0)$ about the initial point x_0. In a practical problem the interval of convergence is of vital concern and should be determined, if possible. Another matter of great importance in a practical problem is the determination of the number of terms which have to be found in order to be certain of a sufficient degree of accuracy. We shall not consider these matters here. Our primary purpose has been merely to explain the details of the methods. We refer the reader to more advanced treatises for discussions of the importance questions of convergence and accuracy.

Exercises

Obtain a power series solution in powers of x of each of the initial-value problems in Exercises 1–7 by (a) the Taylor series method and (b) the method of undetermined coefficients.

1. $\dfrac{dy}{dx} = x + y, \qquad y(0) = 1.$

2. $\dfrac{dy}{dx} = x^2 + 2y^2, \qquad y(0) = 4.$

3. $\dfrac{dy}{dx} = 1 + xy^2, \qquad y(0) = 2.$

4. $\dfrac{dy}{dx} = x^3 + y^3, \qquad y(0) = 3.$

5. $\dfrac{dy}{dx} = x + \sin y, \qquad y(0) = 0.$

6. $\dfrac{dy}{dx} = 1 + x \sin y, \qquad y(0) = 0.$

7. $\dfrac{dy}{dx} = e^x + x \cos y, \qquad y(0) = 0.$

Obtain a power series solution in powers of $x - 1$ of each of the initial-value problems in Exercises 8–12 by (a) the Taylor series method and (b) the method of undetermined coefficients.

8. $\dfrac{dy}{dx} = x^2 + y^2, \qquad y(1) = 4.$

9. $\dfrac{dy}{dx} = x^3 + y^2, \qquad y(1) = 1.$

10. $\dfrac{dy}{dx} = x + y + y^2, \qquad y(1) = 1.$

11. $\dfrac{dy}{dx} = x + \cos y, \qquad y(1) = \pi.$

12. $\dfrac{dy}{dx} = x^2 + x \sin y, \qquad y(1) = \dfrac{\pi}{2}.$

8.3 THE METHOD OF SUCCESSIVE APPROXIMATIONS

A. The Method

We again consider the initial-value problem consisting of the differential equation

$$\frac{dy}{dx} = f(x, y) \tag{8.1}$$

and the initial condition

$$y(x_0) = y_0. \tag{8.6}$$

We now outline the Picard method of successive approximations for finding a solution of this problem which is valid on some interval that includes the initial point x_0.

The first step of the Picard method is quite unlike anything that we have done before and at first glance it appears to be rather fruitless. For the first step actually consists of making a guess at the solution! That is, we choose a function ϕ_0 and call it a "zeroth approximation" to the actual solution. How do we make this guess? In other words, what function do we choose? Actually, many different choices could be made. The only thing that we know about the actual solution is that in order to satisfy the initial condition (8.6) it must assume the value y_0 at $x = x_0$. Therefore it would seem reasonable to choose for ϕ_0 a function that assumes this value y_0 at $x = x_0$. Although this requirement is not essential, it certainly seems as sensible as anything else. In particular, it is often convenient to choose for ϕ_0 the constant function that has the value y_0 for all x. While this choice is certainly not essential, it is often the simplest, most reasonable choice that quickly comes to mind.

In summary, then, the first step of the Picard method is to choose a function ϕ_0 which will serve as a zeroth approximation.

Having thus chosen a zeroth approximation ϕ_0, we now determine a first approximation ϕ_1 in the following manner. We determine $\phi_1(x)$ so that (1) it satisfies the differential equation obtained from (8.1) by replacing y in $f(x, y)$ by $\phi_0(x)$, and (2) it satisfies the initial condition (8.6). Thus ϕ_1 is determined such that

$$\frac{d}{dx}[\phi_1(x)] = f[x, \phi_0(x)] \tag{8.37}$$

and

$$\phi_1(x_0) = y_0. \tag{8.38}$$

We now assume that $f[x, \phi_0(x)]$ is continuous. Then ϕ_1 satisfies (8.37) and (8.38) if and only if

$$\phi_1(x) = y_0 + \int_{x_0}^{x} f[t, \phi_0(t)]\,dt. \tag{8.39}$$

From this equation the first approximation ϕ_1 is determined.

We now determine the second approximation ϕ_2 in a similar manner. The function ϕ_2 is determined such that

$$\frac{d}{dx}[\phi_2(x)] = f[x, \phi_1(x)] \tag{8.40}$$

and

$$\phi_2(x_0) = y_0. \tag{8.41}$$

Assuming that $f[x, \phi_1(x)]$ is continuous, then ϕ_2 satisfies (8.40) and (8.41) if and only if

$$\phi_2(x) = y_0 + \int_{x_0}^{x} f[t, \phi_1(t)]\,dt. \tag{8.42}$$

From this equation the second approximation ϕ_2 is determined.

We now proceed in like manner to determine a third approximation ϕ_3, a fourth approximation ϕ_4, and so on. The nth approximation ϕ_n is determined from

$$\phi_n(x) = y_0 + \int_{x_0}^{x} f[t, \phi_{n-1}(t)]\,dt, \tag{8.43}$$

where ϕ_{n-1} is the $(n-1)$st approximation. We thus obtain a sequence of functions

$$\phi_0, \phi_1, \phi_2, \ldots, \phi_n, \ldots,$$

where ϕ_0 is chosen, ϕ_1 is determined from (8.39), ϕ_2 is determined from (8.42),..., and in general ϕ_n is determined from (8.43) for $n \geq 1$.

Now just how does this sequence of functions relate to the actual solution of the initial-value problem under consideration? It can be proved, under certain general conditions and for x restricted to a sufficiently small interval about the initial point x_0, that (1) as $n \to \infty$ the sequence of functions ϕ_n defined by (8.43) for $n \geq 1$ approaches a limit function ϕ, and (2) this limit function ϕ satisfies both the differential equation (8.1) and the initial condition (8.6). That is, under suitable restrictions the function ϕ defined by

$$\phi = \lim_{n \to \infty} \phi_n$$

is the exact solution of the initial-value problem under consideration. Furthermore, the error in approximating the exact solution ϕ by the nth approximation ϕ_n will be arbitrarily small provided that n is sufficiently large and that x is sufficiently close to the initial point x_0.

B. An Example; Remarks on The Method

We illustrate the Picard method of successive approximations by applying it to the initial-value problem of Examples 8.4 and 8.5.

▶ **Example 8.6**

Use the method of successive approximations to find a sequence of functions that approaches the solution of the initial-value problem

$$\frac{dy}{dx} = x^2 + y^2, \tag{8.10}$$

$$y(0) = 1. \tag{8.11}$$

Solution. Our first step is to choose a function for the zeroth approximation ϕ_0. Since the initial value of y is 1, it would seem reasonable to choose for ϕ_0 the constant function that has the value 1 for all x. Thus, we let ϕ_0 be such that

$$\phi_0(x) = 1$$

for all x. The nth approximation ϕ_n for $n \geq 1$ is given by formula (8.43). Since

$$f(x, y) = x^2 + y^2$$

in the differential equation (8.10), the formula (8.43) becomes in this case

$$\phi_n(x) = 1 + \int_0^x \{t^2 + [\phi_{n-1}(t)]^2\} \, dt, \qquad n \geq 1.$$

Using this formula for $n = 1, 2, 3, \ldots$, we obtain successively

$$\phi_1(x) = 1 + \int_0^x \{t^2 + [\phi_0(t)]^2\}\, dt$$

$$= 1 + \int_0^x (t^2 + 1)\, dt = 1 + x + \frac{x^3}{3},$$

$$\phi_2(x) = 1 + \int_0^x \{t^2 + [\phi_1(t)]^2\}\, dt = 1 + \int_0^x \left[t^2 + \left(1 + t + \frac{t^3}{3}\right)^2 \right] dt$$

$$= 1 + \int_0^x \left(1 + 2t + 2t^2 + \frac{2t^3}{3} + \frac{2t^4}{3} + \frac{t^6}{9} \right) dt$$

$$= 1 + x + x^2 + \frac{2x^3}{3} + \frac{x^4}{6} + \frac{2x^5}{15} + \frac{x^7}{63},$$

$$\phi_3(x) = 1 + \int_0^x \{t^2 + [\phi_2(t)]^2\}\, dt$$

$$= 1 + \int_0^x \left[t^2 + \left(1 + t + t^2 + \frac{2t^3}{3} + \frac{t^4}{6} + \frac{2t^5}{15} + \frac{t^7}{63} \right)^2 \right] dt$$

$$= 1 + \int_0^x \left[1 + 2t + 4t^2 + \frac{10t^3}{3} + \frac{8t^4}{3} + \frac{29t^5}{15} + \frac{47t^6}{45} + \frac{164t^7}{315} \right.$$

$$\left. + \frac{299t^8}{1260} + \frac{8t^9}{105} + \frac{184t^{10}}{4725} + \frac{t^{11}}{189} + \frac{4t^{12}}{945} + \frac{t^{14}}{3969} \right] dt$$

$$= 1 + x + x^2 + \frac{4x^3}{3} + \frac{5x^4}{6} + \frac{8x^5}{15} + \frac{29x^6}{90} + \frac{47x^7}{315} + \frac{41x^8}{630} + \frac{299x^9}{11{,}340}$$

$$+ \frac{4x^{10}}{525} + \frac{184x^{11}}{51{,}975} + \frac{x^{12}}{2268} + \frac{4x^{13}}{12{,}285} + \frac{x^{15}}{59{,}535}.$$

$$\vdots$$

We have chosen the zeroth approximation ϕ_0 and then found the next three approximations ϕ_1, ϕ_2, and ϕ_3 explicitly. We could proceed in like manner to find ϕ_4, $\phi_5, \ldots$ explicitly and thus determine successively the members of the sequence $\{\phi_n\}$ that approaches the exact solution of the initial-value problem under consideration. However, we believe that the procedure is now clear, and for rather obvious reasons we shall not proceed further with this problem.

Remarks. The greatest disadvantage of the method of successive approximations is that it leads to tedious, involved, and sometimes impossible calculations. This is amply illustrated in Example 8.6. At best the calculations are usually very complicated, and in general it is impossible to carry through more than a few of the successive integrations exactly. Nevertheless, the method is of practical importance, for the first few approximations alone are sometimes quite accurate.

However, the principal use of the method of successive approximations is in proving existence theorems. Concerning this, we refer the reader to Chapter 10.

Exercises

For each of the initial-value problems in Exercises 1–8 use the method of successive approximations to find the first three members ϕ_1, ϕ_2, ϕ_3 of a sequence of functions that approaches the exact solution of the problem.

1. $\dfrac{dy}{dx} = xy$, $\quad y(0) = 1$.

2. $\dfrac{dy}{dx} = x + y$, $\quad y(0) = 1$.

3. $\dfrac{dy}{dx} = x + y^2$, $\quad y(0) = 0$.

4. $\dfrac{dy}{dx} = 1 + xy^2$, $\quad y(0) = 0$.

5. $\dfrac{dy}{dx} = e^x + y^2$, $\quad y(0) = 0$.

6. $\dfrac{dy}{dx} = \sin x + y^2$, $\quad y(0) = 0$.

7. $\dfrac{dy}{dx} = 2x + y^3$, $\quad y(0) = 0$.

8. $\dfrac{dy}{dx} = 1 + 6xy^4$, $\quad y(0) = 0$.

8.4 NUMERICAL METHODS

A. Introduction; A Problem for Illustration

In this section we introduce certain basic numerical methods for approximating the solution of the initial-value problem consisting of the differential equation

$$\frac{dy}{dx} = f(x, y) \tag{8.1}$$

and the initial condition

$$y(x_0) = y_0. \tag{8.6}$$

Numerical methods employ the differential equation (8.1) and the condition (8.6) to obtain approximations to the values of the solution corresponding to various, selected values of x. To be more specific, let y denote the solution of the problem and let h denote a positive *increment* in x. The initial condition (8.6) tells us that $y = y_0$ at $x = x_0$. A numerical method will employ the differential equation (8.1) and the condition (8.6) to approximate successively the values of y at $x_1 = x_0 + h, x_2 = x_1 + h, x_3 = x_2 + h, \ldots$.

Let us denote these approximate values of y by $y_1, y_2, y_3, \ldots$, respectively. That is, we let y_n denote the approximate value of y at $x = x_n = x_0 + nh, n = 1, 2, 3, \ldots$. Now all that we know about y before starting is that $y = y_0$ at $x = x_0$. In order to get started we need a method that requires only the value y_n in order to obtain the next value y_{n+1}. For we can apply such a method with $n = 0$ and use the initial value y_0 to obtain the next value y_1. A method that uses only y_n to find y_{n+1}, and that therefore enables us to get started, is called a *starting method*. Once we have used a starting method to find y_1, we can repeat it with $n = 1$ to find y_2, with $n = 2$ to find y_3, and so forth. However, once we have several calculated values at our disposal, it is often convenient to change over to a method that uses both y_n and one or more *preceding* values $y_{n-1}, y_{n-2}, \ldots$ to find the next value y_{n+1}. Such a method, which enables us to continue once we have got sufficiently well started, is called a *continuing method*. Most of our attention in this text will be devoted to starting methods.

Our principal objective in this section is to present the actual details of certain basic numerical methods for solving first-order initial-value problems. In general, we shall not consider the theoretical justifications of these methods, nor shall we enter into detailed discussions of matters such as accuracy and error.

Before turning to the details of the numerical methods to be considered, we introduce a simple initial-value problem that we shall use for purposes of illustration throughout this section. We consider the problem

$$\frac{dy}{dx} = 2x + y, \tag{8.2}$$

$$y(0) = 1. \tag{8.44}$$

We have already employed the differential equation (8.2) to illustrate the graphical methods of Section 8.1. We note at once that it is a linear differential equation and hence it can be solved exactly. Using the methods of Section 2.3, we find at once that its general solution is

$$y = -2(x + 1) + ce^x, \tag{8.45}$$

where c is an arbitrary constant. Applying the initial condition (8.44) to (8.45), we find that the exact solution of the initial-value problem consisting of (8.2) and (8.44) is

$$y = -2(x + 1) + 3e^x. \tag{8.46}$$

We have chosen the problem consisting of (8.2) and (8.44) for illustrative purposes for two reasons. First, the differential equation (8.2) is so simple that numerical methods may be applied to it without introducing involved computations that might obscure the main steps of the method to a beginner. Second, since the exact solution (8.46) of the problem has been found, we can compare approximate solution values obtained numerically with this exact solution and thereby gain some insight into the accuracy of numerical methods.

Of course in practice we would not solve a simple linear differential equation such as (8.2) by a numerical method. The methods of this section are actually designed for equations that can *not* be solved exactly and for equations that, although solvable exactly, have exact solutions so unwieldy that they are practically useless.

B. The Euler Method

The Euler method is very simple but not very practical. An understanding of it, however, paves the way for an understanding of the more practical (but also more complicated) methods which follow.

Let y denote the exact solution of the initial-value problem that consists of the differential equation

$$\frac{dy}{dx} = f(x, y) \tag{8.1}$$

and the initial condition

$$y(x_0) = y_0. \tag{8.6}$$

Let h denote a positive increment in x and let $x_1 = x_0 + h$. Then

$$\int_{x_0}^{x_1} f(x, y)\, dx = \int_{x_0}^{x_1} \frac{dy}{dx}\, dx = y(x_1) - y(x_0).$$

Since y_0 denotes the value $y(x_0)$ of the exact solution y at x_0, we have

$$y(x_1) = y_0 + \int_{x_0}^{x_1} f(x, y)\, dx. \qquad (8.47)$$

If we assume that $f(x, y)$ varies slowly on the interval $x_0 \le x \le x_1$, then we can approximate $f(x, y)$ in (8.47) by its value $f(x_0, y_0)$ at the left endpoint x_0. Then

$$y(x_1) \approx y_0 + \int_{x_0}^{x_1} f(x_0, y_0)\, dx.$$

But

$$\int_{x_0}^{x_1} f(x_0, y_0)\, dx = f(x_0, y_0)(x_1 - x_0) = hf(x_0, y_0).$$

Thus

$$y(x_1) \approx y_0 + hf(x_0, y_0).$$

Thus we obtain the approximate value y_1 of y at $x_1 = x_0 + h$ by the formula

$$y_1 = y_0 + hf(x_0, y_0). \qquad (8.48)$$

Having obtained y_1 by formula (8.48), we proceed in like manner to obtain y_2 by the formula $y_2 = y_1 + hf(x_1, y_1)$, y_3 by the formula $y_3 = y_2 + hf(x_2, y_2)$, and so forth. In general we find y_{n+1} in terms of y_n by the formula

$$y_{n+1} = y_n + hf(x_n, y_n). \qquad (8.49)$$

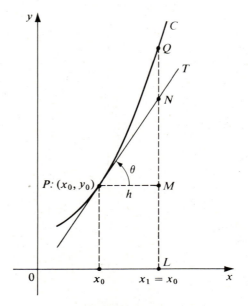

Figure 8.5

Before illustrating the method, we give a useful geometric interpretation. The graph of the exact solution y is a curve C in the xy plane (see Figure 8.5). Let P denote the initial point (x_0, y_0) and let T be the tangent to C at P. Let Q be the point at which the line $x = x_1$ intersects C and let N be the point at which this line intersects T. Then the exact value of y at x_1 is represented by LQ. The approximate value y_1 is represented by LN, since $LN = LM + MN = y_0 + PM \tan \theta = y_0 + hf(x_0, y_0)$. The error in approximating the exact value of y at x_1 by y_1 is thus represented by NQ. The figure suggests that if h is sufficiently small, then this error NQ will also be small and hence that the approximation will be good.

It is clear that the Euler method is indeed simple, but it should also be apparent from our discussion why it is not very practical. If the increment h is *not* very small, then the errors in the approximations generally will not be small and thus the method will lead to quite inaccurate results. If the increment h *is* very small, then the computations will be more lengthy and so the method will involve tedious and time-consuming labor.

▶ **Example 8.7**

Apply the Euler method to the initial-value problem

$$\frac{dy}{dx} = 2x + y, \tag{8.2}$$

$$y(0) = 1. \tag{8.44}$$

Employ the method to approximate the value of the solution y at $x = 0.2, 0.4, 0.6, 0.8$, and 1.0 using $(1)\,h = 0.2$, and $(2)\,h = 0.1$. Obtain results to three figures after the decimal point. Compare with the exact value determined from (8.46).

Solution. 1. We use formula (8.49) with $f(x, y) = 2x + y$ and $h = 0.2$. From the initial condition (8.44), we have $x_0 = 0$, $y_0 = 1$. We now proceed with the calculations.

(a) $x_1 = x_0 + h = 0.2$, $f(x_0, y_0) = f(0, 1) = 1.000$,
$y_1 = y_0 + hf(x_0, y_0) = 1.000 + 0.2(1.000) = 1.200$

(b) $x_2 = x_1 + h = 0.4$, $f(x_1, y_1) = f(0.2, 1.200) = 1.600$,
$y_2 = y_1 + hf(x_1, y_1) = 1.200 + 0.2(1.600) = 1.520$.

(c) $x_3 = x_2 + h = 0.6$, $f(x_2, y_2) = f(0.4, 1.520) = 2.320$,
$y_3 = y_2 + hf(x_2, y_2) = 1.520 + 0.2(2.320) = 1.984$.

(d) $x_4 = x_3 + h = 0.8$, $f(x_3, y_3) = f(0.6, 1.984) = 3.184$,
$y_4 = y_3 + hf(x_3, y_3) = 1.984 + 0.2(3.184) = 2.621$.

(e) $x_5 = x_4 + h = 1.0$, $f(x_4, y_4) = f(0.8, 2.621) = 4.221$,
$y_5 = y_4 + hf(x_4, y_4) = 2.621 + 0.2(4.221) = 3.465$.

These results, corresponding to the various values of x_n, are collected in the second column of Table 8.2.

2. We now use formula (8.49) with $f(x, y) = 2x + y$ and $h = 0.1$. Of course we again have $x_0 = 0$, $y_0 = 1$. The calculations follow.

(a) $x_1 = x_0 + h = 0.1$, $f(x_0, y_0) = f(0, 1) = 1.000$,
$y_1 = y_0 + hf(x_0, y_0) = 1.000 + 0.1(1.000) = 1.100$.

TABLE 8.2

x_n	y_n using $h = 0.2$	y_n using $h = 0.1$	y
0.0	1.000	1.000	1.000
0.1	—	1.100	1.116
0.2	1.200	1.230	1.264
0.3	—	1.393	1.450
0.4	1.520	1.592	1.675
0.5	—	1.831	1.946
0.6	1.984	2.114	2.266
0.7	—	2.445	2.641
0.8	2.621	2.830	3.076
0.9	—	3.273	3.579
1.0	3.465	3.780	4.155

exact value (handwritten annotation)

(b) $x_2 = x_1 + h = 0.2$, $f(x_1, y_1) = f(0.1, 1.100) = 1.300$,
$y_2 = y_1 + hf(x_1, y_1) = 1.100 + 0.1(1.300) = 1.230$.

(c) $x_3 = x_2 + h = 0.3$, $f(x_2, y_2) = f(0.2, 1.230) = 1.630$,
$y_3 = y_2 + hf(x_2, y_2) = 1.230 + 0.1(1.630) = 1.393$.

(d) $x_4 = x_3 + h = 0.4$, $f(x_3, y_3) = f(0.3, 1.393) = 1.993$,
$y_4 = y_3 + hf(x_3, y_3) = 1.393 + 0.1(1.993) = 1.592$.

(e) $x_5 = x_4 + h = 0.5$, $f(x_4, y_4) = f(0.4, 1.592) = 2.392$,
$y_5 = y_4 + hf(x_4, y_4) = 1.592 + 0.1(2.392) = 1.831$.

(f) $x_6 = x_5 + h = 0.6$, $f(x_5, y_5) = f(0.5, 1.831) = 2.831$,
$y_6 = y_5 + hf(x_5, y_5) = 1.831 + 0.1(2.831) = 2.114$.

(g) $x_7 = x_6 + h = 0.7$, $f(x_6, y_6) = f(0.6, 2.114) = 3.314$,
$y_7 = y_6 + hf(x_6, y_6) = 2.114 + 0.1(3.314) = 2.445$.

(h) $x_8 = x_7 + h = 0.8$, $f(x_7, y_7) = f(0.7, 2.445) = 3.845$,
$y_8 = y_7 + hf(x_7, y_7) = 2.445 + 0.1(3.845) = 2.830$.

(i) $x_9 = x_8 + h = 0.9$, $f(x_8, y_8) = f(0.8, 2.830) = 4.430$,
$y_9 = y_8 + hf(x_8, y_8) = 2.830 + 0.1(4.430) = 3.273$.

(j) $x_{10} = x_9 + h = 1.0$, $f(x_9, y_9) = f(0.9, 3.273) = 5.073$,
$y_{10} = y_9 + hf(x_9, y_9) = 3.273 + 0.1(5.073) = 3.780$.

These results are collected in the third column of Table 8.2. The values of the exact solution y, computed from (8.46) to three figures after the decimal point, are listed in the fourth column of Table 8.2. From this table we compute the errors involved in both approximations at $x = 0.2, 0.4, 0.6, 0.8$, and 1.0. These errors are tabulated in Table 8.3.

A study of these tables illustrates two important facts concerning the Euler method. First, for a fixed value of h, the error becomes greater and greater as we proceed over a larger and larger range away from the initial point. Second, for a fixed value of x_n, the error is smaller if the value of h is smaller.

TABLE 8.3

x_n	Error using $h = 0.2$	Error using $h = 0.1$
0.2	0.064	0.034
0.4	0.155	0.083
0.6	0.282	0.152
0.8	0.455	0.246
1.0	0.690	0.375

Exercises

1. Consider the initial-value problem

$$\frac{dy}{dx} = x - 2y,$$

$$y(0) = 1.$$

(a) Apply the Euler method to approximate the values of the solution y at $x = 0.1, 0.2, 0.3,$ and $0.4,$ using $h = 0.1$. Obtain results to three figures after the decimal point.

(b) Proceed as in part (a) using $h = 0.05$.

(c) Find the exact solution of the problem and determine its values at $x = 0.1, 0.2, 0.3,$ and 0.4 (to three figures after the decimal point).

(d) Compare the results obtained in parts (a), (b), and (c). Tabulate errors as in Table 8.3.

2. Proceed as in Exercise 1 for the initial-value problem

$$\frac{dy}{dx} = x + y,$$

$$y(0) = 2.$$

C. The Modified Euler Method

In Section B we observed that the value $y(x_1)$ of the exact solution y of the initial value problem

$$\frac{dy}{dx} = f(x, y), \tag{8.1}$$

$$y(x_0) = y_0, \tag{8.6}$$

at $x_1 = x_0 + h$ is given by

$$y(x_1) = y_0 + \int_{x_0}^{x_1} f(x, y) \, dx. \tag{8.47}$$

In the Euler method we approximated $f(x, y)$ in (8.47) by its value $f(x_0, y_0)$ at the left endpoint of the interval $x_0 \le x \le x_1$ and thereby obtained the approximation

$$y_1 = y_0 + hf(x_0, y_0) \tag{8.48}$$

for y at x_1. It seems reasonable that a more accurate value would be obtained if we were to approximate $f(x, y)$ by the *average* of its values at the left and right endpoints of $x_0 \le x \le x_1$, instead of simply by its value at the left endpoint x_0. This is essentially what is done in the modified Euler method, which we shall now explain.

In order to approximate $f(x, y)$ by the average of its values at x_0 and x_1, we need to know its value $f[x_1, y(x_1)]$ at x_1. However, we do not know the value $y(x_1)$ of y at x_1. We must find a first approximation $y_1^{(1)}$ for $y(x_1)$, and to do this we proceed just as we did at the start of the basic Euler method. That is, we take

$$y_1^{(1)} = y_0 + hf(x_0, y_0) \tag{8.50}$$

as the *first* approximation to the value of y at x_1. Then we approximate $f[x_1, y(x_1)]$ by $f(x_1, y_1^{(1)})$, using the value $y_1^{(1)}$ found by (8.50). From this we obtain

$$\frac{f(x_0, y_0) + f(x_1, y_1^{(1)})}{2}, \tag{8.51}$$

which is approximately the average of the values of $f(x, y)$ at the endpoints x_0 and x_1. We now replace $f(x, y)$ in (8.47) by (8.51) and thereby obtain

$$y_1^{(2)} = y_0 + \frac{f(x_0, y_0) + f(x_1, y_1^{(1)})}{2} h \tag{8.52}$$

as the *second* approximation to the value of y at x_1.

We now use the second approximation $y_1^{(2)}$ to obtain a second approximation $f(x_1, y_1^{(2)})$ for the value of $f(x, y)$ at x_1. From this we proceed to obtain

$$y_1^{(3)} = y_0 + \frac{f(x_0, y_0) + f(x_1, y_1^{(2)})}{2} h \tag{8.53}$$

as the *third* approximation to the value of y at x_1. Proceeding in this way we obtain a sequence of approximations

$$y_1^{(1)}, y_1^{(2)}, y_1^{(3)}, \ldots$$

to the value of the exact solution y at x_1. We proceed to compute the successive members of the sequence until we encounter two consecutive members that have the same value to the number of decimal places required. We take the common value of these two consecutive members as our approximation to the value of the solution y at x_1 and denote it by y_1.

Having finally approximated y at x_1 by y_1, we now move on and proceed to approximate y at $x_2 = x_1 + h$. We proceed in exactly the same way as we did in finding y_1. We find successively

$$y_2^{(1)} = y_1 + hf(x_1, y_1),$$

$$y_2^{(2)} = y_1 + \frac{f(x_1, y_1) + f(x_2, y_2^{(1)})}{2} h,$$

$$y_2^{(3)} = y_1 + \frac{f(x_1, y_1) + f(x_2, y_2^{(2)})}{2} h, \tag{8.54}$$

$$\vdots$$

until two consecutive members of this sequence agree, thereby obtaining an approximation y_2 to the value of y at x_2.

Proceeding further in like manner one obtains an approximation y_3 to the value of y at x_3, and so forth.

▶ **Example 8.8**

Apply the modified Euler method to the initial-value problem

$$\frac{dy}{dx} = 2x + y, \tag{8.2}$$

$$y(0) = 1. \tag{8.44}$$

Employ the method to approximate the value of the solution y at $x = 0.2$ and $x = 0.4$ using $h = 0.2$. Obtain results to three figures after the decimal point. Compare with the results obtained using the basic Euler method with $h = 0.1$ and with the exact values (Example 8.7, Table 8.2).

Solution. Here $f(x, y) = 2x + y$, $x_0 = 0$, and $y_0 = 1$, and we are to use $h = 0.2$. We begin by approximating the value of y at $x_1 = x_0 + h = 0.2$. A first approximation $y_1^{(1)}$ to this value is found using formula (8.50). Since $f(x_0, y_0) = f(0, 1) = 1.000$, we have

$$y_1^{(1)} = y_0 + hf(x_0, y_0) = 1.000 + 0.2(1.000) = 1.200.$$

We now use (8.52) to find a second approximation $y_1^{(2)}$ to the desired value. Since $f(x_1, y_1^{(1)}) = f(0.2, 1.200) = 1.600$, we have

$$y_1^{(2)} = y_0 + \frac{f(x_0, y_0) + f(x_1, y_1^{(1)})}{2} h = 1.000 + \frac{1.000 + 1.600}{2}(0.2) = 1.260.$$

We next employ (8.53) to find a third approximation $y_1^{(3)}$. Since $f(x_1, y_1^{(2)}) = f(0.2, 1.260) = 1.660$, we find

$$y_1^{(3)} = y_0 + \frac{f(x_0, y_0) + f(x_1, y_1^{(2)})}{2} h = 1.000 + \frac{1.000 + 1.660}{2}(0.2) = 1.266.$$

Proceeding in like manner, we obtain fourth and fifth approximations $y_1^{(4)}$ and $y_1^{(5)}$, respectively. We find

$$y_1^{(4)} = y_0 + \frac{f(x_0, y_0) + f(x_1, y_1^{(3)})}{2} h = 1.000 + \frac{1.000 + 1.666}{2}(0.2) = 1.267$$

and

$$y_1^{(5)} = y_0 + \frac{f(x_0, y_0) + f(x_1, y_1^{(4)})}{2} h = 1.000 + \frac{1.000 + 1.667}{2}(0.2) = 1.267.$$

Since the approximations $y_1^{(4)}$ and $y_1^{(5)}$ are the same to the number of decimal places required, we take their common value as the approximation y_1 to the value of the solution y at $x_1 = 0.2$. That is, we take

$$y_1 = 1.267. \tag{8.55}$$

We now proceed to approximate the value of y at $x_2 = x_1 + h = 0.4$. For this purpose we employ the formulas (8.54), using $y_1 = 1.267$. We find successively

$$y_2^{(1)} = y_1 + hf(x_1, y_1) = 1.267 + 0.2(1.667) = 1.600,$$

$$y_2^{(2)} = y_1 + \frac{f(x_1, y_1) + f(x_2, y_2^{(1)})}{2} h = 1.267 + \frac{1.667 + 2.400}{2} (0.2) = 1.674,$$

$$y_2^{(3)} = y_1 + \frac{f(x_1, y_1) + f(x_2, y_2^{(2)})}{2} h = 1.267 + \frac{1.667 + 2.474}{2} (0.2) = 1.681,$$

$$y_2^{(4)} = y_1 + \frac{f(x_1, y_1) + f(x_2, y_2^{(3)})}{2} h = 1.267 + \frac{1.667 + 2.481}{2} (0.2) = 1.682,$$

$$y_2^{(5)} = y_1 + \frac{f(x_1, y_1) + f(x_2, y_2^{(4)})}{2} h = 1.267 + \frac{1.667 + 2.482}{2} (0.2) = 1.682.$$

Since the approximations $y_2^{(4)}$ and $y_2^{(5)}$ are both the same to the required number of decimal places, we take their common value as the approximation y_2 to the value of the solution y at $x_2 = 0.4$. That is, we take

$$y_2 = 1.682. \tag{8.56}$$

We compare the results (8.55) and (8.56) with those obtained using the basic Euler method with $h = 0.1$ and with the exact values. For this purpose the various results and the corresponding errors are listed in Table 8.4.

The principal advantage of the modified Euler method over the basic Euler method is immediately apparent from a study of Table 8.4. The modified method is much more accurate. At $x = 0.4$ the error using the modified method with $h = 0.02$ is 0.007. The corresponding error 0.083 using the basic method with $h = 0.1$ is nearly twelve times as great, despite the fact that a smaller value of h was used in this case. Of course at each step the modified method involves more lengthy and complicated calculations than the basic method. We note, however, that the basic method would require many individual steps to give a result as accurate as that which the modified method can provide in a single step.

TABLE 8.4

x_n	Exact value of y (to three decimal places)	Using basic Euler method with $h = 0.1$		Using modified Euler with $h = 0.2$	
		Approximation y_n	Error	Approximation y_n	Error
0.2	1.264	1.230	0.034	1.267	0.003
0.4	1.675	1.592	0.083	1.682	0.007

Exercises

1. Consider the initial-value problem

$$\frac{dy}{dx} = 3x + 2y, \qquad y(0) = 1.$$

(a) Apply the modified Euler method to approximate the values of the solution y at $x = 0.1, 0.2$, and 0.3 using $h = 0.1$. Obtain results to three figures after the decimal point.

(b) Proceed as in part (a) using $h = 0.05$.

(c) Find the exact solution of the problem and determine its values at $x = 0.1, 0.2$, and 0.3 (to three figures after the decimal point).

(d) Compare the results obtained in parts (a), (b), and (c), and tabulate errors.

2. Proceed as in Exercise 1 for the initial-value problem

$$\frac{dy}{dx} = 2x - y,$$

$$y(0) = 3.$$

3. Consider the initial-value problem

$$\frac{dy}{dx} = x^2 + y^2,$$

$$y(0) = 1.$$

(a) Apply the modified Euler method to approximate the values of the solution y at $x = 0.1, 0.2$, and 0.3, using $h = 0.1$. Obtain results to three figures after the decimal point.

(b) Apply the Euler method to approximate the values of the solution y at $x = 0.1, 0.2$, and 0.3, using $h = 0.1$. Obtain results to three figures after the decimal point.

(c) Compare the results obtained in parts (a) and (b) and tabulate errors.

D. The Runge–Kutta Method

We now consider the so-called Runge–Kutta method for approximating the values of the solution of the initial-value problem

$$\frac{dy}{dx} = f(x, y), \tag{8.1}$$

$$y(x_0) = y_0, \tag{8.6}$$

at $x_1 = x_0 + h$, $x_2 = x_1 + h$, and so forth. This method gives surprisingly accurate results without the need of using extremely small values of the interval h. We shall give no justification for the method but small merely list the several formulas involved and explain how they are used.*

To approximate the value of the solution of the initial-value problem under consideration at $x_1 = x_0 + h$ by the Runge–Kutta method, we proceed in the following way. We calculate successively the numbers k_1, k_2, k_3, k_4, and K_0 defined by the

---

* See F. Hildebrand, *Introduction to Numerical Analysis*, 2nd ed. (McGraw-Hill, New York, 1974).

formulas

$$k_1 = hf(x_0, y_0),$$

$$k_2 = hf\left(x_0 + \frac{h}{2}, y_0 + \frac{k_1}{2}\right),$$

$$k_3 = hf\left(x_0 + \frac{h}{2}, y_0 + \frac{k_2}{2}\right), \tag{8.57}$$

$$k_4 = hf(x_0 + h, y_0 + k_3),$$

and

$$K_0 = \tfrac{1}{6}(k_1 + 2k_2 + 2k_3 + k_4).$$

Then we set

$$y_1 = y_0 + K_0 \tag{8.58}$$

and take this as the approximate value of the exact solution at $x_1 = x_0 + h$.

Having thus determined y_1, we proceed to approximate the value of the solution at $x_2 = x_1 + h$ in an exactly similar manner. Using $x_1 = x_0 + h$ and y_1 as determined by (8.58), we calculate successively the numbers k_1, k_2, k_3, k_4, and K_1 defined by

$$k_1 = hf(x_1, y_1),$$

$$k_2 = hf\left(x_1 + \frac{h}{2}, y_1 + \frac{k_1}{2}\right),$$

$$k_3 = hf\left(x_1 + \frac{h}{2}, y_1 + \frac{k_2}{2}\right), \tag{8.59}$$

$$k_4 = hf(x_1 + h, y_1 + k_3),$$

and

$$K_1 = \tfrac{1}{6}(k_1 + 2k_2 + 2k_3 + k_4).$$

Then we set

$$y_2 = y_1 + K_1 \tag{8.60}$$

and take this as the approximate value of the exact solution at $x_2 = x_1 + h$.

We proceed to approximate the value of the solution at $x_3 = x_2 + h$, $x_4 = x_3 + h$, and so forth, in an exactly similar manner. Letting y_n denote the approximate value obtained for the solution at $x_n = x_0 + nh$, we calculate successively k_1, k_2, k_3, k_4, and K_n defined by

$$k_1 = hf(x_n, y_n),$$

$$k_2 = hf\left(x_n + \frac{h}{2}, y_n + \frac{k_1}{2}\right),$$

$$k_3 = hf\left(x_n + \frac{h}{2}, y_n + \frac{k_2}{2}\right),$$

$$k_4 = hf(x_n + h, y_n + k_3),$$

and

$$K_n = \tfrac{1}{6}(k_1 + 2k_2 + 2k_3 + k_4).$$

Then we set

$$y_{n+1} = y_n + K_n$$

and take this as the approximate value of the exact solution at $x_{n+1} = x_n + h$.

▶ **Example 8.9**

Apply the Runge–Kutta method to the initial-value problem

$$\frac{dy}{dx} = 2x + y, \qquad (8.2)$$

$$y(0) = 1. \qquad (8.44)$$

Employ the method to approximate the value of the solution y at $x = 0.2$ and $x = 0.4$ using $h = 0.2$. Carry the intermediate calculations in each step to five figures after the decimal point, and round off the final results of each step to four such places. Compare with the exact value.

Solution. Here $f(x, y) = 2x + y$, $x_0 = 0$, $y_0 = 1$, and we are to use $h = 0.2$. Using these quantities we calculate successively k_1, k_2, k_3, k_4, and K_0 defined by (8.57). We first find

$$k_1 = hf(x_0, y_0) = 0.2f(0, 1) = 0.2(1) = 0.20000.$$

Then since

$$x_0 + \frac{h}{2} = 0 + \frac{1}{2}(0.2) = 0.1$$

and

$$y_0 + \frac{k_1}{2} = 1.00000 + \frac{1}{2}(0.20000) = 1.10000,$$

we find

$$k_2 = hf\left(x_0 + \frac{h}{2}, y_0 + \frac{k_1}{2}\right) = 0.2f(0.1, 1.10000)$$

$$= 0.2(1.30000) = 0.26000.$$

Next, since

$$y_0 + \frac{k_2}{2} = 1.00000 + \frac{1}{2}(0.26000) = 1.13000,$$

we find

$$k_3 = hf\left(x_0 + \frac{h}{2}, y_0 + \frac{k_2}{2}\right) = 0.2f(0.1, 1.13000)$$

$$= 0.2(1.33000) = 0.26600.$$

Since $x_0 + h = 0.2$ and $y_0 + k_3 = 1.00000 + 0.26600 = 1.26600$, we obtain

$$k_4 = hf(x_0 + h, y_0 + k_3) = 0.2f(0.2, 1.26600)$$
$$= 0.2(1.66600) = 0.33320.$$

Finally, we find

$$K_0 = \tfrac{1}{6}(k_1 + 2k_2 + 2k_3 + k_4) = \tfrac{1}{6}(0.20000 + 0.52000 + 0.53200 + 0.33320)$$
$$= 0.26420.$$

Then by (8.58) the approximate value of the solution at $x_1 = 0.2$ is

$$y_1 = 1 + 0.2642 = 1.2642. \tag{8.61}$$

Now using y_1 as given by (8.61), we calculate successively k_1, k_2, k_3, k_4, and K_1 defined by (8.59). We first find

$$k_1 = hf(x_1, y_1) = 0.2f(0.2, 1.2642) = 0.2(1.6642) = 0.33284.$$

Then since

$$x_1 + \frac{h}{2} = 0.2 + \frac{1}{2}(0.2) = 0.3$$

and

$$y_1 + \frac{k_1}{2} = 1.26420 + \frac{1}{2}(0.33284) = 1.43062,$$

we find

$$k_2 = hf\left(x_1 + \frac{h}{2}, y_1 + \frac{k_1}{2}\right) = 0.2f(0.3, 1.43062)$$
$$= 0.2(2.03062) = 0.40612.$$

Next, since

$$y_1 + \frac{k_2}{2} = 1.26420 + \frac{1}{2}(0.40612) = 1.46726,$$

we find

$$k_3 = hf\left(x_1 + \frac{h}{2}, y_1 + \frac{k_2}{2}\right) = 0.2f(0.3, 1.46726)$$
$$= 0.2(2.06726) = 0.41345.$$

Since $x_1 + h = 0.4$ and $y_1 + k_3 = 1.26420 + 0.41345 = 1.67765$, we obtain

$$k_4 = hf(x_1 + h, y_1 + k_3) = 0.2f(0.4, 1.67765)$$
$$= 0.2(2.47765) = 0.49553.$$

Finally, we find

$$K_1 = \tfrac{1}{6}(k_1 + 2k_2 + 2k_3 + k_4)$$
$$= \tfrac{1}{6}(0.33284 + 0.81224 + 0.82690 + 0.49553) = 0.41125.$$

We round off K_1 according to the well-known rule of rounding off so as to leave the last digit retained an even one. Then by (8.60) we see that the approximate value of the solution at $x_2 = 0.4$ is

$$y_2 = 1.2642 + 0.4112 = 1.6754. \qquad (8.62)$$

As before, the exact values are determined using (8.46). Rounded off to four places after the decimal point, the exact values at $x = 0.2$ and $x = 0.4$ are 1.2642 and 1.6754, respectively. The approximate value at $x = 0.2$ as given by (8.61) is therefore correct to four places after the decimal point, and the approximate value at $x = 0.4$ as given by (8.62) is likewise correct to four places!

The remarkable accuracy of the Runge–Kutta method in this problem is certainly apparent. In fact, if we employ the method to approximate the solution at $x = 0.4$ using $h = 0.4$ (that is, in only one step), we obtain the value 1.6752, which differs from the exact value 1.6754 by merely 0.0002.

Exercises

1. Consider the initial-value problem

$$\frac{dy}{dx} = 3x + 2y,$$

$$y(0) = 1.$$

(a) Apply the Runge–Kutta method to approximate the values of the solution y at $x = 0.1, 0.2,$ and 0.3, using $h = 0.1$. Carry the intermediate calculations in each step to five figures after the decimal point, and round off the final results of each step to four such places.

(b) Find the exact solution of the problem and compare the results obtained in part (a) with the exact values.

2. Proceed as in Exercise 1 for the initial-value problem

$$\frac{dy}{dx} = 2x - y,$$

$$y(0) = 3.$$

3. Proceed as in part (a) of Exercise 1 for initial-value problem

$$\frac{dy}{dx} = x^2 + y^2,$$

$$y(0) = 1.$$

E. The Milne Method

The Euler method, the modified Euler method, and the Runge–Kutta method are all *starting* methods for the numerical solution of an initial-value problem. As we have already pointed out, a starting method is a method that can be used to start the solution. In contrast, a *continuing* method is one that cannot be used to start the solution but that can be used to continue it, once it is sufficiently well started. In this section we consider briefly a useful continuing method, the so-called Milne method. We shall give no justification for this method but shall merely list the formulas involved and indicate how they are employed.*

The Milne method can be used to approximate the value of the solution of the initial-value problem

$$\frac{dy}{dx} = f(x, y), \tag{8.1}$$

$$y(x_0) = y_0 \tag{8.6}$$

at $x_{n+1} = x_0 + (n+1)h$, provided the values at the four previous points $x_{n-3}, x_{n-2}, x_{n-1}$, and x_n have been determined. We assume that these four previous values have been found and denote them by $y_{n-3}, y_{n-2}, y_{n-1}$, and y_n, respectively. Then we can use (8.1) to determine dy/dx at x_{n-2}, x_{n-1}, and x_n. That is, we can determine $y'_{n-2} = f(x_{n-2}, y_{n-2})$, $y'_{n-1} = f(x_{n-1}, y_{n-1})$, and $y'_n = f(x_n, y_n)$. These various numbers being determined, the Milne method proceeds as follows:

We first determine the number $y_{n+1}^{(1)}$ given by the formula

$$y_{n+1}^{(1)} = y_{n-3} + \frac{4h}{3}(2y'_n - y'_{n-1} + 2y'_{n-2}). \tag{8.63}$$

Having thus determined $y_{n+1}^{(1)}$, we next determine the number $y'^{(1)}_{n+1}$ given by

$$y'^{(1)}_{n+1} = f(x_{n+1}, y_{n+1}^{(1)}). \tag{8.64}$$

Finally, having determined $y'^{(1)}_{n+1}$, we proceed to determine the number $y_{n+1}^{(2)}$ given by

$$y_{n+1}^{(2)} = y_{n-1} + \frac{h}{3}(y'^{(1)}_{n+1} + 4y'_n + y'_{n-1}). \tag{8.65}$$

If the numbers $y_{n+1}^{(1)}$ determined from (8.63) and $y_{n+1}^{(2)}$ determined from (8.65) are the same to the number of decimal places required, then we take this common value to be the approximate value of the solution at x_{n+1} and denote it by y_{n+1}.

If the numbers $y_{n+1}^{(1)}$ and $y_{n+1}^{(2)}$ so determined do not agree to the number of decimal places required and all of the calculations have been checked and appear to be correct, then we proceed in the following way. We calculate the number

$$E = \frac{y_{n+1}^{(2)} - y_{n+1}^{(1)}}{29},$$

which is the principal part of the error in the formula (8.65). If E is negligible with respect to the number of decimal places required, then we take the number $y_{n+1}^{(2)}$ given by (8.65) as the approximate value of the solution at x_{n+1} and denote it by y_{n+1}. On the other hand, if E is so large that it is not negligible with respect to the number of decimal places required, then the value of h employed is too large and a smaller value must be used.

We observe that once the values y_0, y_1, y_2, and y_3 have been determined, we can use the Milne formulas with $n = 3$ to determine y_4. Then when y_4 has been determined by the formulas, we can use them with $n = 4$ to determine y_5. Then proceeding in like manner, we can successively determine $y_6, y_7, \ldots$. But we must have y_0, y_1, y_2, and y_3 in order to start the Milne method. Of course y_0 is given exactly by the initial condition

---

* See J. Scarborough, *Numerical Mathematical Analysis*, 6th ed. (Johns Hopkins, Baltimore, 1966).

(8.6), and we can find y_1, y_2, and y_3 by one of the previously explained starting methods (for example, by the Runge–Kutta method).

▶ **Example 8.10**

Apply the Milne method to approximate the value at $x = 0.4$ of the solution y of the initial-value problem

$$\frac{dy}{dx} = 2x + y, \tag{8.2}$$

$$y(0) = 1, \tag{8.44}$$

assuming that the values at $0.1, 0.2$, and 0.3 are $1.1155, 1.2642$, and 1.4496, respectively.

Solution. We apply the formulas (8.63), (8.64), and (8.65) with $n = 3$. We set

$$x_0 = 0, \qquad y_0 = 1.0000,$$
$$x_1 = 0.1, \qquad y_1 = 1.1155,$$
$$x_2 = 0.2, \qquad y_2 = 1.2642,$$
$$x_3 = 0.3, \qquad y_3 = 1.4496,$$

and $x_4 = 0.4$. Then using $f(x, y) = 2x + y$, we find

$$y_1' = f(x_1, y_1) = f(0.1, 1.1155) = 1.3155,$$
$$y_2' = f(x_2, y_2) = f(0.2, 1.2642) = 1.6642,$$
$$y_3' = f(x_3, y_3) = f(0.3, 1.4496) = 2.0496.$$

We now use (8.63) with $n = 3$ and $h = 0.1$ to determine $y_4^{(1)}$. We have

$$y_4^{(1)} = y_0 + \frac{4(0.1)}{3} (2y_3' - y_2' + 2y_1')$$

$$= 1.000 + \frac{0.4}{3} (4.0992 - 1.6642 + 2.6310) = 1.6755.$$

Having thus determined $y_4^{(1)}$, we use (8.64) with $n = 3$ to determine $y_4'^{(1)}$. We find

$$y_4'^{(1)} = f(x_4, y_4^{(1)}) = f(0.4, 1.6755) = 2.4755.$$

Finally, having determined $y_4'^{(1)}$, we use (8.65) with $n = 3$ and $h = 0.1$ to determine $y_4^{(2)}$. We obtain

$$y_4^{(2)} = y_2 + \frac{0.1}{3} (y_4'^{(1)} + 4y_3' + y_2')$$

$$= 1.2642 + \frac{0.1}{3} (2.4755 + 8.1984 + 1.6642)$$

$$= 1.6755.$$

Since the numbers $y_4^{(1)}$ and $y_4^{(2)}$ agree to four decimal places, we take their common value as the approximate value of the solution at $x_4 = 0.4$ and denote it by y_4. That is, we set

$$y_4 = 1.6755.$$

Using (8.46), the exact value at $x = 0.4$, rounded off to four places after the decimal point, is found to be 1.6754.

Exercises

1. Apply the Milne method to approximate the value at $x = 0.4$ of the solution y of the initial-value problem

$$\frac{dy}{dx} = 3x + 2y,$$

$$y(0) = 1,$$

assuming that the values at 0.1, 0.2, and 0.3 are 1.2375, 1.5607, and 1.9887, respectively.

2. Apply the Milne method to approximate the value at $x = 0.4$ of the solution y of the initial-value problem

$$\frac{dy}{dx} = 2x - y,$$

$$y(0) = 3,$$

assuming that the values at 0.1, 0.2, and 0.3 are 2.7242, 2.4937, and 2.3041, respectively.

CHAPTER NINE

The Laplace Transform

In this chapter we shall introduce a concept that is especially useful in the solution of initial-value problems. This concept is the so-called Laplace transform, which transforms a suitable function f of a real variable t into a related function F of a real variable s. When this transform is applied in connection with an initial-value problem involving a linear differential equation in an "unknown" function of t, it transforms the given initial-value problem into an algebraic problem involving the variable s. In Section 9.3 we shall indicate just how this transformation is accomplished and how the resulting algebraic problem is then employed to find the solution of the given initial-value problem. First, however, in Section 9.1 we shall introduce the Laplace transform itself and develop certain of its most basic and useful properties.

9.1 DEFINITION, EXISTENCE, AND BASIC PROPERTIES OF THE LAPLACE TRANSFORM

A. Definition and Existence

DEFINITION

Let f be a real-valued function of the real variable t, defined for $t > 0$. Let s be a variable that we shall assume to be real, and consider the function F defined by

$$F(s) = \int_0^\infty e^{-st} f(t)\, dt, \tag{9.1}$$

for all values of s for which this integral exists. The function F defined by the integral (9.1) is called the Laplace transform *of the function f. We shall denote the Laplace transform F of f by $\mathscr{L}\{f\}$ and shall denote $F(s)$ by $\mathscr{L}\{f(t)\}$.*

In order to be certain that the integral (9.1) does exist for some range of values of s, we must impose suitable restrictions upon the function f under consideration. We shall do this shortly; however, first we shall directly determine the Laplace transforms of a few simple functions.

▶ **Example 9.1**

Consider the function f defined by

$$f(t) = 1, \qquad \text{for } t > 0.$$

Then

$$\mathscr{L}\{1\} = \int_0^\infty e^{-st} \cdot 1 \, dt = \lim_{R \to \infty} \int_0^R e^{-st} \cdot 1 \, dt = \lim_{R \to \infty} \left[\frac{-e^{-st}}{s} \right]_0^R$$

$$= \lim_{R \to \infty} \left[\frac{1}{s} - \frac{e^{-sR}}{s} \right] = \frac{1}{s}$$

for all $s > 0$. Thus we have

$$\mathscr{L}\{1\} = \frac{1}{s} \qquad (s > 0). \tag{9.2}$$

▶ **Example 9.2**

Consider the function f defined by

$$f(t) = t, \qquad \text{for } t > 0.$$

Then

$$\mathscr{L}\{t\} = \int_0^\infty e^{-st} \cdot t \, dt = \lim_{R \to \infty} \int_0^R e^{-st} \cdot t \, dt = \lim_{R \to \infty} \left[-\frac{e^{-st}}{s^2}(st + 1) \right]_0^R$$

$$= \lim_{R \to \infty} \left[\frac{1}{s^2} - \frac{e^{-sR}}{s^2}(sR + 1) \right] = \frac{1}{s^2}$$

for all $s > 0$. Thus

$$\mathscr{L}\{t\} = \frac{1}{s^2} \qquad (s > 0). \tag{9.3}$$

▶ **Example 9.3**

Consider the function f defined by

$$f(t) = e^{at}, \qquad \text{for } t > 0.$$

$$\mathscr{L}\{e^{at}\} = \int_0^\infty e^{-st}e^{at} \, dt = \lim_{R \to \infty} \int_0^R e^{(a-s)t} \, dt = \lim_{R \to \infty} \left[\frac{e^{(a-s)t}}{a - s} \right]_0^R$$

$$= \lim_{R \to \infty} \left[\frac{e^{(a-s)R}}{a - s} - \frac{1}{a - s} \right] = -\frac{1}{a - s} = \frac{1}{s - a} \qquad \text{for all } s > a.$$

Thus

$$\mathcal{L}\{e^{at}\} = \frac{1}{s-a} \qquad (s > a). \tag{9.4}$$

▶ **Example 9.4**

Consider the function f defined by

$$f(t) = \sin bt \qquad \text{for } t > 0.$$

$$\mathcal{L}\{\sin bt\} = \int_0^\infty e^{-st} \cdot \sin bt \, dt = \lim_{R \to \infty} \int_0^R e^{-st} \cdot \sin bt \, dt$$

$$= \lim_{R \to \infty} \left[-\frac{e^{-st}}{s^2 + b^2} (s \sin bt + b \cos bt) \right]_0^R$$

$$= \lim_{R \to \infty} \left[\frac{b}{s^2 + b^2} - \frac{e^{-sR}}{s^2 + b^2} (s \sin bR + b \cos bR) \right]$$

$$= \frac{b}{s^2 + b^2} \qquad \text{for all } s > 0.$$

Thus

$$\mathcal{L}\{\sin bt\} = \frac{b}{s^2 + b^2} \qquad (s > 0). \tag{9.5}$$

▶ **Example 9.5**

Consider the function f defined by

$$f(t) = \cos bt \qquad \text{for } t > 0.$$

$$\mathcal{L}\{\cos bt\} = \int_0^\infty e^{-st} \cdot \cos bt \, dt = \lim_{R \to \infty} \int_0^R e^{-st} \cos bt \, dt$$

$$= \lim_{R \to \infty} \left[\frac{e^{-st}}{s^2 + b^2} (-s \cos bt + b \sin bt) \right]_0^R$$

$$= \lim_{R \to \infty} \left[\frac{e^{-sR}}{s^2 + b^2} (-s \cos bR + b \sin bR) + \frac{s}{s^2 + b^2} \right]$$

$$= \frac{s}{s^2 + b^2} \qquad \text{for all } s > 0.$$

Thus

$$\mathcal{L}\{\cos bt\} = \frac{s}{s^2 + b^2} \qquad (s > 0). \tag{9.6}$$

In each of the above examples we have seen directly that the integral (9.1) actually does exist for some range of values of s. We shall now determine a class of functions f for which this is always the case. To do so we first consider certain properties of functions.

DEFINITION

A function f is said to be piecewise continuous (*or sectionally continuous*) *on a finite interval a ≤ t ≤ b if this interval can be divided into a finite number of subintervals such that (1) f is continuous in the interior of each of these subintervals, and (2) f(t) approaches finite limits as t approaches either endpoint of each of the subintervals from its interior.*

Suppose f is piecewise continuous on $a \le t \le b$, and t_0, $a < t_0 < b$, is an endpoint of one of the subintervals of the above definition. Then the finite limit approached by $f(t)$ as t approaches t_0 from the left (that is, through smaller values of t) is called the *left-hand limit* of $f(t)$ as t approaches t_0, denoted by $\lim_{t \to t_0-} f(t)$ or by $f(t_0 -)$. In like manner, the finite limit approached by $f(t)$ as t approaches t_0 from the right (through larger values) is called the *right-hand limit* of $f(t)$ as t approaches t_0, denoted by $\lim_{t \to t_0+} f(t)$ or $f(t_0 +)$. We emphasize that at such a point t_0, both $f(t_0 -)$ and $f(t_0 +)$ are finite but they are not in general equal.

We point out that if f is continuous on $a \le t \le b$ it is necessarily piecewise continuous on this interval. Also, we note that if f is piecewise continuous on $a \le t \le b$, then f is integrable on $a \le t \le b$.

▶ **Example 9.6**

Consider the function f defined by

$$f(t) = \begin{cases} -1, & 0 < t < 2, \\ 1, & t > 2. \end{cases}$$

f is piecewise continuous on every finite interval $0 \le t \le b$, for every positive number b. At $t = 2$, we have

$$f(2-) = \lim_{t \to 2-} f(t) = -1,$$

$$f(2+) = \lim_{t \to 2+} f(t) = +1.$$

The graph of f is shown in Figure 9.1.

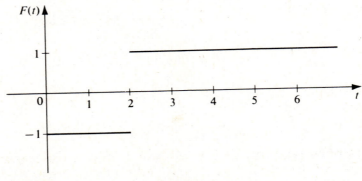

Figure 9.1

DEFINITION

A function f is said to be of exponential order *if there exists a constant α and positive constants t_0 and M such that*

$$e^{-\alpha t}|f(t)| < M \tag{9.7}$$

for all $t > t_0$ at which $f(t)$ is defined. More explicitly, if f is of exponential order corresponding to some definite constant α in (9.7), then we say that f is of exponential order $e^{\alpha t}$.

In other words, we say that f is of exponential order if a constant α exists such that the product $e^{-\alpha t}|f(t)|$ is bounded for all sufficiently large values of t. From (9.7) we have

$$|f(t)| < Me^{\alpha t} \tag{9.8}$$

for all $t > t_0$ at which $f(t)$ is defined. Thus if f is of exponential order and the values $f(t)$ of f become infinite as $t \to \infty$, these values cannot become infinite more rapidly than a multiple M of the corresponding values $e^{\alpha t}$ of some exponential function. We note that if f is of exponential order $e^{\alpha t}$, then f is also of exponential order $e^{\beta t}$ for any $\beta > \alpha$.

▶ **Example 9.7**

Every bounded function is of exponential order, with the constant $\alpha = 0$. Thus, for example, $\sin bt$ and $\cos bt$ are of exponential order.

▶ **Example 9.8**

The function f such that $f(t) = e^{at} \sin bt$ is of exponential order, with the constant $\alpha = a$. For we then have

$$e^{-\alpha t}|f(t)| = e^{-at}e^{at}|\sin bt| = |\sin bt|,$$

which is bounded for all t.

▶ **Example 9.9**

Consider the function f such that $f(t) = t^n$, where $n > 0$. Then $e^{-\alpha t}|f(t)|$ is $e^{-\alpha t}t^n$. For any $\alpha > 0$, $\lim_{t \to \infty} e^{-\alpha t}t^n = 0$. Thus there exists $M > 0$ and $t_0 > 0$ such that

$$e^{-\alpha t}|f(t)| = e^{-\alpha t}t^n < M$$

for $t > t_0$. Hence $f(t) = t^n$ is of exponential order, with the constant α equal to any positive number.

▶ **Example 9.10**

The function f such that $f(t) = e^{t^2}$ is *not* of exponential order, for in this case $e^{-\alpha t}|f(t)|$ is $e^{t^2 - \alpha t}$ and this becomes unbounded as $t \to \infty$, no matter what is the value of α.

We shall now proceed to obtain a theorem giving conditions on f that are sufficient for the integral (9.1) to exist. To obtain the desired result we shall need the following two theorems from advanced calculus, which we state without proof.

THEOREM A   Comparison Test for Improper Integrals

Hypothesis

1. Let g and G be real functions such that

$$0 \le g(t) \le G(t) \qquad on \quad a \le t < \infty.$$

2. Suppose $\int_a^\infty G(t)\, dt$ exists.
3. Suppose g is integrable on every finite closed subinterval of $a \le t < \infty$.

Conclusion.   Then $\int_a^\infty g(t)\, dt$ exists.

THEOREM B

Hypothesis

1. Suppose the real function g is integrable on every finite closed subinterval of $a \le t \le \infty$.
2. Suppose $\int_a^\infty |g(t)|\, dt$ exists.

Conclusion.   Then $\int_a^\infty g(t)\, dt$ exists.

We now state and prove an existence theorem for Laplace transforms.

THEOREM 9.1

Hypothesis.   Let f be a real function that has the following properties:
1. f is piecewise continuous in every finite closed interval $0 \le t \le b\ (b > 0)$.
2. f is of exponential order; that is, there exists α, $M > 0$, and $t_0 > 0$ such that

$$e^{-\alpha t}|f(t)| < M \qquad for \quad t > t_0.$$

Conclusion.   The Laplace transform

$$\int_0^\infty e^{-st} f(t)\, dt$$

of f exists for $s > \alpha$.

Proof.   We have

$$\int_0^\infty e^{-st} f(t)\, dt = \int_0^{t_0} e^{-st} f(t)\, dt + \int_{t_0}^\infty e^{-st} f(t)\, dt.$$

By Hypothesis 1, the first integral of the right member exists. By Hypothesis 2,

$$e^{-st}|f(t)| < e^{-st} M e^{\alpha t} = M e^{-(s-\alpha)t}$$

for $t > t_0$. Also

$$\int_{t_0}^{\infty} Me^{-(s-\alpha)t}\, dt = \lim_{R \to \infty} \int_{t_0}^{R} Me^{-(s-\alpha)t}\, dt = \lim_{R \to \infty} \left[-\frac{Me^{-(s-\alpha)t}}{s-\alpha} \right]_{t_0}^{R}$$

$$= \lim_{R \to \infty} \left[\frac{M}{s-\alpha} \right] [e^{-(s-\alpha)t_0} - e^{-(s-\alpha)R}]$$

$$= \left[\frac{M}{s-\alpha} \right] e^{-(s-\alpha)t_0} \qquad \text{if} \quad s > \alpha.$$

Thus

$$\int_{t_0}^{\infty} Me^{-(s-\alpha)t}\, dt \qquad \text{exists for} \quad s > \alpha.$$

Finally, by Hypothesis 1, $e^{-st}\,|f(t)|$ is integrable on every finite closed subinterval of $t_0 \leq t < \infty$. Thus, applying Theorem A with $g(t) = e^{-st}\,|f(t)|$ and $G(t) = Me^{-(s-\alpha)t}$, we see that

$$\int_{t_0}^{\infty} e^{-st}\,|f(t)|\, dt \qquad \text{exists if} \quad s > \alpha.$$

In other words,

$$\int_{t_0}^{\infty} |e^{-st} f(t)|\, dt \qquad \text{exists if} \quad s > \alpha,$$

and so by Theorem B

$$\int_{t_0}^{\infty} e^{-st} f(t)\, dt$$

also exists if $s > \alpha$. Thus the Laplace transform of f exists for $s > \alpha$. $\qquad$ Q.E.D.

Let us look back at this proof for a moment. Actually we showed that if f satisfies the hypotheses stated, then

$$\int_{t_0}^{\infty} e^{-st}|f(t)|\, dt \qquad \text{exists if} \quad s > \alpha.$$

Further, Hypothesis 1 shows that

$$\int_{0}^{t_0} e^{-st}|f(t)|\, dt \qquad \text{exists.}$$

Thus

$$\int_{0}^{\infty} e^{-st}|f(t)|\, dt \qquad \text{exists if} \quad s > \alpha.$$

In other words, if f satisfies the hypotheses of Theorem 9.1, then not only does $\mathcal{L}\{f\}$ exist for $s > \alpha$, but also $\mathcal{L}\{|f|\}$ exists for $s > \alpha$. That is,

$$\int_{0}^{\infty} e^{-st} f(t)\, dt \qquad \text{is absolutely convergent for} \quad s > \alpha.$$

We point out that the conditions on f described in the hypothesis of Theorem 9.1 are not necessary for the existence of $\mathscr{L}\{f\}$. In other words, there exist functions f that do *not* satisfy the hypotheses of Theorem 9.1, but for which $\mathscr{L}\{f\}$ exists. For instance, suppose we replace Hypothesis 1 by the following less restrictive condition. Let us suppose that f is piecewise continuous in every finite closed interval $a \le t \le b$, where $a > 0$, and is such that $|t^n f(t)|$ remains bounded as $t \to 0^+$ for some number n, where $0 < n < 1$. Then, provided Hypothesis 2 remains satisfied, it can be shown that $\mathscr{L}\{f\}$ still exists. Thus for example, if $f(t) = t^{-1/3}, t > 0, \mathscr{L}\{f\}$ exists. For although f does *not* satisfy the hypothesis of Theorem 9.1 $[f(t) \to \infty$ as $t \to 0^+]$, it *does* satisfy the less restrictive requirement stated above (take $n = \frac{2}{3}$), and f is of exponential order.

Exercises

In each of Exercises 1–6, use the definition of the Laplace transform to find $\mathscr{L}\{f(t)\}$ for the given $f(t)$.

1. $f(t) = t^2$.

2. $f(t) = \sinh t$.

3. $f(t) = \begin{cases} 5, & 0 < t < 2, \\ 0, & t > 2. \end{cases}$

4. $f(t) = \begin{cases} 4, & 0 < t < 3, \\ 2, & t > 3. \end{cases}$

5. $f(t) = \begin{cases} t, & 0 < t < 2, \\ 3, & t > 2. \end{cases}$

6. $f(t) = \begin{cases} 0, & 0 < t < 1, \\ t, & 1 < t < 2, \\ 1, & t > 2. \end{cases}$

B. Basic Properties of the Laplace Transform

THEOREM 9.2   The Linear Property

Let f_1 and f_2 be functions whose Laplace transforms exist, and let c_1 and c_2 be constants. Then

$$\mathscr{L}\{c_1 f_1(t) + c_2 f_2(t)\} = c_1 \mathscr{L}\{f_1(t)\} + c_2 \mathscr{L}\{f_2(t)\}. \tag{9.9}$$

Proof.   This follows directly from the definition (9.1).

▶ **Example 9.11**

Use Theorem 9.2 to find $\mathscr{L}\{\sin^2 at\}$. Since $\sin^2 at = (1 - \cos 2at)/2$, we have

$$\mathscr{L}\{\sin^2 at\} = \mathscr{L}\{\tfrac{1}{2} - \tfrac{1}{2}\cos 2at\}.$$

By Theorem 9.2,

$$\mathscr{L}\{\tfrac{1}{2} - \tfrac{1}{2}\cos 2at\} = \tfrac{1}{2}\mathscr{L}\{1\} - \tfrac{1}{2}\mathscr{L}\{\cos 2at\}.$$

By (9.2), $\mathscr{L}\{1\} = 1/s$, and by (9.6), $\mathscr{L}\{\cos 2at\} = s/(s^2 + 4a^2)$. Thus

$$\mathscr{L}\{\sin^2 at\} = \frac{1}{2}\cdot\frac{1}{s} - \frac{1}{2}\cdot\frac{s}{s^2 + 4a^2} = \frac{2a^2}{s(s^2 + 4a^2)}. \tag{9.10}$$

THEOREM 9.3

Hypothesis

1. Let f be a real function that is continuous for $t \geq 0$ and of exponential order $e^{\alpha t}$.
2. Let f' (the derivative of f) be piecewise continuous in every finite closed interval
$0 \leq t \leq b.$

Conclusion. *Then $\mathscr{L}\{f'\}$ exists for $s > \alpha$; and*

$$\mathscr{L}\{f'(t)\} = s\mathscr{L}\{f(t)\} - f(0). \tag{9.11}$$

Proof. By definition of the Laplace transform,

$$\mathscr{L}\{f'(t)\} = \lim_{R\to\infty} \int_0^R e^{-st}f'(t)\,dt,$$

provided this limit exists. In any closed interval $0 \leq t \leq R$, $f'(t)$ has at most a finite number of discontinuities; denote these by $t_1, t_2, \ldots, t_n$, where

$$0 \leq t_1 < t_2 < \cdots < t_n \leq R.$$

Then we may write

$$\int_0^R e^{-st}f'(t)\,dt = \int_0^{t_1} e^{-st}f'(t)\,dt + \int_{t_1}^{t_2} e^{-st}f'(t)\,dt + \cdots + \int_{t_n}^R e^{-st}f'(t)\,dt.$$

Now the integrand of each of the integrals on the right is continuous. We may therefore integrate each by parts. Doing so, we obtain

$$\int_0^R e^{-st}f'(t)\,dt = \left[e^{-st}f(t)\right]_0^{t_1-} + s\int_0^{t_1} e^{-st}f(t)\,dt + \left[e^{-st}f(t)\right]_{t_1+}^{t_2-}$$

$$+ s\int_{t_1}^{t_2} e^{-st}f(t)\,dt + \cdots + \left[e^{-st}f(t)\right]_{t_n+}^{R-} + s\int_{t_n}^R e^{-st}f(t)\,dt.$$

By Hypothesis 1, f is continuous for $t \geq 0$. Thus

$$f(t_1-) = f(t_1+), f(t_2-) = f(t_2+), \ldots, f(t_n-) = f(t_n+).$$

Thus all of the integrated "pieces" add out, except for $e^{-st}f(t)|_{t=0}$ and $e^{-st}f(t)|_{t=R-}$, and there remains only

$$\int_0^R e^{-st}f'(t)\,dt = -f(0) + e^{-sR}f(R) + s\int_0^R e^{-st}f(t)\,dt.$$

But by Hypothesis 1 f is of exponential order $e^{\alpha t}$. Thus there exists $M > 0$ and $t_0 > 0$

such that $e^{-\alpha t}|f(t)| < M$ for $t > t_0$. Thus $|e^{-sR}f(R)| < Me^{-(s-\alpha)R}$ for $R > t_0$. Thus if $s > \alpha$,

$$\lim_{R \to \infty} e^{-sR}f(R) = 0.$$

Further,

$$\lim_{R \to \infty} s \int_0^R e^{-st}f(t)\, dt = s\mathcal{L}\{f(t)\}.$$

Thus, we have

$$\lim_{R \to \infty} \int_0^R e^{-st}f'(t)\, dt = -f(0) + s\mathcal{L}\{f(t)\},$$

and so $\mathcal{L}\{f'(t)\}$ exists for $s > \alpha$ and is given by (9.11).   Q.E.D

▶ **Example 9.12**

Consider the function defined by $f(t) = \sin^2 at$. This function satisfies the hypotheses of Theorem 9.3. Since $f'(t) = 2a \sin at \cos at$ and $f(0) = 0$, Equation (9.11) gives

$$\mathcal{L}\{2a \sin at \cos at\} = s\mathcal{L}\{\sin^2 at\}.$$

By (9.10),

$$\mathcal{L}\{\sin^2 at\} = \frac{2a^2}{s(s^2 + 4a^2)}.$$

Thus,

$$\mathcal{L}\{2a \sin at \cos at\} = \frac{2a^2}{s^2 + 4a^2}.$$

Since $2a \sin at \cos at = a \sin 2at$, we also have

$$\mathcal{L}\{\sin 2at\} = \frac{2a}{s^2 + 4a^2}.$$

Observe that this is the result (9.5), obtained in Example 9.4, with $b = 2a$.

We now generalize Theorem 9.3 and obtain the following result:

THEOREM 9.4

Hypothesis

1. *Let f be a real function having a continuous $(n - 1)$st derivative $f^{(n-1)}$ (and hence f, $f', \ldots, f^{(n-2)}$ are also continuous) for $t \geq 0$; and assume that $f, f', \ldots, f^{(n-1)}$ are all of exponential order $e^{\alpha t}$.*
2. *Suppose $f^{(n)}$ is piecewise continuous in every finite closed interval $0 \leq t \leq b$.*

Conclusion. $\mathscr{L}\{f^{(n)}\}$ exists for $s > \alpha$ and

$$\mathscr{L}\{f^{(n)}(t)\} = s^n\mathscr{L}\{f(t)\} - s^{n-1}f(0) - s^{n-2}f'(0) - s^{n-3}f''(0) - \cdots - f^{(n-1)}(0).$$
(9.12)

Outline of Proof. One first proceeds as in the proof of Theorem 9.3 to show that $\mathscr{L}\{f^{(n)}\}$ exists for $s > \alpha$ and is given by

$$\mathscr{L}\{f^{(n)}\} = s\mathscr{L}\{f^{(n-1)}\} - f^{(n-1)}(0).$$

One then completes the proof by mathematical induction.

▶ **Example 9.13**

We apply Theorem 9.4, with $n = 2$, to find $\mathscr{L}\{\sin bt\}$, which we have already found directly and given by (9.5). Clearly the function f defined by $f(t) = \sin bt$ satisfies the hypotheses of the theorem with $\alpha = 0$. For $n = 2$, Equation (9.12) becomes

$$\mathscr{L}\{f''(t)\} = s^2\mathscr{L}\{f(t)\} - sf(0) - f'(0).$$
(9.13)

We have $f'(t) = b\cos bt$, $f''(t) = -b^2\sin bt$, $f(0) = 0$, $f'(0) = b$. Substituting into Equation (9.13) we find

$$\mathscr{L}\{-b^2\sin bt\} = s^2\mathscr{L}\{\sin bt\} - b,$$

and so

$$(s^2 + b^2)\mathscr{L}\{\sin bt\} = b.$$

Thus,

$$\mathscr{L}\{\sin bt\} = \frac{b}{s^2 + b^2} \qquad (s > 0),$$

which is the result (9.5), already found directly.

THEOREM 9.5 Translation Property

Hypothesis. *Suppose f is such that $\mathscr{L}\{f\}$ exists for $s > \alpha$.*

Conclusion. *For any constant a,*

$$\mathscr{L}\{e^{at}f(t)\} = F(s - a)$$
(9.14)

for $s > \alpha + a$, where $F(s)$ denotes $\mathscr{L}\{f(t)\}$.

Proof. $F(s) = \mathscr{L}\{f(t)\} = \int_0^\infty e^{-st}f(t)\,dt$. Replacing s by $s - a$, we have

$$F(s - a) = \int_0^\infty e^{-(s-a)t}f(t)\,dt = \int_0^\infty e^{-st}[e^{at}f(t)]\,dt = \mathscr{L}\{e^{at}f(t)\}.$$

Q.E.D

▶ **Example 9.14**

Find $\mathcal{L}\{e^{at}t\}$. We apply Theorem 9.5 with $f(t) = t$.
$$\mathcal{L}\{e^{at}t\} = F(s - a),$$
where $F(s) = \mathcal{L}\{f(t)\} = \mathcal{L}\{t\}$. By (9.3), $\mathcal{L}\{t\} = 1/s^2$ $(s > 0)$. That is, $F(s) = 1/s^2$ and so $F(s - a) = 1/(s - a)^2$. Thus

$$\mathcal{L}\{e^{at}t\} = \frac{1}{(s - a)^2} \qquad (s > a). \tag{9.15}$$

▶ **Example 9.15**

Find $\mathcal{L}\{e^{at} \sin bt\}$. We let $f(t) = \sin bt$. Then $\mathcal{L}\{e^{at} \sin bt\} = F(s - a)$, where

$$F(s) = \mathcal{L}\{\sin bt\} = \frac{b}{s^2 + b^2} \qquad (s > 0).$$

Thus

$$F(s - a) = \frac{b}{(s - a)^2 + b^2}$$

and so

$$\mathcal{L}\{e^{at} \sin bt\} = \frac{b}{(s - a)^2 + b^2} \qquad (s > a). \tag{9.16}$$

THEOREM 9.6

Hypothesis. *Suppose f is a function satisfying the hypotheses of Theorem 9.1, with Laplace transform F, where*

$$F(s) = \int_0^\infty e^{-st} f(t)\, dt. \tag{9.17}$$

Conclusion

$$\mathcal{L}\{t^n f(t)\} = (-1)^n \frac{d^n}{ds^n} [F(s)]. \tag{9.18}$$

Proof. Differentiate both sides of Equation (9.17) n times with respect to s. This differentiation is justified in the present case and yields

$$F'(s) = (-1)^1 \int_0^\infty e^{-st} t f(t)\, dt,$$

$$F''(s) = (-1)^2 \int_0^\infty e^{-st} t^2 f(t)\, dt,$$

$$\vdots$$

$$F^{(n)}(s) = (-1)^n \int_0^\infty e^{-st} t^n f(t)\, dt,$$

from which the conclusion (9.18) is at once apparent.   *Q.E.D*

▶   **Example 9.16**

Find $\mathcal{L}\{t^2 \sin bt\}$. By Theorem 9.6,

$$\mathcal{L}\{t^2 \sin bt\} = (-1)^2 \frac{d^2}{ds^2} [F(s)],$$

where

$$F(s) = \mathcal{L}\{\sin bt\} = \frac{b}{s^2 + b^2}$$

(using (9.5)). From this,

$$\frac{d}{ds} [F(s)] = -\frac{2bs}{(s^2 + b^2)^2}$$

and

$$\frac{d^2}{ds^2} [F(s)] = \frac{6bs^2 - 2b^3}{(s^2 + b^2)^3}.$$

Thus,

$$\mathcal{L}\{t^2 \sin bt\} = \frac{6bs^2 - 2b^3}{(s^2 + b^2)^3}.$$

Exercises

1.   Use Theorem 9.2 to find $\mathcal{L}\{\cos^2 at\}$.

2.   Use Theorem 9.2 to find $\mathcal{L}\{\sin at \sin bt\}$.

3.   Use Theorem 9.2 to find $\mathcal{L}\{\sin^3 at\}$ and then employ Theorem 9.3 to obtain $\mathcal{L}\{\sin^2 at \cos at\}$.

4.   Use Theorem 9.2 to find $\mathcal{L}\{\cos^3 at\}$ and then employ Theorem 9.3 to obtain $\mathcal{L}\{\cos^2 at \sin at\}$.

5.   If $\mathcal{L}\{t^2\} = 2/s^3$, use Theorem 9.3 to find $\mathcal{L}\{t^3\}$.

6.   If $\mathcal{L}\{t^2\} = 2/s^3$, use Theorem 9.4 to find $\mathcal{L}\{t^4\}$.

7.   Use (9.11) and (9.13) to find $\mathcal{L}\{f(t)\}$ if
$$f''(t) + 3f'(t) + 2f(t) = 0, \ f(0) = 1, \text{ and } f'(0) = 2.$$

8.   Use (9.11) and (9.13) to find $\mathcal{L}\{f(t)\}$ if
$$f''(t) + 4f'(t) - 8f(t) = 0, \ f(0) = 3, \ f'(0) = -1.$$

9.   Use Theorem 9.5 to find $\mathcal{L}\{e^{at}t^2\}$.

10.   Use Theorem 9.5 to find $\mathcal{L}\{e^{at} \sin^2 bt\}$.

11.   Use Theorem 9.6 to find $\mathcal{L}\{t^2 \cos bt\}$.

12.   Use Theorem 9.6 to find $\mathcal{L}\{t^3 \sin bt\}$.

13.   Use Theorem 9.6 to find $\mathscr{L}\{t^3 e^{at}\}$.

14.   Use Theorem 9.6 to find $\mathscr{L}\{t^4 e^{at}\}$.

C. Step Functions, Translated Functions, and Periodic Functions

In the application of the Laplace transform to certain differential equations problems, we shall need to find the transform of a function having one or more finite discontinuities. In dealing with these functions, we shall find the concept of the so-called unit step function to be very useful.

For each real number $a \geq 0$, the *unit step function* u_a is defined for nonnegative t by

$$u_a(t) = \begin{cases} 0, & t < a, \\ 1, & t > a \end{cases} \tag{9.19}$$

(see Figure 9.2a). In particular, if $a = 0$, this formally becomes

$$u_0(t) = \begin{cases} 0, & t < 0, \\ 1, & t > 0; \end{cases}$$

but since we have defined u_a in (9.19) only for nonnegative t, this reduces to

$$u_0(t) = 1 \quad \text{for } t > 0 \tag{9.20}$$

(see Figure 9.2b).

The function u_a so defined satisfies the hypotheses of Theorem 9.1, so $\mathscr{L}\{u_a(t)\}$ exists. Using the definition of the Laplace transform, we find

$$\mathscr{L}\{u_a(t)\} = \int_0^\infty e^{-st} u_a(t)\, dt = \int_0^a e^{-st}(0)\, dt + \int_a^\infty e^{-st}(1)\, dt$$

$$= 0 + \lim_{R \to \infty} \int_a^R e^{-st}\, dt = \lim_{R \to \infty} \left[\frac{-e^{-st}}{s} \right]_a^R$$

$$= \lim_{R \to \infty} \frac{-e^{-sR} + e^{-sa}}{s} = \frac{e^{-as}}{s} \quad \text{for } s > 0.$$

Thus we have

$$\mathscr{L}\{u_a(t)\} = \frac{e^{-as}}{s} \quad (s > 0). \tag{9.21}$$

A variety of so-called *step functions* can be expressed as suitable linear combinations of the unit step function u_a. Then, using Theorem 9.2 (the linear property), and $\mathscr{L}\{u_a(t)\}$, we can readily obtain the Laplace transform of such step functions.

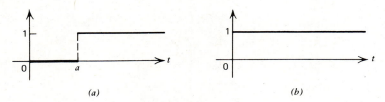

(a)   (b)

Figure 9.2

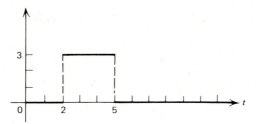

Figure 9.3

► **Example 9.17**

Consider the step function defined by

$$f(t) = \begin{cases} 0, & 0 < t < 2, \\ 3, & 2 < t < 5, \\ 0, & t > 5. \end{cases}$$

The graph of f is shown in Figure 9.3. We may express the values of f in the form

$$f(t) = \begin{cases} 0 - 0, & 0 < t < 2, \\ 3 - 0, & 2 < t < 5, \\ 3 - 3, & t > 5. \end{cases}$$

Hence we see that f is the function with values given by

$$\begin{cases} 0, & 0 < t < 2, \\ 3, & t > 2, \end{cases}$$

minus the function with values given by

$$\begin{cases} 0, & 0 < t < 5, \\ 3, & t > 5. \end{cases}$$

Thus $f(t)$ can be expressed as the linear combination

$$3u_2(t) - 3u_5(t)$$

of the unit step functions u_2 and u_5. Then using Theorem 9.2 and formula (9.21), we find

$$\mathscr{L}\{f(t)\} = \mathscr{L}\{3u_2(t) - 3u_5(t)\} = \frac{3e^{-2s}}{s} - \frac{3e^{-5s}}{s} = \frac{3}{s}[e^{-2s} - e^{-5s}].$$

Another useful property of the unit step function in connection with Laplace transforms is concerned with the translation of a given function a given distance in the positive direction. Specifically, consider the function f with values $f(t)$ defined for $t > 0$ (see Figure 9.4a). Suppose we consider the new function that results from translating the given function f a distance of a units in the positive direction (that is, to the right) and then assigning the value 0 to the new function for $t < a$. Then this new function is defined by

$$\begin{cases} 0, & 0 < t < a, \\ f(t - a), & t > a \end{cases} \tag{9.22}$$

(see Figure 9.4*b*). Then since the unit step function u_a is defined by

$$u_a(t) = \begin{cases} 0, & 0 < t < a, \\ 1, & t > a, \end{cases}$$

we see that the function defined by (9.22) is $u_a(t)f(t - a)$. That is,

$$u_a(t)f(t - a) = \begin{cases} 0, & 0 < t < a, \\ f(t - a), & t > a \end{cases} \tag{9.23}$$

(note Figure 9.4*b* again).

Concerning the Laplace transform of this function we have the following theorem.

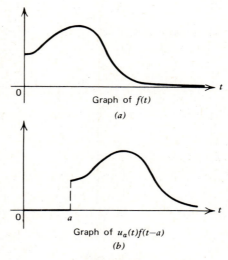

Graph of *f(t)*

(*a*)

Graph of $u_a(t)f(t-a)$

(*b*)

Figure 9.4

THEOREM 9.7

Hypothesis. *Suppose f is a function satisfying the hypotheses of Theorem 9.1 with Laplace transform F so that*

$$F(s) = \int_0^\infty e^{-st}f(t) \, dt;$$

and consider the translated function defined by

$$u_a(t)f(t - a) = \begin{cases} 0, & 0 < t < a, \\ f(t - a), & t > a. \end{cases} \tag{9.24}$$

Conclusion. *Then,*

$$\mathscr{L}\{u_a(t)f(t - a)\} = e^{-as}\mathscr{L}\{f(t)\}$$

that is, $\tag{9.25}$

$$\mathscr{L}\{u_a(t)f(t - a)\} = e^{-as}F(s).$$

Proof

$$\mathscr{L}\{u_a(t)f(t-a)\} = \int_0^\infty e^{-st}u_a(t)f(t-a)\,dt$$

$$= \int_0^a e^{-st}\cdot 0\,dt + \int_a^\infty e^{-st}f(t-a)\,dt$$

$$= \int_0^\infty e^{-st}f(t-a)\,dt.$$

Letting $t - a = \tau$, we obtain

$$\int_a^\infty e^{-st}f(t-a)\,dt = \int_0^\infty e^{-s(\tau+a)}f(\tau)\,d\tau$$

$$= e^{-as}\int_0^\infty e^{-s\tau}f(\tau)\,d\tau = e^{-as}\mathscr{L}\{f(t)\}.$$

Thus

$$\mathscr{L}\{u_a(t)f(t-a)\} = e^{-as}\mathscr{L}\{f(t)\} = e^{-as}F(s). \qquad\qquad Q.E.D$$

In applying Theorem 9.7 to a translated function of the form (9.24), one must be certain that the functional values for $t > a$ are indeed expressed in terms of $t - a$. In general, this will not be so; and if it is not, one must first express these functional values for $t > a$ so that this *is* so. We shall illustrate this in each of Examples 9.18 and 9.19.

▶ **Example 9.18**

Find the Laplace transform of

$$g(t) = \begin{cases} 0 & 0 < t < 5, \\ t-3, & t > 5. \end{cases}$$

Before we can apply Theorem 9.7 to this translated function, we must express the functional values $t - 3$ for $t > 5$ in terms of $t - 5$, as required by (9.24). That is, we express $t - 3$ as $(t - 5) + 2$ and write

$$g(t) = \begin{cases} 0, & 0 < t < 5, \\ (t-5)+2, & t > 5. \end{cases}$$

This is now of the form (9.24), and we recognize it as

$$u_5(t)f(t-5) = \begin{cases} 0, & 0 < t < 5, \\ (t-5)+2, & t > 5, \end{cases}$$

where $f(t) = t + 2$, $t > 0$. Hence we apply Theorem 9.7 with $f(t) = t + 2$. Using Theorem 9.2 (the Linear Property) and formulas (9.2) and (9.3), we find

$$F(s) = \mathscr{L}\{t+2\} = \mathscr{L}\{t\} + 2\mathscr{L}\{1\} = \frac{1}{s^2} + \frac{2}{s}.$$

Then by Theorem 9.7, with $a = 5$, we obtain

$$\mathscr{L}\{u_5(t)f(t-5)\} = e^{-5s}F(s) = e^{-5s}\left(\frac{1}{s^2} + \frac{2}{s}\right).$$

This then is the Laplace transform of the given function $g(t)$.

▶ **Example 9.19**

Find the Laplace transform of

$$g(t) = \begin{cases} 0, & 0 < t < \dfrac{\pi}{2}, \\[2mm] \sin t, & t > \dfrac{\pi}{2}. \end{cases}$$

Before we can apply Theorem 9.7, we must express $\sin t$ in terms of $t - \pi/2$, as required by (9.24). We observe that $\sin t = \cos(t - \pi/2)$ for all t, and hence write

$$g(t) = \begin{cases} 0, & 0 < t < \dfrac{\pi}{2}, \\[2mm] \cos\left(t - \dfrac{\pi}{2}\right), & t > \dfrac{\pi}{2}. \end{cases}$$

This is now of the form (9.24), and we recognize it as

$$u_{\pi/2}(t)f(t - \pi/2) = \begin{cases} 0, & 0 < t < \dfrac{\pi}{2}, \\[2mm] \cos\left(t - \dfrac{\pi}{2}\right), & t > \dfrac{\pi}{2}, \end{cases}$$

where $f(t) = \cos t$, $t > 0$. Hence we apply Theorem 9.7 with $f(t) = \cos t$. Using formula (9.6) with $b = 1$, we obtain

$$F(s) = \mathscr{L}\{\cos t\} = \frac{s}{s^2 + 1}.$$

Then by Theorem 9.7, with $a = \pi/2$, we obtain

$$\mathscr{L}\{g(t)\} = \mathscr{L}\{u_{\pi/2}(t)f(t - \pi/2)\} = \frac{se^{-(\pi/2)s}}{s^2 + 1}.$$

We next obtain a result concerning the Laplace transform of a periodic function. A function f is *periodic* of period P, where $P > 0$, if $f(t + P) = f(t)$ for every t for which f is defined. For example, the functions defined by $\sin bt$ and $\cos bt$ are periodic of period $2\pi/b$.

THEOREM 9.8

Hypothesis. *Suppose f is a periodic function of period P which satisfies the hypotheses of Theorem 9.1.*

Conclusion. *Then*

$$\mathscr{L}\{f(t)\} = \frac{\int_0^P e^{-st}f(t)\,dt}{1 - e^{-Ps}}. \tag{9.26}$$

Proof. By definition of the Laplace transform,

$$\mathscr{L}\{f(t)\} = \int_0^\infty e^{-st}f(t)\,dt. \tag{9.27}$$

The integral on the right can be broken up into the infinite series of integrals

$$\int_0^P e^{-st}f(t)\,dt + \int_P^{2P} e^{-st}f(t)\,dt + \int_{2P}^{3P} e^{-st}f(t)\,dt + \cdots$$

$$+ \int_{nP}^{(n+1)P} e^{-st}f(t)\,dt \cdots. \tag{9.28}$$

We now transform each integral in this series. For each $n = 0, 1, 2, \ldots$, let $t = u + nP$ in the corresponding integral

$$\int_{nP}^{(n+1)P} e^{-st}f(t)\,dt.$$

Then for each $n = 0, 1, 2, \ldots$, this becomes

$$\int_0^P e^{-s(u+nP)}f(u + nP)\,du. \tag{9.29}$$

But by hypothesis, f is periodic of period P. Thus $f(u) = f(u + P) = f(u + 2P) = \cdots = f(u + nP)$ for all u for which f is defined. Also $e^{-s(u+nP)} = e^{-su}e^{-nPs}$, where the factor e^{-nPs} is independent of the variable of integration u in (9.29). Thus for each $n = 0, 1, 2, \ldots$, the integral in (9.29) becomes

$$e^{-nPs}\int_0^P e^{-su}f(u)\,du.$$

Hence the infinite series (9.28) takes the form

$$\int_0^P e^{-su}f(u)\,du + e^{-Ps}\int_0^P e^{-su}f(u)\,du$$

$$+ e^{-2Ps}\int_0^P e^{-su}f(u)\,du + \cdots + e^{-nPs}\int_0^P e^{-su}f(u)\,du + \cdots$$

$$= [1 + e^{-Ps} + e^{-2Ps} + \cdots + e^{-nPs} + \cdots]\int_0^P e^{-su}f(u)\,du. \tag{9.30}$$

Now observe that the infinite series in brackets is a geometric series of first term 1 and common ratio $r = e^{-Ps} < 1$. Such a series converges to $1/(1 - r)$, and hence the series in brackets converges to $1/(1 - e^{-Ps})$. Therefore the right member of (9.30), and hence that of (9.28), reduces to

$$\frac{\int_0^P e^{-su}f(u)\,du}{1 - e^{-Ps}}.$$

Then, since this is the right member of (9.27), upon replacing the dummy variable u by t, we have

$$\mathscr{L}\{f(t)\} = \frac{\int_0^P e^{-st} f(t)\, dt}{1 - e^{-Ps}} \qquad\qquad Q.E.D.$$

▶ **Example 9.20**

Find the Laplace transform of f defined on $0 \le t < 4$ by

$$f(t) = \begin{cases} 1, & 0 \le t < 2, \\ -1, & 2 \le t < 4, \end{cases}$$

and for all other positive t by the periodicity condition

$$f(t + 4) = f(t).$$

The graph of f is shown in Figure 9.5. Clearly this function f is periodic of period $P = 4$. Applying formula (9.26) of Theorem 9.8, we find

$$\mathscr{L}\{f(t)\} = \frac{\int_0^4 e^{-st} f(t)\, dt}{1 - e^{-4s}}$$

$$= \frac{1}{1 - e^{-4s}} \left[\int_0^2 e^{-st}(1)\, dt + \int_2^4 e^{-st}(-1)\, dt \right]$$

$$= \frac{1}{1 - e^{-4s}} \left[\frac{-e^{-st}}{s}\Big|_0^2 + \frac{e^{-st}}{s}\Big|_2^4 \right]$$

$$= \frac{1}{1 - e^{-4s}} \left(\frac{1}{s} \right) [-e^{-2s} + 1 + e^{-4s} - e^{-2s}]$$

$$= \frac{1 - 2e^{-2s} + e^{-4s}}{s(1 - e^{-4s})} = \frac{(1 - e^{-2s})^2}{s(1 - e^{-2s})(1 + e^{-2s})}$$

$$= \frac{1 - e^{-2s}}{s(1 + e^{-2s})}.$$

Exercises

Find $\mathscr{L}\{f(t)\}$ for each of the functions f defined in Exercises 1–18.

1. $f(t) = \begin{cases} 0, & 0 < t < 6, \\ 5, & t > 6. \end{cases}$

2. $f(t) = \begin{cases} 0, & 0 < t < 10, \\ -3, & t > 10. \end{cases}$

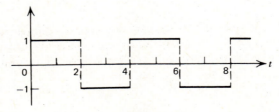

Figure 9.5

3. $f(t) = \begin{cases} 4, & 0 < t < 6, \\ 0, & t > 6. \end{cases}$

4. $f(t) = \begin{cases} 2, & 0 < t < 5, \\ 0, & t > 5. \end{cases}$

5. $f(t) = \begin{cases} 0, & 0 < t < 5, \\ 2, & 5 < t < 7, \\ 0, & t > 7. \end{cases}$

6. $f(t) = \begin{cases} 0, & 0 < t < 3, \\ -6, & 3 < t < 9, \\ 0, & t > 9. \end{cases}$

7. $f(t) = \begin{cases} 1, & 0 < t < 2, \\ 2, & 2 < t < 4, \\ 3, & 4 < t < 6, \\ 0, & t > 6. \end{cases}$

8. $f(t) = \begin{cases} 9, & 0 < t < 5, \\ 6, & 5 < t < 10, \\ 3, & 10 < t < 15, \\ 0, & t > 10. \end{cases}$

9. $f(t) = \begin{cases} 2, & 0 < t < 3, \\ 0, & 3 < t < 6, \\ 2, & t > 6. \end{cases}$

10. $f(t) = \begin{cases} 4, & 0 < t < 5, \\ 0, & 5 < t < 10, \\ 3, & t > 10. \end{cases}$

11. $f(t) = \begin{cases} 0, & 0 < t < 2, \\ t, & t > 2. \end{cases}$

12. $f(t) = \begin{cases} 0, & 0 < t < 4, \\ 3t, & t > 4. \end{cases}$

13. $f(t) = \begin{cases} t, & 0 < t < 3, \\ 3, & t > 3. \end{cases}$

14. $f(t) = \begin{cases} 2t, & 0 < t < 5, \\ 10, & t > 5. \end{cases}$

15. $f(t) = \begin{cases} 0, & 0 < t < \pi/2, \\ \cos t, & t > \pi/2. \end{cases}$

16. $f(t) = \begin{cases} 0, & 0 < t < 2, \\ e^{-t}, & t > 2. \end{cases}$

17. $f(t) = \begin{cases} 0, & 0 < t < 4. \\ t - 4 & 4 < t < 7, \\ 3, & t > 7. \end{cases}$

18. $f(t) = \begin{cases} 6, & 0 < t < 1, \\ 8 - 2t, & 1 < t < 3, \\ 2, & t > 3. \end{cases}$

9.2 THE INVERSE TRANSFORM AND THE CONVOLUTION

A. The Inverse Transform

Thus far in this chapter we have been concerned with the following problem: Given a function f, defined for $t > 0$, to find its Laplace transform, which we denoted by $\mathscr{L}\{f\}$ or F. Now consider the inverse problem: Given a function F, to find a function f whose Laplace transform is the given F. We introduce the notation $\mathscr{L}^{-1}\{F\}$ to denote such a function f, denote $\mathscr{L}^{-1}\{F(s)\}$ by $f(t)$, and call such a function an *inverse transform* of F. That is,

$$f(t) = \mathscr{L}^{-1}\{F(s)\}$$

means that $f(t)$ is such that

$$\mathscr{L}\{f(t)\} = F(s).$$

Three questions arise at once:

1. Given a function F, does an inverse transform of F exist?
2. Assuming F does have an inverse transform, is this inverse transform unique?
3. How is an inverse transform found?

432 THE LAPLACE TRANSFORM

In answer to question 1 we shall say "not necessarily," for there exist functions F that are not Laplace transforms of any function f. In order for F to be a transform it must possess certain continuity properties and also behave suitably as $s \to \infty$. To reassure the reader in a practical way we note that inverse transforms corresponding to numerous functions F have been determined and tabulated.

Now let us consider question 2. Assuming that F is a function that *does have* an inverse transform, in what sense, if any, is this inverse transform unique? We answer this question in a manner that is adequate for our purposes by stating without proof the following theorem.

THEOREM 9.9

Hypothesis. *Let f and g be two functions that are continuous for $t \geq 0$ and that have the same Laplace transform F.*

Conclusion. $f(t) = g(t)$ *for all $t \geq 0$.*

Thus if it is known that a given function F has a *continuous* inverse transform f, then f is the *only* continuous inverse transform of F. Let us consider the following example.

▶ **Example 9.21**

By Equation (9.2), $\mathscr{L}\{1\} = 1/s$. Thus an inverse transform of the function F defined by $F(s) = 1/s$ is the *continuous* function f defined for all t by $f(t) = 1$. Thus by Theorem 9.9 there is no other *continuous* inverse transform of the function F such that $F(s) = 1/s$. However, discontinuous inverse transforms of this function F exist. For example, consider the function g defined as follows:

$$g(t) = \begin{cases} 1, & 0 < t < 3, \\ 2, & t = 3, \\ 1, & t > 3. \end{cases}$$

Then

$$\mathscr{L}\{g(t)\} = \int_0^\infty e^{-st} g(t)\, dt = \int_0^3 e^{-st}\, dt + \int_3^\infty e^{-st}\, dt$$

$$= \left[-\frac{e^{-st}}{s} \right]_0^3 + \lim_{R \to \infty} \left[-\frac{e^{-st}}{s} \right]_3^R = \frac{1}{s} \quad \text{if } s > 0.$$

Thus this discontinuous function g is also an inverse transform of F defined by $F(s) = 1/s$. However, we again emphasize that the only *continuous* inverse transform of F defined by $F(s) = 1/s$ is f defined for all t by $f(t) = 1$. Indeed we write

$$\mathscr{L}^{-1}\left\{ \frac{1}{s} \right\} = 1,$$

with the understanding that f defined for all t by $f(t) = 1$ is the *unique continuous* inverse transform of F defined by $F(s) = 1/s$.

Finally, let us consider question 3. Assuming a unique continuous inverse transform of F exists, how is it actually found? The direct determination of inverse transforms will not be considered in this book. Our primary means of finding the inverse transform of a given F will be to make use of a table of transforms. As already indicated, extensive tables of transforms have been prepared. A short table of this kind appears on page 434.

In using a table of transforms to find the inverse transform of a given F, certain preliminary manipulations often have to be performed in order to put the given $F(s)$ in a form to which the various entries in the table apply. Among the various techniques available, the method of partial fractions is often very useful. We shall illustrate its use in Example 9.23.

▶ **Example 9.22**

Using Table 9.1, find $\mathscr{L}^{-1}\left\{\dfrac{1}{s^2 + 6s + 13}\right\}$.

Solution. Looking in the $F(s)$ column of Table 9.1 we would first look for $F(s) = \dfrac{1}{as^2 + bs + c}$. However, we find no such $F(s)$; but we do find $F(s) = \dfrac{b}{(s + a)^2 + b^2}$ (number 11). We can put the given expression $\dfrac{1}{s^2 + 6s + 13}$ in this form as follows:

$$\frac{1}{s^2 + 6s + 13} = \frac{1}{(s + 3)^2 + 4} = \frac{1}{2} \cdot \frac{2}{(s + 3)^2 + 2^2}.$$

Thus, using number 11 of Table 9.1, we have

$$\mathscr{L}^{-1}\left\{\frac{1}{s^2 + 6s + 13}\right\} = \frac{1}{2}\mathscr{L}^{-1}\left\{\frac{2}{(s + 3)^2 + 2^2}\right\} = \frac{1}{2}e^{-3t}\sin 2t.$$

▶ **Example 9.23**

Using Table 9.1, find $\mathscr{L}^{-1}\left\{\dfrac{1}{s(s^2 + 1)}\right\}$.

Solution. No entry of this form appears in the $F(s)$ column of Table 9.1. We employ the method of partial fractions. We have

$$\frac{1}{s(s^2 + 1)} = \frac{A}{s} + \frac{Bs + C}{s^2 + 1}$$

and hence

$$1 = (A + B)s^2 + Cs + A.$$

Thus

$$A + B = 0, \qquad C = 0, \quad \text{and} \quad A = 1.$$

TABLE 9.1 LAPLACE TRANSFORMS

	$f(t) = \mathscr{L}^{-1}\{F(s)\}$	$F(s) = \mathscr{L}\{f(t)\}$
1	1	$\dfrac{1}{s}$
2	e^{at}	$\dfrac{1}{s-a}$
3	$\sin bt$	$\dfrac{b}{s^2+b^2}$
4	$\cos bt$	$\dfrac{s}{s^2+b^2}$
5	$\sinh bt$	$\dfrac{b}{s^2-b^2}$
6	$\cosh bt$	$\dfrac{s}{s^2-b^2}$
7	$t^n (n=1,2,\ldots)$	$\dfrac{n!}{s^{n+1}}$
8	$t^n e^{at} (n=1,2,\ldots)$	$\dfrac{n!}{(s-a)^{n+1}}$
9	$t \sin bt$	$\dfrac{2bs}{(s^2+b^2)^2}$
10	$t \cos bt$	$\dfrac{s^2-b^2}{(s^2+b^2)^2}$
11	$e^{-at}\sin bt$	$\dfrac{b}{(s+a)^2+b^2}$
12	$e^{-at}\cos bt$	$\dfrac{s+a}{(s+a)^2+b^2}$
13	$\dfrac{\sin bt - bt\cos bt}{2b^3}$	$\dfrac{1}{(s^2+b^2)^2}$
14	$\dfrac{t\sin bt}{2b}$	$\dfrac{s}{(s^2+b^2)^2}$
15	$u_a(t)$ [see equations (9.19) and (9.21)]	$\dfrac{e^{-as}}{s}$
16	$u_a(t)f(t-a)$ [see Theorem 9.7]	$e^{-as}F(s)$

From these equations, we have the partial fractions decomposition

$$\frac{1}{s(s^2 + 1)} = \frac{1}{s} - \frac{s}{s^2 + 1}.$$

Thus

$$\mathscr{L}^{-1}\left\{\frac{1}{s(s^2 + 1)}\right\} = \mathscr{L}^{-1}\left\{\frac{1}{s}\right\} - \mathscr{L}^{-1}\left\{\frac{s}{s^2 + 1}\right\}.$$

By number 1 of Table 9.1, $\mathscr{L}^{-1}\{1/s\} = 1$ and by number 4, $\mathscr{L}^{-1}\{s/(s^2 + 1)\} = \cos t$. Thus

$$\mathscr{L}^{-1}\left\{\frac{1}{s(s^2 + 1)}\right\} = 1 - \cos t.$$

We now give two examples of finding the inverse transform of a function that involves one or more terms of the form $e^{-as} F(s)$.

▶ **Example 9.24**

Find

$$\mathscr{L}^{-1}\left\{\frac{5}{s} - \frac{3e^{-3s}}{s} - \frac{2e^{-7s}}{s}\right\}.$$

Solution.   By number 1 of Table 9.1, we at once have

$$\mathscr{L}^{-1}\left\{\frac{1}{s}\right\} = 1.$$

By number 15, we see that

$$\mathscr{L}^{-1}\left\{\frac{e^{-as}}{s}\right\} = u_a(t). \tag{9.31}$$

Here u_a is the unit step function [see Equations (9.19) and following] defined for $a > 0$ by

$$u_a(t) = \begin{cases} 0, & 0 < t < a, \\ 1, & t > a, \end{cases} \tag{9.32}$$

and for $a = 0$ by

$$u_0(t) = 1 \quad \text{for } t > 0.$$

Applying (9.31) and (9.32) with $a = 3$ and $a = 7$, respectively, we have

$$\mathscr{L}^{-1}\left\{\frac{e^{-3s}}{s}\right\} = u_3(t) = \begin{cases} 0, & 0 < t < 3, \\ 1, & t > 3, \end{cases} \tag{9.33}$$

and

$$\mathscr{L}^{-1}\left\{\frac{e^{-7s}}{s}\right\} = u_7(t) = \begin{cases} 0, & 0 < t < 7, \\ 1, & t > 7. \end{cases} \tag{9.34}$$

Thus we obtain

$$\mathscr{L}^{-1}\left\{\frac{5}{s} - \frac{3e^{-3s}}{s} - \frac{2e^{-7s}}{s}\right\} = 5 - 3u_3(t) - 2u_7(t).$$

Now using (9.33) and (9.34), we see that this equals

$$\begin{cases} 5 - 0 - 0, & 0 < t < 3, \\ 5 - 3 - 0, & 3 < t < 7, \\ 5 - 3 - 2, & t > 7; \end{cases}$$

and hence

$$\mathscr{L}^{-1}\left\{\frac{5}{s} - \frac{3e^{-3s}}{s} - \frac{2e^{-7s}}{s}\right\} = \begin{cases} 5, & 0 < t < 3, \\ 2, & 3 < t < 7, \\ 0, & t > 7. \end{cases}$$

▶ **Example 9.25**

Find

$$\mathscr{L}^{-1}\left\{e^{-4s}\left(\frac{2}{s^2} + \frac{5}{s}\right)\right\}.$$

Solution.   This is of the form $\mathscr{L}^{-1}\{e^{-as}F(s)\}$, where $a = 4$ and $F(s) = 2/s^2 + 5/s$. By number 16 of Table 9.1, we see that

$$\mathscr{L}^{-1}\{e^{-as}F(s)\} = u_a(t)f(t - a). \tag{9.35}$$

Here u_a is the unit step function defined for $a > 0$ by (9.32) and $f(t) = \mathscr{L}^{-1}\{F(s)\}$ [see Theorem 9.7]. By number 1 of Table 9.1, we again find $\mathscr{L}^{-1}\{1/s\} = 1$; and by number 7 with $n = 1$, we obtain $\mathscr{L}^{-1}\{1/s^2\} = t$. Thus

$$f(t) = \mathscr{L}^{-1}\{F(s)\} = \mathscr{L}^{-1}\left\{\frac{2}{s^2} + \frac{5}{s}\right\} = 2t + 5,$$

and so $f(t - 4) = 2(t - 4) + 5 = 2t - 3$. Then by (9.35), with $a = 4$,

$$\mathscr{L}^{-1}\{e^{-4s}F(s)\} = u_4(t)f(t - 4);$$

that is,

$$\mathscr{L}^{-1}\left\{e^{-4s}\left(\frac{2}{s^2} + \frac{5}{s}\right)\right\} = u_4(t)[2t - 3] = \begin{cases} 0, & 0 < t < 4, \\ 2t - 3, & t > 4. \end{cases}$$

Exercises

Use Table 9.1 to find $\mathscr{L}^{-1}\{F(s)\}$ for each of the functions F defined in Exercises 1–28:

1. $F(s) = \dfrac{2}{s^2 + 9}.$

2. $F(s) = \dfrac{3s}{s^2 - 4}.$

3. $F(s) = \dfrac{5}{(s - 2)^4}.$

4. $F(s) = \dfrac{5s}{s^2 + 4s + 4}.$

5. $F(s) = \dfrac{s+2}{s^2 + 4s + 7}.$

6. $F(s) = \dfrac{s+10}{s^2 + 8s + 20}.$

7. $F(s) = \dfrac{1}{s^3 + 4s^2 + 3s}.$

8. $F(s) = \dfrac{s+1}{s^3 + 2s}.$

9. $F(s) = \dfrac{s+3}{(s^2 + 4)^2}.$

10. $F(s) = \dfrac{s+5}{s^4 + 3s^3 + 2s^2}.$

11. $F(s) = \dfrac{5}{(s+2)^5}.$

12. $F(s) = \dfrac{7}{(2s+1)^3}.$

13. $F(s) = \dfrac{2s+7}{(s+3)^4}.$

14. $F(s) = \dfrac{8(s+1)}{(2s-1)^3}.$

15. $F(s) = \dfrac{s-2}{s^2 + 5s + 6}.$

16. $F(s) = \dfrac{2s+6}{8s^2 - 2s - 3}.$

17. $F(s) = \dfrac{7s+12}{s^2 + 9}.$

18. $F(s) = \dfrac{5s+8}{s^2 + 3s - 10}.$

19. $F(s) = \dfrac{5s+6}{s^2 + 9} e^{-\pi s}.$

20. $F(s) = \dfrac{s+10}{s^2 + 2s - 8} e^{-2s}.$

21. $F(s) = \dfrac{s+8}{s^2 + 4s + 13} e^{-(\pi s)/2}.$

22. $F(s) = \dfrac{2s+9}{s^2 + 4s + 13} e^{-3s}.$

23. $F(s) = \dfrac{e^{-4s} - e^{-7s}}{s^2}.$

24. $F(s) = \dfrac{e^{-3s} - e^{-8s}}{s^3}.$

25. $F(s) = \dfrac{1 + e^{-\pi s}}{s^2 + 4}.$

26. $F(s) = \dfrac{2 - e^{-3s}}{s^2 + 9}.$

27. $F(s) = \dfrac{2[1 + e^{-(\pi s)/2}]}{s^2 - 2s + 5}.$

28. $F(s) = \dfrac{4(e^{-2s} - 1)}{s(s^2 + 4)}.$

B. The Convolution

Another important procedure in connection with the use of tables of transforms is that furnished by the so-called convolution theorem which we shall state below. We first define the convolution of two functions f and g.

DEFINITION

Let f and g be two functions that are piecewise continuous on every finite closed interval $0 \le t \le b$ and of exponential order. The function denoted by $f * g$ and defined by

$$f(t) * g(t) = \int_0^t f(\tau)g(t - \tau)\, d\tau \qquad (9.36)$$

is called the convolution *of the functions f and g.*

Let us change the variable of integration in (9.36) by means of the substitution $u = t - \tau$. We have

$$f(t) * g(t) = \int_0^t f(\tau) g(t - \tau)\, d\tau = -\int_t^0 f(t - u) g(u)\, du$$

$$= \int_0^t g(u) f(t - u)\, du = g(t) * f(t).$$

Thus we have shown that

$$f * g = g * f \tag{9.37}$$

Suppose that both f and g are piecewise continuous on every finite closed interval $0 \le t \le b$ and of exponential order e^{at}. Then it can be shown that $f * g$ is also piecewise continuous on every finite closed interval $0 \le t \le b$ and of exponential order $e^{(a + \epsilon)t}$, where ϵ is any positive number. Thus $\mathscr{L}\{f * g\}$ exists for s sufficiently large. More explicitly, it can be shown that $\mathscr{L}\{f * g\}$ exists for $s > a$.

We now prove the following important theorem concerning $\mathscr{L}\{f * g\}$.

THEOREM 9.10

Hypothesis. *Let the functions f and g be piecewise continuous on every finite closed interval $0 \le t \le b$ and of exponential order e^{at}.*

Conclusion

$$\mathscr{L}\{f * g\} = \mathscr{L}\{f\} \mathscr{L}\{g\} \tag{9.38}$$

for $s > a$.

Proof.   By definition of the Laplace transform, $\mathscr{L}\{f * g\}$ is the function defined by

$$\int_0^\infty e^{-st}\left[\int_0^t f(\tau) g(t - \tau)\, d\tau \right] dt. \tag{9.39}$$

The integral (9.39) may be expressed as the iterated integral

$$\int_0^\infty \int_0^t e^{-st} f(\tau) g(t - \tau)\, d\tau\, dt. \tag{9.40}$$

Further, the iterated integral (9.40) is equal to the double integral

$$\iint_{R_1} e^{-st} f(\tau) g(t - \tau)\, d\tau\, dt, \tag{9.41}$$

where R_1 is the 45° wedge bounded by the lines $\tau = 0$ and $t = \tau$ (see Figure 9.6).
We now make the change of variable

$$u = t - \tau,$$

$$v = \tau, \tag{9.42}$$

to transform the double integral (9.41). The change of variables (9.42) has Jacobian 1 and transforms the region R_1 in the τ, t plane into the first quadrant of the u, v plane.

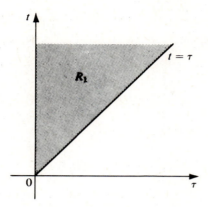

Figure 9.6

Thus the double integral (9.41) transforms into the double integral

$$\iint\limits_{R_2} e^{-s(u+v)} f(v)g(u)\ du\ dv, \tag{9.43}$$

where R_2 is the quarter plane defined by $u > 0$, $v > 0$ (see Figure 9.7). The double integral (9.43) is equal to the iterated integral

$$\int_0^\infty \int_0^\infty e^{-s(u+v)} f(v)g(u)\ du\ dv. \tag{9.44}$$

But the iterated integral (9.44) can be expressed in the form

$$\int_0^\infty e^{-sv} f(v)\ dv \int_0^\infty e^{-su} g(u)\ du. \tag{9.45}$$

But the left-hand integral in (9.45) defines $\mathscr{L}\{f\}$ and the right-hand integral defines $\mathscr{L}\{g\}$. Therefore the expression (9.45) is precisely $\mathscr{L}\{f\}\mathscr{L}\{g\}$.

We note that since the integrals involved are absolutely convergent for $s > a$, the operations performed are indeed legitimate for $s > a$. Therefore we have shown that

$$\mathscr{L}\{f * g\} = \mathscr{L}\{f\}\mathscr{L}\{g\} \qquad \text{for } s > a. \qquad Q.E.D.$$

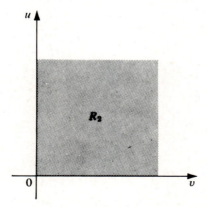

Figure 9.7

Denoting $\mathscr{L}\{f\}$ by F and $\mathscr{L}\{g\}$ by G, we may write the conclusion (9.38) in the form

$$\mathscr{L}\{f(t)*g(t)\} = F(s)G(s).$$

Hence, we have

$$\mathscr{L}^{-1}\{F(s)G(s)\} = f(t)*g(t) = \int_0^t f(\tau)g(t-\tau)\,d\tau, \tag{9.46}$$

and using (9.37), we also have

$$\mathscr{L}^{-1}\{F(s)G(s)\} = g(t)*f(t) = \int_0^t g(\tau)f(t-\tau)\,d\tau. \tag{9.47}$$

Suppose we are given a function H and are required to determine $\mathscr{L}^{-1}\{H(s)\}$. If we can express $H(s)$ as a product $F(s)G(s)$, where $\mathscr{L}^{-1}\{F(s)\} = f(t)$ and $\mathscr{L}^{-1}\{G(s)\} = g(t)$ are known, then we can apply either (9.46) or (9.47) to determine $\mathscr{L}^{-1}\{H(s)\}$.

▶ **Example 9.26**

Find $\mathscr{L}^{-1}\left\{\dfrac{1}{s(s^2+1)}\right\}$ using the convolution and Table 9.1.

Solution. We write $1/s(s^2+1)$ as the product $F(s)G(s)$, where $F(s) = 1/s$ and $G(s) = 1/(s^2+1)$. By Table 9.1, number 1, $f(t) = \mathscr{L}^{-1}\{1/s\} = 1$, and by number 3, $g(t) = \mathscr{L}^{-1}\{1/(s^2+1)\} = \sin t$. Thus by (9.46),

$$\mathscr{L}^{-1}\left\{\frac{1}{s(s^2+1)}\right\} = f(t)*g(t) = \int_0^t 1 \cdot \sin(t-\tau)\,d\tau,$$

and by (9.47),

$$\mathscr{L}^{-1}\left\{\frac{1}{s(s^2+1)}\right\} = g(t)*f(t) = \int_0^t \sin\tau \cdot 1\,d\tau.$$

The second of these two integrals is slightly more simple. Evaluating it, we have

$$\mathscr{L}^{-1}\left\{\frac{1}{s(s^2+1)}\right\} = 1 - \cos t.$$

Observe that we obtained this result in Example 9.23 by means of partial fractions.

Exercises

In each of Exercises 1–6 find $\mathscr{L}^{-1}\{H(s)\}$ using the convolution and Table 9.1.

1. $H(s) = \dfrac{1}{s^2+5s+6}.$

2. $H(s) = \dfrac{1}{s^2+3s-4}.$

3. $H(s) = \dfrac{1}{s(s^2+9)}.$

4. $H(s) = \dfrac{1}{s(s^2+4s+13)}.$

5. $H(s) = \dfrac{1}{s^2(s+3)}.$

6. $H(s) = \dfrac{1}{(s+2)(s^2+1)}.$

9.3 LAPLACE TRANSFORM SOLUTION OF LINEAR DIFFERENTIAL EQUATIONS WITH CONSTANT COEFFICIENTS

A. The Method

We now consider how the Laplace transform may be applied to solve the initial-value problem consisting of the nth-order linear differential equation with constant coefficients

$$a_0 \frac{d^n y}{dt^n} + a_1 \frac{d^{n-1} y}{dt^{n-1}} + \cdots + a_{n-1} \frac{dy}{dt} + a_n y = b(t), \qquad (9.48)$$

plus the initial conditions

$$y(0) = c_0, \, y'(0) = c_1, \ldots, \, y^{(n-1)}(0) = c_{n-1}. \qquad (9.49)$$

Theorem 4.1 (Chapter 4) assures us that this problem has a unique solution.

We now take the Laplace transform of both members of Equation (9.48). By Theorem 9.2, we have

$$a_0 \mathscr{L}\left\{\frac{d^n y}{dt^n}\right\} + a_1 \mathscr{L}\left\{\frac{d^{n-1} y}{dt^{n-1}}\right\} + \cdots + a_{n-1} \mathscr{L}\left\{\frac{dy}{dt}\right\} + a_n \mathscr{L}\{y(t)\} = \mathscr{L}\{b(t)\}. \quad (9.50)$$

We now apply Theorem 9.4 to

$$\mathscr{L}\left\{\frac{d^n y}{dt^n}\right\}, \, \mathscr{L}\left\{\frac{d^{n-1} y}{dt^{n-1}}\right\}, \ldots, \, \mathscr{L}\left\{\frac{dy}{dt}\right\}$$

in the left member of Equation (9.50). Using the initial conditions (9.49), we have

$$\mathscr{L}\left\{\frac{d^n y}{dt^n}\right\} = s^n \mathscr{L}\{y(t)\} - s^{n-1} y(0) - s^{n-2} y'(0) - \cdots - y^{(n-1)}(0)$$

$$= s^n \mathscr{L}\{y(t)\} - c_0 s^{n-1} - c_1 s^{n-2} - \cdots - c_{n-1},$$

$$\mathscr{L}\left\{\frac{d^{n-1}}{dt^{n-1}}\right\} = s^{n-1} \mathscr{L}\{y(t)\} - s^{n-2} y(0) - s^{n-3} y'(0) - \cdots - y^{(n-2)}(0)$$

$$= s^{n-1} \mathscr{L}\{y(t)\} - c_0 s^{n-2} - c_1 s^{n-3} - \cdots - c_{n-2},$$

$$\vdots$$

$$\mathscr{L}\left\{\frac{dy}{dt}\right\} = s \mathscr{L}\{y(t)\} - y(0) = s \mathscr{L}\{y(t)\} - c_0.$$

Thus, letting $Y(s)$ denote $\mathscr{L}\{y(t)\}$ and $B(s)$ denote $\mathscr{L}\{b(t)\}$, Equation (9.50) becomes

$$[a_0 s^n + a_1 s^{n-1} + \cdots + a_{n-1} s + a_n] Y(s)$$

$$- c_0[a_0 s^{n-1} + a_1 s^{n-2} + \cdots + a_{n-1}]$$
$$- c_1[a_0 s^{n-2} + a_1 s^{n-3} + \cdots + a_{n-2}]$$
$$- \cdots - c_{n-2}[a_0 s + a_1] - c_{n-1} a_0 = B(s). \quad (9.51)$$

Since b is a known function of t, then B, assuming it exists and can be determined, is a known function of s. Thus Equation (9.51) is an algebraic equation in the "unknown" $Y(s)$. We now solve the algebraic equation (9.51) to determine $Y(s)$. Once $Y(s)$ has been found, we then find the unique solution

$$y(t) = \mathscr{L}^{-1}\{Y(s)\}$$

of the given initial-value problem using the table of transforms.
We summarize this procedure as follows:

1. Take the Laplace transform of both sides of the differential equation (9.48), applying Theorem 9.4 and using the initial conditions (9.49) in the process, and equate the results to obtain the algebraic equation (9.51) in the "unknown" $Y(s)$.
2. Solve the algebraic equation (9.51) thus obtained to determine $Y(s)$.
3. Having found $Y(s)$, employ the table of transforms to determine the solution $y(t) = \mathscr{L}^{-1}\{Y(s)\}$ of the given initial-value problem.

B. Examples

We shall now consider several detailed examples that will illustrate the procedure outlined above.

▶ **Example 9.27**

Solve the initial-value problem

$$\frac{dy}{dt} - 2y = e^{5t}, \tag{9.52}$$

$$y(0) = 3 \tag{9.53}$$

Step 1. Taking the Laplace transform of both sides of the differential equation (9.52), we have

$$\mathscr{L}\left\{\frac{dy}{dt}\right\} - 2\mathscr{L}\{y(t)\} = \mathscr{L}\{e^{5t}\}. \tag{9.54}$$

Using Theorem 9.4 with $n = 1$ (or Theorem 9.3) and denoting $\mathscr{L}\{y(t)\}$ by $Y(s)$, we may express $\mathscr{L}\{dy/dt\}$ in terms of $Y(s)$ and $y(0)$ as follows:

$$\mathscr{L}\left\{\frac{dy}{dt}\right\} = sY(s) - y(0).$$

Applying the initial condition (9.53), this becomes

$$\mathscr{L}\left\{\frac{dy}{dt}\right\} = sY(s) - 3.$$

Using this, the left member of Equation (9.54) becomes $sY(s) - 3 - 2Y(s)$. From Table 9.1, number 2, $\mathscr{L}\{e^{5t}\} = 1/(s - 5)$. Thus Equation (9.54) reduces to the algebraic equation

$$[s - 2]Y(s) - 3 = \frac{1}{s - 5} \tag{9.55}$$

in the unknown $Y(s)$.

Step 2. We now solve Equation (9.55) for $Y(s)$. We have

$$[s - 2]Y(s) = \frac{3s - 14}{s - 5}$$

and so

$$Y(s) = \frac{3s - 14}{(s - 2)(s - 5)}.$$

Step 3. We must now determine

$$\mathcal{L}^{-1}\left\{\frac{3s - 14}{(s - 2)(s - 5)}\right\}.$$

We employ partial fractions. We have

$$\frac{3s - 14}{(s - 2)(s - 5)} = \frac{A}{s - 2} + \frac{B}{s - 5},$$

and so $3s - 14 = A(s - 5) + B(s - 2)$. From this we find that

$$A = \tfrac{8}{3} \quad \text{and} \quad B = \tfrac{1}{3},$$

and so

$$\mathcal{L}^{-1}\left\{\frac{3s - 14}{(s - 2)(s - 5)}\right\} = \frac{8}{3}\,\mathcal{L}^{-1}\left\{\frac{1}{s - 2}\right\} + \frac{1}{3}\,\mathcal{L}^{-1}\left\{\frac{1}{s - 5}\right\}.$$

Using number 2 of Table 9.1,

$$\mathcal{L}^{-1}\left\{\frac{1}{s - 2}\right\} = e^{2t} \quad \text{and} \quad \mathcal{L}^{-1}\left\{\frac{1}{s - 5}\right\} = e^{5t}.$$

Thus the solution of the given initial-value problem is

$$y = \tfrac{8}{3}e^{2t} + \tfrac{1}{3}e^{5t}.$$

▶ **Example 9.28**

Solve the initial-value problem

$$\frac{d^2y}{dt^2} - 2\frac{dy}{dt} - 8y = 0, \tag{9.56}$$

$$y(0) = 3, \tag{9.57}$$

$$y'(0) = 6. \tag{9.58}$$

Step 1. Taking the Laplace transform of both sides of the differential equation (9.56), we have

$$\mathcal{L}\left\{\frac{d^2y}{dt^2}\right\} - 2\mathcal{L}\left\{\frac{dy}{dt}\right\} - 8\mathcal{L}\{y(t)\} = \mathcal{L}\{0\}. \tag{9.59}$$

Since $\mathscr{L}\{0\} = 0$, the right member of Equation (9.59) is simply 0. Denote $\mathscr{L}\{y(t)\}$ by $Y(s)$. Then, applying Theorem 9.4, we have the following expressions for $\mathscr{L}\{d^2y/dt^2\}$ and $\mathscr{L}\{dy/dt\}$ in terms of $Y(s)$, $y(0)$, and $y'(0)$:

$$\mathscr{L}\left\{\frac{d^2y}{dt^2}\right\} = s^2 Y(s) - s y(0) - y'(0),$$

$$\mathscr{L}\left\{\frac{dy}{dt}\right\} = s Y(s) - y(0).$$

Applying the initial conditions (9.57) and (9.58) to these expressions, they become:

$$\mathscr{L}\left\{\frac{d^2y}{dt^2}\right\} = s^2 Y(s) - 3s - 6,$$

$$\mathscr{L}\left\{\frac{dy}{dt}\right\} = s Y(s) - 3.$$

Now, using these expressions, Equation (9.59) becomes

$$s^2 Y(s) - 3s - 6 - 2s Y(s) + 6 - 8 Y(s) = 0$$

or

$$[s^2 - 2s - 8] Y(s) - 3s = 0. \tag{9.60}$$

Step 2. We now solve Equation (9.60) for $Y(s)$. We have at once

$$Y(s) = \frac{3s}{(s - 4)(s + 2)}.$$

Step 3. We must now determine

$$\mathscr{L}^{-1}\left\{\frac{3s}{(s - 4)(s + 2)}\right\}.$$

We shall again employ partial fractions. From

$$\frac{3s}{(s - 4)(s + 2)} = \frac{A}{s - 4} + \frac{B}{s + 2}$$

we find that $A = 2$, $B = 1$. Thus

$$\mathscr{L}^{-1}\left\{\frac{3s}{(s - 4)(s + 2)}\right\} = 2\mathscr{L}^{-1}\left\{\frac{1}{s - 4}\right\} + \mathscr{L}^{-1}\left\{\frac{1}{s + 2}\right\}.$$

By Table 9.1, number 2, we find

$$\mathscr{L}^{-1}\left\{\frac{1}{s - 4}\right\} = e^{4t} \quad \text{and} \quad \mathscr{L}^{-1}\left\{\frac{1}{s + 2}\right\} = e^{-2t}.$$

Thus the solution of the given initial-value problem is

$$y = 2e^{4t} + e^{-2t}.$$

▶ **Example 9.29**

Solve the initial-value problem

$$\frac{d^2 y}{dt^2} + y = e^{-2t} \sin t, \tag{9.61}$$

$$y(0) = 0, \tag{9.62}$$

$$y'(0) = 0. \tag{9.63}$$

Step 1. Taking the Laplace transform of both sides of the differential equation (9.61), we have

$$\mathscr{L}\left\{\frac{d^2 y}{dt^2}\right\} + \mathscr{L}\{y(t)\} = \mathscr{L}\{e^{-2t} \sin t\}. \tag{9.64}$$

Denoting $\mathscr{L}\{y(t)\}$ by $Y(s)$ and applying Theorem 9.4, we express $\mathscr{L}\{d^2 y/dt^2\}$ in terms of $Y(s)$, $y(0)$, and $y'(0)$ as follows:

$$\mathscr{L}\left\{\frac{d^2 y}{dt^2}\right\} = s^2 Y(s) - s y(0) - y'(0).$$

Applying the initial conditions (9.62) and (9.63) to this expression, it becomes simply

$$\mathscr{L}\left\{\frac{d^2 y}{dt^2}\right\} = s^2 Y(s);$$

and thus the left member of Equation (9.64) becomes $s^2 Y(s) + Y(s)$. By number 11, Table 9.1, the right member of Equation (9.64) becomes

$$\frac{1}{(s+2)^2 + 1}.$$

Thus Equation (9.64) reduces to the algebraic equation

$$(s^2 + 1) Y(s) = \frac{1}{(s+2)^2 + 1} \tag{9.65}$$

in the unknown $Y(s)$.

Step 2. Solving Equation (9.65) for $Y(s)$, we have

$$Y(s) = \frac{1}{(s^2 + 1)[(s+2)^2 + 1]}.$$

Step 3. We must now determine

$$\mathscr{L}^{-1}\left\{\frac{1}{(s^2 + 1)[(s+2)^2 + 1]}\right\}.$$

We may use either partial fractions or the convolution. We shall illustrate both methods.

1. Use of Partial Fractions. We have

$$\frac{1}{(s^2 + 1)(s^2 + 4s + 5)} = \frac{As + B}{s^2 + 1} + \frac{Cs + D}{s^2 + 4s + 5}.$$

From this we find

$$1 = (As + B)(s^2 + 4s + 5) + (Cs + D)(s^2 + 1)$$
$$= (A + C)s^3 + (4A + B + D)s^2 + (5A + 4B + C)s + (5B + D).$$

Thus we obtain the equations

$$A + C = 0,$$
$$4A + B + D = 0,$$
$$5A + 4B + C = 0,$$
$$5B + D = 1.$$

From these equations we find that

$$A = -\tfrac{1}{8}, \qquad B = \tfrac{1}{8}, \qquad C = \tfrac{1}{8}, \qquad D = \tfrac{3}{8},$$

and so

$$\mathscr{L}^{-1}\left\{\frac{1}{(s^2 + 1)(s^2 + 4s + 5)}\right\} = -\frac{1}{8}\mathscr{L}^{-1}\left\{\frac{s}{s^2 + 1}\right\} + \frac{1}{8}\mathscr{L}^{-1}\left\{\frac{1}{s^2 + 1}\right\}$$
$$+ \frac{1}{8}\mathscr{L}^{-1}\left\{\frac{s}{s^2 + 4s + 5}\right\}$$
$$+ \frac{3}{8}\mathscr{L}^{-1}\left\{\frac{1}{s^2 + 4s + 5}\right\}. \tag{9.66}$$

In order to determine

$$\frac{1}{8}\mathscr{L}^{-1}\left\{\frac{s}{s^2 + 4s + 5}\right\} + \frac{3}{8}\mathscr{L}^{-1}\left\{\frac{1}{s^2 + 4s + 5}\right\}, \tag{9.67}$$

we write

$$\frac{s}{s^2 + 4s + 5} = \frac{s + 2}{(s + 2)^2 + 1} - \frac{2}{(s + 2)^2 + 1}.$$

Thus the expression (9.67) becomes

$$\frac{1}{8}\mathscr{L}^{-1}\left\{\frac{s + 2}{(s + 2)^2 + 1}\right\} + \frac{1}{8}\mathscr{L}^{-1}\left\{\frac{1}{(s + 2)^2 + 1}\right\},$$

and so (9.66) may be written

$$\mathscr{L}^{-1}\left\{\frac{1}{(s^2 + 1)(s^2 + 4s + 5)}\right\} = -\frac{1}{8}\mathscr{L}^{-1}\left\{\frac{s}{s^2 + 1}\right\} + \frac{1}{8}\mathscr{L}^{-1}\left\{\frac{1}{s^2 + 1}\right\}$$
$$+ \frac{1}{8}\mathscr{L}^{-1}\left\{\frac{s + 2}{(s + 2)^2 + 1}\right\}$$
$$+ \frac{1}{8}\mathscr{L}^{-1}\left\{\frac{1}{(s + 2)^2 + 1}\right\}.$$

Now using Table 9.1, numbers 4, 3, 12, and 11, respectively, we obtain the solution

$$y(t) = -\tfrac{1}{8}\cos t + \tfrac{1}{8}\sin t + \tfrac{1}{8}e^{-2t}\cos t + \tfrac{1}{8}e^{-2t}\sin t$$

or

$$y(t) = \frac{1}{8}(\sin t - \cos t) + \frac{e^{-2t}}{8}(\sin t + \cos t). \tag{9.68}$$

2. *Use of the Convolution.* We write $\dfrac{1}{(s^2 + 1)[(s + 2)^2 + 1]}$ as the product

$F(s)G(s)$, where $F(s) = \dfrac{1}{s^2 + 1}$ and $G(s) = \dfrac{1}{(s + 2)^2 + 1}$. By Table 9.1, number 3,

$f(t) = \mathscr{L}^{-1}\left\{\dfrac{1}{s^2 + 1}\right\} = \sin t$, and by number 11, $g(t) = \mathscr{L}^{-1}\left\{\dfrac{1}{(s + 2)^2 + 1}\right\} =$

$e^{-2t}\sin t$. Thus by Theorem 9.10 using (9.46) or (9.47), we have, respectively,

$$\mathscr{L}^{-1}\left\{\frac{1}{(s^2 + 1)[(s + 2)^2 + 1]}\right\} = f(t) * g(t) = \int_0^t \sin \tau \cdot e^{-2(t-\tau)}\sin(t - \tau)\,d\tau$$

or

$$\mathscr{L}^{-1}\left\{\frac{1}{(s^2 + 1)[(s + 2)^2 + 1]}\right\} = g(t) * f(t) = \int_0^t e^{-2\tau}\sin \tau \cdot \sin(t - \tau)\,d\tau.$$

The second of these integrals is slightly more simple; it reduces to

$$(\sin t)\int_0^t e^{-2\tau}\sin \tau \cos \tau\,d\tau - (\cos t)\int_0^t e^{-2\tau}\sin^2 \tau\,d\tau.$$

Introducing double-angle formulas this becomes

$$\frac{\sin t}{2}\int_0^t e^{-2\tau}\sin 2\tau\,d\tau - \frac{\cos t}{2}\int_0^t e^{-2\tau}\,d\tau + \frac{\cos t}{2}\int_0^t e^{-2\tau}\cos 2\tau\,d\tau.$$

Carrying out the indicated integrations we find that this becomes

$$-\sin t\left[\frac{e^{-2t}}{8}(\sin 2\tau + \cos 2\tau)\right]_0^t + \frac{\cos t}{4}\left[e^{-2\tau}\right]_0^t + \cos t\left[\frac{e^{-2t}}{8}(\sin 2\tau - \cos 2\tau)\right]_0^t$$

$$= -\frac{e^{-2t}}{8}(\sin t \sin 2t + \sin t \cos 2t) + \frac{\sin t}{8} + \frac{e^{-2t}\cos t}{4} - \frac{\cos t}{4}$$

$$+ \frac{e^{-2t}}{8}(\cos t \sin 2t - \cos t \cos 2t) + \frac{\cos t}{8}.$$

Using double-angle formulas and simplifying, this reduces to

$$\frac{1}{8}(\sin t - \cos t) + \frac{e^{-2t}}{8}(\sin t + \cos t),$$

which is the solution (9.68) obtained above using partial fractions.

▶   **Example 9.30**

Solve the initial-value problem

$$\frac{d^3y}{dt^3} + 4\frac{d^2y}{dt^2} + 5\frac{dy}{dt} + 2y = 10\cos t, \tag{9.69}$$

$$y(0) = 0, \tag{9.70}$$

$$y'(0) = 0, \tag{9.71}$$

$$y''(0) = 3. \tag{9.72}$$

Step 1.   Taking the Laplace transform of both sides of the differential equation (9.69), we have

$$\mathscr{L}\left\{\frac{d^3y}{dt^3}\right\} + 4\mathscr{L}\left\{\frac{d^2y}{dt^2}\right\} + 5\mathscr{L}\left\{\frac{dy}{dt}\right\} + 2\mathscr{L}\{y(t)\} = 10\mathscr{L}\{\cos t\}. \tag{9.73}$$

We denote $\mathscr{L}\{y(t)\}$ by $Y(s)$ and then apply Theorem 9.4 to express

$$\mathscr{L}\left\{\frac{d^3y}{dt^3}\right\}, \qquad \mathscr{L}\left\{\frac{d^2y}{dt^2}\right\}, \quad \text{and} \quad \mathscr{L}\left\{\frac{dy}{dt}\right\}$$

in terms of $Y(s)$, $y(0)$, $y'(0)$, and $y''(0)$. We thus obtain

$$\mathscr{L}\left\{\frac{d^3y}{dt^3}\right\} = s^3Y(s) - s^2y(0) - sy'(0) - y''(0),$$

$$\mathscr{L}\left\{\frac{d^2y}{dt^2}\right\} = s^2Y(s) - sy(0) - y'(0),$$

$$\mathscr{L}\left\{\frac{dy}{dt}\right\} = sY(s) - y(0).$$

Applying the initial conditions (9.70), (9.71), and (9.72), these expressions become

$$\mathscr{L}\left\{\frac{d^3y}{dt^3}\right\} = s^3Y(s) - 3,$$

$$\mathscr{L}\left\{\frac{d^2y}{dt^2}\right\} = s^2Y(s),$$

$$\mathscr{L}\left\{\frac{dy}{dt}\right\} = sY(s).$$

Thus the left member of Equation (9.73) becomes

$$s^3Y(s) - 3 + 4s^2Y(s) + 5sY(s) + 2Y(s)$$

or

$$[s^3 + 4s^2 + 5s + 2]Y(s) - 3.$$

By number 4, Table 9.1,

$$10\mathscr{L}\{\cos t\} = \frac{10s}{s^2 + 1}.$$

Thus Equation (9.73) reduces to the algebraic equation

$$(s^3 + 4s^2 + 5s + 2)Y(s) - 3 = \frac{10s}{s^2 + 1} \qquad (9.74)$$

in the unknown $Y(s)$.

Step 2. We now solve Equation (9.74) for $Y(s)$. We have

$$(s^3 + 4s^2 + 5s + 2)Y(s) = \frac{3s^2 + 10s + 3}{s^2 + 1}$$

or

$$Y(s) = \frac{3s^2 + 10s + 3}{(s^2 + 1)(s^3 + 4s^2 + 5s + 2)}.$$

Step 3. We must now determine

$$\mathscr{L}^{-1}\left\{\frac{3s^2 + 10s + 3}{(s^2 + 1)(s^3 + 4s^2 + 5s + 2)}\right\}.$$

Let us not despair! We can again employ partial fractions to put the expression for $Y(s)$ into a form where Table 9.1 can be used, but the work will be rather involved. We proceed by writing

$$\frac{3s^2 + 10s + 3}{(s^2 + 1)(s^3 + 4s^2 + 5s + 2)} = \frac{3s^2 + 10s + 3}{(s^2 + 1)(s + 1)^2(s + 2)}$$

$$= \frac{A}{s + 2} + \frac{B}{s + 1} + \frac{C}{(s + 1)^2} + \frac{Ds + E}{s^2 + 1}. \qquad (9.75)$$

From this we find

$$3s^2 + 10s + 3 = A(s + 1)^2(s^2 + 1) + B(s + 2)(s + 1)(s^2 + 1)$$
$$+ C(s + 2)(s^2 + 1) + (Ds + E)(s + 2)(s + 1)^2, \qquad (9.76)$$

or

$$3s^2 + 10s + 3 = (A + B + D)s^4 + (2A + 3B + C + 4D + E)s^3$$
$$+ (2A + 3B + 2C + 5D + 4E)s^2$$
$$+ (2A + 3B + C + 2D + 5E)s + (A + 2B + 2C + 2E).$$

From this we obtain the system of equations

$$A + B + D = 0,$$
$$2A + 3B + C + 4D + E = 0,$$
$$2A + 3B + 2C + 5D + 4E = 3, \qquad (9.77)$$
$$2A + 3B + C + 2D + 5E = 10,$$
$$A + 2B + 2C + 2E = 3.$$

Letting $s = -1$ in Equation (9.76), we find that $C = -2$; and letting $s = -2$ in this same equation results in $A = -1$. Using these values for A and C we find from the system (9.77) that

$$B = 2, \qquad D = -1, \quad \text{and} \quad E = 2.$$

Substituting these values thus found for A, B, C, D, and E into Equation (9.75), we see that

$$\mathscr{L}^{-1}\left\{\frac{3s^2 + 10s + 3}{(s^2 + 1)(s^3 + 4s^2 + 5s + 2)}\right\} = -\mathscr{L}^{-1}\left\{\frac{1}{s + 2}\right\} + 2\mathscr{L}^{-1}\left\{\frac{1}{s + 1}\right\}$$

$$-2\mathscr{L}^{-1}\left\{\frac{1}{(s + 1)^2}\right\} - \mathscr{L}^{-1}\left\{\frac{s}{s^2 + 1}\right\}$$

$$+ 2\mathscr{L}^{-1}\left\{\frac{1}{s^2 + 1}\right\}.$$

Using Table 9.1, numbers 2, 2, 8, 4, and 3, respectively, we obtain the solution

$$y(t) = -e^{-2t} + 2e^{-t} - 2te^{-t} - \cos t + 2\sin t.$$

▶ **Example 9.31**

Solve the initial-value problem

$$\frac{d^2y}{dt^2} + 2\frac{dy}{dt} + 5y = h(t), \tag{9.78}$$

where

$$h(t) = \begin{cases} 1, & 0 < t < \pi, \\ 0, & t > \pi, \end{cases} \tag{9.79}$$

$$y(0) = 0, \tag{9.80}$$

$$y'(0) = 0. \tag{9.81}$$

Step 1. We take the Laplace transform of both sides of the differential equation (9.78) to obtain

$$\mathscr{L}\left\{\frac{d^2y}{dt^2}\right\} + 2\mathscr{L}\left\{\frac{dy}{dt}\right\} + 5\mathscr{L}\{y(t)\} = \mathscr{L}\{h(t)\}. \tag{9.82}$$

Denoting $\mathscr{L}\{y(t)\}$ by $Y(s)$, using Theorem 9.4 as in the previous examples, and then applying the initial conditions (9.80) and (9.81), we see that the left member of Equation (9.82) becomes $[s^2 + 2s + 5]Y(s)$. By the definition of the Laplace transform, from (9.79) we have

$$\mathscr{L}\{h(t)\} = \int_0^\infty e^{-st}h(t)\, dt = \int_0^\pi e^{-st}\, dt = \left[\frac{e^{-st}}{-s}\right]_0^\pi = \frac{1 - e^{-\pi s}}{s}.$$

Thus Equation (9.82) becomes

$$[s^2 + 2s + 5]Y(s) = \frac{1 - e^{-\pi s}}{s}. \tag{9.83}$$

Step 2. We solve the algebraic equation (9.83) for $Y(s)$ to obtain

$$Y(s) = \frac{1 - e^{-\pi s}}{s(s^2 + 2s + 5)}.$$

Step 3. We must now determine

$$\mathscr{L}^{-1}\left\{\frac{1 - e^{-\pi s}}{s(s^2 + 2s + 5)}\right\}.$$

Let us write this as

$$\mathscr{L}^{-1}\left\{\frac{1}{s(s^2 + 2s + 5)}\right\} - \mathscr{L}^{-1}\left\{\frac{e^{-\pi s}}{s(s^2 + 2s + 5)}\right\},$$

and apply partial fractions to determine the first of these two inverse transforms. Writing

$$\frac{1}{s(s^2 + 2s + 5)} = \frac{A}{s} + \frac{Bs + C}{s^2 + 2s + 5},$$

we find at once that $A = \frac{1}{5}$, $B = -\frac{1}{5}$, $C = -\frac{2}{5}$. Thus

$$\mathscr{L}^{-1}\left\{\frac{1}{s(s^2 + 2s + 5)}\right\} = \frac{1}{5}\mathscr{L}^{-1}\left\{\frac{1}{s}\right\} - \frac{1}{5}\mathscr{L}^{-1}\left\{\frac{s + 2}{(s + 1)^2 + 4}\right\}$$

$$= \frac{1}{5}\mathscr{L}^{-1}\left\{\frac{1}{s}\right\} - \frac{1}{5}\mathscr{L}^{-1}\left\{\frac{s + 1}{(s + 1)^2 + 4}\right\}$$

$$- \frac{1}{10}\mathscr{L}^{-1}\left\{\frac{2}{(s + 1)^2 + 4}\right\}$$

$$= \frac{1}{5} - \frac{1}{5}e^{-t}\cos 2t - \frac{1}{10}e^{-t}\sin 2t,$$

using Table 9.1, numbers 1, 12, and 11, respectively. Letting

$$F(s) = \frac{1}{s(s^2 + 2s + 5)}$$

and

$$f(t) = \tfrac{1}{5} - \tfrac{1}{5}e^{-t}\cos 2t - \tfrac{1}{10}e^{-t}\sin 2t,$$

we thus have

$$\mathscr{L}^{-1}\{F(s)\} = f(t).$$

We now consider

$$\mathscr{L}^{-1}\left\{\frac{e^{-\pi s}}{s(s^2 + 2s + 5)}\right\} = \mathscr{L}^{-1}\{e^{-\pi s}F(s)\}.$$

By Theorem 9.7,

$$\mathscr{L}^{-1}\{e^{-\pi s}F(s)\} = u_\pi(t)f(t - \pi),$$

where

$$u_\pi(t)f(t - \pi) = \begin{cases} 0, & 0 < t < \pi, \\ f(t - \pi), & t > \pi. \end{cases}$$

Thus,

$$\mathcal{L}^{-1}\left\{\frac{e^{-\pi s}}{s(s^2 + 2s + 5)}\right\}$$

$$= \begin{cases} 0, & 0 < t < \pi, \\ \frac{1}{5} - \frac{1}{5}e^{-(t-\pi)} \cos 2(t - \pi) - \frac{1}{10}e^{-(t-\pi)} \sin 2(t - \pi), & t > \pi \end{cases}$$

$$= \begin{cases} 0, & 0 < t < \pi, \\ \frac{1}{5} - \frac{1}{5}e^{-(t-\pi)} \cos 2t - \frac{1}{10}e^{-(t-\pi)} \sin 2t, & t > \pi. \end{cases}$$

Thus the solution is given by

$$y(t) = \mathcal{L}^{-1}\left\{\frac{1 - e^{-\pi s}}{s(s^2 + 2s + 5)}\right\} = f(t) - u_\pi(t)f(t - \pi)$$

$$= \begin{cases} \frac{1}{5} - \frac{1}{5}e^{-t} \cos 2t - \frac{1}{10}e^{-t} \sin 2t - 0, & 0 < t < \pi, \\ \frac{1}{5} - \frac{1}{5}e^{-t} \cos 2t - \frac{1}{10}e^{-t} \sin 2t - \frac{1}{5} + \frac{1}{5}e^{-(t-\pi)} \cos 2t + \frac{1}{10}e^{-(t-\pi)} \sin 2t, & t > \pi, \end{cases}$$

or

$$y(t) = \begin{cases} \frac{1}{5}\left[1 - e^{-t}\left(\cos 2t + \frac{1}{2}\sin 2t\right)\right], & 0 < t < \pi, \\ \frac{e^{-t}}{5}\left[(e^\pi - 1)\cos 2t + \left(\frac{e^\pi - 1}{2}\right)\sin 2t\right], & t > \pi. \end{cases}$$

Exercises

Use Laplace transforms to solve each of the initial-value problems in Exercises 1–18:

1. $\dfrac{dy}{dt} - y = e^{3t}$,　$y(0) = 2$.

2. $\dfrac{dy}{dt} + y = 2 \sin t$,　$y(0) = -1$.

3. $\dfrac{d^2y}{dt^2} - 5\dfrac{dy}{dt} + 6y = 0$,

　$y(0) = 1$,　$y'(0) = 2$.

4. $\dfrac{d^2y}{dt^2} + \dfrac{dy}{dt} - 12y = 0$,

　$y(0) = 4$,　$y'(0) = -1$.

5. $\dfrac{d^2y}{dt^2} + 4y = 8$,

　$y(0) = 0$,　$y'(0) = 6$.

6. $\dfrac{d^2y}{dt^2} + 2\dfrac{dy}{dt} + 5y = 0$,

　$y(0) = 2$,　$y'(0) = 4$.

7. $\dfrac{d^2y}{dt^2} - \dfrac{dy}{dt} - 2y = 18e^{-t} \sin 3t$,

　$y(0) = 0$,　$y'(0) = 3$.

8. $\dfrac{d^2y}{dt^2} + 2\dfrac{dy}{dt} + y = te^{-2t}$,

　$y(0) = 1$,　$y'(0) = 0$.

9. $\dfrac{d^2y}{dt^2} + 7\dfrac{dy}{dt} + 10y = 4te^{-3t}$,

$y(0) = 0,\qquad y'(0) = -1.$

10. $\dfrac{d^2y}{dt^2} - 8\dfrac{dy}{dt} + 15y = 9te^{2t}$,

$y(0) = 5,\qquad y'(0) = 10.$

11. $\dfrac{d^3y}{dt^3} - 5\dfrac{d^2y}{dt^2} + 7\dfrac{dy}{dt} - 3y = 20\sin t,$

$y(0) = 0,\qquad y'(0) = 0,\qquad y''(0) = -2.$

12. $\dfrac{d^3y}{dt^3} - 6\dfrac{d^2y}{dt^2} + 11\dfrac{dy}{dt} - 6y = 36te^{4t}$,

$y(0) = -1,\qquad y'(0) = 0,\qquad y''(0) = -6.$

13. $\dfrac{d^2y}{dt^2} - 3\dfrac{dy}{dt} + 2y = h(t),$

where $h(t) = \begin{cases} 2, & 0 < t < 4, \\ 0, & t > 4, \end{cases}$

$y(0) = 0,\qquad y'(0) = 0.$

14. $\dfrac{d^2y}{dt^2} + 5\dfrac{dy}{dt} + 6y = h(t),$

where $h(t) = \begin{cases} 6, & 0 < t < 2, \\ 0, & t > 2, \end{cases}$

$y(0) = 0,\qquad y'(0) = 0.$

15. $\dfrac{d^2y}{dt^2} + 4\dfrac{dy}{dt} + 5y = h(t),$

where $h(t) = \begin{cases} 1, & 0 < t < \pi/2, \\ 0, & t > \pi/2, \end{cases}$

$y(0) = 0,\qquad y'(0) = 1.$

16. $\dfrac{d^2y}{dt^2} + 6\dfrac{dy}{dt} + 8y = h(t),$

where $h(t) = \begin{cases} 3, & 0 < t < 2\pi, \\ 0, & t > 2\pi, \end{cases}$

$y(0) = 1,\qquad y'(0) = -1.$

17. $\dfrac{d^2y}{dt^2} + 4y = h(t),$

where $h(t) = \begin{cases} -4t + 8\pi, & 0 < t < 2\pi, \\ 0, & t > 2\pi, \end{cases}$

$y(0) = 2,\qquad y'(0) = 0.$

18. $\dfrac{d^2y}{dt^2} + y = h(t),$

where $h(t) = \begin{cases} t, & 0 < t < \pi, \\ \pi, & t > \pi, \end{cases}$

$y(0) = 2,\qquad y'(0) = 3.$

9.4 LAPLACE TRANSFORM SOLUTION OF LINEAR SYSTEMS

A. The Method

We apply the Laplace transform method to find the solution of a first-order system

$$a_1\frac{dx}{dt} + a_2\frac{dy}{dt} + a_3x + a_4y = \beta_1(t),$$

$$b_1\frac{dx}{dt} + b_2\frac{dy}{dt} + b_3x + b_4y = \beta_2(t),\qquad (9.84)$$

where $a_1, a_2, a_3, a_4, b_1, b_2, b_3,$ and b_4 are constants and β_1 and β_2 are known functions, that satisfies the initial conditions

$$x(0) = c_1 \quad \text{and} \quad y(0) = c_2, \tag{9.85}$$

where c_1 and c_2 are constants.

The procedure is a straightforward extension of the method outlined in Section 9.3. Let $X(s)$ denote $\mathscr{L}\{x(t)\}$ and let $Y(s)$ denote $\mathscr{L}\{y(t)\}$. Then proceed as follows:

1. For each of the two equations of the system (9.84), take the Laplace transform of both sides of the equation, apply Theorem 9.3 and the initial conditions (9.85), and equate the results to obtain a linear algebraic equation in the two "unknowns" $X(s)$ and $Y(s)$.
2. Solve the linear system of two algebraic equations in the two unknowns $X(s)$ and $Y(s)$ thus obtained in Step 1 to explicitly determine $X(s)$ and $Y(s)$.
3. Having found $X(s)$ and $Y(s)$, employ the table of transforms to determine the solution $x(t) = \mathscr{L}^{-1}\{X(s)\}$ and $y(t) = \mathscr{L}^{-1}\{Y(s)\}$ of the given initial-value problem.

B. An Example

▶ **Example 9.32**

Use Laplace transforms to find the solution of the system

$$\frac{dx}{dt} - 6x + 3y = 8e^t,$$

$$\frac{dy}{dt} - 2x - y = 4e^t, \tag{9.86}$$

that satisfies the initial conditions

$$x(0) = -1,$$

$$y(0) = 0. \tag{9.87}$$

Step 1.   Taking the Laplace transform of both sides of each differential equation of system (9.86), we have

$$\mathscr{L}\left\{\frac{dx}{dt}\right\} - 6\mathscr{L}\{x(t)\} + 3\mathscr{L}\{y(t)\} = \mathscr{L}\{8e^t\},$$

$$\mathscr{L}\left\{\frac{dy}{dt}\right\} - 2\mathscr{L}\{x(t)\} - \mathscr{L}\{y(t)\} = \mathscr{L}\{4e^t\}. \tag{9.88}$$

Denote $\mathscr{L}\{x(t)\}$ by $X(s)$ and $\mathscr{L}\{y(t)\}$ by $Y(s)$. Then applying Theorem 9.3 and the initial conditions (9.87), we have

$$\mathscr{L}\left\{\frac{dx}{dt}\right\} = sX(s) - x(0) = sX(s) + 1,$$

$$\mathscr{L}\left\{\frac{dy}{dt}\right\} = sY(s) - y(0) = sY(s). \tag{9.89}$$

Using Table 9.1, number 2, we find

$$\mathscr{L}\{8e^t\} = \frac{8}{s-1} \quad \text{and} \quad \mathscr{L}\{4e^t\} = \frac{4}{s-1}. \tag{9.90}$$

Thus, from (9.89) and (9.90), we see that Equations (9.88) become

$$sX(s) + 1 - 6X(s) + 3Y(s) = \frac{8}{s-1},$$

$$sY(s) - 2X(s) - Y(s) = \frac{4}{s-1},$$

which simplify to the form

$$(s-6)X(s) + 3Y(s) = \frac{8}{s-1} - 1,$$

$$-2X(s) + (s-1)Y(s) = \frac{4}{s-1},$$

or

$$(s-6)X(s) + 3Y(s) = \frac{-s+9}{s-1},$$

$$-2X(s) + (s-1)Y(s) = \frac{4}{s-1}. \tag{9.91}$$

Step 2. We solve the linear algebraic system of the two equations (9.91) in the two "unknowns" $X(s)$ and $Y(s)$. We have

$$(s-1)(s-6)X(s) + 3(s-1)Y(s) = -s+9,$$

$$-6X(s) + 3(s-1)Y(s) = \frac{12}{s-1}.$$

Subtracting we obtain

$$(s^2 - 7s + 12)X(s) = -s + 9 - \frac{12}{s-1},$$

from which we find

$$X(s) = \frac{-s^2 + 10s - 21}{(s-1)(s-3)(s-4)} = \frac{-s+7}{(s-1)(s-4)}.$$

In like manner, we find

$$Y(s) = \frac{2s - 6}{(s-1)(s-3)(s-4)} = \frac{2}{(s-1)(s-4)}.$$

Step 3. We must now determine

$$x(t) = \mathscr{L}^{-1}\{X(s)\} = \mathscr{L}^{-1}\left\{\frac{-s+7}{(s-1)(s-4)}\right\}$$

and

$$y(t) = \mathscr{L}^{-1}\{Y(s)\} = \mathscr{L}^{-1}\left\{\frac{2}{(s-1)(s-4)}\right\}.$$

We first find $x(t)$. We use partial fractions and write

$$\frac{-s+7}{(s-1)(s-4)} = \frac{A}{s-1} + \frac{B}{s-4}.$$

From this we find

$$A = -2 \quad \text{and} \quad B = 1.$$

Thus

$$x(t) = -2\mathscr{L}^{-1}\left\{\frac{1}{s-1}\right\} + \mathscr{L}^{-1}\left\{\frac{1}{s-4}\right\},$$

and using Table 9.1, number 2, we obtain

$$x(t) = -2e^t + e^{4t}. \tag{9.92}$$

In like manner, we find $y(t)$. Doing so, we obtain

$$y(t) = -\tfrac{2}{3}e^t + \tfrac{2}{3}e^{4t}. \tag{9.93}$$

The pair defined by (9.92) and (9.93) constitute the solution of the given system (9.86) that satisfies the given initial conditions (9.87).

Exercises

In each of the following exercises, use the Laplace transform to find the solution of the given linear system that satisfies the given initial conditions.

1. $\dfrac{dx}{dt} + y = 3e^{2t},$

    $\dfrac{dy}{dt} + x = 0,$

    $x(0) = 2, \qquad y(0) = 0.$

2. $\dfrac{dx}{dt} - 2y = 0,$

    $\dfrac{dy}{dt} + x - 3y = 2,$

    $x(0) = 3, \qquad y(0) = 0.$

3. $\dfrac{dx}{dt} - 5x + 2y = 3e^{4t},$

    $\dfrac{dy}{dt} - 4x + y = 0,$

    $x(0) = 3, \qquad y(0) = 0.$

4. $\dfrac{dx}{dt} - 2x - 3y = 0,$

    $\dfrac{dy}{dt} + x + 2y = t,$

    $x(0) = -1, \qquad y(0) = 0.$

5. $\dfrac{dx}{dt} - 4x + 2y = 2t,$

    $\dfrac{dy}{dt} - 8x + 4y = 1,$

    $x(0) = 3, \qquad y(0) = 5.$

6. $\dfrac{dx}{dt} + x + y = 5e^{2t},$

    $\dfrac{dy}{dt} - 5x - y = -3e^{2t},$

    $x(0) = 3, \qquad y(0) = 2.$

7. $2\dfrac{dx}{dt} + \dfrac{dy}{dt} - x - y = e^{-t}$,

   $\dfrac{dx}{dt} + \dfrac{dy}{dt} + 2x + y = e^{t}$,

   $x(0) = 2, \qquad y(0) = 1$.

8. $2\dfrac{dx}{dt} + \dfrac{dy}{dt} + x + 5y = 4t$,

   $\dfrac{dx}{dt} + \dfrac{dy}{dt} + 2x + 2y = 2$,

   $x(0) = 3, \qquad y(0) = -4$.

9. $2\dfrac{dx}{dt} + 4\dfrac{dy}{dt} + x - y = 3e^{t}$,

   $\dfrac{dx}{dt} + \dfrac{dy}{dt} + 2x + 2y = e^{t}$,

   $x(0) = 1, \qquad y(0) = 0$.

10. $\dfrac{d^{2}x}{dt^{2}} - 3\dfrac{dx}{dt} + \dfrac{dy}{dt} + 2x - y = 0$,

   $\dfrac{dx}{dt} + \dfrac{dy}{dt} - 2x + y = 0$,

   $x(0) = 0, \qquad y(0) = -1, \qquad x'(0) = 0$.

PART TWO
FUNDAMENTAL THEORY AND FURTHER METHODS

CHAPTER TEN

Existence And Uniqueness Theory

In certain of the preceding chapters we stated and illustrated several of the basic theorems of the theory of differential equations. However, in these earlier chapters our primary interest was in methods of solution, and our interest in theory was secondary. Now that we have become familiar with basic methods, we shall direct our attention to the more theoretical aspects of differential equations. In particular, in this chapter we shall state, prove, and illustrate existence and uniqueness theorems for ordinary differential equations. In order to fully understand the proofs of these theorems, we shall need to be familiar with certain concepts of real function theory. Since the reader may not be familiar with all of these topics, our first section will be devoted to a brief survey of some of them.

10.1 SOME CONCEPTS FROM REAL FUNCTION THEORY

A. Uniform Convergence

We assume at once that the reader is familiar with the most basic concepts and results concerning functions of one real variable. To be specific, he should be familiar with continuous functions and their properties, and also with the basic theory of differentiation and integration.

This much assumed, let us recall the definition of convergence of a sequence $\{c_n\}$ of real numbers.

DEFINITION A

A sequence $\{c_n\}$ of real numbers is said to converge *to the limit c if, given any $\epsilon > 0$, there exists a positive number N such that*

$$|c_n - c| < \epsilon$$

for all $n > N$. We indicate this by writing $\lim_{n \to \infty} c_n = c.$

▶ **Example 10.1**

Let $c_n = n/(n + 1)$, $n = 1, 2, 3, \ldots$.

Then $\lim\limits_{n \to \infty} \dfrac{n}{n + 1} = 1$.   For, given $\epsilon > 0$,

$$|c_n - c| = \left| \frac{n}{n + 1} - 1 \right| = \frac{1}{n + 1} < \epsilon$$

provided $n > (1/\epsilon) - 1$. Thus, given $\epsilon > 0$, there exists $N = (1/\epsilon) - 1$ such that

$$\left| \frac{n}{n + 1} - 1 \right| < \epsilon \quad \text{for all } n > N.$$

Hence we conclude that the given sequence converges to 1.

Let us now consider a sequence $\{f_n\}$ of real functions, each of which is defined on a real interval $a \leq x \leq b$. Let x be a particular real number belonging to this interval. Then $\{f_n(x)\}$ is simply a sequence of real numbers. If this sequence $\{f_n(x)\}$ of real numbers converges to a limit $f(x)$, then we say that the sequence $\{f_n\}$ of functions converges at this particular x. If the sequence $\{f_n\}$ converges at *every* x such that $a \leq x \leq b$, then we say that the given sequence of functions *converges pointwise* on the interval $a \leq x \leq b$. More precisely, we have:

DEFINITION B

Let $\{f_n\}$ be a sequence of real functions, each function of which is defined for all x on a real interval $a \leq x \leq b$. For each particular x such that $a \leq x \leq b$ consider the corresponding sequence of real numbers $\{f_n(x)\}$. Suppose that the sequence $\{f_n(x)\}$ converges for every x such that $a \leq x \leq b$, and let $f(x) = \lim\limits_{n \to \infty} f_n(x)$ for every $x \in [a, b]$.

Then we say that the sequence of functions $\{f_n\}$ converges pointwise *on the interval $a \leq x \leq b$, and the function f thus defined is called the* limit function *of the sequence $\{f_n\}$.*

▶ **Example 10.2**

Consider the sequence of functions $\{f_n\}$ defined for all x on the real interval $0 \leq x \leq 1$ by

$$f_n(x) = \frac{nx}{nx + 1}, \qquad 0 \leq x \leq 1 \quad (n = 1, 2, 3, \ldots).$$

The first three terms of this sequence are the functions f_1, f_2, and f_3 defined, respectively, by

$$f_1(x) = \frac{x}{x + 1}, \qquad f_2(x) = \frac{2x}{2x + 1}, \qquad f_3(x) = \frac{3x}{3x + 1}.$$

For $x = 0$, the corresponding sequence $\{f_n(0)\}$ of real numbers is simply $0, 0, 0, \ldots$, which obviously converges to the limit 0. For every x such that $0 < x \leq 1$, the corresponding sequence of real numbers is

$$f_n(x) = \frac{nx}{nx + 1}.$$

Since

$$\lim_{n \to \infty} \frac{nx}{nx + 1} = 1$$

for every such x, we see that the sequence $\{f_n\}$ converges to f such that $f(x) = 1$ for $0 < x \le 1$. Thus the sequence $\{f_n\}$ converges pointwise on $0 \le x \le 1$, and the limit function is the function f defined by

$$f(0) = 0,$$

$$f(x) = 1, \quad 0 < x \le 1.$$

The graphs of the functions f_1, f_5, f_{10}, defined, respectively, by

$$f_1(x) = \frac{x}{x + 1}, \qquad f_5(x) = \frac{5x}{5x + 1}, \quad \text{and} \quad f_{10}(x) = \frac{10x}{10x + 1},$$

as well as that of the limit function f, appear in Figure 10.1.

Now suppose that each term of a sequence $\{f_n\}$ of real functions which converges pointwise on $a \le x \le b$ is a *continuous* function on $a \le x \le b$. Can we conclude from this that the limit function f will also be continuous on $a \le x \le b$? The answer to this equation is "no," for under such circumstances f might be continuous or it might not be. For instance, in Example 10.2, each f_n is continuous on $0 \le x \le 1$, and the sequence $\{f_n\}$ converges pointwise on $0 \le x \le 1$. However, the limit function f is such that $f(0) = 0$ but $f(x) = 1$ for $0 < x \le 1$; thus this limit function is certainly *not* continuous

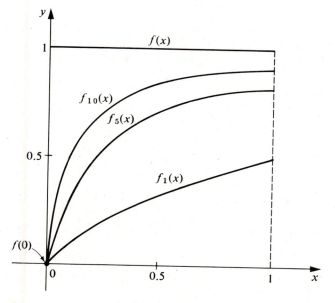

Figure 10.1

on $0 \leq x \leq 1$. To be certain that the limit function of a sequence of continuous functions is itself continuous we shall need a "stronger" type of convergence than mere pointwise convergence. One such type of convergence is *uniform* convergence, which we shall now introduce.

In order to introduce the concept of uniform convergence, let us turn back to the definition of pointwise convergence and recast it in the more basic language of Definition A. We have:

DEFINITION B′

Let $\{f_n\}$ *be a sequence of real functions, each of which is defined on the real interval* $a \leq x \leq b$. *The sequence* $\{f_n\}$ *is said to* converge pointwise *to* f *on* $a \leq x \leq b$, *if, given any* $\epsilon > 0$, *for each* x *such that* $a \leq x \leq b$ *there exists a positive number* N (*in general depending both on* ϵ *and on* x) *such that*

$$|f_n(x) - f(x)| < \epsilon$$

for every $n > N$.

Now observe carefully that in general the number N of this definition depends not only upon ϵ but also on x. For a given ϵ, different numbers $N(x)$ are in general required for different points x. However, if, given $\epsilon > 0$, there exists *one single* N such that

$$|f_n(x) - f(x)| < \epsilon$$

for *every* x such that $a \leq x \leq b$ for every $n > N$, then we say that the convergence is *uniform* on $a \leq x \leq b$. More precisely, we state:

DEFINITION C

Let $\{f_n\}$ *be a sequence of real functions, each of which is defined on the real interval* $a \leq x \leq b$. *The sequence* $\{f_n\}$ *is said to* converge uniformly *to* f *on* $a \leq x \leq b$ *if, given any* $\epsilon > 0$, *there exists a positive number* N (*which depends* only *upon* ϵ) *such that*

$$|f_n(x) - f(x)| < \epsilon$$

for all $n > N$ *for every* x *such that* $a \leq x \leq b$.

Geometrically, this means that given $\epsilon > 0$, the graphs of $y = f_n(x)$ for all $n > N$ all lie between the graphs of $y = f(x) + \epsilon$ and $y = f(x) - \epsilon$ for $a \leq x \leq b$ (see Figure 10.2).

▶ **Example 10.3**

Consider the sequence of functions $\{f_n\}$ defined for all x on the real interval $0 \leq x \leq 1$ by

$$f_n(x) = \frac{nx^2}{nx + 1}, \qquad 0 \leq x \leq 1 \qquad (n = 1, 2, 3, \ldots).$$

For $x = 0$, the corresponding sequence $\{f_n(0)\}$ of real numbers in $0, 0, 0, \ldots$, and

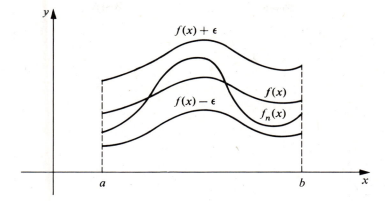

Figure 10.2

$\lim\limits_{n \to \infty} f_n(0) = 0$. For every x such that $0 < x \leq 1$, we have

$$f_n(x) = \frac{nx^2}{nx + 1} \quad \text{and} \quad \lim\limits_{n \to \infty} f_n(x) = x.$$

Thus the sequence $\{f_n\}$ converges pointwise on $0 \leq x \leq 1$ to the limit function f defined by $f(x) = x$, $0 \leq x \leq 1$. Further, the convergence is uniform on $0 \leq x \leq 1$. To see this, consider

$$|f_n(x) - f(x)| = \left| \frac{nx^2}{nx + 1} - x \right| = \frac{x}{nx + 1}.$$

Given $\epsilon > 0$, we have

$$\frac{x}{nx + 1} < \epsilon$$

provided

$$n > \frac{1}{\epsilon} - \frac{1}{x}.$$

But for x such that $0 \leq x \leq 1$, we have

$$\frac{1}{\epsilon} - \frac{1}{x} \leq \frac{1}{\epsilon} - 1.$$

Thus if we choose $N = (1/\epsilon) - 1$, we have

$$n > \frac{1}{\epsilon} - \frac{1}{x}$$

for all $n > N$. Hence, given $\epsilon > 0$, there exists $N = (1/\epsilon) - 1$ (depending only upon ϵ and *not* on x) such that $|f_n(x) - f(x)| < \epsilon$ for all $n > N$ for *every* x such that $0 \leq x \leq 1$. In other words, the convergence is uniform on $0 \leq x \leq 1$. The graphs of the functions $f_1, f_5,$ and f_{10} defined, respectively, by

$$f_1(x) = \frac{x^2}{x + 1}, \quad f_5(x) = \frac{5x^2}{5x + 1}, \quad \text{and} \quad f_{10}(x) = \frac{10x^2}{10x + 1},$$

as well as that of f, appear in Figure 10.3.

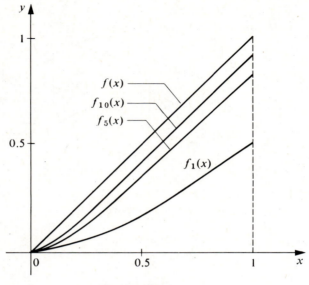

Figure 10.3

We now state and illustrate two important theorems on uniformly convergent sequences, which we shall use in the existence proof in Section 10.2 We shall not prove these theorems here, since the proofs may be found in most texts on advanced calculus and real analysis.

THEOREM A

Hypothesis

1. Let $\{f_n\}$ be a sequence of real functions which converges uniformly to f on the interval $a \leq x \leq b$.

2. Suppose each function f_n ($n = 1, 2, 3, \ldots$) is continuous on $a \leq x \leq b$.

Conclusion.   *The limit function f is continuous on $a \leq x \leq b$.*

▶ **Example 10.4**

In Example 10.3 we saw that the sequence of functions $\{f_n\}$ defined on the real interval $0 \leq x \leq 1$ by

$$f_n(x) = \frac{nx^2}{nx + 1} \qquad (n = 1, 2, 3, \ldots)$$

converges uniformly to a limit function f on $0 \leq x \leq 1$. Further, each function f_n ($n = 1, 2, 3, \ldots$) is continuous on $0 \leq x \leq 1$. By Theorem A we conclude at once that the limit function f is also continuous on $0 \leq x \leq 1$. Indeed, in this example, the limit function f is that defined by $f(x) = x, 0 \leq x \leq 1$, and clearly this function f is continuous on $0 \leq x \leq .1$.

THEOREM B

Hypothesis

1. Let $\{f_n\}$ be a sequence of real functions which converges uniformly to f on the interval $a \leq x \leq b$.

2. Suppose each function $f_n(n = 1, 2, 3, \ldots)$ is continuous on $a \leq x \leq b$.

Conclusion. *Then for every α and β such that $a \leq \alpha < \beta \leq b$,*

$$\lim_{n \to \infty} \int_{\alpha}^{\beta} f_n(x)\, dx = \int_{\alpha}^{\beta} \lim_{n \to \infty} f_n(x)\, dx.$$

► **Example 10.5**

We illustrate this theorem by again considering the sequence of functions $\{f_n\}$ discussed in Examples 10.3 and 10.4 and defined by

$$f_n(x) = \frac{nx^2}{nx + 1}, \qquad 0 \leq x \leq 1 \qquad (n = 1, 2, 3, \ldots).$$

The hypothesis of Theorem B is identical with that of Theorem A, and we have already noted in Example 10.4 that the sequence under consideration satisfies this hypothesis on $0 \leq x \leq 1$. Thus we can conclude that

$$\lim_{n \to \infty} \int_0^1 f_n(x)\, dx = \int_0^1 \lim_{n \to \infty} f_n(x)\, dx.$$

Since $\lim_{n \to \infty} f_n(x) = f(x) = x$ in this example, our conclusion here is that

$$\lim_{n \to \infty} \int_0^1 \frac{nx^2}{nx + 1}\, dx = \int_0^1 x\, dx.$$

We shall verify this directly.

$$\int_0^1 \frac{nx^2}{nx + 1}\, dx = \int_0^1 \left[x - \frac{1}{n} + \frac{1}{n(nx + 1)} \right] dx$$

$$= \frac{x^2}{2} - \frac{x}{n} + \frac{1}{n^2} \ln(nx + 1) \Big|_0^1 = \frac{1}{2} - \frac{1}{n} + \frac{\ln(n + 1)}{n^2}.$$

Thus

$$\lim_{n \to \infty} \int_0^1 \frac{nx^2}{nx + 1}\, dx = \lim_{n \to \infty} \left[\frac{1}{2} - \frac{1}{n} + \frac{\ln(n + 1)}{n^2} \right] = \frac{1}{2}.$$

Clearly $\int_0^1 x\, dx = \frac{1}{2}$ also, and our conclusion is thus verified.

Let us now consider briefly the uniform convergence of an infinite *series* of real functions, each of which is defined on a real interval $a \leq x \leq b$.

DEFINITION D

Consider the infinite series $\sum_{n=1}^{\infty} u_n$ of real functions u_n ($n = 1, 2, 3, \ldots$), each of which is defined on a real interval $a \leq x \leq b$. Consider the sequence $\{f_n\}$ of so-called partial sums *of this series, defined as follows:*

$$f_1 = u_1,$$
$$f_2 = u_1 + u_2,$$
$$f_3 = u_1 + u_2 + u_3,$$
$$\vdots$$
$$f_n = u_1 + u_2 + u_3 + \cdots + u_n,$$
$$\vdots$$

The infinite series $\sum_{n=1}^{\infty} u_n$ is said to converge uniformly *to f on $a \leq x \leq b$ if its sequence of partial sums $\{f_n\}$ converges uniformly to f on $a \leq x \leq b$.*

The following theorem gives a very useful test for uniform convergence of series.

THEOREM C   Weierstrass M-Test

Hypothesis

1. Let $\{M_n\}$ be a sequence of positive constants such that the series of constants $\sum_{n=1}^{\infty} M_n$ converges.

2. Let $\sum_{n=1}^{\infty} u_n$ be a series of real functions such that $|u_n(x)| \leq M_n$ for all x such that $a \leq x \leq b$ and for each $n = 1, 2, 3, \ldots$.

Conclusion.   *The series $\sum_{n=1}^{\infty} u_n$ converges uniformly on $a \leq x \leq b$.*

▶ **Example 10.6**

Consider the series $\sum_{n=1}^{\infty} \dfrac{\sin nx}{n^2}$ on the interval $0 \leq x \leq 1$. The sequence $\{1/n^2\}$ is a sequence of positive constants which is such that the series $\sum_{n=1}^{\infty} \dfrac{1}{n^2}$ is convergent. Let us take $M_n = 1/n^2$ and observe that

$$|u_n(x)| = \left| \frac{\sin nx}{n^2} \right| \leq \frac{1}{n^2} = M_n$$

for all x such that $0 \leq x \leq 1$ and for each $n = 1, 2, 3, \ldots$. Thus by Theorem C the series $\displaystyle\sum_{n=1}^{\infty} \frac{\sin nx}{n^2}$ converges uniformly on $0 \leq x \leq 1$.

B. Functions of Two Real Variables; the Lipschitz Condition

DEFINITION E

1. A set of points A of the xy plane will be called connected *if any two points of A can be joined by a continuous curve which lies entirely in A.*

2. A set of points A of the xy plane is called open *if each point of A is the center of a circle whose interior lies entirely in A.*

3. An open, connected set in the xy plane is called a domain.

4. A point P is called a boundary point *of a domain D if every circle about P contains both points in D and points not in D.*

5. A domain plus its boundary points will be called a closed domain.

▶ **Example 10.7**

The set of all points (x, y) lying *within* the ellipse $x^2 + 2y^2 = 1$ and characterized by $x^2 + 2y^2 < 1$ is a domain D. The boundary points of D are the points of the ellipse itself. The set of points (x, y) such that $x^2 + 2y^2 \leq 1$ *within and on* the ellipse is a closed domain.

We assume that the reader is at least somewhat familiar with functions f of two real variables x and y, defined on a domain of the xy plane or on such a domain plus its boundary. Let us recall a few useful concepts and results.

DEFINITION F

Let f be a real function defined on a domain D of the xy plane, and let (x_0, y_0) be an (interior) point of D. The function f is said to be continuous *at (x_0, y_0) if, given any $\epsilon > 0$, there exists a $\delta > 0$ such that*

$$|f(x, y) - f(x_0, y_0)| < \epsilon$$

for all $(x, y) \in D$ such that

$$|x - x_0| < \delta \quad and \quad |y - y_0| < \delta.$$

DEFINITION G

Let f be a real function defined on D, where D is either a domain or a closed domain of the xy plane. The function f is said to be bounded *on D if there exists a positive number M such that $|f(x, y)| \leq M$ for all (x, y) in D.*

THEOREM D

Hypothesis. *Let f be defined and continuous on a closed rectangle R: $a \leq x \leq b$, $c \leq y \leq d$.*

Conclusion. *Then the function f is bounded on R.*

▶ **Example 10.8**

The function f defined by $f(x, y) = x + y^2$ is continuous on the closed rectangle $R: 0 \leq x \leq 1, 0 \leq y \leq 2$. Thus by Theorem D, the function f is bounded on R. In fact, we have $|f(x, y)| = |x + y^2| \leq 5$ for all $(x, y) \in R$.

Having disposed of these preliminaries, we now introduce a concept which will be of paramount importance in the existence and uniqueness proof of Section 10.2.

DEFINITION H

Let f be defined on D, where D is either a domain or a closed domain of the xy plane. The function f is said to satisfy a Lipschitz condition *(with respect to y) in D if there exists a constant k > 0 such that*

$$|f(x, y_1) - f(x, y_2)| \leq k|y_1 - y_2| \tag{10.1}$$

for every (x, y_1) and (x, y_2) which belong to D. The constant k is called the Lipschitz constant.

Let us investigate the geometric significance of the Lipschitz condition. Let (x, y_1) and (x, y_2) be any two points in D having the same abscissa x. Consider the corresponding points

$$P_1:[x, y_1, f(x, y_1)] \quad \text{and} \quad P_2:[x, y_2, f(x, y_2)]$$

on the surface $z = f(x, y)$, and let α $(0 \leq \alpha \leq \pi/2)$ denote the angle which the chord joining P_1 and P_2 makes with the xy plane. Then if condition (10.1) holds in D, $\tan \alpha$ is bounded in absolute value and so the chord joining P_1 and P_2 is bounded away from being perpendicular to the xy plane. Further, this bound is independent of x for points (x, y_1) and (x, y_2) belonging to D.

If f satisfies condition (10.1) in D, then for each fixed x the resulting function of y is a continuous function of y for (x, y) belonging to D. Note, however, that condition (10.1) implies nothing at all concerning the continuity of f with respect to x. For example, the function f defined by

$$f(x, y) = y + [x],$$

where $[x]$ denotes the greatest integer less than or equal to x, satisfies a Lipschitz condition in any bounded domain D. For each fixed x, the resulting function of y is continuous. However, this function f is discontinuous with respect to x for every integral value of x.

Now suppose the function f is such that $\dfrac{\partial f}{\partial y}$ exists and is bounded for all $(x, y) \in D$. Then by the mean-value theorem of differential calculus, for any (x, y_1) and $(x, y_2) \in D$ there exists ξ, where ξ is between y_1 and y_2, such that

$$f(x, y_1) - f(x, y_2) = (y_1 - y_2) \frac{\partial f(x, \xi)}{\partial y}.$$

Thus

$$|f(x, y_1) - f(x, y_2)| = |y_1 - y_2| \left| \frac{\partial f(x, \xi)}{\partial y} \right| \leq \operatorname*{lub}_{(x,y)\in D} \left| \frac{\partial f(x, y)}{\partial y} \right| |y_1 - y_2|,$$

provided $(x, \xi) \in D$. Thus we have obtained the following very useful result.

THEOREM E

Hypothesis. *Let f be such that $\dfrac{\partial f}{\partial y}$ exists and is bounded for all $(x, y) \in D$, where D is a domain or closed domain such that the line segment joining any two points of D lies entirely within D.*

Conclusion. *Then f satisfies a Lipschitz condition (with respect to y) in D, where the Lipschitz constant is given by*

$$k = \operatorname*{lub}_{(x,y)\in D} \left| \frac{\partial f(x, y)}{\partial y} \right|.$$

▶ **Example 10.9**

Consider the function f defined by $f(x, y) = y^2$, where D is the rectangle defined by $|x| \leq a, |y| \leq b$. Then $\dfrac{\partial f(x, y)}{\partial y} = 2y$, and so $\dfrac{\partial f}{\partial y}$ exists and is bounded for all $(x, y) \in D$. Thus by Theorem E, the function f satisfies a Lipschitz condition in D, where the Lipschitz constant k is given by $2b$. If we directly apply the definition of Lipschitz condition instead of Theorem E, we find that

$$|f(x, y_1) - f(x, y_2)| = |y_1^2 - y_2^2| = |y_1 + y_2||y_1 - y_2| \leq 2b|y_1 - y_2|$$

for all $(x, y_1), (x, y_2) \in D$.

Note that the sufficient condition of Theorem E is not necessary for f to satisfy a Lipschitz condition in D. That is, there exist functions f such that f satisfies a Lipschitz condition (with respect to y) in D but such that the hypothesis of Theorem E is not satisfied.

▶ **Example 10.10**

Consider the function f defined by $f(x, y) = x|y|$, where D is the rectangle defined by $|x| \leq a, |y| \leq b$. We note that

$$|f(x, y_1) - f(x, y_2)| = |x|y_1| - x|y_2|| \leq |x||y_1 - y_2| \leq a|y_1 - y_2|$$

for all $(x, y_1), (x, y_2) \in D$. Thus f satisfies a Lipschitz condition (with respect to y) in D. However, the partial derivative $\dfrac{\partial f}{\partial y}$ does not exist at any point $(x, 0) \in D$ for which $x \neq 0$.

Exercises

1. Consider the following sequences of functions $\{f_n\}$ which are defined for all x on the real interval $0 \leq x \leq 1$. Show that each of these sequences converges uniformly on $0 \leq x \leq 1$.

   (a) $f_n(x) = \dfrac{1}{x + n}$, $0 \leq x \leq 1$ $(n = 1, 2, 3, \ldots)$.

   (b) $f_n(x) = x - \dfrac{x^n}{n}$, $0 \leq x \leq 1$ $(n = 1, 2, 3, \ldots)$.

2. Consider the sequence of functions $\{f_n\}$ defined for all x on the real interval $0 \leq x \leq 1$ by

   $$f_n(x) = \frac{1}{nx + 1},   0 \leq x \leq 1   (n = 1, 2, 3, \ldots).$$

   (a) Show that the sequence $\{f_n\}$ converges pointwise to a function f on $0 \leq x \leq 1$.

   (b) Does f_n converge uniformly to f on $0 \leq x \leq 1$?

   (c) Does f_n converge uniformly to f on $\frac{1}{2} \leq x \leq 1$?

3. (a) Prove the following theorem:

   Hypothesis. Let $\{f_n\}$ be a sequence of real functions defined on $a \leq x \leq b$, and suppose f_n converges pointwise to a function f on $a \leq x \leq b$. Let $\{\phi_n\}$ be a sequence of positive real numbers such that $\lim\limits_{n \to \infty} \phi_n = 0$. Finally, suppose that for each $n = 1, 2, 3, \ldots, |f(x) - f_n(x)| \leq \phi_n$ for all $x \in [a, b]$.

   Conclusion. Then $\{f_n\}$ converges uniformly to f on $a \leq x \leq b$.

   (b) Use the theorem of part (a) to show that the sequence of functions $\{f_n\}$ defined by $f_n(x) = (\sin nx)/n$ converges uniformly on $0 \leq x \leq 1$.

4. Prove Theorem A.

5. Show that each of the following infinite series converges uniformly on every interval $-a \leq x \leq a$, where a is a positive number.

   (a) $\displaystyle\sum_{n=1}^{\infty} \frac{x^n}{n!}$.

   (b) $\displaystyle\sum_{n=1}^{\infty} \frac{\cos nx}{n^2}$.

   (c) $\displaystyle\sum_{n=1}^{\infty} \frac{\sin 2nx}{(2n - 1)(2n + 1)}$.

6. Show that each of the functions defined as follows satisfies a Lipschitz condition in the rectangle D defined by $|x| \le a, |y| \le b$.
   (a) $f(x, y) = x^2 + y^2$.
   (b) $f(x, y) = x \sin y + y \cos x$.
   (c) $f(x, y) = x^2 e^{x+y}$.
   (d) $f(x, y) = A(x)y^2 + B(x)y + C(x)$, where A, B, and C are continuous on $|x| \le a$.

7. Show that neither of the functions defined in (a) and (b) satisfies a Lipschitz condition throughout any domain which includes the line $y = 0$.
   (a) $f(x, y) = y^{2/3}$.
   (b) $f(x, y) = \sqrt{|y|}$.

8. Consider the function f defined by $f(x, y) = x|y|$ for $(x, y) \in D$, where D is the rectangle defined by $|x| \le a, |y| \le b$. In Example 10.10 it is stated that $\dfrac{\partial f}{\partial y}$ does not exist for $(x, 0) \in D$, where $x \ne 0$. Show that this statement is valid.

9. Consider the function f defined by

$$f(x, y) = \begin{cases} \dfrac{4x^3 y}{x^4 + y^2}, & (x, y) \ne (0, 0), \\ 0, & (x, y) = (0, 0). \end{cases}$$

   Show that f does not satisfy a Lipschitz condition throughout any domain which includes $(0, 0)$.
   [*Hint*: Let a and b be real constants and consider $|f(x, y_1) - f(x, y_2)|$ with $(x, y_1) = (x, ax^2)$ and $(x, y_2) = (x, bx^2)$.]

10.2 THE FUNDAMENTAL EXISTENCE AND UNIQUENESS THEOREM

A. The Basic Problem and a Basic Lemma

We now formulate the basic problem with which this chapter is primarily concerned.

DEFINITION

Let D be a domain in the xy plane and let (x_0, y_0) be an (interior) point of D. Consider the differential equation

$$\frac{dy}{dx} = f(x, y), \tag{10.2}$$

where f is a continuous real function defined on D. Consider the following problem. We wish to determine:

   1. an interval $\alpha \le x \le \beta$ of the real x axis such that $\alpha < x_0 < \beta$, and

2. *a differentiable real function ϕ defined on this interval $[\alpha, \beta]$ and satisfying the following three requirements:*

(i) $[x, \phi(x)] \in D$, *and thus* $f[x, \phi(x)]$ *is defined, for all* $x \in [\alpha, \beta]$.

(ii) $\dfrac{d\phi(x)}{dx} = f[x, \phi(x)]$, *and thus* ϕ *satisfies the differential equation* (10.2), *for all*
$x \in [\alpha, \beta]$.

(iii) $\phi(x_0) = y_0$.

We shall call this problem the initial-value problem *associated with the differential equation* (10.2) *and the point* (x_0, y_0). *We shall denote it briefly by*

$$\frac{dy}{dx} = f(x, y),$$

$$y(x_0) = y_0,$$

(10.3)

and call a function ϕ satisfying the above requirements on an interval $[\alpha, \beta]$ a solution *of the problem on the interval $[\alpha, \beta]$.*

If ϕ is a solution of the problem on $[\alpha, \beta]$, then requirement (ii) shows that ϕ has a continuous first derivative ϕ' on $[\alpha, \beta]$.

In order to investigate this problem we shall need the following basic lemma.

LEMMA

Hypothesis

1. *Let f be a continuous function defined on a domain D of the xy plane.*
2. *Let ϕ be a continuous function defined on a real interval $\alpha \leq x \leq \beta$ and such that $[x, \phi(x)] \in D$ for all $x \in [\alpha, \beta]$.*
3. *Let x_0 be any real number such that $\alpha < x_0 < \beta$.*

Conclusion. *Then ϕ is a solution of the differential equation $dy/dx = f(x, y)$ on $[\alpha, \beta]$ and is such that $\phi(x_0) = y_0$ if and only if ϕ satisfies the* integral equation

$$\phi(x) = y_0 + \int_{x_0}^{x} f[t, \phi(t)] \, dt$$

(10.4)

for all $x \in [\alpha, \beta]$.

Proof. If ϕ satisfies the differential equation $dy/dx = f(x, y)$ on $[\alpha, \beta]$, then $d\phi(x)/dx = f[x, \phi(x)]$ for all $x \in [\alpha, \beta]$ and integration yields at once

$$\phi(x) = \int_{x_0}^{x} f[t, \phi(t)] \, dt + c.$$

If also $\phi(x_0) = y_0$, then we have $c = y_0$ and so ϕ satisfies the integral equation (10.4) for all $x \in [\alpha, \beta]$.

Conversely, if ϕ satisfies the integral equation (10.4) for all $x \in [\alpha, \beta]$, then differentiation yields

$$\frac{d\phi(x)}{dx} = f[x, \phi(x)]$$

for all $x \in [\alpha, \beta]$ and so ϕ satisfies the differential equation $dy/dx = f(x, y)$ on $[\alpha, \beta]$; also Equation (10.4) shows that $\phi(x_0) = y_0$.

Q.E.D

B. The Existence and Uniqueness Theorem and Its Proof

We shall now state the main theorem of this chapter.

THEOREM 10.1

Hypothesis

1. Let D be a domain of the xy plane, and let f be a real function satisfying the following two requirements:

(i) f is continuous in D;
(ii) f satisfies a Lipschitz condition (with respect to y) in D; that is, there exists a constant k > 0 such that

$$|f(x, y_1) - f(x, y_2)| \leq k|y_1 - y_2| \tag{10.1}$$

for all $(x, y_1), (x, y_2) \in D$.

2. Let (x_0, y_0) be an (interior) point of D; let a and b be such that the rectangle $R: |x - x_0| \leq a, |y - y_0| \leq b$, lies in D; let $M = \max |f(x, y)|$ for $(x, y) \in R$; and let $h = \min(a, b/M)$.

Conclusion. *There exists a unique solution ϕ of the initial-value problem*

$$\frac{dy}{dx} = f(x, y),$$

$$y(x_0) = y_0, \tag{10.3}$$

on the interval $|x - x_0| \leq h$. That is, there exists a unique differentiable real function ϕ defined on the interval $|x - x_0| \leq h$ which is such that

(i) $\dfrac{d\phi(x)}{dx} = f[x, \phi(x)]$ for all x on this interval;

(ii) $\phi(x_0) = y_0$.

Remarks. Since R lies in D, f satisfies the requirements (i) and (ii) of Hypothesis 1 in R. In particular, since f is thus continuous in the rectangular closed domain R, the constant M defined in Hypothesis 2 actually does exist. Let us examine more closely the number h defined in Hypothesis 2. If $a < b/M$, then $h = a$ and the solution ϕ is defined for all x in the interval $|x - x_0| \leq a$ used in defining the rectangle R. If, however, $b/M < a$, then $h = b/M < a$ and so the solution ϕ is assured only for all x in the

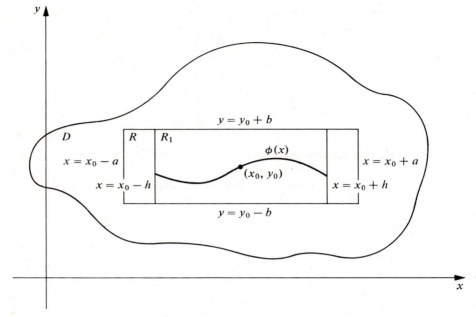

Figure 10.4 The case in which $\dfrac{b}{M} < a$ so that $h = \min\left(a, \dfrac{b}{M}\right) = \dfrac{b}{M}$.

smaller interval $|x - x_0| \leq h < a$ associated with the smaller rectangle R_1 defined by $|x - x_0| \leq h < a, |y - y_0| \leq b$ (see Figure 10.4).

Method of Proof. We shall prove this theorem by the method of *successive approximations*. Let x be such that $|x - x_0| \leq h$. We define a sequence of functions

$$\phi_1, \phi_2, \phi_3, \ldots, \phi_n, \ldots$$

called the *successive approximations* (or *Picard iterants*) as follows:

$$\phi_1(x) = y_0 + \int_{x_0}^{x} f[t, y_0]\, dt,$$

$$\phi_2(x) = y_0 + \int_{x_0}^{x} f[t, \phi_1(t)]\, dt,$$

$$\phi_3(x) = y_0 + \int_{x_0}^{x} f[t, \phi_2(t)]\, dt, \tag{10.5}$$

$$\vdots$$

$$\phi_n(x) = y_0 + \int_{x_0}^{x} f[t, \phi_{n-1}(t)]\, dt.$$

We shall divide the proof into five main steps in which we shall show the following.

1. The functions $\{\phi_n\}$ defined by (10.5) actually exist, have continuous derivatives, and satisfy the inequality $|\phi_n(x) - y_0| \leq b$ on $|x - x_0| \leq h$; and thus $f[x, \phi_n(x)]$ is defined on this interval.

2. The functions $\{\phi_n\}$ satisfy the inequality

$$|\phi_n(x) - \phi_{n-1}(x)| \leq \frac{M}{k} \cdot \frac{(kh)^n}{n!} \quad \text{on} \quad |x - x_0| \leq h.$$

3. As $n \to \infty$, the sequence of functions $\{\phi_n\}$ converges uniformly to a continuous function ϕ on $|x - x_0| \leq h$.
4. The limit function ϕ satisfies the differential equation $dy/dx = f(x, y)$ on $|x - x_0| \leq h$ and is such that $\phi(x_0) = y_0$.
5. This function ϕ is the only differentiable function on $|x - x_0| \leq h$ which satisfies the differential equation $dy/dx = f(x, y)$ and is such that $\phi(x_0) = y_0$.

The Proof. Throughout the entire proof we shall consider the interval $[x_0, x_0 + h]$; similar arguments hold for the interval $[x_0 - h, x_0]$.

1. In this first step of the proof we use mathematical induction. Assume ϕ_{n-1} exists, has a continuous derivative, and is such that $|\phi_{n-1}(x) - y_0| \leq b$ for all x such that $x_0 \leq x \leq x_0 + h$. Thus $[x, \phi_{n-1}(x)]$ lies in the rectangle R and so $f[x, \phi_{n-1}(x)]$ is defined and continuous and satisfies

$$|f[x, \phi_{n-1}(x)]| \leq M \quad \text{on} \quad [x_0, x_0 + h].$$

Thus since

$$\phi_n(x) = y_0 + \int_{x_0}^{x} f[t, \phi_{n-1}(t)] \, dt,$$

we see that ϕ_n also exists and has a continuous derivative on $[x_0, x_0 + h]$. Also,

$$|\phi_n(x) - y_0| = \left| \int_{x_0}^{x} f[t, \phi_{n-1}(t)] \, dt \right|$$

$$\leq \int_{x_0}^{x} |f[t, \phi_{n-1}(t)]| \, dt \leq \int_{x_0}^{x} M \, dt$$

$$= M(x - x_0) \leq Mh \leq b.$$

Thus, also, $[x, \phi_n(x)]$ lies in R and hence $f[x, \phi_n(x)]$ is defined and continuous on $[x_0, x_0 + h]$. Clearly ϕ_1 defined by

$$\phi_1(x) = y_0 + \int_{x_0}^{x} f[t, y_0] \, dt$$

exists and has a continuous derivative on this interval. Also,

$$|\phi_1(x) - y_0| \leq \int_{x_0}^{x} |f[t, y_0]| \, dt \leq M(x - x_0) \leq b$$

and so $f[x, \phi_1(x)]$ is defined and continuous on the interval under consideration. Thus, by mathematical induction, each function ϕ_n of the sequence (10.5) possesses these desired properties on $[x_0, x_0 + h]$.

2. In this step we again employ mathematical induction. We assume that

$$|\phi_{n-1}(x) - \phi_{n-2}(x)| \leq \frac{Mk^{n-2}}{(n-1)!} (x - x_0)^{n-1} \quad \text{on} \quad [x_0, x_0 + h]. \tag{10.6}$$

Then

$$|\phi_n(x) - \phi_{n-1}(x)| = \left| \int_{x_0}^{x} \{ f[t, \phi_{n-1}(t)] - f[t, \phi_{n-2}(t)] \} \, dt \right|$$

$$\leq \int_{x_0}^{x} |f[t, \phi_{n-1}(t)] - f[t, \phi_{n-2}(t)]| \, dt.$$

Since by Step 1, $|\phi_n(x) - y_0| \leq b$ for all n on $[x_0, x_0 + h]$, the Lipschitz condition (10.1) applies with $y_1 = \phi_{n-1}(x)$ and $y_2 = \phi_{n-2}(x)$, and we have

$$|f[t, \phi_{n-1}(x)] - f[t, \phi_{n-2}(x)]| \leq k |\phi_{n-1}(x) - \phi_{n-2}(x)|.$$

Thus we have

$$|\phi_n(x) - \phi_{n-1}(x)| \leq \int_{x_0}^{x} k |\phi_{n-1}(t) - \phi_{n-2}(t)| \, dt.$$

Now using the assumption (10.6), we have

$$|\phi_n(x) - \phi_{n-1}(x)| \leq k \int_{x_0}^{x} \frac{Mk^{n-2}}{(n-1)!} (t - x_0)^{n-1} \, dt$$

$$= \frac{Mk^{n-1}}{(n-1)!} \int_{x_0}^{x} (t - x_0)^{n-1} \, dt$$

$$= \frac{Mk^{n-1}}{(n-1)!} \left[\frac{(t - x_0)^n}{n} \right] \Bigg|_{x_0}^{x} = \frac{Mk^{n-1}}{n!} (x - x_0)^n.$$

That is, we now have

$$|\phi_n(x) - \phi_{n-1}(x)| \leq \frac{Mk^{n-1}}{n!} (x - x_0)^n, \tag{10.7}$$

which is precisely inequality (10.6) with $(n-1)$ replaced by n. When $n = 1$, we have as in Step 1:

$$|\phi_1(x) - y_0| \leq M(x - x_0).$$

This is inequality (10.7) when $n = 1$. Thus by induction the inequality (10.7) is satisfied on $[x_0, x_0 + h]$ for all n.

Since

$$\frac{Mk^{n-1}}{n!} (x - x_0)^n \leq \frac{Mk^{n-1}}{n!} h^n = \frac{M}{k} \frac{(kh)^n}{n!},$$

we have

$$|\phi_n(x) - \phi_{n-1}(x)| \leq \frac{M}{k} \frac{(kh)^n}{n!} \tag{10.8}$$

for $n = 1, 2, 3, \ldots$ on $[x_0, x_0 + h]$.

3. Now the series of positive constants

$$\frac{M}{k} \sum_{n=1}^{\infty} \frac{(kh)^n}{n!} = \frac{M}{k} \frac{kh}{1!} + \frac{M}{k} \frac{(kh)^2}{2!} + \frac{M}{k} \frac{(kh)^3}{3!} + \cdots$$

converges to

$$\frac{M}{k}[e^{kh} - 1].$$

Also, the series

$$\sum_{i=1}^{\infty} [\phi_i(x) - \phi_{i-1}(x)]$$

is such that (10.8) is satisfied for all x on the interval $x_0 \leq x \leq x_0 + h$, for each $n = 1, 2, 3, \ldots$. Thus by the Weierstrass M-test (Theorem C), the series

$$y_0 + \sum_{i=1}^{\infty} [\phi_i(x) - \phi_{i-1}(x)]$$

converges uniformly on $[x_0, x_0 + h]$. Therefore its sequence of partial sums $\{S_n\}$ converges uniformly to a limit function ϕ on $[x_0, x_0 + h]$. But

$$S_n(x) = y_0 + \sum_{i=1}^{n} [\phi_i(x) - \phi_{i-1}(x)] = \phi_n(x).$$

In other words, the sequence ϕ_n converges uniformly to ϕ on $[x_0, x_0 + h]$. Thus, since each ϕ_n is continuous on $[x_0, x_0 + h]$, Theorem A shows that the limit function ϕ is also continuous on $[x_0, x_0 + h]$.

4. Since each ϕ_n satisfies $|\phi_n(x) - y_0| \leq b$ on $[x_0, x_0 + h]$, we also have $|\phi(x) - y_0| \leq b$ on $[x_0, x_0 + h]$. Thus $f[x, \phi(x)]$ is defined on this interval and we can further apply the Lipschitz condition (10.1) with $y_1 = \phi(x)$ and $y_2 = \phi_n(x)$. Doing so, we obtain

$$|f[x, \phi(x)] - f[x, \phi_n(x)]| \leq k |\phi(x) - \phi_n(x)| \qquad (10.9)$$

for $x \in [x_0, x_0 + h]$. By Step 3, given $\epsilon > 0$, there exists $N > 0$ such that $|\phi(x) - \phi_n(x)| < \epsilon/k$ for all $n > N$ and all x on $[x_0, x_0 + h]$. Thus

$$k |\phi(x) - \phi_n(x)| < k\left(\frac{\epsilon}{k}\right) = \epsilon \qquad (10.10)$$

for all $n > N$ and all x on the interval under consideration. Thus from (10.9) and (10.10) we see that the sequence of functions defined for $n = 1, 2, 3, \ldots$ by $f[x, \phi_n(x)]$ converges uniformly to the function defined by $f[x, \phi(x)]$ on $[x_0, x_0 + h]$. Also, each function defined for $n = 1, 2, 3, \ldots$ by $f[x, \phi_n(x)]$ is continuous on this interval. Thus Theorem B applies and

$$\phi(x) = \lim_{n \to \infty} \phi_{n+1}(x) = y_0 + \lim_{n \to \infty} \int_{x_0}^{x} f[t, \phi_n(t)] \, dt$$

$$= y_0 + \int_{x_0}^{x} \lim_{n \to \infty} f[t, \phi_n(t)] \, dt = y_0 + \int_{x_0}^{x} f[t, \phi(t)] \, dt.$$

In short, the limit function ϕ satisfies the integral equation

$$\phi(x) = y_0 + \int_{x_0}^{x} f[t, \phi(t)] \, dt$$

on $[x_0, x_0 + h]$. Thus by the basic lemma, the limit function ϕ satisfies the differential equation $dy/dx = f(x, y)$ on $[x_0, x_0 + h]$ and is such that $\phi(x_0) = y_0$. We have thus proved the *existence* of a solution of the basic initial-value problem (10.3) on the interval $[x_0, x_0 + h]$.

5. We now prove that the solution ϕ is *unique*. Assume that ψ is *another* differentiable function defined on $[x_0, x_0 + h]$ such that

$$\frac{d\psi(x)}{dx} = f[x, \psi(x)]$$

and $\psi(x_0) = y_0$. Then certainly

$$|\psi(x) - y_0| < b \tag{10.11}$$

on some interval $[x_0, x_0 + \delta]$. Let x_1 be such that $|\psi(x) - y_0| < b$ for $x_0 \leq x < x_1$ and $|\psi(x_1) - y_0| = b$. Suppose $x_1 < x_0 + h$. Then

$$M_1 = \left| \frac{\psi(x_1) - y_0}{x_1 - x_0} \right| = \frac{b}{x_1 - x_0} > \frac{b}{h} \geq M.$$

But by the mean-value theorem there exists ξ, where $x_0 < \xi < x_1$, such that

$$M_1 = |\psi'(\xi)| = |f[\xi, \psi(\xi)]| \leq M,$$

a contradiction. Thus $x_1 \geq x_0 + h$ and the inequality (10.11) holds for $x_0 \leq x < x_0 + h$, and so

$$|\psi(x) - y_0| \leq b \tag{10.12}$$

on the interval $x_0 \leq x \leq x_0 + h$.

Since ψ is a solution of $dy/dx = f(x, y)$ on $[x_0, x_0 + h]$ such that $\psi(x_0) = y_0$, from the basic lemma we see that ψ satisfies the integral equation

$$\psi(x) = y_0 + \int_{x_0}^{x} f[t, \psi(t)] \, dt \tag{10.13}$$

on $[x_0, x_0 + h]$. We shall now prove by mathematical induction that

$$|\psi(x) - \phi_n(x)| \leq \frac{k^n b (x - x_0)^n}{n!} \tag{10.14}$$

on $[x_0, x_0 + h]$. We thus assume that

$$|\psi(x) - \phi_{n-1}(x)| \leq \frac{k^{n-1} b (x - x_0)^{n-1}}{(n-1)!} \tag{10.15}$$

on $[x_0, x_0 + h]$. Then from (10.5) and (10.13) we have

$$|\psi(x) - \phi_n(x)| = \left| \int_{x_0}^{x} \{ f[t, \psi(t)] - f[t, \phi_{n-1}(t)] \} \, dt \right|$$

$$\leq \int_{x_0}^{x} |f[t, \psi(t)] - f[t, \phi_{n-1}(t)]| \, dt.$$

By Equation (10.12) and Step 1, respectively, the Lipschitz condition (10.1) applies with $y_1 = \psi(x)$ and $y_2 = \phi_{n-1}(x)$. Thus we have

$$|\psi(x) - \phi_n(x)| \leq \int_{x_0}^{x} k |\psi(t) - \phi_{n-1}(t)| \, dt.$$

By the assumption (10.15) we then have

$$|\psi(x) - \phi_n(x)| \leq k \int_{x_0}^{x} \frac{k^{n-1} b(t - x_0)^{n-1}}{(n-1)!} \, dt$$

$$\leq \frac{k^n b}{(n-1)!} \left[\frac{(t - x_0)^n}{n} \right]\Bigg|_{x_0}^{x} = \frac{k^n b (x - x_0)^n}{n!},$$

which is (10.15) with $(n-1)$ replaced by n. When $n = 1$, we have

$$|\psi(x) - \phi_1(x)| \leq \int_{x_0}^{x} |f[t, \psi(t)] - f[t, y_0]| \, dt$$

$$\leq k \int_{x_0}^{x} |\psi(t) - y_0| \, dt \leq kb(x - x_0),$$

which is (10.14) for $n = 1$. Thus by induction the inequality (10.14) holds for all n on $[x_0, x_0 + h]$. Hence we have

$$|\psi(x) - \phi_n(x)| \leq b \frac{(kh)^n}{n!} \tag{10.16}$$

for $n = 1, 2, 3, \ldots$ on $[x_0, x_0 + h]$.

Now the series $\sum_{n=0}^{\infty} b \frac{(kh)^n}{n!}$ converges, and so $\lim_{n \to \infty} b \frac{(kh)^n}{n!} = 0$. Thus from (10.16), $\psi(x) = \lim_{n \to \infty} \phi_n(x)$ on $[x_0, x_0 + h]$. But recall that $\phi(x) = \lim_{n \to \infty} \phi_n(x)$ on this interval. Thus,

$$\psi(x) = \phi(x)$$

on $[x_0, x_0 + h]$. Thus the solution ϕ of the basic initial-value problem is unique on $[x_0, x_0 + h]$.

We have thus proved that the basic initial-value problem has a unique solution on $[x_0, x_0 + h]$. As we pointed out at the start of the proof, we can carry through similar arguments on the interval $[x_0 - h, x_0]$. We thus conclude that the differential equation $dy/dx = f(x, y)$ has a unique solution ϕ such that $\phi(x_0) = y_0$ on $|x - x_0| \leq h$.

Q.E.D.

C. Remarks and Examples

Notice carefully that Theorem 10.1 is both an existence theorem and uniqueness theorem. It tells us that

if   (i) f is continuous, and
    (ii) f satisfies a Lipschitz condition (with respect to y) in the rectangle R,
then (a)   there *exists* a solution ϕ of $dy/dx = f(x, y)$, defined on $|x - x_0| \leq h$, which is such that $\phi(x_0) = y_0$; and
    (b)   this solution ϕ is the *unique* solution satisfying these conditions.

Let us first concern ourselves with the *existence* aspect, Conclusion (a). Although the above proof made use of both Hypotheses (i) and (ii) in obtaining this first conclusion, other methods of proof can be employed to obtain conclusion (a) under Hypothesis (i) alone. In other words, the continuity of f in R is a sufficient condition for the *existence* of solutions as described in conclusion (a).

We now turn to the *uniqueness* aspect, conclusion (b). The continuity of f alone is *not* sufficient to assure us of a unique solution of the problem. That is, the function f must satisfy some additional condition besides the continuity requirement in order to obtain the uniqueness conclusion (b). One such additional condition which is sufficient for the uniqueness is of course the Lipschitz condition. Other less restrictive conditions are also known, however. For example, the Lipschitz condition (ii) could be replaced by the condition

$$|f(x, y_1) - f(x, y_2)| < k|y_1 - y_2| \ln \frac{1}{|y_1 - y_2|}$$

for all (x, y_1) and (x, y_2) in a sufficiently small rectangle about (x_0, y_0), and uniqueness would still be assured.

▶ **Example 10.11**

Consider the initial-value problem

$$\frac{dy}{dx} = y^{1/3}, \tag{10.17}$$

$$y(0) = 0.$$

Here $f(x, y) = y^{1/3}$ is continuous in the rectangle $R{:}|x| \le a, |y| \le b$ about the origin. Thus there *exists* at least one solution of $dy/dx = y^{1/3}$ on $|x| \le h$ such that $y(0) = 0$.

Existence of a solution of problem (10.17) being thus assured, let us examine the uniqueness aspect. If f satisfies a Lipschitz condition in R, then Theorem 10.1 will apply and uniqueness will also be assured. We have

$$\left| \frac{f(x, y_1) - f(x, y_2)}{y_1 - y_2} \right| = \left| \frac{y_1^{1/3} - y_2^{1/3}}{y_1 - y_2} \right|.$$

If we choose $y_1 = \delta > 0$ and $y_2 = -\delta$, this becomes

$$\left| \frac{\delta^{1/3} - (-\delta)^{1/3}}{\delta - (-\delta)} \right| = \frac{1}{\delta^{2/3}}.$$

Since this becomes unbounded as δ approaches zero, we see that f does *not* satisfy a Lipschitz condition throughout any domain containing the line $y = 0$, and hence not in R. Thus we can *not* apply Theorem 10.1 to obtain a uniqueness conclusion here. On the other hand, we must not conclude that uniqueness is impossible simply because a Lipschitz condition is not satisfied. The simple truth is that at this point we can draw no conclusion one way or the other about uniqueness in this problem.

In fact, the problem does *not* have a *unique* solution; for we can actually exhibit *two* solutions. Indeed, the functions ϕ_1 and ϕ_2 defined, respectively, by $\phi_1(x) = 0$ for all x, and

$$\phi_2(x) = \begin{cases} (\tfrac{2}{3}x)^{3/2}, & x \ge 0, \\ 0, & x \le 0, \end{cases}$$

are both solutions of problem (10.17) on the interval $-\infty < x < \infty$.

We further observe that Theorem 10.1 is a "local" existence theorem or an existence theorem "in the small." For it states that if f satisfies the given hypothesis in a domain D and if (x_0, y_0) is a point of D, then there exists a solution ϕ of

$$\frac{dy}{dx} = f(x, y),$$

$$y(x_0) = y_0,$$

defined on an interval $|x - x_0| \leq h$, where h is *sufficiently small*. It does *not* assert that ϕ is defined for *all* x, even if f satisfies the given hypotheses for all (x, y). Existence "in the large" cannot be asserted unless additional, very specialized restrictions are placed upon f. If f is linear in y, then an existence theorem in the large may be proved.

▶ **Example 10.12**

Consider the initial-value problem

$$\frac{dy}{dx} = y^2, \tag{10.18}$$

$$y(1) = -1.$$

Here $f(x, y) = y^2$ and $\dfrac{\partial f(x, y)}{\partial y} = 2y$ are both continuous for all (x, y). Thus using Theorem E we observe that f satisfies the hypothesis of Theorem 10.1 in every rectangle R,

$$|x - 1| \leq a, \qquad |y + 1| \leq b,$$

about the point $(1, -1)$ (see Figure 10.5).

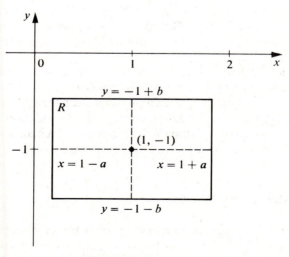

Figure 10.5

As in Theorem 10.1, let $M = \max |f(x, y)|$ for $(x, y) \in R$, and $h = \min (a, b/M)$. Then Theorem 10.1 asserts that the initial-value problem (10.18) possesses a unique solution defined on $|x - 1| \leq h$.

Now in this case $M = (-1 - b)^2 = (b + 1)^2$ and so

$$h = \min\left[a, \frac{b}{(b + 1)^2} \right].$$

Now consider

$$F(b) = \frac{b}{(b + 1)^2}.$$

From

$$F'(b) = \frac{1 - b}{(b + 1)^3},$$

we see that the maximum value of $F(b)$ for $b > 0$ occurs at $b = 1$; and we find $F(1) = \frac{1}{4}$. Thus if $a \geq \frac{1}{4}$,

$$\frac{b}{(b + 1)^2} \leq a$$

for all $b > 0$ and so

$$h = \frac{b}{(b + 1)^2} \leq \frac{1}{4},$$

regardless of the value of a. If, however, $a < \frac{1}{4}$, then certainly $h < \frac{1}{4}$. Thus in any case $h \leq \frac{1}{4}$. For $b = 1$, $a \geq \frac{1}{4}$,

$$h = \min\left[a, \frac{b}{(b + 1)^2} \right] = \min (a, \tfrac{1}{4}) = \tfrac{1}{4}.$$

This is the "best possible" h, according to the theorem. That is, at best Theorem 10.1 assures us that the initial-value problem (10.18) possesses a unique solution on the interval $\frac{3}{4} \leq x \leq \frac{5}{4}$. Thus, although the hypotheses of Theorem 10.1 are satisfied for *all* (x, y), the theorem only assures us of a solution to our problem on the "small" interval $|x - 1| \leq \frac{1}{4}$.

On the other hand, this does *not* necessarily mean that the actual solution of problem (10.18) is defined only on this small interval and nowhere outside of it. It may actually be defined on a much larger interval which includes the "small" interval $|x - 1| \leq \frac{1}{4}$ on which it is guaranteed by the theorem. Indeed, the solution of problem (10.18) is readily found to be $y = -(1/x)$, and this is actually defined and possesses a continuous derivative on the interval $0 < x < \infty$ (see Figure 10.6).

D. Continuation of Solutions

Example 10.12 brings up another point worthy of consideration. In this problem, the solution $y = -(1/x)$ was actually defined on the interval $0 < x < \infty$, while Theorem 10.1 guaranteed its existence only for $\frac{3}{4} \leq x \leq \frac{5}{4}$. We might thus inquire

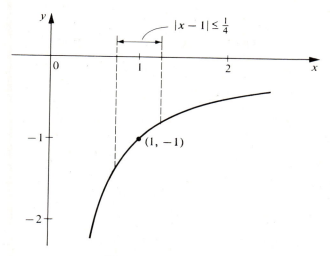

Figure 10.6

whether there is any way in which our theory could be applied to assure us that a solution is actually defined over a larger interval than that guaranteed by the basic application of Theorem 10.1.

Suppose then that f satisfies the hypotheses of Theorem 10.1 in D and that $(x_0, y_0) \in D$. Let $R: |x - x_0| \le a, |y - y_0| \le b$ be a rectangle lying in D which gives rise to the "best possible" h of the conclusion of Theorem 10.1. Then a single application of the Theorem 10.1 asserts that the initial-value problem

$$\frac{dy}{dx} = f(x, y),$$

$$(10.19)$$

$$y(x_0) = y_0,$$

possesses a unique solution ϕ_0 on $|x - x_0| \le h$, but nothing is implied about ϕ_0 *outside* this interval. Now let us consider the extreme right-hand point for which we are thus assured that ϕ_0 is defined. This is the point (x_1, y_1), where $x_1 = x_0 + h$, $y_1 = \phi_0(x_1)$. Since this point is a point of R, it is certainly a point of the domain D in which the hypotheses of Theorem 10.1 are satisfied. Thus we can reapply Theorem 10.1 at the point (x_1, y_1) to conclude that the differential equation $dy/dx = f(x, y)$ possesses a unique solution ϕ_1 such that $\phi_1(x_1) = y_1$, which is defined on some interval $x_1 \le x \le x_1 + h_1$, where $h_1 > 0$.

Now let us define ϕ as follows:

$$\phi(x) = \begin{cases} \phi_0(x), & x_0 - h \le x \le x_0 + h = x_1, \\ \phi_1(x), & x_1 \le x \le x_1 + h_1. \end{cases}$$

We now assert that ϕ is a solution of problem (10.19) on the extended interval $x_0 - h \le x \le x_1 + h_1$ (see Figure 10.7). The function ϕ is continuous on this interval and is such that $\phi(x_0) = y_0$. For $x_0 - h \le x \le x_0 + h$ we have

$$\phi_0(x) = y_0 + \int_{x_0}^{x} f[t, \phi_0(t)] \, dt$$

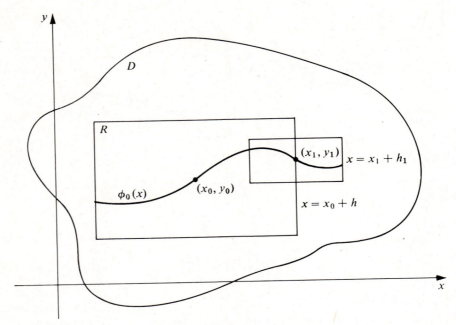

Figure 10.7

and hence

$$\phi(x) = y_0 + \int_{x_0}^{x} f[t, \phi(t)] \, dt \tag{10.20}$$

on this interval. On the interval $x_0 + h < x \leq x_1 + h_1$, we have

$$\phi_1(x) = y_1 + \int_{x_1}^{x} f[t, \phi_1(t)] \, dt,$$

or

$$\phi(x) = y_1 + \int_{x_1}^{x} f[t, \phi(t)] \, dt.$$

Since $y_1 = \phi_0(x_1) = y_0 + \int_{x_0}^{x_1} f[t, \phi(t)] \, dt$, we thus have

$$\phi(x) = y_0 + \int_{x_0}^{x} f[t, \phi(t)] \, dt \tag{10.21}$$

on the interval $x_0 + h < x \leq x_1 + h_1$. Thus, combining the results of (10.20) and (10.21) we see that ϕ satisfies the integral equation (10.21) on the extended interval $x_0 - h \leq x \leq x_1 + h_1$. Since ϕ is continuous on this interval, so is $f[x, \phi(x)]$. Thus,

$$\frac{d\phi(x)}{dx} = f[x, \phi(x)]$$

on $[x_0 - h, x_1 + h_1]$. Therefore ϕ is a solution of problem (10.19) on this larger interval.

The function ϕ so defined is called a *continuation* of the solution ϕ_0 to the interval $[x_0 - h, x_1 + h_1]$. If we now apply Theorem 10.1 again at the point $[x_1 + h_1, \phi(x_1 + h_1)]$, we may thus obtain the continuation over the still longer interval $x_0 - h \le x \le x_2 + h_2$ where $x_2 = x_1 + h_1$ and h_2 is positive. Repeating this process further, we may continue the solution over successively longer intervals $x_0 - h \le x \le x_n + h_n$ extending farther and farther to the right of $x_0 + h$. Also, in like manner, it may be continued over successively longer intervals extending farther and farther to the left of $x_0 - h$.

Thus repeating the process indefinitely on both the left and the right, we continue the solution to successively longer intervals $[a_n, b_n]$, where

$$[x_0 - h, x_0 + h] = [a_0, b_0] \subset [a_1, b_1] \subset [a_2, b_2] \subset \cdots \subset [a_n, b_n] \subset \cdots .$$

Let $a = \lim_{n \to \infty} a_n$ and $b = \lim_{n \to \infty} b_n$.

We thus obtain a largest open interval $a < x < b$ over which the solution ϕ such that $\phi(x_0) = y_0$ may be defined. It is clear that two cases are possible:

1. $a = -\infty$ and $b = +\infty$, in which case ϕ is defined for all x, $-\infty < x < +\infty$.
2. Either a is finite or b is finite or both.

Note carefully that even if f is continuous and satisfies a Lipschitz condition in every *bounded* domain D, we still cannot conclude that (1) will necessarily be the case. The initial-value problem $dy/dx = y^2$, $y(1) = -1$ of Example 10.12 illustrates this. The solution of this problem (given by $y = -[1/x]$) is defined for $0 < x < \infty$ and so $a = 0$, despite the fact that f such that $f(x, y) = y^2$ is continuous and satisfies a Lipschitz condition in every *bounded* domain. Note, however, that this f does *not* satisfy a Lipschitz condition in any *unbounded* domain defined by $a < x < b$, $-\infty < y < +\infty$.

On the other hand, suppose that for the initial-value problem $dy/dx = f(x, y)$, $y(x_0) = y_0$, the function f is continuous and *does* satisfy a Lipschitz condition in an *unbounded* domain $a < x < b$, $-\infty < y < +\infty$ which includes (x_0, y_0). Then we can be more definite concerning the largest open interval over which the solution of this initial-value problem is defined. In this connection we state without proof the following theorem.

THEOREM 10.2

Hypothesis

1. Let f be continuous in the unbounded *domain D*: $a < x < b$, $-\infty < y < +\infty$.
2. Let f satisfy a Lipschitz condition (with respect to y) in this unbounded domain. That is, assume there exists $k > 0$ such that

$$|f(x, y_1) - f(x, y_2)| \le k|y_1 - y_2|$$

for all $(x, y_1), (x, y_2) \in D$.

Conclusion. *A solution ϕ of $dy/dx = f(x, y)$ such that $\phi(x_0) = y_0$, where (x_0, y_0) is any point of D, is defined on the entire open interval $a < x < b$. In particular, if $a = -\infty$ and $b = +\infty$, then ϕ is defined for all x, $-\infty < x < +\infty$.*

For example, a solution of the initial-value problem

$$\frac{dy}{dx} = F(x)y, \qquad y(x_0) = y_0,$$

where F is continuous for $-\infty < x < +\infty$, is defined for all x, $-\infty < x < +\infty$.

Exercises

1. Consider the initial-value problem

$$\frac{dy}{dx} = y^{4/3}, \qquad y(x_0) = y_0.$$

    (a) Discuss the existence of a solution of this problem.
    (b) Discuss the uniqueness of a solution of this problem.

2. For each of the following initial-value problems show that there exists a unique solution of the problem if $y_0 \neq 0$. In each case discuss the existence and uniqueness of a solution if $y_0 = 0$.

    (a) $\dfrac{dy}{dx} = y^{2/3}, \qquad y(x_0) = y_0.$

    (b) $\dfrac{dy}{dx} = \sqrt{|y|}, \qquad y(x_0) = y_0.$

3. For each of the following initial-value problems find the largest interval $|x| \leq h$ on which Theorem 10.1 guarantees the existence of a unique solution. In each case find the unique solution and show that it actually exists over a larger interval than that guaranteed by the theorem.

    (a) $\dfrac{dy}{dx} = 1 + y^2, \qquad y(0) = 0.$

    (b) $\dfrac{dy}{dx} = e^{2y}, \qquad y(0) = 0.$

4. Show that Theorem 10.1 guarantees the existence of a unique solution of the initial-value problem

$$\frac{dy}{dx} = x^2 + y^2, \qquad y(0) = 0$$

on the interval $|x| \leq \sqrt{2}/2$.

10.3 DEPENDENCE OF SOLUTIONS ON INITIAL CONDITIONS AND ON THE FUNCTION f

A. Dependence on Initial Conditions

We now consider how the solution of the differential equation $dy/dx = f(x, y)$ depends upon a slight change in the initial conditions or upon a slight change in the function f. It would seem that such slight changes would cause only slight changes in the solution. We shall show that under suitable restrictions this is indeed the case.

We first consider the result of a slight change in the initial condition $y(x_0) = y_0$. Let f be continuous and satisfy a Lipschitz condition with respect to y in a domain D, and let (x_0, y_0) be a fixed point of D. Then by Theorem 10.1 the initial-value problem

$$\frac{dy}{dx} = f(x, y),$$

$$y(x_0) = y_0,$$

has a unique solution ϕ defined on some sufficiently small interval $|x - x_0| \le h_0$. Now suppose the initial y value is changed from y_0 to Y_0. Our first concern is whether or not the new initial-value problem

$$\frac{dy}{dx} = f(x, y),$$

$$y(x_0) = Y_0, \tag{10.22}$$

also has a unique solution on some sufficiently small interval $|x - x_0| \le h_1$. If Y_0 is such that $|Y_0 - y_0|$ is sufficiently small, then we can be certain that the problem (10.22) does possess a unique solution on some such interval $|x - x_0| \le h_1$. In fact, let the rectangle $R: |x - x_0| \le a, |y - y_0| \le b$, lie in D and let Y_0 be such that $|Y_0 - y_0| \le b/2$. Then an application of Theorem 10.1 to problem (10.22) shows that this problem has a unique solution ψ which is defined and contained in R for $|x - x_0| \le h_1$, where $h_1 = \min(a, b/2M)$ and $M = \max |f(x, y)|$ for $(x, y) \in R$. Thus we may assume that there exists $\delta > 0$ and $h > 0$ such that for each Y_0 satisfying $|Y_0 - y_0| \le \delta$, problem (10.22) possesses a unique solution $\phi(x, Y_0)$ on $|x - x_0| \le h$ (see Figure 10.8).

We are now in a position to state the basic theorem concerning the dependence of solutions on initial conditions.

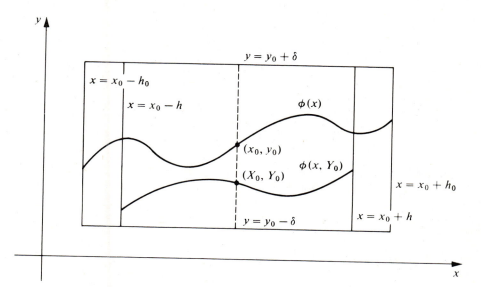

Figure 10.8

THEOREM 10.3

Hypothesis

1. Let f be continuous and satisfy a Lipschitz condition with respect to y, with Lipschitz constant k, in a domain D of the xy plane; and let (x_0, y_0) be a fixed point of D.
2. Assume there exists $\delta > 0$ and $h > 0$ such that for each Y_0 satisfying $|Y_0 - y_0| \leq \delta$ the initial-value problem

$$\frac{dy}{dx} = f(x, y),$$

$$y(x_0) = Y_0, \tag{10.22}$$

possesses a unique solution $\phi(x, Y_0)$ defined and contained in D on $|x - x_0| \leq h$.

Conclusion.   *If ϕ denotes the unique solution of (10.22) when $Y_0 = y_0$, and $\tilde{\phi}$ denotes the unique solution of (10.22) when $Y_0 = \tilde{y}_0$, where $|\tilde{y}_0 - y_0| = \delta_1 \leq \delta$, then*

$$|\tilde{\phi}(x) - \phi(x)| \leq \delta_1 e^{kh} \quad on \quad |x - x_0| \leq h.$$

Thus the solution $\phi(x, Y_0)$ of problem (10.22) is a continuous function of the initial value Y_0 at $Y_0 = y_0$.

Proof.   From Theorem 10.1 we know that

$$\phi = \lim_{n \to \infty} \phi_n,$$

where

$$\phi_n(x) = y_0 + \int_{x_0}^{x} f[t, \phi_{n-1}(t)] \, dt \quad (n = 1, 2, 3, \ldots),$$

and $\phi_0(x) = y_0$; $|x - x_0| \leq h$.
  In like manner,

$$\tilde{\phi} = \lim_{n \to \infty} \tilde{\phi}_n,$$

where

$$\tilde{\phi}_n(x) = \tilde{y}_0 + \int_{x_0}^{x} f[t, \tilde{\phi}_{n-1}(t)] \, dt \quad (n = 1, 2, 3, \ldots),$$

and $\tilde{\phi}_0(x) = \tilde{y}_0$; $|x - x_0| \leq h$.
  We shall show by induction that

$$|\tilde{\phi}_n(x) - \phi_n(x)| \leq \delta_1 \sum_{j=0}^{n} \frac{k^j (x - x_0)^j}{j!} \tag{10.23}$$

on $[x_0, x_0 + h]$, where k is the Lipschitz constant. We thus assume that on $[x_0, x_0 + h]$,

$$|\tilde{\phi}_{n-1}(x) - \phi_{n-1}(x)| \leq \delta_1 \sum_{j=0}^{n-1} \frac{k^j (x - x_0)^j}{j!}. \tag{10.24}$$

Then

$$\left|\tilde{\phi}_n(x) - \phi_n(x)\right| = \left|\tilde{y}_0 + \int_{x_0}^{x} f[t, \tilde{\phi}_{n-1}(t)]\, dt - y_0 - \int_{x_0}^{x} f[t, \phi_{n-1}(t)]\, dt\right|$$

$$\leq |\tilde{y}_0 - y_0| + \int_{x_0}^{x} \left|f[t, \tilde{\phi}_{n-1}(t)] - f[t, \phi_{n-1}(t)]\right|\, dt.$$

Applying the Lipschitz condition, we have

$$\left|f[x, \tilde{\phi}_{n-1}(x)] - f[x, \phi_{n-1}(x)]\right| \leq k \left|\tilde{\phi}_{n-1}(x) - \phi_{n-1}(x)\right|;$$

and so, since $|\tilde{y}_0 - y_0| = \delta_1$,

$$\left|\tilde{\phi}_n(x) - \phi_n(x)\right| \leq \delta_1 + k \int_{x_0}^{x} \left|\tilde{\phi}_{n-1}(t) - \phi_{n-1}(t)\right|\, dt.$$

Using the assumption (10.24), we have

$$\left|\tilde{\phi}_n(x) - \phi_n(x)\right| \leq \delta_1 + k \int_{x_0}^{x} \delta_1 \sum_{j=0}^{n-1} \frac{k^j(t - x_0)^j}{j!}\, dt$$

$$= \delta_1 + k\delta_1 \sum_{j=0}^{n-1} \frac{k^j}{j!} \int_{x_0}^{x} (t - x_0)^j\, dt = \delta_1 \left[1 + \sum_{j=0}^{n-1} \frac{k^{j+1}(x - x_0)^{j+1}}{(j + 1)!}\right].$$

Since

$$\delta_1 \left[1 + \sum_{j=0}^{n-1} \frac{k^{j+1}(x - x_0)^{j+1}}{(j + 1)!}\right] = \delta_1 \sum_{j=0}^{n} \frac{k^j(x - x_0)^j}{j!},$$

we have

$$\left|\tilde{\phi}_n(x) - \phi_n(x)\right| \leq \delta_1 \sum_{j=0}^{n} \frac{k^j(x - x_0)^j}{j!},$$

which is (10.24) with $(n - 1)$ replaced by n.

Also, on $[x_0, x_0 + h]$, we have

$$\left|\tilde{\phi}_1(x) - \phi_1(x)\right| = \left|\tilde{y}_0 + \int_{x_0}^{x} f[t, \tilde{y}_0]\, dt - y_0 - \int_{x_0}^{x} f[t, y_0]\, dt\right|$$

$$\leq |\tilde{y}_0 - y_0| + \int_{x_0}^{x} \left|f[t, \tilde{y}_0] - f[t, y_0]\right|\, dt$$

$$\leq \delta_1 + \int_{x_0}^{x} k|\tilde{y}_0 - y_0|\, dt = \delta_1 + k\delta_1(x - x_0).$$

Thus (10.23) holds for $n = 1$. Hence the induction is complete and (10.23) holds on $[x_0, x_0 + h]$. Using similar arguments on $[x_0 - h, x_0]$, we have

$$\left|\tilde{\phi}_n(x) - \phi_n(x)\right| \leq \delta_1 \sum_{j=0}^{n} \frac{k^j|x - x_0|^j}{j!} \leq \delta_1 \sum_{j=0}^{n} \frac{(kh)^j}{j!}$$

for all x on $|x - x_0| \leq h$, $n = 1, 2, 3, \ldots$. Letting $n \to \infty$, we have

$$\left|\tilde{\phi}(x) - \phi(x)\right| \leq \delta_1 \sum_{j=0}^{\infty} \frac{(kh)^j}{j!}.$$

But $\displaystyle\sum_{j=0}^{\infty} \frac{(kh)^j}{j!} = e^{kh}$; and so we have the desired inequality

$$|\tilde{\phi}(x) - \phi(x)| \le \delta_1 e^{kh} \quad \text{on} \quad |x - x_0| \le h.$$

The remaining statement of the conclusion is now obvious.   Q.E.D.

Thus under the conditions stated, if the initial values of the two solutions ϕ and $\tilde{\phi}$ differ by a sufficiently small amount, then their values will differ by an arbitrarily small amount at every point of $|x - x_0| \le h$. Geometrically, this means that if the corresponding integral curves are sufficiently close to each other initially, then they will be arbitrarily close to each other for all x such that $|x - x_0| \le h$.

B. Dependence on the Function f

We now consider how the solution of $dy/dx = f(x, y)$ will change if the function f is slightly changed. In this connection we have the following theorem.

THEOREM 10.4

Hypothesis

1. *In a domain D of the xy plane, assume that*
   (i) *f is continuous and satisfies a Lipschitz condition with respect to y, with Lipschitz constant k.*
   (ii) *F is continuous.*
   (iii) *$|F(x, y) - f(x, y)| \le \epsilon$ for $(x, y) \in D$.*
2. *Let (x_0, y_0) be a point of D; and let*
   (i) *ϕ be the solution of the initial-value problem*

$$\frac{dy}{dx} = f(x, y),$$

$$y(x_0) = y_0.$$

   (ii) *ψ be a solution of the initial-value problem*

$$\frac{dy}{dx} = F(x, y),$$

$$y(x_0) = y_0,$$

   (iii) *$[x, \phi(x)]$ and $[x, \psi(x)] \in D$ for $|x - x_0| \le h$.*

Conclusion.   *Then*

$$|\phi(x) - \psi(x)| \le \frac{\epsilon}{k}(e^{kh} - 1) \quad on \quad |x - x_0| \le h.$$

Proof.   Let $\bar{\phi}_0(x) = \psi(x)$ and define a sequence of functions $\{\bar{\phi}_n\}$ by

$$\bar{\phi}_n(x) = y_0 + \int_{x_0}^{x} f[t, \bar{\phi}_{n-1}(t)]\, dt, \quad |x - x_0| \le h \quad (n = 1, 2, 3, \ldots).$$

Then $\bar{\phi} = \displaystyle\lim_{n \to \infty} \bar{\phi}_n$ is a solution of $dy/dx = f(x, y)$ such that $\bar{\phi}(x_0) = y_0$ on $|x - x_0| \le h$.

By Hypothesis 1(i) the initial-value problem $dy/dx = f(x, y)$, $y(x_0) = y_0$, has a *unique* solution on $|x - x_0| \le h$. Thus from Hypothesis 2(i) $\bar{\phi}(x) = \phi(x)$ on $|x - x_0| \le h$, and so $\lim\limits_{n \to \infty} \bar{\phi}_n = \phi$.

From Hypothesis 2(ii) we have

$$\psi(x) = y_0 + \int_{x_0}^{x} F[t, \psi(t)]\, dt, \qquad |x - x_0| \le h.$$

We shall show by induction that

$$|\bar{\phi}_n(x) - \psi(x)| \le \epsilon \sum_{j=1}^{n} \frac{k^{j-1}(x - x_0)^j}{j!} \tag{10.25}$$

on $[x_0, x_0 + h]$. We thus assume that on this interval

$$|\bar{\phi}_{n-1}(x) - \psi(x)| \le \epsilon \sum_{j=1}^{n-1} \frac{k^{j-1}(x - x_0)^j}{j!}. \tag{10.26}$$

Then

$$|\bar{\phi}_n(x) - \psi(x)| = \left| y_0 + \int_{x_0}^{x} f[t, \bar{\phi}_{n-1}(t)]\, dt - y_0 - \int_{x_0}^{x} F[t, \psi(t)]\, dt \right|$$

$$\le \int_{x_0}^{x} |f[t, \bar{\phi}_{n-1}(t)] - F[t, \psi(t)]|\, dt.$$

We now write $F(x, y) = f(x, y) + \delta(x, y)$. Then

$$|\bar{\phi}_n(x) - \psi(x)| \le \int_{x_0}^{x} |f[t, \bar{\phi}_{n-1}(t)] - f[t, \psi(t)] - \delta[t, \psi(t)]|\, dt.$$

Applying the inequality $|A - B| \le |A| + |B|$ and then the Lipschitz condition satisfied by f, we have

$$|\bar{\phi}_n(x) - \psi(x)| \le \int_{x_0}^{x} |f[t, \bar{\phi}_{n-1}(t)] - f[t, \psi(t)]|\, dt + \int_{x_0}^{x} |\delta[t, \psi(t)]|\, dt$$

$$\le k \int_{x_0}^{x} |\bar{\phi}_{n-1}(t) - \psi(t)|\, dt + \int_{x_0}^{x} |\delta[t, \psi(t)]|\, dt.$$

Now using the assumption (10.26) and the fact that

$$|\delta(x, y)| = |F(x, y) - f(x, y)| \le \epsilon,$$

we obtain

$$|\bar{\phi}_n(x) - \psi(x)| \le k\epsilon \sum_{j=1}^{n-1} \frac{k^{j-1}}{j!} \int_{x_0}^{x} (t - x_0)^j\, dt + \int_{x_0}^{x} \epsilon\, dt$$

$$= \epsilon \sum_{j=1}^{n-1} \frac{k^j(x - x_0)^{j+1}}{(j+1)!} + \epsilon(x - x_0) = \epsilon \sum_{j=1}^{n} \frac{k^{j-1}(x - x_0)^j}{j!}.$$

Thus (10.26) holds with $(n-1)$ replaced by n. Also Hypothesis 1 (iii) shows that

$$|\bar{\phi}_1(x) - \psi(x)| \le \int_{x_0}^{x} |f[t, \psi(t)] - F[t, \psi(t)]|\, dt \le \int_{x_0}^{x} \epsilon\, dt = \epsilon(x - x_0)$$

on $[x_0, x_0 + h]$. Thus (10.25) holds for $n = 1$. Thus the induction is complete and so (10.25) holds on $[x_0, x_0 + h]$ for $n = 1, 2, 3, \ldots$.

Using similar arguments on $[x_0 - h, x_0]$, we thus have

$$|\bar{\phi}_n(x) - \psi(x)| \leq \epsilon \sum_{j=1}^{n} \frac{k^{j-1}|x - x_0|^j}{j!} \leq \frac{\epsilon}{k} \sum_{j=1}^{n} \frac{(kh)^j}{j!}$$

for all x on $|x - x_0| \leq h$, $n = 1, 2, 3, \ldots$.

Letting $n \to \infty$, we obtain

$$|\phi(x) - \psi(x)| \leq \frac{\epsilon}{k} \sum_{j=1}^{\infty} \frac{(kh)^j}{j!}.$$

But $\sum_{j=1}^{\infty} \frac{(kh)^j}{j!} = e^{kh} - 1$. Thus we obtain the desired inequality

$$|\phi(x) - \psi(x)| \leq \frac{\epsilon}{k}(e^{kh} - 1) \quad \text{on} \quad |x - x_0| \leq h. \qquad Q.E.D$$

Thus, under the hypotheses stated, if ϵ is sufficiently small, the difference between the solutions ϕ and ψ will be arbitrarily small on $|x - x_0| \leq h$. The following example illustrates how this result can be used to advantage.

▶ **Example 10.13**

Consider the initial-value problem

$$\frac{dy}{dx} = x^2 + y^2 + y + 1,$$

$$y(0) = 0. \qquad (10.27)$$

The differential equation of this problem cannot be solved explicitly by any of the methods which we have studied, but the differential equation $dy/dx = y + 1$ can be. If x and y are sufficiently small, the difference $|(x^2 + y^2 + y + 1) - (y + 1)| = |x^2 + y^2|$ will be less than or equal to any given $\epsilon > 0$. Thus the solution of problem (10.27) will differ from that of the problem

$$\frac{dy}{dx} = y + 1,$$

$$y(0) = 0 \qquad (10.28)$$

by an arbitrarily small amount if x and y are sufficiently small. We can thus use the explicit solution of problem (10.28) to obtain information about the solution of problem (10.27) in a sufficiently small neighborhood of $(0, 0)$.

Exercises

1. The differential equation $dy/dx = \sin(xy)$ cannot be solved explicitly by any of our methods, but the differential equation $dy/dx = xy$ can be. Explain why and how we can use the explicit solution of the initial-value problem

$$\frac{dy}{dx} = xy, \qquad y(0) = 0,$$

to obtain information about the solution of the initial-value problem

$$\frac{dy}{dx} = \sin(xy), \qquad y(0) = 0,$$

if x and y are sufficiently small.

10.4 EXISTENCE AND UNIQUENESS THEOREMS FOR SYSTEMS AND HIGHER-ORDER EQUATIONS

A. The General Case

Our first concern is the basic existence and uniqueness theorem for an initial-value problem involving a system of n first-order differential equations of the form

$$\frac{dy_1}{dx} = f_1(x, y_1, y_2, \ldots, y_n),$$

$$\frac{dy_2}{dx} = f_2(x, y_1, y_2, \ldots, y_n), \qquad (10.29)$$

$$\vdots$$

$$\frac{dy_n}{dx} = f_n(x, y_1, y_2, \ldots, y_n),$$

in the n unknowns $y_1, y_2, \ldots, y_n$, where $f_1, f_2, \ldots, f_n$ are n continuous real functions defined in some domain D of real $(n + 1)$-dimensional $x, y_1, y_2, \ldots, y_n$ space.

DEFINITION

By a solution of the system (10.29) we shall mean an ordered set of n continuously differentiable real functions

$$\phi_1, \phi_2, \ldots, \phi_n$$

defined on some real x interval $a \le x \le b$ such that

$$[x, \phi_1(x), \phi_2(x), \ldots, \phi_n(x)] \in D$$

and

$$\frac{d\phi_1(x)}{dx} = f_1[x, \phi_1(x), \phi_2(x), \ldots, \phi_n(x)],$$

$$\frac{d\phi_2(x)}{dx} = f_2[x, \phi_1(x), \phi_2(x), \ldots, \phi_n(x)],$$

$$\vdots$$

$$\frac{d\phi_n(x)}{dx} = f_n[x, \phi_1(x), \phi_2(x), \ldots, \phi_n(x)],$$

for all x such that $a \le x \le b$.

Corresponding to Theorem 10.1 we have following theorem dealing with the system (10.29). Since its proof parallels that of Theorem 10.1 we shall merely outline the major steps of the proof and omit the details.

THEOREM 10.5

Hypothesis

1. Let the functions $f_1, f_2, \ldots, f_n$ be continuous in the $(n + 1)$-dimensional rectangle R defined by

$$|x - x_0| \le a, \qquad |y_1 - c_1| \le b_1, \ldots, |y_n - c_n| \le b_n,$$

where $(x_0, c_1, \ldots, c_n)$ is a point of real $(n + 1)$-dimensional $(x, y_1, \ldots, y_n)$ space and a, $b_1, \ldots, b_n$ are positive constants.
  Let M be such that

$$|f_i(x, y_1, y_2, \ldots, y_n)| \le M \qquad for\ i = 1, 2, \ldots, n \quad for\ all\ (x, y_1, y_2, \ldots, y_n) \in R.$$

Let $h = \min\left(a, \dfrac{b_1}{M}, \dfrac{b_2}{M}, \ldots, \dfrac{b_n}{M}\right)$.

2. Let the functions $f_i (i = 1, 2, \ldots, n)$ satisfy a Lipschitz condition with Lipschitz constant k in R. That is, assume there exists a constant $k > 0$ such that

$$|f_i(x, \bar{y}_1, \bar{y}_2, \ldots, \bar{y}_n) - f_i(x, \tilde{y}_1, \tilde{y}_2, \ldots, \tilde{y}_n)|$$
$$\le k(|\bar{y}_1 - \tilde{y}_1| + |\bar{y}_2 - \tilde{y}_2| + \cdots + |\bar{y}_n - \tilde{y}_n|) \quad (10.30)$$

for any two points $(x, \bar{y}_1, \bar{y}_2, \ldots, \bar{y}_n), (x, \tilde{y}_1, \tilde{y}_2, \ldots, \tilde{y}_n) \in R$, and for $i = 1, 2, \ldots, n$.

Conclusion.   *There exists a unique solution*

$$\phi_1, \phi_2, \ldots, \phi_n$$

of the system

$$\frac{dy_1}{dx} = f_1(x, y_1, y_2, \ldots, y_n),$$

$$\frac{dy_2}{dx} = f_2(x, y_1, y_2, \ldots, y_n), \qquad (10.29)$$

$$\vdots$$

$$\frac{dy_n}{dx} = f_n(x, y_1, y_2, \ldots, y_n),$$

such that

$$\phi_1(x_0) = c_1, \qquad \phi_2(x_0) = c_2, \ldots, \phi_n(x_0) = c_n,$$

defined for $|x - x_0| \le h$.

Outline of Proof.   We first define functions $\phi_{i, j}$ by

$$\phi_{i, 0}(x) = c_i \qquad (i = 1, 2, \ldots, n)$$

and

$$\phi_{i,j}(x) = c_i + \int_{x_0}^{x} f_i[t, \phi_{1,j-1}(t), \ldots, \phi_{n,j-1}(t)] \, dt \tag{10.31}$$

$(i = 1, 2, \ldots, n; \, j = 1, 2, 3, \ldots)$.

We then prove by mathematical induction that all of the function $\phi_{i,j}$ so defined are continuous and satisfy the relations

$$|\phi_{i,j}(x) - c_i| \le b_i \qquad (i = 1, 2, \ldots, n; \, j = 1, 2, 3, \ldots)$$

on $|x - x_0| \le h$. This much assured, the formulas (10.31) and the Lipschitz condition (10.30) can be used to obtain the following inequality by another induction:

$$|\phi_{i,j}(x) - \phi_{i,j-1}(x)| \le \frac{M(kn)^{j-1}|x - x_0|^j}{j!}$$

$(i = 1, 2, \ldots, n; \, j = 1, 2, 3, \ldots; |x - x_0| \le h)$. Thus also

$$|\phi_{i,j}(x) - \phi_{i,j-1}(x)| \le \frac{M}{kn} \frac{(knh)^j}{j!} \tag{10.32}$$

$(i = 1, 2, \ldots, n; \, j = 1, 2, 3, \ldots)$. This enables us to conclude that for each $i = 1, 2, \ldots, n$, the sequence $\{\phi_{i,j}\}$ defined by

$$\phi_{i,j}(x) = \phi_{i,0}(x) + \sum_{p=1}^{j} [\phi_{i,p}(x) - \phi_{i,p-1}(x)] \qquad (j = 1, 2, 3, \ldots)$$

converges uniformly to a continuous function ϕ_i. We may then show that each ϕ_i $(i = 1, 2, \ldots, n)$ satisfies the integral equation

$$\phi_i(x) = c_i + \int_{x_0}^{x} f_i[t, \phi_1(t), \ldots, \phi_n(t)] \, dt,$$

on $|x - x_0| \le h$. From this we have at once that

$$\frac{d\phi_i(x)}{dx} = f_i[x, \phi_1(x), \ldots, \phi_n(x)]$$

on $|x - x_0| \le h$ and $\phi_i(x_0) = c_i (i = 1, 2, \ldots, n)$.

Our outline of the existence proof is thus completed. It is clear that it paralles that for the case $n = 1$ given for Theorem 10.1. The proof of the uniqueness in the present case also parallels that of Theorem 10.1, and we shall leave it to the reader to make the necessary changes and complete the present outline.

Theorem 10.5 can be used to obtain an existence and uniqueness theorem for the basic initial-value problem associated with an nth-order differential equation of the form

$$\frac{d^n y}{dx^n} = f\left[x, y, \frac{dy}{dx}, \ldots, \frac{d^{n-1} y}{dx^{n-1}}\right]. \tag{10.33}$$

DEFINITION

Consider the differential equation

$$\frac{d^n y}{dx^n} = f\left[x, y, \frac{dy}{dx}, \ldots, \frac{d^{n-1} y}{dx^{n-1}}\right], \tag{10.33}$$

where f is a continuous real function defined in a domain D of real $(n + 1)$-dimensional $(x, y, y', \ldots, y^{(n-1)})$-space. By a solution of Equation (10.33) we mean a real function ϕ having a continuous nth derivative (and hence all lower-ordered derivatives) on a real interval $a \le x \le b$ such that

$$[x, \phi(x), \phi'(x), \ldots, \phi^{(n-1)}(x)] \in D$$

and

$$\phi^{(n)}(x) = f[x, \phi(x), \phi'(x), \ldots, \phi^{(n-1)}(x)]$$

for all x such that $a \le x \le b$.

THEOREM 10.6

Hypothesis. *Consider the differential equation*

$$\frac{d^n y}{dx^n} = f\left[x, y, \frac{dy}{dx}, \ldots, \frac{d^{n-1} y}{dx^{n-1}}\right], \tag{10.33}$$

where the function f is continuous and satisfies a Lipschitz condition of the form (10.30) in a domain D of real $(n + 1)$-dimensional $(x, y, y', \ldots, y^{(n-1)})$-space. Let $(x_0, c_0, c_1, \ldots, c_{n-1})$ be a point of D.

Conclusion. *There exists a unique solution ϕ of the nth-order differential equation such that*

$$\phi(x_0) = c_0, \phi'(x_0) = c_1, \ldots, \phi^{(n-1)}(x_0) = c_{n-1}, \tag{10.34}$$

defined on some interval $|x - x_0| \le h$ about $x = x_0$.

Proof

$$\text{Let } y_1 = y, \qquad y_2 = \frac{dy}{dx}, \ldots, \qquad y_n = \frac{d^{n-1} y}{dx^{n-1}}.$$

Then the nth-order differential equation (10.33) is equivalent to the system of n first-order equations

$$\frac{dy_1}{dx} = y_2,$$

$$\frac{dy_2}{dx} = y_3,$$

$$\vdots \tag{10.35}$$

$$\frac{dy_{n-1}}{dx} = y_n,$$

$$\frac{dy_n}{dx} = f[x, y_1, y_2, \ldots, y_n].$$

If ϕ is a solution of (10.33) which satisfies the conditions (10.34), then the ordered set of functions $\phi_1, \phi_2, \ldots, \phi_n$, where $\phi_1 = \phi, \phi_2 = \phi', \ldots, \phi_n = \phi^{(n-1)}$, is a solution of the system (10.35) which satisfies the conditions

$$\phi_1(x_0) = c_0, \; \phi_2(x_0) = c_1, \ldots, \; \phi_n(x_0) = c_{n-1}. \tag{10.36}$$

Conversely, if $\phi_1, \phi_2, \ldots, \phi_n$ is a solution of (10.35) which satisfies conditions (10.36), then the function ϕ, where $\phi = \phi_1$, is a solution of the differential equation (10.33) which satisfies condition (10.34).

Since $(x_0, c_0, c_1, \ldots, c_{n-1})$ is a point of D, there exists an $(n + 1)$-dimensional rectangle R about this point such that the function f satisfies the stated hypotheses in R. Thus the system (10.35) satisfies all the hypotheses of Theorem 10.5 in R. Hence there exists a unique solution $\phi_1, \phi_2, \cdots, \phi_n$ of system (10.35) which satisfies the conditions (10.36) and is defined on some sufficiently small interval $|x - x_0| \leq h$ about $x = x_0$. Thus if we set $\phi = \phi_1$, the above-noted equivalence of (10.33) and (10.34) with (10.35) and (10.36) gives the desired conclusion.

Q.E.D

B. The Linear Case

We now consider the linear system

$$\frac{dy_1}{dx} = a_{11}(x)y_1 + a_{12}(x)y_2 + \cdots + a_{1n}(x)y_n + F_1(x),$$

$$\frac{dy_2}{dx} = a_{21}(x)y_1 + a_{22}(x)y_2 + \cdots + a_{2n}(x)y_n + F_2(x), \tag{10.37}$$

$$\vdots$$

$$\frac{dy_n}{dx} = a_{n1}(x)y_1 + a_{n2}(x)y_2 + \cdots + a_{nn}(x)y_n + F_n(x),$$

where the coefficients a_{ij} and the functions F_i are continuous on the interval $a \leq x \leq b$. We first prove a lemma.

LEMMA

Hypothesis. *Let the functions a_{ij} and $F_i (1, 2, \ldots, n; \; j = 1, 2, \ldots, n)$ be continuous on the interval $a \leq x \leq b$.*

Conclusion. *Then the functions f_i defined by*

$$f_i(x, y_1, y_2, \ldots, y_n) = a_{i1}(x)y_1 + a_{i2}(x)y_2 + \cdots + a_{in}(x)y_n + F_i(x)$$

($i = 1, 2, \ldots, n$), satisfy a Lipschitz condition on

$$a \leq x \leq b, \quad -\infty < y_i < +\infty \quad (i = 1, 2, \ldots, n).$$

That is, there exists a constant $k > 0$ such that

$$|f_i(x, \bar{y}_1, \bar{y}_2, \ldots, \bar{y}_n) - f_i(x, \tilde{y}_1, \tilde{y}_2, \ldots, \tilde{y}_n)|$$

$$\leq k(|\bar{y}_1 - \tilde{y}_1| + |\bar{y}_2 - \tilde{y}_2| + \cdots + |\bar{y}_n - \tilde{y}_n|)$$

for all x such that $a \leq x \leq b$ and any two sets of real numbers $\bar{y}_1, \bar{y}_2, \ldots, \bar{y}_n$ and $\tilde{y}_1, \tilde{y}_2, \ldots, \tilde{y}_n$ ($i = 1, 2, 3, \ldots, n$).

Proof. Since each of the functions a_{ij} is continuous on $a \le x \le b$, corresponding to each of these functions there exists a constant k_{ij} such that $|a_{ij}(x)| \le k_{ij}$ for $x \in [a, b]$ $(i = 1, 2, \ldots, n; j = 1, 2, \ldots, n)$. Let $k = \max\{k_{ij}\}$ for $i = 1, 2, \ldots, n; j = 1, 2, \ldots, n$. Then $|a_{ij}(x)| \le k$ for $x \in [a, b]$. Then for every $x \in [a, b]$ and any two sets of real numbers $\bar{y}_1, \bar{y}_2, \ldots, \bar{y}_n$ and $\tilde{y}_1, \tilde{y}_2, \ldots, \tilde{y}_n$, we have

$$
\begin{aligned}
|f_i(x, \bar{y}_1, \bar{y}_2, \ldots, \bar{y}_n) &- f_i(x, \tilde{y}_1, \tilde{y}_2, \ldots, \tilde{y}_n)| \\
&= |a_{i1}(x)\bar{y}_1 + a_{i2}(x)\bar{y}_2 + \cdots + a_{in}(x)\bar{y}_n + F_i(x) \\
&\quad - a_{i1}(x)\tilde{y}_1 - a_{i2}(x)\tilde{y}_2 - \cdots - a_{in}(x)\tilde{y}_n - F_i(x)| \\
&= |a_{i1}(x)[\bar{y}_1 - \tilde{y}_1] + a_{i2}(x)[\bar{y}_2 - \tilde{y}_2] + \cdots + a_{in}(x)[\bar{y}_n - \tilde{y}_n]| \\
&\le |a_{i1}(x)||\bar{y}_1 - \tilde{y}_1| + |a_{i2}(x)||\bar{y}_2 - \tilde{y}_2| + \cdots + |a_{in}(x)||\bar{y}_n - \tilde{y}_n| \\
&\le k(|\bar{y}_1 - \tilde{y}_1| + |\bar{y}_2 - \tilde{y}_2| + \cdots + |\bar{y}_n - \tilde{y}_n|). \qquad Q.E.D.
\end{aligned}
$$

We now obtain the basic existence theorem concerning the linear system (10.37).

THEOREM 10.7

Hypothesis

1. *Let the coefficients a_{ij} and the functions $F_i (i, j = 1, 2, \ldots, n)$ in the linear system*

$$
\frac{dy_1}{dx} = a_{11}(x)y_1 + a_{12}(x)y_2 + \cdots + a_{1n}(x)y_n + F_1(x),
$$

$$
\frac{dy_2}{dx} = a_{21}(x)y_1 + a_{22}(x)y_2 + \cdots + a_{2n}(x)y_n + F_2(x),
$$

$$
\vdots \qquad\qquad (10.37)
$$

$$
\frac{dy_n}{dx} = a_{n1}(x)y_1 + a_{n2}(x)y_2 + \cdots + a_{nn}(x)y_n + F_n(x),
$$

be continuous on the real interval $a \le x \le b$.

2. *Let x_0 be a point of the interval $a \le x \le b$, and let $c_1, c_2, \ldots, c_n$ be a set of n real constants.*

Conclusion. *There exists a unique solution*

$$
\phi_1, \phi_2, \ldots, \phi_n
$$

of the system (10.37) such that

$$
\phi_1(x_0) = c_1, \phi_2(x_0) = c_2, \ldots, \phi_n(x_0) = c_n,
$$

and this solution is defined on the entire interval $a \le x \le b$.

Outline of Proof. The system (10.37) is a special case of the system (10.29) with which Theorem 10.5 is concerned, and the present outline of proof parallels that given for Theorem 10.5. We first define functions $\phi_{i,j}$ by

$$
\phi_{i,0}(x) = c_i \qquad (i = 1, 2, \ldots, n)
$$

and

$$\phi_{i,j}(x) = c_i + \int_{x_0}^{x} [a_{i1}(t)\phi_{1,j-1}(t) + \cdots + a_{in}(t)\phi_{n,j-1}(t) + F_i(t)]\, dt \quad (10.38)$$

$(i = 1, 2, \ldots, n; j = 1, 2, 3, \ldots)$ on $a \leq x \leq b$. The functions $\phi_{i,j}$ so defined are continuous on the entire interval $a \leq x \leq b$. Also, by hypothesis there exists $M > 0$ such that $|a_{i1}(x)c_1 + \cdots + a_{in}(x)c_n + F_i(x)| \leq M$ $(i = 1, 2, \ldots, n)$, $a \leq x \leq b$.

By the lemma the functions defined by

$$a_{i1}(x)y_1 + a_{i2}(x)y_2 + \cdots + a_{in}(x)y_n + F_i(x)$$

satisfy a Lipschitz condition on $a \leq x \leq b$. We can thus use the formulas (10.38) and this Lipschitz condition to obtain by induction the inequality

$$|\phi_{i,j}(x) - \phi_{i,j-1}(x)| \leq \frac{M(kn)^{j-1}|x - x_0|^j}{j!},$$

$(i = 1, 2, \ldots, n; j = 1, 2, 3, \ldots)$ on the entire interval $a \leq x \leq b$. Thus also

$$|\phi_{i,j}(x) - \phi_{i,j-1}(x)| \leq \frac{M}{kn} \frac{(knH)^j}{j!} \quad (10.39)$$

$(i = 1, 2, \ldots, n; j = 1, 2, 3, \ldots)$, $a \leq x \leq b$, where $H = \max(|a - x_0|, |b - x_0|)$. The inequality (10.39) here corresponds to the inequality (10.32) in the proof of Theorem 10.5. The remainder of the proof outlined for Theorem 10.5 now carries over to the present case for $a \leq x \leq b$ and we obtain the desired conclusion.

Now we are in a position to obtain the basic existence theorem for the initial-value problem associated with a single nth-order linear differential equation.

THEOREM 10.8

Hypothesis

1. Consider the differential equation

$$a_0(x)\frac{d^n y}{dx^n} + a_1(x)\frac{d^{n-1}y}{dx^{n-1}} + \cdots + a_{n-1}(x)\frac{dy}{dx} + a_n(x)y = F(x), \quad (10.40)$$

where $a_0, a_1, \ldots, a_{n-1}, a_n$, and F are continuous on the interval $a \leq x \leq b$ and $a_0(x) \neq 0$ on $a \leq x \leq b$.

2. Let x_0 be a point of the interval $a \leq x \leq b$, and let $c_0, c_1, \ldots, c_{n-1}$ be a set of n real constants.

Conclusion. *There exists a unique solution ϕ of (10.40) such that*

$$\phi(x_0) = c_0, \phi'(x_0) = c_1, \ldots, \phi^{(n-1)}(x_0) = c_{n-1}, \quad (10.41)$$

and this solution is defined over the entire interval $a \leq x \leq b$.

Proof. As in the proof of Theorem 10.6, we let

$$y_1 = y, \quad y_2 = \frac{dy}{dx}, \ldots, y_n = \frac{d^{n-1}y}{dx^{n-1}}.$$

Then the nth-order linear differential equation (10.40) is equivalent to the linear system

$$\frac{dy_1}{dx} = y_2,$$

$$\frac{dy_2}{dx} = y_3,$$

$$\vdots \tag{10.42}$$

$$\frac{dy_{n-1}}{dx} = y_n,$$

$$\frac{dy_n}{dx} = -\frac{a_n(x)}{a_0(x)}\,y_1 - \frac{a_{n-1}(x)}{a_0(x)}\,y_2 - \cdots - \frac{a_1(x)}{a_0(x)}\,y_n + \frac{F(x)}{a_0(x)}.$$

If ϕ is a solution of (10.40) which satisfies the conditions (10.41), then the ordered set of functions $\phi_1, \phi_2, \ldots, \phi_n$, where $\phi_1 = \phi$, $\phi_2 = \phi', \ldots, \phi_n = \phi^{(n-1)}$, is a solution of the linear system (10.42) which satisfies the conditions

$$\phi_1(x_0) = c_0, \qquad \phi_2(x_0) = c_1, \ldots, \phi_n(x_0) = c_{n-1}. \tag{10.43}$$

Conversely, if $\phi_1, \ldots, \phi_n$ is a solution of (10.42) which satisfies (10.43), then the function $\phi = \phi_1$ is a solution of the differential equation (10.40) which satisfies conditions (10.41).

The system (10.42) is simply a special case of the linear system (10.37) to which Theorem 10.7 applies. Thus the system (10.42) possesses a unique solution $\phi_1, \ldots, \phi_n$ defined on the entire interval $a \le x \le b$ which satisfies the conditions (10.43). Thus if we set $\phi = \phi_1$, the above-noted equivalence of (10.40) and (10.41) with (10.42) and (10.43) gives the desired conclusion.   *Q.E.D*

Observe that this is the basic existence theorem which was first stated without proof at the beginning of Chapter 4 (Theorem 4.1). Two examples illustrating it were given in that chapter (page 104) and it has been tacitly employed in the intervening pages whenever initial-value problems were solved. We given one further example.

▶   **Example 10.14**

Consider the initial-value problem:

$$(x^2 - x - 6)\frac{d^2y}{dx^2} + (x^2 + 4)\frac{dy}{dx} + \frac{1}{2x + 3}\,y = e^{-x},$$

$$y(2) = 0,$$

$$y'(2) = 4.$$

The coefficient of y is continuous except at $x = -\frac{3}{2}$. The remaining coefficients and the nonhomogeneous term are continuous for all values of x, $-\infty < x < \infty$. The leading coefficient $(x^2 - x - 6)$ equal zero at $x = -2$ and $x = 3$. Thus the hypothesis of Theorem 10.8 is satisfied in every closed interval $a \le x \le b$ such that $-\frac{3}{2} < a < x_0 = 2 < b < 3$. Therefore the given initial-value problem has a unique solution, and we are assured that this solution is defined over every such closed interval $a \le x \le b$.

In Chapter 4 we stated an important corollary to this theorem concerning the homogeneous equation

$$a_0(x)\frac{d^n y}{dx^n} + a_1(x)\frac{d^{n-1}y}{dx^{n-1}} + \cdots + a_{n-1}(x)\frac{dy}{dx} + a_n(x)y = 0. \tag{10.44}$$

We now restate this corollary, which will soon be very useful to us, and give its proof.

COROLLARY

Hypothesis. *The function ϕ is a solution of the homogeneous equation (10.44) such that*

$$\phi(x_0) = 0, \ \phi'(x_0) = 0, \ldots, \ \phi^{(n-1)}(x_0) = 0, \tag{10.45}$$

where x_0 is a point of an interval $a \le x \le b$ on which the coefficients $a_0, a_1, \ldots, a_n$ are all continuous and $a_0(x) \ne 0$.

Conclusion. $\phi(x) = 0$ *for all x such that $a \le x \le b$.*

Proof. First note that ϕ such that $\phi(x) = 0$ for all $x \in [a, b]$ is indeed *a* solution of the differential equation (10.44) which satisfies the initial conditions (10.45). But by Theorem 10.8 the initial-value problem composed of Equation (10.44) and conditions (10.45) has a *unique* solution on $a \le x \le b$. Hence the stated conclusion follows.

Exercises

1. Consider the third-order differential equation

$$\frac{d^3 y}{dx^3} = x^2 + y\frac{dy}{dx} + \left(\frac{d^2 y}{dx^2}\right)^2$$

of the form (10.33) of the text.

(a) Does there exist a unique solution ϕ of the given equation such that

$$\phi(0) = 1, \qquad \phi'(0) = -3, \qquad \phi''(0) = 0?$$

Explain precisely why or why not.

(b) Find the system of three first-order equations of the form (10.35) to which the given third-order equation is equivalent.

2. Does there exist a solution of the initial-value problem

$$(x^2 - 4)\frac{d^4 y}{dx^4} + 2x\frac{d^2 y}{dx^2} + (\sin x)y = 0,$$

$$y(0) = 0, \qquad y'(0) = 1, \qquad y''(0) = 1, \qquad y'''(0) = -1?$$

If so, is the solution unique and over what interval are we assured that it is defined? Explain precisely.

3. Given that each of the functions f_1 and f_2 defined for all x by

    $$f_1(x) = \sin x \quad \text{and} \quad f_2(x) = \sum_{n=1}^{\infty} \frac{(-1)^{n+1} x^{2n-1}}{(2n-1)!}$$

    are solutions of the initial-value problem

    $$\frac{d^2 y}{dx^2} + y = 0,$$

    $$y(0) = 0, \qquad y'(0) = 1,$$

    for all x, $-\infty < x < \infty$, what theorem enables us to conclude that $f_1(x) = f_2(x)$ for all x, $-\infty < x < \infty$? Explain.

4. Consider the differential equation

    $$a_0(x) \frac{d^2 y}{dx^2} + a_1(x) \frac{dy}{dx} + a_2(x) y = 0, \tag{A}$$

    where a_0, a_1, and a_2 are continuous for all x, $-\infty < x < \infty$, and $a_0(x) \neq 0$ for all values of x.

    (a) Let f be a nontrivial solution of differential equation (A), let f' denote the derivative of f, and let $x_0 \in [a, b]$. Prove that if $f(x_0) = 0$, then $f'(x_0) \neq 0$.

    (b) Let f and g be two distinct nontrivial solutions of differential equation (A), and suppose there exists $x_0 \in [a, b]$ such that $f(x_0) = g(x_0) = 0$. Prove that there exists a constant c such that $f = cg$.

    [*Hint*: Observe that the function h defined by $h(x) = Af(x) - Bg(x)$, where $A = g'(x_0)$ and $B = f'(x_0)$, is also a solution of differential equation (A).]

5. Complete the details of the proof of Theorem 10.5.

6. Complete the details of the proof of Theorem 10.7.

CHAPTER ELEVEN

The Theory of Linear Differential Equations

We stated and illustrated certain basic theorems on higher-order linear differential equations in Chapter 4 and on linear systems in Chapter 7. Our main purpose in introducing them in these earlier chapters was to have them available as background material for the methods which followed. In this chapter it is the theory itself that is of primary concern. Here we shall restate and also prove the various theorems previously introduced and shall also proceed to obtain and illustrate further theorems on higher-order linear equations and linear systems. We shall begin by developing the basic theory of linear systems.

11.1 INTRODUCTION

We consider the normal form of linear system of n first-order differential equations in n unknown functions $x_1, x_2, \ldots, x_n$. As noted in Section 7.1A, this system is of the form

$$\frac{dx_1}{dt} = a_{11}(t)x_1 + a_{12}(t)x_2 + \cdots + a_{1n}(t)x_n + F_1(t),$$

$$\frac{dx_2}{dt} = a_{21}(t)x_1 + a_{22}(t)x_2 + \cdots + a_{2n}(t)x_n + F_2(t), \qquad (11.1)$$

$$\vdots$$

$$\frac{dx_n}{dt} = a_{n1}(t)x_1 + a_{n2}(t)x_2 + \cdots + a_{nn}(t)x_n + F_n(t).$$

We shall assume that all of the functions defined by $a_{ij}(t)$, $i = 1, 2, \ldots, n$, $j = 1, 2, \ldots, n$, and $F_i(t)$, $i = 1, 2, \ldots, n$, are continuous on a real interval $a \le t \le b$. If all $F_i(t) = 0$, $i = 1, 2, \ldots, n$, for all t, then the system (11.1) is called *homogeneous*. Otherwise, the system is called *nonhomogeneous*.

▶ **Example 11.1**

The system

$$\frac{dx_1}{dt} = 7x_1 - x_2 + 6x_3,$$

$$\frac{dx_2}{dt} = -10x_1 + 4x_2 - 12x_3, \qquad (11.2)$$

$$\frac{dx_3}{dt} = -2x_1 + x_2 - x_3,$$

is a homogeneous linear system of the type (11.1) with $n = 3$ and having constant coefficients. The system

$$\frac{dx_1}{dt} = 7x_1 - x_2 + 6x_3 - 5t - 6,$$

$$\frac{dx_2}{dt} = -10x_1 + 4x_2 - 12x_3 - 4t + 23, \qquad (11.3)$$

$$\frac{dx_3}{dt} = -2x_1 + x_2 - x_3 + 2,$$

is a nonhomogeneous linear system of the type (11.1) with $n = 3$, the nonhomogeneous terms being $-5t - 6$, $-4t + 23$, and 2, respectively.

We note that the system (11.1) can be written more compactly as

$$\frac{dx_i}{dt} = \sum_{j=1}^{n} a_{ij}(t)x_j + F_i(t) \qquad (i = 1, 2, \ldots, n).$$

We shall now proceed to express the system in an even more compact manner using vectors and matrices. We introduce the matrix $\mathbf{A}$ defined by

$$\mathbf{A}(t) = \begin{pmatrix} a_{11}(t) & a_{12}(t) & \cdots & a_{1n}(t) \\ a_{21}(t) & a_{22}(t) & \cdots & a_{2n}(t) \\ \vdots & \vdots & & \vdots \\ a_{n1}(t) & a_{n2}(t) & \cdots & a_{nn}(t) \end{pmatrix} \qquad (11.4)$$

and the vectors $\mathbf{F}$ and $\mathbf{x}$ defined respectively by

$$\mathbf{F}(t) = \begin{pmatrix} F_1(t) \\ F_2(t) \\ \vdots \\ F_n(t) \end{pmatrix} \quad \text{and} \quad \mathbf{x} = \begin{pmatrix} x_1 \\ x_2 \\ \vdots \\ x_n \end{pmatrix}. \qquad (11.5)$$

Then (1) by definition of the derivative of a vector, and (2) by multiplication of a matrix by a vector followed by addition of vectors, we have respectively

$$\frac{d\mathbf{x}}{dt} = \begin{pmatrix} \dfrac{dx_1}{dt} \\[2mm] \dfrac{dx_2}{dt} \\[1mm] \vdots \\[1mm] \dfrac{dx_n}{dt} \end{pmatrix}$$

and

$$\mathbf{A}(t)\mathbf{x} + \mathbf{F}(t) = \begin{pmatrix} a_{11}(t) & a_{12}(t) & \cdots & a_{1n}(t) \\ a_{21}(t) & a_{22}(t) & \cdots & a_{2n}(t) \\ \vdots & \vdots & & \vdots \\ a_{n1}(t) & a_{n2}(t) & \cdots & a_{nn}(t) \end{pmatrix} \begin{pmatrix} x_1 \\ x_2 \\ \vdots \\ x_n \end{pmatrix} + \begin{pmatrix} F_1(t) \\ F_2(t) \\ \vdots \\ F_n(t) \end{pmatrix}$$

$$= \begin{pmatrix} a_{11}(t)x_1 + a_{12}(t)x_2 + \cdots + a_{1n}(t)x_n + F_1(t) \\ a_{21}(t)x_1 + a_{22}(t)x_2 + \cdots + a_{2n}(t)x_n + F_2(t) \\ \vdots \\ a_{n1}(t)x_1 + a_{n2}(t)x_2 + \cdots + a_{nn}(t)x_n + F_n(t) \end{pmatrix}.$$

Comparing the components of $d\mathbf{x}/dt$ with the left members of (11.1) and the components of $\mathbf{A}(t)\mathbf{x} + \mathbf{F}(t)$ with the right members of (11.1), we see that system (11.1) can be expressed as the linear *vector* differential equation

$$\frac{d\mathbf{x}}{dt} = \mathbf{A}(t)\mathbf{x} + \mathbf{F}(t). \tag{11.6}$$

Conversely, if $\mathbf{A}(t)$ is given by (11.4) and $\mathbf{F}(t)$ and $\mathbf{x}$ are given by (11.5), then we see that the vector differential equation (11.6) can be expressed as the system (11.1). Thus the system (11.1) and the vector differential equation (11.6) both express the same relations and so are equivalent to one another. We refer to (11.6) as the *vector differential equation corresponding to the system* (11.1), and we shall sometimes call the system (11.1) the *scalar form of the vector differential equation* (11.6). Henceforth throughout this section, we shall usually write the system (11.1) as the corresponding vector differential equation (11.6).

▶ **Example 11.2**

The vector differential equation corresponding to the nonhomogeneous system (11.3) of Example 11.1 is

$$\frac{d\mathbf{x}}{dt} = \mathbf{A}(t)\mathbf{x} + \mathbf{F}(t),$$

where

$$\mathbf{A}(t) = \begin{pmatrix} 7 & -1 & 6 \\ -10 & 4 & -12 \\ -2 & 1 & -1 \end{pmatrix}, \quad \mathbf{x} = \begin{pmatrix} x_1 \\ x_2 \\ x_3 \end{pmatrix}, \quad \text{and} \quad \mathbf{F}(t) = \begin{pmatrix} -5t - 6 \\ -4t + 23 \\ 2 \end{pmatrix}.$$

Thus we can write this vector differential equation as

$$\frac{d\mathbf{x}}{dt} = \begin{pmatrix} 7 & -1 & 6 \\ -10 & 4 & -12 \\ -2 & 1 & -1 \end{pmatrix} \mathbf{x} + \begin{pmatrix} -5t - 6 \\ -4t + 23 \\ 2 \end{pmatrix},$$

where $\mathbf{x}$ is the vector with components x_1, x_2, x_3, as given above.

DEFINITION

By a solution *of the vector differential equation (11.6) we mean an $n \times 1$ column vector function*

$$\boldsymbol{\phi} = \begin{pmatrix} \phi_1 \\ \phi_2 \\ \vdots \\ \phi_n \end{pmatrix}, \tag{11.7}$$

whose components $\phi_1, \phi_2, \ldots, \phi_n$ each have a continuous derivative on the real interval $a \leq t \leq b$, which is such that

$$\frac{d\boldsymbol{\phi}(t)}{dt} = \mathbf{A}(t)\boldsymbol{\phi}(t) + \mathbf{F}(t) \tag{11.8}$$

for all t such that $a \leq t \leq b$. In other words, $\mathbf{x} = \boldsymbol{\phi}(t)$ satisfies the vector differential equation (11.6) identically on $a \leq t \leq b$. That is, the components $\phi_1, \phi_2, \ldots, \phi_n$ of $\boldsymbol{\phi}$ are such that

$$x_1 = \phi_1(t),$$
$$x_2 = \phi_2(t),$$
$$\vdots \tag{11.9}$$
$$x_n = \phi_n(t),$$

simultaneously satisfy all n equations of the scalar form (11.1) of the vector differential equation (11.6) for $a \leq t \leq b$. Hence we say that a solution of the system (11.1) is an ordered set of n real functions $\phi_1, \phi_2, \ldots, \phi_n$, each having continuous derivatives on $a \leq t \leq b$, such that

$$x_1 = \phi_1(t),$$
$$x_2 = \phi_2(t),$$
$$\vdots \tag{11.9}$$
$$x_n = \phi_n(t),$$

simultaneously satisfy all n equations of the system (11.1) for $a \leq t \leq b$.

▶ **Example 11.3**

The vector differential equation corresponding to the homogeneous linear system

$$\frac{dx_1}{dt} = 7x_1 - x_2 + 6x_3,$$

$$\frac{dx_2}{dt} = -10x_1 + 4x_2 - 12x_3, \qquad (11.2)$$

$$\frac{dx_3}{dt} = -2x_1 + x_2 - x_3,$$

is

$$\frac{d\mathbf{x}}{dt} = \begin{pmatrix} 7 & -1 & 6 \\ -10 & 4 & -12 \\ -2 & 1 & -1 \end{pmatrix} \mathbf{x}, \quad \text{where } \mathbf{x} = \begin{pmatrix} x_1 \\ x_2 \\ x_3 \end{pmatrix}. \qquad (11.10)$$

The column vector function $\boldsymbol{\phi}$ defined by

$$\boldsymbol{\phi}(t) = \begin{pmatrix} e^{3t} \\ -2e^{3t} \\ -e^{3t} \end{pmatrix}$$

is a solution of the vector differential equation (11.10) on every real interval $a \leq t \leq b$; for $\mathbf{x} = \boldsymbol{\phi}(t)$ satisfies (11.10) identically on $a \leq t \leq b$, that is,

$$\begin{pmatrix} 3e^{3t} \\ -6e^{3t} \\ -3e^{3t} \end{pmatrix} = \begin{pmatrix} 7 & -1 & 6 \\ -10 & 4 & -12 \\ -2 & 1 & -1 \end{pmatrix} \begin{pmatrix} e^{3t} \\ -2e^{3t} \\ -e^{3t} \end{pmatrix}.$$

Thus

$$x_1 = e^{3t},$$

$$x_2 = -2e^{3t}, \qquad (11.11)$$

$$x_3 = -e^{3t},$$

simultaneously satisfy all three equations of the system (11.2) for $a \leq t \leq b$, and so we call (11.11) a solution of the system.

The linear system (10.37) of Theorem 10.7 and the linear system (11.1) of this chapter are identical, except for obvious changes in notation. Thus we see that Theorem 10.7 is in fact the basic existence and uniqueness theorem dealing with the linear system (11.1). Rewriting this theorem in the notation of vectors and matrices, we thus have the following basic existence and uniqueness theorem dealing with the vector differential equation (11.6).

THEOREM 11.1

Consider the vector differential equation

$$\frac{d\mathbf{x}}{dt} = \mathbf{A}(t)\mathbf{x} + \mathbf{F}(t) \qquad (11.6)$$

corresponding to the linear system (11.1) of n equations in n unknown functions. Let the components $a_{ij}(t)$, $i = 1, 2, \ldots, n$, $j = 1, 2, \ldots, n$, of the matrix $\mathbf{A}(t)$ and the components $F_i(t)$, $i = 1, 2, \ldots, n$, of the vector $\mathbf{F}(t)$ all be continuous on the real interval $a \le t \le b$. Let t_0 be any point of the interval $a \le t \le b$, and let

$$\mathbf{c} = \begin{pmatrix} c_1 \\ c_2 \\ \vdots \\ c_n \end{pmatrix}$$

be an $n \times 1$ column vector of any n numbers $c_1, c_2, \ldots, c_n$.

Then there exists a unique solution

$$\boldsymbol{\phi} = \begin{pmatrix} \phi_1 \\ \phi_2 \\ \vdots \\ \phi_n \end{pmatrix}$$

of the vector differential equation (11.6) such that

$$\boldsymbol{\phi}(t_0) = \mathbf{c}; \tag{11.12}$$

that is,

$$\phi_1(t_0) = c_1,$$
$$\phi_2(t_0) = c_2,$$
$$\vdots$$
$$\phi_n(t_0) = c_n, \tag{11.13}$$

and this solution is defined on the entire interval $a \le t \le b$.

11.2 BASIC THEORY OF THE HOMOGENEOUS LINEAR SYSTEM

We now assume that all $F_i(t) = 0$, $i = 1, 2, \ldots, n$, for all t in the linear system (11.1) and consider the resulting *homogeneous* linear system

$$\frac{dx_1}{dt} = a_{11}(t)x_1 + a_{12}(t)x_2 + \cdots + a_{1n}(t)x_n,$$

$$\frac{dx_2}{dt} = a_{21}(t)x_1 + a_{22}(t)x_2 + \cdots + a_{2n}(t)x_n, \tag{11.14}$$

$$\vdots$$

$$\frac{dx_n}{dt} = a_{n1}(t)x_1 + a_{n2}(t)x_2 + \cdots + a_{nn}(t)x_n.$$

The corresponding homogeneous vector equation is the equation of the form (11.6) for which $\mathbf{F}(t) = 0$ for all t and hence is

$$\frac{d\mathbf{x}}{dt} = \mathbf{A}(t)\mathbf{x}. \tag{11.15}$$

Throughout Sections 11.2 and 11.3 we shall always make the following assumption whenever we write or refer to the homogeneous vector differential equation (11.15): We shall *always* assume that (11.15) is the vector differential equation corresponding to the homogeneous linear system (11.14) of n equations in n unknown functions and that *the components $a_{ij}(t)$, $i = 1, 2, \ldots, n$, $j = 1, 2, \ldots, n$, of the $n \times n$ matrix $\mathbf{A}(t)$ are all continuous on the real interval $a \le t \le b$.* Our first result concerning Equation (11.15) is an immediate consequence of Theorem 11.1.

COROLLARY TO THEOREM 11.1

Consider the homogeneous vector differential equation

$$\frac{d\mathbf{x}}{dt} = \mathbf{A}(t)\mathbf{x}. \tag{11.15}$$

Let t_0 be any point of $a \le t \le b$; and let

$$\boldsymbol{\phi} = \begin{pmatrix} \phi_1 \\ \phi_2 \\ \vdots \\ \phi_n \end{pmatrix}$$

be a solution of (11.15) such that $\boldsymbol{\phi}(t_0) = \mathbf{0}$, that is, such that

$$\phi_1(t_0) = \phi_2(t_0) = \cdots = \phi_n(t_0) = 0. \tag{11.16}$$

Then $\boldsymbol{\phi}(t) = \mathbf{0}$ for all t on $a \le t \le b$; that is,

$$\phi_1(t) = \phi_2(t) = \cdots = \phi_n(t) = 0$$

for all t on $a \le t \le b$.

Proof. Obviously $\boldsymbol{\phi}$ defined by $\boldsymbol{\phi}(t) = \mathbf{0}$ for *all* t on $a \le t \le b$ is *a* solution of the vector differential equation (11.15) that satisfies conditions (11.16). These conditions are of the form (11.13) where $c_1 = c_2 = \cdots = c_n = 0$; and by Theorem 11.1 there is a *unique* solution of the differential equation satisfying such a set of conditions. Thus $\boldsymbol{\phi}$ such that $\boldsymbol{\phi}(t) = \mathbf{0}$ for *all* t on $a \le t \le b$ is the *only* solution of (11.15) such that $\boldsymbol{\phi}(t_0) = \mathbf{0}$.

Q.E.D.

THEOREM 11.2

A linear combination of m solutions of the homogeneous vector differential equation

$$\frac{d\mathbf{x}}{dt} = \mathbf{A}(t)\mathbf{x} \tag{11.15}$$

is also a solution of (11.15). That is, if the vector functions $\boldsymbol{\phi}_1, \boldsymbol{\phi}_2, \ldots, \boldsymbol{\phi}_m$ are m solutions of (11.15) and $c_1, c_2, \ldots, c_m$ are m numbers, then the vector function

$$\boldsymbol{\phi} = \sum_{k=1}^{m} c_k \boldsymbol{\phi}_k$$

is also a solution of (11.15).

Proof. We have

$$\frac{d}{dt}\left[\sum_{k=1}^{m} c_k \boldsymbol{\phi}_k(t)\right] = \sum_{k=1}^{m}\left[\frac{d}{dt} c_k \boldsymbol{\phi}_k(t)\right] = \sum_{k=1}^{m} c_k\left[\frac{d\boldsymbol{\phi}_k(t)}{dt}\right].$$

Now since each $\boldsymbol{\phi}_k$ is a solution of (11.15),

$$\frac{d\boldsymbol{\phi}_k(t)}{dt} = \mathbf{A}(t)\boldsymbol{\phi}_k(t) \quad \text{for } k = 1, 2, \ldots, m.$$

Thus we have

$$\frac{d}{dt}\left[\sum_{k=1}^{m} c_k \boldsymbol{\phi}_k(t)\right] = \sum_{k=1}^{m} c_k \mathbf{A}(t)\boldsymbol{\phi}_k(t).$$

We now use Results A and B of Section 7.5A. First applying Result B to each term in the right member above, and then applying Result A $(m-1)$ times, we obtain

$$\sum_{k=1}^{m} c_k \mathbf{A}(t)\boldsymbol{\phi}_k(t) = \sum_{k=1}^{m} \mathbf{A}(t)[c_k\boldsymbol{\phi}_k(t)] = \mathbf{A}(t)\sum_{k=1}^{m} c_k\boldsymbol{\phi}_k(t).$$

Thus we have

$$\frac{d}{dt}\left[\sum_{k=1}^{m} c_k \boldsymbol{\phi}_k(t)\right] = \mathbf{A}(t)\left[\sum_{k=1}^{m} c_k\boldsymbol{\phi}_k(t)\right];$$

that is,

$$\frac{d\boldsymbol{\phi}(t)}{dt} = \mathbf{A}(t)\boldsymbol{\phi}(t),$$

for all t on $a \le t \le b$. Thus the linear combination

$$\boldsymbol{\phi} = \sum_{k=1}^{m} c_k\boldsymbol{\phi}_k$$

is a solution of (11.15).   *Q.E.D.*

Before proceeding, the student should return to Section 7.5C and review the concepts of linear dependence and linear independence of vector functions.

In each of the next four theorems we shall be concerned with n vector functions, and we shall use the following common notation for the n vector functions of each of these theorems. We let $\boldsymbol{\phi}_1, \boldsymbol{\phi}_2, \ldots, \boldsymbol{\phi}_n$ be the n vector functions defined respectively by

$$\boldsymbol{\phi}_1(t) = \begin{pmatrix} \phi_{11}(t) \\ \phi_{21}(t) \\ \vdots \\ \phi_{n1}(t) \end{pmatrix}, \boldsymbol{\phi}_2(t) = \begin{pmatrix} \phi_{12}(t) \\ \phi_{22}(t) \\ \vdots \\ \phi_{n2}(t) \end{pmatrix}, \ldots, \boldsymbol{\phi}_n(t) = \begin{pmatrix} \phi_{1n}(t) \\ \phi_{2n}(t) \\ \vdots \\ \phi_{nn}(t) \end{pmatrix}. \tag{11.17}$$

Carefully observe the notation scheme. For each vector, the first subscript of a component indicates the row of that component in the vector, whereas the second subscript indicates the vector of which the component is an element. For instance, ϕ_{35} would be the component occupying the third row of the vector $\boldsymbol{\phi}_5$.

DEFINITION

The $n \times n$ determinant

$$\begin{vmatrix} \phi_{11} & \phi_{12} & \cdots & \phi_{1n} \\ \phi_{21} & \phi_{22} & \cdots & \phi_{2n} \\ \vdots & \vdots & & \vdots \\ \phi_{n1} & \phi_{n2} & \cdots & \phi_{nn} \end{vmatrix} \qquad (11.18)$$

is called the Wronskian *of the n vector functions* $\phi_1, \phi_2, \ldots, \phi_n$ *defined by (11.17). We will denote it by* $W(\phi_1, \phi_2, \ldots, \phi_n)$ *and its value at t by* $W(\phi_1, \phi_2, \ldots, \phi_n)(t)$.

THEOREM 11.3

If the n vector functions $\phi_1, \phi_2, \ldots, \phi_n$ *defined by (11.17) are linearly dependent on $a \le t \le b$, then their Wronskian $W(\phi_1, \phi_2, \ldots, \phi_n)(t)$ equals zero for all t on $a \le t \le b$.*

Proof. We begin by employing the definition of linear dependence of vector functions on an interval: Since $\phi_1, \phi_2, \ldots, \phi_n$ are linearly dependent on the interval $a \le t \le b$, there exist n numbers $c_1, c_2, \ldots, c_n$, not all zero, such that

$$c_1 \phi_1(t) + c_2 \phi_2(t) + \cdots + c_n \phi_n(t) = \mathbf{0}$$

for all $t \in [a, b]$. Now using the definition (11.17) of $\phi_1, \phi_2, \ldots, \phi_n$, and writing the preceding vector relation in the form of the n equivalent relations involving corresponding components, we have

$$c_1 \phi_{11}(t) + c_2 \phi_{12}(t) + \cdots + c_n \phi_{1n}(t) = 0,$$
$$c_1 \phi_{21}(t) + c_2 \phi_{22}(t) + \cdots + c_n \phi_{2n}(t) = 0,$$
$$\vdots$$
$$c_1 \phi_{n1}(t) + c_2 \phi_{n2}(t) + \cdots + c_n \phi_{nn}(t) = 0,$$

for all $t \in [a, b]$. Thus, in particular, these must hold at an *arbitrary* point $t_0 \in [a, b]$. Thus, letting $t = t_0$ in the preceding n relations, we obtain the homogeneous linear algebraic system

$$\phi_{11}(t_0)c_1 + \phi_{12}(t_0)c_2 + \cdots + \phi_{1n}(t_0)c_n = 0,$$
$$\phi_{21}(t_0)c_1 + \phi_{22}(t_0)c_2 + \cdots + \phi_{2n}(t_0)c_n = 0,$$
$$\vdots$$
$$\phi_{n1}(t_0)c_1 + \phi_{n2}(t_0)c_2 + \cdots + \phi_{nn}(t_0)c_n = 0,$$

in the n unknowns $c_1, c_2, \ldots, c_n$. Since $c_1, c_2, \ldots, c_n$ are not all all zero, the determinant of coefficients of the preceding system must be zero, by Theorem A of Section 7.5C. That is, we must have

$$\begin{vmatrix} \phi_{11}(t_0) & \phi_{12}(t_0) & \cdots & \phi_{1n}(t_0) \\ \phi_{21}(t_0) & \phi_{22}(t_0) & \cdots & \phi_{2n}(t_0) \\ \vdots & \vdots & & \vdots \\ \phi_{n1}(t_0) & \phi_{n2}(t_0) & \cdots & \phi_{nn}(t_0) \end{vmatrix} = 0.$$

But the left member of this is the Wronskian $W(\phi_1, \phi_2, \ldots, \phi_n)(t_0)$. Thus we have

$$W(\phi_1, \phi_2, \ldots, \phi_n)(t_0) = 0.$$

Since t_0 is an arbitrary point of $[a, b]$, we must have

$$W(\phi_1, \phi_2, \ldots, \phi_n)(t) = 0$$

for *all* t on $a \le t \le b$. *Q.E.D.*

Examine the proof of Theorem 11.3 and observe that it makes absolutely no use of the properties of solutions of differential equations. Thus it holds for arbitrary vector functions, whether they are solutions of a vector differential equation of the form (11.15) or not.

▶ **Example 11.4**

In Example 7.31 of Section 7.5C we saw that the three vector functions ϕ_1, ϕ_2, and ϕ_3 defined respectively by

$$\phi_1(t) = \begin{pmatrix} e^{2t} \\ 2e^{2t} \\ 5e^{2t} \end{pmatrix}, \qquad \phi_2(t) = \begin{pmatrix} e^{2t} \\ 4e^{2t} \\ 11e^{2t} \end{pmatrix}, \quad \text{and} \quad \phi_3(t) = \begin{pmatrix} e^{2t} \\ e^{2t} \\ 2e^{2t} \end{pmatrix}$$

are linearly dependent on any interval $a \le t \le b$. Therefore, by Theorem 11.3 their Wronskian must equal zero for all t on $a \le t \le b$. Indeed, we find

$$W(\phi_1, \phi_2, \phi_3)(t) = \begin{vmatrix} e^{2t} & e^{2t} & e^{2t} \\ 2e^{2t} & 4e^{2t} & e^{2t} \\ 5e^{2t} & 11e^{2t} & 2e^{2t} \end{vmatrix} = 0 \quad \text{for all } t.$$

THEOREM 11.4

Let the vector functions $\phi, \phi_2, \ldots, \phi_n$ defined by (11.17) be n solutions of the homogeneous linear vector differential equation

$$\frac{d\mathbf{x}}{dt} = \mathbf{A}(t)\mathbf{x}. \tag{11.15}$$

If the Wronskian $W(\phi_1, \phi_2, \ldots, \phi_n)(t_0) = 0$ at some $t_0 \in [a, b]$, then $\phi_1, \phi_2, \ldots, \phi_n$ are linearly dependent on $a \le t \le b$.

Proof. Consider the linear algebraic system

$$c_1 \phi_{11}(t_0) + c_2 \phi_{12}(t_0) + \cdots + c_n \phi_{1n}(t_0) = 0,$$
$$c_1 \phi_{21}(t_0) + c_2 \phi_{22}(t_0) + \cdots + c_n \phi_{2n}(t_0) = 0,$$
$$\vdots \tag{11.19}$$
$$c_1 \phi_{n1}(t_0) + c_2 \phi_{n2}(t_0) + \cdots + c_n \phi_{nn}(t_0) = 0,$$

in the n unknowns $c_1, c_2, \ldots, c_n$. Since the determinant of coefficients is

$$W(\phi_1, \phi_2, \ldots, \phi_n)(t_0) \quad \text{and} \quad W(\phi_1, \phi_2, \ldots, \phi_n)(t_0) = 0$$

by hypothesis, this system has a nontrivial solution by Theorem A of Section 7.5C. That is, there exist numbers $c_1, c_2, \ldots, c_n$, not all zero, that satisfy all n equations of system (11.19). These n equations are the n corresponding component relations equivalent to the one vector relation

$$c_1 \boldsymbol{\phi}_1(t_0) + c_2 \boldsymbol{\phi}_2(t_0) + \cdots + c_n \boldsymbol{\phi}_n(t_0) = \mathbf{0}. \tag{11.20}$$

Thus there exist numbers $c_1, c_2, \ldots, c_n$, not all zero, such that (11.20) holds.

Now consider the vector function $\boldsymbol{\phi}$ defined by

$$\boldsymbol{\phi}(t) = c_1 \boldsymbol{\phi}_1(t) + c_2 \boldsymbol{\phi}_2(t) + \cdots + c_n \boldsymbol{\phi}_n(t) \tag{11.21}$$

for all $t \in [a, b]$. Since $\boldsymbol{\phi}_1, \boldsymbol{\phi}_2, \ldots, \boldsymbol{\phi}_n$ are solutions of the differential equation (11.15) by Theorem 11.2, the linear combination $\boldsymbol{\phi}$ defined by (11.21) is also a solution of (11.15). Now from (11.20), we see that this solution $\boldsymbol{\phi}$ is such that $\boldsymbol{\phi}(t_0) = \mathbf{0}$. Thus by the corollary to Theorem 11.1, we must have $\boldsymbol{\phi}(t) = \mathbf{0}$ for *all* $t \in [a, b]$. That is, using the definition (11.21).

$$c_1 \boldsymbol{\phi}_1(t) + c_2 \boldsymbol{\phi}_2(t) + \cdots + c_n \boldsymbol{\phi}_n(t) = \mathbf{0}$$

for *all* $t \in [a, b]$, where $c_1, c_2, \ldots, c_n$ are not all zero. Thus, by definition, $\boldsymbol{\phi}_1, \boldsymbol{\phi}_2, \ldots, \boldsymbol{\phi}_n$ are linearly dependent on $a \leq t \leq b$.

Q.E.D.

▶   **Example 11.5**

Consider the vector functions $\boldsymbol{\phi}_1, \boldsymbol{\phi}_2$, and $\boldsymbol{\phi}_3$ defined respectively by

$$\boldsymbol{\phi}_1(t) = \begin{pmatrix} e^{3t} \\ -2e^{3t} \\ -e^{3t} \end{pmatrix}, \qquad \boldsymbol{\phi}_2(t) = \begin{pmatrix} 2e^{3t} \\ -4e^{3t} \\ -2e^{3t} \end{pmatrix}, \quad \text{and} \quad \boldsymbol{\phi}_3(t) = \begin{pmatrix} -3e^{3t} \\ 6e^{3t} \\ 3e^{3t} \end{pmatrix}.$$

It is easy to verify that $\boldsymbol{\phi}_1, \boldsymbol{\phi}_2$, and $\boldsymbol{\phi}_3$ are all solutions of the homogeneous linear vector differential equation

$$\frac{d\mathbf{x}}{dt} = \begin{pmatrix} 7 & -1 & 6 \\ -10 & 4 & -12 \\ -2 & 1 & -1 \end{pmatrix} \mathbf{x}, \qquad \text{where} \quad \mathbf{x} = \begin{pmatrix} x_1 \\ x_2 \\ x_3 \end{pmatrix} \tag{11.10}$$

on every real interval $a \leq t \leq b$ (see Example 11.3). Thus, in particular, $\boldsymbol{\phi}_1, \boldsymbol{\phi}_2$, and $\boldsymbol{\phi}_3$ are solutions of (11.10) on every interval $[a, b]$ containing $t_0 = 0$. It is easy to see that

$$W(\boldsymbol{\phi}_1, \boldsymbol{\phi}_2, \boldsymbol{\phi}_3)(0) = \begin{vmatrix} 1 & 2 & -3 \\ -2 & -4 & 6 \\ -1 & -2 & 3 \end{vmatrix} = 0.$$

Thus by Theorem 11.4, $\boldsymbol{\phi}_1, \boldsymbol{\phi}_2$, and $\boldsymbol{\phi}_3$ are linearly dependent on every $[a, b]$ containing 0. Indeed, note that

$$\boldsymbol{\phi}_1(t) + \boldsymbol{\phi}_2(t) + \boldsymbol{\phi}_3(t) = \mathbf{0}$$

for all t on every interval $[a, b]$, and recall the definition of linear dependence.

Note. Theorem 11.4 is *not* true for vector functions $\boldsymbol{\phi}_1, \boldsymbol{\phi}_2, \ldots, \boldsymbol{\phi}_n$ that are *not* solutions of a homogeneous linear vector differential equation (11.15). For example, consider the vector functions $\boldsymbol{\phi}_1$ and $\boldsymbol{\phi}_2$ defined respectively by

$$\boldsymbol{\phi}_1(t) = \begin{pmatrix} t \\ 0 \end{pmatrix} \quad \text{and} \quad \boldsymbol{\phi}_2(t) = \begin{pmatrix} t^2 \\ 0 \end{pmatrix}.$$

It can be shown that $\boldsymbol{\phi}_1$ and $\boldsymbol{\phi}_2$ are *not solutions* of any differential equation of the form (11.15) for which $n = 2$ [see I. G. Petrovski, *Ordinary Differential Equations* (Prentice-Hall, Englewood Cliffs, N.J., 1966), Theorem, pages 110–111, for the method of doing so]. Clearly

$$W(\boldsymbol{\phi}_1, \boldsymbol{\phi}_2)(t_0) = \begin{vmatrix} t_0 & t_0^2 \\ 0 & 0 \end{vmatrix} = 0$$

for *all* t_0 in *every* interval $a \le t \le b$. However, $\boldsymbol{\phi}_1$ and $\boldsymbol{\phi}_2$ are *not* linearly dependent. To show this, proceed as in Example 7.32; see Exercise 6 at the end of Section 7.5C.

THEOREM 11.5

Let the vector functions $\boldsymbol{\phi}_1, \boldsymbol{\phi}_2, \ldots, \boldsymbol{\phi}_n$ *defined by (11.17) be n solutions of the homogeneous linear vector differential equation*

$$\frac{d\mathbf{x}}{dt} = \mathbf{A}(t)\mathbf{x} \tag{11.15}$$

on the real interval $[a, b]$. *Then*

$$\begin{aligned} \text{either} \quad & W(\boldsymbol{\phi}_1, \boldsymbol{\phi}_2, \ldots, \boldsymbol{\phi}_n)(t) = 0 && \text{for all } t \in [a, b], \\ \text{or} \quad & W(\boldsymbol{\phi}_1, \boldsymbol{\phi}_2, \ldots, \boldsymbol{\phi}_n)(t) = 0 && \text{for no } t \in [a, b]. \end{aligned}$$

Proof.
$$\begin{aligned} \text{Either} \quad & W(\boldsymbol{\phi}_1, \boldsymbol{\phi}_2, \ldots, \boldsymbol{\phi}_n)(t) = 0 && \text{for some } t \in [a, b] \\ \text{or} \quad & W(\boldsymbol{\phi}_1, \boldsymbol{\phi}_2, \ldots, \boldsymbol{\phi}_n)(t) = 0 && \text{for no } t \in [a, b]. \end{aligned}$$

If $W(\boldsymbol{\phi}_1, \boldsymbol{\phi}_2, \ldots, \boldsymbol{\phi}_n)(t) = 0$ for *some* $t \in [a, b]$, then by Theorem 11.4, the solutions $\boldsymbol{\phi}_1, \boldsymbol{\phi}_2, \ldots, \boldsymbol{\phi}_n$ are linearly dependent on $[a, b]$; and then by Theorem 11.3, $W(\boldsymbol{\phi}_1, \boldsymbol{\phi}_2, \ldots, \boldsymbol{\phi}_n)(t) = 0$ for *all* $t \in [a, b]$. Thus the Wronksian of $\boldsymbol{\phi}_1, \boldsymbol{\phi}_2, \ldots, \boldsymbol{\phi}_n$ either equals zero for *all* $t \in [a, b]$ or equals zero for *no* $t \in [a, b]$.   Q.E.D.

THEOREM 11.6

Let the vector functions $\boldsymbol{\phi}_1, \boldsymbol{\phi}_2, \ldots, \boldsymbol{\phi}_n$ *defined by (11.17) be n solutions of the homogeneous linear vector differential equation*

$$\frac{d\mathbf{x}}{dt} = \mathbf{A}(t)\mathbf{x} \tag{11.15}$$

on the real interval $[a, b]$. *These n solutions* $\boldsymbol{\phi}_1, \boldsymbol{\phi}_2, \ldots, \boldsymbol{\phi}_n$ *of (11.15) are linearly independent on* $[a, b]$ *if and only if*

$$W(\boldsymbol{\phi}_1, \boldsymbol{\phi}_2, \ldots, \boldsymbol{\phi}_n)(t) \ne 0$$

for all $t \in [a, b]$.

Proof. By Theorems 11.3 and 11.4, the solutions $\phi_1, \phi_2, \ldots, \phi_n$ are linearly *dependent* on $[a, b]$ if and only if $W(\phi_1, \phi_2, \ldots, \phi_n)(t) = 0$ for *all* $t \in [a, b]$. Hence, $\phi_1, \phi_2, \ldots, \phi_n$ are linearly *independent* on $[a, b]$ if and only if $W(\phi_1, \phi_2, \ldots, \phi_n)(t_0) \neq 0$ for *some* $t_0 \in [a, b]$. Then by Theorem 11.5, $W(\phi_1, \phi_2, \ldots, \phi_n)(t_0) \neq 0$ for *some* $t_0 \in [a, b]$ if and only if $W(\phi_1, \phi_2, \ldots, \phi_n)(t) \neq 0$ for *all* $t \in [a, b]$. $\qquad$ *Q.E.D.*

▶ **Example 11.6**

Consider the vector functions ϕ_1, ϕ_2, and ϕ_3 defined respectively by

$$\phi_1(t) = \begin{pmatrix} e^{2t} \\ -e^{2t} \\ -e^{2t} \end{pmatrix}, \qquad \phi_2(t) = \begin{pmatrix} e^{3t} \\ -2e^{3t} \\ -e^{3t} \end{pmatrix}, \quad \text{and} \quad \phi_3(t) = \begin{pmatrix} 3e^{5t} \\ -6e^{5t} \\ -2e^{5t} \end{pmatrix}. \quad (11.22)$$

It is easy to verify that ϕ_1, ϕ_2, and ϕ_3 are all solutions of the homogeneous linear vector differential equation

$$\frac{d\mathbf{x}}{dt} = \begin{pmatrix} 7 & -1 & 6 \\ -10 & 4 & -12 \\ -2 & 1 & -1 \end{pmatrix} \mathbf{x}, \qquad \text{where} \quad \mathbf{x} = \begin{pmatrix} x_1 \\ x_2 \\ x_3 \end{pmatrix}, \quad (11.10)$$

on every real interval $a \leq t \leq b$. We evaluate

$$W(\phi_1, \phi_2, \phi_3)(t) = \begin{vmatrix} e^{2t} & e^{3t} & 3e^{5t} \\ -e^{2t} & -2e^{3t} & -6e^{5t} \\ -e^{2t} & -e^{3t} & -2e^{5t} \end{vmatrix} = -e^{10t} \neq 0$$

for all real t. Thus by Theorem 11.6, the solutions ϕ_1, ϕ_2, and ϕ_3 of (11.10) defined by (11.22) are linearly independent on every real interval $[a, b]$.

DEFINITIONS

Consider the homogeneous linear vector differential equation

$$\frac{d\mathbf{x}}{dt} = \mathbf{A}(t)\mathbf{x}, \quad (11.15)$$

where $\mathbf{x}$ *is an* $n \times 1$ *column vector.*

1. A set of n linearly independent solutions of (11.15) is called a fundamental set of solutions *of (11.15).*

2. A matrix whose individual columns consist of a fundamental set of solutions of (11.15) is called a fundamental matrix *of (11.15). That is, if the vector functions* $\phi_1, \phi_2, \ldots, \phi_n$ *defined by (11.17) make up a fundamental set of solutions of (11.15), then the* $n \times n$ *square matrix*

$$\begin{pmatrix} \phi_{11}(t) & \phi_{12}(t) & \cdots & \phi_{1n}(t) \\ \phi_{21}(t) & \phi_{22}(t) & \cdots & \phi_{2n}(t) \\ \vdots & \vdots & & \vdots \\ \phi_{n1}(t) & \phi_{n2}(t) & \cdots & \phi_{nn}(t) \end{pmatrix}$$

is a fundamental matrix of (11.15).

▶ **Example 11.7**

In Example 11.6 we saw that the three vector functions ϕ_1, ϕ_2, and ϕ_3 defined respectively by

$$\phi_1(t) = \begin{pmatrix} e^{2t} \\ -e^{2t} \\ -e^{2t} \end{pmatrix}, \quad \phi_2(t) = \begin{pmatrix} e^{3t} \\ -2e^{3t} \\ -e^{3t} \end{pmatrix}, \quad \text{and} \quad \phi_3(t) = \begin{pmatrix} 3e^{5t} \\ -6e^{5t} \\ -2e^{5t} \end{pmatrix} \tag{11.22}$$

are linearly independent solutions of the differential equation

$$\frac{d\mathbf{x}}{dt} = \begin{pmatrix} 7 & -1 & 6 \\ -10 & 4 & -12 \\ -2 & 1 & -1 \end{pmatrix} \mathbf{x}, \quad \text{where} \quad \mathbf{x} = \begin{pmatrix} x_1 \\ x_2 \\ x_3 \end{pmatrix}, \tag{11.10}$$

on every real interval $[a, b]$. Thus these three solutions ϕ_1, ϕ_2, and ϕ_3 form a fundamental set of differential equation (11.10), and a fundamental matrix of the differential equation is

$$\begin{pmatrix} e^{2t} & e^{3t} & 3e^{5t} \\ -e^{2t} & -2e^{3t} & -6e^{5t} \\ -e^{2t} & -e^{3t} & -2e^{5t} \end{pmatrix}.$$

   We know that the differential equation (11.10) of Examples 11.6 and 11.7 has the fundamental set of solutions ϕ_1, ϕ_2, ϕ_3 defined by (11.22). We now show that every vector differential equation (11.15) has fundamental sets of solutions.

THEOREM 11.7

There exist fundamental sets of solutions of the homogeneous linear vector differential equation

$$\frac{d\mathbf{x}}{dt} = \mathbf{A}(t)\mathbf{x}. \tag{11.15}$$

   Proof. We begin by defining a special set of constant vectors $\mathbf{u}_1, \mathbf{u}_2, \ldots, \mathbf{u}_n$. We define

$$\mathbf{u}_1 = \begin{pmatrix} 1 \\ 0 \\ 0 \\ \vdots \\ 0 \\ 0 \end{pmatrix}, \mathbf{u}_2 = \begin{pmatrix} 0 \\ 1 \\ 0 \\ \vdots \\ 0 \\ 0 \end{pmatrix}, \ldots, \mathbf{u}_n = \begin{pmatrix} 0 \\ 0 \\ 0 \\ \vdots \\ 0 \\ 1 \end{pmatrix}.$$

That is, in general, for each $i = 1, 2, \ldots, n$, $\mathbf{u}_i$ has ith component one and all other components zero. Now let $\phi_1, \phi_2, \ldots, \phi_n$ be the n solutions of (11.15) that satisfy the conditions

$$\phi_i(t_0) = \mathbf{u}_i \qquad (i = 1, 2, \ldots, n),$$

that is

$$\phi_1(t_0) = \mathbf{u}_1, \phi_2(t_0) = \mathbf{u}_2, \ldots, \phi_n(t_0) = \mathbf{u}_n,$$

where t_0 is an arbitrary (but fixed) point of $[a, b]$. Note that these solutions exist and are unique by Theorem 11.1. We now find

$$W(\phi_1, \phi_2, \ldots, \phi_n)(t_0) = W(\mathbf{u}_1, \mathbf{u}_2, \ldots, \mathbf{u}_n) = \begin{vmatrix} 1 & 0 & \cdots & 0 \\ 0 & 1 & \cdots & 0 \\ \vdots & \vdots & & \vdots \\ 0 & 0 & \cdots & 1 \end{vmatrix} = 1 \neq 0.$$

Then by Theorem 11.5, $W(\phi_1, \phi_2, \ldots, \phi_n)(t) \neq 0$ for all $t \in [a, b]$; and so by Theorem 11.6 solutions $\phi_1, \phi_2, \ldots, \phi_n$ are linearly independent on $[a, b]$. Thus, by definition, $\phi_1, \phi_2, \ldots, \phi_n$ form a fundamental set of differential equation (11.15).

Q.E.D

THEOREM 11.8

Let $\phi_1, \phi_2, \ldots, \phi_n$ defined by (11.17) be a fundamental set of solutions of the homogeneous linear vector differential equation

$$\frac{d\mathbf{x}}{dt} = \mathbf{A}(t)\mathbf{x} \tag{11.15}$$

and let ϕ be an arbitrary solution of (11.15) on the real interval $[a, b]$. Then ϕ can be represented as a suitable linear combination of $\phi_1, \phi_2, \ldots, \phi_n$; that is, there exist numbers $c_1, c_2, \ldots, c_n$ such that

$$\phi = c_1\phi_1 + c_2\phi_2 + \cdots + c_n\phi_n$$

on $[a, b]$.

Proof. Suppose $\phi(t_0) = \mathbf{u}_0$, where $t_0 \in [a, b]$ and

$$\mathbf{u}_0 = \begin{pmatrix} u_{10} \\ u_{20} \\ \vdots \\ u_{n0} \end{pmatrix}$$

is a constant vector. Consider the linear algebraic system

$$c_1\phi_{11}(t_0) + c_2\phi_{12}(t_0) + \cdots + c_n\phi_{1n}(t_0) = u_{10},$$
$$c_1\phi_{21}(t_0) + c_2\phi_{22}(t_0) + \cdots + c_n\phi_{2n}(t_0) = u_{20},$$
$$\vdots \tag{11.23}$$
$$c_1\phi_{n1}(t_0) + c_2\phi_{n2}(t_0) + \cdots + c_n\phi_{nn}(t_0) = u_{n0},$$

of n equations in the n unknowns $c_1, c_2, \ldots, c_n$. Since $\phi_1, \phi_2, \ldots, \phi_n$ is a fundamental set of solutions of (11.15) on $[a, b]$, we know that $\phi_1, \phi_2, \ldots, \phi_n$ are linearly independent solutions on $[a, b]$ and hence, by Theorem 11.6, $W(\phi_1, \phi_2, \ldots, \phi_n)(t_0) \neq 0$. Now observe that $W(\phi_1, \phi_2, \ldots, \phi_n)(t_0)$ is the determinant of coefficients of system (11.23) and so this determinant of coefficients is unequal to zero. Thus by Theorem B of Section 7.5C, the system (11.23) has a unique solution for $c_1, c_2, \ldots, c_n$. That is, there exists a unique set of numbers $c_1, c_2, \ldots, c_n$ such that

$$c_1 \phi_1(t_0) + c_2 \phi_2(t_0) + \cdots + c_n \phi_n(t_0) = \mathbf{u}_0,$$

and hence such that

$$\phi(t_0) = \mathbf{u}_0 = \sum_{k=1}^{n} c_k \phi_k(t_0). \tag{11.24}$$

Now consider the vector function ψ defined by

$$\psi(t) = \sum_{k=1}^{n} c_k \phi_k(t).$$

By Theorem 11.2, the vector function ψ is also a solution of the vector differential equation (11.15). Now note that

$$\psi(t_0) = \sum_{k=1}^{n} c_k \phi_k(t_0).$$

Hence by (11.24), we obtain $\psi(t_0) = \phi(t_0)$. Thus by Theorem 11.1 we must have $\psi(t) = \phi(t)$ for *all* $t \in [a, b]$. That is

$$\phi(t) = \sum_{k=1}^{n} c_k \phi_k(t)$$

for all $t \in [a, b]$. Thus ϕ is expressed as the linear combination

$$\phi = c_1 \phi_1 + c_2 \phi_2 + \cdots + c_n \phi_n$$

of $\phi_1, \phi_2, \ldots, \phi_n$ where $c_1, c_2, \ldots, c_n$ is the *unique* solution of system (11.23).   Q.E.D.

As a result of Theorem 11.8, we are led to make the following definition.

DEFINITION

Consider the homogeneous linear vector differential equation

$$\frac{d\mathbf{x}}{dt} = \mathbf{A}(t)\mathbf{x}, \tag{11.15}$$

where $\mathbf{x}$ is an $n \times 1$ column vector. By a general solution *of (11.15), we mean a solution of the form*

$$c_1 \phi_1 + c_2 \phi_2 + \cdots + c_n \phi_n,$$

where $c_1, c_2, \ldots, c_n$ are n arbitrary numbers and $\phi_1, \phi_2, \ldots, \phi_n$ is a fundamental set of solutions of (11.15).

▶ **Example 11.8**

Consider the differential equation

$$\frac{d\mathbf{x}}{dt} = \begin{pmatrix} 7 & -1 & 6 \\ -10 & 4 & -12 \\ -2 & 1 & -1 \end{pmatrix} \mathbf{x}, \qquad \text{where} \quad \mathbf{x} = \begin{pmatrix} x_1 \\ x_2 \\ x_3 \end{pmatrix}. \qquad (11.10)$$

In Example 11.7 we saw that the three vector functions $\boldsymbol{\phi}_1$, $\boldsymbol{\phi}_2$, and $\boldsymbol{\phi}_3$ defined respectively by

$$\boldsymbol{\phi}_1(t) = \begin{pmatrix} e^{2t} \\ -e^{2t} \\ -e^{2t} \end{pmatrix}, \qquad \boldsymbol{\phi}_2(t) = \begin{pmatrix} e^{3t} \\ -2e^{3t} \\ -e^{3t} \end{pmatrix}, \quad \text{and} \quad \boldsymbol{\phi}_3(t) = \begin{pmatrix} 3e^{5t} \\ -6e^{5t} \\ -2e^{5t} \end{pmatrix} \qquad (11.22)$$

form a fundamental set of differential equation (11.10). Thus by Theorem 11.8 if $\boldsymbol{\phi}$ is an arbitrary solution of (11.10), then $\boldsymbol{\phi}$ can be represented as a suitable linear combination of these three linearly independent solutions $\boldsymbol{\phi}_1, \boldsymbol{\phi}_2$, and $\boldsymbol{\phi}_3$ of (11.10). Further, if c_1, c_2, and c_3 are arbitrary numbers, we see from the definition that $c_1\boldsymbol{\phi}_1 + c_2\boldsymbol{\phi}_2 + c_3\boldsymbol{\phi}_3$ is a general solution of (11.10). That is, a general solution of (11.10) is defined by

$$c_1 \begin{pmatrix} e^{2t} \\ -e^{2t} \\ -e^{2t} \end{pmatrix} + c_2 \begin{pmatrix} e^{3t} \\ -2e^{3t} \\ -e^{3t} \end{pmatrix} + c_3 \begin{pmatrix} 3e^{5t} \\ -6e^{5t} \\ -2e^{5t} \end{pmatrix}$$

and can be written as

$$x_1 = c_1 e^{2t} + c_2 e^{3t} + 3c_3 e^{5t},$$
$$x_2 = -c_1 e^{2t} - 2c_2 e^{3t} - 6c_3 e^{5t},$$
$$x_3 = -c_1 e^{2t} - c_2 e^{3t} - 2c_3 e^{5t},$$

where c_1, c_2, and c_3 are arbitrary numbers.

Exercises

In each of Exercises 1–6, determine whether or not the matrix in Column I is a fundamental matrix of the corresponding linear system in Column II.

I

II

1.
$$\begin{pmatrix} e^t & e^{2t} & e^{-3t} \\ e^t & 2e^{2t} & 7e^{-3t} \\ e^t & e^{2t} & 11e^{-3t} \end{pmatrix},$$
$$\frac{d\mathbf{x}}{dt} = \begin{pmatrix} 1 & 1 & -1 \\ 2 & 3 & -4 \\ 4 & 1 & -4 \end{pmatrix} \mathbf{x}.$$

2.
$$\begin{pmatrix} e^{2t} & e^{3t} & e^{-2t} \\ 0 & -e^{3t} & -e^{-2t} \\ -e^{2t} & -e^{3t} & 4e^{-2t} \end{pmatrix},$$
$$\frac{d\mathbf{x}}{dt} = \begin{pmatrix} 1 & -1 & -1 \\ 1 & 3 & 1 \\ -3 & 1 & -1 \end{pmatrix} \mathbf{x}.$$

3.
$$\begin{pmatrix} e^{4t} & 0 & 2e^{4t} \\ 2e^{4t} & 3e^t & 4e^{4t} \\ e^{4t} & e^t & 2e^{4t} \end{pmatrix},$$
$$\frac{d\mathbf{x}}{dt} = \begin{pmatrix} 1 & -3 & 9 \\ 0 & -5 & 18 \\ 0 & -3 & 10 \end{pmatrix} \mathbf{x}.$$

4. $\begin{pmatrix} e^t & 0 & e^{4t} \\ 0 & 3e^t & 2e^{4t} \\ 0 & e^t & e^{4t} \end{pmatrix}$,
$\qquad \dfrac{d\mathbf{x}}{dt} = \begin{pmatrix} 1 & -3 & 9 \\ 0 & -5 & 18 \\ 0 & -3 & 10 \end{pmatrix} \mathbf{x}.$

5. $\begin{pmatrix} e^t & e^{2t} & e^{2t} \\ e^t & -e^{2t} & 0 \\ 3e^t & 0 & e^{2t} \end{pmatrix}$,
$\qquad \dfrac{d\mathbf{x}}{dt} = \begin{pmatrix} 3 & 1 & -1 \\ 1 & 3 & -1 \\ 3 & 3 & -1 \end{pmatrix} \mathbf{x}.$

6. $\begin{pmatrix} e^t & te^t & e^{2t} \\ e^t & (t+1)e^t & 2e^{2t} \\ e^t & (t+2)e^t & 4e^{2t} \end{pmatrix}$,
$\qquad \dfrac{d\mathbf{x}}{dt} = \begin{pmatrix} 0 & 1 & 0 \\ 0 & 0 & 1 \\ 2 & -5 & 4 \end{pmatrix} \mathbf{x}.$

11.3 FURTHER THEORY OF THE HOMOGENEOUS LINEAR SYSTEM

THEOREM 11.9

Let

$$\Phi = \begin{pmatrix} \phi_{11} & \phi_{12} & \cdots & \phi_{1n} \\ \phi_{21} & \phi_{22} & \cdots & \phi_{2n} \\ \vdots & \vdots & & \vdots \\ \phi_{n1} & \phi_{n2} & \cdots & \phi_{nn} \end{pmatrix}$$

be a fundamental matrix of the homogeneous linear vector differential equation.

$$\frac{d\mathbf{x}}{dt} = \mathbf{A}(t)\mathbf{x}, \tag{11.15}$$

and let $\boldsymbol{\phi}$ be an arbitrary solution of (11.15) on the real interval $[a, b]$. Then there exists a suitable constant vector

$$\mathbf{c} = \begin{pmatrix} c_1 \\ c_2 \\ \vdots \\ c_n \end{pmatrix}$$

such that

$$\boldsymbol{\phi} = \Phi \mathbf{c} \tag{11.25}$$

on $[a, b]$.

Proof. Since Φ is a fundamental matrix of differential equation (11.15), its individual columns

$$\boldsymbol{\phi}_k = \begin{pmatrix} \phi_{1k} \\ \phi_{2k} \\ \vdots \\ \phi_{nk} \end{pmatrix},$$

$(k = 1, 2, \ldots, n)$, constitute a fundamental set of solutions of (11.15). Thus by

Theorem 11.8 there exist numbers $c_1, c_2, \ldots, c_n$ such that

$$\boldsymbol{\phi} = c_1 \boldsymbol{\phi}_1 + c_2 \boldsymbol{\phi}_2 + \cdots + c_n \boldsymbol{\phi}_n$$

on $[a, b]$. In component form, this is

$$\phi_1 = c_1 \phi_{11} + c_2 \phi_{12} + \cdots + c_n \phi_{1n}$$
$$\phi_2 = c_1 \phi_{21} + c_2 \phi_{22} + \cdots + c_n \phi_{2n}$$
$$\vdots$$
$$\phi_n = c_1 \phi_{n1} + c_2 \phi_{n2} + \cdots + c_n \phi_{nn}.$$

Now observe that

$$\begin{pmatrix} c_1 \phi_{11} + c_2 \phi_{12} + \cdots + c_n \phi_{1n} \\ c_1 \phi_{21} + c_2 \phi_{22} + \cdots + c_n \phi_{2n} \\ \vdots \\ c_1 \phi_{n1} + c_2 \phi_{n2} + \cdots + c_n \phi_{nn} \end{pmatrix} = \begin{pmatrix} \phi_{11} & \phi_{12} & \cdots & \phi_{1n} \\ \phi_{21} & \phi_{22} & \cdots & \phi_{2n} \\ \vdots & \vdots & & \vdots \\ \phi_{n1} & \phi_{n2} & \cdots & \phi_{nn} \end{pmatrix} \begin{pmatrix} c_1 \\ c_2 \\ \vdots \\ c_n \end{pmatrix}.$$

In vector form this is precisely

$$\boldsymbol{\phi} = \boldsymbol{\Phi} \mathbf{c}, \tag{11.25}$$

where

$$\mathbf{c} = \begin{pmatrix} c_1 \\ c_2 \\ \vdots \\ c_n \end{pmatrix}.$$

$$Q.E.D.$$

Note. In terms of the concept of general solution, we see that the result of Theorem 11.9 can be expressed as follows: The general solution of the homogeneous linear vector differential equation (11.15) can be written in the form $\boldsymbol{\Phi} \mathbf{c}$, where $\boldsymbol{\Phi}$ is a fundamental matrix of (11.15) and $\mathbf{c}$ is an arbitrary n-rowed constant vector.

▶ **Example 11.9**

As noted in Examples 11.7 and 11.8 we have seen that the three vector functions defined by (11.22) form a fundamental set of differential equation (11.10). It follows immediately that

$$\boldsymbol{\Phi}(t) = \begin{pmatrix} e^{2t} & e^{3t} & 3e^{5t} \\ -e^{2t} & -2e^{3t} & -6e^{5t} \\ -e^{2t} & -e^{3t} & -2e^{5t} \end{pmatrix}$$

is a fundamental matrix of differential equation (11.10). Thus an arbitrary solution $\boldsymbol{\phi}(t)$ of (11.10) can be written in the form

$$\boldsymbol{\phi}(t) = \boldsymbol{\Phi}(t) \mathbf{c},$$

where $\mathbf{c} = \begin{pmatrix} c_1 \\ c_2 \\ c_3 \end{pmatrix}$ is a suitable constant vector.

THEOREM 11.10

The unique solution $\boldsymbol{\phi}$ *of the homogeneous linear vector differential equation*

$$\frac{d\mathbf{x}}{dt} = \mathbf{A}(t)\mathbf{x} \qquad (11.15)$$

that satisfies the initial condition

$$\boldsymbol{\phi}(t_0) = \mathbf{x}_0, \qquad (11.26)$$

where $t_0 \in [a, b]$, *can be expressed in the form*

$$\boldsymbol{\phi}(t) = \boldsymbol{\Phi}(t)\boldsymbol{\Phi}^{-1}(t_0)\mathbf{x}_0, \qquad (11.27)$$

where $\boldsymbol{\Phi}(t)$ *is an arbitrary fundamental matrix of differential equation (11.15).*

   Proof.   By Theorem 11.9 there exists a constant vector $\mathbf{c}$ such that $\boldsymbol{\phi}(t) = \boldsymbol{\Phi}(t)\mathbf{c}$ for all $t \in [a, b]$. Thus, in particular, $\boldsymbol{\phi}(t_0) = \boldsymbol{\Phi}(t_0)\mathbf{c}$; and applying the initial condition (11.26), we have $\mathbf{x}_0 = \boldsymbol{\Phi}(t_0)\mathbf{c}$. The determinant $|\boldsymbol{\Phi}(t)|$ is the Wronskian of the n linearly independent solutions of (11.15) that constitute the individual columns of the fundamental matrix $\boldsymbol{\Phi}(t)$. Thus by Theorem 11.6 we have $|\boldsymbol{\Phi}(t_0)| \neq 0$; and so $\boldsymbol{\Phi}(t_0)$ is nonsingular and its inverse $\boldsymbol{\Phi}^{-1}(t_0)$ exists. Thus we obtain

$$\boldsymbol{\Phi}^{-1}(t_0)\mathbf{x}_0 = \boldsymbol{\Phi}^{-1}(t_0)\boldsymbol{\Phi}(t_0)\mathbf{c} = \mathbf{I}\mathbf{c} = \mathbf{c}.$$

Now, using this value of c, we find

$$\boldsymbol{\phi}(t) = \boldsymbol{\Phi}(t)\boldsymbol{\Phi}^{-1}(t_0)\mathbf{x}_0. \qquad Q.E.D.$$

▶ **Example 11.10**

Suppose we seek the unique solution $\boldsymbol{\phi}$ of the differential equation

$$\frac{d\mathbf{x}}{dt} = \begin{pmatrix} 7 & -1 & 6 \\ -10 & 4 & -12 \\ -2 & 1 & -1 \end{pmatrix}\mathbf{x}, \quad \text{where } \mathbf{x} = \begin{pmatrix} x_1 \\ x_2 \\ x_3 \end{pmatrix}, \qquad (11.10)$$

that satisfies the initial condition

$$\boldsymbol{\phi}(0) = x_0, \quad \text{where } \mathbf{x}_0 = \begin{pmatrix} -1 \\ 4 \\ 2 \end{pmatrix}. \qquad (11.28)$$

By Theorem 11.10, this solution is given by

$$\boldsymbol{\phi}(t) = \boldsymbol{\Phi}(t)\boldsymbol{\Phi}^{-1}(0)\mathbf{x}_0, \qquad (11.29)$$

where $\boldsymbol{\Phi}(t)$ is a fundamental matrix of differential equation (11.10). In Example 11.9 we observed that

$$\boldsymbol{\Phi}(t) = \begin{pmatrix} e^{2t} & e^{3t} & 3e^{5t} \\ -e^{2t} & -2e^{3t} & -6e^{5t} \\ -e^{2t} & -e^{3t} & -2e^{5t} \end{pmatrix}$$

is such a fundamental matrix. After performing the required calculations (see Exercise 1(a) at the end of this section), we find

$$
\Phi^{-1}(t) = \begin{pmatrix} 2e^{-2t} & e^{-2t} & 0 \\ -4e^{-3t} & -e^{-3t} & -3e^{-3t} \\ e^{-5t} & 0 & e^{-5t} \end{pmatrix},
$$

and hence

$$
\Phi^{-1}(0) = \begin{pmatrix} 2 & 1 & 0 \\ -4 & -1 & -3 \\ 1 & 0 & 1 \end{pmatrix}.
$$

Thus, using (11.29), the solution of the problem is

$$
\phi(t) = \begin{pmatrix} e^{2t} & e^{3t} & 3e^{5t} \\ -e^{2t} & -2e^{3t} & -6e^{5t} \\ -e^{2t} & -e^{3t} & -2e^{5t} \end{pmatrix} \begin{pmatrix} 2 & 1 & 0 \\ -4 & -1 & -3 \\ 1 & 0 & 1 \end{pmatrix} \begin{pmatrix} -1 \\ 4 \\ 2 \end{pmatrix}
$$

$$
= \begin{pmatrix} e^{2t} & e^{3t} & 3e^{5t} \\ -e^{2t} & -2e^{3t} & -6e^{5t} \\ -e^{2t} & -e^{3t} & -2e^{5t} \end{pmatrix} \begin{pmatrix} 2 \\ -6 \\ 1 \end{pmatrix}
$$

$$
= \begin{pmatrix} 2e^{2t} - 6e^{3t} + 3e^{5t} \\ -2e^{2t} + 12e^{3t} - 6e^{5t} \\ -2e^{2t} + 6e^{3t} - 2e^{5t} \end{pmatrix}.
$$

We write this as

$$
\begin{aligned}
x_1 &= 2e^{2t} - 6e^{3t} + 3e^{5t}, \\
x_2 &= -2e^{2t} + 12e^{3t} - 6e^{5t}, \\
x_3 &= -2e^{2t} + 6e^{3t} - 2e^{5t},
\end{aligned}
\tag{11.30}
$$

We should note that the determination of $\Phi^{-1}(t)$ requires some calculations (we again refer to the aforementioned Exercise 1(a)). For this reason, the following alternate procedure for solving the problem is preferable.

In Example 11.8, we noted that the general solution of the differential equation (11.10) can be written as

$$
\begin{aligned}
x_1 &= c_1 e^{2t} + c_2 e^{3t} + 3c_3 e^{5t}, \\
x_2 &= -c_1 e^{2t} - 2c_2 e^{3t} - 6c_3 e^{5t}, \\
x_3 &= -c_1 e^{2t} - c_2 e^{3t} - 2c_3 e^{5t}.
\end{aligned}
\tag{11.31}
$$

In terms of components, the given initial condition (11.28) is

$$
x_1(0) = -1, \quad x_2(0) = 4, \quad x_3(0) = 2.
$$

Applying these conditions to (11.31) we at once have

$$
\begin{aligned}
c_1 + c_2 + 3c_3 &= -1, \\
-c_1 - 2c_2 - 6c_3 &= 4, \\
-c_1 - c_2 - 2c_3 &= 2.
\end{aligned}
\tag{11.32}
$$

This system of three linear algebraic equations in the three unknowns c_1, c_2, c_3 has the unique solution

$$c_1 = 2, \quad c_2 = -6, \quad c_3 = 1.$$

Substituting these values back into (11.31) we again obtain the unique solution (11.30) of the given initial-value problem.

Comparing this alternate procedure for solving the given initial-value problem with that which uses formula (11.29) we note that obtaining the solution of the system (11.32) also requires some calculations (we here refer to Exercise 1(b) at the end of this section). Nonetheless, the alternate procedure seems more practical.

THEOREM 11.11

If the n vector functions

$$\boldsymbol{\phi}_k = \begin{pmatrix} \phi_{1k} \\ \phi_{2k} \\ \vdots \\ \phi_{nk} \end{pmatrix} \quad (k = 1, 2, \dots, n),$$

satisfy the homogeneous linear vector differential equation

$$\frac{d\mathbf{x}}{dt} = \mathbf{A}(t)\mathbf{x} \tag{11.15}$$

on the real interval $[a, b]$, then the matrix function

$$\boldsymbol{\Phi} = \begin{pmatrix} \phi_{11} & \phi_{12} & \cdots & \phi_{1n} \\ \phi_{21} & \phi_{22} & \cdots & \phi_{2n} \\ \vdots & \vdots & & \vdots \\ \phi_{n1} & \phi_{n2} & \cdots & \phi_{nn} \end{pmatrix}$$

is such that

$$\frac{d\boldsymbol{\Phi}(t)}{dt} = \mathbf{A}(t)\boldsymbol{\Phi}(t) \tag{11.33}$$

for all $t \in [a, b]$; and conversely.

In particular, if $\boldsymbol{\Phi}$ is a fundamental matrix of (11.15) on $[a, b]$, then $\boldsymbol{\Phi}$ satisfies (11.33) on $[a, b]$.

Proof. Since $\boldsymbol{\phi}_k$ satisfies $d\mathbf{x}/dt = \mathbf{A}(t)\mathbf{x}$ on $[a, b]$, we have $\boldsymbol{\phi}'_k(t) = \mathbf{A}(t)\boldsymbol{\phi}_k(t)$ for all $t \in [a, b]$ $(k = 1, 2, \dots, n)$. That is,

$$\begin{pmatrix} \phi'_{1k}(t) \\ \vdots \\ \phi'_{jk}(t) \\ \vdots \\ \phi'_{nk}(t) \end{pmatrix} = \begin{pmatrix} a_{11}(t) & a_{12}(t) & \cdots & a_{1n}(t) \\ \vdots & \vdots & & \vdots \\ a_{j1}(t) & a_{j2}(t) & \cdots & a_{jn}(t) \\ \vdots & \vdots & & \vdots \\ a_{n1}(t) & a_{n2}(t) & \cdots & a_{nn}(t) \end{pmatrix} \begin{pmatrix} \phi_{1k}(t) \\ \phi_{2k}(t) \\ \vdots \\ \phi_{nk}(t) \end{pmatrix},$$

and hence

$$\phi'_{jk}(t) = \sum_{i=1}^{n} a_{ji}(t)\phi_{ik}(t),$$

for all $t \in [a, b]$ ($j = 1, 2, \ldots, n$; $k = 1, 2, \ldots, n$). Now observe that this derivative $\phi'_{jk}(t)$ is the element in jth row and kth column of the derivative $\mathbf{\Phi}'(t)$ of $\mathbf{\Phi}(t)$. On the other hand, we see that the element in the jth row and the kth column of the product

$$\mathbf{A}(t)\mathbf{\Phi}(t) = \begin{pmatrix} a_{11}(t) & a_{12}(t) & \cdots & a_{1n}(t) \\ \vdots & \vdots & & \vdots \\ a_{j1}(t) & a_{j2}(t) & \cdots & a_{jn}(t) \\ \vdots & \vdots & & \vdots \\ a_{n1}(t) & a_{n2}(t) & \cdots & a_{nn}(t) \end{pmatrix} \begin{pmatrix} \phi_{11}(t) & \cdots & \phi_{1k}(t) & \cdots & \phi_{1n}(t) \\ \phi_{21}(t) & \cdots & \phi_{2k}(t) & \cdots & \phi_{2n}(t) \\ \vdots & & \vdots & & \vdots \\ a_{n1}(t) & \cdots & \phi_{nk}(t) & \cdots & \phi_{nn}(t) \end{pmatrix}$$

is also

$$\sum_{i=1}^{n} a_{ji}(t)\phi_{ik}(t),$$

for all $t \in [a, b]$ ($j = 1, 2, \ldots, n$; $k = 1, 2, \ldots, n$). Therefore, we have

$$\mathbf{\Phi}'(t) = \mathbf{A}(t)\mathbf{\Phi}(t)$$

for all $t \in [a, b]$.

Conversely, suppose $\mathbf{\Phi}'(t) = \mathbf{A}(t)\mathbf{\Phi}(t)$ for all $t \in [a, b]$. Then equating corresponding columns of $\mathbf{\Phi}'(t)$ and $\mathbf{A}(t)\mathbf{\Phi}(t)$, we at once have

$$\mathbf{\phi}'_k(t) = \mathbf{A}(t)\mathbf{\phi}_k(t),$$

$t \in [a, b]$ ($k = 1, 2, \ldots, n$). Thus each $\mathbf{\phi}_k$ ($k = 1, 2, \ldots, n$) satisfies differential equation (11.15) on $[a, b]$.

Finally, the statement concerning a fundamental matrix of (11.15) is now obvious

Q.E.D.

▶ **Example 11.11**

Referring back to Example 11.10 and applying Theorem 11.11, we see that the fundamental matrix

$$\mathbf{\Phi}(t) = \begin{pmatrix} e^{2t} & e^{3t} & 3e^{5t} \\ -e^{2t} & -2e^{3t} & -6e^{5t} \\ -e^{2t} & -e^{3t} & -2e^{5t} \end{pmatrix}$$

of the differential equation

$$\frac{d\mathbf{x}}{dt} = \begin{pmatrix} 7 & -1 & 6 \\ -10 & 4 & -12 \\ -2 & 1 & -1 \end{pmatrix}\mathbf{x} \tag{11.10}$$

is such that

$$\frac{d\mathbf{\phi}(t)}{dt} = \begin{pmatrix} 7 & -1 & 6 \\ -10 & 4 & -12 \\ -2 & 1 & -1 \end{pmatrix}\mathbf{\Phi}(t)$$

for all real t. That is, for all real t, we have

$$
\begin{pmatrix}
2e^{2t} & 3e^{3t} & 15e^{5t} \\
-2e^{2t} & -6e^{3t} & -30e^{5t} \\
-2e^{2t} & -3e^{3t} & -10e^{5t}
\end{pmatrix}
=
\begin{pmatrix}
7 & -1 & 6 \\
-10 & 4 & -12 \\
-2 & 1 & -1
\end{pmatrix}
\begin{pmatrix}
e^{2t} & e^{3t} & 3e^{5t} \\
-e^{2t} & -2e^{3t} & -6e^{5t} \\
-e^{2t} & -e^{3t} & -2e^{5t}
\end{pmatrix}.
$$

The student should carry out the indicated multiplication in the right member and thus directly verify the validity of this equality.

DEFINITION

The trace *of an $n \times n$ matrix $\mathbf{A}$ is given by the formula* $\operatorname{tr} \mathbf{A} = \sum\limits_{j=1}^{n} a_{jj}$. *That is, the trace of $\mathbf{A}$ is the sum of its main diagonal elements.*

THEOREM 11.12   Abel–Liouville Formula

Let the functions

$$
\boldsymbol{\phi}_k =
\begin{pmatrix}
\phi_{1k} \\
\phi_{2k} \\
\vdots \\
\phi_{nk}
\end{pmatrix},
\quad (k = 1, 2, \ldots, n),
$$

be n solutions of the homogeneous linear vector differential equation

$$
\frac{d\mathbf{x}}{dt} = \mathbf{A}(t)\mathbf{x} \tag{11.15}
$$

on the real interval $[a, b]$; let t_0 be any point of $[a, b]$; and let W denote the Wronskian of $\phi_1, \phi_2, \ldots, \phi_n$. Then

$$
W(t) = W(t_0) \exp\left[\int_{t_0}^{t} \operatorname{tr} \mathbf{A}(s)\, ds \right] \tag{11.34}
$$

for all $t \in [a, b]$.

Proof.   We differentiate the Wronskian determinant

$$
W =
\begin{vmatrix}
\phi_{11} & \phi_{12} & \cdots & \phi_{1n} \\
\phi_{21} & \phi_{22} & \cdots & \phi_{2n} \\
\vdots & \vdots & & \vdots \\
\phi_{i1} & \phi_{i2} & \cdots & \phi_{in} \\
\vdots & \vdots & & \vdots \\
\phi_{n1} & \phi_{n2} & \cdots & \phi_{nn}
\end{vmatrix}
$$

to obtain

$$
W' = \begin{vmatrix} \phi'_{11} & \phi'_{12} & \cdots & \phi'_{1n} \\ \phi_{21} & \phi_{22} & \cdots & \phi_{2n} \\ \vdots & \vdots & & \vdots \\ \phi_{i1} & \phi_{i2} & \cdots & \phi_{in} \\ \vdots & \vdots & & \vdots \\ \phi_{n1} & \phi_{n2} & \cdots & \phi_{nn} \end{vmatrix} + \begin{vmatrix} \phi_{11} & \phi_{12} & \cdots & \phi_{1n} \\ \phi'_{21} & \phi'_{22} & \cdots & \phi'_{2n} \\ \vdots & \vdots & & \vdots \\ \phi_{i1} & \phi_{i2} & \cdots & \phi_{in} \\ \vdots & \vdots & & \vdots \\ \phi_{n1} & \phi_{n2} & \cdots & \phi_{nn} \end{vmatrix} + \cdots
$$

$$
+ \begin{vmatrix} \phi_{11} & \phi_{12} & \cdots & \phi_{1n} \\ \phi_{21} & \phi_{22} & \cdots & \phi_{2n} \\ \vdots & \vdots & & \vdots \\ \phi'_{i1} & \phi'_{i2} & \cdots & \phi'_{in} \\ \vdots & \vdots & & \vdots \\ \phi_{n1} & \phi_{n2} & \cdots & \phi_{nn} \end{vmatrix} + \cdots + \begin{vmatrix} \phi_{11} & \phi_{12} & \cdots & \phi_{1n} \\ \phi_{21} & \phi_{22} & \cdots & \phi_{2n} \\ \vdots & \vdots & & \vdots \\ \phi_{i1} & \phi_{i2} & \cdots & \phi_{in} \\ \vdots & \vdots & & \vdots \\ \phi'_{n1} & \phi'_{n2} & \cdots & \phi'_{nn} \end{vmatrix}. \tag{11.35}
$$

Here, and throughout the rest of the proof, primes denote derivatives with respect to t. Thus W' is the sum of n determinants, in each of which the elements of precisely one row are differentiated. Since $\boldsymbol{\phi}_k$, $(k = 1, 2, \ldots, n)$, satisfies the vector differential equation (11.15), we have $\boldsymbol{\phi}'_k = \mathbf{A}\boldsymbol{\phi}_k$, $(k = 1, 2, \ldots, n)$, and so $\phi'_{ik} = \sum_{j=1}^{n} a_{ij}\phi_{jk}$, $(i = 1, 2, \ldots, n; j = 1, 2, \ldots, n)$. Substitute these for the indicated derivatives in each of the n determinants in the preceding expression (11.35) for W'. Then the ith determinant, $(i = 1, 2, \ldots, n)$ in (11.35) becomes

$$
\begin{vmatrix} \phi_{11} & \phi_{12} & \cdots & \phi_{1n} \\ \phi_{21} & \phi_{22} & \cdots & \phi_{2n} \\ \vdots & \vdots & & \vdots \\ \phi'_{i1} & \phi'_{i2} & \cdots & \phi'_{in} \\ \vdots & \vdots & & \vdots \\ \phi_{n1} & \phi_{n2} & \cdots & \phi_{nn} \end{vmatrix} = \begin{vmatrix} \phi_{11} & \phi_{12} & \cdots & \phi_{1n} \\ \phi_{21} & \phi_{22} & \cdots & \phi_{2n} \\ \vdots & \vdots & & \vdots \\ \sum_{j=1}^{n} a_{ij}\phi_{j1} & \sum_{j=1}^{n} a_{ij}\phi_{j2} & \cdots & \sum_{j=1}^{n} a_{ij}\phi_{jn} \\ \vdots & \vdots & & \vdots \\ \phi_{n1} & \phi_{n2} & \cdots & \phi_{nn} \end{vmatrix}
$$

Writing out each of the indicated sums in the preceding determinant and using fundamental properties of determinants, we see that it breaks up into the following sum of n determinants:

$$
a_{i1} \begin{vmatrix} \phi_{11} & \phi_{12} & \cdots & \phi_{1n} \\ \phi_{21} & \phi_{22} & \cdots & \phi_{2n} \\ \vdots & \vdots & & \vdots \\ \phi_{11} & \phi_{12} & \cdots & \phi_{1n} \\ \vdots & \vdots & & \vdots \\ \phi_{n1} & \phi_{n2} & \cdots & \phi_{nn} \end{vmatrix} + a_{i2} \begin{vmatrix} \phi_{11} & \phi_{12} & \cdots & \phi_{1n} \\ \phi_{21} & \phi_{22} & \cdots & \phi_{2n} \\ \vdots & \vdots & & \vdots \\ \phi_{21} & \phi_{22} & \cdots & \phi_{2n} \\ \vdots & \vdots & & \vdots \\ \phi_{n1} & \phi_{n2} & \cdots & \phi_{nn} \end{vmatrix} + \cdots
$$

$$
+ a_{ii} \begin{vmatrix} \phi_{11} & \phi_{12} & \cdots & \phi_{1n} \\ \phi_{21} & \phi_{22} & \cdots & \phi_{2n} \\ \vdots & \vdots & & \vdots \\ \phi_{i1} & \phi_{i2} & \cdots & \phi_{in} \\ \vdots & \vdots & & \vdots \\ \phi_{n1} & \phi_{n2} & \cdots & \phi_{nn} \end{vmatrix} + \cdots + a_{in} \begin{vmatrix} \phi_{11} & \phi_{12} & \cdots & \phi_{1n} \\ \phi_{21} & \phi_{22} & \cdots & \phi_{2n} \\ \vdots & \vdots & & \vdots \\ \phi_{n1} & \phi_{n2} & \cdots & \phi_{nn} \\ \vdots & \vdots & & \vdots \\ \phi_{n1} & \phi_{n2} & \cdots & \phi_{nn} \end{vmatrix}.
$$

Each of these n determinants has two equal rows, except the ith one, and the coefficient of this exceptional one is a_{ii}. Since a determinant having two equal rows is zero, this leaves only the single exceptional determinant having the coefficient a_{ii}. Thus we have

$$\begin{vmatrix} \phi_{11} & \phi_{12} & \cdots & \phi_{1n} \\ \phi_{21} & \phi_{22} & \cdots & \phi_{2n} \\ \vdots & \vdots & & \vdots \\ \phi'_{i1} & \phi'_{i2} & \cdots & \phi'_{in} \\ \vdots & \vdots & & \vdots \\ \phi_{n1} & \phi_{n2} & \cdots & \phi_{nn} \end{vmatrix} = a_{ii} \begin{vmatrix} \phi_{11} & \phi_{12} & \cdots & \phi_{1n} \\ \phi_{21} & \phi_{22} & \cdots & \phi_{2n} \\ \vdots & \vdots & & \vdots \\ \phi_{i1} & \phi_{i2} & \cdots & \phi_{in} \\ \vdots & \vdots & & \vdots \\ \phi_{n1} & \phi_{n2} & \cdots & \phi_{nn} \end{vmatrix} \quad \text{for each} \quad i = 1, 2, \ldots, n.$$

Using this identity with $i = 1, 2, \ldots, n$, we replace each of the n determinants in (11.35) accordingly. Doing so, (11.35) then takes the form

$$W' = a_{11} \begin{vmatrix} \phi_{11} & \phi_{12} & \cdots & \phi_{1n} \\ \phi_{21} & \phi_{22} & \cdots & \phi_{2n} \\ \vdots & \vdots & & \vdots \\ \phi_{i1} & \phi_{i2} & \cdots & \phi_{in} \\ \vdots & \vdots & & \vdots \\ \phi_{n1} & \phi_{n2} & \cdots & \phi_{nn} \end{vmatrix} + a_{22} \begin{vmatrix} \phi_{11} & \phi_{12} & \cdots & \phi_{1n} \\ \phi_{21} & \phi_{22} & \cdots & \phi_{2n} \\ \vdots & \vdots & & \vdots \\ \phi_{i1} & \phi_{i2} & \cdots & \phi_{in} \\ \vdots & \vdots & & \vdots \\ \phi_{n1} & \phi_{n2} & \cdots & \phi_{nn} \end{vmatrix} + \cdots$$

$$+ a_{ii} \begin{vmatrix} \phi_{11} & \phi_{12} & \cdots & \phi_{1n} \\ \phi_{21} & \phi_{22} & \cdots & \phi_{2n} \\ \vdots & \vdots & & \vdots \\ \phi_{i1} & \phi_{i2} & \cdots & \phi_{in} \\ \vdots & \vdots & & \vdots \\ \phi_{n1} & \phi_{n2} & \cdots & \phi_{nn} \end{vmatrix} + \cdots + a_{nn} \begin{vmatrix} \phi_{11} & \phi_{12} & \cdots & \phi_{1n} \\ \phi_{21} & \phi_{22} & \cdots & \phi_{2n} \\ \vdots & \vdots & & \vdots \\ \phi_{i1} & \phi_{i2} & \cdots & \phi_{in} \\ \vdots & \vdots & & \vdots \\ \phi_{n1} & \phi_{n2} & \cdots & \phi_{nn} \end{vmatrix}.$$

That is, $W' = \left[\sum_{j=1}^{n} a_{jj} \right] W$, and so

$$W' = (\text{tr } \mathbf{A})W. \tag{11.36}$$

That is, (11.36) states that W satisfies the first-order scalar homogeneous linear differential equation

$$\frac{dW(t)}{dt} = [\text{tr } \mathbf{A}(t)] W(t).$$

Integrating this, we at once obtain

$$W(t) = c \exp\left[\int_{t_0}^{t} \text{tr } \mathbf{A}(s) \, ds \right].$$

Letting $t = t_0$, we find that $c = W(t_0)$, and hence we obtain the stated Abel–Liouville formula

$$W(t) = W(t_0) \exp\left[\int_{t_0}^{t} \text{tr } \mathbf{A}(s) \, ds \right]. \tag{11.34}$$

$Q.E.D.$

Exercises

1. This exercise deals with the verification of certain algebraic calculations referred to in Example 11.10.

   (a) Find the inverse of the matrix

   $$\begin{pmatrix} e^{2t} & e^{3t} & 3e^{5t} \\ -e^{2t} & -2e^{3t} & -6e^{5t} \\ -e^{2t} & -e^{3t} & -2e^{5t} \end{pmatrix}.$$

   (b) Find the solution of the linear algebraic system

   $$c_1 + c_2 + 3c_2 = -1,$$
   $$-c_1 - 2c_2 - 6c_3 = 4,$$
   $$-c_1 - c_2 - 2c_3 = 2.$$

2. (a) Show that

   $$\Phi(t) = \begin{pmatrix} e^{5t} & e^{3t} \\ e^{5t} & 3e^{3t} \end{pmatrix}$$

   is a fundamental matrix of the linear system

   $$\frac{dx}{dt} = \begin{pmatrix} 6 & -1 \\ 3 & 2 \end{pmatrix} x.$$

   (b) Use Formula (11.27) of Theorem 11.10 to find the unique solution ϕ of this linear system that satisfies the initial condition

   $$\phi(0) = \begin{pmatrix} 2 \\ 0 \end{pmatrix}.$$

   (c) Verify directly that Φ defined in part (a) satisfies Formula (11.33) of Theorem 11.11.

3. (a) Show that

   $$\Phi(t) = \begin{pmatrix} e^{3t} & e^{-3t} \\ 5e^{3t} & -3e^{-3t} \end{pmatrix}$$

   is a fundamental matrix of the linear system

   $$\frac{dx}{dt} = \begin{pmatrix} -2 & 1 \\ 5 & 2 \end{pmatrix} x.$$

   (b) Use Formula (11.27) of Theorem 11.10 to find the unique solution ϕ of this linear system that satisfies the initial condition

   $$\phi(0) = \begin{pmatrix} 5 \\ 7 \end{pmatrix}.$$

   (c) Verify directly that Φ defined in part (a) satisfies Formula (11.33) of Theorem 11.11.

4. (a) Show that

$$\Phi(t) = \begin{pmatrix} e^t & e^{5t} \\ -2e^t & 2e^{5t} \end{pmatrix}$$

is a fundamental matrix of the linear system

$$\frac{d\mathbf{x}}{dt} = \begin{pmatrix} 3 & 1 \\ 4 & 3 \end{pmatrix} \mathbf{x}.$$

(b) Use Formula (11.27) of Theorem 11.10 to find the unique solution ϕ of this linear system that satisfies the initial condition

$$\phi(0) = \begin{pmatrix} 1 \\ -6 \end{pmatrix}.$$

(c) Verify directly that Φ defined in part (a) satisfies Formula (11.33) of Theorem 11.11.

5. (a) Show that

$$\Phi(t) = \begin{pmatrix} 2e^t & e^{2t} & 0 \\ 2e^t & 2e^{2t} & 3e^{5t} \\ e^t & e^{2t} & e^{5t} \end{pmatrix}$$

is a fundamental matrix of the linear system

$$\frac{d\mathbf{x}}{dt} = \begin{pmatrix} 0 & -2 & 6 \\ -2 & 9 & -12 \\ -1 & 2 & -1 \end{pmatrix} \mathbf{x}.$$

(b) Use Formula (11.27) of Theorem 11.10 to find the unique solution ϕ of this linear system that satisfies the initial condition

$$\phi(0) = \begin{pmatrix} 6 \\ 3 \\ 2 \end{pmatrix}.$$

(c) Verify directly that Φ defined in part (a) satisfies Formula (11.33) of Theorem 11.11.

6. (a) Show that

$$\Phi(t) = \begin{pmatrix} e^t & e^{3t} & 0 \\ e^t & e^{3t} & e^{-2t} \\ 3e^t & 2e^{3t} & e^{-2t} \end{pmatrix}$$

is a fundamental matrix of the linear system

$$\frac{d\mathbf{x}}{dt} = \begin{pmatrix} 5 & 2 & -2 \\ 7 & 0 & -2 \\ 11 & 1 & -3 \end{pmatrix} \mathbf{x}.$$

(b) Use Formula (11.27) of Theorem 11.10 to find the unique solution ϕ of this linear system that satisfies the initial condition

$$\phi(0) = \begin{pmatrix} 1 \\ 0 \\ 4 \end{pmatrix}.$$

(c) Verify directly that Φ defined in part (a) satisfies Formula (11.33) of Theorem 11.11.

7. Let Φ be a fundamental matrix of $dx/dt = A(t)x$ on $[a, b]$, and let C be a constant nonsingular matrix.
    (a) Show that ΦC is also a fundamental matrix of $dx/dt = A(t)x$ on $[a, b]$.
    (b) Show that in general $C\Phi$ is not a fundamental matrix of $dx/dt = A(t)x$ on $[a, b]$.
    (c) Determine B such that $C\Phi$ is a fundamental matrix of $dx/dt = B(t)x$ on $[a, b]$.

11.4 THE NONHOMOGENEOUS LINEAR SYSTEM

We return briefly to the nonhomogeneous linear vector differential equation

$$\frac{dx}{dt} = A(t)x + F(t), \tag{11.6}$$

where $A(t)$ is given by (11.4) and $F(t)$ and x are given by (11.5). We shall see the solutions of this nonhomogeneous equation are closely related to those of the corresponding homogeneous equation

$$\frac{dx}{dt} = A(t)x. \tag{11.15}$$

THEOREM 11.13

Let ϕ_0 be any solution of the nonhomogeneous linear vector differential equation

$$\frac{dx}{dt} = A(t)x + F(t); \tag{11.6}$$

let $\phi_1, \phi_2, \ldots, \phi_n$ be a fundamental set of solutions of the corresponding homogeneous differential equation

$$\frac{dx}{dt} = A(t)x; \tag{11.15}$$

and let $c_1, c_2, \ldots, c_n$ be n numbers.
    Then: (1) the vector function

$$\phi_0 + \sum_{k=1}^{n} c_k \phi_k \tag{11.37}$$

is also a solution of the nonhomogeneous differential equation (11.6) for every choice of $c_1, c_2, \ldots, c_n$; *and*

(2) *an arbitrary solution* ϕ *of the nonhomogeneous differential equation (11.6) is of the form (11.37) for a suitable choice of* $c_1, c_2, \ldots, c_n$.

Proof. (1) We show that (11.37) satisfies (11.6) for all choices of $c_1, c_2, \ldots, c_n$. We have

$$\frac{d}{dt}\left[\phi_0(t) + \sum_{k=1}^{n} c_k \phi_k(t)\right] = \frac{d\phi_0(t)}{dt} + \frac{d}{dt}\left[\sum_{k=1}^{n} c_k \phi_k(t)\right].$$

Now since ϕ_0 satisfies (11.6), we have

$$\frac{d\phi_0(t)}{dt} = A(t)\phi_0(t) + F(t);$$

and since by Theorem 11.2, $\sum_{k=1}^{n} c_k \phi_k$ satisfies (11.15), we also have

$$\frac{d}{dt}\left[\sum_{k=1}^{n} c_k \phi_k(t)\right] = A(t)\left[\sum_{k=1}^{n} c_k \phi_k(t)\right].$$

Thus

$$\frac{d}{dt}\left[\phi_0(t) + \sum_{k=1}^{n} c_k \phi_k(t)\right] = A(t)\phi_0(t) + F(t) + A(t)\left[\sum_{k=1}^{n} c_k \phi_k(t)\right]$$

$$= A(t)\left[\phi_0(t) + \sum_{k=1}^{n} c_k \phi_k(t)\right] + F(t).$$

That is,

$$\frac{d\psi(t)}{dt} = A(t)\psi(t) + F(t)$$

where

$$\psi = \phi_0 + \sum_{k=1}^{n} c_k \phi_k;$$

and so

$$\psi = \phi_0 + \sum_{k=1}^{n} c_k \phi_k$$

is a solution of (11.6) for every choice of $c_1, c_2, \ldots, c_n$.

(2) Now consider an arbitrary solution ϕ of (11.6) and evaluate the derivative of the difference $\phi - \phi_0$. We have

$$\frac{d}{dt}\left[\phi(t) - \phi_0(t)\right] = \frac{d\phi(t)}{dt} - \frac{d\phi_0(t)}{dt}.$$

Since both ϕ and ϕ_0 satisfy (11.6) we have respectively

$$\frac{d\phi(t)}{dt} = A(t)\phi(t) + F(t),$$

$$\frac{d\phi_0(t)}{dt} = A(t)\phi_0(t) + F(t).$$

Thus we obtain

$$\frac{d}{dt}[\boldsymbol{\phi}(t) - \boldsymbol{\phi}_0(t)] = [\mathbf{A}(t)\boldsymbol{\phi}(t) + \mathbf{F}(t)] - [\mathbf{A}(t)\boldsymbol{\phi}_0(t) + \mathbf{F}(t)],$$

which at once reduces to

$$\frac{d}{dt}[\boldsymbol{\phi}(t) - \boldsymbol{\phi}_0(t)] = \mathbf{A}(t)[\boldsymbol{\phi}(t) - \boldsymbol{\phi}_0(t)].$$

Thus $\boldsymbol{\phi} - \boldsymbol{\phi}_0$ satisfies the *homogeneous* differential equation (11.15). Hence by Theorem 11.8, there exists a suitable choice of numbers $c_1, c_2, \ldots, c_n$ such that

$$\boldsymbol{\phi} - \boldsymbol{\phi}_0 = \sum_{k=1}^{n} c_k \boldsymbol{\phi}_k.$$

Thus the arbitrary solution $\boldsymbol{\phi}$ of (11.6) is of the form

$$\boldsymbol{\phi} = \boldsymbol{\phi}_0 + \sum_{k=1}^{n} c_k \boldsymbol{\phi}_k \tag{11.37}$$

for a suitable choice of $c_1, c_2, \ldots, c_n$.

Q.E.D.

DEFINITION

Consider the nonhomogeneous linear vector differential equation (11.6) and the corresponding homogeneous linear vector differential equation (11.15). By a general solution of (11.6), we mean a solution of the form

$$c_1 \boldsymbol{\phi}_1 + c_2 \boldsymbol{\phi}_2 + \cdots + c_n \boldsymbol{\phi}_n + \boldsymbol{\phi}_0,$$

where $c_1, c_2, \ldots, c_n$ are n arbitrary numbers, $\boldsymbol{\phi}_1, \boldsymbol{\phi}_2, \ldots, \boldsymbol{\phi}_n$ is a fundamental set of solutions of (11.15), and $\boldsymbol{\phi}_0$ is any solution of (11.6).

▶ **Example 11.12**

Consider the nonhomogeneous differential equation

$$\frac{d\mathbf{x}}{dt} = \begin{pmatrix} 7 & -1 & 6 \\ -10 & 4 & -12 \\ -2 & 1 & -1 \end{pmatrix} \mathbf{x} + \begin{pmatrix} -5t - 6 \\ -4t + 23 \\ 2 \end{pmatrix} \tag{11.3}$$

and the corresponding homogeneous differential equation

$$\frac{d\mathbf{x}}{dt} = \begin{pmatrix} 7 & -1 & 6 \\ -10 & 4 & -12 \\ -2 & 1 & -1 \end{pmatrix} \mathbf{x}, \qquad \text{where} \quad \mathbf{x} = \begin{pmatrix} x_1 \\ x_2 \\ x_3 \end{pmatrix}. \tag{11.2}$$

These were introduced in Example 11.1 where they were written out in component form; and (11.2) has been used in Example 11.8 and several other examples as well. In

Example 11.8 we observed that ϕ_1, ϕ_2, ϕ_3 defined respectively by

$$\phi_1(t) = \begin{pmatrix} e^{2t} \\ -e^{2t} \\ -e^{2t} \end{pmatrix}, \qquad \phi_2(t) = \begin{pmatrix} e^{3t} \\ -2e^{3t} \\ -e^{3t} \end{pmatrix}, \quad \text{and} \quad \phi_3(t) = \begin{pmatrix} 3e^{5t} \\ -6e^{5t} \\ -2e^{5t} \end{pmatrix} \qquad (11.22)$$

form a fundamental set of the homogeneous differential equation (11.2) [or (11.10) as it is numbered there]. Now observe that the vector function ϕ_0 defined by

$$\phi_0(t) = \begin{pmatrix} 2t \\ 3t - 2 \\ -t + 1 \end{pmatrix}$$

is a solution of the nonhomogeneous differential equation (11.3). Thus a general solution of (11.3) is given by

$$\mathbf{x} = c_1\phi_1(t) + c_2\phi_2(t) + c_3\phi_3(t) + \phi_0(t),$$

that is,

$$\mathbf{x} = c_1 \begin{pmatrix} e^{2t} \\ -e^{2t} \\ -e^{2t} \end{pmatrix} + c_2 \begin{pmatrix} e^{3t} \\ -2e^{3t} \\ -e^{3t} \end{pmatrix} + c_3 \begin{pmatrix} 3e^{5t} \\ -6e^{5t} \\ -2e^{5t} \end{pmatrix} + \begin{pmatrix} 2t \\ 3t - 2 \\ -t + 1 \end{pmatrix},$$

where $c_1, c_2,$ and c_3 are arbitrary numbers. Thus a general solution of (11.3) can be written as

$$x_1 = c_1e^{2t} + c_2e^{3t} + 3c_3e^{5t} + 2t,$$
$$x_2 = -c_1e^{2t} - 2c_2e^{3t} - 6c_3e^{5t} + 3t - 2,$$
$$x_3 = -c_1e^{2t} - c_2e^{3t} - 2c_3e^{5t} - t + 1,$$

where $c_1, c_2,$ and c_3 are arbitrary numbers.

We now develop a systematic procedure for finding a solution of the nonhomogeneous linear vector differential equation

$$\frac{d\mathbf{x}}{dt} = \mathbf{A}(t)\mathbf{x} + \mathbf{F}(t), \qquad (11.6)$$

assuming that we know a fundamental matrix of the corresponding homogeneous linear vector differential equation

$$\frac{d\mathbf{x}}{dt} = \mathbf{A}(t)\mathbf{x}. \qquad (11.15)$$

Specifically, let $\mathbf{\Phi}(t)$ be a fundamental matrix of (11.15) on $[a, b]$. Then by the note immediately following Theorem 11.9, the general solution of (11.15) can be expressed in the form

$$\mathbf{x}(t) = \mathbf{\Phi}(t)\mathbf{c}, \qquad (11.38)$$

where $\mathbf{c}$ is an arbitrary n-rowed constant vector. We shall now proceed as we did in the analogous case of the single nth-order nonhomogeneous linear differential equation in

Section 4.4, that is, by *variation of parameters*. We thus replace the constant vector **c** in (11.38) by the vector *function* **v**(*t*), thereby obtaining

$$\mathbf{x}(t) = \mathbf{\Phi}(t)\mathbf{v}(t), \tag{11.39}$$

where **v**(*t*) will be determined so that (11.39) will be a solution of the nonhomogeneous differential equation (11.6).

Differentiating (11.39), we obtain

$$\mathbf{x}'(t) = \mathbf{\Phi}'(t)\mathbf{v}(t) + \mathbf{\Phi}(t)\mathbf{v}'(t); \tag{11.40}$$

and substituting (11.39) and (11.40) into (11.6) we find

$$\mathbf{\Phi}'(t)\mathbf{v}(t) + \mathbf{\Phi}(t)\mathbf{v}'(t) = \mathbf{A}(t)[\mathbf{\Phi}(t)\mathbf{v}(t)] + \mathbf{F}(t). \tag{11.41}$$

Since $\mathbf{\Phi}(t)$ is a fundamental matrix for the homogeneous differential equation (11.15), by Theorem 11.11, $\mathbf{\Phi}'(t) = \mathbf{A}(t)\mathbf{\Phi}(t)$. Hence $\mathbf{\Phi}'(t)\mathbf{v}(t) = \mathbf{A}(t)\mathbf{\Phi}(t)\mathbf{v}(t)$, and so (11.41) reduces to

$$\mathbf{\Phi}(t)\mathbf{v}'(t) = \mathbf{F}(t).$$

Since $\mathbf{\Phi}(t)$ is a fundamental matrix of (11.6), by Theorem 11.6 we know that $|\mathbf{\Phi}(t)| \neq 0$ on $[a, b]$ and hence $\mathbf{\Phi}^{-1}(t)$ exists and is unique on this interval. Hence, we at once have

$$\mathbf{v}'(t) = \mathbf{\Phi}^{-1}(t)\mathbf{F}(t);$$

and integrating, we obtain

$$\mathbf{v}(t) = \int_{t_0}^{t} \mathbf{\Phi}^{-1}(u)\mathbf{F}(u)\, du,$$

where t_0 and $t \in [a, b]$. Thus a solution of the assumed form (11.39) is given by

$$\mathbf{x}(t) = \mathbf{\Phi}(t) \int_{t_0}^{t} \mathbf{\Phi}^{-1}(u)\mathbf{F}(u)\, du.$$

We shall now show that every function of this form is indeed a solution of the nonhomogeneous differential equation (11.6).

THEOREM 11.14

Let $\mathbf{\Phi}$ be a fundamental matrix of the homogeneous linear vector differential equation $d\mathbf{x}/dt = \mathbf{A}(t)\mathbf{x}$ on $[a, b]$. Then $\mathbf{\Phi}_0$ defined by

$$\mathbf{\phi}_0(t) = \mathbf{\Phi}(t) \int_{t_0}^{t} \mathbf{\Phi}^{-1}(u)\mathbf{F}(u)\, du, \tag{11.42}$$

where $t_0 \in [a, b]$, is a solution of the nonhomogeneous linear vector differential equation

$$\frac{d\mathbf{x}}{dt} = \mathbf{A}(t)\mathbf{x} + \mathbf{F}(t) \tag{11.6}$$

on $[a, b]$.

Proof.   Differentiating (11.42), we find

$$\phi_0'(t) = \Phi'(t) \int_{t_0}^{t} \Phi^{-1}(u)F(u)\,du + \Phi(t)\Phi^{-1}(t)F(t)$$

$$= \Phi'(t) \int_{t_0}^{t} \Phi^{-1}(u)F(u)\,du + F(t).$$

Since $\Phi(t)$ is a fundamental matrix of the homogeneous differential equation $dx/dt = A(t)x$, by Theorem 11.11 we have $\Phi'(t) = A(t)\Phi(t)$. Thus we find that

$$\phi_0'(t) = A(t)\Phi(t) \int_{t_0}^{t} \Phi^{-1}(u)F(u)\,du + F(t). \qquad (11.43)$$

Also from (11.42), we find

$$A(t)\phi_0(t) + F(t) = A(t)\Phi(t) \int_{t_0}^{t} \Phi^{-1}(u)F(u)\,du + F(t). \qquad (11.44)$$

Comparing (11.43) and (11.44), we see that $\phi_0(t)$ defined by (11.42) is such that

$$\phi_0'(t) = A(t)\phi_0(t) + F(t);$$

that is, every function $\phi_0(t)$ defined by (11.42) satisfies the nonhomogeneous differential equation (11.6).   Q.E.D.

▶ **Example 11.13**

Consider the nonhomogeneous differential equation

$$\frac{dx}{dt} = \begin{pmatrix} 6 & -3 \\ 2 & 1 \end{pmatrix} x + \begin{pmatrix} e^{5t} \\ 4 \end{pmatrix}. \qquad (11.45)$$

Clearly this is of the form (11.6), where

$$A(t) = \begin{pmatrix} 6 & -3 \\ 2 & 1 \end{pmatrix}, \quad F(t) = \begin{pmatrix} e^{5t} \\ 4 \end{pmatrix}, \quad \text{and } x = \begin{pmatrix} x_1 \\ x_2 \end{pmatrix}.$$

The corresponding homogeneous differential equation is

$$\frac{dx}{dt} = \begin{pmatrix} 6 & -3 \\ 2 & 1 \end{pmatrix} x. \qquad (11.46)$$

This is the system (7.79) of Example 7.17 (in different notation). In that example, we found that

$$\phi_1(t) = \begin{pmatrix} e^{3t} \\ e^{3t} \end{pmatrix} \quad \text{and} \quad \phi_2(t) = \begin{pmatrix} 3e^{4t} \\ 2e^{4t} \end{pmatrix} \qquad (11.47)$$

constitute a fundamental set (pair of linearly independent solutions) of (11.46). Thus a fundamental matrix of (11.46) is given by

$$\Phi(t) = \begin{pmatrix} e^{3t} & 3e^{4t} \\ e^{3t} & 2e^{4t} \end{pmatrix}. \qquad (11.48)$$

By Theorem 11.14 we know that

$$\boldsymbol{\phi}_0(t) = \boldsymbol{\Phi}(t) \int_{t_0}^{t} \boldsymbol{\Phi}^{-1}(u)\mathbf{F}(u)\,du$$

is a solution of (11.45) for every real number t_0. For convenience, we choose $t_0 = 0$. We thus proceed to evaluate

$$\boldsymbol{\phi}_0(t) = \boldsymbol{\Phi}(t) \int_{0}^{t} \boldsymbol{\Phi}^{-1}(u)\mathbf{F}(u)\,du, \tag{11.49}$$

where $\boldsymbol{\Phi}(t)$ is given by (11.48) and $\mathbf{F}(u) = \begin{pmatrix} e^{5u} \\ 4 \end{pmatrix}$. We find that

$$\boldsymbol{\Phi}^{-1}(u) = \begin{pmatrix} -2e^{-3u} & 3e^{-3u} \\ e^{-4u} & -e^{-4u} \end{pmatrix}.$$

Thus (11.49) becomes

$$\begin{aligned}
\boldsymbol{\phi}_0(t) &= \begin{pmatrix} e^{3t} & 3e^{4t} \\ e^{3t} & 2e^{4t} \end{pmatrix} \int_{0}^{t} \begin{pmatrix} -2e^{-3u} & 3e^{-3u} \\ e^{-4u} & -e^{-4u} \end{pmatrix} \begin{pmatrix} e^{5u} \\ 4 \end{pmatrix} du \\
&= \begin{pmatrix} e^{3t} & 3e^{4t} \\ e^{3t} & 2e^{4t} \end{pmatrix} \int_{0}^{t} \begin{pmatrix} -2e^{2u} + 12e^{-3u} \\ e^{u} - 4e^{-4u} \end{pmatrix} du \\
&= \begin{pmatrix} e^{3t} & 3e^{4t} \\ e^{3t} & 2e^{4t} \end{pmatrix} \begin{pmatrix} -e^{2t} - 4e^{-3t} + 5 \\ e^{t} + e^{-4t} - 2 \end{pmatrix} \\
&= \begin{pmatrix} 2e^{5t} - 1 + 5e^{3t} - 6e^{4t} \\ e^{5t} - 2 + 5e^{3t} - 4e^{4t} \end{pmatrix}.
\end{aligned}$$

This, then, is a solution of the nonhomogeneous differential equation (11.45). In component form it is

$$x_1 = 5e^{3t} - 6e^{4t} + 2e^{5t} - 1,$$
$$x_2 = 5e^{3t} - 4e^{4t} + e^{5t} - 2.$$

Observe that it can also be expressed in the form

$$\boldsymbol{\phi}_0(t) = 5\boldsymbol{\phi}_1(t) - 2\boldsymbol{\phi}_2(t) + \boldsymbol{\phi}_0^*(t), \tag{11.50}$$

where $\boldsymbol{\phi}_1$ and $\boldsymbol{\phi}_2$ are the fundamental set of homogeneous differential equation (11.46) defined by (11.47) and $\boldsymbol{\phi}_0^*(t)$ is given by

$$\boldsymbol{\phi}_0^*(t) = \begin{pmatrix} 2e^{5t} - 1 \\ e^{5t} - 2 \end{pmatrix}.$$

The reader should verify that both $\boldsymbol{\phi}_0(t)$ and $\boldsymbol{\phi}_0^*(t)$ satisfy the nonhomogeneous differential equation (11.45) and he should then observe how (11.50) thus illustrates Theorem 11.13 conclusion (1).

THEOREM 11.15

The unique solution $\boldsymbol{\phi}$ *of the nonhomogeneous linear vector differential equation*

$$\frac{d\mathbf{x}}{dt} = \mathbf{A}(t)\mathbf{x} + \mathbf{F}(t) \tag{11.6}$$

that satisfies the initial condition

$$\phi(t_0) = \mathbf{x}_0, \tag{11.51}$$

where $t_0 \in [a, b]$, can be expressed in the form

$$\phi(t) = \Phi(t)\Phi^{-1}(t_0)\mathbf{x}_0 + \Phi(t) \int_{t_0}^{t} \Phi^{-1}(u)\mathbf{F}(u)\, du, \tag{11.52}$$

where $\Phi(t)$ is an arbitrary fundamental matrix of the corresponding homogeneous vector differential equation

$$\frac{d\mathbf{x}}{dt} = \mathbf{A}(t)\mathbf{x} \tag{11.15}$$

Proof. By Theorem 11.13, Conclusion (2), the solution ϕ can be expressed in the form

$$\phi(t) = \sum_{k=1}^{n} c_k \phi_k(t) + \phi_0(t), \tag{11.53}$$

where the c_k are suitably chosen constants, the ϕ_k are a fundamental set of (11.15), and ϕ_0 is any solution of (11.6). The proof of Theorem 11.9 shows that the linear combination $\sum_{k=1}^{n} c_k \phi_k$ in (11.53) can be written as $\Phi\mathbf{c}$, where Φ is the fundamental matrix of (11.15) having the ϕ_k as its individual columns. Also, by Theorem 11.14 a solution ϕ_0 of (11.6) is given by

$$\phi_0(t) = \Phi(t) \int_{t_0}^{t} \Phi^{-1}(u)\mathbf{F}(u)\, du,$$

where Φ is this same fundamental matrix of (11.15). Thus the solution ϕ given by (11.53) takes the form

$$\phi(t) = \Phi(t)\mathbf{c} + \Phi(t) \int_{t_0}^{t} \Phi^{-1}(u)\mathbf{F}(u)\, du. \tag{11.54}$$

Now apply the initial condition $\phi(t_0) = \mathbf{x}_0$, Equation (11.51). Letting $t = t_0$ in (11.54) we obtain

$$\phi(t_0) = \Phi(t_0)\mathbf{c} + \mathbf{0}$$

(since the integral from t_0 to t_0 is zero); and hence application of the initial condition gives

$$\mathbf{x}_0 = \Phi(t_0)\mathbf{c}.$$

Since $|\Phi(t_0)| \neq 0$, $\Phi^{-1}(t_0)$ exists and we find

$$\Phi^{-1}(t_0)\mathbf{x}_0 = \Phi^{-1}(t_0)\Phi(t_0)\mathbf{c} = \mathbf{I}\mathbf{c} = \mathbf{c}.$$

Thus substituting this value of $\mathbf{c}$ back into (11.54) we obtain the solution in the form

$$\phi(t) = \Phi(t)\Phi^{-1}(t_0)\mathbf{x}_0 + \Phi(t) \int_{t_0}^{t} \Phi^{-1}(u)\mathbf{F}(u)\, du. \tag{11.52}$$

Q.E.D.

▶ **Example 11.14**

Find the unique solution ϕ of the nonhomogeneous differential equation

$$\frac{d\mathbf{x}}{dt} = \begin{pmatrix} 6 & -3 \\ 2 & 1 \end{pmatrix}\mathbf{x} + \begin{pmatrix} e^{5t} \\ 4 \end{pmatrix}$$ (11.45)

that satisfies the initial condition

$$\phi(0) = \begin{pmatrix} 9 \\ 4 \end{pmatrix}.$$ (11.55)

By Theorem 11.15, we know that the desired solution can be expressed in the form (11.52), where $\Phi(t)$ is a fundamental matrix of the corresponding homogeneous differential equation

$$\frac{d\mathbf{x}}{dt} = \begin{pmatrix} 6 & -3 \\ 2 & 1 \end{pmatrix}\mathbf{x},$$ (11.46)

[handwritten: $(6-\lambda)(1-\lambda)+6$, $\lambda^2-7\lambda+12$, $(\lambda-3)(\lambda-4)$, $J=\begin{pmatrix}3&0\\0&4\end{pmatrix}$, $tJ=\begin{pmatrix}e^{3t}&0\\0&e^{4t}\end{pmatrix}$]

and where

$$t_0 = 0, \quad \mathbf{x}_0 = \begin{pmatrix} 9 \\ 4 \end{pmatrix}, \quad \text{and} \quad \mathbf{F}(u) = \begin{pmatrix} e^{5u} \\ 4 \end{pmatrix}.$$

From Example 11.13 we know that a fundamental matrix of (11.46) is given by

$$\Phi(t) = \begin{pmatrix} e^{3t} & 3e^{4t} \\ e^{3t} & 2e^{4t} \end{pmatrix}.$$ (11.48)

[handwritten: $3-3$, $2-2$, $v_1=v_2$; $2-3$, $2v_1=3v_2$; $\begin{array}{c}2-3\\2-3\end{array}$]

In that example we also found that

$$\Phi^{-1}(t) = \begin{pmatrix} -2e^{-3t} & 3e^{-3t} \\ e^{-4t} & -e^{-4t} \end{pmatrix},$$

and from this we at once obtain

$$\Phi^{-1}(0) = \begin{pmatrix} -2 & 3 \\ 1 & -1 \end{pmatrix}.$$

Hence, applying (11.52) with these assorted matrices, vectors, and values, we see that the desired solution is given by

$$\phi(t) = \begin{pmatrix} e^{3t} & 3e^{4t} \\ e^{3t} & 2e^{4t} \end{pmatrix}\begin{pmatrix} -2 & 3 \\ 1 & -1 \end{pmatrix}\begin{pmatrix} 9 \\ 4 \end{pmatrix} + \begin{pmatrix} e^{3t} & 3e^{4t} \\ e^{3t} & 2e^{4t} \end{pmatrix}\int_0^t \begin{pmatrix} -2e^{-3u} & 3e^{-3u} \\ e^{-4u} & -e^{-4u} \end{pmatrix}\begin{pmatrix} e^{5u} \\ 4 \end{pmatrix}du.$$ (11.56)

The first term on the right quickly reduces to

$$\begin{pmatrix} e^{3t} & 3e^{4t} \\ e^{3t} & 2e^{4t} \end{pmatrix}\begin{pmatrix} -6 \\ 5 \end{pmatrix} = \begin{pmatrix} -6e^{3t} + 15e^{4t} \\ -6e^{3t} + 10e^{4t} \end{pmatrix}.$$ (11.57)

The second term on the right is precisely the solution $\phi_0(t)$ of Example 11.13, and this was evaluated there and found to be

$$\begin{pmatrix} 5e^{3t} - 6e^{4t} + 2e^{5t} - 1 \\ 5e^{3t} - 4e^{4t} + e^{5t} - 2 \end{pmatrix}.$$ (11.58)

Adding these two constituent parts (11.57) and (11.58) of (11.56), we find the desired solution

$$\phi(t) = \begin{pmatrix} -e^{3t} + 9e^{4t} + 2e^{5t} - 1 \\ -e^{3t} + 6e^{4t} + e^{5t} - 2 \end{pmatrix}.$$

In component form this is

$$x_1 = -e^{3t} + 9e^{4t} + 2e^{5t} - 1,$$
$$x_2 = -e^{3t} + 6e^{4t} + e^{5t} - 2.$$

Exercises

For each of the nonhomogeneous linear systems in Exercises 1–6,
(a)   find a fundamental matrix of the corresponding homogeneous system, and
(b)   find a solution of the given nonhomogeneous system.

1. $\dfrac{d\mathbf{x}}{dt} = \begin{pmatrix} 6 & -3 \\ 2 & 1 \end{pmatrix} \mathbf{x} + \begin{pmatrix} e^{2t} \\ -e^{2t} \end{pmatrix}.$

2. $\dfrac{d\mathbf{x}}{dt} = \begin{pmatrix} 2 & 3 \\ -1 & -2 \end{pmatrix} \mathbf{x} + \begin{pmatrix} 4e^{3t} \\ -e^{3t} \end{pmatrix}.$

3. $\dfrac{d\mathbf{x}}{dt} = \begin{pmatrix} 1 & 2 \\ 3 & 2 \end{pmatrix} \mathbf{x} + \begin{pmatrix} 6e^{t} \\ -6e^{2t} \end{pmatrix}.$

4. $\dfrac{d\mathbf{x}}{dt} = \begin{pmatrix} 3 & -1 \\ 4 & -1 \end{pmatrix} \mathbf{x} + \begin{pmatrix} 2e^{2t} \\ -2 \end{pmatrix}.$

5. $\dfrac{d\mathbf{x}}{dt} = \begin{pmatrix} 3 & 1 \\ 4 & 3 \end{pmatrix} \mathbf{x} + \begin{pmatrix} -2 \sin t \\ 6 \cos t \end{pmatrix}.$

6. $\dfrac{d\mathbf{x}}{dt} = \begin{pmatrix} 1 & -5 \\ 2 & -1 \end{pmatrix} \mathbf{x} + \begin{pmatrix} 5 \cos 2t \\ -5 \sin 2t \end{pmatrix}.$

7.   Find the unique solution ϕ of the nonhomogeneous linear system

$$\frac{d\mathbf{x}}{dt} = \begin{pmatrix} -10 & 6 \\ -12 & 7 \end{pmatrix} \mathbf{x} + \begin{pmatrix} 10e^{-3t} \\ 18e^{-3t} \end{pmatrix}.$$

that satisfies the initial condition $\phi(0) = \begin{pmatrix} 1 \\ -2 \end{pmatrix}.$

8.   Find the unique solution ϕ of the nonhomogeneous linear system

$$\frac{d\mathbf{x}}{dt} = \begin{pmatrix} -1 & 1 \\ -12 & 6 \end{pmatrix} \mathbf{x} + \begin{pmatrix} 3e^{4t} \\ 8e^{4t} \end{pmatrix}$$

that satisfies the initial condition $\phi(0) = \begin{pmatrix} 2 \\ 4 \end{pmatrix}.$

11.5 BASIC THEORY OF THE *n*th-ORDER HOMOGENEOUS LINEAR DIFFERENTIAL EQUATION

A. Fundamental Results

Throughout Sections 11.5 and 11.6 we shall be concerned with the single *n*th-order homogeneous linear scalar differential equation

$$a_0(t)\frac{d^n x}{dt^n} + a_1(t)\frac{d^{n-1}x}{dt^{n-1}} + \cdots + a_{n-1}(t)\frac{dx}{dt} + a_n(t)x = 0, \tag{11.59}$$

where $a_0, a_1, \ldots, a_{n-1}, a_n$ are continuous on an interval $a \le t \le b$ and $a_0(t) \ne 0$ on $a \le t \le b$. Let L_n denote the formal *n*th-order linear differential operator defined by

$$L_n = a_0(t)\frac{d^n}{dt^n} + a_1(t)\frac{d^{n-1}}{dt^{n-1}} + \cdots + a_{n-1}(t)\frac{d}{dt} + a_n(t). \tag{11.60}$$

Then differential equation (11.59) may be written

$$L_n x = 0. \tag{11.61}$$

If we divide through by $a_0(t)$ on $a \le t \le b$, we obtain the equation

$$\frac{d^n x}{dt^n} + \frac{a_1(t)}{a_0(t)}\frac{d^{n-1}x}{dt^{n-1}} + \cdots + \frac{a_{n-1}(t)}{a_0(t)}\frac{dx}{dt} + \frac{a_n(t)}{a_0(t)}x = 0. \tag{11.62}$$

This equation, in which the coefficient of $\dfrac{d^n x}{dt^n}$ is 1, is said to be normalized.

The basic existence theorem for the initial-value problem associated with the *n*th-order homogeneous linear scalar differential equation (11.59) is the following special case of Theorem 10.8.

THEOREM 11.16

Hypothesis

1. Consider the differential equation

$$a_0(t)\frac{d^n x}{dt^n} + a_1(t)\frac{d^{n-1}x}{dt^{n-1}} + \cdots + a_{n-1}(t)\frac{dx}{dt} + a_n(t)x = 0, \tag{11.59}$$

where $a_0, a_1, \ldots, a_{n-1}$, and a_n are continuous on the interval $a \le t \le b$, and $a_0(t) \ne 0$ on $a \le t \le b$.

2. Let t_0 be a point of the interval $a \le t \le b$, and let $c_0, c_1, \ldots, c_{n-1}$ be a set of n real constants.

Conclusion. *There exists a unique solution ϕ of (11.59) such that*

$$\phi(t_0) = c_0, \quad \phi'(t_0) = c_1, \ldots, \phi^{(n-1)}(t_0) = c_{n-1}, \tag{11.63}$$

and this solution is defined over the entire interval $a \le t \le b$.

The following important corollary to this theorem will soon be very useful to us.

COROLLARY

Hypothesis. *The function ϕ is a solution of the homogeneous equation (11.59) such that*

$$\phi(t_0) = 0, \quad \phi'(t_0) = 0,\ldots, \phi^{(n-1)}(t_0) = 0, \tag{11.64}$$

where t_0 is a point of an interval $a \le t \le b$ on which the coefficients $a_0, a_1,\ldots, a_n$ are all continuous and $a_0(t) \neq 0$.

Conclusion. *$\phi(t) = 0$ for all t such that $a \le t \le b$.*

Proof. First note that ϕ such that $\phi(t) = 0$ for all $t \in [a, b]$ is indeed a solution of the differential equation (11.59) which satisfies the initial conditions (11.64). But by Theorem 11.16 the initial-value problem composed of Equation (11.59) and conditions (11.64) has a unique solution on $a \le t \le b$. Hence the stated conclusion follows.

In the statements of this theorem and corollary we have explicitly stated the hypothesis that the coefficients $a_0, a_1,\ldots, a_n$ are all continuous on the interval $a \le t \le b$ under consideration and that $a_0(t) \neq 0$ there. In the remainder of this section and in Section 11.6, we shall always make this assumption whenever we refer to the homogeneous linear differential equation (11.59). That is, we always assume that in homogeneous linear differential equation (11.59) *the coefficients $a_0, a_1,\ldots, a_n$ are all continuous on the interval $a \le t \le b$* under consideration and that *the leading coefficient $a_0(t) \neq 0$ on this interval.*

In the proofs of Theorem 10.6 and 10.8 we noted that a single nth-order differential equation is closely related to a certain system of n first-order differential equations. We now investigate this relationship more carefully in the case of the nth-order homogeneous linear scalar differential equation

$$a_0(t)\frac{d^n x}{dt^n} + a_1(t)\frac{d^{n-1}x}{dt^{n-1}} + \cdots + a_{n-1}(t)\frac{dx}{dt} + a_n(t)x = 0. \tag{11.59}$$

Let

$$x_1 = x$$

$$x_2 = \frac{dx}{dt}$$

$$x_3 = \frac{d^2 x}{dt^2}$$

$$\vdots \tag{11.65}$$

$$x_{n-1} = \frac{d^{n-2}x}{dt^{n-2}}$$

$$x_n = \frac{d^{n-1}x}{dt^{n-1}}$$

Differentiating (11.65), we have

$$\frac{dx}{dt} = \frac{dx_1}{dt}, \quad \frac{d^2x}{dt^2} = \frac{dx_2}{dt}, \dots, \quad \frac{d^{n-1}x}{dt^{n-1}} = \frac{dx_{n-1}}{dt}, \quad \frac{d^nx}{dt^n} = \frac{dx_n}{dt}. \tag{11.66}$$

The first $n-1$ equations of (11.66) and the last $n-1$ equations of (11.65) at once give

$$\frac{dx_1}{dt} = x_2, \quad \frac{dx_2}{dt} = x_3, \dots, \quad \frac{dx_{n-1}}{dt} = x_n. \tag{11.67}$$

Now assuming $a_0(t) \neq 0$ on $a \leq t \leq b$, (11.59) is equivalent to

$$\frac{d^nx}{dt^n} = -\frac{a_n(t)}{a_0(t)} x - \frac{a_{n-1}(t)}{a_0(t)} \frac{dx}{dt} - \dots - \frac{a_1(t)}{a_0(t)} \frac{d^{n-1}x}{dt^{n-1}}.$$

Using both (11.65) and (11.66), this becomes

$$\frac{dx_n}{dt} = -\frac{a_n(t)}{a_0(t)} x_1 - \frac{a_{n-1}(t)}{a_0(t)} x_2 - \dots - \frac{a_1(t)}{a_0(t)} x_n. \tag{11.68}$$

Combining (11.67) and (11.68), we have

$$\frac{dx_1}{dt} = x_2$$

$$\frac{dx_2}{dt} = x_3$$

$$\vdots \tag{11.69}$$

$$\frac{dx_{n-1}}{dt} = x_n$$

$$\frac{dx_n}{dt} = -\frac{a_n(t)}{a_0(t)} x_1 - \frac{a_{n-1}(t)}{a_0(t)} x_2 - \dots - \frac{a_1(t)}{a_0(t)} x_n.$$

This is a special homogeneous linear system of the form (11.1). In vector notation, it is the homogeneous linear vector differential equation

$$\frac{d\mathbf{x}}{dt} = \mathbf{A}(t)\mathbf{x}, \tag{11.70}$$

where

$$\mathbf{x} = \begin{pmatrix} x_1 \\ x_2 \\ \vdots \\ x_n \end{pmatrix} \quad \text{and} \quad \mathbf{A} = \begin{pmatrix} 0 & 1 & 0 & \cdots & 0 & 0 \\ 0 & 0 & 1 & \cdots & 0 & 0 \\ \vdots & \vdots & \vdots & \cdots & \vdots & \vdots \\ 0 & 0 & 0 & & 0 & 1 \\ -\dfrac{a_n}{a_0} & -\dfrac{a_{n-1}}{a_0} & -\dfrac{a_{n-2}}{a_0} & \cdots & -\dfrac{a_2}{a_0} & -\dfrac{a_1}{a_0} \end{pmatrix}.$$

The equation (11.70) is sometimes called the *companion vector equation* of *n*th-order scalar equation (11.59).

Now suppose f satisfies the nth-order homogeneous linear differential equation (11.59). Then

$$a_0(t)f^{(n)}(t) + a_1(t)f^{(n-1)}(t) + \cdots + a_{n-1}(t)f'(t) + a_n(t)f(t) = 0, \qquad (11.72)$$

for $t \in [a, b]$. Consider the vector $\boldsymbol{\phi}$ defined by

$$\boldsymbol{\phi}(t) = \begin{pmatrix} \phi_1(t) \\ \phi_2(t) \\ \phi_3(t) \\ \vdots \\ \phi_{n-1}(t) \\ \phi_n(t) \end{pmatrix} = \begin{pmatrix} f(t) \\ f'(t) \\ f''(t) \\ \vdots \\ f^{(n-2)}(t) \\ f^{(n-1)}(t) \end{pmatrix}. \qquad (11.73)$$

From (11.72) and (11.73) we see at once that

$$\phi_1'(t) = \phi_2(t),$$
$$\phi_2'(t) = \phi_3(t),$$
$$\vdots \qquad\qquad (11.74)$$
$$\phi_{n-1}'(t) = \phi_n(t),$$
$$\phi_n'(t) = -\frac{a_n(t)}{a_0(t)}\phi_1(t) - \frac{a_{n-1}(t)}{a_0(t)}\phi_2(t) - \cdots - \frac{a_1(t)}{a_0(t)}\phi_n(t).$$

Comparing this with (11.69), we see that the vector $\boldsymbol{\phi}$ defined by (11.73) satisfies the system (11.69).

Conversely, suppose

$$\boldsymbol{\phi}(t) = \begin{pmatrix} \phi_1(t) \\ \phi_2(t) \\ \vdots \\ \phi_n(t) \end{pmatrix}$$

satisfies system (11.69) on $[a, b]$. Then (11.74) holds for all $t \in [a, b]$. The first $n-1$ equations of (11.74) give

$$\phi_2(t) = \phi_1'(t)$$
$$\phi_3(t) = \phi_2'(t) = \phi_1''(t) \qquad\qquad (11.75)$$
$$\vdots$$
$$\phi_n(t) = \phi_{n-1}'(t) = \phi_{n-2}''(t) = \cdots = \phi_1^{[n-1]}(t),$$

and so $\phi_n'(t) = \phi_1^{[n]}(t)$. The last equation of (11.74) then becomes

$$\phi_1^{[n]}(t) = -\frac{a_n(t)}{a_0(t)}\phi_1(t) - \frac{a_{n-1}(t)}{a_0(t)}\phi_1'(t) - \cdots - \frac{a_1(t)}{a_0(t)}\phi_1^{[n-1]}(t)$$

or

$$a_0(t)\phi_1^{[n]}(t) + a_1(t)\phi_1^{[n-1]}(t) + \cdots + a_{n-1}(t)\phi_1'(t) + a_n(t)\phi_1(t) = 0.$$

Thus ϕ_1 is a solution f of the nth-order homogeneous linear differential equation

(11.59) and moreover (11.75) shows that, in fact,

$$\phi(t) = \begin{pmatrix} f(t) \\ f'(t) \\ f''(t) \\ \vdots \\ f^{(n-1)}(t) \end{pmatrix}.$$

We have thus obtained the following fundamental result.

THEOREM 11.17

Consider the nth-order homogeneous linear differential equation

$$a_0(t)\frac{d^n x}{dt^n} + a_1(t)\frac{d^{n-1}x}{dt^{n-1}} + \cdots + a_{n-1}(t)\frac{dx}{dt} + a_n(t)x = 0, \tag{11.59}$$

and the corresponding homogeneous linear system (11.69). If f is a solution of (11.59) on [a, b], *then*

$$\phi = \begin{pmatrix} f \\ f' \\ f'' \\ \vdots \\ f^{(n-1)} \end{pmatrix} \tag{11.76}$$

is a solution of (11.69) on [a, b]. *Conversely, if*

$$\phi = \begin{pmatrix} \phi_1 \\ \phi_2 \\ \vdots \\ \phi_n \end{pmatrix}$$

is a solution of (11.69) on [a, b], *then its first component ϕ_1 is a solution f of (11.59) on* [a, b] *and ϕ is, in fact, of the form (11.76).*

The following corollary is an immediate consequence of the theorem.

COROLLARY TO THEOREM 11.17

If f is the solution of the nth-order homogeneous linear differential equation (11.59) on [a, b] *satisfying the initial condition*

$$f(t_0) = c_0, \quad f'(t_0) = c_1, \quad f''(t_0) = c_2, \dots, f^{(n-1)}(t_0) = c_{n-1},$$

then

$$\phi = \begin{pmatrix} f \\ f' \\ f'' \\ \vdots \\ f^{(n-1)} \end{pmatrix}$$

is the solution of corresponding homogeneous linear system (11.69) on $[a, b]$ satisfying the initial condition

$$\phi(t_0) = \begin{pmatrix} c_0 \\ c_1 \\ c_2 \\ \vdots \\ c_n \end{pmatrix}.$$

Conversely, if

$$\phi = \begin{pmatrix} \phi_1 \\ \phi_2 \\ \vdots \\ \phi_n \end{pmatrix}$$

is a solution of (11.69) on $[a, b]$ satisfying the initial condition

$$\phi_1(t_0) = c_0, \quad \phi_2(t_0) = c_1, \quad \phi_3(t_0) = c_2, \ldots, \phi_n(t_0) = c_{n-1},$$

then ϕ_1 is the solution f of (11.59) on $[a, b]$ satisfying the initial condition

$$f(t_0) = c_0, \quad f'(t_0) = c_1, \quad f''(t_0) = c_2, \ldots, f^{(n-1)}(t_0) = c_{n-1}.$$

We shall now employ these stated relationships between the solutions of (11.59) and (11.69) to obtain results about the solutions of (11.59) from the corresponding theorems about solutions of (11.69). We first obtain the following basic result.

THEOREM 11.18

A linear combination of m solutions of the nth-order homogeneous linear differential equation

$$a_0(t)\frac{d^n x}{dt^n} + a_1(t)\frac{d^{n-1}x}{dt^{n-1}} + \cdots + a_{n-1}(t)\frac{dx}{dt} + a_n(t)x = 0 \qquad (11.59)$$

is also a solution of (11.59). That is, if the functions $f_1, f_2, \ldots, f_m$ are m solutions of (11.59) and $c_1, c_2, \ldots, c_m$ are m numbers, then the function

$$f = \sum_{k=1}^{m} c_k f_k$$

is also a solution of (11.59).

Proof. By hypothesis, $f_1, f_2, \ldots, f_m$ are solutions of (11.59). Therefore by Theorem 11.17,

$$\phi_1 = \begin{pmatrix} f_1 \\ f'_1 \\ \vdots \\ f_1^{(n-1)} \end{pmatrix}, \quad \phi_2 = \begin{pmatrix} f_2 \\ f'_2 \\ \vdots \\ f_2^{(n-1)} \end{pmatrix}, \ldots, \phi_m = \begin{pmatrix} f_m \\ f'_m \\ \vdots \\ f_m^{(n-1)} \end{pmatrix}$$

are solutions of homogeneous linear system (11.69). Then by Theorem 11.2,

$$\phi = \sum_{k=1}^{m} c_k \phi_k$$

is also a solution of (11.69). That is,

$$\begin{pmatrix} \sum_{k=1}^{m} c_k f_k \\ \sum_{k=1}^{m} c_k f'_k \\ \vdots \\ \sum_{k=1}^{m} c_k f_k^{(n-1)} \end{pmatrix}$$

is a solution of (11.69). Then by the converse part of Theorem 11.18, the first component of this,

$$\sum_{k=1}^{m} c_k f_k,$$

is a solution of (11.59).

Q.E.D.

We now show that the relationship between solutions of (11.59) and (11.69) preserves linear independence. We first review the definitions of linear dependence and independence for solutions of (11.59) and solutions of (11.69).

B. Fundamental Sets of Solutions

DEFINITION

The m solutions $f_1, f_2, \ldots, f_m$ of the nth-order homogeneous linear differential equation (11.59) are linearly dependent *on $a \le t \le b$ if and only if there exist m numbers $c_1, c_2, \ldots, c_m$, not all zero, such that*

$$c_1 f_1(t) + c_2 f_2(t) + \cdots + c_m f_m(t) = 0$$

for all t such that $a \le t \le b$.

DEFINITION

The m solutions $f_1, f_2, \ldots, f_m$ of the nth-order homogeneous linear differential equation (11.59) are linearly independent *on $a \le t \le b$ if and only if they are not linearly dependent on the interval. That is, the solutions $f_1, f_2, \ldots, f_m$ are linearly independent on $a \le t \le b$ if and only if*

$$c_1 f_1(t) + c_2 f_2(t) + \cdots + c_m f_m(t) = 0$$

for all t such that $a \le t \le b$ implies that

$$c_1 = c_2 = \cdots = c_m = 0.$$

DEFINITION

The m solutions $\phi_1, \phi_2, \ldots, \phi_m$ *of the homogeneous linear system* (*11.69*) *are* linearly dependent *on* $a \leq t \leq b$ *if there exists m numbers* $c_1, c_2, \ldots, c_m$, *not all zero, such that*

$$c_1\phi_1(t) + c_2\phi_2(t) + \cdots + c_m\phi_m(t) = 0$$

for all t such that $a \leq t \leq b$.

DEFINITION

The m solutions $\phi_1, \phi_2, \ldots, \phi_m$ *of the homogeneous linear system* (*11.69*) *are* linearly independent *on* $a \leq t \leq b$ *if and only if they are not linearly dependent on this interval. That is, the solutions* $\phi_1, \phi_2, \ldots, \phi_m$ *are linearly independent on* $a \leq t \leq b$ *if and only if the relation*

$$c_1\phi_1(t) + c_2\phi_2(t) + \cdots + c_m\phi_m(t) = 0$$

for all t such that $a \leq t \leq b$ *implies that*

$$c_1 = c_2 = \cdots = c_m = 0.$$

THEOREM 11.19

If the solutions $f_1, f_2, \ldots, f_n$ *of the nth-order homogeneous linear differential equation* (*11.59*) *are linearly independent on* $a \leq t \leq b$, *then the corresponding solutions*

$$\phi_1 = \begin{pmatrix} f_1 \\ f'_1 \\ \vdots \\ f_1^{(n-1)} \end{pmatrix}, \quad \phi_2 = \begin{pmatrix} f_2 \\ f'_2 \\ \vdots \\ f_2^{(n-1)} \end{pmatrix}, \ldots, \phi_n = \begin{pmatrix} f_n \\ f'_n \\ \vdots \\ f_n^{(n-1)} \end{pmatrix}$$

of the homogeneous linear system (*11.69*) *are linearly independent on* $a \leq t \leq b$; *and conversely.*

Proof. If $f_1, f_2, \ldots, f_n$ are linearly independent on $a \leq t \leq b$, then the relation

$$c_1 f_1(t) + c_2 f_2(t) + \cdots + c_n f_n(t) = 0 \tag{11.77}$$

for all $t \in [a, b]$ implies that $c_1 = c_2 = \cdots = c_n = 0$. Now suppose

$$c_1\phi_1(t) + c_2\phi_2(t) + \cdots + c_n\phi_n(t) = 0 \tag{11.78}$$

for all $t \in [a, b]$. The first scalar component relation of the vector relation (11.78) is precisely (11.77). So if (11.78) holds for all $t \in [a, b]$, so does (11.77), and this implies $c_1 = c_2 = \cdots = c_n = 0$. Thus if (11.78) holds for all $t \in [a, b]$, then we must have $c_1 = c_2 = \cdots = c_n = 0$; and therefore $\phi_1, \phi_2, \ldots, \phi_n$ are linearly independent on $a \leq t \leq b$.

Now suppose $f_1, f_2, \ldots, f_n$ are linearly dependent on $a \leq t \leq b$. Then there exist numbers $c_1, c_2, \ldots, c_n$, not all zero, such that

$$c_1 f_1(t) + c_2 f_2(t) + \cdots + c_n f_n(t) = 0 \tag{11.79}$$

for all $t \in [a, b]$. Differentiating this $n - 1$ times we obtain

$$c_1 f'_1(t) + c_2 f'_2(t) + \cdots + c_n f'_n(t) = 0,$$
$$c_1 f''_1(t) + c_2 f''_2(t) + \cdots + c_n f''_n(t) = 0,$$
$$\vdots \qquad\qquad (11.80)$$
$$c_1 f_1^{(n-1)}(t) + c_2 f_2^{(n-1)}(t) + \cdots + c_n f_n^{(n-1)}(t) = 0,$$

all of which hold for all $t \in [a, b]$. Relations (11.79) and (11.80) together are n scalar relations which are equivalent to the single vector relation

$$c_1 \phi_1(t) + c_2 \phi_2(t) + \cdots + c_n \phi_n(t) = 0 \qquad (11.81)$$

for all $t \in [a, b]$. Thus there exist numbers $c_1, c_2, \ldots, c_n$, not all zero, such that (11.81) holds for all $t \in [a, b]$; and so $\phi_1, \phi_2, \ldots, \phi_n$ are linearly dependent on $a \le t \le b$. Thus the linear dependence of $f_1, f_2, \ldots, f_n$ on $a \le t \le b$ implies that of $\phi_1, \phi_2, \ldots, \phi_n$ on the same interval; and so the linear independence of $\phi_1, \phi_2, \ldots, \phi_n$ on $a \le t \le b$ implies that of $f_1, f_2, \ldots, f_n$ on this interval, which is the converse part of the theorem.

Q.E.D.

DEFINITION

Let $f_1, f_2, \ldots, f_n$ be n real functions, each of which has an $(n - 1)$st derivative on $a \le x \le b$. The determinant

$$\begin{vmatrix} f_1 & f_2 & \cdots & f_n \\ f'_1 & f'_2 & \cdots & f'_n \\ \vdots & \vdots & & \vdots \\ f_1^{(n-1)} & f_2^{(n-1)} & & f_n^{(n-1)} \end{vmatrix}$$

where the primes denote derivatives, is called the Wronskian *of the n functions $f_1, f_2, \ldots, f_n$. We denote it by $W(f_1, f_2, \ldots, f_n)$ and denote its value at t by $W(f_1, f_2, \ldots, f_n)(t)$.*

Let $f_1, f_2, \ldots, f_n$ be n solutions of the nth-order homogeneous linear differential equation (11.59) on $a \le t \le b$, and let

$$\phi_1 = \begin{pmatrix} f_1 \\ f'_1 \\ \vdots \\ f_1^{(n-1)} \end{pmatrix}, \quad \phi_2 = \begin{pmatrix} f_2 \\ f'_2 \\ \vdots \\ f_2^{(n-1)} \end{pmatrix}, \ldots, \phi_n = \begin{pmatrix} f_n \\ f'_n \\ \vdots \\ f_n^{(n-1)} \end{pmatrix}$$

be the corresponding solutions of homogeneous linear system (11.69) on $a \le t \le b$. By definition, the Wronskian of the n solutions $f_1, f_2, \ldots, f_n$ of (11.59) is

$$\begin{vmatrix} f_1 & f_2 & \cdots & f_n \\ f'_1 & f'_2 & \cdots & f'_n \\ \vdots & \vdots & & \vdots \\ f_1^{(n-1)} & f_2^{(n-1)} & \cdots & f_n^{(n-1)} \end{vmatrix}.$$

Now note that by definition of the Wronskian of n solutions of (11.69) (Section 11.2), this is also the Wronskian of the n solutions $\phi_1, \phi_2, \ldots, \phi_n$ of (11.69). That is,

$$W(f_1, f_2, \ldots, f_n)(t) = W(\phi_1, \phi_2, \ldots, \phi_n)(t) \qquad (11.82)$$

for all $t \in [a, b]$. By Theorem 11.5, either $W(\phi_1, \phi_2, \ldots, \phi_n)(t) = 0$ for all $t \in [a, b]$ or $W(\phi_1, \phi_2, \ldots, \phi_n)(t) = 0$ for no $t \in [a, b]$. Thus either $W(f_1, f_2, \ldots, f_n)(t) = 0$ for all $t \in [a, b]$ or $W(f_1, f_2, \ldots, f_n)(t) = 0$ for no $t \in [a, b]$. Thus we have obtained the following result.

THEOREM 11.20

Let $f_1, f_2, \ldots, f_n$ be n solutions of the nth-order homogeneous linear differential equation (11.59) on $a \le t \le b$. Then either $W(f_1, f_2, \ldots, f_n)(t) = 0$ for all $t \in [a, b]$ or $W(f_1, f_2, \ldots, f_n)(t) = 0$ for no $t \in [a, b]$.

By Theorem 11.19 the solutions $f_1, f_2, \ldots, f_n$ of (11.59) are linearly independent on $[a, b]$ if and only if the corresponding solutions

$$\phi_1 = \begin{pmatrix} f_1 \\ f'_1 \\ \vdots \\ f_1^{(n-1)} \end{pmatrix}, \quad \phi_2 = \begin{pmatrix} f_2 \\ f'_2 \\ \vdots \\ f_2^{(n-1)} \end{pmatrix}, \ldots, \phi_n = \begin{pmatrix} f_n \\ f'_n \\ \vdots \\ f_n^{(n-1)} \end{pmatrix}$$

of (11.69) are linearly independent on $[a, b]$. By Theorem 11.6, the solutions $\phi_1, \phi_2, \ldots, \phi_n$ of (11.69) are linearly independent on $[a, b]$ if and only if $W(\phi_1, \phi_2, \ldots, \phi_n)(t) \ne 0$ for all $t \in [a, b]$. Thus, using (11.82), $\phi_1, \phi_2, \ldots, \phi_n$ are linearly independent on $[a, b]$ if and only if $W(f_1, f_2, \ldots, f_n)(t) \ne 0$ for all $t \in [a, b]$. Combining this with the first sentence of this paragraph, we see that $f_1, f_2, \ldots, f_n$ are linearly independent on $[a, b]$ if and only if $W(f_1, f_2, \ldots, f_n)(t) \ne 0$ for all $t \in [a, b]$. Thus we have the following results.

THEOREM 11.21

Let $f_1, f_2, \ldots, f_n$ be n solutions of the homogeneous linear differential equation

$$a_0(t) \frac{d^n x}{dt^n} + a_1(t) \frac{d^{n-1} x}{dt^{n-1}} + \cdots + a_{n-1}(t) \frac{dx}{dt} + a_n(t)x = 0 \qquad (11.59)$$

on $[a, b]$. These n solutions $f_1, f_2, \ldots, f_n$ of (11.59) are linearly independent on $[a, b]$ if and only if

$$W(f_1, f_2, \ldots, f_n)(t) \ne 0$$

for all $t \in [a, b]$.

Using this result and that of Theorem 11.20, we at once obtain the following corollary.

COROLLARY TO THEOREM 11.21

Let $f_1, f_2, \ldots, f_n$ be n solutions of the homogeneous linear differential equation (11.59) on $[a, b]$. These n solutions are linearly dependent on $[a, b]$ if and only if $W(f_1, f_2, \ldots, f_n)(t) = 0$ for all $t \in [a, b]$.

▶   **Example 11.15**

Consider the third-order homogeneous linear differential equation

$$t^3 \frac{d^3 x}{dt^3} - 4t^2 \frac{d^2 x}{dt^2} + 8t \frac{dx}{dt} - 8x = 0. \tag{11.83}$$

The overall hypothesis of this section that the coefficients be continuous and the leading coefficient be unequal to zero holds on every interval $a \leq t \leq b$ not including zero. One can easily verify that the functions f_1, f_2, f_3 defined respectively by $f_1(t) = t$, $f_2(t) = t^2$, $f_3(t) = t^4$ are all solutions of (11.83) on any such interval. Evaluating the Wronskian of these solutions, we find

$$W(f_1, f_2, f_3)(t) = W(t, t^2, t^4) = \begin{vmatrix} t^4 & t^2 & t \\ 4t^3 & 2t & 1 \\ 12t^2 & 2 & 0 \end{vmatrix} = -6t^4 \neq 0.$$

Thus by Theorem 11.21, the three solutions f_1, f_2, f_3 are linearly independent on every interval $a \leq t \leq b$, not including zero.

DEFINITION

A set of n linearly independent solutions of the nth-order homogeneous linear differential equation (11.59) is called a fundamental set *of (11.59).*

THEOREM 11.22

There exist fundamental sets of solutions of the nth-order homogeneous linear differential equation

$$a_0(t) \frac{d^n x}{dt^n} + a_1(t) \frac{d^{n-1} x}{dt^{n-1}} + \cdots + a_{n-1}(t) \frac{dx}{dt} + a_n(t)x = 0 \tag{11.59}$$

on $a \leq t \leq b$.

   Proof.   By Theorem 11.7 there exists fundamental sets of the corresponding homogeneous linear system (11.69) on $[a, b]$. Let

$$\Phi_1 = \begin{pmatrix} f_1 \\ f'_1 \\ \vdots \\ f_1^{(n-1)} \end{pmatrix}, \quad \Phi_2 = \begin{pmatrix} f_2 \\ f'_2 \\ \vdots \\ f_2^{(n-1)} \end{pmatrix}, \dots, \Phi_n = \begin{pmatrix} f_n \\ f'_n \\ \vdots \\ f_n^{(n-1)} \end{pmatrix}$$

be such a set. By definition of fundamental set of (11.69), these solutions $\Phi_1, \Phi_2, \dots, \Phi_n$ of (11.69) are linearly independent on $[a, b]$. Then by the converse part of Theorem 11.19, the corresponding solutions $f_1, f_2, \dots, f_n$ of (11.59) are linearly independent on $[a, b]$. Finally, by definition of fundamental set of (11.59) these solutions $f_1, f_2, \dots, f_n$ constitute a fundamental set of (11.59) on $[a, b]$.   *Q.E.D.*

We now exhibit a fundamental set of (11.59). Consider the solutions

$$\mathbf{\Phi}_1 = \begin{pmatrix} f_1 \\ f'_1 \\ \vdots \\ f_1^{(n-1)} \end{pmatrix}, \quad \mathbf{\Phi}_2 = \begin{pmatrix} f_2 \\ f'_2 \\ \vdots \\ f_2^{(n-1)} \end{pmatrix}, \dots, \mathbf{\Phi}_n = \begin{pmatrix} f_n \\ f'_n \\ \vdots \\ f_n^{(n-1)} \end{pmatrix}$$

of corresponding homogeneous linear system (11.69) defined respectively by

$$\mathbf{\Phi}_1(t_0) = \begin{pmatrix} 1 \\ 0 \\ 0 \\ \vdots \\ 0 \\ 0 \end{pmatrix}, \quad \mathbf{\Phi}_2(t_0) = \begin{pmatrix} 0 \\ 1 \\ 0 \\ \vdots \\ 0 \\ 0 \end{pmatrix}, \dots, \mathbf{\Phi}_n(t_0) = \begin{pmatrix} 0 \\ 0 \\ 0 \\ \vdots \\ 0 \\ 1 \end{pmatrix},$$

that is, in general, by

$$f_i^{(j)}(t_0) = \begin{cases} 0, & j \neq i - 1, \\ 1, & j = i - 1, \end{cases} \quad (i = 1, 2, \dots, n; \, j = 0, 1, \dots, n - 1). \tag{11.84}$$

By the Corollary to Theorem 11.17, $f_1, f_2, \dots, f_n$ is a set of solutions of (11.59) which satisfy the same initial condition (11.84). That is, $f_1, f_2, \dots, f_n$ are n solutions of (11.59) such that

$$
\begin{array}{lll}
f_1(t_0) = 1 & f_2(t_0) = 0 & f_n(t_0) = 0 \\
f'_1(t_0) = 0 & f'_2(t_0) = 1 & f'_n(t_0) = 0 \\
f''_1(t_0) = 0 & f''_2(t_0) = 0 & f''_n(t_0) = 0 \\
\quad \vdots & \quad \vdots \quad \cdots & \quad \vdots \\
f_1^{(n-2)}(t_0) = 0 & f_2^{(n-2)}(t_0) = 0 & f_n^{(n-2)}(t_0) = 0 \\
f_1^{(n-1)}(t_0) = 0 & f_2^{(n-1)}(t_0) = 0 & f_n^{(n-1)}(t_0) = 1.
\end{array}
\tag{11.84}
$$

Thus the Wronskian $W(f_1, f_2, \dots, f_n)(t_0) = 1$, that is, is not zero; and so by Theorem 11.20, $W(f_1, f_2, \dots, f_n)(t) \neq 0$ for all $t \in [a, b]$. Thus by Theorem 11.21, the solutions $f_1, f_2, \dots, f_n$ of (11.59) so defined form a fundamental set of (11.59) on $[a, b]$.

We thus see that any n solutions $f_1, f_2, \dots, f_n$ of Equation (11.59) which satisfy conditions of the form (11.84) at some point t_0 of $a \leq t \leq b$ are linearly independent on $a \leq t \leq b$. However, there are also sets of n linearly independent solutions of Equation (11.59) which do not satisfy conditions of the form (11.84) at any one point t_0 of $a \leq t \leq b$. For example, the functions f_1 and f_2 defined respectively by $f_1(t) = \sin t$, $f_2(t) = \sin t + \cos t$ are a set of linearly independent solutions of

$$\frac{d^2 x}{dt^2} + x = 0$$

on every closed interval $a \leq t \leq b$, but there is no point t_0 such that $f_1(t_0) = 1$, $f'_1(t_0) = 0$, $f_2(t_0) = 0$, $f'_2(t_0) = 1$.

THEOREM 11.23

Let $f_1, f_2, \ldots, f_n$ be a fundamental set of solutions of the nth-order homogeneous linear differential equation

$$a_0(t)\frac{d^n x}{dt^n} + a_1(t)\frac{d^{n-1} x}{dt^{n-1}} + \cdots + a_{n-1}(t)\frac{dx}{dt} + a_n(t)x = 0, \qquad (11.59)$$

and let f be an arbitrary solution of (11.59) on $[a, b]$. Then f can be represented as a suitable linear combination of $f_1, f_2, \ldots, f_n$. That is, there exist numbers $c_1, c_2, \ldots, c_n$ such that

$$f = c_1 f_1 + c_2 f_2 + \cdots + c_n f_n \qquad (11.85)$$

on $[a, b]$.

Proof. Let

$$\Phi = \begin{pmatrix} f \\ f' \\ \vdots \\ f^{(n-1)} \end{pmatrix}, \quad \Phi_1 = \begin{pmatrix} f_1 \\ f_1' \\ \vdots \\ f_1^{(n-1)} \end{pmatrix}, \quad \Phi_2 = \begin{pmatrix} f_2 \\ f_2' \\ \vdots \\ f_2^{(n-1)} \end{pmatrix}, \ldots, \Phi_n = \begin{pmatrix} f_n \\ f_n' \\ \vdots \\ f_n^{(n-1)} \end{pmatrix}$$

be the solutions of the related homogeneous linear system (11.69) corresponding respectively to the solutions $f, f_1, f_2, \ldots, f_n$ of (11.59). Since $f_1, f_2, \ldots, f_n$ are a fundamental set of (11.59) on $[a, b]$, they are linearly independent on $[a, b]$. Then by Theorem 11.19, $\Phi_1, \Phi_2, \ldots, \Phi_n$ are linearly independent on $[a, b]$; that is, they constitute a fundamental set of (11.69) on $[a, b]$. Thus by Theorem 11.8 there exist numbers $c_1, c_2, \ldots, c_n$ such that

$$\Phi = c_1 \Phi_1 + c_2 \Phi_2 + \cdots + c_n \Phi_n.$$

The first component of this vector relation is

$$f = c_1 f_1 + c_2 f_2 + \cdots + c_n f_n. \qquad (11.85)$$

Thus there exist numbers $c_1, c_2, \ldots, c_n$ such that (11.85) holds. *Q.E.D.*

▶ **Example 11.16**

Every solution f of

$$\frac{d^2 x}{dt^2} + x = 0$$

can be expressed as a linear combination of any two linearly independent solutions f_1 and f_2. For example, if $f(t) = 2 \sin t + 4 \cos t$, $f_1(t) = \sin t$, and $f_2(t) = \cos t$, we have $f(t) = 2f_1(t) + 4f_2(t)$; while if $f_1(t) = \sin t$ and $f_2(t) = \sin t + \cos t$, we have $f(t) = -2f_1(t) + 4f_2(t)$.

Exercises

1. Consider the third-order homogeneous linear differential equation

    $$\frac{d^3x}{dt^3} - \frac{d^2x}{dt^2} + \frac{dx}{dt} - x = 0.$$

    (a) Find the three linearly independent solutions f_1, f_2, f_3 of this equation which are such that

    $$f_1(0) = 1, \qquad f_2(0) = 0, \qquad f_3(0) = 0,$$
    $$f'_1(0) = 0, \qquad f'_2(0) = 1, \qquad f'_3(0) = 0,$$
    $$f''_1(0) = 0, \qquad f''_2(0) = 0, \qquad f''_3(0) = 1,$$

    respectively.

    (b) Express the solution

    $$2 \sin t + 3 \cos t + 5e^t$$

    as a linear combination of the three linearly independent solutions f_1, f_2, and f_3 defined in part (a).

2. (a) Show that e^t, $e^t(t - 1)$, and $2e^t - e^{2t}$ are linearly independent solutions of

    $$\frac{d^3x}{dt^3} - 4\frac{d^2x}{dt^2} + 5\frac{dx}{dt} - 2x = 0.$$

    (b) Express the solution $2e^t + 3te^t + 4e^{2t}$ as a linear combination of the three linearly independent solutions of part (a).

    (c) Find another set of three linearly independent solutions of the given equation.

3. (a) Show that $\sin(t^3)$ and $\cos(t^3)$ is a fundamental set of the differential equation

    $$t\frac{d^2x}{dt^2} - 2\frac{dx}{dt} + 9t^5x = 0$$

    on every closed interval $[a, b]$, where $0 < a < b < \infty$.

    (b) Show that t, t^2, and t^3 is a fundamental set of the differential equation

    $$t^3\frac{d^3x}{dt^3} - 3t^2\frac{d^2x}{dt^2} + 6t\frac{dx}{dt} - 6x = 0$$

    on every closed interval $[a, b]$, where $0 < a < b < \infty$.

4. Consider the second-order homogeneous linear differential equation

    $$a_0(t)\frac{d^2x}{dt^2} + a_1(t)\frac{dx}{dt} + a_2(t)x = 0, \qquad \text{(A)}$$

    where a_0, a_1, and a_2 are continuous on an interval $a \le t \le b$, and $a_0(t) \ne 0$ on this interval. Let f and g be distinct solutions of differential equation (A) on $a \le t \le b$, and suppose $g(t) \ne 0$ for all $t \in [a, b]$.

    (a) Show that

    $$\frac{d}{dt}\left[\frac{f(t)}{g(t)}\right] = -\frac{W[f(t), g(t)]}{[g(t)]^2} \qquad \text{for all } t \in [a, b].$$

(b)   Use the result of part (a) to show that if $W[f(t), g(t)] = 0$ for all $t \in [a, b]$, then the solutions f and g are linearly dependent on this interval.

(c)   Suppose the solutions f and g are linearly independent on $a \leq t \leq b$, and let h be the function defined by $h(t) = f(t)/g(t)$, $a \leq t \leq b$. Show that h is a monotonic function on $a \leq t \leq b$.

5.   Let f and g be two solutions of the second-order homogeneous linear differential equation (A) of Exercise 4.

(a)   Show that if f and g have a common zero at $t_0 \in [a, b]$, then f and g are linearly dependent on $a \leq t \leq b$.

(b)   Show that if f and g have relative maxima at a common point $t_0 \in [a, b]$, then f and g are linearly dependent on $a \leq t \leq b$.

6.   Consider the second-order homogeneous linear differential equation (A) of Exercise 4. Let f and g be two solutions of this equation. Show that if f and g are linearly independent on $a \leq t \leq b$ and A_1, A_2, B_1, and B_2 are constants such that $A_1 B_2 - A_2 B_1 \neq 0$, then the solutions $A_1 f + A_2 g$ and $B_1 f + B_2 g$ of Equation (A) are also linearly independent on $a \leq t \leq b$.

7.   Let f and g be two solutions of the second-order homogeneous linear differential equation (A) of Exercise 4. Show that if f and g are linearly independent on $a \leq t \leq b$ and are such that $f''(t_0) = g''(t_0) = 0$ at some point $t_0 \in [a, b]$, then $a_1(t_0) = a_2(t_0) = 0$.

8.   (a)   Show that the substitution

$$x = y \exp\left[-\tfrac{1}{2} \int P_1(t)\, dt \right]$$

transforms the second-order homogeneous linear differential equation

$$\frac{d^2 x}{dt^2} + P_1(t)\, \frac{dx}{dt} + P_2(t)x = 0 \tag{A}$$

into the second-order homogeneous linear differential equation

$$\frac{d^2 y}{dt^2} + \left\{ P_2(t) - \frac{1}{2} P_1'(t) - \frac{1}{4} [P_1(t)]^2 \right\} y = 0 \tag{B}$$

in which the first-derivative term is missing.

(b)   If $g_1(t)$ and $g_2(t)$ is a fundamental set of Equation (B) on an interval $a \leq t \leq b$, show that

$$f_1(t) = g_1(t) \exp\left[-\tfrac{1}{2} \int P_1(t)\, dt \right]$$

and

$$f_2(t) = g_2(t) \exp\left[-\tfrac{1}{2} \int P_1(t)\, dt \right]$$

is a fundamental set of Equation (A) on this interval.

9. Use the substitution given in Exercise 8(a) to transform the differential equation

$$\frac{d^2 x}{dt^2} + (2t + 1)\frac{dx}{dt} + \left(t^2 + t + \frac{1}{4}\right)x = 0 \qquad \text{(C)}$$

into an equation (D) of the form (B) (of Exercise 8) in which the first-derivative term is missing. Obtain a fundamental set of the equation (D) which thus results, and use this fundamental set to obtain a fundamental set of the given equation (C).

11.6 FURTHER PROPERTIES OF THE nth-ORDER HOMOGENEOUS LINEAR DIFFERENTIAL EQUATION

A. More Wronskian Theory

We first use Theorem 11.12, the Abel–Liouville formula for homogeneous linear systems, to obtain the corresponding result for the single nth-order homogeneous linear scalar differential equation

$$a_0(t)\frac{d^n x}{dt^n} + a_1(t)\frac{d^{n-1}x}{dt^{n-1}} + \cdots + a_{n-1}(t)\frac{dx}{dt} + a_n(t)x = 0. \qquad (11.59)$$

THEOREM 11.24   Abel–Liouville Formula

Let $f_1, f_2, \ldots, f_n$ be n solutions of the homogeneous linear differential equation (11.59) on $a \le t \le b$; and let $t_0 \in [a, b]$. Let W denote the Wronskian of $f_1, f_2, \ldots, f_n$. Then

$$W(t) = W(t_0)\exp\left[-\int_{t_0}^{t}\frac{a_1(s)}{a_0(s)}\,ds\right] \qquad (11.86)$$

for all $t \in [a, b]$.

Proof. Since $f_1, f_2, \ldots, f_n$ are solutions of (11.59), by Theorem 11.17,

$$\phi_1 = \begin{pmatrix} f_1 \\ f'_1 \\ f''_1 \\ \vdots \\ f_1^{(n-1)} \end{pmatrix}, \quad \phi_2 = \begin{pmatrix} f_2 \\ f'_2 \\ f''_2 \\ \vdots \\ f_2^{(n-1)} \end{pmatrix}, \ldots, \phi_n = \begin{pmatrix} f_n \\ f'_n \\ f''_n \\ \vdots \\ f_n^{(n-1)} \end{pmatrix}$$

are n solutions of the homogeneous linear system (11.69) on $[a, b]$. Let W^* denote the Wronskian of $\phi_1, \phi_2, \ldots, \phi_n$. Then by Theorem 11.12, the Abel–Liouville formula for systems,

$$W^*(t) = W^*(t_0)\exp\left[\int_{t_0}^{t}\text{tr } \mathbf{A}(s)\,ds\right] \qquad (11.87)$$

where $\mathbf{A}$ is the coefficient matrix of system (11.69), holds for all $t \in [a, b]$. By formula (11.82), $W^*(t) = W(t)$ for all $t \in [a, b]$. Also, the coefficient matrix $\mathbf{A}$ of system (11.69) is given by (11.71); and from this we see that tr $\mathbf{A}(s)$ in (11.87) is simply $-a_1(s)/a_0(s)$. Thus (11.87) reduces at once to

$$W(t) = W(t_0) \exp\left[-\int_{t_0}^{t} \frac{a_1(s)}{a_0(s)} \, ds \right].$$

Q.E.D.

▶ **Example 11.17**

The coefficients $t^2 + 1$ and $2t$ in the differential equation

$$(t^2 + 1)\frac{d^2x}{dt^2} + 2t\frac{dx}{dt} = 0 \tag{11.88}$$

are continuous for all t, $-\infty < t < \infty$, and $t^2 + 1 \neq 0$ for all real t. The reader can verify that f_1 and f_2 defined, respectively, by

$$f_1(t) = 1 \quad \text{and} \quad f_2(t) = \arctan t$$

are solutions of Equation (11.88) on $-\infty < t < \infty$. Since

$$W(f_1, f_2)(t) = \begin{vmatrix} 1 & \arctan t \\ 0 & \dfrac{1}{1 + t^2} \end{vmatrix} = \frac{1}{1 + t^2} \neq 0$$

on every finite closed interval $a \leq t \leq b$, we conclude that f_1 and f_2 are a set of linearly independent solutions of Equation (11.88) on every such interval. In this case the Abel–Liouville formula

$$W(t) = W(t_0) \exp\left[-\int_{t_0}^{t} \frac{a_1(s)}{a_0(s)} \, ds \right]$$

becomes

$$\frac{1}{1 + t^2} = \frac{1}{1 + t_0^2} \exp\left[-\int_{t_0}^{t} \frac{2s}{s^2 + 1} \, ds \right]$$

for all $t \in [a, b]$.

In order to obtain the next theorem, we shall need the following result from linear algebra, which we state without proof.

THEOREM A

Suppose $m > n$. Then a system of n homogeneous linear algebraic equations in m unknowns always has nontrivial solutions.

THEOREM 11.25

Suppose $m > n$; and let $f_1, f_2, \ldots, f_m$ be a set of m solutions of the nth-order homogeneous linear differential equation (11.59) on $a \leq t \leq b$. Then $f_1, f_2, \ldots, f_m$ are linearly dependent on $a \leq t \leq b$.

Proof. By hypothesis, $f_1, f_2, \ldots, f_m$ are m solutions of the nth-order equation (11.59) on $a \le t \le b$, where $m > n$. Let t_0 be a point of the interval $a \le t \le b$ and consider the following system of n homogeneous linear algebraic equations in the m unknowns $k_1, k_2, \ldots, k_m$.

$$k_1 f_1(t_0) + k_2 f_2(t_0) + \cdots + k_m f_m(t_0) = 0,$$
$$k_1 f'_1(t_0) + k_2 f'_2(t_0) + \cdots + k_m f'_m(t_0) = 0,$$
$$\vdots \tag{11.89}$$
$$k_1 f_1^{(n-1)}(t_0) + k_2 f_2^{(n-1)}(t_0) + \cdots + k_m f_m^{(n-1)}(t_0) = 0.$$

By Theorem A this system has a solution

$$k_1 = c_1, k_2 = c_2, \ldots, k_m = c_m,$$

where $c_1, c_2, \ldots, c_m$ are *not all zero*.

Now consider the function f defined by

$$f(t) = c_1 f_1(t) + c_2 f_2(t) + \cdots + c_m f_m(t) \tag{11.90}$$

for all t on $a \le t \le b$. The function f is a linear combination of the solutions $f_1, f_2, \ldots, f_m$. Hence by Theorem 11.18 the function f is also a solution of Equation (11.59) on $a \le t \le b$.

Now differentiate (11.90) $(n-1)$ times and let $t = t_0$ in (11.90) and the $(n-1)$ derivatives so obtained. We have

$$f(t_0) = c_1 f_1(t_0) + c_2 f_2(t_0) + \cdots + c_m f_m(t_0),$$
$$f'(t_0) = c_1 f'_1(t_0) + c_2 f'_2(t_0) + \cdots + c_m f'_m(t_0),$$
$$\vdots$$
$$f^{(n-1)}(t_0) = c_1 f_1^{(n-1)}(t_0) + c_2 f_2^{(n-1)}(t_0) + \cdots + c_m f_m^{(n-1)}(t_0).$$

The right members of these expressions are precisely the left members of Equations (11.89) with the unknowns k_i replaced by c_i $(i = 1, 2, \ldots, m)$. Since $c_1, c_2, \ldots, c_m$ is a solution of the system (11.89), we see at once that the solution f of Equation (11.59) is such that

$$f(t_0) = f'(t_0) = \cdots = f^{(n-1)}(t_0) = 0.$$

Thus by the Corollary to Theorem 11.16 we conclude that $f(t) = 0$ for all t such that $a \le t \le b$. Thus from (11.90) we have

$$c_1 f_1(t) + c_2 f_2(t) + \cdots + c_m f_m(t) = 0,$$

$a \le t \le b$, where $c_1, c_2, \ldots, c_m$ are *not all zero*. Thus the solutions $f_1, f_2, \ldots, f_m$ are linearly dependent on $a \le t \le b$. Q.E.D.

The Corollary to Theorem 11.21 states a necessary and sufficient condition that n solutions of the nth-order homogeneous linear differential equation

$$a_0(t) \frac{d^n x}{dt^n} + a_1(t) \frac{d^{n-1} x}{dt^{n-1}} + \cdots + a_{n-1}(t) \frac{dx}{dt} + a_n(t) x = 0 \tag{11.59}$$

be linearly dependent on $a \leq t \leq b$. This condition is that the value of the Wronskian of the n solutions be zero for all t on $a \leq t \leq b$. We now inquire whether or not this condition remains both necessary and sufficient if the n functions involved are not necessarily solutions of an nth-order homogeneous linear differential equation but are merely functions having a continuous $(n-1)$st derivative on $a \leq t \leq b$. The following theorem tells us that the condition is still necessary.

THEOREM 11.26

Hypothesis

1. Let each of the functions $f_1, f_2, \ldots, f_n$ *have a continuous* $(n-1)$st *derivative on* $a \leq t \leq b$.

2. Let the functions $f_1, f_2, \ldots, f_n$ *be linearly dependent on* $a \leq t \leq b$.

Conclusion. $W(f_1, f_2, \ldots, f_n)(t) = 0$ *for all* t *on* $a \leq t \leq b$.

Proof. Since $f_1, f_2, \ldots, f_n$ are linearly dependent on $a \leq t \leq b$, there exist constants $c_1, c_2, \ldots, c_n$, *not all zero*, such that

$$c_1 f_1(t) + c_2 f_2(t) + \cdots + c_n f_n(t) = 0 \tag{11.91}$$

for all t on $a \leq t \leq b$. From (11.91) we obtain

$$c_1 f'_1(t) + c_2 f'_2(t) + \cdots + c_n f'_n(t) = 0,$$
$$\vdots \tag{11.92}$$
$$c_1 f_1^{(n-1)}(t) + c_2 f_2^{(n-1)}(t) + \cdots + c_n f_n^{(n-1)}(t) = 0$$

for all $t \in [a, b]$. Now let $t = t_0$ be an arbitrary point of $a \leq t \leq b$. Then (11.91) and (11.92) hold at $t = t_0$. That is, the relations

$$c_1 f_1(t_0) + c_2 f_2(t_0) + \cdots + c_n f_n(t_0) = 0$$
$$c_1 f'_1(t_0) + c_2 f'_2(t_0) + \cdots + c_n f'_n(t_0) = 0$$
$$\vdots$$
$$c_1 f_1^{(n-1)}(t_0) + c_2 f_2^{(n-1)}(t_0) + \cdots + c_n f_n^{(n-1)}(t_0) = 0.$$

hold, where $c_1, c_2, \ldots, c_n$ are not all zero. Thus by Chapter 7, Section 7.5, Theorem A, we have

$$\begin{vmatrix} f_1(t_0) & f_2(t_0) & \cdots & f_n(t_0) \\ f'_1(t_0) & f'_2(t_0) & \cdots & f'_n(t_0) \\ f_1^{(n-1)}(t_0) & f_2^{(n-1)}(t_0) & \cdots & f_n^{(n-1)}(t_0) \end{vmatrix} = 0.$$

But this determinant is the value of the Wronskian of $f_1, f_2, \ldots, f_n$ at $t = t_0$, and t_0 is an *arbitrary* point of $[a, b]$. Hence

$$W(f_1, f_2, \ldots, f_n)(t) = 0$$

for all $t \in [a, b]$.

Q.E.D.

We now consider whether or not the condition $W(f_1, f_2, \ldots, f_n)(t) = 0$ for all t on $a \le t \le b$ is still sufficient for linear dependence in the present case. The following example shows that it is *not*.

▶ **Example 11.18**

Consider the functions f_1, f_2, and f_3 defined, respectively, by

$$f_1(t) = t^3, \qquad f_2(t) = 1, \qquad f_3(t) = |t|^3.$$

Each of these functions has a continuous second derivative on every interval $a \le t \le b$. Further,

$$W(f_1, f_2, f_3)(t) = \begin{vmatrix} t^3 & 1 & |t|^3 \\ 3t^2 & 0 & 3t|t| \\ 6t & 0 & 6|t| \end{vmatrix} = 0, \qquad a \le t \le b.$$

But the functions f_1, f_2, and f_3 are *not* linearly dependent on any interval $a \le t \le b$ which includes $t = 0$ (see Exercise 3 at the end of this section).

Thus we see that the condition $W(f_1, f_2, \ldots, f_n)(t) = 0$ for all t on $a \le t \le b$ is not sufficient for linear dependence in general. However, we shall obtain the following result.

THEOREM 11.27

Hypothesis

1. Let each of the functions $f_1, f_2, \ldots, f_n$ have a continuous $(n-1)$st derivative on $a \le t \le b$.
2. Suppose $W(f_1, f_2, \ldots, f_n)(t) = 0$ for all t on $a \le t \le b$.
3. Suppose the Wronskian of some $(n-1)$ of the functions $f_1, f_2, \ldots, f_n$ is unequal to zero for all t on $a \le t \le b$. That is, without loss in generality, suppose that $W(f_1, f_2, \ldots, f_{n-1})(t) \ne 0$ for all t on $a \le t \le b$.

Conclusion. *Then the functions $f_1, f_2, \ldots, f_n$ are linearly dependent on $a \le t \le b$.*

Proof. The equation

$$W[f_1(t), f_2(t), \ldots, f_{n-1}(t), x] = 0;$$

that is,

$$\begin{vmatrix} f_1(t) & f_2(t) & \cdots & f_{n-1}(t) & x \\ f_1'(t) & f_2'(t) & \cdots & f_{n-1}'(t) & x' \\ \vdots & \vdots & & \vdots & \vdots \\ f_1^{(n-1)}(t) & f_2^{(n-1)}(t) & \cdots & f_{n-1}^{(n-1)}(t) & x^{(n-1)} \end{vmatrix} = 0 \qquad (11.93)$$

is a homogeneous linear differential equation of order $(n-1)$. By Hypothesis 1 the coefficients are continuous on $a \leq t \leq b$. The coefficient of $\dfrac{d^{n-1}x}{dt^{n-1}}$ is $W(f_1, f_2, \ldots, f_{n-1})(t)$, and by Hypothesis 3 this is never zero on $a \leq t \leq b$. If any one of $f_1(t), f_2(t), \ldots, f_{n-1}(t)$ is substituted for x in Equation (11.93), the resulting determinant will have two identical columns. Thus each of the functions $f_1, f_2, \ldots, f_{n-1}$ is a solution of the differential equation (11.93) on $a \leq t \leq b$. By Hypothesis 2 the function f_n is also a solution of Equation (11.93) on $a \leq t \leq b$. Thus on $a \leq t \leq b$ the n functions $f_1, f_2, \ldots, f_{n-1}, f_n$ are all solutions of the differential equation (11.93). Since this differential equation is of order $(n-1)$, we know from Theorem 11.25 that no set of n solutions of it can be linearly independent. Thus the functions $f_1, f_2, \ldots, f_n$ are linearly dependent on $a \leq t \leq b$.

Q.E.D.

▶ **Example 11.19**

Although the functions f_1, f_2, f_3 of Example 11.18 are *not* linearly dependent on any interval $a \leq t \leq b$, *including* $t = 0$, it is clear that they *are* linearly dependent on every interval $a \leq t \leq b$ which does *not* include $t = 0$. Observe that on any such interval not including $t = 0$ the hypotheses of Theorem 11.27 are indeed satisfied. In particular, note that Hypothesis 3 holds, since

$$W(f_1, f_2)(t) = \begin{vmatrix} t^3 & 1 \\ 3t^2 & 0 \end{vmatrix} = -3t^2 \neq 0 \qquad \text{for } t \neq 0.$$

Theorem 11.22 showed that every nth-order homogeneous linear differential equation (11.59) possesses sets of n linearly independent solutions on $a \leq t \leq b$, and Theorem 11.21 showed that the value of the Wronskian of such a set is unequal to zero on $a \leq t \leq b$. Now, given a set of n functions having continuous nth derivatives and a nonzero Wronskian on $a \leq t \leq b$, does there exist an nth-order homogeneous linear differential equation for which this set of n functions constitutes a fundamental set (set of n linearly independent solutions) on $a \leq t \leq b$? In answer to this we shall obtain the following theorem.

THEOREM 11.28

Hypothesis

1. Let $f_1, f_2, \ldots, f_n$ be a set of n functions each of which has a continuous nth derivative on $a \leq t \leq b$.

2. Suppose $W(f_1, f_2, \ldots, f_n)(t) \neq 0$ for all t on $a \leq t \leq b$.

Conclusion.   *There exists a unique normalized (coefficient of $\dfrac{d^n x}{dt^n}$ is unity) homogeneous linear differential equation of order n (with continuous coefficients) which has $f_1, f_2, \ldots, f_n$ as a fundamental set on $a \leq t \leq b$. This equation is*

$$\frac{W[f_1(t), f_2(t), \ldots, f_n(t), x]}{W[f_1(t), f_2(t), \ldots, f_n(t)]} = 0. \tag{11.94}$$

Proof. The differential equation (11.94) is actually

$$\frac{\begin{vmatrix} f_1(t) & f_2(t) & \cdots & f_n(t) & x \\ f'_1(t) & f'_2(t) & \cdots & f'_n(t) & x' \\ \vdots & \vdots & & \vdots & \vdots \\ f_1^{(n)}(t) & f_2^{(n)}(t) & \cdots & f_n^{(n)}(t) & x^{(n)} \end{vmatrix}}{\begin{vmatrix} f_1(t) & f_2(t) & \cdots & f_n(t) \\ f'_1(t) & f'_2(t) & \cdots & f'_n(t) \\ \vdots & \vdots & & \vdots \\ f_1^{(n-1)}(t) & f_2^{(n-1)}(t) & \cdots & f_n^{(n-1)}(t) \end{vmatrix}} = 0. \tag{11.95}$$

This is of the form

$$\frac{d^n x}{dt^n} + p_1(t)\frac{d^{n-1}x}{dt^{n-1}} + \cdots + p_{n-1}(t)\frac{dx}{dt} + p_n(t)x = 0,$$

and so is a normalized homogeneous linear differential equation of order n. Also, by Hypotheses 1 and 2 the coefficients $p_i(i = 1, 2, \ldots, n)$ are continuous on $a \le t \le b$. If any one of $f_1(t), f_2(t), \ldots, f_n(t)$ is substituted for x in Equation (11.95), the resulting determinant in the numerator will have two identical columns. Thus each of the functions $f_1, f_2, \ldots, f_n$ is a solution of Equation (11.94) on $a \le t \le b$; and by Theorem 11.21 we see from Hypothesis 2 that these solutions are linearly independent on $a \le t \le b$. Thus Equation (11.94) is indeed an equation of the required type having $f_1, f_2, \ldots, f_n$ as a fundamental set.

We must now show that Equation (11.94) is the *only* normalized nth-order homogeneous linear differential equation with continuous coefficients which has this property. Suppose there are two such equations

$$\frac{d^n x}{dt^n} + q_1(t)\frac{d^{n-1}x}{dt^{n-1}} + \cdots + q_{n-1}(t)\frac{dx}{dt} + q_n(t)x = 0,$$

$$\frac{d^n x}{dt^n} + r_1(t)\frac{d^{n-1}x}{dt^{n-1}} + \cdots + r_{n-1}(t)\frac{dx}{dt} + r_n(t)x = 0. \tag{11.96}$$

Then the equation

$$[q_1(t) - r_1(t)]\frac{d^{n-1}x}{dt^{n-1}} + \cdots + [q_n(t) - r_n(t)]x = 0 \tag{11.97}$$

is a homogeneous linear differential equation of order at most $(n - 1)$, and the coefficients in Equation (11.97) are continuous on $a \le t \le b$. Further, since $f_1, f_2, \ldots, f_n$ satisfy both of Equations (11.96) these n functions are all solutions of Equation (11.97) on $a \le t \le b$.

We shall show that $q_1(t) - r_1(t) = 0$ for all t on $a \le t \le b$. To do so, let t_0 be a point of the interval $a \le t \le b$ and suppose that $q_1(t_0) - r_1(t_0) \ne 0$. Then there exists a subinterval I, $\alpha \le t \le \beta$, of $a \le t \le b$ containing t_0 such that $q_1(t) - r_1(t) \ne 0$ on I. Since the n solutions $f_1, f_2, \ldots, f_n$ of Equation (11.97) are linearly independent on $a \le t \le b$, they are also linearly independent on I. Thus on I, Equation (11.97) of order at most $(n - 1)$, has a set of n linearly independent solutions. But this contradicts Theorem 11.25. Thus there exists no $t_0 \in [a, b]$ such that $q_1(t_0) - r_1(t_0) \ne 0$. In other words, $q_1(t) - r_1(t) = 0$ for all t on $a \le t \le b$.

In like manner, one shows that $q_k(t) - r_k(t) = 0, k = 2, 3, \ldots, n$, for all t on $a \leq t \leq b$. Thus Equations (11.96) are identical on $a \leq t \leq b$ and the uniqueness is proved.

Q.E.D.

► **Example 11.20**

Consider the functions f_1 and f_2 defined, respectively, by $f_1(t) = t$ and $f_2(t) = te^t$. We note that

$$W(f_1, f_2)(t) = \begin{vmatrix} t & te^t \\ 1 & te^t + e^t \end{vmatrix} = t^2 e^t \neq 0 \qquad \text{for } t \neq 0.$$

Thus by Theorem 11.28 on every closed interval $a \leq t \leq b$ not including $t = 0$ there exists a unique normalized second-order homogeneous linear differential equation with continuous coefficients which has f_1 and f_2 as a fundamental set. Indeed, Theorem 11.28 states that this equation is

$$\frac{W[t, te^t, x]}{W[t, te^t]} = 0.$$

Writing out the two Wronskians involved, this becomes

$$\frac{\begin{vmatrix} t & te^t & x \\ 1 & te^t + e^t & x' \\ 0 & te^t + 2e^t & x'' \end{vmatrix}}{\begin{vmatrix} t & te^t \\ 1 & te^t + e^t \end{vmatrix}} = 0.$$

Since

$$\begin{vmatrix} t & t & x \\ 1 & t+1 & x' \\ 0 & t+2 & x'' \end{vmatrix} = t^2 \frac{d^2 x}{dt^2} - t(t+2)\frac{dx}{dt} + (t+2)x \quad \text{and} \quad \begin{vmatrix} t & t \\ 1 & t+1 \end{vmatrix} = t^2,$$

we see that this differential equation is

$$\frac{d^2 x}{dt^2} - \left(\frac{t+2}{t}\right)\frac{dx}{dt} + \left(\frac{t+2}{t^2}\right)x = 0.$$

B. Reduction of Order

In Chapter 4 we introduced the process of reduction of order of an nth-order homogeneous linear differential equation. There we stated the basic theorem without proof and went on to consider the second-order case in greater detail. We employed the process in Chapter 4 and later in Chapter 6. We shall now restate the basic theorem, give its proof, and indicate further how it may be employed.

THEOREM 11.29

Hypothesis. *Let f be a nontrivial solution of the nth-order homogeneous linear differential equation*

$$a_0(t)\frac{d^n x}{dt^n} + a_1(t)\frac{d^{n-1}x}{dt^{n-1}} + \cdots + a_{n-1}(t)\frac{dx}{dt} + a_n(t)x = 0. \tag{11.59}$$

Conclusion. *The transformation*

$$x = f(t)v \tag{11.98}$$

reduces Equation (11.59) to an (n − 1)st-order homogeneous linear differential equation in the dependent variable w, where w = dv/dt.

Proof. Let $x = f(t)v$. Then

$$\frac{dx}{dt} = f(t)\frac{dv}{dt} + f'(t)v,$$

$$\frac{d^2 x}{dt^2} = f(t)\frac{d^2 v}{dt^2} + 2f'(t)\frac{dv}{dt} + f''(t)v,$$

$$\vdots$$

$$\frac{d^n x}{dt^n} = f(t)\frac{d^n v}{dt^n} + nf'(t)\frac{d^{n-1}v}{dt^{n-1}} + \frac{n(n-1)}{2!}f''(t)\frac{d^{n-2}v}{dt^{n-2}} + \cdots + f^{(n)}(t)v.$$

Substituting these expressions into the differential equation (11.59), we have

$$a_0(t)\left[f(t)\frac{d^n v}{dt^n} + nf'(t)\frac{d^{n-1}v}{dt^{n-1}} + \cdots + f^{(n)}(t)v \right]$$

$$+ a_1(t)\left[f(t)\frac{d^{n-1}v}{dt^{n-1}} + (n-1)f'(t)\frac{d^{n-2}v}{dt^{n-2}} + \cdots + f^{(n-1)}(t)v \right]$$

$$+ \cdots + a_{n-1}(t)\left[f(t)\frac{dv}{dt} + f'(t)v \right] + a_n(t)f(t)v = 0$$

or

$$a_0(t)f(t)\frac{d^n v}{dt^n} + [na_0(t)f'(t) + a_1(t)f(t)]\frac{d^{n-1}v}{dt^{n-1}} + \cdots$$

$$+ [na_0(t)f^{(n-1)}(t) + \cdots + a_{n-1}(t)f(t)]\frac{dv}{dt}$$

$$+ [a_0(t)f^{(n)}(t) + a_1(t)f^{(n-1)}(t) + \cdots + a_{n-1}(t)f'(t) + a_n(t)f(t)]v = 0. \tag{11.99}$$

Now since f is a solution of Equation (11.59), the coefficient of v is zero. Then, letting $w = dv/dt$, Equation (11.99) reduces to the $(n-1)$st-order equation in w,

$$A_0(t)\frac{d^{n-1}w}{dt^{n-1}} + A_1(t)\frac{d^{n-2}w}{dt^{n-2}} + \cdots + A_{n-1}(t)w = 0, \tag{11.100}$$

where

$$A_0(t) = a_0(t)f(t),$$

$$A_1(t) = na_0(t)f'(t) + a_1(t)f(t), \ldots,$$

$$A_{n-1}(t) = na_0(t)f^{(n-1)}(t) + \cdots + a_{n-1}(t)f(t).$$

Q.E.D.

Now suppose that $w_1, w_2, \ldots, w_{n-1}$ is a known fundamental set of equation (11.100). Then $v_1, v_2, \ldots, v_{n-1}$ defined by

$$v_1(t) = \int w_1(t)\,dt, \quad v_2(t) = \int w_2(t)\,dt, \ldots, v_{n-1}(t) = \int w_{n-1}(t)\,dt$$

is a set of $(n-1)$ solutions of Equation (11.99). Also, the function v_n such that $v_n(t) = 1$ for all t is a solution of Equation (11.99). These n solutions $v_1, v_2, \ldots, v_n$ of Equation (11.99) are linearly independent. Then, using (11.98) we obtain n solutions f_i, where $f_i(t) = f(t)v_i(t)$ $(i = 1, 2, \ldots, n)$ of the original nth-order equation (11.59). The n solutions f_i so defined are also linearly independent and thus constitute a fundamental set of Equation (11.59).

One may extend Theorem 11.29 to show that if m (where $m < n$) linearly independent solutions of Equation (11.59) are known, then Equation (11.59) may be reduced to a homogeneous linear equation of order $(n - m)$.

Exercises

1. In each case, verify the truth of the Abel–Liouville formula (Theorem 11.24) for the given fundamental set of the given differential equation.

    (a) Fundamental set: $\sin(t^3)$, $\cos(t^3)$.
    Differential equation:

    $$t\frac{d^2x}{dt^2} - 2\frac{dx}{dt} + 9t^5x = 0.$$

    (b) Fundamental set: t, t^2, t^3.
    Differential equation:

    $$t^3\frac{d^3x}{dt^3} - 3t^2\frac{d^2x}{dt^2} + 6t\frac{dx}{dt} - 6x = 0.$$

2. Let f and g be defined, respectively, by

    $$f(t) = \begin{cases} t^2, & t \geq 0, \\ 0, & t < 0 \end{cases} \quad \text{and} \quad g(t) = \begin{cases} 0, & t \geq 0, \\ t^2, & t < 0. \end{cases}$$

    Show that $W(f, g)(t) = 0$ for all t but that f and g are not linearly dependent on any interval $a \leq t \leq b$ which includes $t = 0$.

3. Consider the functions considered in Example 11.18, namely, f_1, f_2, and f_3 defined respectively by

    $$f_1(t) = t^3, \qquad f_2(t) = 1, \qquad f_3(t) = |t|^3.$$

    (a) Show that each of these functions has a continuous second derivative on every interval $a \leq t \leq b$.

(b) Show that these three functions are *not* linearly dependent on any interval which includes $t = 0$.

4. In each case find the unique normalized homogeneous linear differential equation of order 3 (with continuous coefficients) which has the given set as a fundamental set on some interval $a \le t \le b$.

   (a) t, t^2, e^t.

   (b) $t, \sin t, \cos t$.

   (c) t, t^3, t^5.

   (d) $e^t, t^2, t^2 e^t$.

5. The differential equation

$$(t^3 - 2t^2) \frac{d^2 x}{dt^2} - (t^3 + 2t^2 - 6t) \frac{dx}{dt} + (3t^2 - 6)x = 0$$

has a solution of the form t^n, where n is an integer.

   (a) Find this solution of the form t^n.

   (b) Using the solution found in part (a) reduce the order and find the general solution of the given differential equation.

6. Given that the differential equation

$$t^3 \frac{d^3 x}{dt^3} - (t + 3)t^2 \frac{d^2 x}{dt^2} + 2t(t + 3) \frac{dx}{dt} - 2(t + 3)x = 0$$

has two linearly independent solutions of the form t^n, where n is an integer, find the general solution.

7. Let $\{f, g\}$ and $\{F, G\}$ be two fundamental sets of the differential equation

$$a_0(t) \frac{d^2 x}{dt^2} + a_1(t) \frac{dx}{dt} + a_2(t)x = 0$$

on $a \le t \le b$. Use the Abel–Liouville formula to show there exists a constant $c \neq 0$ such that $W(f, g)(t) = cW(F, G)(t)$ for all $t \in [a, b]$.

8. Consider the normalized nth-order homogeneous linear differential equation

$$\frac{d^n x}{dt^n} + P_1(t) \frac{d^{n-1} x}{dt^{n-1}} + \cdots + P_{n-1}(t) \frac{dx}{dt} + P_n(t)x = 0, \qquad \text{(A)}$$

where $P_i(i = 1, 2, \ldots, n)$ are continuous on $a \le t \le b$. Let $f_1, f_2, \ldots, f_n$ be a fundamental set of this equation on $a \le t \le b$. By the uniqueness result of Theorem 11.28 the differential equation (A) may be written

$$\frac{W[f_1(t), f_2(t), \ldots, f_n(t), x]}{W[f_1(t), f_2(t), \ldots, f_n(t)]} = 0, \qquad \text{(B)}$$

and the coefficients of $\dfrac{d^{n-1} x}{dt^{n-1}}$ in the two expressions (A) and (B) for the differential equation must be identical. Equate these coefficients to show that

$$W(f_1, f_2, \ldots, f_n)$$

satisfies the first-order differential equation

$$\frac{dW}{dt} + P_1(t)W = 0.$$

Note that this paves the way for the proof of the Abel–Liouville formula (Theorem 11.24) in the case of the normalized nth-order equation.

11.7   THE *n*th-ORDER NONHOMOGENEOUS LINEAR EQUATION

In this section we consider briefly the nth-order nonhomogeneous linear scalar differential equation

$$a_0(t)\frac{d^n x}{dt^n} + a_1(t)\frac{d^{n-1} x}{dt^{n-1}} + \cdots + a_{n-1}(t)\frac{dx}{dt} + a_n(t)x = F(t). \qquad (11.101)$$

Using the operator notation already introduced in Section 11.5 of this chapter, we may write this as

$$L_n x = F(t), \qquad (11.102)$$

where, as before,

$$L_n = a_0(t)\frac{d^n}{dt^n} + a_1(t)\frac{d^{n-1}}{dt^{n-1}} + \cdots + a_{n-1}(t)\frac{d}{dt} + a_n(t).$$

We now prove the basic theorem dealing with Equation (11.102).

THEOREM 11.30

Hypothesis

1. *Let v be any solution of the nonhomogeneous equation (11.102).*
2. *Let u be any solution of the corresponding homogeneous equation.*

$$L_n x = 0. \qquad (11.103)$$

Conclusion.   *Then u + v is also a solution of the nonhomogeneous equation (11.102).*

Proof.   We have

$$L_n[u(t) + v(t)]$$

$$= a_0(t)\frac{d^n}{dt^n}[u(t) + v(t)] + a_1(t)\frac{d^{n-1}}{dt^{n-1}}[u(t) + v(t)]$$

$$+ \cdots + a_{n-1}(t)\frac{d}{dt}[u(t) + v(t)] + a_n(t)[u(t) + v(t)]$$

$$= a_0(t)\frac{d^n}{dt^n}u(t) + a_1(t)\frac{d^{n-1}}{dt^{n-1}}u(t) + \cdots + a_{n-1}(t)\frac{d}{dt}u(t) + a_n(t)u(t)$$

$$+ a_0(t)\frac{d^n}{dt^n}v(t) + a_1(t)\frac{d^{n-1}}{dt^{n-1}}v(t) + \cdots + a_{n-1}(t)\frac{d}{dt}v(t) + a_n(t)v(t)$$

$$= L_n[u(t)] + L_n[v(t)].$$

Now by Hypothesis 1, $L_n[v(t)] = F(t)$; and by Hypothesis 2, $L_n[u(t)] = 0$. Thus $L_n[u(t) + v(t)] = F(t)$; that is, $u + v$ is a solution of Equation (11.102).   *Q.E.D.*

In particular, if $f_1, f_2, \ldots, f_n$ is a fundamental set of the homogeneous equation (11.103) and v is any particular solution of the nonhomogeneous equation (11.102) then

$$c_1 f_1 + c_2 f_2 + \cdots + c_n f_n + v$$

is also a solution of the nonhomogeneous equation (11.102). If a fundamental set $f_1, f_2, \ldots, f_n$ of the corresponding homogeneous equation (11.103) is known, then a particular solution of the nonhomogeneous equation (11.102) can always be found by the method of *variation of parameters*. This method was introduced in Chapter 4, Section 4.4. In that section we gave a detailed development of the method for the case of the second-order equation and considered several illustrative examples. We now state and prove the relevant theorem for the general nth-order equation.

THEOREM 11.31

Hypothesis

1. Let $f_1, f_2, \ldots, f_n$ be a fundamental set of the homogeneous equation

$$L_n x = 0 \tag{11.103}$$

on the interval $a \leq t \leq b$.
2. Let v be a function defined by

$$v(t) = v_1(t) f_1(t) + v_2(t) f_2(t) + \cdots + v_n(t) f_n(t), \tag{11.104}$$

$a \leq t \leq b$, where $v_1, v_2, \ldots, v_n$ are functions such that

$$v_1'(t) f_1(t) + v_2'(t) f_2(t) + \cdots + v_n'(t) f_n(t) = 0,$$

$$v_1'(t) f_1'(t) + v_2'(t) f_2'(t) + \cdots + v_n'(t) f_n'(t) = 0,$$

$$\vdots \tag{11.105}$$

$$v_1'(t) f_1^{(n-2)}(t) + v_2'(t) f_2^{(n-2)}(t) + \cdots + v_n'(t) f_n^{(n-2)}(t) = 0,$$

$$v_1'(t) f_1^{(n-1)}(t) + v_2'(t) f_2^{(n-1)}(t) + \cdots + v_n'(t) f_n^{(n-1)}(t) = \frac{F(t)}{a_0(t)},$$

for $a \leq t \leq b$.

Conclusion.   *Each such function v is a particular solution of the nonhomogeneous equation*

$$L_n x = F(t) \tag{11.102}$$

on $a \leq t \leq b$.

Proof.   We first consider the linear algebraic system (11.105) in the n unknowns $v_1'(t), v_2'(t), \ldots, v_n'(t)$. The determinant of coefficients of this system is

$$W(f_1, f_2, \ldots, f_n)(t).$$

Since $f_1, f_2, \ldots, f_n$ is a fundamental set of the homogeneous equation (11.103) on $a \le t \le b$, this determinant is different from zero for all $t \in [a, b]$. Therefore the system (11.105) uniquely determines $v'_1(t), v'_2(t), \ldots, v'_n(t), a \le t \le b$, and hence $v_1(t), v_2(t), \ldots, v_n(t)$ may be determined by antidifferentiation.

In this manner functions v are indeed defined by (11.104). We must now show that any such function v satisfies the nonhomogeneous equation (11.102).

Differentiating (11.104), we obtain

$$v'(t) = v_1(t)f'_1(t) + v_2(t)f'_2(t) + \cdots + v_n(t)f'_n(t)$$
$$+ v'_1(t)f_1(t) + v'_2(t)f_2(t) + \cdots + v'_n(t)f_n(t);$$

and using the first equation of (11.105) this reduces to

$$v'(t) = v_1(t)f'_1(t) + v_2(t)f'_2(t) + \cdots + v_n(t)f'_n(t). \tag{11.106}$$

Differentiating (11.106), we obtain

$$v''(t) = v_1(t)f''_1(t) + v_2(t)f''_2(t) + \cdots + v_n(t)f''_n(t)$$
$$+ v'_1(t)f'_1(t) + v'_2(t)f'_2(t) + \cdots + v'_n(t)f'_n(t);$$

and using the second equation of (11.105) this reduces to

$$v''(t) = v_1(t)f''_1(t) + v_2(t)f''_2(t) + \cdots + v_n(t)f''_n(t). \tag{11.107}$$

Proceeding in like manner, we also obtain

$$v'''(t) = v_1(t)f'''_1(t) + v_2(t)f'''_2(t) + \cdots + v_n(t)f'''_n(t),$$
$$\vdots$$
$$v^{(n-1)}(t) = v_1(t)f_1^{(n-1)}(t) + v_2(t)f_2^{(n-1)}(t) + \cdots + v_n(t)f_n^{(n-1)}(t), \tag{11.108}$$

$$v^{(n)}(t) = v_1(t)f_1^{(n)}(t) + v_2(t)f_2^{(n)}(t) + \cdots + v_n(t)f_n^{(n)}(t) + \frac{F(t)}{a_0(t)}.$$

We now substitute (11.104), (11.106), (11.107), and (11.108) into the nonhomogeneous differential equation (11.102). We obtain

$$a_0(t)\left[v_1(t)f_1^{(n)}(t) + v_2(t)f_2^{(n)}(t) + \cdots + v_n(t)f_n^{(n)}(t) + \frac{F(t)}{a_0(t)} \right]$$
$$+ a_1(t)[v_1(t)f_1^{(n-1)}(t) + v_2(t)f_2^{(n-1)}(t) + \cdots + v_n(t)f_n^{(n-1)}(t)]$$
$$+ \cdots + a_n(t)[v_1(t)f_1(t) + v_2(t)f_2(t) + \cdots + v_n(t)f_n(t)] = F(t)$$

or

$$v_1(t)[a_0(t)f_1^{(n)}(t) + a_1(t)f_1^{(n-1)}(t) + \cdots + a_n(t)f_1(t)]$$
$$+ v_2(t)[a_0(t)f_2^{(n)}(t) + a_1(t)f_2^{(n-1)}(t) + \cdots + a_n(t)f_2(t)]$$
$$+ \cdots + v_n(t)[a_0(t)f_n^{(n)}(t) + a_1(t)f_n^{(n-1)}(t) + \cdots + a_n(t)f_n(t)]$$
$$+ F(t) = F(t). \tag{11.109}$$

Since $f_1, f_2, \ldots, f_n$ are solutions of the homogeneous equation (11.103) on $a \le t \le b$, each of the expressions in brackets in (11.109) is equal to zero for all $t \in [a, b]$. Therefore (11.109) reduces to $F(t) = F(t)$ and thus the function v satisfies the nonhomogeneous differential equation (11.102) on $a \le t \le b$. Q.E.D.

We now prove the so-called superposition principle for particular solutions of Equation (11.102).

THEOREM 11.32

Hypothesis

1. Let $L_n x = F_i(t)$, $i = 1, 2, \ldots, m$, be m different nonhomogeneous linear equations, each of which has the same left member

$$L_n x = a_0(t) \frac{d^n x}{dt^n} + a_1(t) \frac{d^{n-1} x}{dt^{n-1}} + \cdots + a_n(t)x.$$

2. Let f_i be a particular solution of $L_n x = F_i(t)$, $i = 1, 2, \ldots, m$.

Conclusion. *Then $\sum_{i=1}^{m} k_i f_i$ is a particular solution of the equation*

$$L_n x = \sum_{i=1}^{m} k_i F_i(t),$$

where $k_1, k_2, \ldots, k_m$ are constants.

Proof. We have

$$L_n \left[\sum_{i=1}^{m} k_i f_i(t) \right] = a_0(t) \frac{d^n}{dt^n} \left[\sum_{i=1}^{m} k_i f_i(t) \right] + \cdots + a_n(t) \left[\sum_{i=1}^{m} k_i f_i(t) \right]$$

$$= k_1 \left[a_0(t) \frac{d^n}{dt^n} f_1(t) + \cdots + a_n(t) f_1(t) \right] + \cdots$$

$$+ k_m \left[a_0(t) \frac{d^n}{dt^n} f_m(t) + \cdots + a_n(t) f_m(t) \right]$$

$$= k_1 L_n[f_1(t)] + \cdots + k_m L_n[f_m(t)].$$

By Hypothesis 2, f_i satisfies $L_n x = F_i(t)$, $i = 1, 2, \ldots, m$. Thus $L_n[f_1(t)] = F_1(t), \ldots,$ $L_n[f_m(t)] = F_m(t)$, and so $L_n \left[\sum_{i=1}^{m} k_i f_i(t) \right] = k_1 F_1(t) + \cdots + k_m F_m(t)$. That is $\sum_{i=1}^{m} k_i f_i$ is a solution of $L_n x = \sum_{i=1}^{m} k_i F_i(t)$. Q.E.D.

Exercises

1. Show that Theorem 11.30 does not in general apply to nonlinear equations by showing that although f such that $f(t) = t$ is solution of

$$\frac{d^2 x}{dt^2} - t^2 \frac{dx}{dt} + x^2 = 0$$

and g such that $g(t) = 1$ is a solution of

$$\frac{d^2 x}{dt^2} - t^2 \frac{dx}{dt} + x^2 = 1,$$

the sum $f + g$ is *not* a solution of

$$\frac{d^2x}{dt^2} - t^2\frac{dx}{dt} + x^2 = 1.$$

2. Apply the superposition principle for particular solutions (Theorem 11.32) to find a particular solution of

$$\frac{d^2x}{dt^2} - x = \sum_{k=1}^{m} \frac{\sin kt}{k^2}.$$

11.8 STURM THEORY

A. Self-Adjoint Equations of the Second Order

In Chapter 14 we shall present a basic method for obtaining a formal solution of a type of problem which involves a partial differential equation and various supplementary conditions. Many problems of this type actually arise in the application of differential equations to certain problems of mathematical physics and engineering. In solving such problems we encounter boundary-value problems involving second-order ordinary differential equations; many of these second-order ordinary differential equations are, or can be transformed into, so-called self-adjoint form. In this section we shall consider some basic properties of these second-order self-adjoint ordinary differential equations. We begin by introducing the adjoint of a second-order homogeneous linear differential equation.

DEFINITION

Consider the second-order homogeneous linear differential equation

$$a_0(t)\frac{d^2x}{dt^2} + a_1(t)\frac{dx}{dt} + a_2(t)x = 0, \qquad (11.110)$$

where a_0 has a continuous second derivative, a_1 has a continuous first derivative, a_2 is continuous, and $a_0(t) \neq 0$ on $a \leq t \leq b$. The adjoint equation to Equation (11.110) is

$$\frac{d^2}{dt^2}[a_0(t)x] - \frac{d}{dt}[a_1(t)x] + a_2(t)x = 0,$$

that is, after taking the indicated derivatives,

$$a_0(t)\frac{d^2x}{dt^2} + [2a_0'(t) - a_1(t)]\frac{dx}{dt} + [a_0''(t) - a_1'(t) + a_2(t)]x = 0, \qquad (11.111)$$

where the primes denote differentiation with respect to t.

Note. We assume the additional hypothesis on a_0 and a_1 stated in the definition in order that each of the coefficients in the adjoint equation (11.111) be continuous on $a \leq t \leq b$.

▶ **Example 11.21**

Consider

$$t^2 \frac{d^2x}{dt^2} + 7t \frac{dx}{dt} + 8x = 0.$$

Here $a_0(t) = t^2$, $a_1(t) = 7t$, $a_2(t) = 8$. By (11.111), the adjoint equation to this equation is

$$t^2 \frac{d^2x}{dt^2} + [4t - 7t] \frac{dx}{dt} + [2 - 7 + 8]x = 0$$

or simply

$$t^2 \frac{d^2x}{dt^2} - 3t \frac{dx}{dt} + 3x = 0.$$

The adjoint equation of the adjoint equation of Equation (11.110) is always the original equation (11.110) itself (see Exercise 2 at the end of this section). We now turn to the special situation in which the adjoint equation (11.111) of Equation (11.110) is also Equation (11.110) itself.

DEFINITION

The second-order homogeneous linear differential equation

$$a_0(t) \frac{d^2x}{dt^2} + a_1(t) \frac{dx}{dt} + a_2(t)x = 0 \qquad (11.110)$$

is called self-adjoint *if it is identical with its adjoint Equation (11.111).*

THEOREM 11.33

Consider the second-order linear differential equation

$$a_0(t) \frac{d^2x}{dt^2} + a_1(t) \frac{dx}{dt} + a_2(t)x = 0, \qquad (11.110)$$

where a_0 has a continuous second derivative, a_1 has a continuous first derivative, a_2 is continuous, and $a_0(t) \neq 0$ on $a \leq t \leq b$. A necessary and sufficient condition that Equation (11.110) be self-adjoint is that

$$\frac{d}{dt}[a_0(t)] = a_1(t) \qquad (11.112)$$

on $a \leq t \leq b$.

Proof.   By the definition, the adjoint equation to Equation (11.110) is

$$a_0(t) \frac{d^2x}{dt^2} + [2a_0'(t) - a_1(t)] \frac{dx}{dt} + [a_0''(t) - a_1'(t) + a_2(t)]x = 0. \qquad (11.111)$$

If condition (11.112) is satisfied, then

$$2a_0'(t) - a_1(t) = a_1(t)$$

and

$$a_0''(t) - a_1'(t) + a_2(t) = a_2(t).$$

These relations show that Equations (11.110) and (11.111) are identical; that is, equation (11.110) is self-adjoint.

Conversely, if Equations (11.110) and (11.111) are identical then

$$2a_0'(t) - a_1(t) = a_1(t)$$

and

$$a_0''(t) - a_1'(t) + a_2(t) = a_2(t).$$

The second of these conditions shows that $a_0'(t) = a_1(t) + c$, where c is an arbitrary constant. From the first condition we see that $a_0'(t) = a_1(t)$. Thus $c = 0$ and we have the condition (11.112).

Q.E.D.

COROLLARY

Hypothesis. *Suppose the second-order equation*

$$a_0(t)\frac{d^2x}{dt^2} + a_1(t)\frac{dx}{dt} + a_2(t)x = 0 \tag{11.110}$$

is self-adjoint.

Conclusion. *Then Equation (11.110) can be written in the form*

$$\frac{d}{dt}\left[a_0(t)\frac{dx}{dt}\right] + a_2(t)x = 0. \tag{11.113}$$

Proof. Since Equation (11.110) is self-adjoint, condition (11.112) is satisfied. Thus Equation (11.110) may be written

$$a_0(t)\frac{d^2x}{dt^2} + a_0'(t)\frac{dx}{dt} + a_2(t)x = 0$$

or

$$\frac{d}{dt}\left[a_0(t)\frac{dx}{dt}\right] + a_2(t)x = 0.$$

Q.E.D.

▶ **Example 11.22 Legendre's equation**

$$(1 - t^2)\frac{d^2x}{dt^2} - 2t\frac{dx}{dt} + n(n + 1)x = 0$$

is self-adjoint, for $a_0(t) = 1 - t^2$, $a_1(t) = -2t$ and $(d/dt)(1 - t^2) = -2t$. Written in the form (11.113), it is

$$\frac{d}{dt}\left[(1 - t^2)\frac{dx}{dt}\right] + n(n + 1)x = 0.$$

THEOREM 11.34

Hypothesis.   *The coefficients a_0, a_1, and a_2 in the differential equation*

$$a_0(t)\frac{d^2x}{dt^2} + a_1(t)\frac{dx}{dt} + a_2(t)x = 0 \tag{11.110}$$

are continuous on $a \leq t \leq b$, and $a_0(t) \neq 0$ on $a \leq t \leq b$.

Conclusion.   *Equation (11.110) can be transformed into the equivalent self-adjoint equation*

$$\frac{d}{dt}\left[P(t)\frac{dx}{dt}\right] + Q(t)x = 0 \tag{11.114}$$

on $a \leq t \leq b$, where

$$P(t) = \exp\left[\int \frac{a_1(t)}{a_0(t)}\,dt\right], \qquad Q(t) = \frac{a_2(t)}{a_0(t)}\exp\left[\int \frac{a_1(t)}{a_0(t)}\,dt\right], \tag{11.115}$$

by multiplication throughout by the factor

$$\frac{1}{a_0(t)}\exp\left[\int \frac{a_1(t)}{a_0(t)}\,dt\right]. \tag{11.116}$$

Proof.   Multiplying Equation (11.110) through by the factor (11.116) we obtain

$$\exp\left[\int \frac{a_1(t)}{a_0(t)}\,dt\right]\frac{d^2x}{dt^2} + \frac{a_1(t)}{a_0(t)}\exp\left[\int \frac{a_1(t)}{a_0(t)}\,dt\right]\frac{dx}{dt} + \frac{a_2(t)}{a_0(t)}\exp\left[\int \frac{a_1(t)}{a_0(t)}\,dt\right]x = 0.$$

This equation is clearly self-adjoint and may be written as

$$\frac{d}{dt}\left\{\exp\left[\int \frac{a_1(t)}{a_0(t)}\,dt\right]\frac{dx}{dt}\right\} + \frac{a_2(t)}{a_0(t)}\exp\left[\int \frac{a_1(t)}{a_0(t)}\,dt\right]x = 0$$

or

$$\frac{d}{dt}\left[P(t)\frac{dx}{dt}\right] + Q(t)x = 0,$$

where P and Q are given by (11.115).   *Q.E.D.*

▶ **Example 11.23**

Consider the equation

$$t^2\frac{d^2x}{dt^2} - 2t\frac{dx}{dt} + 2x = 0. \tag{11.117}$$

Here $a_0(t) = t^2, a_1(t) = -2t, a_2(t) = 2$. Since $a_0'(t) = 2t \neq -2t = a_1(t)$, Equation (11.117) is *not* self-adjoint. Let us form the factor (11.116) for this equation. We have

$$\frac{1}{a_0(t)} \exp\left[\int \frac{a_1(t)}{a_0(t)} \, dt\right] = \frac{1}{t^2} \exp\left[\int \frac{-2t}{t^2} \, dt\right] = \frac{1}{t^4}.$$

Multiplying Equation (11.117) by $1/t^4$ on any interval $a \leq t \leq b$ which does not include $t = 0$, we obtain

$$\frac{1}{t^2} \frac{d^2 x}{dt^2} - \frac{2}{t^3} \frac{dx}{dt} + \frac{2}{t^4} x = 0.$$

Since

$$\frac{d}{dt}\left(\frac{1}{t^2}\right) = -\frac{2}{t^3},$$

this equation is self-adjoint and may be written in the form

$$\frac{d}{dt}\left[\frac{1}{t^2} \frac{dx}{dt}\right] + \frac{2}{t^4} x = 0.$$

B. Some Basic Results of Sturm Theory

Henceforth we write the self-adjoint second-order equation in the form

$$\frac{d}{dt}\left[P(t) \frac{dx}{dt}\right] + Q(t)x = 0, \tag{11.118}$$

where P has a continuous derivative, Q is continuous, and $P(t)$ is > 0 on $a \leq t \leq b$.

In order to obtain our first result we shall need a well-known theorem on point sets which is known as the Bolzano–Weierstrass theorem. Suppose E is a set of points on the t axis. A point t_0 is called a *limit point* of E if there exists a sequence of distinct points $t_1, t_2, t_3, \ldots$ of E such that $\lim_{n \to \infty} t_n = t_0$. The Bolzano–Weierstrass theorem states that every bounded infinite set E has at least one limit point.

We now proceed to study Equation (11.118).

THEOREM 11.35

Hypothesis

1. Let f be a solution of

$$\frac{d}{dt}\left[P(t) \frac{dx}{dt}\right] + Q(t)x = 0, \tag{11.118}$$

having first derivative f' on $a \leq t \leq b$.

2. Suppose f has an infinite number of zeros on $a \leq t \leq b$.

Conclusion. *Then $f(t) = 0$ for all t on $a \leq t \leq b$.*

Proof. Since f has an infinite number of zeros on $[a, b]$, by the Bolzano–Weierstrass theorem the set of zeros has a limit point t_0, where $t_0 \in [a, b]$. Thus there exists a sequence $\{t_n\}$ of zeros which converges to t_0 (where $t_n \neq t_0$). Since f is continuous, $\lim_{t \to t_0} f(t) = f(t_0)$, where $t \to t_0$ through any sequence of points on $[a, b]$. Let $t \to t_0$ through the sequence of zeros $\{t_n\}$. Then

$$\lim_{t \to t_0} f(t) = 0 = f(t_0).$$

Now since $f'(t_0)$ exists,

$$f'(t_0) = \lim_{t \to t_0} \frac{f(t) - f(t_0)}{t - t_0},$$

where $t \to t_0$ through the sequence $\{t_n\}$. For such points

$$\frac{f(t) - f(t_0)}{t - t_0} = 0 \quad \text{and thus} \quad f'(t_0) = 0.$$

Thus f is a solution of Equation (11.118) such that $f(t_0) = f'(t_0) = 0$. Hence by the Corollary to Theorem 11.16 we conclude that

$$f(t) = 0 \quad \text{for all } t \text{ on} \quad a \leq t \leq b.$$

$$Q.E.D$$

THEOREM 11.36   Abel's Formula

Hypothesis.   *Let f and g be any two solutions of*

$$\frac{d}{dt}\left[P(t)\frac{dx}{dt}\right] + Q(t)x = 0 \tag{11.118}$$

on the interval $a \leq t \leq b$.

Conclusion.   *Then for all t on $a \leq t \leq b$,*
$$P(t)[f(t)g'(t) - f'(t)g(t)] = k, \tag{11.119}$$

where k is a constant.

Proof.   Since f and g are solutions of (11.118) on $a \leq t \leq b$, we have

$$\frac{d}{dt}[P(t)f'(t)] + Q(t)f(t) = 0 \tag{11.120}$$

and

$$\frac{d}{dt}[P(t)g'(t)] + Q(t)g(t) = 0 \tag{11.121}$$

for all $t \in [a, b]$. Multiply (11.120) by $-g(t)$ and (11.121) by $f(t)$ and add to obtain

$$f(t)\frac{d}{dt}[P(t)g'(t)] - g(t)\frac{d}{dt}[P(t)f'(t)] = 0. \tag{11.122}$$

Now integrate both sides of (11.122) from a to t. Using integration by parts, we obtain

$$f(s)P(s)g'(s)\Big|_a^t - \int_a^t P(s)g'(s)f'(s)\,ds - g(s)P(s)f'(s)\Big|_a^t + \int_a^t P(s)f'(s)g'(s)\,ds = 0,$$

which reduces to

$$P(t)[f(t)g'(t) - f'(t)g(t)] = P(a)[f(a)g'(a) - f'(a)g(a)].$$

The right member is a constant k, and thus we have Abel's formula (11.119).

Q.E.D.

THEOREM 11.37

A. Hypothesis. *Let f and g be two solutions of*

$$\frac{d}{dt}\left[P(t)\frac{dx}{dt}\right] + Q(t)x = 0 \tag{11.118}$$

such that f and g have a common zero on $a \le t \le b$.

Conclusion. *Then f and g are linearly dependent on $a \le t \le b$.*

B. Hypothesis. *Let f and g be nontrivial linearly dependent solutions of Equation (11.118) on $a \le t \le b$, and suppose $f(t_0) = 0$, where t_0 is such that $a \le t_0 \le b$.*

Conclusion. *Then $g(t_0) = 0$.*

Proof

A. We apply Abel's formula

$$P(t)[f(t)g'(t) - f'(t)g(t)] = k. \tag{11.119}$$

Let $t_0 \in [a, b]$ be the common zero of f and g. Letting $t = t_0$ in the formula (11.119) we obtain $k = 0$. Thus

$$P(t)[f(t)g'(t) - f'(t)g(t)] = 0 \qquad \text{for all } t \in [a, b].$$

Since we have assumed throughout that $P(t) > 0$ on $a \le t \le b$, the quantity in brackets above must be zero for all t on $a \le t \le b$. But this quantity is $W(f, g)(t)$. Thus by the corollary to Theorem 11.21, the solutions f and g are linearly dependent on $a \le t \le b$.

B. Since f and g are linearly dependent on $a \le t \le b$, there exist constants c_1 and c_2, not both zero, such that

$$c_1 f(t) + c_2 g(t) = 0, \tag{11.123}$$

for all t on $a \le t \le b$. Now by hypothesis neither f nor g is zero for *all* t on $a \le t \le b$. If $c_1 = 0$, then $c_2 g(t) = 0$ for *all* t on $a \le t \le b$. Since g is not zero for *all* t on $[a, b]$, we must have $c_2 = 0$, which is a contradiction. Thus $c_1 \ne 0$, and likewise $c_2 \ne 0$. Thus neither c_1 nor c_2 in (11.123) is zero. Since $f(t_0) = 0$, letting $t = t_0$ in (11.123), we have $c_2 g(t_0) = 0$. Thus $g(t_0) = 0$.

Q.E.D.

▶ **Example 11.24**

The equation

$$\frac{d^2x}{dt^2} + x = 0$$

is of the type (11.118), where $P(t) = Q(t) = 1$ on every interval $a \le t \le b$. The linearly dependent solutions $A \sin t$ and $B \sin t$ have the common zeros $t = \pm n\pi$ ($n = 0, 1, 2, \ldots$) and no other zeros.

C. The Separation and Comparison Theorems

THEOREM 11.38   Sturm Separation Theorem

Hypothesis.   *Let f and g be real linearly independent solutions of*

$$\frac{d}{dt}\left[P(t)\frac{dx}{dt}\right] + Q(t)x = 0 \tag{11.118}$$

on the interval $a \le t \le b$.

Conclusion.   *Between any two consecutive zeros of f there is precisely one zero of g.*

Proof.   Let t_1 and t_2 be two consecutive zeros of f on $[a, b]$. Then by Theorem 11.37, Part A, we know that $g(t_1) \ne 0$ and $g(t_2) \ne 0$. Now assume that g has *no* zero in the open interval $t_1 < t < t_2$. Then since the solutions f and g have continuous derivatives on $[a, b]$, the quotient f/g has a continuous derivative on the interval $t_1 \le t \le t_2$. Further, $f(t)/g(t)$ is zero at the endpoints of this interval. Thus by Rolle's theorem there exists ξ, where $t_1 < \xi < t_2$, such that

$$\frac{d}{dt}\left[\frac{f(t)}{g(t)}\right]\bigg|_{t=\xi} = 0.$$

But

$$\frac{d}{dt}\left[\frac{f(t)}{g(t)}\right] = \frac{W(g, f)(t)}{[g(t)]^2};$$

and thus since f and g are linearly independent on $a \le t \le b$,

$$\frac{d}{dt}\left[\frac{f(t)}{g(t)}\right] \ne 0 \quad \text{on} \quad t_1 < t < t_2.$$

This contradiction shows that g has *at least one* zero in $t_1 < t < t_2$.

Now suppose g has *more than one* zero in $t_1 < t < t_2$, and let t_3 and t_4 be two such consecutive zeros of g. Then interchanging f and g, the preceding paragraph shows that f must have at least one zero t_5 in the open interval $t_3 < t < t_4$. Then $t_1 < t_5 < t_2$, and so t_1 and t_2 would not be *consecutive* zeros of f, contrary to our assumption concerning t_1 and t_2. Thus g has *precisely one* zero in the open interval $t_1 < t < t_2$.
                                                                            Q.E.D.

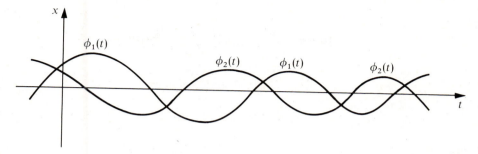

Figure 11.1

We may restate Theorem 11.38 in the following form: The zeros of one of two real linearly independent solutions of Equation (11.118) separate the zeros of the other solution (see Figure 11.1).

▶ **Example 11.25**

We have already observed that the equation

$$\frac{d^2x}{dt^2} + x = 0$$

is of the type (11.118). The functions f and g defined, respectively, by $f(t) = \sin t$ and $g(t) = \cos t$ are linearly independent solutions of this equation. Between any two consecutive zeros of one of these two linearly independent solutions there is indeed precisely one zero of the other solution.

THEOREM 11.39   Sturm's Fundamental Comparison Theorem

Hypothesis.   *On the interval $a \le t \le b$,*

1. Let ϕ_1 be a real solution of

$$\frac{d}{dt}\left[P(t)\frac{dx}{dt} \right] + Q_1(t)x = 0. \tag{11.124}$$

2. Let ϕ_2 be a real solution of

$$\frac{d}{dt}\left[P(t)\frac{dx}{dt} \right] + Q_2(t)x = 0. \tag{11.125}$$

3. Let P have a continuous derivative and be such that $P(t) > 0$, and let Q_1 and Q_2 be continuous and such that $Q_2(t) > Q_1(t)$.

Conclusion.   *If t_1 and t_2 are successive zeros of ϕ_1 on $[a, b]$, then ϕ_2 has at least one zero at some point of the open interval $t_1 < t < t_2$.*

Proof.   Assume that ϕ_2 does *not* have a zero on the open interval $t_1 < t < t_2$. Then without loss in generality we can assume that $\phi_1(t) > 0$ and $\phi_2(t) > 0$ on $t_1 < t < t_2$. By hypothesis, we have

$$\frac{d}{dt}[P(t)\phi_1'(t)] + Q_1(t)\phi_1(t) = 0, \tag{11.126}$$

$$\frac{d}{dt}[P(t)\phi_2'(t)] + Q_2(t)\phi_2(t) = 0, \tag{11.127}$$

for all $t \in [a, b]$. Multiply (11.126) by $\phi_2(t)$ and (11.127) by $\phi_1(t)$, and subtract, to obtain

$$\phi_2(t)\frac{d}{dt}[P(t)\phi_1'(t)] - \phi_1(t)\frac{d}{dt}[P(t)\phi_2'(t)] = [Q_2(t) - Q_1(t)]\phi_1(t)\phi_2(t). \tag{11.128}$$

Since

$$\phi_2(t)\frac{d}{dt}[P(t)\phi_1'(t)] - \phi_1(t)\frac{d}{dt}[P(t)\phi_2'(t)] = \frac{d}{dt}\{P(t)[\phi_1'(t)\phi_2(t) - \phi_1(t)\phi_2'(t)]\},$$

the identity (11.128) reduces to

$$\frac{d}{dt}\{P(t)[\phi_1'(t)\phi_2(t) - \phi_1(t)\phi_2'(t)]\} = [Q_2(t) - Q_1(t)]\phi_1(t)\phi_2(t).$$

We now integrate this from t_1 to t_2 to obtain

$$\int_{t_1}^{t_2} \frac{d}{dt}\{P(t)[\phi_1'(t)\phi_2(t) - \phi_1(t)\phi_2'(t)]\}\, dt = \int_{t_1}^{t_2}[Q_2(t) - Q_1(t)]\phi_1(t)\phi_2(t)\, dt$$

or

$$P(t)[\phi_1'(t)\phi_2(t) - \phi_1(t)\phi_2'(t)]\Big|_{t_1}^{t_2} = \int_{t_1}^{t_2}[Q_2(t) - Q_1(t)]\phi_1(t)\phi_2(t)\, dt. \tag{11.129}$$

Since $\phi_1(t_1) = \phi_1(t_2) = 0$, the equality (11.129) becomes

$$P(t_2)\phi_1'(t_2)\phi_2(t_2) - P(t_1)\phi_1'(t_1)\phi_2(t_1) = \int_{t_1}^{t_2}[Q_2(t) - Q_1(t)]\phi_1(t)\phi_2(t)\, dt. \tag{11.130}$$

By hypothesis, $P(t_2) > 0$. Since $\phi_1(t_2) = 0$ and $\phi_1(t) > 0$ on $t_1 < t < t_2$, we have $\phi_1'(t_2) < 0$. Since $\phi_2(t) > 0$ on $t_1 < t < t_2$, we have $\phi_2(t_2) \geq 0$. Thus $P(t_2)\phi_1'(t_2)\phi_2(t_2) \leq 0$. In like manner, we have $P(t_1)\phi_1'(t_1)\phi_2(t_1) \geq 0$. Thus, the left member of (11.130) is *not* positive.

But by hypothesis $Q_2(t) - Q_1(t) > 0$ on $t_1 \leq t \leq t_2$, and so the right member of (11.130) *is* positive. Thus the assumption that ϕ_2 does *not* have a zero on the open interval $t_1 < t < t_2$ leads to a contradiction, and so ϕ_2 has a zero at some point of this open interval.   Q.E.D.

As a particular case of importance, suppose that the hypotheses of Theorem 11.39 are satisfied and that t_1 is a zero of *both* ϕ_1 and ϕ_2. Then if t_2 and ξ are the "next" zeros of ϕ_1 and ϕ_2, respectively, we must have $\xi < t_2$ (see Figure 11.2).

▶ **Example 11.26**

Consider the equations

$$\frac{d^2x}{dt^2} + A^2x = 0 \tag{11.131}$$

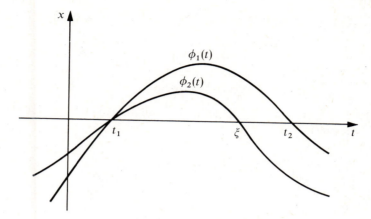

Figure 11.2

and

$$\frac{d^2x}{dt^2} + B^2x = 0, \tag{11.132}$$

where A and B are constants such that $B > A > 0$. The functions ϕ_1 and ϕ_2 defined respectively by $\phi_1(t) = \sin At$ and $\phi_2(t) = \sin Bt$ are real solutions of these respective equations. Consecutive zeros of $\sin At$ are

$$\frac{n\pi}{A} \quad \text{and} \quad \frac{(n+1)\pi}{A} \qquad (n = 0, \pm 1, \pm 2, \dots).$$

Theorem 11.39 applies, and so we are assured that $\sin Bt$ has at least one zero ξ_n such that

$$\frac{n\pi}{A} < \xi_n < \frac{(n+1)\pi}{A} \qquad (n = 0, \pm 1, \pm 2, \dots).$$

In particular, $t = 0$ is a zero of both $\sin At$ and $\sin Bt$. The "next" zero of $\sin At$ is π/A, while the "next" zero of $\sin Bt$ is π/B; and clearly $\pi/B < \pi/A$.

Exercises

1. Find the adjoint equation to each of the following equations:

    (a) $t^2 \dfrac{d^2x}{dt^2} + 3t \dfrac{dx}{dt} + 3x = 0.$

    (b) $(2t + 1) \dfrac{d^2x}{dt^2} + t^3 \dfrac{dx}{dt} + x = 0.$

    (c) $t^2 \dfrac{d^2x}{dt^2} + (2t^3 + 7t) \dfrac{dx}{dt} + (8t^2 + 8)x = 0.$

    (d) $t^3 \dfrac{d^2x}{dt^2} - (t^3 + 2t^2 - t) \dfrac{dx}{dt} + (t^2 + t - 1)x = 0.$

2. Show that the adjoint equation of the adjoint equation (11.111) of the equation

$$a_0(t)\frac{d^2x}{dt^2} + a_1(t)\frac{dx}{dt} + a_2(t)x = 0 \qquad (11.110)$$

is the original equation (11.110) itself.

3. Show that each of the following equations is self-adjoint and write each in the form (11.113).

(a) $t^3\dfrac{d^2x}{dt^2} + 3t^2\dfrac{dx}{dt} + x = 0.$

(b) $\sin t\,\dfrac{d^2x}{dt^2} + \cos t\,\dfrac{dx}{dt} + 2x = 0.$

(c) $\left(\dfrac{t+1}{t}\right)\dfrac{d^2x}{dt^2} - \dfrac{1}{t^2}\dfrac{dx}{dt} + \dfrac{1}{t^3}x = 0.$

4. Transform each of the following equations into an equivalent self-adjoint equation:

(a) $t^2\dfrac{d^2x}{dt^2} + t\dfrac{dx}{dt} + x = 0.$

(b) $(t^4 + t^2)\dfrac{d^2x}{dt^2} + 2t^3\dfrac{dx}{dt} + 3x = 0.$

(c) $\dfrac{d^2x}{dt^2} - \tan t\,\dfrac{dx}{dt} + x = 0.$

(d) $f(t)\dfrac{d^2x}{dt^2} + g(t)x = 0.$

5. (a) A first-order differential equation of the form

$$\frac{dx}{dt} + a(t)x + b(t)x^2 + c(t) = 0$$

is called a Riccati equation. Show that the transformation

$$u = \frac{P(t)\dfrac{dx}{dt}}{x}$$

transforms the self-adjoint second-order equation

$$\frac{d}{dt}\left[P(t)\frac{dx}{dt}\right] + Q(t)x = 0$$

into the special Riccati equation

$$\frac{du}{dt} + \frac{1}{P(t)}u^2 + Q(t) = 0.$$

(b) Use the result of part (a) to transform the self-adjoint equation

$$\frac{d}{dt}\left[t\frac{dx}{dt}\right] + (1 - t)x = 0 \tag{A}$$

into a Riccati equation. For this Riccati equation find a solution of the form ct^n. Then employ the transformation of part (a) to find a solution of Equation (A). Finally, use the method of reduction of order to obtain the general solution of Equation (A).

6. Use the Sturm separation theorem to show that between any two consecutive zeros of $\sin 2t + \cos 2t$ there is precisely one zero of $\sin 2t - \cos 2t$.

7. (a) Show that every real solution of

$$\frac{d^2x}{dt^2} + (t + 1)x = 0$$

has an infinite number of positive zeros.

(b) More generally, show that if q is continuous and such that $q(t) > 0$ for $t > 0$ and k is a positive constant, then every real solution of

$$\frac{d^2x}{dt^2} + [q(t) + k^2]x = 0$$

has an infinite number of positive zeros.

8. Show that if q is continuous and $q(t) < 0$ throughout $a \le t \le b$, then no nontrivial solution of

$$\frac{d^2x}{dt^2} + q(t)x = 0$$

can have more than one zero on $a \le t \le b$.

9. Consider the equation

$$\frac{d^2x}{dt^2} + q(t)x = 0, \tag{A}$$

where q is continuous on $a \le t \le b$ and such that $0 < m < q(t) < M$. Let ϕ_1 be a solution of Equation (A) having consecutive zeros at t_1 and t_2 (where $a \le t_1 < t_2 \le b$). Show that

$$\frac{\pi}{\sqrt{M}} < t_2 - t_1 < \frac{\pi}{\sqrt{m}}.$$

[*Hint:* Consider a solution ϕ_2 of $\dfrac{d^2x}{dt^2} + Mx = 0$ which is such that $\phi_2(t_1) = 0$, and apply Theorem 11.39.]

10. Consider the differential equation

$$\frac{d^2x}{dt^2} + q(t)x = 0, \tag{A}$$

where q is continuous and $q(t) > 0$ on $a \le t \le b$. Let q_m denote the minimum value of $q(t)$ on $a \leq t \leq b$. Show that if $q_m > k^2\pi^2/(b-a)^2$, then every real solution of Equation (A) has at least k zeros on $a \leq t \leq b$.

[*Hint*: Consider the zeros of the real solutions of

$$\frac{d^2x}{dt^2} + \frac{k^2\pi^2}{(b-a)^2}x = 0$$

and apply Theorem 11.39.]

11. Consider the Bessel differential equation

$$t^2\frac{d^2x}{dt^2} + t\frac{dx}{dt} + (t^2 - p^2)x = 0, \tag{A}$$

where p is a parameter. Suppose $t > 0$; and let $[a, a + \pi]$, where $a > 0$, be an arbitrary interval of length π on the positive t axis. Prove the following result:

  If $p = 0$, every such interval $[a, a + \pi]$ contains at least one zero of any solution of the Bessel differential equation (A) of order zero; and if $p > \frac{1}{2}$, every such interval $[a, a + \pi]$ contains at most one zero of any nontrivial solution of the Bessel differential equation of order p.

[*Hint*: According to the result stated in Exercise 2 of Section 6.3 the transformation

$$y = \frac{u(t)}{\sqrt{t}}$$

reduces Equation (A) to the form

$$\frac{d^2u}{dt^2} + \left[1 + \left(\frac{1}{4} - p^2\right)\frac{1}{t^2}\right]u = 0$$

or, equivalently,

$$\frac{d^2u}{dt^2} + \left[1 - \frac{4p^2 - 1}{4t^2}\right]u = 0. \tag{B}$$

Note that if a solution f of an equation of the form (B) has a zero at $t_0 > 0$, then the solution g such that $g(t) = f(t)/\sqrt{t}$ of the corresponding Bessel equation (A) also has a zero at t_0. Then apply Theorem 11.39 to compare the number of zeros of solutions of Equation (B) to the number of zeros of solutions of

$$\frac{d^2u}{dt^2} + u = 0.]$$

12. Consider the differential equations

$$\frac{d}{dt}\left[P(t)\frac{dx}{dt}\right] + Q_1(t)x = 0 \tag{A}$$

and

$$\frac{d}{dt}\left[P(t)\frac{dx}{dt}\right] + Q_2(t)x = 0, \tag{B}$$

where P has a continuous derivative and is such that $P(t) > 0$, and Q_1 and Q_2 are continuous and such that $Q_2(t) > Q_1(t)$ on $a \leq t \leq b$. Let t_0 and t_1 be such that $a \leq t_0 < t_1 \leq b$. Let ϕ_1 be the solution of Equation (A) such that $\phi_1(t_0) = c_0$, $\phi_1'(t_0) = c_1$, and let ϕ_2 be the solution of Equation (B) such that $\phi_2(t_0) = c_0$ and $\phi_2'(t_0) = c_1$, where $c_0 \geq 0$ and $c_1 > 0$ if $c_0 = 0$. Suppose that $\phi_2(t) > 0$ for $t_0 < t < t_1$.

Show that $\phi_1(t) > \phi_2(t)$ for $t_0 < t < t_1$.

[*Hint:* Proceed as in the proof of Theorem 11.39 to show that

$$P(t)[\phi_1'(t)\phi_2(t) - \phi_1(t)\phi_2'(t)] = \int_{t_0}^{t} [Q_2(s) - Q_1(s)]\phi_1(s)\phi_2(s)\, ds$$

for all $t \in [a, b]$. Thus show that the function h such that $h(t) = \phi_1(t)/\phi_2(t)$ is increasing for $t_0 < t < t_1$, and observe the value $h(t_0)$.]

CHAPTER TWELVE

Sturm–Liouville Boundary-Value Problems and
Fourier Series

In Chapter 1 we encountered boundary-value problems consisting of a second-order linear differential equation and two supplementary conditions which the solution of the equation must satisfy. In this chapter we shall consider a special kind of boundary-value problem known as a *Sturm–Liouville problem.* Our study of this type of problem will introduce us to several important concepts including *characteristic function,* *orthogonality,* and *Fourier series.* These concepts are frequently employed in the applications of differential equations to physics and engineering. In Chapter 14 we shall use them to obtain solutions of boundary-value problems which involve partial differential equations.

12.1 STURM–LIOUVILLE PROBLEMS

A. Definition and Examples

Our first concern in this chapter is a study of the special type of two-point boundary-value problem given in the following definition:

DEFINITION

We consider a boundary-value problem which consists of
   1. a second-order homogeneous linear differential equation of the form

$$\frac{d}{dx}\left[p(x)\frac{dy}{dx}\right] + [q(x) + \lambda r(x)]y = 0, \tag{12.1}$$

where p, q, and r are real functions such that p has a continuous derivative, q and r are

588

continuous, and p(x) > 0 and r(x) > 0 for all x on a real interval $a \leq x \leq b$; and λ is a parameter independent of x; and

    2. *two supplementary conditions*

$$A_1 y(a) + A_2 y'(a) = 0,$$

$$B_1 y(b) + B_2 y'(b) = 0, \qquad (12.2)$$

where A_1, A_2, B_1, and B_2 are real constants such that A_1 and A_2 are not both zero and B_1 and B_2 are not both zero.

This type of boundary-value problem is called a Sturm–Liouville problem (*or* Sturm–Liouville system).

Two important special cases are those in which the supplementary conditions (12.2) are either of the form

$$y(a) = 0, \qquad y(b) = 0 \qquad (12.3)$$

or of the form

$$y'(a) = 0, \qquad y'(b) = 0. \qquad (12.4)$$

▶ **Example 12.1**

The boundary-value problem

$$\frac{d^2 y}{dx^2} + \lambda y = 0, \qquad (12.5)$$

$$y(0) = 0, \qquad y(\pi) = 0 \qquad (12.6)$$

is a Sturm–Liouville problem. The differential equation (12.5) may be written

$$\frac{d}{dx}\left[1 \cdot \frac{dy}{dx} \right] + [0 + \lambda \cdot 1]y = 0$$

and hence is of the form (12.1), where $p(x) = 1$, $q(x) = 0$, and $r(x) = 1$. The supplementary conditions (12.6) are of the special form (12.3) of (12.2).

▶ **Example 12.2**

The boundary-value problem

$$\frac{d}{dx}\left[x\,\frac{dy}{dx} \right] + [2x^2 + \lambda x^3]y = 0, \qquad (12.7)$$

$$3y(1) + 4y'(1) = 0,$$

$$5y(2) - 3y'(2) = 0, \qquad (12.8)$$

is a Sturm–Liouville problem. The differential equation (12.7) is of the form (12.1), where $p(x) = x$, $q(x) = 2x^2$, and $r(x) = x^3$. The conditions (12.8) are of the form (12.2), where $a = 1$, $b = 2$, $A_1 = 3$, $A_2 = 4$, $B_1 = 5$, and $B_2 = -3$.

Let us now see what is involved in solving a Sturm–Liouville problem. We must find a function f which satisfies both the differential equation (12.1) and the two supplementary conditions (12.2). Clearly one solution of *any* problem of this type is the *trivial* solution ϕ such that $\phi(x) = 0$ for all values of x. Equally clear is the fact that this trivial solution is not very useful. We shall therefore focus our attention on the search for *nontrivial* solutions of the problem. That is, we shall attempt to find functions, *not identically zero*, which satisfy both the differential equation (12.1) and the two conditions (12.2). We shall see that the existence of such nontrivial solutions depends upon the value of the parameter λ in the differential equation (12.1). To illustrate this, let us return to the Sturm–Liouville problem of Example 12.1 and attempt to find nontrivial solutions.

▶ **Example 12.3**

Find nontrivial solutions of the Sturm–Liouville problem

$$\frac{d^2y}{dx^2} + \lambda y = 0, \tag{12.5}$$

$$y(0) = 0, \qquad y(\pi) = 0. \tag{12.6}$$

Solution.   We shall consider separately the three cases $\lambda = 0$, $\lambda < 0$, and $\lambda > 0$. In each case we shall first find the general solution of the differential equation (12.5). We shall then attempt to determine the two arbitrary constants in this solution so that the supplementary conditions (12.6) will also be satisfied.

Case 1: $\lambda = 0$.   In this case the differential equation (12.5) reduces at once to

$$\frac{d^2y}{dx^2} = 0$$

and so the general solution is

$$y = c_1 + c_2 x. \tag{12.9}$$

We now apply the conditions (12.6) to the solution (12.9). Applying the first condition $y(0) = 0$, we obtain $c_1 = 0$. Applying the second condition $y(\pi) = 0$, we find that $c_1 + c_2\pi = 0$. Hence, since $c_1 = 0$, we must also have $c_2 = 0$. Thus in order for the solution (12.9) to satisfy the conditions (12.6), we must have $c_1 = c_2 = 0$. But then the solution (12.9) becomes the solution y such that $y(x) = 0$ for all values of x. Thus if the parameter $\lambda = 0$, the only solution of the given problem is the trivial solution.

Case 2: $\lambda < 0$.   The auxiliary equation of the differential equation (12.5) is $m^2 + \lambda = 0$ and has the roots $\pm\sqrt{-\lambda}$. Since in this case $\lambda - 0$, these roots are real and unequal. Denoting $\sqrt{-\lambda}$ by α, we see that for $\lambda < 0$ the general solution of (12.5) is of the form

$$y = c_1 e^{\alpha x} + c_2 e^{-\alpha x}. \tag{12.10}$$

We now apply the conditions (12.6) to the solution (12.10). Applying the first condition $y(0) = 0$, we obtain

$$c_1 + c_2 = 0. \tag{12.11}$$

Applying the second condition $y(\pi) = 0$, we find that

$$c_1 e^{\alpha\pi} + c_2 e^{-\alpha\pi} = 0. \tag{12.12}$$

We must thus determine c_1 and c_2 such that the system consisting of (12.11) and (12.12) is satisfied. Thus in order for the solution (12.10) to satisfy the conditions (12.6), the constants c_1 and c_2 must satisfy the system of Equations (12.11) and (12.12). Obviously $c_1 = c_2 = 0$ is a solution of this system; but these values of c_1 and c_2 would only give the trivial solution of the given problem. We must therefore seek nonzero values of c_1 and c_2 which satisfy (12.11) and (12.12). By Chapter 7, Section 7.5C, Theorem A, this system has nonzero solutions only if the determinant of coefficients is zero. Therefore we must have

$$\begin{vmatrix} 1 & 1 \\ e^{\alpha\pi} & e^{-\alpha\pi} \end{vmatrix} = 0.$$

But this implies that $e^{\alpha\pi} = e^{-\alpha\pi}$ and hence that $\alpha = 0$. Thus in order for a nontrivial function of the form (12.10) to satisfy the conditions (12.6) we must have $\alpha = 0$. Since $\alpha = \sqrt{-\lambda}$, we must then have $\lambda = 0$. But $\lambda < 0$ in this case. Thus there are no non-trivial solutions of the given problem in the case $\lambda < 0$.

Case 3: $\lambda > 0$.   Since $\lambda > 0$ here, the roots $\pm\sqrt{-\lambda}$ of the auxiliary equation of (12.5) are the conjugate-complex numbers $\pm\sqrt{\lambda}i$. Thus in this case the general solution of (12.5) is of the form

$$y = c_1 \sin\sqrt{\lambda}x + c_2 \cos\sqrt{\lambda}x. \tag{12.13}$$

We now apply the conditions (12.6) to this general solution. Applying the first condition $y(0) = 0$, we obtain

$$c_1 \sin 0 + c_2 \cos 0 = 0$$

and hence $c_2 = 0$. Applying the second condition $y(\pi) = 0$, we find that

$$c_1 \sin\sqrt{\lambda}\pi + c_2 \cos\sqrt{\lambda}\pi = 0.$$

Since $c_2 = 0$, this reduces at once to

$$c_1 \sin\sqrt{\lambda}\pi = 0 \tag{12.14}$$

We must therefore satisfy (12.14). At first glance it appears that we can do this in either of two ways: we can set $c_1 = 0$ or we can set $\sin\sqrt{\lambda}\pi = 0$. However, if we set $c_1 = 0$, then (since $c_2 = 0$ also) the solution (12.13) reduces immediately to the unwanted trivial solution. Thus to obtain a *nontrivial* solution we can *not* set $c_1 = 0$ but rather we *must* set

$$\sin\sqrt{\lambda}\pi = 0. \tag{12.15}$$

If $k > 0$, then $\sin k\pi = 0$ only if k is a positive integer $n = 1, 2, 3, \ldots$. Thus in order to satisfy (12.15), we must have $\sqrt{\lambda} = n$, where $n = 1, 2, 3, \ldots$. Therefore, in order that the differential equation (12.5) have a nontrivial solution of the form (12.13) satisfying the conditions (12.6), we must have

$$\lambda = n^2, \qquad \text{where } n = 1, 2, 3, \ldots. \tag{12.16}$$

In other words, the parameter λ in (12.5) must be a member of the infinite sequence

$$1, 4, 9, 16, \ldots, n^2, \ldots.$$

Summary. If $\lambda \leq 0$, the Sturm–Liouville problem consisting of (12.5) and (12.6) does *not* have a nontrivial solution; if $\lambda > 0$, a nontrivial solution can exist only if λ is one of the values given by (12.16). We now note that if λ *is* one of the values (12.16), then the problem *does have* nontrivial solutions. Indeed, from (12.13) we see that nontrivial solutions corresponding to $\lambda = n^2$ ($n = 1, 2, 3, \ldots$) are given by

$$y = c_n \sin nx \qquad (n = 1, 2, 3, \ldots), \tag{12.17}$$

where c_n ($n = 1, 2, 3, \ldots$) is an arbitrary nonzero constant. That is, the functions defined by $c_1 \sin x, c_2 \sin 2x, c_3 \sin 3x, \ldots$, where $c_1, c_2, c_3, \ldots$ are arbitrary nonzero constants, are nontrivial solutions of the given problem.

B. Characteristic Values and Characteristic Functions

Example 12.3 shows that the existence of nontrivial solutions of a Sturm–Liouville problem does indeed depend upon the value of the parameter λ in the differential equation of the problem. Those values of the parameter for which nontrivial solutions do exist, as well as the corresponding nontrivial solutions themselves, are singled out by the following definition:

DEFINITION

Consider the Sturm–Liouville problem consisting of the differential equation (12.1) and the supplementary conditions (12.2). The values of the parameter λ in (12.1) for which there exist nontrivial solutions of the problem are called the characteristic values *of the problem. The corresponding nontrivial solutions themselves are called the* characteristic functions *of the problem.**

▶ **Example 12.4**

Consider again the Sturm–Liouville problem

$$\frac{d^2 y}{dx^2} + \lambda y = 0, \tag{12.5}$$

$$y(0) = 0, \qquad y(\pi) = 0. \tag{12.6}$$

In Example 12.3 we found that the values of λ in (12.5) for which there exist nontrivial solutions of this problem are the values

$$\lambda = n^2, \qquad \text{where } n = 1, 2, 3, \ldots. \tag{12.16}$$

These then are the characteristic values of the problem under consideration. The characteristic functions of the problem are the corresponding nontrivial solutions

$$y = c_n \sin nx \qquad (n = 1, 2, 3, \ldots), \tag{12.17}$$

where c_n ($n = 1, 2, 3, \ldots$) is an arbitrary nonzero constant.

---

* The characteristic values are also called *eigenvalues*; and the characteristic functions are also called *eigenfunctions*.

▶ **Example 12.5**

Find the characteristic values and characteristic functions of the Sturm–Liouville problem

$$\frac{d}{dx}\left[x\frac{dy}{dx}\right] + \frac{\lambda}{x}y = 0. \tag{12.18}$$

$$y'(1) = 0, \qquad y'(e^{2\pi}) = 0, \tag{12.19}$$

where we assume that the parameter λ in (12.18) is nonnegative.

Solution. We consider separately the cases $\lambda = 0$ and $\lambda > 0$. If $\lambda = 0$, the differential equation (12.18) reduces to

$$\frac{d}{dx}\left[x\frac{dy}{dx}\right] = 0.$$

The general solution of this differential equation is

$$y = C \ln |x| + C_0,$$

where C and C_0 are arbitrary constants. If we apply the conditions (12.19) to this general solution, we find that both of them require that $C = 0$ but neither of them imposes any restriction upon C_0. Thus for $\lambda = 0$ we obtain the solutions $y = C_0$, where C_0 is an arbitrary constant. These are nontrivial solutions for all choices of $C_0 \neq 0$. Thus $\lambda = 0$ is a characteristic value and the corresponding characteristic functions are given by $y = C_0$, where C_0 is an arbitrary nonzero constant.

If $\lambda > 0$, we see that for $x \neq 0$ this equation is equivalent to the Cauchy-Euler equation

$$x^2\frac{d^2y}{dx^2} + x\frac{dy}{dx} + \lambda y = 0. \tag{12.20}$$

Letting $x = e^t$, Equation (12.20) transforms into

$$\frac{d^2y}{dt^2} + \lambda y = 0. \tag{12.21}$$

Since $\lambda > 0$, the general solution of (12.21) is of the form

$$y = c_1 \sin\sqrt{\lambda}t + c_2 \cos\sqrt{\lambda}t.$$

Thus for $\lambda > 0$ and $x > 0$ the general solution of (12.18) may be written

$$y = c_1 \sin(\sqrt{\lambda}\ln x) + c_2 \cos(\sqrt{\lambda}\ln x). \tag{12.22}$$

We now apply the supplementary conditions (12.19). From (12.22) we find that

$$\frac{dy}{dx} = \frac{c_2\sqrt{\lambda}}{x}\cos(\sqrt{\lambda}\ln x) - \frac{c_2\sqrt{\lambda}}{x}\sin(\sqrt{\lambda}\ln x) \tag{12.23}$$

for $x > 0$. Applying the first condition $y'(1) = 0$ of (12.19) to (12.23), we have

$$c_1\sqrt{\lambda}\cos(\sqrt{\lambda}\ln 1) - c_2\sqrt{\lambda}\sin(\sqrt{\lambda}\ln 1) = 0$$

or simply $c_1\sqrt{\lambda} = 0$. Thus we must have

$$c_1 = 0. \tag{12.24}$$

Applying the second condition $y'(e^{2\pi}) = 0$ of (12.19) to (12.23), we obtain

$$c_1\sqrt{\lambda}e^{-2\pi}\cos(\sqrt{\lambda}\ln e^{2\pi}) - c_2\sqrt{\lambda}e^{-2\pi}\sin(\sqrt{\lambda}\ln e^{2\pi}) = 0.$$

Since $c_1 = 0$ by (12.24) and $\ln e^{2\pi} = 2\pi$, this reduces at once to

$$c_2\sqrt{\lambda}e^{-2\pi}\sin(2\pi\sqrt{\lambda}) = 0.$$

Since $c_1 = 0$, the choice $c_2 = 0$ would lead to the trivial solution.

We must have $\sin(2\pi\sqrt{\lambda}) = 0$ and hence $2\pi\sqrt{\lambda} = n\pi$, where $n = 1, 2, 3, \ldots$. Thus in order to satisfy the second condition (12.19) nontrivially we must have

$$\lambda = \frac{n^2}{4} \qquad (n = 1, 2, 3, \ldots). \tag{12.25}$$

Corresponding to these values of λ we obtain for $x > 0$ the nontrivial solutions

$$y = C_n \cos\left(\frac{n \ln x}{2}\right) \qquad (n = 1, 2, 3, \ldots), \tag{12.26}$$

where the $C_n (n = 1, 2, 3, \ldots)$ are arbitrary nonzero constants.

Thus the values

$$\lambda = 0, \frac{1}{4}, 1, \frac{9}{4}, 4, \frac{25}{4}, \ldots, \frac{n^2}{4}, \ldots,$$

given by (12.25) for $n \geq 0$, are the characteristic values of the given problem. The functions

$$C_0, C_1\cos\left(\frac{\ln x}{2}\right), C_2\cos(\ln x), C_3\cos\left(\frac{3\ln x}{2}\right), \ldots,$$

given by (12.26) for $n \geq 0$, where $C_0, C_1, C_2, C_3, \ldots$ are arbitrary nonzero constants, are the corresponding characteristic functions.

For each of the Sturm–Liouville problems of Examples 12.3 and 12.5 we found an infinite number of characteristic values. We observe that in each of these problems the infinite set of characteristic values thus found can be arranged in a monotonic increasing sequence*

$$\lambda_1 < \lambda_2 < \lambda_3 < \cdots$$

such that $\lambda_n \to +\infty$ as $n \to +\infty$. For example, the characteristic values of the problem of Example 12.3 can be arranged in the monotonic increasing sequence

$$1 < 4 < 9 < 16 < \cdots \tag{12.27}$$

such that $\lambda_n = n^2 \to +\infty$ as $n \to +\infty$. We also note that in each problem there is a one-parameter family of characteristic functions corresponding to each characteristic value, and any two characteristic functions corresponding to the same characteristic value are merely nonzero constant multiples of each other. For example, in the problem of Example 12.3 the one-parameter family of characteristic functions corresponding to the characteristic value n^2 is $c_n \sin nx$, where $c_n \neq 0$ is the parameter.

---

* An infinite sequence $\{x_n\}$ is said to be monotonic increasing if $x_{n+1} \geq x_n$ for every n.

We might now inquire whether or not all Sturm–Liouville problems of the type under consideration possess characteristic values and characteristic functions having the properties noted in the preceding paragraph. We can answer in the affirmative by stating the following important theorem.

THEOREM 12.1

Hypothesis. *Consider the Sturm–Liouville problem consisting of*

1. the differential equation

$$\frac{d}{dx}\left[p(x)\frac{dy}{dx} \right] + [q(x) + \lambda r(x)]\, y = 0, \tag{12.1}$$

where p, q, and r are real functions such that p has a continuous derivative, q and r are continuous, and p(x) > 0 and r(x) > 0 for all x on a real interval $a \le x \le b$; and λ is a parameter independent of x; and

2. the conditions

$$A_1 y(a) + A_2 y'(a) = 0,$$
$$B_1 y(b) + B_2 y'(b) = 0, \tag{12.2}$$

where A_1, A_2, B_1, and B_2 are real constants such that A_1 and A_2 are not both zero and B_1 and B_2 are not both zero.

Conclusions

1. There exists an infinite number of characteristic values λ_n of the given problem. These characteristic values λ_n can be arranged in a monotonic increasing sequence

$$\lambda_1 < \lambda_2 < \lambda_3 < \cdots$$

such that $\lambda_n \to +\infty$ as $n \to +\infty$.

2. Corresponding to each characteristic value λ_n there exists a one-parameter family of characteristic functions ϕ_n. Each of these characteristic functions is defined on $a \le x \le b$, and any two characteristic functions corresponding to the same characteristic value are nonzero constant multiples of each other.

3. Each characteristic function ϕ_n corresponding to the characteristic value λ_n (n = 1, 2, 3,...) has exactly (n − 1) zeros in the open *interval $a < x < b$.*

We regard the proof of this theorem as beyond the scope of this book; therefore we omit it.

▶ **Example 12.6**

Consider again the Sturm–Liouville problem of Examples 12.3 and 12.4,

$$\frac{d^2y}{dx^2} + \lambda y = 0, \tag{12.5}$$

$$y(0) = 0, \qquad y(\pi) = 0. \tag{12.6}$$

We have already noted the validity of Conclusions 1 and 2 of Theorem 12.1 for this problem. The infinite number of characteristic values $\lambda_n = n^2$ $(n = 1, 2, 3,\ldots)$ can be arranged in the unbounded monotonic increasing sequence indicated by (12.27); and the characteristic functions $c_n \sin nx (c_n \neq 0)$, corresponding to $\lambda_n = n^2$, possess the properties stated.

We now illustrate Conclusion 3 by showing that each characteristic function $c_n \sin nx$ corresponding to $\lambda_n = n^2$ has exactly $(n-1)$ zeros in the open interval $0 < x < \pi$. We know that $\sin nx = 0$ if and only if $nx = k\pi$, where k is an integer. Thus the zero of $c_n \sin nx$ are given by

$$x = \frac{k\pi}{n} \qquad (k = 0, \pm 1, \pm 2,\ldots).\tag{12.28}$$

The zeros (12.28) which lie in the open interval $0 < x < \pi$ are precisely those for which $k = 1, 2, 3,\ldots, n-1$. Thus, just as Conclusion 3 asserts, each characteristic function $c_n \sin nx$ has precisely $(n-1)$ zeros in the open interval $0 < x < \pi$.

Exercises

Find the characteristic values and characteristic functions of each of the following Sturm–Liouville problems.

1. $\dfrac{d^2y}{dx^2} + \lambda y = 0,$ $y(0) = 0,$ $y\left(\dfrac{\pi}{2}\right) = 0.$

2. $\dfrac{d^2y}{dx^2} + \lambda y = 0,$ $y(0) = 0,$ $y'(\pi) = 0.$

3. $\dfrac{d^2y}{dx^2} + \lambda y = 0,$ $y(0) = 0,$ $y(L) = 0,$ where $L > 0.$

4. $\dfrac{d^2y}{dx^2} + \lambda y = 0,$ $y'(0) = 0,$ $y'(L) = 0,$ where $L > 0.$

5. $\dfrac{d^2y}{dx^2} + \lambda y = 0,$ $y(0) = 0.$ $y(\pi) - y'(\pi) = 0.$

6. $\dfrac{d^2y}{dx^2} + \lambda y = 0,$ $y(0) - y'(0) = 0,$ $y(\pi) - y'(\pi) = 0.$

7. $\dfrac{d}{dx}\left[x\dfrac{dy}{dx}\right] + \dfrac{\lambda}{x}y = 0,$ $y(1) = 0,$ $y(e^{\pi}) = 0.$

8. $\dfrac{d}{dx}\left[x\dfrac{dy}{dx}\right] + \dfrac{\lambda}{x}y = 0,$ $y(1) = 0,$ $y'(e^{\pi}) = 0.$

9. $\dfrac{d}{dx}\left[(x^2 + 1)\dfrac{dy}{dx}\right] + \dfrac{\lambda}{x^2 + 1}y = 0,$ $y(0) = 0,$ $y(1) = 0.$

    [*Hint:*   Let $x = \tan t.$]

10. $\dfrac{d}{dx}\left[\dfrac{1}{3x^2 + 1}\dfrac{dy}{dx}\right] + \lambda(3x^2 + 1)y = 0,$ $y(0) = 0,$ $y(\pi) = 0.$

    [*Hint:*   Let $t = x^3 + x.$]

12.2 ORTHOGONALITY OF CHARACTERISTIC FUNCTIONS

A. Orthogonality

We now introduce the concept of orthogonality of functions in the following definition.

DEFINITION

Two functions f and g are called orthogonal *with respect to the weight function r on the interval $a \leq x \leq b$ if and only if*

$$\int_a^b f(x)g(x)r(x)\, dx = 0.$$

▶ **Example 12.7**

The functions $\sin x$ and $\sin 2x$ are orthogonal with respect to the weight function having the constant value 1 on the interval $0 \leq x \leq \pi$, for

$$\int_0^\pi (\sin x)(\sin 2x)(1)\, dx = \frac{2\sin^3 x}{3}\bigg|_0^\pi = 0.$$

DEFINITION

Let $\{\phi_n\}$, $n = 1, 2, 3, \ldots$, be an infinite set of functions defined on the interval $a \leq x \leq b$. The set $\{\phi_n\}$ is called an orthogonal system *with respect to the weight function r on $a \leq x \leq b$ if every two distinct functions of the set are orthogonal with respect to r on $a \leq x \leq b$. That is, the set $\{\phi_n\}$ is orthogonal with respect to r on $a \leq x \leq b$ if*

$$\int_a^b \phi_m(x)\phi_n(x)r(x)\, dx = 0 \qquad \text{for } m \neq n.$$

▶ **Example 12.8**

Consider the infinite set of functions $\{\phi_n\}$, where $\phi_n(x) = \sin nx$ $(n = 1, 2, 3, \ldots)$, on the interval $0 \leq x \leq \pi$. The set $\{\phi_n\}$ is an orthogonal system with respect to the weight function having the constant value 1 on the interval $0 \leq x \leq \pi$, for

$$\int_0^\pi (\sin mx)(\sin nx)(1)\, dx = \frac{\sin(m-n)x}{2(m-n)} - \frac{\sin(m+n)x}{2(m+n)}\bigg|_0^\pi = 0 \qquad \text{for } m \neq n.$$

Note. In Examples 12.7 and 12.8 and also in all later examples in this chapter which involve orthogonality, the weight function *r* is the function having the constant value 1 on the interval of the problem. Of course one should *not* conclude from this that the weight function r is *always* $r(x) = 1$ for all x under consideration in *every* problem involving orthogonality. For instance, see Exercise 3 at the end of this section and also the orthogonality relation involving the Bessel function of the first kind of order zero in Section 14.2D.

B. Orthogonality of Characteristic Functions

We now state and prove a basic theorem concerning the orthogonality of characteristic functions of a Sturm–Liouville problem.

THEOREM 12.2

Hypothesis.   *Consider the Sturm–Liouville problem consisting of*

1. the differential equation

$$\frac{d}{dx}\left[p(x)\frac{dy}{dx} \right] + [q(x) + \lambda r(x)]y = 0, \tag{12.1}$$

where p, q, and r are real functions such that p has a continuous derivative, q and r are continuous, and $p(x) > 0$ and $r(x) > 0$ for all x on a real interval $a \le x \le b$; and λ is a parameter independent of x; and
  2. the conditions

$$A_1 y(a) + A_2 y'(a) = 0,$$
$$B_1 y(b) + B_2 y'(b) = 0, \tag{12.2}$$

where A_1, A_2, B_1, and B_2 are real constants such that A_1 and A_2 are not both zero and B_1 and B_2 are not both zero.
  Let λ_m and λ_n be any two distinct characteristic values of this problem. Let ϕ_m be a characteristic function corresponding to λ_m and let ϕ_n be a characteristic function corresponding to λ_n.

Conclusion.   *The characteristic functions ϕ_m and ϕ_n are orthogonal with respect to the weight function r on the interval $a \le x \le b$.*

Proof.   Since ϕ_m is a characteristic function corresponding to λ_m, the function ϕ_m satisfies the differential equation (12.1) with $\lambda = \lambda_m$; and since ϕ_n is a characteristic function corresponding to λ_n, the function ϕ_n satisfies the differential equation (12.1) with $\lambda = \lambda_n$. Thus, denoting the derivatives of ϕ_m and ϕ_n by ϕ'_m and ϕ'_n, respectively, we have

$$\frac{d}{dx}[p(x)\phi'_m(x)] + [q(x) + \lambda_m r(x)]\phi_m(x) = 0, \tag{12.29}$$

$$\frac{d}{dx}[p(x)\phi'_n(x)] + [q(x) + \lambda_n r(x)]\phi_n(x) = 0, \tag{12.30}$$

for all x such that $a \le x \le b$. Multiplying both sides of (12.29) by $\phi_n(x)$ and both sides of

(12.30) by $\phi_m(x)$ and then subtracting the results we obtain

$$\phi_n(x)\frac{d}{dx}\left[p(x)\phi'_m(x)\right] + \lambda_m\phi_m(x)\phi_n(x)r(x) - \phi_m(x)\frac{d}{dx}\left[p(x)\phi'_n(x)\right]$$

$$- \lambda_n\phi_m(x)\phi_n(x)r(x) = 0,$$

and hence

$$(\lambda_m - \lambda_n)\phi_m(x)\phi_n(x)r(x) = \phi_m(x)\frac{d}{dx}\left[p(x)\phi'_n(x)\right] - \phi_n(x)\frac{d}{dx}\left[p(x)\phi'_m(x)\right].$$

We now integrate both members of this identity from a to b to obtain

$$(\lambda_m - \lambda_n)\int_a^b \phi_m(x)\phi_n(x)r(x)\,dx = \int_a^b \phi_m(x)\frac{d}{dx}\left[p(x)\phi'_n(x)\right]\,dx$$

$$- \int_a^b \phi_n(x)\frac{d}{dx}\left[p(x)\phi'_m(x)\right]\,dx. \qquad (12.31)$$

Applying integration by parts to each integral in the right member of (12.31), this right member becomes

$$\phi_m(x)p(x)\phi'_n(x)\Big|_a^b - \int_a^b p(x)\phi'_n(x)\phi'_m(x)\,dx - \phi_n(x)p(x)\phi'_m(x)\Big|_a^b + \int_a^b p(x)\phi'_m(x)\phi'_n(x)\,dx$$

or simply

$$\left\{p(x)\left[\phi_m(x)\phi'_n(x) - \phi_n(x)\phi'_m(x)\right]\right\}\Big|_a^b.$$

Therefore the identity (12.31) becomes

$$(\lambda_m - \lambda_n)\int_a^b \phi_m(x)\phi_n(x)r(x)\,dx = p(b)[\phi_m(b)\phi'_n(b) - \phi_n(b)\phi'_m(b)]$$

$$- p(a)[\phi_m(a)\phi'_n(a) - \phi_n(a)\phi'_m(a)]. \qquad (12.32)$$

Since ϕ_m and ϕ_n are characteristic functions of the problem under consideration, they satisfy the supplementary conditions (12.2) of the problem. If $A_2 = B_2 = 0$ in (12.2), these conditions reduce to $y(a) = 0$, $y(b) = 0$. Then in this case $\phi_m(a) = 0$, $\phi_m(b) = 0$, $\phi_n(a) = 0$, and $\phi_n(b) = 0$, and so the right member of (12.32) is equal to zero.

If $A_2 = 0$ but $B_2 \neq 0$ in (12.2), these conditions reduce to $y(a) = 0$, $\beta y(b) + y'(b) = 0$, where $\beta = B_1/B_2$. Then the second bracket in the right member of (12.32) is again equal to zero. Also, the first bracket in this member may then be written

$$[\beta\phi_n(b) + \phi'_n(b)]\phi_m(b) - [\beta\phi_m(b) + \phi'_m(b)]\phi_n(b),$$

and so it is also equal to zero. Thus in this case the right member of (12.32) is equal to zero.

In like manner, if either $A_2 \neq 0$, $B_2 = 0$ or $A_2 \neq 0$, $B_2 \neq 0$ in (12.2), then the right member of (12.32) is equal to zero. Thus in all cases the right member of (12.32) is equal to zero and so

$$(\lambda_m - \lambda_n)\int_a^b \phi_m(x)\phi_n(x)r(x)\,dx = 0.$$

Since λ_m and λ_n are *distinct* characteristic values, their difference $\lambda_m - \lambda_n \neq 0$. Therefore we must have

$$\int_a^b \phi_m(x)\phi_n(x)r(x)\,dx = 0,$$

and so ϕ_m and ϕ_n are orthogonal with respect to r on $a \leq x \leq b$.   *Q.E.D.*

Let $\{\lambda_n\}$ be the infinite set of characteristic values of a Sturm–Liouville problem, arranged in a monotonic increasing sequence $\lambda_1 < \lambda_2 < \lambda_3 < \cdots$. For each $n = 1, 2, 3, \ldots$, let ϕ_n be one of the characteristic functions corresponding to the characteristic value λ_n. Then Theorem 12.2 implies at once that the infinite set of characteristic functions $\phi_1, \phi_2, \phi_3, \ldots$ is an orthogonal system with respect to the weight function r on $a \leq x \leq b$.

▶ **Example 12.9**

Consider once more the Sturm–Liouville problem

$$\frac{d^2y}{dx^2} + \lambda y = 0, \tag{12.5}$$

$$y(0) = 0, \qquad y(\pi) = 0, \tag{12.6}$$

which we have already investigated in Examples 12.3, 12.4, and 12.6. Corresponding to each characteristic value $\lambda_n = n^2$ $(n = 1, 2, 3, \ldots)$ we found the characteristic functions $c_n \sin nx$ $(n = 1, 2, 3, \ldots)$, where c_n $(n = 1, 2, 3, \ldots)$ is an arbitrary nonzero constant. Let $\{\phi_n\}$ denote the infinite set of characteristic functions for which $c_n = 1$ $(n = 1, 2, 3, \ldots)$. That is,

$$\phi_n(x) = \sin nx \qquad (0 \leq x \leq \pi; \quad n = 1, 2, 3, \ldots).$$

Then by Theorem 12.2, the infinite set $\{\phi_n\}$ is an orthogonal system with respect to the weight function r, where $r(x) = 1$ for all x, on the interval $0 \leq x \leq \pi$. That is, Theorem 12.2 shows that

$$\int_0^\pi (\sin mx)(\sin nx)(1)\,dx = 0 \tag{12.33}$$

for $m = 1, 2, 3, \ldots$; $n = 1, 2, 3, \ldots$; $m \neq n$. We have already noted the validity of (12.33) in Example 12.8.

Exercises

Directly verify the validity of the conclusion of Theorem 12.2 for the characteristic functions of each of the Sturm–Liouville problems in Exercises 1, 2, and 3.

1. $\dfrac{d^2y}{dx^2} + \lambda y = 0,$ $y(0) = 0,$ $y(\pi/2) = 0.$

(See Exercise 1 at the end of Section 12.1.)

2. $\dfrac{d^2y}{dx^2} + \lambda y = 0, \qquad y'(0) = 0, \qquad y'(L) = 0,$

where $L > 0$. (See Exercise 4 at the end of Section 12.1.)

3. $\dfrac{d}{dx}\left[x\,\dfrac{dy}{dx}\right] + \dfrac{\lambda}{x}\,y = 0, \qquad y'(1) = 0, \qquad y'(e^{2\pi}) = 0.$

(See Example 12.5 of the text.)

4. Let $\{P_n\}$, $n = 0, 1, 2, \dots$, be an infinite set of polynomial functions such that
    1. P_n is of degree n, $n = 0, 1, 2, \dots$,
    2. $P_n(1) = 1$, $n = 0, 1, 2, \dots$, and
    3.   the set $\{P_n\}$ is an orthogonal system with respect to the weight function r such that $r(x) = 1$ on the interval $-1 \le x \le 1$.

    Construct consecutively the members P_0, P_1, P_2, and P_3 of this set by writing

    $$P_0(x) = a_0,$$
    $$P_1(x) = b_0 x + b_1,$$
    $$P_2(x) = c_0 x^2 + c_1 x + c_2,$$
    $$P_3(x) = d_0 x^3 + d_1 x^2 + d_2 x + d_3,$$

    and determining the constants in each expression so that it has the value 1 at $x = 1$ and is orthogonal to each of the preceding expressions with respect to r on $-1 \le x \le 1$.

12.3   THE EXPANSION OF A FUNCTION IN A SERIES OF ORTHONORMAL FUNCTIONS

A. Orthonormal Systems

We proceed to introduce the concept of an orthonormal system.

DEFINITION

A function f is called normalized *with respect to the weight function r on the interval $a \le x \le b$ if and only if*

$$\int_a^b [f(x)]^2 r(x)\,dx = 1.$$

▶ **Example 12.10**

The function f such that $f(x) = \sqrt{2/\pi}\,\sin x$ is normalized with respect to the weight function having the constant value 1 on the interval $0 \le x \le \pi$, for

$$\int_0^\pi \left(\sqrt{\dfrac{2}{\pi}}\,\sin x\right)^2 (1)\,dx = \dfrac{2}{\pi}\int_0^\pi \sin^2 x\,dx = \dfrac{2}{\pi}\cdot\dfrac{\pi}{2} = 1.$$

DEFINITION

Let $\{\phi_n\}$ ($n = 1, 2, 3, \ldots$) be an infinite set of functions defined on the interval $a \leq x \leq b$. The set $\{\phi_n\}$ is called an orthonormal system *with respect to the weight function r on $a \leq x \leq b$ if (1) it is an orthogonal system with respect to r on $a \leq x \leq b$ and (2) every function of the system is normalized with respect to r on $a \leq x \leq b$. That is, the set $\{\phi_n\}$ is orthonormal with respect to r on $a \leq x \leq b$ if*

$$\int_a^b \phi_m(x)\phi_n(x)r(x)\, dx = \begin{cases} 0 & \text{for } m \neq n, \\ 1 & \text{for } m = n. \end{cases}$$

▶ **Example 12.11**

Consider the infinite set of functions $\{\phi_n\}$, where $\phi_n(x) = \sqrt{2/\pi} \sin nx$ ($n = 1, 2, 3, \ldots$) on the interval $0 \leq x \leq \pi$. The set $\{\phi_n\}$ is an orthogonal system with respect to the weight function having the constant value 1 on the interval $0 \leq x \leq \pi$, for

$$\int_0^\pi \left(\sqrt{\frac{2}{\pi}} \sin mx\right)\left(\sqrt{\frac{2}{\pi}} \sin nx\right)(1)\, dx = 0 \qquad \text{for } m \neq n.$$

Further, every function of the system is normalized with respect to this weight function on $0 \leq x \leq \pi$, for

$$\int_0^\pi \left(\sqrt{\frac{2}{\pi}} \sin nx\right)^2 (1)\, dx = 1.$$

Thus the set $\{\phi_n\}$ is an orthonormal system with respect to the weight function having the constant value 1 on $0 \leq x \leq \pi$.

Now consider the Sturm–Liouville problem consisting of the differential equation (12.1) and the supplementary conditions (12.2). Let $\{\lambda_n\}$ be the infinite set of characteristic values of this problem, arranged in a monotonic increasing sequence $\lambda_1 < \lambda_2 < \lambda_3 < \cdots$. If ϕ_n ($n = 1, 2, 3, \ldots$) is one of the characteristic functions corresponding to the characteristic value λ_n, then we know from Theorem 12.2 that the infinite set of characteristic functions $\phi_1, \phi_2, \phi_3, \ldots$ is an orthogonal system with respect to the weight function r on $a \leq x \leq b$. But this set of characteristic functions is not necessarily orthonormal with respect to r on $a \leq x \leq b$.

Now recall that if ϕ_n is one of the characteristic functions corresponding to λ_n, then $k_n\phi_n$, where k_n is an arbitrary nonzero constant, is also a characteristic function corresponding to λ_n. Thus from the given set of characteristic functions $\phi_1, \phi_2, \phi_3, \ldots$, we can form a set of "new" characteristic functions $k_1\phi_1, k_2\phi_2, k_3\phi_3, \ldots$; and this "new" set is also orthogonal with respect to r on $a \leq x \leq b$. Now if we can choose the constants $k_1, k_2, k_3 \ldots$ in such a way that every characteristic function of the "new" set is also *normalized* with respect to r on $a \leq x \leq b$, then the "new" set of characteristic functions $k_1\phi_1, k_2\phi_2, k_3\phi_3, \ldots$ will be an *orthonormal* system with respect to r on $a \leq x \leq b$.

We now show that the constants $k_1, k_2, k_3, \ldots$ can indeed be chosen so that the set $k_1\phi_1, k_2\phi_2, k_3\phi_3, \ldots$ is orthonormal. Recall that the function r in the differential equation (12.1) is such that $r(x) > 0$ for all x on the interval $a \leq x \leq b$. Also recall that by definition no characteristic function ϕ_n ($n = 1, 2, 3, \ldots$) is identically zero on

$a \leq x \leq b$. Therefore

$$\int_a^b [\phi_n(x)]^2 r(x)\, dx = K_n > 0 \qquad (n = 1, 2, 3, \ldots),$$

and so

$$\int_a^b \left[\frac{1}{\sqrt{K_n}} \phi_n(x) \right]^2 r(x)\, dx = 1 \qquad (n = 1, 2, 3, \ldots).$$

Thus the set

$$\frac{1}{\sqrt{K_1}} \phi_1, \frac{1}{\sqrt{K_2}} \phi_2, \frac{1}{\sqrt{K_3}} \phi_3, \ldots$$

is an orthonormal set with respect to r on $a \leq x \leq b$. We have thus shown that from a given set of orthogonal characteristic functions $\phi_1, \phi_2, \phi_3, \ldots$, we can always form the set of *orthonormal* characteristic functions $k_1\phi_1, k_2\phi_2, k_3\phi_3, \ldots$, where

$$k_n = \frac{1}{\sqrt{K_n}} = \frac{1}{\sqrt{\int_a^b [\phi_n(x)]^2 r(x)\, dx}} \qquad (n = 1, 2, 3, \ldots). \tag{12.34}$$

▶ **Example 12.12**

The Sturm–Liouville problem consisting of the differential equation (12.5) and the conditions (12.6) has the set of orthogonal characteristic functions $\{\phi_n\}$, where $\phi_n(x) = c_n \sin nx$ ($n = 1, 2, 3, \ldots; 0 \leq x \leq \pi$), and c_n ($n = 1, 2, 3, \ldots$) is a nonzero constant. We now form the sequence of *orthonormal* characteristic functions $\{k_n\phi_n\}$, where k_n is given by (12.34). We have

$$K_n = \int_0^\pi (c_n \sin nx)^2 (1)\, dx = \frac{c_n^2 \pi}{2},$$

$$k_n = \frac{1}{\sqrt{K_n}} = \frac{1}{c_n} \sqrt{\frac{2}{\pi}},$$

$$k_n \phi_n(x) = \left(\frac{1}{c_n} \sqrt{\frac{2}{\pi}} \right)(c_n \sin nx) = \sqrt{\frac{2}{\pi}} \sin nx \qquad (n = 1, 2, 3, \ldots).$$

Thus the Sturm–Liouville problem under consideration has the set of orthonormal characteristic functions $\{\psi_n\}$. were $\psi_n(x) = \sqrt{2/\pi} \sin nx$ ($n = 1, 2, 3, \ldots; 0 \leq x \leq \pi$). We observe that this is the set of orthonormal functions considered in Example 12.11.

B. The Expansion Problem

We now consider a problem which has many applications and has led to the development of a vast amount of advanced mathematical analysis. Let $\{\phi_n\}$ be an orthonormal system with respect to a weight function r on an interval $a \leq x \leq b$, and let f be an "arbitrary" function. The basic problem under consideration is to expand the function f in an infinite series of the orthonormal functions $\phi_1, \phi_2, \phi_3, \ldots$.

Let us first assume that such an expansion exists. That is, we assume that

$$f(x) = \sum_{n=1}^{\infty} c_n \phi_n(x) \tag{12.35}$$

for each x in the interval $a \le x \le b$. Now assuming that this expansion does exist, how do we determine the coefficients c_n ($n = 1, 2, 3,\ldots$)? Let us proceed formally for the moment, without considering questions of convergence. We multiply both sides of (12.35) by $\phi_k(x)r(x)$, where r is the weight function and ϕ_k is the kth function of the system $\{\phi_n\}$, thereby obtaining

$$f(x)\phi_k(x)r(x) = \sum_{n=1}^{\infty} c_n \phi_n(x)\phi_k(x)r(x).$$

We now integrate both sides of this equation from a to b and obtain

$$\int_a^b f(x)\phi_k(x)r(x)\,dx = \int_a^b \left[\sum_{n=1}^{\infty} c_n \phi_n(x)\phi_k(x)r(x) \right] dx = \sum_{n=1}^{\infty} \int_a^b c_n \phi_n(x)\phi_k(x)r(x)\,dx,$$

assuming that the integral of the sum in the middle member is the sum of the integrals in the right member. Under this assumption we thus have

$$\int_a^b f(x)\phi_k(x)r(x)\,dx = \sum_{n=1}^{\infty} c_n \int_a^b \phi_n(x)\phi_k(x)r(x)\,dx. \tag{12.36}$$

But $\{\phi_n\}$ is an orthonormal set with respect to r on $a \le x \le b$, and so

$$\int_a^b \phi_n(x)\phi_k(x)r(x)\,dx = \begin{cases} 0 & \text{for } n \ne k, \\ 1 & \text{for } n = k. \end{cases}$$

Therefore every term except the kth term in the series in the right member of (12.36) is zero, and the kth term is simply $c_k \cdot 1 = c_k$. Thus (12.36) reduces to

$$\int_a^b f(x)\phi_k(x)r(x)\,dx = c_k.$$

This is a formula for the kth coefficient ($k = 1, 2, 3,\ldots$) in the assumed series expansion (12.35). Thus under suitable convergence conditions the coefficients c_n in the expansion (12.35) are given by the formula

$$c_n = \int_a^b f(x)\phi_n(x)r(x)\,dx \qquad (n = 1, 2, 3,\ldots). \tag{12.37}$$

In particular, we note that *if* the series $\sum_{n=1}^{\infty} c_n \phi_n(x)$ converges *uniformly* to $f(x)$ on $a \le x \le b$, then the above formal procedure is justified and we are assured that the coefficients c_n are given by the formula (12.37).

Now, given an orthonormal system $\{\phi_n\}$ and a function f, we can form the series

$$\sum_{n=1}^{\infty} c_n \phi_n, \tag{12.38}$$

where c_n ($n = 1, 2, 3,\ldots$) is given by formula (12.37). But note carefully that this formula was obtained under the *assumption* that f *is* representable by an expansion of the form (12.38). Therefore in general the expansion (12.38) obtained using formula (12.37) is indeed a strictly *formal* expansion. That is, in general we have no assurance that the series (12.38) determined by (12.37) converges *pointwise* on $[a, b]$. Further, if this series *does* converge at a point x on $[a, b]$, we have no assurance that it converges to $f(x)$ at this point. To be assured that the formal expansion (12.38) *does* converge to $f(x)$ at every point x of $[a, b]$, the functions f and $\{\phi_n\}$ must satisfy certain restrictive conditions. It is beyond the scope of this book to attempt to give a detailed discussion of the various conditions on f and $\{\phi_n\}$ which are sufficient for the convergence of this expansion. However, we shall state without proof one basic convergence theorem concerned with the case in which the system $\{\phi_n\}$ is the system of orthonormal characteristic functions of a Sturm–Liouville problem.

THEOREM 12.3

Hypothesis

1. Consider the Sturm–Liouville problem

$$\frac{d}{dx}\left[p(x)\frac{dy}{dx} \right] + [q(x) + \lambda r(x)]y = 0,$$

$$A_1 y(a) + A_2 y'(a) = 0,$$

$$B_1 y(b) + B_2 y'(b) = 0,$$

where p, q, and r are real functions such that p has a continuous derivative, q and r are continuous, and $p(x) > 0$ and $r(x) > 0$ for all x on the real interval $a \leq x \leq b$.
Let $\{\lambda_n\}$ ($n = 1, 2, 3, \ldots$) be the infinite set of characteristic values of this problem, arranged in a monotonic increasing sequence $\lambda_1 < \lambda_2 < \lambda_3 < \cdots$. Let $\{\phi_n\}$ ($n = 1, 2, 3, \ldots$) be the corresponding set of orthonormal characteristic functions of the given problem.
2. Let f be a function which is continuous on the interval $a \leq x \leq b$, has a piecewise continuous derivative f' on $a \leq x \leq b$, and is such that $f(a) = 0$ if $\phi_1(a) = 0$ and $f(b) = 0$ if $\phi_1(b) = 0$.

Conclusion. *The series*

$$\sum_{n=1}^{\infty} c_n \phi_n,$$

where

$$c_n = \int_a^b f(x)\phi_n(x)r(x)\,dx \qquad (n = 1, 2, 3, \ldots),$$

converges uniformly and absolutely to f on the interval $a \leq x \leq b$.

▶ Example 12.13

Obtain the formal expansion of the function f, where $f(x) = \pi x - x^2, 0 \leq x \leq \pi$, in the series of orthonormal characteristic functions $\{\phi_n\}$ of the Sturm–Liouville problem

$$\frac{d^2 y}{dx^2} + \lambda y = 0, \tag{12.5}$$

$$y(0) = 0, \qquad y(\pi) = 0. \tag{12.6}$$

Discuss the convergence of this formal expansion.

Solution. In Example 12.12 we found that the orthonormal characteristic functions $\{\phi_n\}$ of the problem under consideration are given by

$$\phi_n(x) = \sqrt{\frac{2}{\pi}} \sin nx, \qquad 0 \le x \le \pi \qquad (n = 1, 2, 3, \ldots).$$

We now form the series

$$\sum_{n=1}^{\infty} c_n \phi_n(x), \tag{12.39}$$

where for each $n = 1, 2, 3, \ldots$,

$$
\begin{aligned}
c_n &= \int_a^b f(x)\phi_n(x)r(x)\,dx \\
&= \int_0^\pi (\pi x - x^2)\left(\sqrt{\frac{2}{\pi}}\sin nx\right)(1)\,dx \\
&= \sqrt{\frac{2}{\pi}}\left[\pi \int_0^\pi x \sin nx\,dx - \int_0^\pi x^2 \sin nx\,dx\right] \\
&= \sqrt{\frac{2}{\pi}}\left[\left(\frac{\pi}{n^2}\sin nx - \frac{\pi x}{n}\cos nx\right)\Big|_0^\pi - \left(\frac{2x}{n^2}\sin nx + \frac{2}{n^3}\cos nx - \frac{x^2}{n}\cos nx\right)\Big|_0^\pi\right] \\
&= \sqrt{\frac{2}{\pi}}\left[\left(-\frac{\pi^2}{n}\cos n\pi\right) - \left(\frac{2}{n^3}\cos n\pi - \frac{\pi^2}{n}\cos n\pi - \frac{2}{n^3}\right)\right] \\
&= \sqrt{\frac{2}{\pi}}\cdot\frac{2}{n^3}(1 - \cos n\pi) \\
&= \begin{cases} \sqrt{\frac{2}{\pi}}\cdot\dfrac{4}{n^3} & \text{if } n \text{ is odd,} \\[2mm] 0 & \text{if } n \text{ is even.} \end{cases}
\end{aligned}
$$

Thus the series (12.39) becomes

$$\sum_{\substack{n=1 \\ (n\,\text{odd})}}^{\infty} \frac{8}{\pi n^3}\sin nx \quad \text{or} \quad \frac{8}{\pi}\sum_{n=1}^{\infty}\frac{\sin(2n-1)x}{(2n-1)^3}. \tag{12.40}$$

We write

$$\pi x - x^2 \sim \frac{8}{\pi}\sum_{n=1}^{\infty}\frac{\sin(2n-1)x}{(2n-1)^3}, \qquad 0 \le x \le \pi,$$

to denote that the series (12.40) is the *formal* expansion of f in terms of the orthonormal characteristic functions $\{\phi_n\}$. We note that the first few terms of this expansion are

$$\frac{8}{\pi}\left[\frac{\sin x}{1^3} + \frac{\sin 3x}{3^3} + \frac{\sin 5x}{5^3} + \cdots\right].$$

Let us now discuss the convergence of this formal expansion. We note first that the coefficients p, q, and r in the differential equation (12.5) satisfy Hypothesis 1 of Theorem 12.3. Let us now see if the function f satisfies Hypothesis 2. Since f is a polynomial function, it certainly satisfies the requirements of continuity and a piecewise continuous derivative on $0 \leq x \leq \pi$. Hypothesis 2 also requires that $f(0) = 0$ if $\phi_1(0) = 0$ and $f(\pi) = 0$ if $\phi_1(\pi) = 0$. These requirements are also fulfilled, for

$$\phi_1(0) = \sqrt{\frac{2}{\pi}} \sin 0 = 0, \qquad f(0) = 0,$$

$$\phi_1(\pi) = \sqrt{\frac{2}{\pi}} \sin \pi = 0, \qquad f(\pi) = \pi^2 - \pi^2 = 0.$$

Thus all the hypotheses of Theorem 12.3 are satisfied and we conclude that the series (12.40) converges uniformly and absolutely to $\pi x - x^2$ on the interval $0 \leq x \leq \pi$. Thus we may write

$$\pi x - x^2 = \frac{8}{\pi} \sum_{n=1}^{\infty} \frac{\sin(2n - 1)x}{(2n - 1)^3}, \qquad 0 \leq x \leq \pi.$$

Exercises

1. Consider the set of functions $\{\phi_n\}$, where

$$\phi_1(x) = \frac{1}{\sqrt{\pi}},$$

$$\phi_{n+1}(x) = \sqrt{\frac{2}{\pi}} \cos nx \qquad (n = 1, 2, 3, \ldots),$$

on the interval $0 \leq x \leq \pi$. Show that this set $\{\phi_n\}$ is an orthonormal system with respect to the weight function having the constant value 1 on $0 \leq x \leq \pi$.

2. Obtain the formal expansion of the function f defined by $f(x) = x (0 \leq x \leq \pi)$, in a series of orthonormal characteristic functions $\{\phi_n\}$ of the Sturm–Liouville problem

$$\frac{d^2y}{dx^2} + \lambda y = 0,$$

$$y(0) = 0,$$

$$y(\pi) = 0.$$

3. Obtain the formal expansion of the function f defined by $f(x) = 1 (1 \leq x \leq e^\pi)$, in a series of orthonormal characteristic functions $\{\phi_n\}$ of the Sturm–Liouville problem

$$\frac{d}{dx}\left[x \frac{dy}{dx}\right] + \frac{\lambda}{x} y = 0,$$

$$y(1) = 0,$$

$$y(e^\pi) = 0.$$

(See Exercise 7 at the end of Section 12.1.)

4. Obtain the formal expansion of the function f defined by $f(x) = \ln x, 1 \leq x \leq e^{2\pi}$, in a series of orthonormal characteristic functions $\{\phi_n\}$ of the Sturm–Liouville

problem

$$\frac{d}{dx}\left[x\frac{dy}{dx}\right] + \frac{\lambda}{x}y = 0,$$

$$y'(1) = 0,$$

$$y'(e^{2\pi}) = 0,$$

where $\lambda \geq 0$ (see Example 12.5 of the text).

12.4 TRIGONOMETRIC FOURIER SERIES

A. Definition of Trigonometric Fourier Series

In Section 12.3 we introduced the problem of expanding a function f in a series of the form

$$\sum_{n=1}^{\infty} c_n \phi_n, \tag{12.38}$$

where $\{\phi_n\}$ $(n = 1, 2, 3, \ldots)$ is an orthonormal system with respect to a weight function r on $a \leq x \leq b$. We first assumed that an expansion for f of the form (12.38) does indeed exist. Then, assuming suitable convergence, we found that the coefficients c_n in (12.38) are given by

$$c_n = \int_a^b f(x)\phi_n(x)r(x)\,dx \qquad (n = 1, 2, 3, \ldots). \tag{12.37}$$

We noted that in general this is a strictly formal expansion and that certain restrictions must be imposed upon f and $\{\phi_n\}$ in order to be assured that the series (12.38) thus formed does indeed converge to f on $a \leq x \leq b$. Nevertheless, assuming that the functions involved in (12.37) are integrable, we can determine the formal expansion (12.38) and give it a name, even though we have no assurance in advance that it actually represents the given function f in any sense. This is essentially what we do in the following definition.

DEFINITION

Let $\{\phi_n\}$ $(n = 1, 2, 3, \ldots)$, be an orthonormal system with respect to a weight function r on $a \leq x \leq b$. Let f be a function such that for each $n = 1, 2, 3, \ldots$, the product $f\phi_n r$ is integrable on $a \leq x \leq b$. Then the series

$$\sum_{n=1}^{\infty} c_n \phi_n, \tag{12.38}$$

where

$$c_n = \int_a^b f(x)\phi_n(x)r(x)\,dx \qquad (n = 1, 2, 3, \ldots). \tag{12.37}$$

is called the Fourier series *of f relative to the system $\{\phi_n\}$; the coefficients c_n are called the*

Fourier coefficients *of f relative to* $\{\phi_n\}$; *and we write*

$$f(x) \sim \sum_{n=1}^{\infty} c_n \phi_n(x), \qquad a \le x \le b.$$

We now proceed to introduce an important special class of Fourier series. For this purpose, consider the system of functions $\{\psi_n\}$ defined by

$$\psi_1(x) = 1,$$

$$\psi_{2n}(x) = \cos\frac{n\pi x}{L} \qquad (n = 1, 2, 3, \dots),$$

$$\psi_{2n+1}(x) = \sin\frac{n\pi x}{L} \qquad (n = 1, 2, 3, \dots),$$

(12.41)

for all x on the interval $-L \le x \le L$, where L is a positive constant. Since

$$\int_{-L}^{L} \cos\frac{m\pi x}{L} \cos\frac{n\pi x}{L}\, dx = 0 \qquad (m, n = 0, 1, 2, \dots; m \ne n),$$

$$\int_{-L}^{L} \sin\frac{m\pi x}{L} \sin\frac{n\pi x}{L}\, dx = 0 \qquad (m, n = 1, 2, 3, \dots; m \ne n),$$

$$\int_{-L}^{L} \cos\frac{m\pi x}{L} \sin\frac{n\pi x}{L}\, dx = 0 \qquad (m = 0, 1, 2, \dots; n = 1, 2, 3, \dots),$$

we see that the system (12.41) is orthogonal with respect to the weight function r defined by $r(x) = 1$ on $-L \le x \le L$. Now note that

$$\int_{-L}^{L} (1)^2\, dx = 2L,$$

$$\int_{-L}^{L} \cos^2\left(\frac{n\pi x}{L}\right) dx = L \qquad (n = 1, 2, 3, \dots),$$

and

$$\int_{-L}^{L} \sin^2\left(\frac{n\pi x}{L}\right) dx = L \qquad (n = 1, 2, 3, \dots).$$

Thus the corresponding *orthonormal* system is the system of functions $\{\phi_n\}$ defined by

$$\phi_1(x) = \frac{1}{\sqrt{2L}}$$

$$\phi_{2n}(x) = \frac{1}{\sqrt{L}}\cos\frac{n\pi x}{L} \qquad (n = 1, 2, 3, \dots),$$

(12.42)

$$\phi_{2n+1}(x) = \frac{1}{\sqrt{L}}\sin\frac{n\pi x}{L} \qquad (n = 1, 2, 3, \dots),$$

on the interval $-L \le x \le L$.

Now suppose f is a function such that the product $f\phi_n$ is integrable for each function ϕ_n of the system (12.42) on the interval $-L \le x \le L$. Then we can form the Fourier series of f relative to the orthonormal system $\{\phi_n\}$ defined by (12.42). This is of the form (12.38), were the coefficients c_n are given by (12.37). Thus using (12.37), with $r(x) = 1$, we find the Fourier coefficients

$$c_1 = \int_a^b f(x)\phi_1(x)\,dx = \frac{1}{\sqrt{2L}} \int_{-L}^L f(x)\,dx,$$

$$c_{2n} = \int_a^b f(x)\phi_{2n}(x)\,dx = \frac{1}{\sqrt{L}} \int_{-L}^L f(x)\cos\frac{n\pi x}{L}\,dx, \tag{12.43}$$

$$c_{2n+1} = \int_a^b f(x)\phi_{2n+1}(x)\,dx = \frac{1}{\sqrt{L}} \int_{-L}^L f(x)\sin\frac{n\pi x}{L}\,dx,$$

$(n = 1, 2, 3,...)$. Thus the Fourier series of f relative to the orthonormal system $\{\phi_n\}$ defined by (12.42) is

$$\sum_{n=1}^\infty c_n\phi_n,$$

where c_n $(n = 1, 2, 3,...)$ is given by (12.43) and ϕ_n $(n = 1, 2, 3,...)$ is, of course, given by (12.42). We now write this series in the form

$$c_1\phi_1 + \sum_{n=1}^\infty (c_{2n}\phi_{2n} + c_{2n+1}\phi_{2n+1}).$$

Substituting c_n $(n = 1, 2, 3,...)$ given by (12.43) and ϕ_n $(n = 1, 2, 3,...)$ given by (12.42) into this series, it takes the form

$$\left[\frac{1}{\sqrt{2L}}\int_{-L}^L f(x)\,dx\right]\left[\frac{1}{\sqrt{2L}}\right] + \sum_{n=1}^\infty \left\{\left[\frac{1}{\sqrt{L}}\int_{-L}^L f(x)\cos\frac{n\pi x}{L}\,dx\right]\left[\frac{1}{\sqrt{L}}\cos\frac{n\pi x}{L}\right]\right.$$

$$\left. + \left[\frac{1}{\sqrt{L}}\int_{-L}^L f(x)\sin\frac{n\pi x}{L}\,dx\right]\left[\frac{1}{\sqrt{L}}\sin\frac{n\pi x}{L}\right]\right\}.$$

Rewriting this slightly, it assumes the form

$$\frac{1}{2}\left[\frac{1}{L}\int_{-L}^L f(x)\,dx\right] + \sum_{n=1}^\infty \left\{\left[\frac{1}{L}\int_{-L}^L f(x)\cos\frac{n\pi x}{L}\,dx\right]\cos\frac{n\pi x}{L}\right.$$

$$\left. + \left[\frac{1}{L}\int_{-L}^L f(x)\sin\frac{n\pi x}{L}\,dx\right]\sin\frac{n\pi x}{L}\right\}.$$

Thus the Fourier series of f relative to system (12.42) may be written as

$$\frac{1}{2}a_0 + \sum_{n=1}^\infty \left(a_n\cos\frac{n\pi x}{L} + b_n\sin\frac{n\pi x}{L}\right),$$

where

$$a_n = \frac{1}{L}\int_{-L}^L f(x)\cos\frac{n\pi x}{L}\,dx \qquad (n = 0, 1, 2, 3,...),$$

$$b_n = \frac{1}{L}\int_{-L}^L f(x)\sin\frac{n\pi x}{L}\,dx \qquad (n = 1, 2, 3,...).$$

This special type of Fourier series is singled out in the following definition.

DEFINITION

Let f be a function which is defined on the interval $-L < x < L$ and is such that the integrals

$$\int_{-L}^{L} f(x) \cos \frac{n\pi x}{L}\, dx \quad and \quad \int_{-L}^{L} f(x) \sin \frac{n\pi x}{L}\, dx$$

$(n = 0, 1, 2, \ldots)$ *exist. Then the series*

$$\frac{1}{2} a_0 + \sum_{n=1}^{\infty} \left(a_n \cos \frac{n\pi x}{L} + b_n \sin \frac{n\pi x}{L} \right), \tag{12.44}$$

where

$$a_n = \frac{1}{L} \int_{-L}^{L} f(x) \cos \frac{n\pi x}{L}\, dx \quad (n = 0, 1, 2, \ldots),$$

$$b_n = \frac{1}{L} \int_{-L}^{L} f(x) \sin \frac{n\pi x}{L}\, dx \quad (n = 1, 2, 3, \ldots), \tag{12.45}$$

is called the trigonometric Fourier series *of f on the interval $-L \le x \le L$. We denote this by writing*

$$f(x) \sim \frac{1}{2} a_0 + \sum_{n=1}^{\infty} \left(a_n \cos \frac{n\pi x}{L} + b_n \sin \frac{n\pi x}{L} \right), \quad -L \le x \le L.$$

The numbers a_n $(n = 0, 1, 2, \ldots)$ and b_n $(n = 1, 2, 3, \ldots)$ defined by (12.45) will be called simply the Fourier coefficients *of f.*

Thus to find the trigonometric Fourier series of a function f having the required integrability properties on $-L \le x \le L$, we determine the Fourier coefficients from (12.45) and substitute these coefficients into the series (12.44). There are two important special cases in which the determination of the Fourier coefficients is considerably simplified. These are the cases in which f is either a so-called *even* function or a so-called *odd* function.

A function F is said to be an *even* function if $F(-x) = F(x)$ for every x. If F is an even function, the graph of $y = F(x)$ is symmetric with respect to the y axis and

$$\int_{-c}^{c} F(x)\, dx = 2 \int_{0}^{c} F(x)\, dx$$

for every $c > 0$. A function F is said to be an *odd* function if $F(-x) = -F(x)$ for every x. If F is an odd function, the graph of $y = F(x)$ is symmetric with respect to the origin and

$$\int_{-c}^{c} F(x)\, dx = 0$$

for every $c > 0$.

Now suppose we wish to find the trigonometric Fourier series of an *even* function f. Then each function F_n defined by

$$F_n(x) = f(x) \cos \frac{n\pi x}{L} \quad (n = 0, 1, 2, \ldots)$$

is also even, since

$$F_n(-x) = f(-x) \cos\left(-\frac{n\pi x}{L}\right) = f(x) \cos \frac{n\pi x}{L} = F_n(x).$$

Therefore

$$\int_{-L}^{L} F_n(x)\, dx = 2 \int_{0}^{L} F_n(x)\, dx \qquad (n = 0, 1, 2, \ldots).$$

and so

$$a_n = \frac{1}{L} \int_{-L}^{L} f(x) \cos \frac{n\pi x}{L}\, dx = \frac{2}{L} \int_{0}^{L} f(x) \cos \frac{n\pi x}{L}\, dx \qquad (n = 0, 1, 2, \ldots).$$

Further, each function G_n defined by

$$G_n(x) = f(x) \sin \frac{n\pi x}{L} \qquad (n = 1, 2, 3, \ldots)$$

is odd, since

$$G_n(-x) = f(-x) \sin\left(-\frac{n\pi x}{L}\right) = -f(x) \sin \frac{n\pi x}{L} = -G_n(x).$$

Therefore

$$\int_{-L}^{L} G_n(x)\, dx = 0 \qquad (n = 1, 2, 3, \ldots),$$

and so

$$b_n = \frac{1}{L} \int_{-L}^{L} f(x) \sin \frac{n\pi x}{L}\, dx = 0 \qquad (n = 1, 2, 3, \ldots).$$

Thus the trigonometric Fourier series of an *even* function f on the interval $-L \le x \le L$ is given by

$$\frac{1}{2} a_0 + \sum_{n=1}^{\infty} a_n \cos \frac{n\pi x}{L}, \qquad\qquad (12.46)$$

where

$$a_n = \frac{2}{L} \int_{0}^{L} f(x) \cos \frac{n\pi x}{L}\, dx \qquad (n = 0, 1, 2, \ldots), \qquad\qquad (12.47)$$

In like manner, one finds that the trigonometric Fourier series of an *odd* function f on the interval $-L \le x \le L$ is given by

$$\sum_{n=1}^{\infty} b_n \sin \frac{n\pi x}{L}, \qquad\qquad (12.48)$$

where

$$b_n = \frac{2}{L} \int_{0}^{L} f(x) \sin \frac{n\pi x}{L}\, dx \qquad (n = 1, 2, 3, \ldots). \qquad\qquad (12.49)$$

We note that in general the trigonometric Fourier series of a function f is a strictly formal expansion, for we have said nothing concerning its convergence. We shall discuss this important matter briefly in Part D of this section. In the meantime, let us gain some facility in finding trigonometric Fourier series by considering the following examples.

B. Examples of Trigonometric Fourier Series

▶ **Example 12.14**

Find the trigonometric Fourier series of the function f defined by

$$f(x) = |x|, \qquad -\pi \le x \le \pi,$$

on the interval $-\pi \le x \le \pi$ (see Figure 12.1).

Solution. In this problem $L = \pi$. Since $f(-x) = |-x| = |x| = f(x)$ for all x, the function f is an *even* function. Therefore the trigonometric Fourier series of f on $-\pi \le x \le \pi$ is given by (12.46) with $L = \pi$, where the coefficients a_n are given by (12.47) with $L = \pi$. Thus, since $|x| = x$ for $0 \le x \le \pi$, we have

$$a_0 = \frac{2}{\pi} \int_0^\pi f(x)\, dx = \frac{2}{\pi} \int_0^\pi x\, dx = \pi,$$

and

$$a_n = \frac{2}{\pi} \int_0^\pi f(x) \cos nx\, dx = \frac{2}{\pi} \int_0^\pi x \cos nx\, dx = \frac{2}{\pi} \left[\frac{\cos nx}{n^2} + \frac{x \sin nx}{n} \right]_0^\pi$$

$$= \frac{2}{\pi} \left[\frac{\cos n\pi - 1}{n^2} \right] = \frac{2}{\pi} \left[\frac{(-1)^n - 1}{n^2} \right]$$

$$= \begin{cases} -\dfrac{4}{\pi n^2}, & n \text{ odd} \\[2mm] 0, & n \text{ even} \end{cases} \qquad (n = 1, 2, 3, \ldots).$$

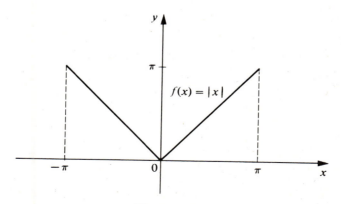

Figure 12.1

Thus the required series is

$$\frac{\pi}{2} - \frac{4}{\pi} \sum_{\substack{n=1 \\ (n \, \text{odd})}}^{\infty} \frac{\cos nx}{n^2} = \frac{\pi}{2} - \frac{4}{\pi} \sum_{n=1}^{\infty} \frac{\cos(2n-1)x}{(2n-1)^2}$$

$$= \frac{\pi}{2} - \frac{4}{\pi} \left[\frac{\cos x}{1^2} + \frac{\cos 3x}{3^2} + \frac{\cos 5x}{5^2} + \cdots \right]$$

and we write

$$|x| \sim \frac{\pi}{2} - \frac{4}{\pi} \sum_{n=1}^{\infty} \frac{\cos(2n-1)x}{(2n-1)^2}, \qquad -\pi \le x \le \pi.$$

▶ **Example 12.15**

Find the trigonometric Fourier series of the function f defined by

$$f(x) = x, \qquad -4 \le x \le 4,$$

on the interval $-4 \le x \le 4$ (see Figure 12.2).

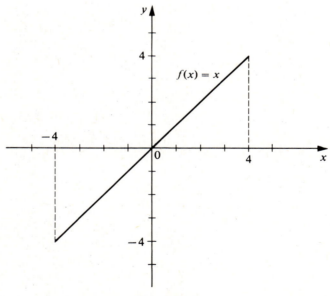

Figure 12.2

Solution. In this problem $L = 4$. Since $f(-x) = -x = -f(x)$ for all x, the function is an *odd* function. Therefore the trigonometric Fourier series of f on $-4 \le x \le 4$ is given by (12.48) with $L = 4$, where the coefficients b_n are given by (12.49) with $L = 4$.

Thus we have

$$b_n = \frac{2}{4} \int_0^4 f(x) \sin \frac{n\pi x}{4} \, dx = \frac{1}{2} \int_0^4 x \sin \frac{n\pi x}{4} \, dx$$

$$= \frac{1}{2} \left[\frac{16}{n^2 \pi^2} \sin \frac{n\pi x}{4} - \frac{4x}{n\pi} \cos \frac{n\pi x}{4} \right]_0^4$$

$$= -\frac{8}{n\pi} \cos n\pi = -\frac{8(-1)^n}{n\pi} = \frac{8(-1)^{n+1}}{n\pi} \qquad (n = 1, 2, 3, \ldots).$$

Thus the required series is

$$\frac{8}{\pi} \sum_{n=1}^{\infty} \frac{(-1)^{n+1}}{n} \sin \frac{n\pi x}{4} = \frac{8}{\pi} \left[\sin \frac{\pi x}{4} - \frac{1}{2} \sin \frac{\pi x}{2} + \frac{1}{3} \sin \frac{3\pi x}{4} - \cdots \right]$$

and we write

$$x \sim \frac{8}{\pi} \sum_{n=1}^{\infty} \frac{(-1)^{n+1}}{n} \sin \frac{n\pi x}{4}, \qquad -4 \le x \le 4.$$

▶ **Example 12.16**

Find the trigonometric Fourier series of the function f defined by

$$f(x) = \begin{cases} \pi, & -\pi \le x < 0, \\ x, & 0 \le x \le \pi, \end{cases} \tag{12.50}$$

on the interval $-\pi \le x \le \pi$ (see Figure 12.3).

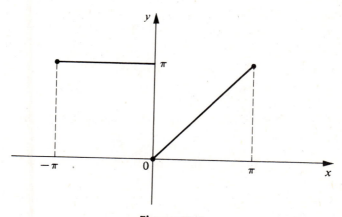

Figure 12.3

Solution. In this problem $L = \pi$. The function f defined by (12.50) is neither even nor odd. Therefore the trigonometric Fourier series of this function on $-\pi \le x \le \pi$ is given by (12.44) with $L = \pi$, where the coefficients a_n and b_n are given by (12.45) with $L = \pi$. We find

$$a_0 = \frac{1}{\pi} \int_{-\pi}^{\pi} f(x) \, dx = \frac{1}{\pi} \left[\int_{-\pi}^{0} \pi \, dx + \int_0^{\pi} x \, dx \right] = \frac{3\pi}{2};$$

$$a_n = \frac{1}{\pi} \int_{-\pi}^{\pi} f(x) \cos nx\, dx = \frac{1}{\pi} \left[\int_{-\pi}^{0} \pi \cos nx\, dx + \int_{0}^{\pi} x \cos nx\, dx \right]$$

$$= \frac{1}{\pi} \left\{ \left[\frac{\pi \sin nx}{n} \right]_{-\pi}^{0} + \left[\frac{\cos nx}{n^2} + \frac{x \sin nx}{n} \right]_{0}^{\pi} \right\}$$

$$= \frac{1}{\pi} \left[\frac{\cos n\pi - 1}{n^2} \right] = \frac{(-1)^n - 1}{\pi n^2} = \begin{cases} -\dfrac{2}{\pi n^2}, & n \text{ odd} \\ 0, & n \text{ even} \end{cases} \qquad (n = 1, 2, 3, \dots);$$

$$b_n = \frac{1}{\pi} \int_{-\pi}^{\pi} f(x) \sin nx\, dx = \frac{1}{\pi} \left[\int_{-\pi}^{0} \pi \sin nx\, dx + \int_{0}^{\pi} x \sin nx\, dx \right]$$

$$= \frac{1}{\pi} \left\{ \left[-\frac{\pi \cos nx}{n} \right]_{-\pi}^{0} + \left[\frac{\sin nx}{n^2} - \frac{x \cos nx}{n} \right]_{0}^{\pi} \right\}$$

$$= \frac{1}{\pi} \left(-\frac{\pi}{n} \right) = -\frac{1}{n} \qquad (n = 1, 2, 3, \dots).$$

Thus the required series is

$$\frac{3\pi}{4} + \sum_{n=1}^{\infty} \left[\frac{(-1)^n - 1}{\pi n^2} \cos nx - \frac{1}{n} \sin nx \right]$$

$$= \frac{3\pi}{4} - \left(\frac{2}{\pi} \cos x + \sin x \right) - \frac{1}{2} \sin 2x - \left(\frac{2}{9\pi} \cos 3x + \frac{1}{3} \sin 3x \right) - \frac{1}{4} \sin 4x - \cdots.$$

Now that we have gained some familiarity with the actual details of finding trigonometric Fourier series, we call attention to a matter which is of considerable importance in this connection. Suppose f is a function defined and integrable on an interval $a \leq x \leq b$, and suppose g is a function such that $g(x) = f(x)$ for all except a finite number of points of the interval $a \leq x \leq b$. Then it is shown in advanced calculus that

$$\int_{a}^{b} g(x)\, dx = \int_{a}^{b} f(x)\, dx.$$

Now suppose f is a function which is defined on the interval $-L \leq x \leq L$ and is such that the integrals

$$\int_{-L}^{L} f(x) \cos \frac{n\pi x}{L}\, dx \quad \text{and} \quad \int_{-L}^{L} f(x) \sin \frac{n\pi x}{L}\, dx$$

$(n = 0, 1, 2, \dots)$ exist, and suppose g is a function such that $g(x) = f(x)$ for all except a finite number of points of $-L \leq x \leq L$. Then

$$\frac{1}{L} \int_{-L}^{L} g(x) \cos \frac{n\pi x}{L}\, dx = \frac{1}{L} \int_{-L}^{L} f(x) \cos \frac{n\pi x}{L}\, dx \qquad (n = 0, 1, 2, \dots).$$

$$\frac{1}{L} \int_{-L}^{L} g(x) \sin \frac{n\pi x}{L}\, dx = \frac{1}{L} \int_{-L}^{L} f(x) \sin \frac{n\pi x}{L}\, dx \qquad (n = 1, 2, 3, \dots).$$

The left members of these two identities are the Fourier coefficients of g, and the right members are the Fourier coefficients of f. Thus f and g have the same Fourier coefficients (12.45) and hence the same trigonometric Fourier series (12.44). We thus see that if f and g are two functions which have the same values at all except a finite number of points of the interval $-L \leq x \leq L$, then f and g have the same trigonometric Fourier series on $-L \leq x \leq L$, assuming that the Fourier coefficients of either of these functions exist.

▶ **Example 12.17**

In Example 12.16 we found the trigonometric Fourier series of the function f defined by

$$f(x) = \begin{cases} \pi, & -\pi \leq x < 0, \\ x, & 0 \leq x \leq \pi, \end{cases} \tag{12.50}$$

on the interval $-\pi \leq x \leq \pi$ (see Figure 12.3 again). The series is

$$\frac{3\pi}{4} + \sum_{n=1}^{\infty} \left[\frac{(-1)^n - 1}{\pi n^2} \cos nx - \frac{1}{n} \sin nx \right]. \tag{12.51}$$

Now consider the function g defined by

$$g(x) = \begin{cases} \pi, & -\pi \leq x < 0, \\ \dfrac{\pi}{2}, & x = 0, \\ x, & 0 < x \leq \pi. \end{cases}$$

(See Figure 12.4 and compare with Figure 12.3.)

The functions f and g have the same values *except at $x = 0$*; for $f(x) = g(x) = \pi$, $-\pi \leq x < 0$, and $f(x) = g(x) = x$, $0 < x \leq \pi$, but $f(0) = 0$ and $g(0) = \pi/2$. Thus, since f and g have the same values at all except a finite number (in this case, one) of points of $-\pi \leq x \leq \pi$, they have the same trigonometric Fourier series on this interval. Thus the trigonometric Fourier series of the function g on $-\pi \leq x \leq \pi$ is also the series (12.51).

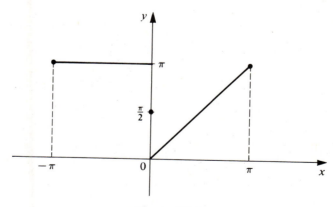

Figure 12.4

Exercises

For each of the functions f defined in Exercises 1–8 find the trigonometric Fourier series of f on $-\pi \leq x \leq \pi$.

1. $f(x) = x, \quad -\pi \leq x \leq \pi.$

2. $f(x) = x^2, \quad -\pi \leq x \leq \pi.$

3. $f(x) = \begin{cases} -2, & -\pi \leq x < 0, \\ 2, & 0 \leq x \leq \pi. \end{cases}$

4. $f(x) = \begin{cases} 0, & -\pi \leq x < 0, \\ 2, & 0 < x \leq \pi. \end{cases}$

5. $f(x) = \begin{cases} 0, & -\pi \leq x < 0, \\ x^2, & 0 \leq x \leq \pi. \end{cases}$

6. $f(x) = \begin{cases} 0, & -\pi \leq x < -\pi/2, \\ 1, & -\pi/2 \leq x < 0, \\ 2, & 0 \leq x < \pi/2, \\ 3, & \pi/2 \leq x \leq \pi. \end{cases}$

7. $f(x) = x|x|, \quad -\pi \leq x \leq \pi.$

8. $f(x) = e^x, \quad -\pi \leq x \leq \pi.$

For each of the functions f defined in Exercises 9–16 find the trigonometric Fourier series of f on the interval stated.

9. $f(x) = |x|, \quad -6 \leq x \leq 6.$

10. $f(x) = 2x + 1, \quad -1 \leq x \leq 1.$

11. $f(x) = \begin{cases} 0, & -3 \leq x < 0, \\ 1, & 0 \leq x \leq 3, \end{cases}$ on the interval $-3 \leq x \leq 3.$

12. $f(x) = \begin{cases} -2, & -4 \leq x < 0, \\ 0, & x = 0, \\ 2, & 0 < x \leq 4, \end{cases}$ on the interval $-4 \leq x \leq 4.$

13. $f(x) = \begin{cases} 0, & -L \leq x < 0, \\ L, & 0 \leq x \leq L, \end{cases}$ on the interval $-L \leq x \leq L.$

14. $f(x) = \begin{cases} 0, & -L \leq x < -L/2, \\ 1, & -L/2 \leq x \leq L, \end{cases}$ on the interval $-L \leq x \leq L.$

15. $f(x) = ax + b, -L \leq x \leq L,$ where a and b are constants.

16. $f(x) = \begin{cases} 1, & -L \leq x < 0, \\ L - x, & 0 \leq x \leq L, \end{cases}$ on the interval $-L \leq x \leq L.$

C. Fourier Sine Series and Fourier Cosine Series

In Example 12.12 we noted that the Sturm–Liouville problem

$$\frac{d^2y}{dx^2} + \lambda y = 0, \tag{12.5}$$

$$y(0) = 0, \qquad y(\pi) = 0 \tag{12.6}$$

has the set of orthonormal characteristic functions $\{\psi_n\}$ defined by

$$\psi_n(x) = \sqrt{\frac{2}{\pi}} \sin nx, \qquad 0 \le x \le \pi \qquad (n = 1, 2, 3, \ldots).$$

If we replace the second condition (12.6) by the more general condition $y(L) = 0$, where $L > 0$; we obtain the problem

$$\frac{d^2 y}{dx^2} + \lambda y = 0,$$

$$y(0) = 0, \qquad y(L) = 0.$$

This problem has the set of orthonormal characteristic functions $\{\phi_n\}$ defined by

$$\phi_n(x) = \sqrt{\frac{2}{L}} \sin \frac{n\pi x}{L}, \qquad 0 \le x \le L \qquad (n = 1, 2, 3, \ldots). \qquad (12.52)$$

Let us now consider the problem of expanding an "arbitrary" function f which is defined on $0 \le x \le L$ in a series of these orthonormal functions $\{\phi_n\}$. In other words, given a function f defined on $0 \le x \le L$, let us find the Fourier series of f relative to the orthonormal system $\{\phi_n\}$ defined by (12.52). From the definition of Fourier series we see that the desired series is of the form

$$\sum_{n=1}^{\infty} c_n \phi_n,$$

where ϕ_n $(n = 1, 2, 3, \ldots)$, is given by (12.52) and

$$c_n = \int_a^b f(x) \phi_n(x) r(x)\, dx = \int_0^L f(x) \sqrt{\frac{2}{L}} \sin \frac{n\pi x}{L}\, dx \qquad (n = 1, 2, 3, \ldots).$$

Thus the series takes the form

$$\sum_{n=1}^{\infty} \left[\sqrt{\frac{2}{L}} \int_0^L f(x) \sin \frac{n\pi x}{L}\, dx \right] \left[\sqrt{\frac{2}{L}} \sin \frac{n\pi x}{L} \right]$$

or

$$\sum_{n=1}^{\infty} b_n \sin \frac{n\pi x}{L},$$

where

$$b_n = \frac{2}{L} \int_0^L f(x) \sin \frac{n\pi x}{L}\, dx \qquad (n = 1, 2, 3, \ldots).$$

This special type of Fourier series is identified in the following definition.

DEFINITION

Let f be a function which is defined on the interval $0 \le x \le L$ and is such that the integrals

$$\int_0^L f(x) \sin \frac{n\pi x}{L}\, dx \qquad (n = 1, 2, 3, \ldots)$$

exist. Then the series

$$\sum_{n=1}^{\infty} b_n \sin \frac{n\pi x}{L}, \tag{12.53}$$

where

$$b_n = \frac{2}{L} \int_0^L f(x) \sin \frac{n\pi x}{L} \, dx \qquad (n = 1, 2, 3, \ldots), \tag{12.54}$$

is called the Fourier sine series *of f on the interval* $0 \leq x \leq L$. *We indicate this by writing*

$$f(x) \sim \sum_{n=1}^{\infty} b_n \sin \frac{n\pi x}{L}, \qquad 0 \leq x \leq L.$$

We thus see that a function f having the required integrability properties on $0 \leq x \leq L$ can be formally expanded into a sine series of the form (12.53) with coefficients given by (12.54). We observe that this expansion is identical with the trigonometric Fourier series (12.48) of the *odd* function which is defined on $-L \leq x \leq L$ and which coincides with f on $0 \leq x \leq L$.

Now suppose that we wish to expand a function f defined on $0 \leq x \leq L$ into a series of cosines instead of a series of sines. If we consider the Sturm–Liouville problem

$$\frac{d^2 y}{dx^2} + \lambda y = 0,$$

$$y'(0) = 0, \qquad y'(L) = 0 \qquad \text{(where } L > 0\text{)},$$

we find that the set of orthonormal characteristic functions of this problem is the set $\{\phi_n\}$ defined by

$$\phi_1(x) = \frac{1}{\sqrt{L}},$$

$$\phi_{n+1}(x) = \sqrt{\frac{2}{L}} \cos \frac{n\pi x}{L}, \qquad 0 \leq x \leq L \qquad (n = 1, 2, 3, \ldots).$$

Using the definition of Fourier series one finds that the Fourier series of a function f relative to this orthonormal system is the series identified in the following definition.

DEFINITION

Let f be a function which is defined on the interval $0 \leq x \leq L$ *and is such that the integrals*

$$\int_0^L f(x) \cos \frac{n\pi x}{L} \, dx \qquad (n = 0, 1, 2, \ldots)$$

exist. Then the series

$$\frac{1}{2} a_0 + \sum_{n=1}^{\infty} a_n \cos \frac{n\pi x}{L}, \tag{12.55}$$

where

$$a_n = \frac{2}{L} \int_0^L f(x) \cos \frac{n\pi x}{L} \, dx \qquad (n = 0, 1, 2, \ldots), \tag{12.56}$$

is called the Fourier cosine series *of f on the interval* $0 \leq x \leq L$. *We indicate this by writing*

$$f(x) \sim \frac{1}{2} a_0 + \sum_{n=1}^{\infty} a_n \cos \frac{n\pi x}{L}, \qquad 0 \leq x \leq L.$$

Note that this expansion is identical with the trigonometric Fourier series (12.46) of the *even* function which is defined on $-L \leq x \leq L$ and which coincides with f on $0 \leq x \leq L$.

▶ **Example 12.18**

Consider the function f defined by

$$f(x) = 2x, \qquad 0 \leq x \leq \pi.$$

(See Figure 12.5.) Find (1) the Fourier sine series of f on $0 \leq x \leq \pi$, and (2) the Fourier cosine series of f on $0 \leq x \leq \pi$.

Solution

1. In this problem $L = \pi$. The Fourier sine series of f on $0 \leq x \leq \pi$ is given by (12.53) with $L = \pi$, where the coefficients b_n are given by (12.54) with $L = \pi$. We find

$$b_n = \frac{2}{\pi} \int_0^\pi f(x) \sin nx \, dx = \frac{2}{\pi} \int_0^\pi 2x \sin nx \, dx$$

$$= \frac{4}{\pi} \left[\frac{\sin nx}{n^2} - \frac{x \cos nx}{n} \right]_0^\pi$$

$$= -\frac{4 \cos n\pi}{n} = \frac{(-1)^{n+1} 4}{n} \qquad (n = 1, 2, 3, \ldots).$$

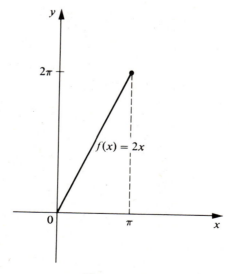

Figure 12.5

Thus the Fourier sine series of f on $0 \le x \le \pi$ is

$$4 \sum_{n=1}^{\infty} \frac{(-1)^{n+1}}{n} \sin nx = 4 \left[\sin x - \frac{1}{2} \sin 2x + \frac{1}{3} \sin 3x - \cdots \right].$$

This expansion is identical with the trigonometric Fourier series (12.48) of the odd function g defined by

$$g(x) = 2x, \qquad -\pi \le x \le \pi,$$

on the interval $-\pi \le x \le \pi$ (see Figure 12.6).

2. The Fourier cosine series of f on $0 \le x \le \pi$ is given by (12.55) with $L = \pi$, where the coefficients a_n are given by (12.56) with $L = \pi$. We find

$$a_0 = \frac{2}{\pi} \int_0^{\pi} f(x)\,dx = \frac{2}{\pi} \int_0^{\pi} 2x\,dx = 2\pi;$$

$$a_n = \frac{2}{\pi} \int_0^{\pi} f(x) \cos nx\,dx = \frac{2}{\pi} \int_0^{\pi} 2x \cos nx\,dx$$

$$= \frac{4}{\pi} \left[\frac{\cos nx}{n^2} + \frac{x \sin nx}{n} \right]_0^{\pi} = \frac{4}{\pi n^2} (\cos n\pi - 1)$$

$$= \frac{4}{\pi n^2} [(-1)^n - 1] = \begin{cases} -\dfrac{8}{\pi n^2}, & n \text{ odd} \\[2mm] 0, & n \text{ even} \end{cases}$$

$(n = 1, 2, 3, \ldots)$. Thus the Fourier cosine series of f on $0 \le x \le \pi$ is

$$\pi - \frac{8}{\pi} \sum_{\substack{n=1 \\ (n\,\text{odd})}}^{\infty} \frac{\cos nx}{n^2} = \pi - \frac{8}{\pi} \sum_{n=1}^{\infty} \frac{\cos(2n-1)x}{(2n-1)^2}$$

$$= \pi - \frac{8}{\pi} \left[\frac{\cos x}{1^2} + \frac{\cos 3x}{3^2} + \frac{\cos 5x}{5^2} + \cdots \right].$$

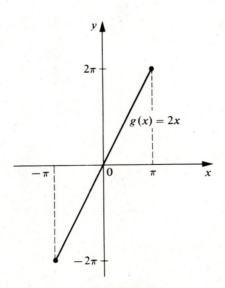

Figure 12.6

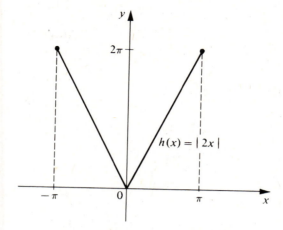

Figure 12.7

This expansion is identical with the trigonometric Fourier series (12.46) of the even function h defined by

$$h(x) = |2x|, \qquad -\pi \le x \le \pi,$$

on the interval $-\pi \le x \le \pi$ (see Figure 12.7).

Exercises

For each of the functions f defined in Exercises 1–6 find (a) the Fourier sine series of f on $0 \le x \le \pi$, and (b) the Fourier cosine series of f on $0 \le x \le \pi$.

1. $f(x) = 1, \quad 0 \le x \le \pi$.

2. $f(x) = x, \quad 0 \le x \le \pi$.

3. $f(x) = \begin{cases} 0, & 0 \le x < \pi/2, \\ 2, & \pi/2 \le x \le \pi. \end{cases}$

4. $f(x) = \begin{cases} x, & 0 \le x < \pi/2, \\ \pi - x, & \pi/2 \le x \le \pi. \end{cases}$

5. $f(x) = \sin x, \quad 0 \le x \le \pi$.

6. $f(x) = e^x, \quad 0 \le x \le \pi$.

For each of the functions f defined in Exercises 7–12 find (a) the Fourier sine series of f on $0 \le x \le L$, and (b) the Fourier cosine series of f on $0 \le x \le L$.

7. $f(x) = 2L, \quad 0 \le x \le L$.

8. $f(x) = 4x, \quad 0 \le x \le L$.

9. $f(x) = \begin{cases} 0, & 0 \le x < L/2, \\ L^2, & L/2 \le x \le L. \end{cases}$

10. $f(x) = \begin{cases} \dfrac{x}{10}, & 0 \le x < L/2, \\[2mm] \dfrac{L-x}{10}, & L/2 \le x \le L. \end{cases}$

11. $f(x) = Lx - x^2, \quad 0 \le x \le L.$

12. $f(x) = \begin{cases} x^2, & 0 \le x < \dfrac{3L}{4}, \\[2mm] \dfrac{9L^2}{4} - \dfrac{9Lx}{4}, & \dfrac{3L}{4} \le x \le L. \end{cases}$

D. Convergence of Trigonometric Fourier Series

In Part A of this section we defined the trigonometric Fourier series of a function f on the interval $-L \le x \le L$. According to our definition one finds the trigonometric Fourier series of f by first determining certain constants (the Fourier coefficients of f) from the formulas

$$
a_n = \frac{1}{L} \int_{-L}^{L} f(x) \cos \frac{n\pi x}{L}\, dx \qquad (n = 0, 1, 2, \ldots),
$$

$$
b_n = \frac{1}{L} \int_{-L}^{L} f(x) \sin \frac{n\pi x}{L}\, dx \qquad (n = 1, 2, 3, \ldots),
$$

(12.45)

and then substituting these constants into the series

$$
\frac{1}{2} a_0 + \sum_{n=1}^{\infty} \left(a_n \cos \frac{n\pi x}{L} + b_n \sin \frac{n\pi x}{L} \right).
$$

(12.44)

We have already pointed out that this development is a formal expansion since no mention of convergence is made. Thus the following equations arise concerning the trigonometric Fourier series (12.44) of the function f.

1. Is the trigonometric Fourier series of f convergent for some or all values of x in the interval $-L \le x \le L$?
2. If the trigonometric Fourier series of f converges at a certain point x of this interval, does it converge to $f(x)$?

In this section we shall state a theorem which will enable us to answer these questions in a manner which is satisfactory for our purposes.

As a first step toward stating this theorem, let us consider a general trigonometric series of the form (12.44). We shall say that a function F is *periodic of period P* if there exists a constant $P > 0$ such that $F(x + P) = F(x)$ for every x for which F is defined. In particular, the functions defined by $\sin(n\pi x/L)$ and $\cos(n\pi x/L)$ are periodic of period $2L/n$. Now observe that if F is periodic of period P, then F is also periodic of period nP, where n is a positive integer. Hence for every positive integer n, the functions defined by $\sin(n\pi x/L)$ and $\cos(n\pi x/L)$ are also periodic of period $2L$. We thus see that every term in the series (12.44) is periodic of period $2L$. Therefore if the series converges for all x, then the function defined by the series must be periodic of period $2L$.

In particular, suppose that the series (12.44) is the trigonometric Fourier series of a function f defined on $-L \leq x \leq L$ and *assume* that this series converges for every x in this interval. Then this series will also converge for every value of $x(-\infty < x < \infty)$ and the function so defined will be periodic of period $2L$. This suggests that a trigonometric Fourier series of the form (12.44) might possibly be employed to represent a periodic function of period $2L$ for *all* values of x. Such a representation is indeed possible for a periodic function which satisfies suitable additional conditions. The convergence theorem which we shall state gives one set of such conditions. In order to state this theorem as simply as possible, we introduce the following definition.

DEFINITION

A function f is said to be piecewise smooth (*or* sectionally smooth) *on a finite interval $a \leq x \leq b$ if this interval can be divided into a finite number of subintervals such that (1) f has a continuous derivative f' in the interior of each of these subintervals, and (2) both $f(x)$ and $f'(x)$ approach finite limits as x approaches either endpoint of each of these subintervals from its interior. In other words, we may say that f is piecewise smooth on $a \leq x \leq b$ if both f and f' are piecewise continuous on $a \leq x \leq b$.*

► **Example 12.19**

The function f defined on the interval $0 \leq x \leq 5$ by

$$f(x) = \begin{cases} x^2, & 0 \leq x < 1, \\ 2 - x, & 1 \leq x < 3, \\ 1, & 3 \leq x < 4, \\ x - 4, & 4 \leq x \leq 5, \end{cases}$$

is piecewise smooth on this interval (see Figure 12.8).

We now state the convergence theorem for trigonometric Fourier series on the interval $-L \leq x \leq L$.

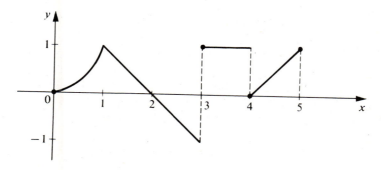

Figure 12.8

THEOREM 12.4

Hypothesis.   *Let f be a function such that*

1. f is periodic of period 2L, and
2. f is piecewise smooth on the interval $-L \leq x \leq L$.

Conclusion.   *The trigonometric Fourier series of f,*

$$\frac{1}{2} a_0 + \sum_{n=1}^{\infty} \left(a_n \cos \frac{n\pi x}{L} + b_n \sin \frac{n\pi x}{L} \right), \tag{12.44}$$

where

$$a_n = \frac{1}{L} \int_{-L}^{L} f(x) \cos \frac{n\pi x}{L} \, dx \qquad (n = 0, 1, 2, \ldots),$$

$$b_n = \frac{1}{L} \int_{-L}^{L} f(x) \sin \frac{n\pi x}{L} \, dx \qquad (n = 1, 2, 3, \ldots), \tag{12.45}$$

converges at every point x to the value

$$\frac{f(x+) + f(x-)}{2}, \tag{12.57}$$

where f(x+) is the right-hand limit of f at x and f(x−) is the left-hand limit of f at x.
In particular, if f is also continuous at x, the value (12.57) reduces to f(x) and the
trigonometric Fourier series of f at x converges to f(x).

We observe that the value (12.57) is the arithmetic mean ("average") of the right-hand and left-hand limits of f at x.

▶ **Example 12.20**

Consider the function f defined for x in the interval $-\pi \leq x \leq \pi$ by

$$f(x) = \begin{cases} \pi, & -\pi \leq x < 0, \\ x, & 0 \leq x \leq \pi, \end{cases} \tag{12.58}$$

and for all other x by the periodicity condition

$$f(x + 2\pi) = f(x) \quad \text{for all } x. \tag{12.59}$$

(See Figure 12.9.)
The condition $f(x + 2\pi) = f(x)$ for all x states that f is periodic of period 2π.
Further, the function f is piecewise smooth on the interval $-\pi \leq x \leq \pi$. Thus the
hypothesis of Theorem 12.4 is satisfied by this function f (with $L = \pi$).
On the interval $-\pi \leq x \leq \pi$ the function f is identical with the function f of
Example 12.16. Thus the trigonometric Fourier series (12.44) of f is the series

$$\frac{3\pi}{4} + \sum_{n=1}^{\infty} \left[\frac{(-1)^n - 1}{\pi n^2} \cos nx - \frac{1}{n} \sin nx \right] \tag{12.51}$$

which was determined in Example 12.16.

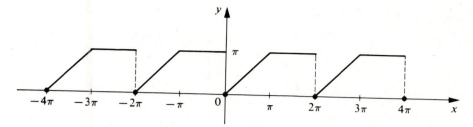

Figure 12.9

The conclusion of Theorem 12.4 assures us that the series (12.51) converges to

$$\frac{f(x+) + f(x-)}{2} \tag{12.57}$$

at every point x, $-\infty < x < \infty$. Further, since f is continuous at every x except $x = \pm 2n\pi$ $(n = 0, 1, 2,\ldots)$, the series (12.51) converges to $f(x)$ at every x except $x = \pm 2n\pi$ $(n = 0, 1, 2,\ldots)$. For example, at $x = \pi$, the series (12.51) is

$$\frac{3\pi}{4} + \sum_{n=1}^{\infty} \frac{1 - (-1)^n}{\pi n^2} = \frac{3\pi}{4} + \frac{2}{\pi} \sum_{n=1}^{\infty} \frac{1}{(2n-1)^2} = \frac{3\pi}{4} + \frac{2}{\pi}\left[\frac{1}{1^2} + \frac{1}{3^2} + \frac{1}{5^2} + \cdots\right]$$

and this converges to $f(\pi) = \pi$. Thus we may write

$$\pi = \frac{3\pi}{4} + \frac{2}{\pi}\left[\frac{1}{1^2} + \frac{1}{3^2} + \frac{1}{5^2} + \cdots\right],$$

from which we obtain the interesting result

$$\frac{\pi^2}{8} = \frac{1}{1^2} + \frac{1}{3^2} + \frac{1}{5^2} + \cdots + \frac{1}{(2n-1)^2} + \cdots. \tag{12.60}$$

At each point $x = \pm 2n\pi$ $(n = 0, 1, 2,\ldots)$, at which f is discontinuous, we find that $f(x+) = f(0+) = 0$, $f(x-) = f(0-) = \pi$, and hence

$$\frac{f(x+) + f(x-)}{2} = \frac{\pi}{2}.$$

Thus at $x = \pm 2n\pi$ $(n = 0, 1, 2,\ldots)$, the series (12.51) converges to the value $\pi/2$. For example, at $x = 0$, the series is

$$\frac{3\pi}{4} + \sum_{n=1}^{\infty} \frac{(-1)^n - 1}{\pi n^2} = \frac{3\pi}{4} - \frac{2}{\pi} \sum_{n=1}^{\infty} \frac{1}{(2n-1)^2}$$

and this converges to $\pi/2$. Thus we may write

$$\frac{\pi}{2} = \frac{3\pi}{4} - \frac{2}{\pi} \sum_{n=1}^{\infty} \frac{1}{(2n-1)^2},$$

and this again leads to the result (12.60).

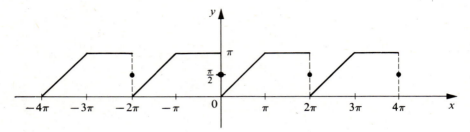

Figure 12.10

In summary, we see that the trigonometric Fourier series (12.51) of the function f defined by (12.58) and (12.59) converges for all x to the function g defined by

$$g(x) = \begin{cases} \pi, & -\pi \le x < 0, \\ \dfrac{\pi}{2}, & x = 0, \\ x, & 0 < x \le \pi, \end{cases} \tag{12.61}$$

$$g(x + 2\pi) = g(x) \quad \text{for all } x.$$

(See Figure 12.10 and compare with Figure 12.9.)

Concerning the convergence of Fourier sine and Fourier cosine series on the interval $0 \le x \le L$, we state the following theorem.

THEOREM 12.5

Hypothesis. *Let f be a function which is piecewise smooth on the interval $0 \le x \le L$.*

Conclusion

1. The Fourier sine series of f,

$$\sum_{n=1}^{\infty} b_n \sin \frac{n\pi x}{L}, \tag{12.53}$$

where

$$b_n = \frac{2}{L} \int_0^L f(x) \sin \frac{n\pi x}{L} \, dx \qquad (n = 1, 2, 3, \ldots), \tag{12.54}$$

converges to the value

$$\frac{f(x+) + f(x-)}{2}$$

for every x such that $0 < x < L$.

In particular, if f is also continuous at x, $0 < x < L$, the Fourier sine series of f at x converges to $f(x)$.

The Fourier sine series of f converges to zero at $x = 0$ and $x = L$.

The Fourier sine series of f converges at every point x to the value

$$\frac{g(x+) + g(x-)}{2},$$

where g is the odd, periodic function of period 2L which coincides with f in the interval $0 < x < L$ and is such that $g(0) = g(L) = 0$.

2. *The Fourier cosine series of f,*

$$\frac{1}{2} a_0 + \sum_{n=1}^{\infty} a_n \cos \frac{n\pi x}{L}, \tag{12.55}$$

where

$$a_n = \frac{2}{L} \int_0^L f(x) \cos \frac{n\pi x}{L} \, dx \qquad (n = 0, 1, 2, \ldots). \tag{12.56}$$

converges to the value

$$\frac{f(x+) + f(x-)}{2}$$

for every x such that $0 < x < L$.

In particular, if f is also continuous at $x, 0 < x < L$, the Fourier cosine series of f at x converges to $f(x)$.

The Fourier cosine series of f converges to $f(0+)$ at $x = 0$ and to $f(L-)$ at $x = L$.

The Fourier cosine series of f converges at every point x to the value

$$\frac{h(x+) + h(x-)}{2},$$

where h is the even, periodic function of period 2L which coincides with f in the interval $0 \le x \le L$.

▶ **Example 12.21**

Consider the function f defined by

$$f(x) = x, \qquad 0 \le x \le \pi.$$

The Fourier sine and cosine series of f on $0 \le x \le \pi$ are readily found to be

$$2 \sum_{n=1}^{\infty} \frac{(-1)^{n+1} \sin nx}{n} \tag{12.62}$$

and

$$\frac{\pi}{2} - \frac{4}{\pi} \sum_{n=1}^{\infty} \frac{\cos(2n-1)x}{(2n-1)^2}, \tag{12.63}$$

respectively.

The function f is piecewise smooth on $0 \le x \le \pi$; in fact, it is continuous on this interval. Therefore, applying Theorem 12.5, we see that the Fourier sine series (12.62) and the Fourier cosine series (12.63) of f both converge to $f(x)$ at every x such that $0 < x < \pi$. Further, the sine series (12.62) converges to zero at $x = 0$ and $x = \pi$; and the cosine series (12.63) converges to $f(0+) = 0$ at $x = 0$ and to $f(\pi-) = \pi$ at $x = \pi$.

The sine series (12.62) converges at every point x to the value

$$\frac{g(x+) + g(x-)}{2},$$

where g is the odd, periodic function of period 2π which coincides with f in the interval $0 < x < \pi$ and is such that $g(0) = g(\pi) = 0$. This function g is defined by

$$g(x) = x, \qquad 0 \le x < \pi,$$
$$g(x) = 0, \qquad x = \pi,$$
$$g(-x) = -g(x) \quad \text{for all } x, \qquad (12.64)$$
$$g(x + 2\pi) = g(x) \quad \text{for all } x.$$

Its graph is shown in Figure 12.11. For this function g we see that

$$\frac{g(x+) + g(x-)}{2} = g(x)$$

for every x. Thus the sine series (12.62) converges at every point x to the value $g(x)$ given by (12.64).

The cosine series (12.63) converges at every point x to the value

$$\frac{h(x+) + h(x-)}{2},$$

where h is the even, periodic function of period 2π which coincides with f in the interval $0 \le x \le \pi$. This function h is defined by

$$h(x) = x, \qquad 0 \le x \le \pi,$$
$$h(-x) = h(x) \quad \text{for all } x, \qquad (12.65)$$
$$h(x + 2\pi) = h(x) \quad \text{for all } x.$$

Its graph is shown in Figure 12.12. For this function h we see that

$$\frac{h(x+) + h(x-)}{2} = h(x)$$

for every x. Thus the cosine series (12.63) converges at every point x to the value $h(x)$ given by (12.65).

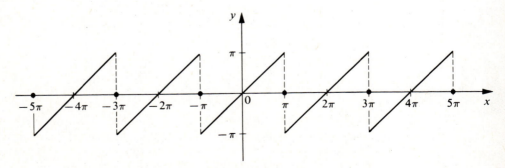

Figure 12.11

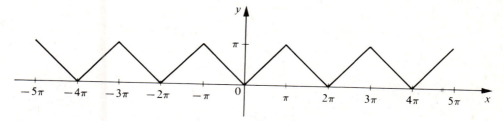

Figure 12.12

Exercises

1. Consider the function f defined for x on the interval $-\pi \le x < \pi$ by

$$f(x) = \begin{cases} 0, & -\pi \le x < 0, \\ x + 1, & 0 \le x < \pi/2, \\ 2x, & \pi/2 \le x < \pi, \end{cases}$$

and for all other x by the periodicity condition

$$f(x + 2\pi) = f(x) \quad \text{for all } x.$$

Discuss the convergence of the trigonometric Fourier series of f. In particular, determine the value to which the series converges at each of the points $x = 0$, $x = \pi/2$, and $x = \pi$.

2. Consider the function f defined on the interval $-\pi \le x < \pi$ by

$$f(x) = \begin{cases} (x + \pi)^2, & -\pi \le x < 0, \\ \pi^2, & 0 \le x < \pi/3, \\ 8, & \pi/3 \le x < 2\pi/3, \\ 5, & 2\pi/3 \le x < \pi, \end{cases}$$

and for all other x by the periodicity condition

$$f(x + 2\pi) = f(x) \quad \text{for all } x.$$

Discuss the convergence of the trigonometric Fourier series of f. Determine those values of x in $-\pi \le x \le \pi$ at which this series does *not* converge to $f(x)$ and determine the value to which the series *does* converge at each of these points.

CHAPTER THIRTEEN

Nonlinear Differential Equations

The mathematical formulation of numerous physical problems results in differential equations which are actually nonlinear. In many cases it is possible to replace such a nonlinear equation by a related linear equation which approximates the actual nonlinear equation closely enough to give useful results. However, such a "linearization" is not always feasible; and when it is not, the original nonlinear equation itself must be considered. While the general theory and methods of linear equations are highly developed, very little of a general character is known about nonlinear equations. The study of nonlinear equations is generally confined to a variety of rather special cases, and one must resort to various methods of approximation. In this chapter we shall give a brief introduction to certain of these methods.

13.1 PHASE PLANE, PATHS, AND CRITICAL POINTS

A. Basic Concepts and Definitions

In this chapter we shall be concerned with second-order nonlinear differential equations of the form

$$\frac{d^2x}{dt^2} = F\left(x, \frac{dx}{dt}\right). \tag{13.1}$$

As a specific example of such an equation we list the important van der Pol equation

$$\frac{d^2x}{dt^2} + \mu(x^2 - 1)\frac{dx}{dt} + x = 0, \tag{13.2}$$

where μ is a positive constant. We shall consider this equation at a later stage of our study. For the time being, we merely note that (13.2) may be put in the form (13.1), where

$$F\left(x, \frac{dx}{dt}\right) = -\mu(x^2 - 1)\frac{dx}{dt} - x.$$

Let us suppose that the differential equation (13.1) describes a certain dynamical system having one degree of freedom. The state of this system at time t is determined by the values of x (position) and dx/dt (velocity). The plane of the variables x and dx/dt is called a *phase plane*.

If we let $y = dx/dt$, we can replace the second-order equation (13.1) by the equivalent system

$$\frac{dx}{dt} = y,$$

$$\frac{dy}{dt} = F(x, y). \tag{13.3}$$

We can determine information about Equation (13.1) from a study of the system (13.3). In particular we shall be interested in the configurations formed by the curves which the solutions of (13.3) define. We shall regard t as a parameter so that these curves will appear in the xy plane. Since $y = dx/dt$, this xy plane is simply the $x, dx/dt$ phase plane mentioned in the preceding paragraph.

More generally, we shall consider systems of the form

$$\frac{dx}{dt} = P(x, y),$$

$$\frac{dy}{dt} = Q(x, y), \tag{13.4}$$

where P and Q have continuous first partial derivatives for all (x, y). Such a system, in which the independent variable t appears only in the differentials dt of the left members and not explicitly in the functions P and Q on the right, is called an *autonomous system*. We shall now proceed to study the configurations formed in the xy phase plane by the curves which are defined by the solutions of (13.4).

From Chapter 10, Section 10.4A, Theorem 10.5, it follows that, given any number t_0 and any pair (x_0, y_0) of real numbers, there exists a unique solution

$$x = f(t),$$

$$y = g(t), \tag{13.5}$$

of the system (13.4) such that

$$f(t_0) = x_0,$$

$$g(t_0) = y_0.$$

If f and g are not both constant functions, then (13.5) defines a curve in the xy plane which we shall call a *path* (or *orbit* or *trajectory*) of the system (13.4).

If the ordered pair of functions defined by (13.5) is a solution of (13.4) and t_1 is any real number, then it is easy to see that the ordered pair of functions defined by

$$x = f(t - t_1),$$

$$y = g(t - t_1), \tag{13.6}$$

is also a solution of (13.4). Assuming that f and g in (13.5) are not both constant functions and that $t_1 \neq 0$, the solutions defined by (13.5) and (13.6) are two *different solutions* of (13.4). However, these two *different solutions* are simply different parametrizations of the *same path*. We thus observe that the terms *solution* and *path* are not synonymous. On the one hand, a *solution* of (13.4) is an ordered *pair of functions* (f, g) such that $x = f(t)$, $y = g(t)$ simultaneously satisfy the two equations of the system (13.4) identically; on the other hand, a *path* of (13.4) is a *curve* in the xy phase plane, which may be defined parametrically by more than one solution of (13.4).

Through any point of the xy phase plane there passes at most one path of (13.4). Let C be a path of (13.4) and consider the totality of different solutions of (13.4) which define this path C parametrically. For each of these defining solutions, C is traced out in the *same direction* as the parameter t increases. Thus with each path C there is associated a definite direction, the direction of increase of the parameter t in the various possible parametric representations of C by the corresponding solutions of the system. In our figures we shall use arrows to indicate this direction associated with a path.

Eliminating t between the two equations of the system (13.4), we obtain the equation

$$\frac{dy}{dx} = \frac{Q(x, y)}{P(x, y)}. \tag{13.7}$$

This equation gives the slope of the tangent to the path of (13.4) passing through the point (x, y), provided the functions P and Q are not both zero at this point. The one-parameter family of solutions of (13.7) thus provides the one-parameter family of paths of (13.4). However, the description (13.7) does not indicate the directions associated with these paths.

At a point (x_0, y_0) at which both P and Q are zero, the slope of the tangent to the path, as defined by (13.7), is indeterminate. Such points are singled out in the following definition.

DEFINITION

Given the autonomous system

$$\frac{dx}{dt} = P(x, y),$$

$$\frac{dy}{dt} = Q(x, y), \tag{13.4}$$

a point (x_0, y_0) *at which both*

$$P(x_0, y_0) = 0 \quad and \quad Q(x_0, y_0) = 0$$

is called a critical point* *of* (13.4).

---

* The terms *equilibrium point* and *singular point* are also used.

▶ **Example 13.1**

Consider the linear autonomous system

$$\frac{dx}{dt} = y,$$

$$\frac{dy}{dt} = -x. \tag{13.8}$$

Using the methods developed in Chapter 7, we find that the general solution of this system may be written

$$x = c_1 \sin t - c_2 \cos t,$$

$$y = c_1 \cos t + c_2 \sin t,$$

where c_1 and c_2 are arbitrary constants. The solution satisfying the conditions $x(0) = 0$, $y(0) = 1$ is readily found to be

$$x = \sin t,$$

$$y = \cos t. \tag{13.9}$$

This solution defines a path C_1 in the xy plane. The solution satisfying the conditions $x(0) = -1$, $y(0) = 0$ is

$$x = \sin(t - \pi/2),$$

$$y = \cos(t - \pi/2). \tag{13.10}$$

The solution (13.10) is different from the solution (13.9), but (13.10) also defines the same path C_1. That is, the ordered pairs of functions defined by (13.9) and (13.10) are two *different solutions* of (13.8) which are different parametrizations of the *same path C_1*. Eliminating t from either (13.9) or (13.10) we obtain the equation $x^2 + y^2 = 1$ of the path C_1 in the xy phase plane. Thus the path C_1 is the circle with center at $(0, 0)$ and radius 1. From either (13.9) or (13.10) we see that the direction associated with C_1 is the *clockwise* direction.

Eliminating t between the equations of the system (13.8) we obtain the differential equation

$$\frac{dy}{dx} = -\frac{x}{y}, \tag{13.11}$$

which gives the slope of the tangent to the path of (13.8) passing through the point (x, y), provided $(x, y) \neq (0, 0)$.

The one-parameter family of solutions

$$x^2 + y^2 = c^2$$

of Equation (13.11) gives the one-parameter family of paths in the xy phase plane. Several of these are shown in Figure 13.1. The path C_1 referred to above is, of course, that for which $c = 1$.

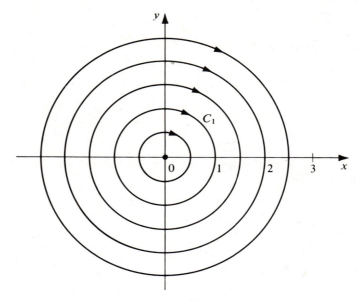

Figure 13.1

Looking back at the system (13.8), we see that $P(x, y) = y$ and $Q(x, y) = -x$. Therefore the only critical point of the system is the origin $(0, 0)$. Given any real number t_0, the solution $x = f(t)$, $y = g(t)$ such that $f(t_0) = g(t_0) = 0$ is simply

$$x = 0,$$
$$y = 0, \qquad \text{for all } t.$$

We can also interpret the autonomous system (13.4) as defining a *velocity vector field* V, where

$$V(x, y) = [P(x, y), Q(x, y)].$$

The x component of this velocity vector at a point (x, y) is given by $P(x, y)$, and the y component there is given by $Q(x, y)$. This vector is the velocity vector of a representative point R describing a path of (13.4) defined parametrically by a solution $x = f(t)$, $y = g(t)$. At a critical point both components of this velocity vector are zero, and hence at a critical point the point R is at rest.

In particular, let us consider the special case (13.3), which arises from a dynamical system described by the differential equation (13.1). At a critical point of (13.3) both dx/dt and dy/dt are zero. Since

$$\frac{dy}{dt} = \frac{d^2x}{dt^2},$$

we thus see that at such a point the velocity and acceleration of the dynamical system described by (13.1) are both zero. Thus the critical points of (13.3) are equilibrium points of the dynamical system described by (13.1).

We now introduce certain basic concepts dealing with critical points and paths.

DEFINITION

A critical point (x_0, y_0) *of the system* (13.4) *is called* isolated *if there exists a circle*

$$(x - x_0)^2 + (y - y_0)^2 = r^2$$

about the point (x_0, y_0) *such that* (x_0, y_0) *is the only critical point of* (13.4) *within this circle.*

In what follows we shall assume that every critical point is isolated.

Note. For convenience, we shall take the critical point (x_0, y_0) to be the origin $(0, 0)$. There is no loss in generality in doing this, for if $(x_0, y_0) \neq (0, 0)$, then the translation of coordinates $\xi = x - x_0, \eta = y - y_0$ transforms (x_0, y_0) into the origin in the $\xi\eta$ plane.

DEFINITION

Let C be a path of the system (13.4), *and let* $x = f(t), y = g(t)$ *be a solution of* (13.4) *which represents C parametrically. Let* $(0, 0)$ *be a critical point of* (13.4). *We shall say that the path C approaches the critical point* $(0, 0)$ *as* $t \to +\infty$ *if*

$$\lim_{t \to +\infty} f(t) = 0, \qquad \lim_{t \to +\infty} g(t) = 0. \tag{13.12}$$

Thus when we say that a path C defined parametrically by $x = f(t), y = g(t)$ approaches the critical point $(0, 0)$ as $t \to +\infty$, we understand the following: a point R tracing out C according to the equations $x = f(t), y = g(t)$ will approach the point $(0, 0)$ as $t \to +\infty$. This approach of a path C to the critical point $(0, 0)$ is independent of the solution actually used to represent C. That is, if C approaches $(0, 0)$ as $t \to +\infty$, then (13.12) is true for all solutions $x = f(t), y = g(t)$ representing C.

In like manner, a path C_1 approaches the critical point $(0, 0)$ as $t \to -\infty$ if

$$\lim_{t \to -\infty} f_1(t) = 0, \qquad \lim_{t \to -\infty} g_1(t) = 0,$$

where $x = f_1(t), y = g_1(t)$ is a solution defining the path C_1.

DEFINITION

Let C be a path of the system (13.4) *which approaches the critical point* $(0, 0)$ *of* (13.4) *as* $t \to +\infty$, *and let* $x = f(t), y = g(t)$ *be a solution of* (13.4) *which represents C parametrically. We say that C enters the critical point* $(0, 0)$ *as* $t \to +\infty$ *if*

$$\lim_{t \to +\infty} \frac{g(t)}{f(t)} \tag{13.13}$$

exists or if the quotient in (13.13) *becomes either positively or negatively infinite as* $t \to +\infty$.

We observe that the quotient $g(t)/f(t)$ in (13.13) represents the slope of the line joining critical point $(0, 0)$ and a point R with coordinates $[f(t), g(t)]$ on C. Thus when we say that a path C enters the critical point $(0, 0)$ as $t \to +\infty$ we mean that the line joining $(0, 0)$ and a point R tracing out C approaches a definite "limiting" direction as $t \to +\infty$.

B. Types of Critical Points

We shall now discuss certain types of critical points which we shall encounter. We shall first give a geometric description of each type of critical point, referring to an appropriate figure as we do so; we shall then follow each such description with a more precise definition.

(1) The critical point $(0, 0)$ of Figure 13.2 is called a *center*. Such a point is surrounded by an infinite family of *closed* paths, members of which are arbitrarily close to $(0, 0)$, but it is not approached by any path either as $t \to +\infty$ or as $t \to -\infty$.
   More precisely, we state:

DEFINITION

The isolated critical point $(0, 0)$ of (13.4) is called a center *if there exists a neighborhood of $(0, 0)$ which contains a countably infinite number of closed paths P_n $(n = 1, 2, 3, \ldots)$, each of which contains $(0, 0)$ in its interior, and which are such that the diameters of the paths approach 0 as $n \to \infty$ [but $(0, 0)$ is not approached by any path either as $t \to +\infty$ or as $t \to -\infty$].*

Note. The preceding definition contains several terms which may be unfamiliar to some readers. Moreover, an explanatory statement concerning the infinite number of closed paths should be made. We proceed to discuss these matters.

1. We define a *neighborhood* of $(0, 0)$ to be the set of all points (x, y) lying within some fixed (positive) distance d of $(0, 0)$.

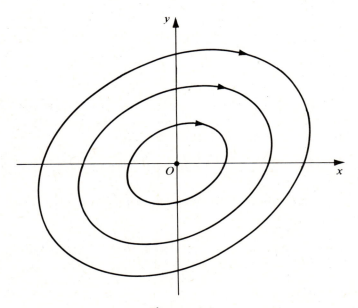

Figure 13.2

2. An infinite set is called *countable* if it can be put into one-to-one correspondence with the set of all positive integers. An example of a countable set is the set of all rational numbers.

3. An infinite set is *uncountable* if it is not countable. An example of an uncountable set is the set of all real numbers.

4. It can be shown that an uncountably infinite set always contains a countably infinite subset. Thus the definition of center does not exclude the possibility of an *uncountably* infinite set of closed paths each of which contains (0, 0) in its interior. Indeed, in the case of a linear system and all of the nonlinear systems which we shall encounter in this text, there always exists such an uncountably infinite set of closed paths containing the center in their interior.

5. We call the *diameter* of a closed curve C the maximum of the distance $d(A, B)$ between points A and B on C for all possible pairs of points A and B on C.

(2) The critical point (0, 0) of Figure 13.3 is called a *saddle point*. Such a point may be characterized as follows:

1. It is approached and entered by two half-line paths $(AO$ and $BO)$ as $t \to +\infty$, these two paths forming the geometric curve AB.

2. It is approached and entered by two half-line paths $(CO$ and $DO)$ as $t \to -\infty$, these two paths forming the geometric curve CD.

3. Between the four half-line paths described in (1) and (2) there are four domains R_1, R_2, R_3, R_4, each containing an infinite family of semi-hyperbolic-like paths which do not approach O as $t \to +\infty$ or as $t \to -\infty$, but which become asymptotic to one or another of the four half-line paths as $t \to +\infty$ and as $t \to -\infty$.

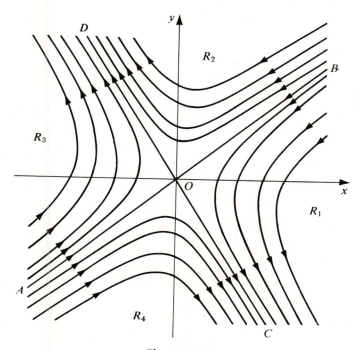

Figure 13.3

More precisely, we state:

DEFINITION

The isolated critical point $(0, 0)$ of (13.4) is called a saddle point *if there exists a neighborhood of $(0, 0)$ in which the following two conditions hold:*

1. There exist two paths which approach and enter $(0,0)$ from a pair of opposite directions as $t \to +\infty$, and there exist two paths which approach and enter $(0,0)$ from a different pair of opposite directions as $t \to -\infty$.

2. In each of the four domains between any two of the four directions in (1) there are infinitely many paths which are arbitrarily close to $(0, 0)$ but which do not approach $(0, 0)$ either as $t \to +\infty$ or as $t \to -\infty$.

(3) The critical point $(0, 0)$ of Figure 13.4 is called a *spiral point* (or *focal point*). Such a point is approached in a spiral-like manner by an infinite family of paths as $t \to +\infty$ (or as $t \to -\infty$). Observe that while the paths *approach* O, they do not *enter* it. That is, a point R tracing such a path C approaches O as $t \to +\infty$ (or as $t \to -\infty$), but the line OR does not tend to a definite direction, since the path constantly winds about O.

More precisely, we state:

DEFINITION

The isolated critical point $(0, 0)$ of (13.4) is called a spiral point *(or focal point) if there exists a neighborhood of $(0, 0)$ such that every path P in this neighborhood has the following properties:*

1. P is defined for all $t > t_0$ (or for all $t < t_0$) for some number t_0;

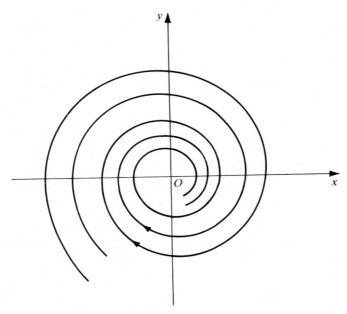

Figure 13.4

2. P approaches $(0, 0)$ as $t \to +\infty$ (or as $t \to -\infty$); and
3. P approaches $(0, 0)$ in a spiral-like manner, winding around $(0, 0)$ an infinite number of times as $t \to +\infty$ (or as $t \to -\infty$).

(4) The critical point $(0, 0)$ of Figure 13.5 is called a *node*. Such a point is not only approached but also *entered* by an infinite family of paths as $t \to +\infty$ (or as $t \to -\infty$). That is, a point R tracing such a path not only approaches O but does so in such a way that the line OR tends to a definite direction as $t \to +\infty$ (or as $t \to -\infty$). For the node shown in Figure 13.5 there are four rectilinear paths, AO, BO, CO, and DO. All other paths are like "semiparabolas." As each of these semiparabolic-like paths approaches O, its slope approaches that of the line AB.

More precisely, we state:

DEFINITION

The isolated critical point $(0, 0)$ of (13.4) is called a node *if there exists a neighborhood of $(0, 0)$ such that every path P in this neighborhood has the following properties:*

1. P is defined for all $t > t_0$ [or for all $t < t_0$] for some number t_0;
2. P approaches $(0, 0)$ as $t \to +\infty$ [or as $t \to -\infty$]; and
3. P enters $(0, 0)$ as $t \to +\infty$ [or as $t \to -\infty$].

C. Stability

We assume that $(0, 0)$ is an isolated critical point of the system

$$\frac{dx}{dt} = P(x, y),$$

$$\frac{dy}{dt} = Q(x, y), \tag{13.4}$$

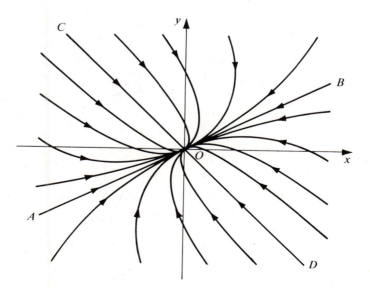

Figure 13.5

and proceed to introduce the concepts of *stability* and *asymptotic stability* for this critical point.

DEFINITION

Assume that (0, 0) *is an isolated critical point of the system* (13.4). *Let C be a path of* (13.4); *let* $x = f(t)$, $y = g(t)$ *be a solution of* (13.4) *defining C parametrically. Let*

$$D(t) = \sqrt{[f(t)]^2 + [g(t)]^2} \tag{13.14}$$

denote the distance between the critical point (0, 0) *and the point* $R: [f(t), g(t)]$ *on C. The critical point* (0, 0) *is called* stable *if for every number* $\epsilon > 0$, *there exists a number* $\delta > 0$ *such that the following is true: Every path C for which*

$$D(t_0) < \delta \qquad \text{for some value } t_0 \tag{13.15}$$

is defined for all $t \geq t_0$ *and is such that*

$$D(t) < \epsilon \qquad \text{for } t_0 \leq t < \infty. \tag{13.16}$$

Let us try to analyze this definition, making use of Figure 13.6 as we do so. The critical point (0, 0) is said to be stable if, corresponding to every positive number ϵ, we can find another positive number δ which does "something" for us. Now what is this "something"? To answer this we must understand what the inequalities in (13.15) and (13.16) mean. According to (13.14), the inequality $D(t_0) < \delta$ for some value t_0 in (13.15) means that the distance between the critical point (0, 0) and the point R on the patch C must be less than δ at $t = t_0$. This means that at $t = t_0$, R lies within the circle K_1 of radius δ about (0, 0) (see Figure 13.6). Likewise the inequality $D(t) < \epsilon$ for $t_0 \leq t < \infty$ in (13.16) means that the distance between (0, 0) and R is less than ϵ for all $t \geq t_0$, and hence that for $t \geq t_0$, R lies within the circle K_2 of radius ϵ about (0. 0). Now we can understand the "something" which the number δ does for us. If (0, 0) is stable, then *every* path C which is inside the circle K_1 of radius δ at $t = t_0$ will remain inside the

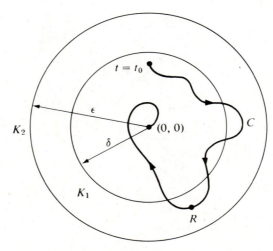

Figure 13.6

circle K_2 of radius $t \geq t_0$. Roughly speaking, if every path C stays as close to $(0, 0)$ as we want it to (that is, within distance ϵ) after it once gets close enough (that is, within distance δ), then $(0, 0)$ is stable.

DEFINITION

Assume that $(0, 0)$ is an isolated critical point of the system (13.4). Let C be a path of (13.4); and let $x = f(t), y = g(t)$ be a solution of (13.4) representing C parametrically. Let

$$D(t) = \sqrt{[f(t)]^2 + [g(t)]^2} \tag{13.14}$$

denote the distance between the critical point $(0, 0)$ and the point R: $[f(t), g(t)]$ on C.

The critical point $(0, 0)$ is called asymptotically stable *if (1) it is stable and (2) there exists a number $\delta_0 > 0$ such that if*

$$D(t_0) < \delta_0 \tag{13.17}$$

for some value t_0, then

$$\lim_{t \to +\infty} f(t) = 0, \qquad \lim_{t \to +\infty} g(t) = 0. \tag{13.18}$$

To analyze this definition, note that condition (1) requires that $(0, 0)$ must be *stable*. That is, every path C will stay as close to $(0, 0)$ as we desire after it once gets sufficiently close. But asymptotic stability is a stronger condition than mere stability. For, in addition to stability, the condition (2) requires that every path that gets sufficiently close to $(0, 0)$ [see (13.17)] ultimately approaches $(0, 0)$ as $t \to +\infty$ [see (13.18)]. Note that the path C of Figure 13.6 has this property.

DEFINITION

A critical point is called unstable *if it is not stable.*

As illustrations of stable critical points, we point out that the center in Figure 13.2, the spiral point in Figure 13.4, and the node in Figure 13.5 are all stable. Of these three, the spiral point and the node are asymptotically stable. If the directions of the paths in Figures 13.4 and 13.5 has been reversed, then the spiral point and the node of these respective figures would have both been unstable. The saddle point of Figure 13.3 is unstable.

Exercises

For each of the autonomous systems in Exercises 1 and 2, (a) find the real critical points of the system, (b) obtain the differential equation which gives the slope of the tangent to the paths of the system, and (c) solve this differential equation to obtain the one-parameter family of paths.

1. $\dfrac{dx}{dt} = x - y^2,$

   $\dfrac{dy}{dt} = x^2 - y.$

2. $\dfrac{dx}{dt} = y,$

   $\dfrac{dy}{dt} = \sin x.$

3. Consider the linear autonomous system

$$\frac{dx}{dt} = x,$$

$$\frac{dy}{dt} = x + y.$$

(a) Find the solution of this system which satisfies the conditions $x(0) = 1$, $y(0) = 3$.

(b) Find the solution of this system which satisfies the conditions $x(4) = e$, $y(4) = 4e$.

(c) Show that the two *different* solutions found in (a) and (b) both represent the *same* path.

(d) Find the differential equation which gives the slope of the tangent to the paths, and solve it to obtain the one-parameter family of paths.

4. Consider the linear autonomous systems

(a) $\dfrac{dx}{dt} = -x,$ (b) $\dfrac{dx}{dt} = 4y,$

$\dfrac{dy}{dt} = -2y.$ $\dfrac{dy}{dt} = -x.$

Show that the critical point $(0, 0)$ of system (a) is asymptotically stable, but that the critical point $(0, 0)$ of system (b) is stable but not asymptotically stable.

13.2 CRITICAL POINTS AND PATHS OF LINEAR SYSTEMS

A. Basic Theorems

Our main purpose in this chapter is the study of nonlinear differential equations and the corresponding nonlinear autonomous systems of the form (13.4). We shall be interested in classifying the critical points of such nonlinear systems. In Section 13.3 we shall see that under appropriate circumstances we may replace a given nonlinear system by a related linear system and then employ this linear system to determine the nature of the critical point of the given system. Thus in this section we shall first investigate the critical points of a linear autonomous system.

We consider the linear system

$$\frac{dx}{dt} = ax + by,$$

$$\frac{dy}{dt} = cx + dy,$$

(13.19)

where $a, b, c,$ and d (in the right member of the second equation) are real constants. The origin (0, 0) is clearly a critical point of (13.19). We assume that

$$\begin{vmatrix} a & b \\ c & d \end{vmatrix} \neq 0, \tag{13.20}$$

and hence (0, 0) is the *only* critical point of (13.19).

Recall that in Chapter 7 we sought and found solutions of (13.19) of the form

$$x = Ae^{\lambda t},$$
$$y = Be^{\lambda t}. \tag{13.21}$$

We saw there that if (13.21) is to be a solution of (13.19), then λ must satisfy the quadratic equation

$$\lambda^2 - (a + d)\lambda + (ad - bc) = 0, \tag{13.22}$$

called the *characteristic equation* of (13.19). Note that by condition (13.20) zero cannot be a root of the equation (13.22) in the problem under discussion.

Let λ_1 and λ_2 be the roots of the characteristic equation (13.22). We shall prove that the nature of the critical point (0, 0) of the system (13.19) depends upon the nature of the roots λ_1 and λ_2. We would expect three possibilities, according as λ_1 and λ_2 are real and distinct, real and equal, or conjugate complex. But actually the situation here is not quite so simple and we must consider the following five cases:

1. λ_1 and λ_2 are real, unequal, and of the same sign.
2. λ_1 and λ_2 are real, unequal, and of opposite sign.
3. λ_1 and λ_2 are real and equal.
4. λ_1 and λ_2 are conjugate complex but not pure imaginary.
5. λ_1 and λ_2 are pure imaginary.

Case 1

THEOREM 13.1

Hypothesis. *The roots λ_1 and λ_2 of the characteristic equation (13.22) are real, unequal, and of the same sign.*

Conclusion. *The critical point (0, 0) of the linear system (13.19) is a* node.

Proof. We first assume that λ_1 and λ_2 are both negative and take $\lambda_1 < \lambda_2 < 0$. By Theorem 7.7 of Chapter 7 the general solution of the system (13.19) may then be written

$$x = c_1 A_1 e^{\lambda_1 t} + c_2 A_2 e^{\lambda_2 t},$$
$$y = c_1 B_1 e^{\lambda_1 t} + c_2 B_2 e^{\lambda_2 t}, \tag{13.23}$$

where A_1, B_1, A_2, and B_2 are definite constants and $A_1 B_2 \neq A_2 B_1$, and where c_1 and c_2 are arbitrary constants. Choosing $c_2 = 0$ we obtain the solutions

$$x = c_1 A_1 e^{\lambda_1 t},$$
$$y = c_1 B_1 e^{\lambda_1 t}; \tag{13.24}$$

choosing $c_1 = 0$ we obtain the solutions

$$x = c_2 A_2 e^{\lambda_2 t},$$
$$y = c_2 B_2 e^{\lambda_2 t}. \tag{13.25}$$

For any $c_1 > 0$, the solutions (13.24) represent a path consisting of "half" of the line $B_1 x = A_1 y$ of slope B_1/A_1. For any $c_1 < 0$, they represent a path consisting of the "other half" of this line. Since $\lambda_1 < 0$, both of these half-line paths approach $(0, 0)$ as $t \to +\infty$. Also, since $y/x = B_1/A_1$, these two paths *enter* $(0, 0)$ with slope B_1/A_1.

In like manner, for any $c_2 > 0$ the solutions (13.25) represent a path consisting of "half" of the line $B_2 x = A_2 y$; while for any $c_2 < 0$, the path so represented consists of the "other half" of this line. These two half-line paths also approach $(0, 0)$ as $t \to +\infty$ and enter it with slope B_2/A_2.

Thus the solutions (13.24) and (13.25) provide us with four half-line paths which all approach and enter $(0, 0)$ as $t \to +\infty$.

If $c_1 \neq 0$ and $c_2 \neq 0$, the general solution (13.23) represents nonrectilinear paths. Again, since $\lambda_1 < \lambda_2 < 0$, all of these paths approach $(0, 0)$ as $t \to +\infty$. Further, since

$$\frac{y}{x} = \frac{c_1 B_1 e^{\lambda_1 t} + c_2 B_2 e^{\lambda_2 t}}{c_1 A_1 e^{\lambda_1 t} + c_2 A_2 e^{\lambda_2 t}} = \frac{(c_1 B_1/c_2)e^{(\lambda_1 - \lambda_2)t} + B_2}{(c_1 A_1/c_2)e^{(\lambda_1 - \lambda_2)t} + A_2},$$

we have

$$\lim_{t \to \infty} \frac{y}{x} = \frac{B_2}{A_2}$$

and so all of these paths *enter* $(0, 0)$ with limiting slope B_2/A_2.

Thus all the paths (both rectilinear and nonrectilinear) *enter* $(0, 0)$ as $t \to +\infty$, and all except the two rectilinear ones defined by (13.24) enter with slope B_2/A_2. According to the definition of Section 13.1B, the critical point $(0, 0)$ is a *node*. Clearly, it is *asymptotically stable*. A qualitative diagram of the paths appears in Figure 13.7.

If now λ_1 and λ_2 are both positive and we take $\lambda_1 > \lambda_2 > 0$, we see that the general solution of (13.19) is still of the form (13.23) and particular solutions of the forms (13.24) and (13.25) still exist. The situation is the same as before, except all the paths approach and enter $(0, 0)$ as $t \to -\infty$. The qualitative diagram of Figure 13.7 is unchanged, except that all the arrows now point in the opposite directions. The critical point $(0, 0)$ is still a *node*, but in this case it is clear that it is *unstable*.

Case 2

THEOREM 13.2

Hypothesis. *The roots λ_1 and λ_2 of the characteristic equation (13.22) are real, unequal, and of opposite sign.*

Conclusion. *The critical point $(0, 0)$ of the linear system (13.19) is a* saddle point.

Proof. We assume that $\lambda_1 < 0 < \lambda_2$. The general solution of the system (13.19) may again be written in the form (13.23) and particular solutions of the forms (13.24) and (13.25) are again present.

For any $c_1 > 0$, the solutions (13.24) again represent a path consisting of "half" the line $B_1 x = A_1 y$; while for any $c_1 < 0$, they again represent a path consisting of the

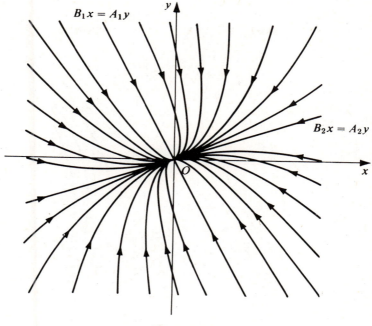

$B_1 x = A_1 y$

$B_2 x = A_2 y$

Figure 13.7

"other half" of this line. Also, since $\lambda_1 < 0$, both of these half-line paths still approach and enter $(0, 0)$ as $t \to +\infty$.

Also, for any $c_2 > 0$, the solutions (13.25) represent a path consisting of "half" the line $B_2 x = A_2 y$; and for any $c_2 < 0$, the path which they represent consists of the "other half" of this line. But in this case, since $\lambda_2 > 0$, both of these half-line paths now approach and enter $(0, 0)$ as $t \to -\infty$.

Once again, if $c_1 \neq 0$ and $c_2 \neq 0$, the general solution (13.23) represents non-rectilinear paths. But here since $\lambda_1 < 0 < \lambda_2$, none of these paths can approach $(0, 0)$ as $t \to +\infty$ or as $t \to -\infty$. Further, none of them pass through $(0, 0)$ for any t_0 such that $-\infty < t_0 < +\infty$. As $t \to +\infty$, we see from (13.23) that each of these nonrectilinear paths becomes asymptotic to one of the half-line paths defined by (13.25). As $t \to -\infty$, each of them becomes asymptotic to one of the paths defined by (13.24).

Thus there are two half-line paths which approach and enter $(0, 0)$ as $t \to +\infty$ and two other half-line paths which approach and enter $(0, 0)$ as $t \to -\infty$. All other paths are nonrectilinear paths which do not approach $(0, 0)$ as $t \to +\infty$ or as $t \to -\infty$, but which become asymptotic to one or another of the four half-line paths as $t \to +\infty$ and as $t \to -\infty$. According to the description and definition of Section 13.1B, the critical point $(0, 0)$ is a *saddle point*. Clearly, it is *unstable*. A qualitative diagram of the paths appears in Figure 13.8.

Case 3

THEOREM 13.3

  Hypothesis. *The roots λ_1 and λ_2 of the characteristic equation (13.22) are real and equal.*

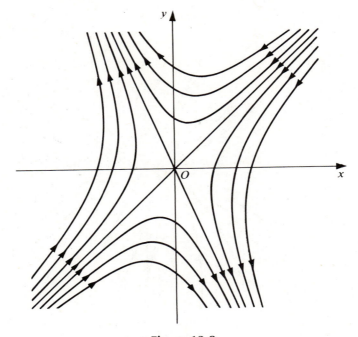

Figure 13.8

Conclusion.   *The critical point* $(0, 0)$ *of the linear system* (13.19) *is a* node.

Proof.   Let us first assume that $\lambda_1 = \lambda_2 = \lambda < 0$. We consider two subcases:

3(a) $a = d \neq 0, b = c = 0.$
3(b) All other possibilities leading to a double root.

We consider first the special case 3(a). The characteristic equation (13.22) becomes

$$\lambda^2 - 2a\lambda + a^2 = 0$$

and hence $\lambda = a = d$. The system (13.19) itself is thus simply

$$\frac{dx}{dt} = \lambda x,$$

$$\frac{dy}{dt} = \lambda y.$$

The general solution of this system is clearly

$$x = c_1 e^{\lambda t},$$
$$y = c_2 e^{\lambda t}, \tag{13.26}$$

where c_1 and c_2 are arbitrary constants. The paths defined by (13.26) for the various values of c_1 and c_2 are half-lines of all possible slopes. Since $\lambda < 0$, we see that each of these half-lines approaches and enters $(0, 0)$ as $t \to +\infty$. That is, all the paths of the system enter $(0, 0)$ as $t \to +\infty$. According to the definition of Section 13.1B, the critical point $(0, 0)$ is a *node*. Clearly, it is *asymptotically stable*. A qualitative diagram of the paths appears in Figure 13.9.

If $\lambda > 0$, the situation is the same except that the paths enter $(0, 0)$ as $t \to -\infty$, the node $(0, 0)$ is *unstable*, and the arrows in Figure 13.9 are all reversed.

We mention that this type of node is sometimes called a *star-shaped node*.

Let us now consider Case 3(b). Here the characteristic equation has the double root $\lambda < 0$, but we exclude the special circumstances of Case 3(a). By Theorem 7.9 of Chapter 7 the general solution of the system (13.19) may in this case be written

$$x = c_1 A e^{\lambda t} + c_2(A_1 t + A_2)e^{\lambda t},$$
$$y = c_1 B e^{\lambda t} + c_2(B_1 t + B_2)e^{\lambda t}, \qquad (13.27)$$

where the A's and B's are definite constants, c_1 and c_2 are arbitrary constants, and $B_1/A_1 = B/A$. Choosing $c_2 = 0$ in (13.27) we obtain solutions

$$x = c_1 A e^{\lambda t},$$
$$y = c_1 B e^{\lambda t}. \qquad (13.28)$$

For any $c_1 > 0$, the solutions (13.28) represent a path consisting of "half" of the line $Bx = Ay$ of slope B/A; for any $c_1 < 0$, they represent a path which consists of the "other half" of this line. Since $\lambda < 0$, both of these half-line paths approach $(0, 0)$ as $t \to +\infty$. Further, since $y/x = B/A$, both paths *enter* $(0, 0)$ with slope B/A.

If $c_2 \neq 0$, the general solution (13.27) represents nonrectilinear paths. Since $\lambda < 0$, we see from (13.27) that

$$\lim_{t \to +\infty} x = 0, \qquad \lim_{t \to +\infty} y = 0.$$

Thus the nonrectilinear paths defined by (13.27) all approach $(0, 0)$ as $t \to +\infty$. Also,

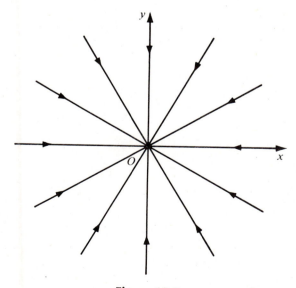

Figure 13.9

since

$$\frac{y}{x} = \frac{c_1 B e^{\lambda t} + c_2 (B_1 t + B_2) e^{\lambda t}}{c_1 A e^{\lambda t} + c_2 (A_1 t + A_2) e^{\lambda t}} = \frac{(c_1 B/c_2) + B_2 + B_1 t}{(c_1 A/c_2) + A_2 + A_1 t},$$

we see that

$$\lim_{t \to +\infty} \frac{y}{x} = \frac{B_1}{A_1} = \frac{B}{A}.$$

Thus all the nonrectilinear paths *enter* (0, 0) with limiting slope B/A.

Thus all the paths (both rectilinear and nonrectilinear) *enter* (0, 0) as $t \to +\infty$ with slope B/A. According to the definition of Section 13.1B, the critical point (0, 0) is a *node*. Clearly, it is *asymptotically stable*. A qualitative diagram of the paths appears in Figure 13.10.

If $\lambda > 0$, the situation is again the same except that the paths enter (0, 0) as $t \to -\infty$, the *node* (0, 0) is *unstable*, and the arrows in Figure 13.10 are reversed.

Case 4

THEOREM 13.4

Hypothesis. *The roots λ_1 and λ_2 of the characteristic equation (13.22) are conjugate complex with real part not zero (that is, not pure imaginary).*

Conclusion. *The critical point (0, 0) of the linear system (13.19) is a spiral point.*

Proof. Since λ_1 and λ_2 are conjugate complex with real part not zero, we may write these roots as $\alpha \pm i\beta$, where α and β are both real and unequal to zero. Then by

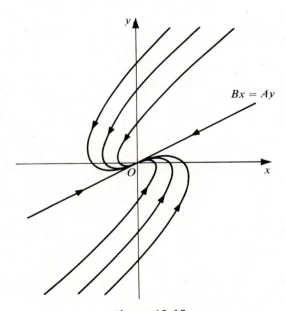

Figure 13.10

Theorem 7.8 of Chapter 7, the general solution of the system (13.19) may be written

$$x = e^{\alpha t}[c_1(A_1 \cos \beta t - A_2 \sin \beta t) + c_2(A_2 \cos \beta t + A_1 \sin \beta t)],$$
$$y = e^{\alpha t}[c_1(B_1 \cos \beta t - B_2 \sin \beta t) + c_2(B_2 \cos \beta t + B_1 \sin \beta t)], \qquad (13.29)$$

where A_1, A_2, B_1, and B_2 are definite real constants and c_1 and c_2 are arbitrary constants.

Let us first assume that $\alpha < 0$. Then from (13.29) we see that

$$\lim_{t \to +\infty} x = 0 \quad \text{and} \quad \lim_{t \to +\infty} y = 0$$

and hence all paths defined by (13.29) approach $(0, 0)$ as $t \to +\infty$. We may rewrite (13.29) in the form

$$x = e^{\alpha t}(c_3 \cos \beta t + c_4 \sin \beta t),$$
$$y = e^{\alpha t}(c_5 \cos \beta t + c_6 \sin \beta t), \qquad (13.30)$$

where $c_3 = c_1 A_1 + c_2 A_2$, $c_4 = c_2 A_1 - c_1 A_2$, $c_5 = c_1 B_1 + c_2 B_2$, and $c_6 = c_2 B_1 - c_1 B_2$. Assuming c_1 and c_2 are real, the solutions (13.30) represent all paths in the real xy phase plane. We may now put these solutions in the form

$$x = K_1 e^{\alpha t} \cos(\beta t + \phi_1),$$
$$y = K_2 e^{\alpha t} \cos(\beta t + \phi_2), \qquad (13.31)$$

where $K_1 = \sqrt{c_3^2 + c_4^2}$, $K_2 = \sqrt{c_5^2 + c_6^2}$ and ϕ_1 and ϕ_2 are defined by the equations

$$\cos \phi_1 = \frac{c_3}{K_1}, \qquad \cos \phi_2 = \frac{c_5}{K_2},$$

$$\sin \phi_1 = -\frac{c_4}{K_1}, \qquad \sin \phi_2 = -\frac{c_6}{K_2}.$$

Let us now consider

$$\frac{y}{x} = \frac{K_2 e^{\alpha t} \cos(\beta t + \phi_2)}{K_1 e^{\alpha t} \cos(\beta t + \phi_1)}. \qquad (13.32)$$

Letting $K = K_2/K_1$ and $\phi_3 = \phi_1 - \phi_2$, this becomes

$$\frac{y}{x} = \frac{K \cos(\beta t + \phi_1 - \phi_3)}{\cos(\beta t + \phi_1)}$$
$$= K\left[\frac{\cos(\beta t + \phi_1) \cos \phi_3 + \sin(\beta t + \phi_1) \sin \phi_3}{\cos(\beta t + \phi_1)}\right] \qquad (13.33)$$
$$= K[\cos \phi_3 + \sin \phi_3 \tan(\beta t + \phi_1)],$$

provided $\cos(\beta t + \phi_1) \neq 0$. As a result of the periodicity of the trigonometric functions involved in (13.32) and (13.33) we conclude from these expressions that $\lim_{t \to +\infty} \frac{y}{x}$ does not exist and so the paths *do not enter* $(0, 0)$. Instead, it follows from (13.32) and (13.33) that the paths approach $(0, 0)$ in a spiral-like manner, winding around $(0, 0)$ an infinite number of times as $t \to +\infty$. According to the definition of Section 13.1B, the critical

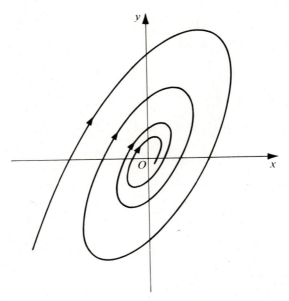

Figure 13.11

point $(0, 0)$ is a *spiral point*. Clearly, it is *asymptotically stable*. A qualitative diagram of the paths appears in Figure 13.11.

If $\alpha > 0$, the situation is the same except that the paths approach $(0, 0)$ as $t \to -\infty$, the *spiral point* $(0, 0)$ is *unstable*, and the arrows in Figure 13.11 are reversed.

Case 5

THEOREM 13.5

Hypothesis. *The roots λ_1 and λ_2 of the characteristic equation (13.22) are pure imaginary.*

Conclusion. *The critical point $(0, 0)$ of the linear system (13.19) is a center.*

Proof. Since λ_1 and λ_2 are pure imaginary we may write them as $\alpha \pm \beta i$, where $\alpha = 0$ but β is real and unequal to zero. Then the general solution of the system (13.19) is of the form (13.29), where $\alpha = 0$. In the notation of (13.31) all real solutions may be written in the form

$$x = K_1 \cos(\beta t + \phi_1),$$
$$y = K_2 \cos(\beta t + \phi_2),$$

(13.34)

where K_1, K_2, ϕ_1, and ϕ_2 are defined as before. The solutions (13.34) define the paths in the real xy phase plane. Since the trigonometric functions in (13.34) oscillate indefinitely between ± 1 as $t \to +\infty$ and as $t \to -\infty$, the paths do not approach $(0, 0)$ as $t \to +\infty$ or as $t \to -\infty$. Rather it is clear from (13.34) that x and y are periodic functions of t and

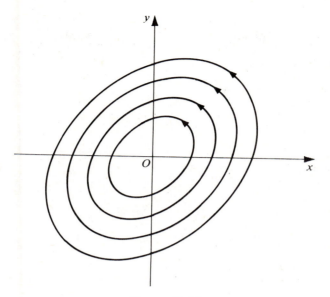

Figure 13.12

hence the paths are closed curves surrounding (0, 0), members of which are arbitrarily close to (0, 0). Indeed they are an infinite family of ellipses with center at (0, 0). According to the definition of Section 13.1B, the critical point (0, 0) is a *center*. Clearly, it is *stable*. However, since the paths do not approach (0, 0), the critical point is *not* asymptotically stable. A qualitative diagram of the paths appears in Figure 13.12.

We summarize our results in Table 13.1. The stability results of column 3 of this table lead at once to Theorem 13.6.

THEOREM 13.6

Consider the linear system

$$\frac{dx}{dt} = ax + by,$$

$$\frac{dy}{dt} = cx + dy,$$ (13.19)

where $ad - bc \neq 0$, *so that* (0, 0) *is the only critical point of the system.*

1. If both roots of the characteristic equation (13.22) are real and negative or conjugate complex with negative real parts, then the critical point (0, 0) *of (13.19) is asymptotically stable.*

2. If the roots of (13.22) are pure imaginary, then the critical point (0, 0) *of (13.19) is stable, but not asymptotically stable.*

3. If either of the roots of (13.22) is real and positive or if the roots are conjugate complex with positive real parts, then the critical point (0, 0) *of (13.19) is unstable.*

TABLE 13.1

Nature of roots λ_1 and λ_2 of characteristic equation $\lambda^2 - (a + d)\lambda + (ad - bc) = 0$	Nature of critical point $(0, 0)$ of linear system $\dfrac{dx}{dt} = ax + by,$ $\dfrac{dy}{dt} = cx + dy$	Stability of critical point $(0, 0)$
real, unequal, and of same sign	node	asymptotically stable if roots are negative; unstable if roots are positive
real, unequal, and of opposite sign	saddle point	unstable
real and equal	node	asymptotically stable if roots are negative; unstable if roots are positive
conjugate complex but not pure imaginary	spiral point	asymptotically stable if real part of roots is negative; unstable if real part of roots is positive
pure imaginary	center	stable, but not asymptotically stable

B. Examples and Applications

▶ **Example 13.2**

Determine the nature of the critical point $(0, 0)$ of the system

$$\frac{dx}{dt} = 2x - 7y,$$

$$\frac{dy}{dt} = 3x - 8y$$

(13.35)

and determine whether or not the point is stable.

Solution. The system (13.35) is of the form (13.19) where $a = 2, b = -7, c = 3$, and $d = -8$. The characteristic equation (13.22) is

$$\lambda^2 + 6\lambda + 5 = 0.$$

Hence the roots of the characteristic equation are $\lambda_1 = -5, \lambda_2 = -1$. Since the roots are real, unequal, and of the same sign (both negative), we conclude that the critical point $(0, 0)$ of (13.35) is a *node*. Since the roots are real and negative, the point is *asymptotically stable*.

▶ **Example 13.3**

Determine the nature of the critical point $(0, 0)$ of the system

$$\frac{dx}{dt} = 2x + 4y,$$

$$\frac{dy}{dt} = -2x + 6y, \tag{13.36}$$

and determine whether or not the point is stable.

Solution. Here $a = 2$, $b = 4$, $c = -2$, and $d = 6$. The characteristic equation is

$$\lambda^2 - 8\lambda + 20 = 0$$

and its roots are $4 \pm 2i$. Since these roots are conjugate complex but not pure imaginary, we conclude that the critical point $(0, 0)$ of (13.36) is a *spiral point*. Since the real part of the conjugate complex roots is positive, the point is *unstable*.

Application to Dynamics. Recall that in Chapter 5 we studied the free vibrations of a mass on a coil spring. The differential equation for the motion was

$$m \frac{d^2x}{dt^2} + \alpha \frac{dx}{dt} + kx = 0, \tag{13.37}$$

where $m > 0$, $\alpha \geq 0$, and $k > 0$ are constants denoting the mass, damping coefficient, and spring constant, respectively. We use α instead of a for the damping coefficient here to avoid confusion with the coefficient a of the system (13.19).

The dynamical equation (13.37) is equivalent to the system

$$\frac{dx}{dt} = y,$$

$$\frac{dy}{dt} = -\frac{k}{m} x - \frac{\alpha}{m} y. \tag{13.38}$$

The solutions of the system (13.38) define the paths (phase trajectories) associated with the dynamical equation (13.37) in the xy phase plane. From (13.38) the differential equation of these paths is

$$\frac{dy}{dx} = -\frac{kx + \alpha y}{my}.$$

We observe that $(0, 0)$ is the only critical point of the system (13.38). The auxiliary equation of the differential equation (13.37) is

$$mr^2 + \alpha r + k = 0, \tag{13.39}$$

while the characteristic equation of the system (13.38) is

$$\lambda^2 + \frac{\alpha}{m} \lambda + \frac{k}{m} = 0. \tag{13.40}$$

TABLE 13.2

Damping factor α	Nature of roots of auxiliary and characteristic equations	Form of solution of dynamical equation $\omega = \sqrt{\dfrac{k}{m}}, \beta = \dfrac{\alpha}{2m}$	Nature of critical point $(0, 0)$ in xy phase plane	Interpretation
$\alpha = 0$ (no damping)	pure imaginary	$x = c\cos(\omega t + \phi)$	(stable) center	oscillatory motion. Displacement and velocity are periodic functions of time
$\alpha < 2\sqrt{km}$ (under-damped)	conjugate complex with negative real parts	$x = ce^{-\beta t}\cos(\omega_1 t + \phi)$, where $\omega_1 = \sqrt{\omega^2 - \beta^2}.$	asymptotically stable spiral point	damped oscillatory motion. Displacement and velocity $\to 0$ through smaller and smaller oscillations
$\alpha = 2\sqrt{km}$ (critically damped)	real, equal, and negative	$x = (c_1 + c_2 t)e^{-\beta t}$	asymptotically stable node	displacement and velocity $\to 0$ without oscillating
$\alpha > 2\sqrt{km}$ (over-damped)	real, unequal, and negative	$x = c_1 e^{r_1 t} + c_2 e^{r_2 t}$, where $r_1 = -\beta + \sqrt{\beta^2 - \omega^2}$, $r_2 = -\beta - \sqrt{\beta^2 - \omega^2}$	asymptotically stable node	displacement and velocity $\to 0$ without oscillating

The two equations (13.39) and (13.40) clearly have the same roots λ_1 and λ_2. We summarize the possible situations in Table 13.2, which gives the form of the solution of the dynamical equation, the phase plane analysis of the critical point, and a brief interpretation.

Exercises

Determine the nature of the critical point $(0, 0)$ of each of the linear autonomous systems in Exercises 1–8. Also, determine whether or not the critical point is stable.

1. $\dfrac{dx}{dt} = x + 3y, \qquad \dfrac{dy}{dt} = 3x + y.$

2. $\dfrac{dx}{dt} = 3x + 2y, \qquad \dfrac{dy}{dt} = x + 2y.$

3. $\dfrac{dx}{dt} = 3x + 4y, \qquad \dfrac{dy}{dt} = 3x + 2y.$

4. $\dfrac{dx}{dt} = 2x + 5y, \qquad \dfrac{dy}{dt} = x - 2y.$

5. $\dfrac{dx}{dt} = 2x - 4y, \qquad \dfrac{dy}{dt} = 2x - 2y.$

6. $\dfrac{dx}{dt} = x - 2y, \qquad \dfrac{dy}{dt} = 4x + 5y.$

7. $\dfrac{dx}{dt} = x - y, \qquad \dfrac{dy}{dt} = x + 5y.$

8. $\dfrac{dx}{dt} = x + 7y, \qquad \dfrac{dy}{dt} = 3x + 5y.$

9. Consider the linear autonomous system

$$\frac{dx}{dt} = x + y,$$

$$\frac{dy}{dt} = 3x - y.$$

(a) Determine the nature of the critical point $(0, 0)$ of this system.

(b) Find the general solution of this system.

(c) Find the differential equation of the paths in the xy plane and obtain the general solution of this differential equation.

(d) Using the equation of the family of paths [the general solution obtained in part (c)], carefully plot the paths which pass through each of the following points: $(1, 1)$, $(-1, -1)$, $(1, -3)$, $(-1, 3)$, $(0, 2)$, $(0, -2)$, $(0, 4)$, $(0, -4)$, $(1, -1)$, $(-1, 1)$, $(2, -2)$, $(-2, 2)$. Does the configuration thus obtained agree with your answer to part (a)? Comment.

10. Consider the linear autonomous system

$$\frac{dx}{dt} = ax + by,$$

$$\frac{dy}{dt} = cx + dy, \qquad \text{(A)}$$

where a, b, c, and d are real constants.

(a) Show that if $a = d$, and b and c are of opposite sign, then the critical point $(0, 0)$ of system (A) is a spiral point.

(b) Show that if $a = d$, and either b or $c = 0$, then the critical point $(0, 0)$ of (A) is a node.

(c)   Show that if $a = d$, and b and c are of the same sign and such that $\sqrt{bc} < |a|$, then the critical point $(0, 0)$ of (A) is a node.

(d)   Show that if $a = d$, and b and c are of the same sign and such that $\sqrt{bc} > |a|$, then the critical point $(0, 0)$ of (A) is a saddle point.

(e)   Show that if $a = -d$ and $ad > bc$, then the critical point $(0, 0)$ of (A) is a center.

(f)   Show that if $a = -d$, and b and c are of the same sign, then the critical point $(0, 0)$ of (A) is a saddle point.

13.3   CRITICAL POINTS AND PATHS OF NONLINEAR SYSTEMS

A. Basic Theorems on Nonlinear Systems

We now consider the nonlinear real autonomous system

$$\frac{dx}{dt} = P(x, y),$$

$$\frac{dy}{dt} = Q(x, y). \tag{13.4}$$

We assume that the system (13.4) has an isolated critical point which we shall choose to be the origin $(0, 0)$. We now assume further that the functions P and Q in the right members of (13.4) are such that $P(x, y)$ and $Q(x, y)$ can be written in the form

$$P(x, y) = ax + by + P_1(x, y),$$

$$Q(x, y) = cx + dy + Q_1(x, y), \tag{13.41}$$

where

1. a, b, c, and d, are real constants, and

$$\begin{vmatrix} a & b \\ c & d \end{vmatrix} \neq 0,$$

and

2. P_1 and Q_1 have continuous first partial derivatives for all (x, y), and are such that

$$\lim_{(x,y)\to(0,0)} \frac{P_1(x, y)}{\sqrt{x^2 + y^2}} = \lim_{(x,y)\to(0,0)} \frac{Q_1(x, y)}{\sqrt{x^2 + y^2}} = 0. \tag{13.42}$$

Thus the system under consideration may be written in the form

$$\frac{dx}{dt} = ax + by + P_1(x, y),$$

$$\frac{dy}{dt} = cx + dy + Q_1(x, y), \tag{13.43}$$

where a, b, c, d, P_1, and Q_1 satisfy the requirements (1) and (2) above.

If $P(x, y)$ and $Q(x, y)$ in (13.4) can be expanded in power series about $(0, 0)$, the system (13.4) takes the form

$$\frac{dx}{dt} = \left[\frac{\partial P}{\partial x}\right]_{(0,0)} x + \left[\frac{\partial P}{\partial y}\right]_{(0,0)} y + a_{12}x^2 + a_{22}xy + a_{21}y^2 + \cdots,$$

$$\frac{dy}{dt} = \left[\frac{\partial Q}{\partial x}\right]_{(0,0)} x + \left[\frac{\partial Q}{\partial y}\right]_{(0,0)} y + b_{12}x^2 + b_{22}xy + b_{21}y^2 + \cdots.$$

(13.44)

This system is of the form (13.43), where $P_1(x, y)$ and $Q_1(x, y)$ are the terms of higher degree in the right members of the equations. The requirements (1) and (2) will be met, provided the Jacobian

$$\left.\frac{\partial(P, Q)}{\partial(x, y)}\right|_{(0,0)} \neq 0.$$

Observe that the constant terms are missing in the expansions in the right members of (13.44), since $P(0, 0) = Q(0, 0) = 0$.

▶ **Example 13.4**

The system

$$\frac{dx}{dt} = x + 2y + x^2,$$

$$\frac{dy}{dt} = -3x - 4y + 2y^2$$

is of the form (13.43) and satisfies the requirements (1) and (2). Here $a = 1$, $b = 2$, $c = -3$, $d = -4$, and

$$\begin{vmatrix} a & b \\ c & d \end{vmatrix} = 2 \neq 0.$$

Further $P_1(x, y) = x^2$, $Q_1(x, y) = 2y^2$, and hence

$$\lim_{(x,y)\to(0,0)} \frac{P_1(x, y)}{\sqrt{x^2 + y^2}} = \lim_{(x,y)\to(0,0)} \frac{x^2}{\sqrt{x^2 + y^2}} = 0$$

and

$$\lim_{(x,y)\to(0,0)} \frac{Q_1(x, y)}{\sqrt{x^2 + y^2}} = \lim_{(x,y)\to(0,0)} \frac{2y^2}{\sqrt{x^2 + y^2}} = 0.$$

By the requirement (2) the nonlinear terms $P_1(x, y)$ and $Q_1(x, y)$ in (13.43) tend to zero more rapidly than the linear terms $ax + by$ and $cx + dy$. Hence one would suspect that the behavior of the paths of the system (13.43) near $(0, 0)$ would be similar to that of the paths of the related linear system

$$\frac{dx}{dt} = ax + by,$$

$$\frac{dy}{dt} = cx + dy,$$

(13.19)

obtained from (13.43) by neglecting the nonlinear terms. In other words, it would seem that the nature of the critical point (0, 0) of the nonlinear system (13.43) should be similar to that of the linear system (13.19). In general this is actually the case. We now state without proof the main theorem regarding this relation.

THEOREM 13.7

Hypothesis.   *Consider the nonlinear system*

$$\frac{dx}{dt} = ax + by + P_1(x, y),$$

$$\frac{dy}{dt} = cx + dy + Q_1(x, y),$$

(13.43)

where a, b, c, d, P_1, and Q_1 satisfy the requirements (1) and (2) above. Consider also the corresponding linear system

$$\frac{dx}{dt} = ax + by,$$

$$\frac{dy}{dt} = cx + dy,$$

(13.19)

obtained from (13.43) by neglecting the nonlinear terms $P_1(x, y)$ and $Q_1(x, y)$. Both systems have an isolated critical point at (0, 0). Let λ_1 and λ_2 be the roots of the characteristic equation

$$\lambda^2 - (a + d)\lambda + (ad - bc) = 0$$

(13.22)

of the linear system (13.19).

Conclusions

1. The critical point (0, 0) of the nonlinear system (13.43) is of the same type as that of the linear system (13.19) in the following cases:

(i)   *If λ_1 and λ_2 are real, unequal, and of the same sign, then not only is (0, 0) a node of (13.19), but also (0, 0) is a node of (13.43).*

(ii)   *If λ_1 and λ_2 are real, unequal, and of opposite sign, then not only is (0, 0) a saddle point of (13.19), but also (0, 0) is a saddle point of (13.43).*

(iii)   *If λ_1 and λ_2 are real and equal and the system (13.19) is not such that $a = d \neq 0$, $b = c = 0$, then not only is (0, 0) a node of (13.19), but also (0, 0) is a node of (13.43).*

(iv)   *If λ_1 and λ_2 are conjugate complex with real part not zero, then not only is (0, 0) a spiral point of (13.19), but also (0, 0) is a spiral point of (13.43).*

2. The critical point (0, 0) of the nonlinear system (13.43) is not necessarily of the same type as that of the linear system (13.19) in the following cases:

(v)   *If λ_1 and λ_2 are real and equal and the system (13.19) is such that $a = d \neq 0$, $b = c = 0$, then although (0, 0) is a node of (13.19), the point (0, 0) may be either a node or a spiral point of (13.43).*

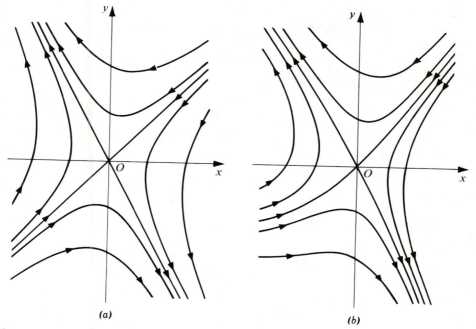

(a)

Figure 13.13a Linear system.

(b)

Figure 13.13b Nonlinear system.

(vi) *If λ_1 and λ_2 are pure imaginary, then although $(0, 0)$ is a center of (13.19), the point $(0, 0)$ may be either a center or a spiral point of (13.43).*

   Although the critical point $(0, 0)$ of the nonlinear system (13.43) is of the same type as that of the linear system (13.19) in cases (i), (ii), (iii), and (iv) of the conclusion, the actual appearance of the paths is somewhat different. For example, if $(0, 0)$ is a saddle point of the linear system (13.19), then we know that there are four half-line paths entering $(0, 0)$, two for $t \to +\infty$ and two for $t \to -\infty$. However, at the saddle point of the nonlinear system (13.43), in general we have four nonrectilinear curves entering $(0, 0)$, two for $t \to +\infty$ and two for $t \to -\infty$, in place of the half-line paths of the linear case (see Figure 13.13).

   Theorem 13.7 deals with the *type* of the critical point $(0, 0)$ of the nonlinear system (13.43). Concerning the *stability* of this point, we state without proof the following theorem of Liapunov.

THEOREM 13.8

Hypothesis.   *Exactly as in Theorem 13.7.*

Conclusions

   1. If both roots of the characteristic equation (13.22) of the linear system (13.19) are real and negative or conjugate complex with negative real parts, then not only is $(0, 0)$ an *asymptotically stable critical point of (13.19) but also* (0, 0) *is an asymptotically stable critical point of (13.43).*

2. *If the roots of (13.22) are pure imaginary, then although* (0, 0) *is a stable critical point of (13.19), it is not necessarily a stable critical point of (13.43). Indeed, the critical point* (0, 0) *of (13.43) may be asymptotically stable, stable but not asymptotically stable, or unstable.*

3. *If either of the roots of (13.22) is real and positive or if the roots are conjugate complex with positive real parts, then not only is* (0, 0) *an unstable critical point of (13.19), but also* (0, 0) *is an unstable critical point of (13.43).*

▶ **Example 13.5**

Consider the nonlinear system

$$\frac{dx}{dt} = x + 4y - x^2,$$

$$\frac{dy}{dt} = 6x - y + 2xy. \tag{13.45}$$

This is of the form (13.43), where $P_1(x, y) = -x^2$ and $Q_1(x, y) = 2xy$. We see at once that the hypotheses of Theorems 13.7 and 13.8 are satisfied. Hence to investigate the critical point (0, 0) of (13.45), we consider the linear system

$$\frac{dx}{dt} = x + 4y,$$

$$\frac{dy}{dt} = 6x - y, \tag{13.46}$$

of the form (13.19). The characteristic equation (13.22) of this system is

$$\lambda^2 - 25 = 0.$$

Hence the roots are $\lambda_1 = 5, \lambda_2 = -5$. Since the roots are real, unequal, and of opposite sign, we see from Conclusion (ii) of Theorem 13.7 that the critical point (0, 0) of the nonlinear system (13.45) is a *saddle point*. From Conclusion (3) of Theorem 13.8 we further conclude that this point is *unstable*.

Eliminating dt from Equations (13.45), we obtain the differential equation

$$\frac{dy}{dx} = \frac{6x - y + 2xy}{x + 4y - x^2}, \tag{13.47}$$

which gives the slope of the paths in the xy phase plane defined by the solutions of (13.45). The first-order equation (13.47) is exact. Its general solution is readily found to be

$$x^2y + 3x^2 - xy - 2y^2 + c = 0, \tag{13.48}$$

where c is an arbitrary constant. Equation (13.48) is the equation of the family of paths in the xy phase plane. Several of these are shown in Figure 13.14.

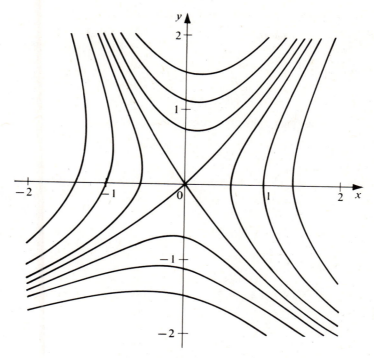

Figure 13.14

▶ **Example 13.6**

Consider the nonlinear system

$$\frac{dx}{dt} = \sin x - 4y,$$

$$\frac{dy}{dt} = \sin 2x - 5y. \tag{13.49}$$

Using the expansion

$$\sin x = x - \frac{x^3}{3!} + \frac{x^5}{5!} - \cdots,$$

we write this system in the form

$$\frac{dx}{dt} = x - 4y - \frac{x^3}{6} + \frac{x^5}{120} + \cdots,$$

$$\frac{dy}{dt} = 2x - 5y - \frac{4x^3}{3} + \frac{4x^5}{15} - \cdots. \tag{13.50}$$

The hypotheses of Theorems 13.7 and 13.8 are satisfied. Thus to investigate the critical point (0, 0) of (13.49) [or (13.50)], we consider the linear system

$$\frac{dx}{dt} = x - 4y,$$

$$\frac{dy}{dt} = 2x - 5y. \tag{13.51}$$

The characteristic equation of this system is

$$\lambda^2 + 4\lambda + 3 = 0.$$

Thus the roots are $\lambda_1 = -3$, $\lambda_2 = -1$. Since the roots are real, unequal, and of the same sign, we see from Conclusion (i) of Theorem 13.7 that the critical point (0,0) of the nonlinear system (13.49) is a *node*. From Conclusion (1) of Theorem 13.8 we conclude that this node is *asymptotically stable*.

▶ **Example 13.7**

In this example we shall find all the real critical points of the nonlinear system

$$\frac{dx}{dt} = 8x - y^2,$$

$$\frac{dy}{dt} = -6y + 6x^2, \tag{13.52}$$

and determine the type and stability of each of these critical points.

Clearly (0, 0) is one critical point of system (13.52). Also observe that (13.52) is of the form (13.43) and that the hypotheses of Theorems 13.7 and 13.8 are satisfied. To determine the type of critical point (0, 0), we consider the linear system

$$\frac{dx}{dt} = 8x,$$

$$\frac{dy}{dt} = -6y,$$

of the form (13.19). The characteristic equation of this linear system is

$$\lambda^2 - 2\lambda - 48 = 0,$$

and thus the roots are $\lambda_1 = 8$ and $\lambda_2 = -6$. Since the roots are real, unequal, and of opposite sign, we see from Conclusion (ii) of Theorem 13.7 that the critical point (0, 0) of the given nonlinear system (13.52) is a *saddle point*. From Conclusion (3) of Theorem 13.8 we conclude that this critical point is *unstable*.

We now proceed to find all other critical points of (13.52). By definition, the critical points of this system must simultaneously satisfy the system of algebraic equations

$$8x - y^2 = 0,$$

$$-6y + 6x^2 = 0. \tag{13.53}$$

From the second equation of this pair, $y = x^2$. Then, substituting this into the first equation of the pair, we obtain

$$8x - x^4 = 0,$$

which factors into

$$x(2 - x)(4 + 2x + x^2) = 0.$$

This equation has only two real roots, $x = 0$ and $x = 2$. These are the abscissas of the real critical points of (13.52); the corresponding ordinates are determined from $y = x^2$. Thus we obtain the two real critical points $(0, 0)$ and $(2, 4)$.

Since we have already considered the critical point $(0, 0)$ and found that it is an (unstable) saddle point of the given system (13.52), we now investigate the type and stability of the other critical point $(2, 4)$. To do this, we make the translation of coordinates

$$\xi = x - 2,$$
$$\eta = y - 4, \tag{13.54}$$

which transforms the critical point $x = 2$, $y = 4$ into the origin $\xi = \eta = 0$ in the $\xi\eta$ plane. We now transform the given system (13.52) into (ξ, η) coordinates. From (13.54), we have

$$x = \xi + 2, \qquad y = \eta + 4;$$

and substituting these into (13.52) and simplifying, we obtain

$$\frac{d\xi}{dt} = 8\xi - 8\eta - \eta^2,$$
$$\frac{d\eta}{dt} = 24\xi - 6\eta + 6\xi^2, \tag{13.55}$$

which is (13.52) in (ξ, η) coordinates. The system (13.55) is of the form (13.43) and the hypotheses of Theorems 13.7 and 13.8 are satisfied in these coordinates. To determine the type of the critical point $\xi = \eta = 0$ of (13.55), we consider the linear system

$$\frac{d\xi}{dt} = 8\xi - 8\eta,$$
$$\frac{d\eta}{dt} = 24\xi - 6\eta.$$

The characteristic equation of this linear system is

$$\lambda^2 - 2\lambda + 144 = 0.$$

The roots of this system are $1 \pm \sqrt{143}\,i$, which are conjugate complex with real part not zero. Thus by Conclusion (iv) of Theorem 13.7, the critical point $\xi = \eta = 0$ of the nonlinear system (13.55) is a *spiral point*. From Conclusion (3) of Theorem 13.8, we conclude that this critical point is *unstable*. But this critical point is the critical point $x = 2$, $y = 4$ of the given system (13.52). Thus the critical point $(2, 4)$ of the given system (13.52) is an unstable spiral point.

In conclusion, the given system (13.52) has two real critical points, namely:

1. critical point $(0, 0)$; a saddle point; unstable;
2. critical point $(2, 4)$; a spiral point; unstable.

▶ **Example 13.8**

Consider the two nonlinear systems

$$\frac{dx}{dt} = -y - x^2,$$

$$\frac{dy}{dt} = x,$$

(13.56)

and

$$\frac{dx}{dt} = -y - x^3,$$

$$\frac{dy}{dt} = x.$$

(13.57)

The point $(0, 0)$ is a critical point for each of these systems. The hypotheses of Theorem 13.7 are satisfied in each case, and in each case the corresponding linear system to be investigated is

$$\frac{dx}{dt} = -y,$$

$$\frac{dy}{dt} = x.$$

(13.58)

The characteristic equation of the system (13.58) is

$$\lambda^2 + 1 = 0$$

with the pure imaginary roots $\pm i$. Thus the critical point $(0, 0)$ of the linear system (13.58) is a center. However, Theorem 13.7 does not give us definite information concerning the nature of this point for either of the nonlinear systems (13.56) or (13.57). Conclusion (vi) of Theorem 13.7 tells us that in each case $(0, 0)$ is *either* a center *or* a spiral point; but this is all that this theorem tells us concerning the two systems under consideration.

Hurewicz (*Lectures on Ordinary Differential Equations,* p. 99) shows that the critical point $(0, 0)$ of the system (13.56) is a *center,* while the same point for the system (13.57) is a *spiral point.*

Suppose we consider a nonlinear real autonomous system of the form

$$\frac{dx}{dt} = P(x, y),$$

$$\frac{dy}{dt} = Q(x, y),$$

where no linear terms are present and P and Q are of the respective forms

$$P(x, y) = Ax^2 + Bxy + Cy^2$$

and

$$Q(x, y) = Dx^2 + Exy + Fy^2.$$

Then we find that the origin $(0, 0)$ will be one of several types of critical point which are of a more complicated nature than any of the simple types that we have previously encountered.

▶ **Example 13.9**

Two systems of the preceding type are given by

$$\frac{dx}{dt} = 2xy,$$

$$\frac{dy}{dt} = 3y^2 - x^2,$$

(13.59)

and

$$\frac{dx}{dt} = x^2,$$

$$\frac{dy}{dt} = 2y^2 - xy.$$

(13.60)

Qualitative portraits of the paths in the neighborhood of the critical point $(0, 0)$ for the systems (13.59) and (13.60) are shown in Figures 13.15 and 13.16, respectively.

The systems of this example and their corresponding phase plane portraits in the neighborhood of their common critical point $(0, 0)$ should be sufficient to suggest that a wide variety of critical points is possible for various types of more general nonlinear systems. Indeed, for the general nonlinear autonomous system

$$\frac{dx}{dt} = P(x, y),$$

$$\frac{dy}{dt} = Q(x, y),$$

(13.4)

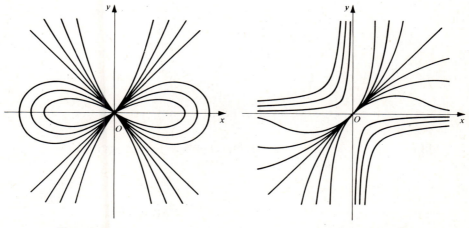

Figure 13.15   **Figure 13.16**

an infinite number of possible types of critical points exists, and, depending upon the nature of the nonlinear terms in (13.4), there are an infinite number of types which are incredibly complicated.

B. Nonlinear Conservative Systems

Let us consider a dynamical system in which the dissipation of energy is so slow that it may be neglected. Thus neglecting this slow dissipation, we assume the law of conservation of energy, namely, that the sum of the kinetic energy and the potential energy is a constant. In other words, the system is regarded as *conservative*.

Specifically, we consider a particle of mass m in rectilinear motion under the action of restoring force F which is a function of the displacement x only. The differential equation of motion is thus

$$m\frac{d^2x}{dt^2} = F(x), \tag{13.61}$$

where we assume that F is analytic for all values of x. The differential equation (13.61) is equivalent to the nonlinear autonomous system

$$\frac{dx}{dt} = y,$$

$$\frac{dy}{dt} = \frac{F(x)}{m}. \tag{13.62}$$

Eliminating dt we obtain the differential equation of the paths defined by the solutions of (13.62) in the xy phase plane:

$$\frac{dy}{dx} = \frac{F(x)}{my}. \tag{13.63}$$

Separating variables in this equation, we obtain

$$my\,dy = F(x)\,dx.$$

Suppose $x = x_0$ and $y\,(=dx/dt) = y_0$ at $t = t_0$. Then, integrating, we find

$$\tfrac{1}{2}my^2 - \tfrac{1}{2}my_0^2 = \int_{x_0}^{x} F(x)\,dx$$

or

$$\tfrac{1}{2}my^2 - \int_{0}^{x} F(x)\,dx = \tfrac{1}{2}my_0^2 - \int_{0}^{x_0} F(x)\,dx. \tag{13.64}$$

But $\tfrac{1}{2}m(dx/dt)^2 = \tfrac{1}{2}my^2$ is the kinetic energy of the system and

$$V(x) = -\int_{0}^{x} F(x)\,dx$$

is the potential energy. Thus Equation (13.64) takes the form

$$\tfrac{1}{2}my^2 + V(x) = h, \tag{13.65}$$

where the constant

$$h = \tfrac{1}{2}my_0^2 - \int_0^{x_0} F(x)\,dx = \tfrac{1}{2}my_0^2 + V(x_0) \geq 0$$

is the total energy of the system.

Since Equation (13.65) was obtained by integrating (13.63), we see that (13.65) gives the family of paths in the xy phase plane. For a given value of h, the path given by (13.65) is a curve of constant energy in this plane. In other words, along a particular path the total energy of the system is a constant; this expresses the law of conservation of energy.

The critical points of the system (13.62) are the points with coordinates $(x_c, 0)$, where x_c are the roots of the equation $F(x) = 0$. These are equilibrium points of the given dynamical system. From the differential equation (13.63) we see that the paths cross the x axis at right angles and have horizontal tangents along the lines $x = x_c$. From Equation (13.65) we observe that the paths are symmetrical with respect to the x axis.

From the equation (13.65) of the paths we find at once

$$y = \pm\sqrt{\frac{2}{m}[h - V(x)]}. \tag{13.66}$$

This may be used to construct the paths in the following convenient way.

1. On a rectangular xY plane construct the graph of $Y = V(x)$ and the lines $Y = h$, for various values of h [see Figure 13.17a, where one such line is shown]. For each fixed value h this shows the difference $h - V(x)$ for each x.
2. Directly below the xY plane draw the xy phase plane, with the y axis on the same vertical line as the Y axis [see Figure 13.17b]. For each x multiply the difference $h - V(x)$ obtained from the graph in Step (1) by the constant factor $2/m$ and use (13.66) to plot the corresponding y value on the phase plane.

We observed that the critical points of the system (13.62) are the points $(x_c, 0)$ such that $F(x_c) = 0$. Now note that $F(x) = -V'(x)$, and recall that $V'(x_c) = 0$ implies that $V(x)$ has either a relative extremum or a horizontal inflection point at $x = x_c$. We thus conclude that all the critical points of (13.62) lie on the x axis and correspond to points where the potential energy $V(x)$ has a relative minimum, a relative maximum, or a horizontal inflection point. We now discuss each of these three possibilities, using the graphical technique just introduced.

1. Suppose $V(x)$ has a *relative minimum* at $x = x_c$ (see Figure 13.18).

Let $V(x_c) = h_0 \geq 0$, and consider the path corresponding to a typical value h_1 of h such that $h_1 > h_0$. For $h = h_1$, the values of y given by (13.66) are real for $a_1 \leq x \leq b_1$ and equal to zero at both $x = a_1$ and $x = b_1$. Hence, using the symmetry of the paths, we see that the path corresponding to $h = h_1$ is a closed ellipse-like curve surrounding the critical point $(x_c, 0)$. This is typical of all paths corresponding to all $h > h_0$ and sufficiently near to the critical point $(x_c, 0)$. Thus, we conclude that if the potential energy $V(x)$ has a relative minimum at $x = x_c$, then the critical point $(x_c, 0)$ is a *center* and so is *stable*.

2. Suppose $V(x)$ has a *relative maximum* at $x = x_c$ (see Figure 13.19).

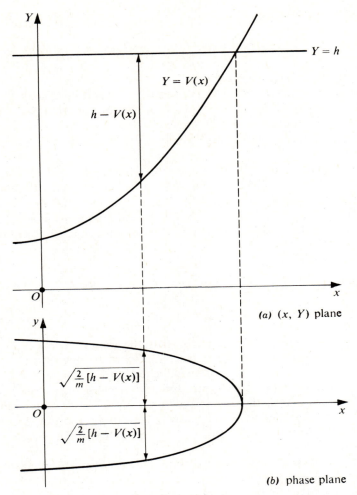

(a) (x, Y) plane

(b) phase plane

Figure 13.17

Let $V(x_c) = h_0 > 0$, and consider the paths corresponding to a typical value h_1 of h such that $h_1 < h_0$. For $h = h_1$, the values of y given by (13.66) are real both for $x \leq a_1$ (and sufficiently close to $x = a_1$) and for $x \geq b_1$ (and sufficiently close to $x = b_1$); the value of y is equal to zero at both $x = a_1$ and $x = b_1$. Thus for $x \leq a_1$ (and sufficiently close to $x = a_1$), the path corresponding to $h = h_1$ is a hyperbolic-like curve opening to the left of $x = a_1$, whereas, for $x \geq b_1$ (and sufficiently close to $x = b_1$) the path corresponding to $h = h_1$ is a hyperbolic-like curve opening to the right of $x = b_1$.

Now consider the paths corresponding to a typical value h_2 of h such that $h_2 > h_0$. For $h = h_2$, the values of y given by (13.66) are real for all x sufficiently close to x_c. Moreover, the positive value of y has a relative minimum (say m) corresponding to $x = x_c$, and the negative value has a relative maximum (which would then be $-m$) corresponding to $x = x_c$. Thus for all x sufficiently close to x_c, the paths corresponding to $h = h_2$ are a pair of hyperbolic-like curves, one opening upward above (x_c, m) and the other opening downward below $(x_c, -m)$.

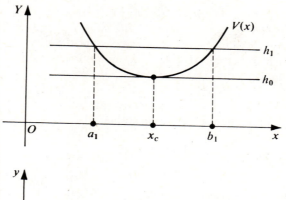

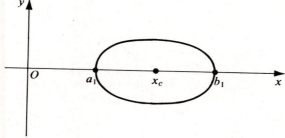

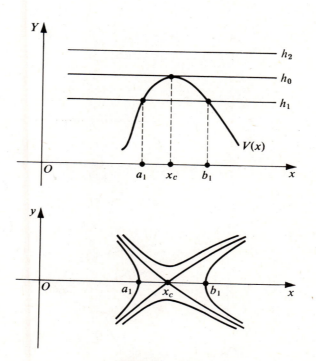

Figure 13.18

Figure 13.19

Finally, consider the paths corresponding to $h = h_0$ itself. For $h = h_0$, the value of y given by (13.66) for $x = x_c$ is zero. The values of y given by (13.66) for $x \neq x_c$ (and sufficiently close to $x = x_c$) are real and of opposite sign. Thus, there are two geometric curves which cross the x axis at $x = x_c$, and these two curves provide four paths which approach and enter the critical point $(x_c, 0)$, two as $t \to +\infty$ and two as $t \to -\infty$. From the discussion in this and the two preceding paragraphs, we conclude the following. If the potential energy $V(x)$ has a relative maximum at $x = x_c$, then the critical point $(x_c, 0)$ is a saddle point and so is unstable.

3.   Suppose $V(x)$ has a *horizontal inflection point* at $x = x_c$ (see Figure 13.20).

To be definite, suppose $V(x)$ is strictly increasing for all $x \neq x_c$ (and sufficiently close to x_c); and let $V(x_c) = h_0 \geq 0$. Then, proceeding as in the discussions of cases 1 and 2, we find that the paths (for x sufficiently close to x_c) have the appearance shown in the phase plane part of Figure 13.20. For typical $h \neq h_0$ (say $h_1 < h_0$ or $h_2 > h_0$), the paths merely have x-axis symmetry and intersect that axis at right angles. For $h = h_0$, the symmetry remains, but the path has a so-called *cusp* as it passes through the critical point $(x_c, 0)$ (see Figure 13.20 again).

The critical point $(x_c, 0)$ is a "degenerate" type of critical point and is unstable. The case in which $V(x)$ is strictly decreasing (instead of increasing) is analogous.

We have presented a graphical treatment of the situation. We point out that an analytic treatment, using principles of elementary calculus, could also be employed to obtain the same results. We summarize these results as the following theorem.

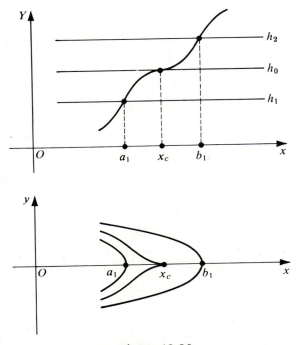

Figure 13.20

THEOREM 13.9

Let

$$m\frac{d^2x}{dt^2} = F(x), \tag{13.61}$$

where F is analytic for all values of x, be the differential equation of a conservative dynamical system. Consider the equivalent autonomous system

$$\frac{dx}{dt} = y,$$

$$\frac{dy}{dt} = \frac{F(x)}{m}, \tag{13.62}$$

and let $(x_c, 0)$ be a critical point of this system. Let V be the potential energy function of the dynamical system having differential equation (13.61); that is, V is the function defined by

$$V(x) = -\int_0^x F(x)\, dx.$$

Then we conclude the following:

1. If the potential energy function has a relative minimum at $x = x_c$, then the critical point $(x_c, 0)$ is a center and is stable.
2. If the potential energy function has a relative maximum at $x = x_c$, then the critical point $(x_c, 0)$ is a saddle point and is unstable.
3. If the potential energy function has a horizontal inflection point at $x = x_c$, then the critical point $(x_c, 0)$ is of a "degenerate" type called a cusp and is unstable.

▶ **Example 13.10**

Suppose $F(x) = 4x^3 - 4x$ and $m = 1$. Then the differential equation (13.61) of motion is

$$\frac{d^2x}{dt^2} = 4x^3 - 4x.$$

The equivalent nonlinear system (13.62) is

$$\frac{dx}{dt} = y,$$

$$\frac{dy}{dt} = 4x^3 - 4x. \tag{13.67}$$

The critical points of this system are $(0, 0)$, $(1, 0)$, and $(-1, 0)$. The differential equation (13.63) of the paths in the xy phase plane is

$$\frac{dy}{dx} = \frac{4x^3 - 4x}{y}.$$

Integrating this we obtain the family of paths

$$\tfrac{1}{2}y^2 - x^4 + 2x^2 = h. \tag{13.68}$$

This is of the form (13.65), where

$$V(x) = -\int_0^x F(x)\,dx = -x^4 + 2x^2 \tag{13.69}$$

is the potential energy.

We now construct the xy-phase-plane diagram by first constructing the curve $Y = V(x)$ and the lines $Y = h$ for various values of h in the xY plane [see Figure 13.21a]. From this xY plane we read off the values $h - V(x) = h + x^4 - 2x^2$ and plot the paths in the xy phase plane directly below [see Figure 13.21b] using

$$y = \pm\sqrt{2(h + x^4 - 2x^2)}.$$

From the formula

$$V(x) = -x^4 + 2x^2$$

for the potential energy, we find

$$V'(x) = -4x^3 + 4x = -4x(x + 1)(x - 1)$$

and

$$V''(x) = -12x^2 + 4.$$

Thus we see that $V'(x) = 0$ at $x = -1$, $x = 0$, and $x = 1$. We also observe that $V''(-1) < 0$, $V''(0) > 0$, $V''(1) < 0$. Thus the potential energy function V has a relative minimum at $x = 0$ and relative maxima at $x = -1$ and $x = 1$. Hence, by Theorem 13.9, the critical point $(0, 0)$ is a *center* and is stable and the critical points $(-1, 0)$ and $(1, 0)$ are *saddle points* and are unstable.

From Figure 13.21 we see that if the total energy h is less than 1 and x is initially between -1 and $+1$, then the corresponding paths are closed and hence the motion is periodic. If, however, $h > 1$, the paths are not closed and the motion is not periodic. The value $h = 1$ clearly separates the motions of one type (periodic) from those of another (nonperiodic). For this reason the corresponding path is called a *separatrix*. We observe the nature of the critical points from the phase plane diagram. The point $(0, 0)$ is clearly a *center*, while the points $(1, 0)$ and $(-1, 0)$ are *saddle points*.

▶ **Example 13.11**

Consider a simple pendulum which is composed of a mass m (the bob) at the end of a straight wire of negligible mass and length l. Suppose this pendulum is suspended from a fixed point S and is free to vibrate in a vertical plane. Let x denote the angle which the straight wire makes with the vertical ray from S downward at time t (see Section 5.2, Exercise 8, and Figure 5.4, where θ there has been replaced by x here). We neglect air resistance and assume that the only forces acting on the mass m are the tension in the wire and the force due to gravity. Then the differential equation of the displacement of the pendulum is

$$\frac{d^2x}{dt^2} = -\frac{g}{l}\sin x,$$

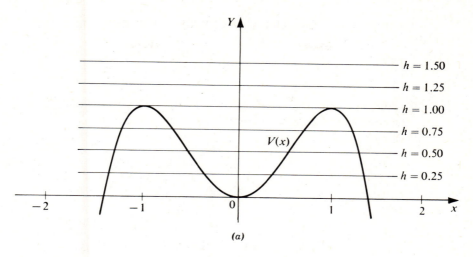

(a)

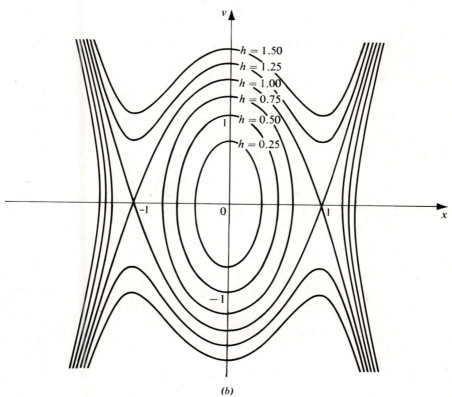

(b)

Figure 13.21

where g is the acceleration due to gravity. For simplicity, we let $g = l = 1$, and consider the special case

$$\frac{d^2x}{dt^2} = -\sin x.$$

This differential equation is of the form (13.61), where $m = 1$ and $F(x) = -\sin x$. The equivalent nonlinear system of the form (13.62) is

$$\frac{dx}{dt} = y,$$

$$\frac{dy}{dt} = -\sin x.$$

The critical points of this system are the points $(x_c, 0)$, where x_c are the roots of $\sin x = 0$. Thus the critical points are the infinite set of points

$$(n\pi, 0), \qquad \text{where } n = 0, \pm 1, \pm 2, \ldots.$$

The differential equation of the paths in the xy phase plane is

$$\frac{dy}{dx} = -\frac{\sin x}{y}.$$

Integrating, we obtain the family of paths

$$\tfrac{1}{2}y^2 = \cos x + c$$

or

$$\tfrac{1}{2}y^2 + (1 - \cos x) = h,$$

where $\tfrac{1}{2}y^2$ is the kinetic energy,

$$V(x) = -\int_0^x F(x)\, dx = \int_0^x \sin x\, dx = 1 - \cos x$$

is the potential energy, and $h = c + 1$ is the total energy of the system.

We may now construct the xy-phase-plane diagram by first constructing the curve $Y = V(x)$ and the lines $Y = h$ for various values of $h \geq 0$ in the xY plane [see Figure 13.22a]. From this xY plane, we read off the values $h - V(x) = h + \cos x - 1$ and plot the paths in the xy phase plane directly below [see Figure 13.22b] using

$$y = \pm\sqrt{2(h + \cos x - 1)}.$$

From the formula

$$V(x) = 1 - \cos x$$

for the potential energy, we find

$$V'(x) = \sin x \quad \text{and} \quad V''(x) = \cos x.$$

Thus we see that $V'(x) = 0$ at $x = n\pi$, where $n = 0, \pm 1, \pm 2, \ldots$. We also observe that $V''(n\pi) > 0$ if n is even and $V''(n\pi) < 0$ if n is odd. Thus the potential energy function V has a relative minimum at $x = 2n\pi$ for every even integer $2n$ $(n = 0, \pm 1, \pm 2, \ldots)$ and a

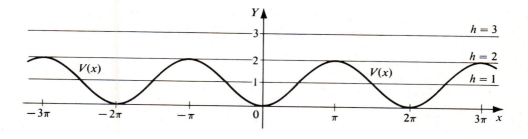

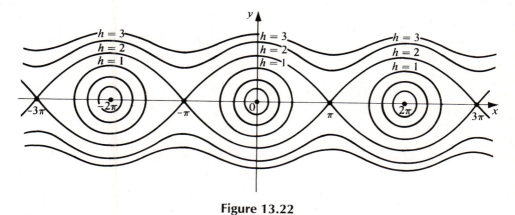

Figure 13.22

relative maximum at $x = (2n + 1)\pi$ for every odd integer $2n + 1$ ($n = 0, \pm 1, \pm 2, \ldots$). Hence, from Theorem 13.9 we conclude the following: The critical points

$$(2n\pi, 0), \qquad n = 0, \pm 1, \pm 2, \ldots, \qquad \text{are } centers$$

and are stable; and the critical points

$$[(2n + 1)\pi, 0], \qquad n = 0, \pm 1, \pm 2, \ldots, \qquad \text{are } saddle\ points$$

and are unstable.

From Figure 13.22 we see that if the total energy h is less than 2, then the corresponding paths are closed. Each of these paths surrounds one of the centers $(2n\pi, 0)$. Physically, all of these (stable) centers correspond to exactly one physical state, namely, the pendulum at rest with the bob in its lowest position (the stable equilibrium position). Thus, each closed path about a center corresponds to a periodic back-and-forth oscillatory motion of the pendulum bob about its lowest position.

On the other hand, if the total energy h is greater than 2, then the corresponding paths are not closed. Clearly $x \to +\infty$ as $t \to +\infty$ if $y = dx/dt > 0$, and $x \to -\infty$ as $t \to +\infty$ if $y = dx/dt < 0$. Thus the motion corresponding to such a path does not define x as a periodic function of t. Nevertheless, physically, the corresponding motion *is* periodic. To see this, we note the following: All of the (unstable) saddle points $[(2n + 1)\pi, 0]$ correspond to exactly one physical state, namely, the pendulum at rest with the bob in its highest position (the unstable equilibrium position); and each of the nonclosed paths corresponds to a physically periodic round-and-round motion of the pendulum about its point of support.

The total energy value $h = 2$ separates the back-and-forth type of motion from the round-and-round type. The paths corresponding to $h = 2$ are the ones which enter the saddle points as $t \to +\infty$ or $t \to -\infty$, and are the *separatrices*.

Dependence on a Parameter. We briefly consider the differential equation of a conservative dynamical system in which the force F depends not only on the displacement x but also on a parameter λ. Specifically, we consider a differential equation of the form

$$\frac{d^2x}{dt^2} = F(x, \lambda),\tag{13.70}$$

where F is analytic for all values of x and λ. The differential equation (13.70) is equivalent to the nonlinear autonomous system

$$\frac{dx}{dt} = y,$$

$$\frac{dy}{dt} = F(x, \lambda).\tag{13.71}$$

For each *fixed* value of the parameter λ, the critical points of (13.71) are the points with coordinates $(x_c, 0)$, where the abscissas x_c are the roots of the equation $F(x, \lambda) = 0$, considered as an equation in the unknown x. In general, as λ varies continuously through a given range of values, the corresponding x_c vary and hence so do the corresponding critical points, paths, and solutions of (13.71). A value of the parameter λ at which two or more critical points coalesce into less than their previous number (or, vice versa, where one or more split up into more than their previous number) is called a *critical value* (or *bifurcation value*) of the parameter. At such a value the nature of the corresponding paths changes abruptly. In determining both the critical values of the parameter and the critical points of the system (13.71), it is often very useful to investigate the graph of the relation $F(x, \lambda) = 0$ in the $x\lambda$ plane.

▶ **Example 13.12**

Consider the differential equation

$$\frac{d^2x}{dt^2} = x^2 - 4x + \lambda\tag{13.72}$$

of the form (13.70), where

$$F(x, \lambda) = x^2 - 4x + \lambda\tag{13.73}$$

and λ is a parameter. The equivalent nonlinear system of the form (13.71) is

$$\frac{dx}{dt} = y,$$

$$\frac{dy}{dt} = x^2 - 4x + \lambda.\tag{13.74}$$

The critical points of this system are the points $(x_1, 0)$ and $(x_2, 0)$, where x_1 and x_2 are the roots of the quadratic equation $F(x, \lambda) = 0$; that is,

$$x^2 - 4x + \lambda = 0, \tag{13.75}$$

in the unknown x. We find

$$x = \frac{4 \pm \sqrt{16 - 4\lambda}}{2} = 2 \pm \sqrt{4 - \lambda}.$$

Thus the critical points of (13.74) are

$$\left(2 + \sqrt{4 - \lambda}, 0\right) \quad \text{and} \quad \left(2 - \sqrt{4 - \lambda}, 0\right). \tag{13.76}$$

For $\lambda < 4$, the roots $2 \pm \sqrt{4 - \lambda}$ of the quadratic equation (13.75) are real and distinct; for $\lambda = 4$, the roots are real and equal, the common value being 2; and for $\lambda > 4$, they are conjugate complex. Thus for $\lambda < 4$, the critical points (13.76) are real and distinct. As $\lambda \to 4-$, the two critical points approach each other; and at $\lambda = 4$, they coalesce into the one single critical point $(2, 0)$. For $\lambda > 4$, there simply are no real critical points. Thus we see that $\lambda = 4$ is the *critical value* of the parameter.

We consider the three cases introduced in the preceding paragraph.

1. $\lambda < 4$.   For each fixed value λ_0 of λ such that $\lambda_0 < 4$, the critical points of (13.74) are the real distinct points with coordinates

$$\left(2 + \sqrt{4 - \lambda_0}, 0\right) \quad \text{and} \quad \left(2 - \sqrt{4 - \lambda_0}, 0\right).$$

For each such λ_0, the potential energy function V is the function of x only defined by

$$V(x, \lambda_0) = -\int_0^x F(x, \lambda_0)\, dx = -\int_0^x (x^2 - 4x + \lambda_0)\, dx = -\frac{x^3}{3} + 2x^2 - \lambda_0 x.$$

Then

$$V'(x, \lambda_0) = -F(x, \lambda_0) = -x^2 + 4x - \lambda_0 \quad \text{and} \quad V''(x, \lambda_0) = -2x + 4.$$

Let $x = 2 - \sqrt{4 - \lambda_0}$. Then

$$V'(2 - \sqrt{4 - \lambda_0}, \lambda_0) = 0 \quad \text{and} \quad V''(2 - \sqrt{4 - \lambda_0}, \lambda_0) > 0.$$

Thus the potential energy $V(x, \lambda_0)$ has a relative minimum at $x = 2 - \sqrt{4 - \lambda_0}$, and so the critical point $(2 - \sqrt{4 - \lambda_0}, 0)$ is a *center* and is stable. In like manner, one finds that the critical point $(2 + \sqrt{4 - \lambda_0}, 0)$ is a *saddle point* and is unstable. As a particular illustration of Case 1, we choose $\lambda_0 = 3 < 4$. Then the two critical points are $(1, 0)$, the center, and $(3, 0)$, the saddle point. A qualitative portrait of the critical points and paths in the xy phase plane for this value $\lambda_0 = 3$ is shown in Figure 13.23.

2. $\lambda = 4$.   As already noted, this is the critical value of the parameter and there is only the one critical point $(2, 0)$ of the system (13.74) corresponding to it. With $x = 2$ and $\lambda_0 = 4$, we quickly find that $V'(2, 4) = 0$ and $V''(2, 4) = 0$. However, it is easy to see that $V'''(x, \lambda_0) = -2$ for all x and λ_0, and so $V'''(2, 4) \neq 0$. Thus the potential energy has a horizontal inflection point at $x = 2$, and so the critical point $(2, 0)$ is a so-called *cusp* and is unstable. A qualitative portrait of the critical point and paths in the xy phase plane is shown in Figure 13.24.

3. $\lambda > 4$.   For each fixed value λ_0 of λ such that $\lambda_0 > 4$, there are no real critical points of the system (13.74). Of course, there are still paths in the xy phase plane and a qualitative portrait of them for $\lambda_0 = 5 > 4$ is shown in Figure 13.25.

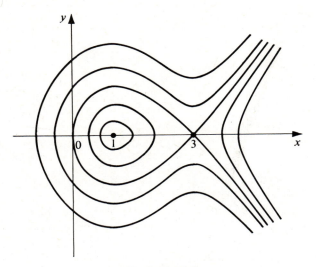

Figure 13.23

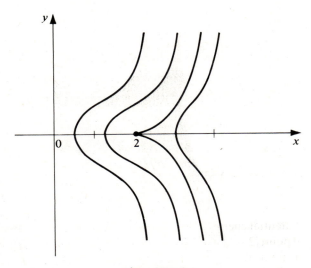

Figure 13.24

   In the general discussion preceding this example, we pointed out that the graph of the relation $F(x, \lambda) = 0$ in the $x\lambda$ plane is often useful in determining both the critical values of the parameter λ and the critical points of the given nonlinear system (13.71). We illustrate this for the particular system (13.74) of this example. From (13.73) we again observe that $F(x, \lambda) = x^2 - 4x + \lambda$ in this example and hence that $F(x, \lambda) = 0$ is $x^2 - 4x + \lambda = 0$. Writing this relation as

$$\lambda = -x^2 + 4x, \tag{13.77}$$

we see at once that its graph in the $x\lambda$ plane is a parabola with vertex at the point (2, 4), axis along the line $x = 2$, and which opens downward (see Figure 13.26).

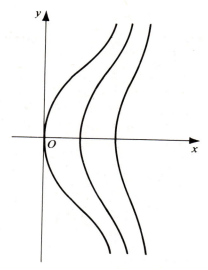

Figure 13.25

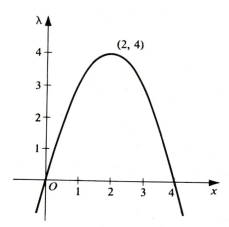

Figure 13.26

For each fixed value of λ, the critical points of system (13.74) are the points in the xy phase plane having coordinates $(x_c, 0)$, where x_c is a root of $\lambda = -x^2 + 4x$, considered as an equation in x. From the graph of $\lambda = -x^2 + 4x$, we at once observe the following results: For each fixed $\lambda_0 < 4$, there exist two real critical points $(x_c, 0)$ of the system (13.74). The abscissas x_c of these two points are the abscissas of the two points in the $x\lambda$ plane where the horizontal line $\lambda = \lambda_0$ intersects the parabola $\lambda = -x^2 + 4x$. For $\lambda_0 = 4$, there exists only one real critical point $(x_c, 0)$ of the system (13.74). The abscissa x_c of this point is the abscissa 2 of the single point where the horizontal line $\lambda_0 = 4$ is tangent to the parabola $\lambda = -x^2 + 4x$ at its vertex. Finally, for $\lambda_0 > 4$, there exist no real critical points $(x_c, 0)$ of the system (13.74), since the horizontal line $\lambda = \lambda_0$ lies above the vertex of the parabola $\lambda = -x^2 + 4x$ and so never intersects the parabola.

C. Liapunov's Direct Method

In Section 13.3B we observed the following concerning an equilibrium point of a conservative dynamical system: If the potential energy has a relative minimum at the equilibrium point, then the equilibrium point is stable; otherwise, it is unstable. This principle was generalized by the Russian mathematician Liapunov to obtain a method for studying the stability of more general autonomous systems. The procedure is known as *Liapunov's direct* (or *second*) *method*.

Consider the nonlinear autonomous system

$$\frac{dx}{dt} = P(x, y),$$

$$\frac{dy}{dt} = Q(x, y). \tag{13.4}$$

Assume that this system has an isolated critical point at the origin $(0, 0)$ and that P and Q have continuous first partial derivatives for all (x, y).

DEFINITIONS

Let $E(x, y)$ have continuous first partial derivatives at all points (x, y) in a domain D containing the origin $(0, 0)$.

1. The function E is called positive definite *in D if $E(0, 0) = 0$ and $E(x, y) > 0$ for all other points (x, y) in D.*

2. The function E is called positive semidefinite *in D if $E(0, 0) = 0$ and $E(x, y) \geq 0$ for all other points (x, y) in D.*

3. The function E is called negative definite *in D if $E(0, 0) = 0$ and $E(x, y) < 0$ for all other points in D.*

4. The function E is called negative semidefinite *in D if $E(0, 0) = 0$ and $E(x, y) \leq 0$ for all other points (x, y) in D.*

▶ **Example 13.13**

The function E defined by $E(x, y) = x^2 + y^2$ is positive definite in every domain D containing $(0, 0)$. Clearly, $E(0, 0) = 0$ and $E(x, y) > 0$ for all $(x, y) \neq (0, 0)$.

The function E defined by $E(x, y) = x^2$ is positive semidefinite in every domain D containing $(0, 0)$. Note that $E(0, 0) = 0$, $E(0, y) = 0$ for all $(0, y)$ such that $y \neq 0$ in D, and $E(x, y) > 0$ for all (x, y) such that $x \neq 0$ in D. There are no other points in D, and so we see that $E(0, 0) = 0$ and $E(x, y) \geq 0$ for all other points in D.

In like manner, we see that the function E defined by $E(x, y) = -x^2 - y^2$ is negative definite in D and that defined by $E(x, y) = -x^2$ is negative semidefinite in D.

DEFINITION

Let $E(x, y)$ have continuous first partial derivatives at all points (x, y) in a domain D containing the origin $(0, 0)$. The derivative of E with respect to the system (13.4) *is the* *function $\dot{E}$ defined by*

$$\dot{E}(x, y) = \frac{\partial E(x, y)}{\partial x} P(x, y) + \frac{\partial E(x, y)}{\partial y} Q(x, y). \tag{13.78}$$

▶ **Example 13.14**

Consider the system

$$\frac{dx}{dt} = -x + y^2,$$

$$\frac{dy}{dt} = -y + x^2, \tag{13.79}$$

and the function E defined by

$$E(x, y) = x^2 + y^2. \tag{13.80}$$

For the system (13.79), $P(x, y) = -x + y^2$, $Q(x, y) = -y + x^2$; and for the function E defined by (13.80),

$$\frac{\partial E(x, y)}{\partial x} = 2x, \qquad \frac{\partial E(x, y)}{\partial y} = 2y.$$

Thus the derivative of E defined by (13.80) with respect to the system (13.79) is given by

$$\dot{E}(x, y) = 2x(-x + y^2) + 2y(-y + x^2) = -2(x^2 + y^2) + 2(x^2 y + xy^2). \tag{13.81}$$

Now let C be a path of system (13.4); let $x = f(t), y = g(t)$ be an arbitrary solution of (13.4) defining C parametrically; and let $E(x, y)$ have continuous first partial derivatives for all (x, y) in a domain containing C. Then E is a composite function of t along C; and using the chain rule, we find that the derivative of E with respect to t along C is

$$\frac{dE[f(t), g(t)]}{dt} = E_x[f(t), g(t)] \frac{df(t)}{dt} + E_y[f(t), g(t)] \frac{dg(t)}{dt}$$

$$= E_x[f(t), g(t)] P[f(t), g(t)] + E_y[f(t), g(t)] Q[f(t), g(t)]$$

$$= \dot{E}[f(t), g(t)]. \tag{13.82}$$

Thus we see that the derivative of $E[f(t), g(t)]$ with respect to t along the path C is equal to the derivative of E with respect to the system (13.4) evaluated at $x = f(t)$, $y = g(t)$.

DEFINITION

Consider the system

$$\frac{dx}{dt} = P(x, y), \qquad \frac{dy}{dt} = Q(x, y). \tag{13.4}$$

Assume that this system has an isolated critical point at the origin $(0, 0)$ and that P and Q

have continuous first partial derivatives for all (x, y). *Let* $E(x, y)$ *be positive definite for all* (x, y) *in a domain D containing the origin and such that the derivative* $\dot{E}(x, y)$ *of E with respect to the system* (13.4) *is negative semidefinite for all* $(x, y) \in D$. *Then E is called a Liapunov function for the system* (13.4) *in D.*

▶ **Example 13.15**

Consider the system

$$\frac{dx}{dt} = -x + y^2,$$

$$\frac{dy}{dt} = -y + x^2, \tag{13.79}$$

and the function E defined by

$$E(x, y) = x^2 + y^2, \tag{13.80}$$

introduced in Example 13.14. Obviously the system (13.79) satisfies all the requirements of the immediately preceding definition in every domain containing the critical point $(0, 0)$. Also, in Example 13.13 we observed that the function E defined by (13.80) is positive definite in every such domain. In Example 13.14, we found the derivative of E with respect to the system (13.79) is given by

$$\dot{E}(x, y) = -2(x^2 + y^2) + 2(x^2y + xy^2) \tag{13.81}$$

for all (x, y). If this is negative semidefinite for all (x, y) in some domain D containing $(0, 0)$, then E defined by (13.80) is a Liapunov function for the system (13.79).

Clearly $\dot{E}(0, 0) = 0$. Now observe the following: If $x < 1$ and $y \neq 0$, then $xy^2 < y^2$; if $y < 1$ and $x \neq 0$, then $x^2y < x^2$. Thus if $x < 1$, $y < 1$, and $(x, y) \neq (0, 0)$, $x^2y + xy^2 < x^2 + y^2$ and hence

$$-(x^2 + y^2) + (x^2y + xy^2) < 0.$$

Thus in every domain D containing $(0, 0)$ and such that $x < 1$ and $y < 1$, $\dot{E}(x, y)$ given by (13.81) is negative definite and hence negative semidefinite. Thus E defined by (13.80) is a Liapunov function for the system (13.79).

We now state and prove two theorems on the stability of the critical point $(0, 0)$ of system (13.4).

THEOREM 13.10

Consider the system

$$\frac{dx}{dt} = P(x, y),$$

$$\frac{dy}{dt} = Q(x, y). \tag{13.4}$$

Assume that this system has an isolated critical point at the origin $(0, 0)$ *and that P and Q have continuous first partial derivatives for all* (x, y). *If there exists a Liapunov function*

E for the system (13.4) in some domain D containing (0, 0), then the critical point (0, 0) of (13.4) is stable.

Proof. Let K_ϵ be a circle of radius $\epsilon > 0$ with center at the critical point $(0, 0)$, where $\epsilon > 0$ is small enough so that this circle K_ϵ lies entirely in the domain D (see Figure 13.27). From a theorem of real analysis, we know that a real-valued function which is continuous on a closed bounded set assumes both a maximum value and a minimum value on that set. Since the circle K_ϵ is a closed bounded set in the plane and E is continuous in D and hence on K_ϵ, the real analysis theorem referred to in the preceding sentence applies to E on K_ϵ and so, in particular, E assumes a minimum value on K_ϵ. Further, since E is also positive definite in D (why?), this minimum value must be positive. Thus E assumes a positive minimum m on the circle K_ϵ. Next observe that since E is continuous at $(0, 0)$ and $E(0, 0) = 0$, there exists a positive number δ satisfying $\delta < \epsilon$ such that $E(x, y) < m$ for all (x, y) within or on the circle K_δ of radius δ and center at $(0, 0)$. (Again see Figure 13.27.)

Now let C be any path of (13.4); let $x = f(t)$, $y = g(t)$ be an arbitrary solution of (13.4) defining C parametrically; and suppose C defined by $[f(t), g(t)]$ is at a point within the "inner" circle K_δ at $t = t_0$. Then

$$E[f(t_0), g(t_0)] < m.$$

Since $\dot{E}$ is negative semidefinite in D (why?), using (13.82), we have

$$\frac{dE[f(t), g(t)]}{dt} \le 0$$

for $[f(t), g(t)] \in D$. Thus $E[f(t), g(t)]$ is a nonincreasing function of t along C. Hence

$$E[f(t), g(t)] \le E[f(t_0), g(t_0)] < m$$

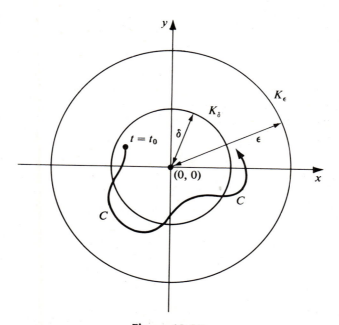

Figure 13.27

for all $t > t_0$. Since $E[f(t), g(t)]$ would have to be $\geq m$ on the "cuter" circle K_ϵ, we see that the path C defined by $x = f(t)$, $y = g(t)$ must remain within K_ϵ for all $t > t_0$. Thus, from the definition of stability of the critical point $(0, 0)$, we see that the critical point $(0, 0)$ of (13.4) is stable.   *Q.E.D.*

THEOREM 13.11

Consider the system

$$\frac{dx}{dt} = P(x, y),$$

$$\frac{dy}{dt} = Q(x, y). \tag{13.4}$$

Assume that this system has an isolated critical point at the origin $(0, 0)$ and that P and Q have continuous first partial derivatives for all (x, y). If there exists a Liapunov function E for the system (13.4) in some domain D containing $(0,0)$ such that E also has the property that $\dot{E}$ defined by (13.78) is negative definite in D, then the critical point $(0, 0)$ of (13.4) is asymptotically stable.

Proof. As in the proof of the previous theorem, let K_ϵ be a circle of radius $\epsilon > 0$ with center at the critical point $(0, 0)$ and lying entirely in D. Also, let C be any path of (13.4); let $x = f(t)$, $y = g(t)$ be an arbitrary solution of (13.4) defining C parametrically; and suppose C defined by $[f(t), g(t)]$ is at a point within K_ϵ at $t = t_0$ (see Figure 13.28).

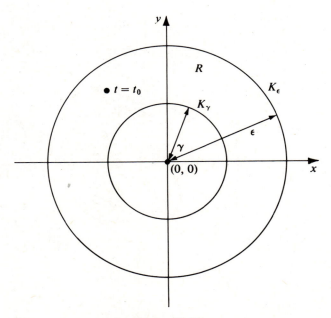

Figure 13.28

Now since $\dot{E}$ is negative definite in D, using (13.82), we have

$$\frac{dE[f(t), g(t)]}{dt} < 0$$

for $[f(t), g(t)] \in D$. Thus $E[f(t), g(t)]$ is a strictly decreasing function of t along C. Since E is positive definite in D, $E[f(t), g(t)] \geq 0$ for $[f(t), g(t)] \in D$. Thus $\lim_{t \to \infty} E[f(t), g(t)]$ exists and is some number $L \geq 0$. We shall show that $L = 0$.

On the contrary, assume that $L > 0$. Since E is positive definite, there exists a positive number γ satisfying $\gamma < \epsilon$ such that $E(x, y) < L$ for all (x, y) within the circle K_γ of radius γ and center at $(0, 0)$ (again see Figure 13.28). Now, we can apply the same real analysis theorem on maximum and minimum values that we used in the proof of the preceding theorem to the continuous function $\dot{E}$ on the closed region R *between and on* the two circles K_ϵ and K_γ. Doing so, since $\dot{E}$ is negative definite in D and hence in this region R which does not include $(0, 0)$, we see that $\dot{E}$ assumes a *negative* maximum $-k$ on R. Since $E[f(t), g(t)]$ is a strictly decreasing function of t along C and

$$\lim_{t \to \infty} E[f(t), g(t)] = L,$$

the path C defined by $x = f(t)$, $y = g(t)$ cannot enter the domain within K_γ for *any* $t > t_0$ and so remains in R for *all* $t \geq t_0$. Thus we have $\dot{E}[f(t), g(t)] \leq -k$ for all $t \geq t_0$. Then by (13.82) we have

$$\frac{dE[f(t), g(t)]}{dt} = \dot{E}[f(t), g(t)] \leq -k \qquad (13.83)$$

for all $t \geq t_0$. Now consider the identity

$$E[f(t), g(t)] - E[f(t_0), g(t_0)] = \int_{t_0}^{t} \frac{dE[f(t), g(t)]}{dt} \, dt. \qquad (13.84)$$

Then (13.83) gives

$$E[f(t), g(t)] - E[f(t_0), g(t_0)] \leq - \int_{t_0}^{t} k \, dt$$

and hence

$$E[f(t), g(t)] \leq E[f(t_0), g(t_0)] - k(t - t_0)$$

for all $t \geq t_0$. Now let $t \to \infty$. Since $-k < 0$, this gives

$$\lim_{t \to \infty} E[f(t), g(t)] = -\infty.$$

But this contradicts the hypothesis that E is positive definite in D and the assumption that

$$\lim_{t \to \infty} E[f(t), g(t)] = L > 0.$$

Thus $L = 0$; that is,

$$\lim_{t \to \infty} E[f(t), g(t)] = 0.$$

Since E is positive definite in D, $E(x, y) = 0$ if and only if $(x, y) = (0, 0)$. Thus,

$$\lim_{t \to \infty} E[f(t), g(t)] = 0$$

if and only if

$$\lim_{t \to \infty} f(t) = 0 \quad \text{and} \quad \lim_{t \to \infty} g(t) = 0.$$

But, from the definition of asymptotic stability of the critical point $(0, 0)$, we see that the critical point $(0, 0)$ of (13.4) is asymptotically stable.

▶ **Example 13.16**

Consider the system

$$\frac{dx}{dt} = -x + y^2,$$

$$\frac{dy}{dt} = -y + x^2, \tag{13.79}$$

and the function E defined by

$$E(x, y) = x^2 + y^2, \tag{13.80}$$

previously studied in Examples 13.14 and 13.15. Before that, in Example 13.13, we noticed that the function E defined by (13.80) is positive definite in every domain containing $(0, 0)$. In Example 13.14, we found the derivative of E with respect to the system (13.79) is given by

$$\dot{E}(x, y) = -2(x^2 + y^2) + 2(x^2 y + xy^2). \tag{13.81}$$

Then, in Example 13.15, we found that $\dot{E}$ defined by (13.81) is negative semidefinite in every domain containing $(0, 0)$ and hence that E defined by (13.80) is a Liapunov function for the system (13.79) in every such domain. Now, applying Theorem 13.10, we see that the critical point $(0, 0)$ of (13.79) is stable.

However, in Example 13.15 we actually showed that $\dot{E}$ defined by (13.81) is *negative definite* in every domain D containing $(0, 0)$. Thus by Theorem 13.11, we see that the critical point $(0, 0)$ of (13.79) is asymptotically stable.

Note that the asymptotic stability of the critical point $(0, 0)$ of system (13.79) could also be determined by applying Theorem 13.8.

Liapunov's direct method is indeed "direct" in the sense that it does not require any previous knowledge about the solutions of the system (13.4) or the type of its critical point $(0, 0)$. Instead, if one can construct a Liapunov function for (13.4), then one can "directly" obtain information about the stability of the critical point $(0, 0)$. However, there is no general method for constructing a Liapunov function, although methods for doing so are available for certain classes of equations.

Exercises

Determine the type and stability of the critical point $(0, 0)$ of each of the nonlinear autonomous systems in Exercises 1–6.

1. $\dfrac{dx}{dt} = x + x^2 - 3xy, \qquad \dfrac{dy}{dt} = -2x + y + 3y^2.$

2. $\dfrac{dx}{dt} = x + 4y + 3x^4$, $\qquad \dfrac{dy}{dt} = 2x - 3y - y^2 + 2x^3$.

3. $\dfrac{dx}{dt} = x + y + x^2 y$, $\qquad \dfrac{dy}{dt} = 3x - y + 2xy^3$,

4. $\dfrac{dx}{dt} = y + \tan x$, $\qquad \dfrac{dy}{dt} = y \cos x$.

5. $\dfrac{dx}{dt} = (y + 1)^2 - \cos x$, $\qquad \dfrac{dy}{dt} = \sin(x + y)$.

6. $\dfrac{dx}{dt} = x - 2y + e^{xy} - \cos^2 x$, $\qquad \dfrac{dy}{dt} = \sin(x + y)$.

7. Consider the autonomous system

$$\frac{dx}{dt} = ye^x,$$

$$\frac{dy}{dt} = e^x - 1.$$

(a) Determine the type of the critical point $(0, 0)$.

(b) Obtain the differential equation of the paths and find its general solution.

(c) Carefully plot several of the paths found in step (b) to obtain a phase plane diagram which clearly exhibits the nature of the critical point $(0, 0)$.

(d) Obtain the differential equation of the paths of the corresponding reduced linear system and find its general solution.

(e) Proceed as in step (c) for the paths found in step (d).

(f) Compare the results of the two phase plane diagrams constructed in steps (c) and (e). Comment.

8. Consider the autonomous system

$$\frac{dx}{dt} = 10x + x^2,$$

$$\frac{dy}{dt} = 20y + x^3,$$

and proceed as in Exercise 7.

9. The differential equation

$$\frac{d^2x}{dt^2} = x^5 - 5x^4 + 5x^3 + 5x^2 - 6x$$

is of the form $\dfrac{d^2x}{dt^2} = F(x)$ and may be considered as the differential equation of a nonlinear conservative dynamical system.

(a) Obtain the equivalent nonlinear autonomous system and find its critical points.

(b)   Obtain the differential equation of the paths in the xy phase plane and show that its general solutions is $\frac{1}{2}y^2 + V(x) = h$, where

$$V(x) = -\int_0^x F(x)\, dx.$$

(c)   Carefully construct the graph of $Y = V(x)$ along with the lines $Y = h$ for several well-chosen values of h in an auxiliary xY plane. Then use this auxiliary diagram to sketch the paths $\frac{1}{2}y^2 + V(x) = h$ in the xy phase plane.

(d)   Use the phase plane diagram constructed in part (c) to determine the type of each of the critical points found in step (a). Also use the diagram to draw conclusions concerning the motion described by the second-order nonlinear equation of the problem.

10.   Proceed as in Exercise 9 for the differential equation

$$\frac{d^2x}{dt^2} = \sin x\,(\cos x - 1).$$

11.   Proceed as in Exercise 9 for the differential equation

$$\frac{d^2x}{dt^2} = x^5 - 3x^4 + x^3 + 3x^2 - 2x.$$

12.   Proceed as in Exercise 9 for the differential equation

$$\frac{d^2x}{dt^2} = \sin x\,(2 \cos x - 1).$$

13.   Consider the nonlinear system

$$\frac{dx}{dt} = 6x - y + x^2,$$

$$\frac{dy}{dt} = \alpha x + 2y + y^2,$$

which depends on a parameter α. Assuming that $\alpha \neq -12$, determine the type of the critical point $(0, 0)$ of the system as a function of α.

14.   For the system

$$\frac{dx}{dt} = 5x - y,$$

$$\frac{dy}{dt} = 3x + y + 2x^2 - x^3,$$

(a)   find all critical points;

(b)   determine the type of each critical point (α, β) for which $\alpha \geq 0$.

15.   Find the real critical points of each of the following systems, and determine the type and stability of each critical point found.

(a) $\dfrac{dx}{dt} = x + 4y,$ $\dfrac{dx}{dt} = x + y - x^2.$

(b) $\dfrac{dx}{dt} = x - y + x^2, \qquad \dfrac{dy}{dt} = 12x - 6y + xy.$

16. Find the real critical points of each of the following systems, and determine the type and stability of each critical point found.

(a) $\dfrac{dx}{dt} = y - x^2, \qquad \dfrac{dy}{dt} = 8x - y^2.$

(b) $\dfrac{dx}{dt} = 2x - y + x^2, \qquad \dfrac{dy}{dt} = -12x + y + y^2.$

17. Consider the differential equation

$$\frac{d^2x}{dt^2} = x^2 - 6x + \lambda,$$

where λ is a real parameter independent of x.
(a) Find the critical value λ_0 of the parameter λ.
(b) Let $\lambda = 5$. Find the critical points of the related autonomous system, make a qualitative sketch of the paths in the xy phase plane, and determine the type of each critical point.
(c) Let $\lambda = \lambda_0$ [the critical value determined in part (a)]. Proceed as in part (b).

18. Consider the differential equation

$$\frac{d^2x}{dt^2} = x^2 - \lambda x + 9,$$

where λ is a real parameter independent of x.
(a) Find the critical values of the parameter λ.
(b) Let $\lambda = 10$. Find the critical points of the related autonomous system, make a qualitative sketch of the paths in the xy phase plane, and determine the type of each critical point.
(c) Let $\lambda = \lambda_0$ [the positive critical value determined in part (a)]. Proceed as in part (b).

19. For each of the following systems, construct a Liapunov function of the form $Ax^2 + By^2$, where A and B are constants, and use the function to determine whether the critical point $(0, 0)$ of the system is asymptotically stable or at least stable.

(a) $\dfrac{dx}{dt} = -x + 2x^2 + y^2, \qquad \dfrac{dy}{dt} = -y + xy.$

(b) $\dfrac{dx}{dt} = -x + y - x^3 - xy^2, \qquad \dfrac{dy}{dt} = -x - y - x^2y - y^3.$

(c) $\dfrac{dx}{dt} = -x - y - x^3, \qquad \dfrac{dy}{dt} = x - y - y^3.$

(d) $\dfrac{dx}{dt} = -3x + x^3 + 2xy^2, \qquad \dfrac{dy}{dt} = -2y + \tfrac{2}{3}y^3.$

13.4 LIMIT CYCLES AND PERIODIC SOLUTIONS

A. Limit Cycles

We have already encountered autonomous systems having closed paths. For example, the system (13.67) of Section 13.3B has a center at $(0, 0)$ and in the neighborhood of this center there is an infinite family of closed paths resembling ellipses (see Figure 13.21). In this example the closed paths about $(0, 0)$ form a continuous family in the sense that arbitrarily near to any one of the closed paths of this family there is always another closed path of the family. Now we shall consider systems having closed paths which are isolated in the sense that there are no other closed paths of the system arbitrarily near to a given closed path of the system.

What is the significance of a closed path? Given an autonomous system

$$\frac{dx}{dt} = P(x, y),$$

$$\frac{dy}{dt} = Q(x, y), \tag{13.4}$$

one is often most interested in determining the existence of *periodic* solutions of this system. It is easy to see that periodic solutions and closed paths of (13.4) are very closely related. For, in the first place, if $x = f_1(t)$, $y = g_1(t)$, where f_1 and g_1 are not both constant functions, is a periodic solution of (13.4), then the path which this solution defines is a *closed* path. On the other hand, let C be a closed path of (13.4) defined by a solution $x = f(t)$, $y = g(t)$ of (13.4), and suppose $f(t_0) = x_0$, $g(t_0) = y_0$. Since C is closed, there exists a value $t_1 = t_0 + T$, where $T > 0$, such that $f(t_1) = x_0, g(t_1) = y_0$. Now the pair $x = f(t + T)$, $y = g(t + T)$ is also a solution of (13.4). At $t = t_0$, this latter solution also assumes the values $x = x_0$, $y = y_0$. Thus by Chapter 10, Section 10.4A, Theorem 10.5, the two solutions $x = f(t)$, $y = g(t)$ and $x = f(t + T)$, $y = g(t + T)$ are identical for all t. In other words, $f(t + T) = f(t), g(t + T) = g(t)$ for all t, and so the solution $x = f(t)$, $y = g(t)$ defining the closed path C is a *periodic* solution. Thus, the search for periodic solutions of (13.4) falls back on the search for closed paths.

Now suppose the system (13.4) has a closed path C. Further, suppose (13.4) possesses a nonclosed path C_1 defined by a solution $x = f(t)$, $y = g(t)$ of (13.4) and having the following property: As a point R traces out C_1 according to the equations $x = f(t)$, $y = g(t)$, the path C_1 spirals and the distance between R and the nearest point on the closed path C approaches zero either as $t \to +\infty$ or as $t \to -\infty$. In other words, the nonclosed path C_1 spirals closer and closer around the closed path C either from the inside of C or from the outside either as $t \to +\infty$ or as $t \to -\infty$ (see Figure 13.29 where C_1 approaches C from the outside).

In such a case we call the closed path C a *limit cycle*, according to the following definition:

DEFINITION

A closed path C of the system (13.4) which is approached spirally from either the inside or the outside by a nonclosed path C_1 of (13.4) either as $t \to +\infty$ or as $t \to -\infty$ is called a limit cycle of (13.4).

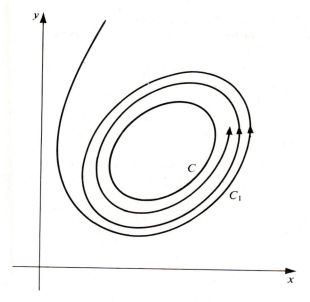

Figure 13.29

▶ **Example 13.17**

The following well-known example of a system having a limit cycle will illustrate the above discussion and definition.

$$\frac{dx}{dt} = y + x(1 - x^2 - y^2),$$

$$\frac{dy}{dt} = -x + y(1 - x^2 - y^2). \qquad (13.85)$$

To study this system we shall introduce polar coordinates (r, θ), where

$$x = r \cos \theta,$$

$$y = r \sin \theta. \qquad (13.86)$$

From these relations we find that

$$x\frac{dx}{dt} + y\frac{dy}{dt} = r\frac{dr}{dt},$$

$$x\frac{dy}{dt} - y\frac{dx}{dt} = r^2\frac{d\theta}{dt}. \qquad (13.87)$$

Now, multiplying the first equation of (13.85) by x and the second by y and adding, we obtain

$$x\frac{dx}{dt} + y\frac{dy}{dt} = (x^2 + y^2)(1 - x^2 - y^2).$$

Introducing the polar coordinates defined by (13.86), and making use of (13.87), this becomes

$$r \frac{dr}{dt} = r^2(1 - r^2).$$

For $r \neq 0$, we may thus write

$$\frac{dr}{dt} = r(1 - r^2).$$

Now multiplying the first equation of (13.85) by y and the second by x and subtracting, we obtain

$$y \frac{dx}{dt} - x \frac{dy}{dt} = y^2 + x^2.$$

Again using (13.87), this becomes

$$-r^2 \frac{d\theta}{dt} = r^2,$$

and so for $r \neq 0$ we may write

$$\frac{d\theta}{dt} = -1.$$

Thus in polar coordinates the system (13.85) becomes

$$\frac{dr}{dt} = r(1 - r^2),$$

$$\frac{d\theta}{dt} = -1.$$

(13.88)

From the second of these equations we find at once that

$$\theta = -t + t_0,$$

where t_0 is an arbitrary constant. The first of the equations (13.88) is separable. Separating variables, we have

$$\frac{dr}{r(1 - r^2)} = dt,$$

and an integration using partial fractions yields

$$\ln r^2 - \ln|1 - r^2| = 2t + \ln|c_0|.$$

After some manipulations we obtain

$$r^2 = \frac{c_0 e^{2t}}{1 + c_0 e^{2t}}.$$

Thus we may write

$$r = \frac{1}{\sqrt{1 + c e^{-2t}}}, \qquad \text{where} \quad c = \frac{1}{c_0}.$$

Thus, the solution of the system (13.88) may be written

$$r = \frac{1}{\sqrt{1 + ce^{-2t}}},$$

$$\theta = -t + t_0,$$

where c and t_0 are arbitrary constants. We may choose $t_0 = 0$. Then $\theta = -t$; using (13.86), the solution of the system (13.85) becomes

$$x = \frac{\cos t}{\sqrt{1 + ce^{-2t}}},$$

$$y = -\frac{\sin t}{\sqrt{1 + ce^{-2t}}}. \qquad (13.89)$$

The solutions (13.89) of (13.85) define the paths of (13.85) in the xy plane. Examining these paths for various values of c, we note the following conclusions:

1. If $c = 0$, the path defined by (13.89) is the circle $x^2 + y^2 = 1$, described in the clockwise direction.
2. If $c \neq 0$, the paths defined by (13.89) are *not* closed paths but rather paths having a spiral behavior. If $c > 0$, the paths are spirals lying inside the circle $x^2 + y^2 = 1$. As $t \to +\infty$, they approach this circle; while as $t \to -\infty$, they approach the critical point $(0, 0)$ of (13.85). If $c < 0$, the paths lie outside the circle $x^2 + y^2 = 1$. These "outer" paths also approach this circle as $t \to +\infty$; while as $t \to \ln\sqrt{|c|}$, both $|x|$ and $|y|$ becomes infinite.

Since the closed path $x^2 + y^2 = 1$ is approached spirally from both the inside and the outside by nonclosed paths as $t \to +\infty$, we conclude that this circle is a limit cycle of the system (13.85). (See Figure 13.30).

B. Existence and Nonexistence of Limit cycles

In Example 13.17 the existence of a limit cycle was ascertained by actually finding this limit cycle. In general such a procedure is, of course, impossible. Given the autonomous system (13.4) we need a theorem giving sufficient conditions for the existence of a limit cycle of (13.4). One of the few general theorems of this nature is the Poincaré-Bendixson theorem, which we shall state below (Theorem 13.13). First, however, we shall state and prove a theorem on the *nonexistence* of closed paths of the system (13.4)

THEOREM 13.12   Bendixson's Nonexistence Criterion

Hypothesis.   *Let D be a domain in the xy plane. Consider the autonomous system*

$$\frac{dx}{dt} = P(x, y),$$

$$\frac{dy}{dt} = Q(x, y), \qquad (13.4)$$

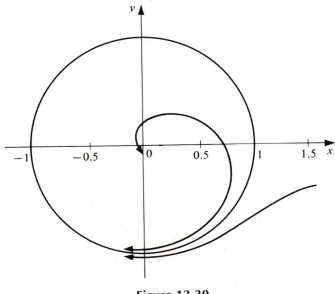

Figure 13.30

where P and Q have continuous first partial derivatives in D. Suppose that

$\dfrac{\partial P(x, y)}{\partial x} + \dfrac{\partial Q(x, y)}{\partial y}$ has the same sign throughout D.

Conclusion.   The system (13.4) has no closed path in the domain D.

Proof.   Let C be a closed curve in D; let R be the region bounded by C; and apply Green's theorem in the plane. We have

$$\int_C [P(x, y)\, dy - Q(x, y)\, dx] = \iint_R \left[\frac{\partial P(x, y)}{\partial x} + \frac{\partial Q(x, y)}{\partial y} \right] ds,$$

where the line integral is taken in the positive sense. Now assume that C is a closed path of (13.4); let $x = f(t)$, $y = g(t)$ be an arbitrary solution of (13.4) defining C parametrically; and let T denote the period of this solution. Then

$$\frac{df(t)}{dt} = P[f(t), g(t)],$$

$$\frac{dg(t)}{dt} = Q[f(t), g(t)],$$

along C and we have

$$\int_C [P(x, y)\, dy - Q(x, y)\, dx]$$

$$= \int_0^T \left\{ P[f(t), g(t)] \frac{dg(t)}{dt} - Q[f(t), g(t)] \frac{df(t)}{dt} \right\} dt$$

$$= \int_0^T \{P[f(t), g(t)]Q[f(t), g(t)] - Q[f(t), g(t)]P[f(t), g(t)]\}\, dt$$

$$= 0.$$

Thus

$$\iint_R \left[\frac{\partial P(x, y)}{\partial x} + \frac{\partial Q(x, y)}{\partial y} \right] ds = 0.$$

But this double integral can be zero only if $\dfrac{\partial P(x, y)}{\partial x} + \dfrac{\partial Q(x, y)}{\partial y}$ changes sign. This is a contradiction. Thus C is not a *path* of (13.4) and hence (13.4) possesses no *closed* path in D.

Q.E.D.

▶ **Example 13.18**

$$\frac{dx}{dt} = 2x + y + x^3,$$

$$\frac{dy}{dt} = 3x - y + y^3. \tag{13.90}$$

Here

$$P(x, y) = 2x + y + x^3,$$

$$Q(x, y) = 3x - y + y^3,$$

and

$$\frac{\partial P(x, y)}{\partial x} + \frac{\partial Q(x, y)}{\partial y} = 3(x^2 + y^2) + 1.$$

Since this expression is positive throughout every domain D in the xy plane, the system (13.90) has no closed path in any such domain. In particular, then, the system (13.90) has no limit cycles and hence no periodic solutions.

Having considered this nonexistence result, we now turn to the Poincaré-Bendixson existence theorem, the proof of which is outside the scope of this book. We shall merely state the theorem and indicate its significance. In order to make the statement of the theorem less involved, we introduce two further concepts.

DEFINITION

Let C be a path of the system (13.4) and let $x = f(t)$, $y = g(t)$ be a solution of (13.4) defining C. Then we shall call the set of all points of C for $t \geq t_0$, where t_0 is some value of t, a half-path of (13.4). In other words, by a half-path of (13.4) we mean the set of all points with coordinates $[f(t), g(t)]$ for $t_0 \leq t < +\infty$. We denote a half-path of (13.4) by C^+.

DEFINITION

Let C^+ be a half-path of (13.4) defined by $x = f(t)$, $y = g(t)$ for $t \geq t_0$. Let (x_1, y_1) be a point in the xy plane. If there exists a sequence of real numbers $\{t_n\}$, $n = 1, 2, \ldots$, such that $t_n \to +\infty$ and $[f(t_n), g(t_n)] \to (x_1, y_1)$ as $n \to +\infty$, then we call (x_1, y_1) a limit point of C^+. The set of all limit points of a half-path C^+ will be called the limit set *of C^+ and will be denoted by $L(C^+)$.*

▶   **Example 13.19**

The paths of the system (13.85) are given by Equations (13.89). Letting $c = 1$ we obtain the path C defined by

$$x = \frac{\cos t}{\sqrt{1 + e^{-2t}}},$$

$$y = -\frac{\sin t}{\sqrt{1 + e^{-2t}}}. \tag{13.91}$$

The set of all points of C for $t \geq 0$ is a half-path C^+. That is, C^+ is the set of all points with coordinates

$$\left[\frac{\cos t}{\sqrt{1 + e^{-2t}}}, -\frac{\sin t}{\sqrt{1 + e^{-2t}}} \right], \qquad 0 \leq t < +\infty.$$

Consider the sequence 0, 2π, $4\pi, \ldots$, $2n\pi, \ldots$, tending to $+\infty$ as $n \to +\infty$. The corresponding sequence of points on C^+ is

$$\left[\frac{\cos 2n\pi}{\sqrt{1 + e^{-4n\pi}}}, -\frac{\sin 2n\pi}{\sqrt{1 + e^{-4n\pi}}} \right], \qquad (n = 0, 1, 2, \ldots),$$

and this sequence approaches the point $(1, 0)$ as $n \to +\infty$.

Thus $(1, 0)$ is a *limit point* of the half-path C^+.

The set of *all* limit points of C^+ is the set of points such that $x^2 + y^2 = 1$. In other words, the circle $x^2 + y^2 = 1$ is the *limit set* of C^+.

We are now in a position to state the Poincaré-Bendixson theorem.

THEOREM 13.13   Poincaré-Bendixson Theorem; "Strong" Form

Hypothesis

1.   *Consider the autonomous system*

$$\frac{dx}{dt} = P(x, y),$$

$$\frac{dy}{dt} = Q(x, y), \tag{13.4}$$

where P and Q have continuous first partial derivatives in a domain D of the xy plane. Let D_1 be a bounded subdomain of D, and let R denote D_1 plus its boundary.

2. *Let C^+ defined by $x = f(t)$, $y = g(t)$, $t \geq t_0$, be a half-path of (13.4) contained entirely in R. Suppose the limit set $L(C^+)$ of C^+ contains no critical points of (13.4).*

Conclusion. *Either (1) the half-path C^+ is itself a closed path [in this case C^+ and $L(C^+)$ are identical], or (2) $L(C^+)$ is a closed path which C^+ approaches spirally from either the inside or the outside [in this case $L(C^+)$ is a limit cycle]. Thus in either case, there exists a closed path of (13.4) in R.*

A slightly weaker but somewhat more practical form of this theorem may be seen at once. If the region R of Hypothesis 1 contains no critical points of (13.4), then the limit set $L(C^+)$ will contain no critical points of (13.4) and so the second statement of Hypothesis 2 will automatically be satisfied. Thus we may state:

THEOREM 13.13A Poincaré-Bendixson Theorem; "Weak" Form

Hypothesis

1. *Exactly as in Theorem 13.13.*
2. *Suppose R contains no critical points of (13.4).*

Conclusion. *If R contains a half-path of (13.4), then R also contains a closed path of (13.4).*

Let us indicate how this theorem may be applied to determine the existence of a closed path of (13.4). Suppose the continuity requirements concerning the derivatives of $P(x, y)$ and $Q(x, y)$ are satisfied for all (x, y). Further suppose that (13.4) has a critical point at (x_0, y_0) but no other critical points within some circle

$$K : (x - x_0)^2 + (y - y_0)^2 = r^2$$

about (x_0, y_0) (see Figure 13.31). Then an annular region whose boundary consists of two smaller circles $K_1 : (x - x_0)^2 + (y - y_0)^2 = r_1^2$ and $K_2 : (x - x_0)^2 + (y - y_0)^2 = r_2^2$, where $0 < r_1 < r_2 < r$, about (x_0, y_0) may be taken as a region R containing no critical points of (13.4). If we can then show that a half-path C^+ of (13.4) (for $t \geq$ some t_0) is entirely contained in this annular region R, then we can conclude at once that a closed path C_0 of (13.4) is also contained in R.

The difficulty in applying Theorem 13.13A usually comes in being able to show that a half-path C^+ is entirely contained in R. If one can show that the vector $[P(x, y), Q(x, y)]$ determined by (13.4) points *into R* at *every* point of the boundary of R, then a path C entering R at $t = t_0$ will remain in R for $t \geq t_0$ and hence provide the needed half-path C^+.

▶ **Example 13.20**

Consider again the system (13.85) with critical point $(0, 0)$. The annular region R bounded by $x^2 + y^2 = \frac{1}{4}$ and $x^2 + y^2 = 4$ contains no critical points of (13.85). If we can show that R contains a half-path of (13.85), the Poincaré-Bendixson theorem ("weak" form) will apply.

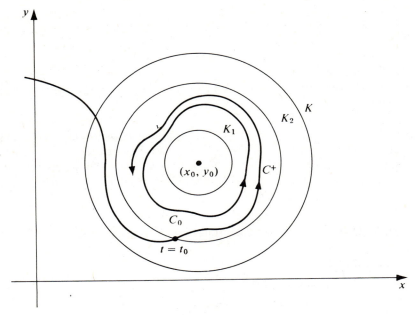

Figure 13.31

In our previous study of this system we found that

$$\frac{dr}{dt} = r(1 - r^2)$$

for $r > 0$, where $r = \sqrt{x^2 + y^2}$. On the circle $x^2 + y^2 = \frac{1}{4}$, $dr/dt > 0$ and hence $r = \sqrt{x^2 + y^2}$ is increasing. Thus the vector $[P(x, y), Q(x, y)]$ points *into* R at *every* point of this inner circle. On the circle $x^2 + y^2 = 4$, $dr/dt < 0$ and hence $r = \sqrt{x^2 + y^2}$ is decreasing. Thus the vector $[P(x, y), Q(x, y)]$ also points *into* R at *every* point of this outer circle. (See Figure 13.32, in which this situation is illustrated qualitatively.) Hence a path C entering R at $t = t_0$ will remain in R for $t \geq t_0$, and this provides us with the needed half-path contained in R.

Thus by the Poincaré-Bendixson theorem ("weak" form), we know that R contains a closed path C_0. We have already seen that the circle $x^2 + y^2 = 1$ is indeed such a closed path of (13.85).

C. The Index of a Critical Point

We again consider the system (13.4), where P and Q have continuous first partial derivatives for all (x, y), and assume that all of the critical points of (13.4) are isolated. Now consider a simple closed curve* C [not necessarily a path of (13.4)] which passes

---

* By a simple closed curve we mean a closed curve having no double points; for example, a circle is a simple closed curve, but a figure eight is not.

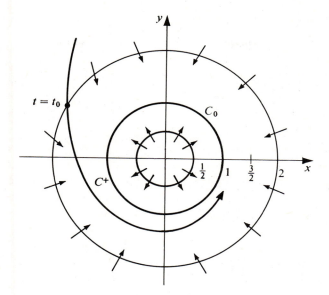

Figure 13.32

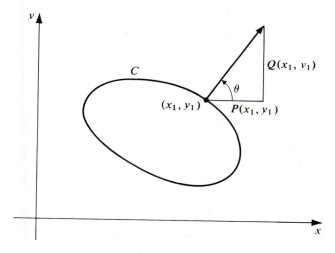

Figure 13.33

through no critical points of (13.4). Consider a point (x_1, y_1) on C and the vector $[P(x_1, y_1), Q(x_1, y_1)]$ defined by (13.4) at the point (x_1, y_1). Let θ denote the angle from the positive x direction to this vector (see Figure 13.33).

Now let (x_1, y_1) describe the curve C once in the counterclockwise direction and return to its original position. As (x_1, y_1) describes C, the vector $[P(x_1, y_1), Q(x_1, y_1)]$ changes continuously, and consequently the angle θ also varies continuously. When (x_1, y_1) reaches its original position, the angle θ will have changed by an amount $\Delta\theta$. We are now prepared to define the index of the curve C.

DEFINITION

Let θ denote the angle from the positive x direction to the vector $[P(x_1, y_1), Q(x_1, y_1)]$ defined by (13.4) at (x_1, y_1). Let $\Delta\theta$ denote the total change in θ as (x_1, y_1) describes the simple closed curve C once in the counterclockwise direction. We call the number

$$I = \frac{\Delta\theta}{2\pi}$$

the index *of the curve C with respect to the system (13.4).*

Clearly $\Delta\theta$ is either equal to zero or a positive or negative integral multiple of 2π and hence I is either zero or a positive or negative integer. If $[P(x_1, y_1), Q(x_1, y_1)]$ merely oscillates but does not make a complete rotation as (x_1, y_1) describes C, then I is zero. If the net change $\Delta\theta$ in θ is a decrease, then I is negative.

Now let (x_0, y_0) be an isolated critical point of (13.4). It can be shown that all simple closed curves enclosing (x_0, y_0) but containing no other critical point of (13.4) have the same index. This leads us to make the following definition.

DEFINITION

By the index *of an isolated critical point (x_0, y_0) of (13.4) we mean the index of a simple closed curve C which encloses (x_0, y_0) but no other critical points of (13.4).*

From an examination of Figure 13.34 we may reach the following conclusion intuitively: The index of a node, a center, or a spiral point is $+1$, while the index of a saddle point is -1.

We now list several interesting results concerning the index of a simple closed curve C and then point out several important consequences of these results. Most of these results are intuitively obvious, but few are easy to prove rigorously. In each case when we say index we shall mean the index with respect to the system (13.4) where $P(x, y)$ and $Q(x, y)$ have continuous first partial derivatives for all (x, y) and (13.4) has only isolated critical points.

1. The index of a simple closed curve which neither passes through a critical point of (13.4) nor has a critical point of (13.4) in its interior is zero.
2. The index of a simple closed curve which surrounds a finite number of critical points of (13.4) is equal to the sum of the indices of these critical points.
3. The index of a closed path of (13.4) is $+1$.

From these results the following conclusions follow as once.

(a) A closed path of (13.4) contains at least one critical point of (13.4) in its interior [for otherwise, by (1), the index of such a closed path would be zero; and this would contradict (3)].

(b) A closed path of (13.4) may contain in its interior a finite number of critical points of (13.4), the sum of the indices of which is $+1$ [this follows at once from (2) and (3)].

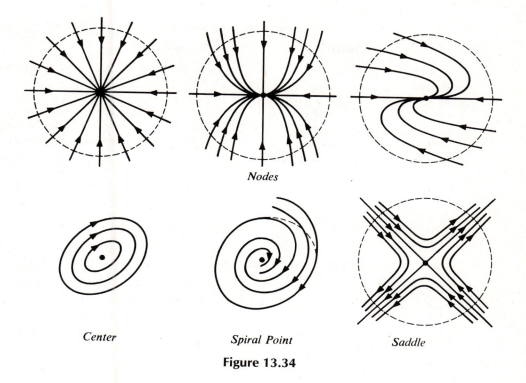

Nodes

Center   Spiral Point   Saddle

Figure 13.34

D. The Lienard-Levinson-Smith Theorem and the van der Pol Equation

Throughout this chapter we have stated a number of important results without proof. We have done this because we feel that every serious student of differential equations should become aware of these results as soon as possible, even though their proofs are definitely beyond our scope and properly belong to a more advanced study of our subject. In keeping with this philosophy, we close this section by stating without proof an important theorem dealing with the existence of periodic solutions for a class of second-order nonlinear equations. We shall then apply this theorem to the famous van der Pol equation already introduced at the beginning of the chapter.

THEOREM 13.14   Lienard-Levinson-Smith

Hypothesis.   *Consider the differential equation*

$$\frac{d^2x}{dt^2} + f(x)\frac{dx}{dt} + g(x) = 0, \tag{13.92}$$

where f, g, F defined by $F(x) = \int_0^x f(u)\,du$, and G defined by $G(x) = \int_0^x g(u)\,du$ are real functions having the following properties:

1. *f is even and is continuous for all x.*
2. *There exists a number $x_0 > 0$ such that $F(x) < 0$ for $0 < x < x_0$ and $F(x) > 0$ and monotonic increasing for $x > x_0$. Further, $F(x) \to \infty$ as $x \to \infty$.*

3. g is odd, has a continuous derivative for all x, and is such that g(x) > 0 for x > 0.
4. G(x) → ∞ as x → ∞.

Conclusion. *Equation (13.92) possesses an essentially unique nontrivial periodic solution.*

Remark. By "essentially unique" in the above conclusion we mean that if $x = \phi(t)$ is a nontrivial periodic solution of (13.92), then all other nontrivial periodic solutions of (13.92) are of the form $x = \phi(t - t_1)$, where t_1 is a real number. In other words, the equivalent autonomous system

$$\frac{dx}{dt} = y,$$

$$\frac{dy}{dt} = -f(x)y - g(x),$$

has a unique closed path in the xy plane.

One of the most important examples of an equation of the form (13.92) which satisfies the hypotheses of Theorem 13.14 is the van der Pol equation

$$\frac{d^2x}{dt^2} + \mu(x^2 - 1)\frac{dx}{dt} + x = 0, \tag{13.2}$$

where μ is a positive constant. Here $f(x) = \mu(x^2 - 1)$, $g(x) = x$,

$$F(x) = \int_0^x \mu(u^2 - 1)\, du = \mu\left(\frac{x^3}{3} - x\right),$$

and

$$G(x) = \int_0^x u\, du = \frac{x^2}{2}.$$

We check that the hypotheses of Theorem 13.14 are indeed satisfied:

1. Since $f(-x) = \mu(x^2 - 1) = f(x)$, the function f is even. Clearly it is continuous for all x.
2. $F(x) = \mu(x^3/3 - x)$ is negative for $0 < x < \sqrt{3}$. For $x > \sqrt{3}$, $F(x)$ is positive and monotonic increasing (it is, in fact, monotonic increasing for $x > 1$). Clearly $F(x) \to \infty$ as $x \to \infty$.
3. Since $g(-x) = -x = -g(x)$, the function g is odd. Since $g'(x) = 1$, the derivative of g is continuous for all x. Obviously $g(x) > 0$ for $x > 0$.
4. Obviously $G(x) \to \infty$ as $x \to \infty$.

Thus the conclusion of Theorem 13.14 is valid, and we conclude that the van der Pol equation (13.2) has an essentially unique nontrivial periodic solution. In other words, the equivalent autonomous system

$$\frac{dx}{dt} = y,$$

$$\frac{dy}{dt} = \mu(1 - x^2)y - x, \tag{13.93}$$

has a unique closed path in the xy plane.

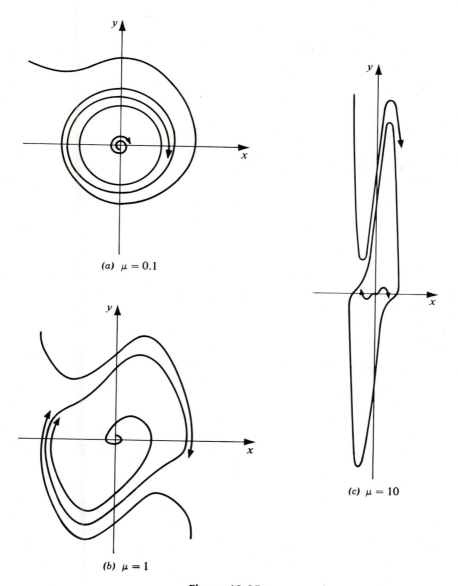

(a) $\mu = 0.1$

(b) $\mu = 1$

(c) $\mu = 10$

Figure 13.35

The differential equation of the paths of the system (13.93) is

$$\frac{dy}{dx} = \frac{\mu(1 - x^2)y - x}{y}.\tag{13.94}$$

Using the method of isoclines (Section 8.1B) one can obtain the paths defined by (13.94) in the xy plane. The results for $\mu = 0.1$, $\mu = 1$, and $\mu = 10$ are shown in Figure 13.35a, 13.35b, and 13.35c, respectively. The limit cycle C in each of these figures is the unique closed path whose existence we have already ascertained on the basis of Theorem 13.14. For $\mu = 0.1$ we note that this limit cycle is very nearly a circle of radius 2. For $\mu = 1$, it has lost its circle-like form and appears rather "baggy," while for $\mu = 10$, it is very long and narrow.

Exercises

1. Consider the nonlinear autonomous system

    $$\frac{dx}{dt} = 4x - 4y - x(x^2 + y^2),$$

    $$\frac{dy}{dt} = 4x + 4y - y(x^2 + y^2).$$

    (a) Introduce polar coordinates (r, θ) and transform the system to polar coordinate form.

    (b) Apply the Poincaré-Bendixson theorem to show that there exists a limit cycle between the circles

    $$x^2 + y^2 = \tfrac{1}{4} \quad \text{and} \quad x^2 + y^2 = 16.$$

    (c) Find the explicit solutions $x = f(t)$, $y = g(t)$ of the original system. In particular, find "the" periodic solution corresponding to the limit cycle whose existence was ascertained in step (b).

    (d) Sketch the limit cycle and at least one of the nonclosed paths.

2. Proceed as in Exercise 1 for the system

    $$\frac{dx}{dt} = y + \frac{x}{\sqrt{x^2 + y^2}} [1 - (x^2 + y^2)],$$

    $$\frac{dy}{dt} = -x + \frac{y}{\sqrt{x^2 + y^2}} [1 - (x^2 + y^2)].$$

Consider the nonlinear differential equations in Exercises 3–7. In each case apply one of the theorems of this section to determine whether or not the differential equation has periodic solutions.

3. $\dfrac{d^2x}{dt^2} + x^4 \dfrac{dx}{dt} - \dfrac{dx}{dt} + x = 0.$

4. $\dfrac{d^2x}{dt^2} + \dfrac{dx}{dt} + \left(\dfrac{dx}{dt}\right)^3 + x = 0.$

5. $\dfrac{d^2x}{dt^2} + (x^4 + x^2) \dfrac{dx}{dt} + (x^3 + x) = 0.$

6. $\dfrac{d^2x}{dt^2} + (5x^4 - 6x^2) \dfrac{dx}{dt} + x^3 = 0.$

7. $\dfrac{d^2x}{dt^2} + (x^2 + 1) \dfrac{dx}{dt} + x^3 = 0.$

8. Consider the nonlinear autonomous system

    $$\frac{dx}{dt} = x - x^2,$$

    $$\frac{dy}{dt} = 2y - y^2.$$

(a)   Locate, classify (as to type), and list the index of each of the critical points.

(b)   Find the equation of the paths of the system in the xy phase plane and sketch several of these paths.

(c)   Determine the index of each of the following simple closed curves with respect to the given system. In each case explain how the index was determined.

(i) $4(x^2 + y^2) = 1$.

(ii) $100(x^2 + y^2) = 441$.

(iii) $x^2 + y^2 = 9$.

(iv) $4(x^2 + y^2) - 8(x + 2y) + 11 = 0$.

13.5   THE METHOD OF KRYLOFF AND BOGOLIUBOFF

A. The First Approximation of Kryloff and Bogoliuboff

In this section we consider a method of finding an approximate solution of an equation of the form

$$\frac{d^2x}{dt^2} + \omega^2 x + \mu f\left(x, \frac{dx}{dt}\right) = 0, \tag{13.95}$$

where μ is a sufficiently small parameter so that the nonlinear term $\mu f(x, dx/dt)$ is relatively small. The method which we shall discuss is due to the Russian scientists Kryloff and Bogoliuboff, and it is basically a method of variation of parameters.

If $\mu = 0$ in Equation (13.95), then Equation (13.95) reduces to the linear equation

$$\frac{d^2x}{dt^2} + \omega^2 x = 0. \tag{13.96}$$

The solution of Equation (13.96) may be written

$$x = a \sin(\omega t + \phi), \tag{13.97}$$

where a and ϕ are constants. The derivative of the solution (13.97) of (13.96) is

$$\frac{dx}{dt} = a\omega \cos(\omega t + \phi). \tag{13.98}$$

If $\mu \neq 0$ but sufficiently small (that is, $|\mu| \ll 1$), one might reasonably assume that the nonlinear equation (13.95) also has a solution of the *form* (13.97) with derivative of the *form* (13.98), provided that a and ϕ now be regarded as functions of t rather than constants. This is precisely what we shall do in applying the Kryloff-Bogoliuboff method. That is, we assume a solution of (13.95) of the form

$$x = a(t) \sin[\omega t + \phi(t)], \tag{13.99}$$

where a and ϕ are functions of t to be determined, such that the derivative of the solution (13.99) is of the form

$$\frac{dx}{dt} = \omega a(t) \cos[\omega t + \phi(t)]. \tag{13.100}$$

Differentiating the assumed solution (13.99), we obtain

$$\frac{dx}{dt} = \omega a(t) \cos[\omega t + \phi(t)] + a(t) \frac{d\phi}{dt} \cos[\omega t + \phi(t)]$$

$$+ \frac{da}{dt} \sin[\omega t + \phi(t)]. \tag{13.101}$$

In order for dx/dt to be of the desired form (13.100), we see from (13.101) that we must have

$$a(t) \frac{d\phi}{dt} \cos[\omega t + \phi(t)] + \frac{da}{dt} \sin[\omega t + \phi(t)] = 0. \tag{13.102}$$

Now differentiating the assumed derivative (13.100), we obtain

$$\frac{d^2x}{dt^2} = -\omega^2 a(t) \sin[\omega t + \phi(t)] - \omega a(t) \frac{d\phi}{dt} \sin[\omega t + \phi(t)]$$

$$+ \omega \frac{da}{dt} \cos[\omega t + \phi(t)]. \tag{13.103}$$

Substituting the assumed solution (13.99), its assumed derivative (13.100), and the second derivative given by (13.103) into the differential equation (13.95), we obtain

$$- \omega^2 a(t) \sin[\omega t + \phi(t)] - \omega a(t) \frac{d\phi}{dt} \sin[\omega t + \phi(t)]$$

$$+ \omega \frac{da}{dt} \cos[\omega t + \phi(t)] + \omega^2 a(t) \sin[\omega t + \phi(t)]$$

$$+ \mu f\{a(t) \sin[\omega t + \phi(t)], \omega a(t) \cos[\omega t + \phi(t)]\} = 0$$

or

$$\omega \frac{da}{dt} \cos[\omega t + \phi(t)] - \omega a(t) \frac{d\phi}{dt} \sin[\omega t + \phi(t)]$$

$$= -\mu f\{a(t)\sin[\omega t + \phi(t)], \omega a(t)\cos[\omega t + \phi(t)]\}. \tag{13.104}$$

Let $\theta(t)$ denote $\omega t + \phi(t)$. Then Equations (13.102) and (13.104) may be written

$$\sin \theta(t) \frac{da}{dt} + a(t) \cos \theta(t) \frac{d\phi}{dt} = 0$$

$$\omega \cos \theta(t) \frac{da}{dt} - \omega a(t) \sin \theta(t) \frac{d\phi}{dt} = -\mu f[a(t) \sin \theta(t), \omega a(t) \cos \theta(t)]. \tag{13.105}$$

Solving the system (13.105) for da/dt and $d\phi/dt$, we obtain the following equations:

$$\frac{da}{dt} = -\frac{\mu}{\omega} f[a(t) \sin \theta(t), \omega a(t) \cos \theta(t)] \cos \theta(t),$$

$$\frac{d\phi}{dt} = \frac{\mu}{\omega a(t)} f[a(t) \sin \theta(t), \omega a(t) \cos \theta(t)] \sin \theta(t). \tag{13.106}$$

These are the *exact* equations for the functions a and ϕ in the solution of the form (13.99) with derivative of the form (13.100). Note that they are nonlinear and nonautonomous and hence are quite complicated, to say the least. In order to make these equations tractable, we now apply the first approximation of Kryloff and Bogoliuboff.

From Equations (13.106) we see that da/dt and $d\phi/dt$ are both proportional to the *small* parameter μ. Hence although a and ϕ *are* functions of t, they are *slowly varying functions* of t during a period $T = 2\pi/\omega$ of t. That is, in a time interval of length $2\pi/\omega$, $a(t)$ and $\phi(t)$ are almost constant, while $\theta(t) = \omega t + \phi(t)$ increases by approximately 2π. Thus in the right members of (13.106) we regard $a(t)$ and $\phi(t)$ as constant during an interval of 2π in θ and then replace the functions of the right members of (13.106) by their mean values over such an interval in θ. This leads to the equations

$$\frac{da}{dt} = -\frac{\mu}{2\pi\omega} \int_0^{2\pi} f(a \sin \theta, \omega a \cos \theta) \cos \theta \, d\theta,$$

$$\frac{d\phi}{dt} = \frac{\mu}{2\pi\omega a} \int_0^{2\pi} f(a \sin \theta, \omega a \cos \theta) \sin \theta \, d\theta. \tag{13.107}$$

These are the equations of the *first approximation* of Kryloff and Bogoliuboff for the functions a and ϕ of the solution (13.99). We hasten to point out that we have not rigorously justified this procedure. Assuming that the procedure is justifiable for the problem under consideration, a first approximation to the solution of Equation (13.95) is thus given by

$$x = a(t) \sin[\omega t + \phi(t)], \tag{13.99}$$

where $a(t)$ and $\phi(t)$ are determined by Equations (13.107).

B. Two Special Cases

Case 1. The term $f(x, dx/dt)$ depends upon x only. If $f(x, dx/dt)$ in Equation (13.95) depends upon x only, then the equation reduces to

$$\frac{d^2x}{dt^2} + \omega^2 x + \mu f(x) = 0. \tag{13.108}$$

Then $f(a \sin \theta, \omega a \cos \theta)$ becomes simply $f(a \sin \theta)$, and the Equations (13.107) reduce at once to

$$\frac{da}{dt} = -\frac{\mu}{2\pi\omega} \int_0^{2\pi} f(a \sin \theta) \cos \theta \, d\theta,$$

$$\frac{d\phi}{dt} = \frac{\mu}{2\pi\omega a} \int_0^{2\pi} f(a \sin \theta) \sin \theta \, d\theta. \tag{13.109}$$

Consider the integral in the right member of the first equation of (13.109). Letting $u = a \sin \theta$, we have

$$\int_0^{2\pi} f(a \sin \theta) \cos \theta \, d\theta = \frac{1}{a} \int_0^0 f(u) \, du = 0.$$

Thus from the first equation of (13.109) we see that $da/dt = 0$, and so the amplitude $a(t)$ is a constant a_0. The second equation of (13.109) now reduces to

$$\frac{d\phi}{dt} = F(a_0), \tag{13.110}$$

where

$$F(a_0) = \frac{\mu}{2\pi\omega a_0} \int_0^{2\pi} f(a_0 \sin \theta) \sin \theta \, d\theta. \tag{13.111}$$

Thus from (13.110), $\phi(t) = F(a_0)t + \phi_0$, where ϕ_0 is a constant of integration.
Hence the first approximation to the solution of (13.108) is given by

$$x = a_0 \sin\{[F(a_0) + \omega]t + \phi_0\}, \tag{13.112}$$

where the constant $F(a_0)$ is given by (13.111).

Therefore in the special case (13.108) in which the nonlinear term $f(x)$ depends upon x only (and not on dx/dt), the first approximation to a solution is a periodic oscillation the period of which depends upon the amplitude.

Case 2.   *The term $f(x, dx/dt)$ depends upon dx/dt only.* In this case Equation (13.95) reduces to

$$\frac{d^2x}{dt^2} + \omega^2 x + \mu f\left(\frac{dx}{dt}\right) = 0. \tag{13.113}$$

Thus $f(a \sin \theta, \omega a \cos \theta)$ becomes simply $f(\omega a \cos \theta)$, and Equations (13.107) now reduce to

$$\frac{da}{dt} = -\frac{\mu}{2\pi\omega} \int_0^{2\pi} f(\omega a \cos \theta) \cos \theta \, d\theta,$$

$$\frac{d\phi}{dt} = \frac{\mu}{2\pi\omega a} \int_0^{2\pi} f(\omega a \cos \theta) \sin \theta \, d\theta. \tag{13.114}$$

Letting $u = \omega a \cos \theta$ in the integral in the right member of the second equation of (13.114), we have

$$\int_0^{2\pi} f(\omega a \cos \theta) \sin \theta \, d\theta = -\frac{1}{\omega a} \int_{\omega a}^{\omega a} f(u) \, du = 0.$$

Thus from the second equation of (13.114) we see that $d\phi/dt = 0$, and so $\phi(t)$ is a constant ϕ_0.

Hence the first approximation to the solution of (13.113) is given by

$$x = a(t) \sin[\omega t + \phi_0], \tag{13.115}$$

where $a(t)$ is given by the first equation of (13.114).

Therefore in the special case (13.113) in which the nonlinear term $f(dx/dt)$ depends upon dx/dt only (and not on x), the first approximation to a solution is a nonperiodic oscillation of variable "amplitude" having the same frequency ($\omega/2\pi$) as that of solution (13.97) of the related linear equation (13.96).

C. Examples

We now apply the equations of the first approximation of Kryloff and Bogoliuboff to several differential equations of the form (13.95).

▶ **Example 13.21**

Consider the differential equation

$$\frac{d^2x}{dt^2} + \omega^2 x + \mu x^3 = 0, \tag{13.116}$$

where μ is a small parameter. Here $f(x, dx/dt) = f(x) = x^3$, and hence Equation (13.116) is of the special form (13.108) in which the nonlinear term depends upon x only. Thus the first approximation to the solution of (13.116) is given by (13.112), where $F(a_0)$ is determined by (13.111). Since $f(a_0 \sin \theta) = a_0^3 \sin^3 \theta$, we see from (13.111) that

$$F(a_0) = \frac{\mu}{2\pi\omega a_0} \int_0^{2\pi} a_0^3 \sin^3 \theta \sin \theta \, d\theta$$

$$= \frac{\mu a_0^2}{2\pi\omega} \int_0^{2\pi} \sin^4 \theta \, d\theta = \frac{\mu a_0^2}{2\pi\omega} \left(\frac{3\pi}{4}\right) = \frac{3\mu a_0^2}{8\omega}.$$

Thus the first approximation to the solution of Equation (13.116) is given by

$$x = a_0 \sin\left[\left(\frac{3\mu a_0^2}{8\omega} + \omega\right)t + \phi_0\right], \tag{13.117}$$

where a_0 (the amplitude) and ϕ_0 are arbitrary constants. Observe that the period

$$\frac{2\pi}{\dfrac{3\mu a_0^2}{8\omega} + \omega}$$

of the oscillation (13.117) is a function of the amplitude.

▶ **Example 13.22**

Consider the differential equation

$$\frac{d^2x}{dt^2} + \omega^2 x + \mu\left[\left(\frac{dx}{dt}\right)^2 + \left(\frac{dx}{dt}\right)^3\right] = 0, \tag{13.118}$$

where μ is a small parameter. Here

$$f\left(x, \frac{dx}{dt}\right) = f\left(\frac{dx}{dt}\right) = \left(\frac{dx}{dt}\right)^2 + \left(\frac{dx}{dt}\right)^3,$$

and hence Equation (13.118) is of the special form (13.113) in which the nonlinear term depends upon dx/dt only. Thus the first approximation to the solution of (13.118) is given by (13.115), where the amplitude $a(t)$ is determined from the first equation of (13.114). Since $f(\omega a \cos \theta) = \omega^2 a^2 \cos^2 \theta + \omega^3 a^3 \cos^3 \theta$, this equation is

$$\frac{da}{dt} = -\frac{\mu}{2\pi\omega} \int_0^{2\pi} (\omega^2 a^2 \cos^2 \theta + \omega^3 a^3 \cos^3 \theta) \cos \theta \, d\theta$$

or

$$\frac{da}{dt} = -\frac{\mu\omega a^2}{2\pi}\left[\int_0^{2\pi} \cos^3 \theta \, d\theta + \omega a \int_0^{2\pi} \cos^4 \theta \, d\theta\right].$$

Since

$$\int_0^{2\pi} \cos^3 \theta \, d\theta = 0 \quad \text{and} \quad \int_0^{2\pi} \cos^4 \theta \, d\theta = \frac{3\pi}{4},$$

this reduces at once to

$$\frac{da}{dt} = -\frac{3\mu\omega^2 a^3}{8}. \tag{13.119}$$

Separating variables in (13.119) and integrating, we obtain

$$\frac{1}{2a^2} = \frac{3\mu\omega^2}{8} t + c,$$

where c is an arbitrary constant. If $a(0) = a_0$, then $c = 1/2a_0^2$. We thus have

$$\frac{1}{2a^2} = \frac{3\mu\omega^2}{8} t + \frac{1}{2a_0^2},$$

and so

$$a = 2\sqrt{\frac{a_0^2}{4 + 3\mu\omega^2 a_0^2 t}}.$$

Thus the first approximation to the solution of Equation (13.118) is given by

$$x = 2\sqrt{\frac{a_0^2}{4 + 3\mu\omega^2 a_0^2 t}} \sin[\omega t + \phi_0].$$

Assuming $\mu > 0$, the amplitude of this oscillation tends to zero as $t \to \infty$.

▶ **Example 13.23**

Consider the van der Pol equation

$$\frac{d^2 x}{dt^2} + \mu(x^2 - 1)\frac{dx}{dt} + x = 0, \tag{13.2}$$

where $0 < \mu \ll 1$. This equation is of neither of the special forms (13.108) or (13.113) considered above but rather of the more general form (13.95), where $\omega^2 = 1$ and

$$f\left(x, \frac{dx}{dt}\right) = (x^2 - 1)\frac{dx}{dt}.$$

Therefore the first approximation to the solution of (13.2) is given by (13.99), where $a(t)$ and $\phi(t)$ are determined by Equations (13.107) and where $\omega = 1$.

Since $f(a \sin \theta, a \cos \theta) = (a^2 \sin^2 \theta - 1)a \cos \theta$, the Equations (13.107) become

$$\frac{da}{dt} = -\frac{\mu}{2\pi} \int_0^{2\pi} (a^2 \sin^2 \theta - 1)a \cos^2 \theta \, d\theta,$$

$$\frac{d\phi}{dt} = \frac{\mu}{2\pi a} \int_0^{2\pi} (a^2 \sin^2 \theta - 1)a \cos \theta \sin \theta \, d\theta.$$

Since

$$\int_0^{2\pi} \sin^2 \theta \cos^2 \theta \, d\theta = \frac{\pi}{4}, \qquad \int_0^{2\pi} \cos^2 \theta \, d\theta = \pi,$$

$$\int_0^{2\pi} \sin^3 \theta \cos \theta \, d\theta = 0, \quad \text{and} \quad \int_0^{2\pi} \cos \theta \sin \theta \, d\theta = 0,$$

these equations reduce at once to

$$\frac{da}{dt} = \frac{\mu a}{2}\left(1 - \frac{a^2}{4}\right),$$

$$\frac{d\phi}{dt} = 0. \tag{13.120}$$

From the second of Equations (13.120) we see at once that $\phi(t) = \phi_0$, a constant. Thus by (13.99) the first approximation to the solution is given by

$$x = a(t) \sin(t + \phi_0),$$

where $a(t)$ is to be found from the first of Equations (13.120). Separating variables in this equation we obtain

$$\frac{da}{a(a^2 - 4)} = -\frac{\mu}{8} \, dt. \tag{13.121}$$

Integrating and simplifying we find that

$$\frac{a^2 - 4}{a^2} = ce^{-\mu t}$$

and hence

$$a^2 = \frac{4}{1 - ce^{-\mu t}}.$$

If $a(0) = a_0 > 0$, then $c = (a_0^2 - 4)/a_0^2$ and hence

$$a^2 = \frac{4}{1 - \left(\dfrac{a_0^2 - 4}{a_0^2}\right)e^{-\mu t}} = \frac{a_0^2 e^{\mu t}}{1 + \dfrac{a_0^2}{4}(e^{\mu t} - 1)}.$$

Thus the first approximation to the solution of the van der Pol equation (13.2) for $0 < \mu \ll 1$ is given by

$$x = \frac{a_0 e^{\mu t/2}}{\sqrt{1 + \dfrac{a_0^2}{4}(e^{\mu t} - 1)}} \sin(t + \phi_0), \tag{13.122}$$

where a_0 is the initial value of the amplitude.

If $a_0 = 2$, the first approximation (13.122) reduces to the *periodic* function defined by

$$x = 2 \sin(t + \phi_0) \tag{13.123}$$

with constant amplitude 2.

If $0 < a_0 < 2$ or if $a_0 > 2$, we see that

$$\lim_{t \to \infty} a(t) = \lim_{t \to \infty} \frac{a_0 e^{\mu t/2}}{\sqrt{1 + \dfrac{a_0^2}{4}(e^{\mu t} - 1)}} = 2.$$

Thus for any finite $a_0 > 0$ except $a_0 = 2$ the nonperiodic oscillation given by (13.122) tends to the periodic oscillation given by (13.123) as $t \to \infty$.

Now we must remember that the formula (13.122) is only a first *approximation* to the exact solution of Equation (13.2). Hence our above observations give only an approximate appraisal of the actual situation. Nevertheless, if $0 < \mu \ll 1$, they indicate the actual situation quite well. Equation (13.2) has only *one* type of nontrivial periodic solutions, and these periodic solutions are given approximately by (13.123). All other nontrivial solutions of Equation (13.2) are nonperiodic oscillations which tend to one of the aforementioned periodic solutions as $t \to \infty$. These nonperiodic solutions are, of course, given approximately by (13.122).

In the previous section of this chapter we studied the phase plane paths associated with the van der Pol equation (13.2). We recall that there is a unique closed path, a limit cycle which all nonclosed paths approach as $t \to \infty$. For $0 < \mu \ll 1$, this limit cycle is approximately a circle of radius 2, and the nonclosed paths approach this nearly circular path spirally as $t \to \infty$. The limit cycle is represented approximately by the periodic oscillations (13.123) above, while the spiral paths are represented approximately by the nonperiodic oscillations (13.122).

Exercises

Use the method of Kryloff and Bogoliuboff to find a first approximation to the solution of each of the differential equations in Exercises 1–6. In each case assume that $0 < \mu \ll 1$ and that the initial value of the "amplitude" $a(t)$ of the "solution" is given by $a_0 > 0$.

1. $\dfrac{d^2x}{dt^2} + \omega^2 x + \mu x^5 = 0.$

2. $\dfrac{d^2x}{dt^2} + x + \mu \dfrac{dx}{dt}\left|\dfrac{dx}{dt}\right| = 0.$

3. $\dfrac{d^2x}{dt^2} + \mu\left[\left(\dfrac{dx}{dt}\right)^3 + x^3\right] + \omega^2 x = 0.$

4. $\dfrac{d^2x}{dt^2} + \omega^2 x + \mu(x^2 + 1)\dfrac{dx}{dt} = 0.$

5. $\dfrac{d^2x}{dt^2} + \omega^2 x + \mu\left(x^3 + \dfrac{dx}{dt}\right) = 0.$

6. $\dfrac{d^2x}{dt^2} + x + \mu\left(\dfrac{8}{9}x^4 - 1\right)\dfrac{dx}{dt} = 0.$

CHAPTER FOURTEEN

Partial Differential Equations

In the previous chapters of this text we have concentrated upon ordinary differential equations. In this final chapter we shall give a brief introduction to partial differential equations. The subject of partial differential equations is a vast one, and an entire book of this size could be devoted to an introduction to it. In this short chapter we shall merely introduce certain basic concepts of partial differential equations and present a basic method of solution which is very useful in many applied problems.

14.1 SOME BASIC CONCEPTS AND EXAMPLES

A. Partial Differential Equations and Their Solutions

Let us first recall that a partial differential equation is a differential equation which involves partial derivatives of one or more dependent variables with respect to one or more independent variables. We shall say that a *solution* of a partial differential equation is an explicit or implicit relation between the variables which does not contain derivatives and which identically satisfies the equation. In certain very simple cases a solution can be obtained immediately.

For example, consider the first-order partial differential equation

$$\frac{\partial u}{\partial x} = x^2 + y^2, \tag{14.1}$$

in which u is the dependent variable and x and y are independent variables. We have already solved equations of this type in Chapter 2. The solution is

$$u = \int (x^2 + y^2) \, \partial x + \phi(y),$$

where $\int (x^2 + y^2)\, \partial x$ indicates a "partial integration" with respect to x, holding y constant, and ϕ is an arbitrary function of y only. Thus the solution of Equation (14.1) is

$$u = \frac{x^3}{3} + xy^2 + \phi(y), \tag{14.2}$$

where ϕ is an arbitrary function of y.

As a second example, consider the second-order partial differential equation

$$\frac{\partial^2 u}{\partial y\, \partial x} = x^3 - y. \tag{14.3}$$

We first write this in the form

$$\frac{\partial}{\partial y}\left(\frac{\partial u}{\partial x}\right) = x^3 - y$$

and integrate partially with respect to y, holding x constant, to obtain

$$\frac{\partial u}{\partial x} = x^3 y - \frac{y^2}{2} + \phi(x),$$

where ϕ is an arbitrary function of x. We now integrate this result partially with respect to x, holding y constant, and so obtain the solution of Equation (14.3) in the form

$$u = \frac{x^4 y}{4} - \frac{xy^2}{2} + f(x) + g(y), \tag{14.4}$$

where f defined by $f(x) = \int \phi(x)\, dx$ is an arbitrary differentiable function of x and g is an arbitrary function of y.

As a result of these two simple examples, we note that whereas ordinary differential equations have solutions which involve arbitrary *constants*, partial differential equations have solutions which involve arbitrary *functions*. In particular, we note that the solution (14.2) of the *first-order* partial differential equation (14.1) contains *one* arbitrary function, and the solution (14.4) of the *second-order* partial differential equation (14.3) contains *two* arbitrary functions.

Let us now consider the nature of a typical mathematical problem which involves a partial differential equation and which originated in the mathematical formulation of some physical problem. Such a problem involves not only the partial differential equation itself, but also certain supplementary conditions (so-called *boundary conditions* or *initial conditions* or both). The number and nature of these conditions depend upon the nature of the physical problem from which the mathematical problem originated. The solution of the problem must satisfy both the partial differential equation and these supplementary conditions. In other words, the solution of the whole problem (differential equation plus conditions) is a *particular* solution of the differential equation of the problem. In Section 2 of this chapter we shall consider a basic method which may be employed to obtain this particular solution in certain cases. First, however, we consider briefly the class of partial differential equations which most frequently occurs in such problems. This is the class of so-called linear partial differential equations of the second order.

B. Linear Partial Differential Equations of the Second Order

The general linear partial differential equation of the second order in two independent variables x and y is an equation of the form

$$A \frac{\partial^2 u}{\partial x^2} + B \frac{\partial^2 u}{\partial x \, \partial y} + C \frac{\partial^2 u}{\partial y^2} + D \frac{\partial u}{\partial x} + E \frac{\partial u}{\partial y} + Fu = G, \tag{14.5}$$

where A, B, C, D, E, F, and G are functions of x and y. If $G(x, y) = 0$ for all (x, y), Equation (14.5) reduces to

$$A \frac{\partial^2 u}{\partial x^2} + B \frac{\partial^2 u}{\partial x \, \partial y} + C \frac{\partial^2 u}{\partial y^2} + D \frac{\partial u}{\partial x} + E \frac{\partial u}{\partial y} + Fu = 0. \tag{14.6}$$

Concerning solutions of Equation (14.6), we state the following basic theorem, which is analogous to Theorem 11.18 for linear ordinary differential equations.

THEOREM 14.1

Hypothesis. *Let $f_1, f_2, \ldots, f_n$ be n solutions of Equation (14.6) in a region R of the xy plane.*

Conclusion. *The linear combination $c_1 f_1 + c_2 f_2 + \cdots + c_n f_n$, where $c_1, c_2, \ldots, c_n$ are arbitrary constants, is also a solution of Equation (14.6) in the region R.*

Extending this, it may be shown that the following theorem is also valid.

THEOREM 14.2

Hypothesis

1. *Let $f_1, f_2, \ldots, f_n, \ldots$ be an infinite set of solutions of Equation (14.6) in a region R of the xy plane.*
2. *Suppose the infinite series*

$$\sum_{n=1}^{\infty} f_n = f_1 + f_2 + \cdots + f_n + \cdots$$

converges to f in R, and suppose this series may be differentiated term by term in R to obtain the various derivatives (of f) which appear in Equation (14.6).

Conclusion. *The function f (defined by $f = \sum\limits_{n=1}^{\infty} f_n$) is also a solution of Equation (14.6) in R.*

We now consider an important special class of second-order linear partial differential equations, the so-called *homogeneous* linear equations of the second order *with constant coefficients*. An equation of this class is of the form

$$a \frac{\partial^2 u}{\partial x^2} + b \frac{\partial^2 u}{\partial x \, \partial y} + c \frac{\partial^2 u}{\partial y^2} = 0, \tag{14.7}$$

where a, b, and c are constants. The word *homogeneous* here refers to the fact that all terms in (14.7) contain derivatives of the *same* order (the second).

We shall seek solutions of (14.7) of the form

$$u = f(y + mx), \tag{14.8}$$

where f is an arbitrary function of its argument (having sufficient derivatives)* and m is a constant. Differentiating (14.8), we obtain

$$\frac{\partial^2 u}{\partial x^2} = m^2 f''(y + mx),$$

$$\frac{\partial^2 u}{\partial x \, \partial y} = m f''(y + mx), \tag{14.9}$$

$$\frac{\partial^2 u}{\partial y^2} = f''(y + mx).$$

Substituting (14.9) into Equation (14.7), we obtain

$$am^2 f''(y + mx) + bm f''(y + mx) + c f''(y + mx) = 0$$

or

$$f''(y + mx)[am^2 + bm + c] = 0.$$

Thus $f(y + mx)$ will be a solution of (14.7) if m satisfies the quadratic equation

$$am^2 + bm + c = 0. \tag{14.10}$$

We now consider the following four cases of Equation (14.7).

1. $a \neq 0$, and the roots of the quadratic equation (14.10) are distinct.
2. $a \neq 0$, and the roots of (14.10) are equal.
3. $a = 0$, $b \neq 0$.
4. $a = 0$, $b = 0$, $c \neq 0$.

In Case 1 let the distinct roots of (14.10) be m_1 and m_2. Then Equation (14.7) has the solutions

$$f(y + m_1 x) \quad \text{and} \quad g(y + m_2 x),$$

where f and g are arbitrary functions of their respective arguments. Applying Theorem 14.1, we see that

$$f(y + m_1 x) + g(y + m_2 x) \tag{14.11}$$

is a solution of Equation (14.7).

In Case 2, let the double root of (14.10) be m_1. Then Equation (14.7) has the solution $f(y + m_1 x)$, where f is an arbitrary function of its argument. Further, it can be shown that in this case Equation (14.7) also has the solution $xg(y + m_1 x)$, where g is an

---

* In this chapter, whenever we state that a solution is an "arbitrary" function of its argument, we shall always mean an arbitrary function having a sufficient number of derivatives to satisfy the differential equation.

arbitrary function of its argument. Hence by Theorem 14.1 we see that

$$f(y + m_1 x) + xg(y + m_1 x) \tag{14.12}$$

is a solution of (14.7)

In Case 3, the quadratic equation (14.10) reduces to $bm + c = 0$ and hence has only one root. Denoting this root by m_1, Equation (14.7) has the solution $f(y + m_1 x)$, where f is an arbitrary function of its argument. Further, it can be verified at once that in this case $g(x)$, where g is an arbitrary function of x only, is also a solution of Equation (14.7). Thus by Theorem 14.1, we see that

$$f(y + m_1 x) + g(x) \tag{14.13}$$

is a solution of (14.7).

Finally, in Case 4, Equation (14.10) reduces to $c = 0$, which is impossible. Thus in this case there exist no solutions of the form (14.8). However, in this case the differential equation (14.7) is simply

$$c \frac{\partial^2 u}{\partial y^2} = 0 \quad \text{or} \quad \frac{\partial}{\partial y}\left(\frac{\partial u}{\partial y}\right) = 0.$$

Integrating partially with respect to y twice in succession, we obtain $u = f(x) + yg(x)$, where f and g are arbitrary functions of x only. Thus in this case

$$f(x) + yg(x) \tag{14.14}$$

is a solution of (14.7).

In each of the four cases under consideration we have obtained a solution of (14.7) which involves two arbitrary functions. Every equation of the form (14.7) with constant coefficients is in one and only one of the four categories covered by Cases 1 through 4. Thus Equation (14.7) always has a solution which involves two arbitrary functions. This solution is given by (14.11) in Case 1, by (14.12) in Case 2, by (14.13) in Case 3, and by (14.14) in Case 4.

We now consider two simple examples.

▶ **Example 14.1**

Find a solution of

$$\frac{\partial^2 u}{\partial x^2} - 5 \frac{\partial^2 u}{\partial x \, \partial y} + 6 \frac{\partial^2 u}{\partial y^2} = 0 \tag{14.15}$$

which contains two arbitrary functions.

Solution. The quadratic equation (14.10) corresponding to the differential equation (14.15) is

$$m^2 - 5m + 6 = 0,$$

and this equation has the distinct roots $m_1 = 2$ and $m_2 = 3$. Therefore this is an example of Case 1. Hence using (14.11) we see that

$$u = f(y + 2x) + g(y + 3x)$$

is a solution of (14.15) which contains two arbitrary functions.

▶ **Example 14.2**

Find a solution of

$$\frac{\partial^2 u}{\partial x^2} - 4\frac{\partial^2 u}{\partial x\,\partial y} + 4\frac{\partial^2 u}{\partial y^2} = 0 \qquad (14.16)$$

which contains two arbitrary functions.

Solution. The quadratic equation (14.10) corresponding to the differential equation (14.16) is

$$m^2 - 4m + 4 = 0,$$

and this equation has the double root $m_1 = 2$. Therefore this is an example of Case 2. Hence using (14.12) we see that

$$u = f(y + 2x) + xg(y + 2x)$$

is a solution of (14.16) which contains two arbitrary functions.

We close this section by classifying equations of the form (14.6) in the special case in which the coefficients A, B, C, D, E, and F are real constants. We shall illustrate this classification with certain of the famous equations of mathematical physics.

DEFINITION

The second-order linear partial differential equation

$$A\frac{\partial^2 u}{\partial x^2} + B\frac{\partial^2 u}{\partial x\,\partial y} + C\frac{\partial^2 u}{\partial y^2} + D\frac{\partial u}{\partial x} + E\frac{\partial u}{\partial y} + Fu = 0, \qquad (14.6)$$

where A, B, C, D, E, and F are real constants, is said to be

1. *hyperbolic if $B^2 - 4AC > 0$;*
2. *parabolic if $B^2 - 4AC = 0$;*
3. *elliptic if $B^2 - 4AC < 0$.*

▶ **Example 14.3**

The equation

$$\frac{\partial^2 u}{\partial x^2} - \frac{\partial^2 u}{\partial y^2} = 0 \qquad (14.17)$$

is hyperbolic, since $A = 1$, $B = 0$, $C = -1$, and $B^2 - 4AC = 4 > 0$. Equation (14.17) is a special case of the so-called one-dimensional *wave equation*, which is satisfied by the small transverse displacements of the points of a vibrating string. Observing that Equation (14.17) is a homogeneous linear equation with constant coefficients, we find at once that it has the solution

$$u = f(y + x) + g(y - x),$$

where f and g are arbitrary functions of their respective arguments.

▶ **Example 14.4**

The equation

$$\frac{\partial^2 u}{\partial x^2} - \frac{\partial u}{\partial y} = 0 \tag{14.18}$$

is parabolic, since $A = 1$, $B = C = 0$, and $B^2 - 4AC = 0$. Equation (14.18) is a special case of the one-dimensional *heat equation* (or *diffusion equation*), which is satisfied by the temperature at a point of a homogeneous rod. We observe that Equation (14.18) is not homogeneous.

▶ **Example 14.5**

The equation

$$\frac{\partial^2 u}{\partial x^2} + \frac{\partial^2 u}{\partial y^2} = 0 \tag{14.19}$$

is elliptic, since $A = 1$, $B = 0$, $C = 1$, and $B^2 - 4AC = -4 < 0$. Equation (14.19) is the so-called two-dimensional *Laplace equation*, which is satisfied by the steady-state temperature at points of a thin rectangular plate. Observing that Equation (14.19) is a homogeneous linear equation with constant coefficients, we find at once that it has the solution

$$u = f(y + ix) + g(y - ix),$$

where f and g are arbitrary functions of their respective arguments.

Exercises

For each of the differential equations in Exercises 1–10 find a solution which contains two arbitrary functions. In each case determine whether the equation is hyperbolic, parabolic, or elliptic.

1. $\dfrac{\partial^2 u}{\partial x^2} - 7\dfrac{\partial^2 u}{\partial x\, \partial y} + 6\dfrac{\partial^2 u}{\partial y^2} = 0.$ 2. $\dfrac{\partial^2 u}{\partial x^2} + \dfrac{\partial^2 u}{\partial x\, \partial y} - 6\dfrac{\partial^2 u}{\partial y^2} = 0.$

3. $\dfrac{\partial^2 u}{\partial x^2} + 6\dfrac{\partial^2 u}{\partial x\, \partial y} + 9\dfrac{\partial^2 u}{\partial y^2} = 0.$ 4. $4\dfrac{\partial^2 u}{\partial x^2} - 4\dfrac{\partial^2 u}{\partial x\, \partial y} + \dfrac{\partial^2 u}{\partial y^2} = 0.$

5. $\dfrac{\partial^2 u}{\partial x^2} + \dfrac{\partial^2 u}{\partial y^2} = 0.$ 6. $2\dfrac{\partial^2 u}{\partial x\, \partial y} + 3\dfrac{\partial^2 u}{\partial y^2} = 0.$

7. $\dfrac{\partial^2 u}{\partial x^2} - 2\dfrac{\partial^2 u}{\partial x\, \partial y} = 0.$ 8. $\dfrac{\partial^2 u}{\partial x^2} + 2\dfrac{\partial^2 u}{\partial x\, \partial y} + 5\dfrac{\partial^2 u}{\partial y^2} = 0.$

9. $8\dfrac{\partial^2 u}{\partial x^2} - 2\dfrac{\partial^2 u}{\partial x\, \partial y} - 3\dfrac{\partial^2 u}{\partial y^2} = 0.$ 10. $2\dfrac{\partial^2 u}{\partial x^2} - 2\dfrac{\partial^2 u}{\partial x\, \partial y} + 5\dfrac{\partial^2 u}{\partial y^2} = 0.$

11. Consider the equation

$$A \frac{\partial^2 u}{\partial x^2} + B \frac{\partial^2 u}{\partial x\, \partial y} + C \frac{\partial^2 u}{\partial y^2} + D \frac{\partial u}{\partial x} + E \frac{\partial u}{\partial y} + Fu = 0.$$

where $A, B, C, D, E,$ and F are functions of x and y. Extending the definition of the text, we say that this equation is

(i)   hyperbolic at all points at which $B^2 - 4AC > 0$;

(ii) parabolic at all points at which $B^2 - 4AC = 0$; and

(iii) elliptic at all points at which $B^2 - 4AC < 0$.

(a)   Show that the equation

$$(x^2 - 1) \frac{\partial^2 u}{\partial x^2} + 2y \frac{\partial^2 u}{\partial x\, \partial y} - \frac{\partial^2 u}{\partial y^2} = 0$$

is hyperbolic for all (x, y) outside the region bounded by the circle $x^2 + y^2 = 1$, parabolic on the boundary of this region, and elliptic for all (x, y) inside this region.

(b)   Determine all points (x, y) at which the equation

$$\frac{\partial^2 u}{\partial x^2} + x \frac{\partial^2 u}{\partial x\, \partial y} + y \frac{\partial^2 u}{\partial y^2} - xy \frac{\partial u}{\partial x} = 0$$

(i)   is hyperbolic;

(ii) is parabolic;

(iii) is elliptic.

14.2   THE METHOD OF SEPARATION OF VARIABLES

A. Introduction

In this section we introduce the so-called *method of separation of variables*. This is a basic method which is very powerful for obtaining solutions of certain problems involving partial differential equations. Although the class of problems to which the method applies is relatively small, it nevertheless includes many problems of great physical interest.

We have already pointed out that the mathematical statement of such a problem involves (1) a partial differential equation and (2) certain supplementary conditions. The solution of the problem is a function which satisfies both the equation and the various conditions, and the task before us is to find such a function. Let us recall a common procedure in solving analogous problems involving *ordinary* differential equations and supplementary conditions. In such a problem we first find the general solution of the ordinary differential equation and then apply the supplementary conditions, one at a time, to obtain the particular solution which satisfies them. Should we attempt to follow a similar procedure in problems in which the differential equation is *partial* instead of ordinary? Generally speaking, the answer is "no." In the first place, it is often impossible to obtain a "general" solution of the partial differential equation; and in the second place, even if such a solution can be found, it is usually impossible to apply the supplementary conditions to it. How then should we proceed? Instead of first finding the most general solution and then specializing it to obtain the particular solution desired, we shall first find particular solutions which satisfy *some* of the supplementary conditions and then combine these particular solutions in some manner to obtain a solution which satisfies *all* of the supplementary conditions. That is, instead of "breaking down" to the desired particular solution, the method of separation of variables in partial differential equations is one of "building up" to it.

In attempting to apply the method of separation of variables to a partial differential equations problem, one makes a certain basic assumption at the outset. If the partial differential equation involves n independent variables $x_1, x_2, \ldots, x_n$, one first *assumes* that the equation possesses product solutions of the form $X_1 X_2 \cdots X_n$, where X_i is a function of x_i only ($i = 1, 2, \ldots, n$). If the method actually does apply to the problem, then this basic assumption will produce n ordinary differential equations, one in each of the unknown functions X_i ($i = 1, 2, \ldots, n$). Certain of the supplementary conditions of the problem lead to supplementary conditions which must be satisfied by certain of the functions X_i. Thus at this stage of the method, one must solve n problems in ordinary differential equations, some of which involve initial or boundary conditions. The solutions of these n problems produce *particular* solutions of the form $X_1 X_2 \cdots X_n$ which satisfy *some* of the supplementary conditions of the original problem. One must now attempt to combine these particular solutions in some manner to produce a solution which satisfies *all* of the original supplementary conditions. One does this by first applying Theorem 14.2 to obtain a series solution and then applying the remaining supplementary conditions to this series solution.

We point out that the procedure outlined in the preceding paragraph is a strictly formal one. We shall illustrate it in considerable detail in parts *B*, *C*, and *D* of this section, and our treatment there will again be of a strictly formal nature. We have not made, and shall not make, any attempt to justify the procedure; but at least we should be aware of some of the questions with which a rigorous treatment would be concerned. In the first place, there is the question of *existence*. Does a solution of the given partial differential equations problem actually exist? In analogous problems involving ordinary differential equations we have seen that existence theorems are available which often enable us to answer questions of this type. Existence theorems for problems in partial differential equations are also known (see Greenspan, *Introduction to Partial Differential Equations*, for example), and these theorems can often be applied to answer the existence equation. But even if such a theorem answers this question in the affirmative, how do we know that a solution can be built up by the method which we have outlined? Even if the method seems to work formally, the various steps require justification. In particular, questions of convergence enter the problem, and a rigorous justification of the method involves a careful study of these convergence questions. Finally, assuming that each step can be justified and that the final result is indeed a solution, the question of *uniqueness* enters the picture. Is the solution obtained the *only* solution of the problem? Uniqueness theorems are known which can often be applied to answer this question.

In summary, our treatment of the method of separation of variables is a *formal* one. In a rigorous treatment one must show (1) that the formal "solution" obtained actually does satisfy both the partial differential equation and all of the supplementary conditions, and (2) that the solution thus justified is the only solution of the problem.

B. Example: The Vibrating String Problem

We now illustrate the method of separation of variables by applying it to obtain a formal solution of the so-called *vibrating string problem*.

The Physical Problem. Consider a tightly stretched elastic string the ends of which are fixed on the x axis at $x = 0$ and $x = L$. Suppose that for each x in the interval $0 < x < L$ the string is displaced into the xy plane and that for each such x the displacement from the x axis is given by $f(x)$, where f is a known function of x (see Figure 14.1).

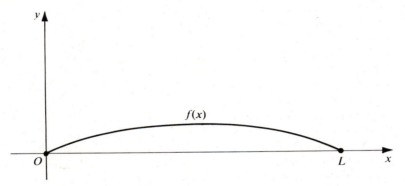

Figure 14.1

Suppose that at $t = 0$ the string is released from the initial position defined by $f(x)$, with an initial velocity given at each point of the interval $0 \le x \le L$ by $g(x)$, where g is a known function of x. Obviously the string will vibrate, and its displacement in the y direction at any point x at any time t will be a function of both x and t. We seek to find this displacement as a function of x and t; we denote it by y or $y(x, t)$.

We now make certain assumptions concerning the string, its vibrations, and its surroundings. To begin with, we assume that the string is perfectly flexible, is of constant linear density ρ, and is of constant tension T at all times. Concerning the vibrations, we assume that the motion is confined to the xy plane and that each point on the string moves on a straight line perpendicular to the x axis as the string vibrates. Further, we assume that the displacement y at each point of the string is small compared to the length L and that the angle between the string and the x axis at each point is also sufficiently small. Finally, we assume that no external forces (such as damping forces, for example) act upon the string.

Although these assumptions are not actually valid in any physical problem, nevertheless they are approximately satisfied in many cases. They are made in order to make the resulting mathematical problem more tractable. With these assumptions, then, the problem is to find the displacement y as a function of x and t.

The Mathematical Problem. Under the assumptions stated it can be shown that the displacement y satisfies the *partial differential equation*,

$$\alpha^2 \frac{\partial^2 y}{\partial x^2} = \frac{\partial^2 y}{\partial t^2}, \tag{14.20}$$

where $\alpha^2 = T/\rho$. This is the one-dimensional wave equation, a special case of which we have already studied in Example 14.3. Since our primary concern here is to illustrate the method of separation of variables, we omit the derivation of the equation.

Since the ends of the string are fixed at $x = 0$ and $x = L$ for all time t, the displacement y must satisfy the *boundary conditions*

$$y(0, t) = 0, \qquad 0 \le t < \infty;$$
$$y(L, t) = 0, \qquad 0 \le t < \infty. \tag{14.21}$$

At $t = 0$ the string is released from the initial position defined by $f(x), 0 \le x \le L$, with initial velocity given by $g(x), 0 \le x \le L$. Thus the displacement y must also satisfy the *initial conditions*

$$y(x, 0) = f(x), \qquad 0 \le x \le L;$$
$$\frac{\partial y(x, 0)}{\partial t} = g(x), \qquad 0 \le x \le L. \tag{14.22}$$

This, then, is our problem. We must find a function y of x and t which satisfies the partial differential equation (14.20), the boundary conditions (14.21), and the initial conditions (14.22).

Solution. We apply the method of separation of variables. We first make the basic assumption that the differential equation (14.20) has product solutions of the form XT, where X is a function of x only and T is a function of t only. To emphasize this, we write

$$y(x, t) = X(x)T(t). \tag{14.23}$$

We now differentiate (14.23) and substitute into the differential equation (14.20). Differentiating, we find

$$\frac{\partial^2 y}{\partial x^2} = T\frac{d^2 X}{dx^2} \quad \text{and} \quad \frac{\partial^2 y}{\partial t^2} = X\frac{d^2 T}{dt^2};$$

substituting, we obtain

$$\alpha^2 T\frac{d^2 X}{dx^2} = X\frac{d^2 T}{dt^2}.$$

From this we obtain at once

$$\alpha^2\frac{\dfrac{d^2 X}{dx^2}}{X} = \frac{\dfrac{d^2 T}{dt^2}}{T}. \tag{14.24}$$

Since X is a function of x only, the left member of (14.24) is also a function of x only and hence is independent of t. Further, since T is a function of t only, the right member of (14.24) is also a function of t only and hence is independent of x. Since one of the two equal expressions in (14.24) is independent of t and the other one is independent of x, both of them must be equal to a constant k. That is, we have

$$\alpha^2\frac{\dfrac{d^2 X}{dx^2}}{X} = k \quad \text{and} \quad \frac{\dfrac{d^2 T}{dt^2}}{T} = k.$$

From this we obtain the two ordinary differential equations

$$\frac{d^2X}{dx^2} - \frac{k}{\alpha^2} X = 0 \tag{14.25}$$

and

$$\frac{d^2T}{dt^2} - kT = 0. \tag{14.26}$$

Let us now consider the boundary conditions (14.21). Since $y(x, t) = X(x)T(t)$, we see that $y(0, t) = X(0)T(t)$ and $y(L, t) = X(L)T(t)$. Thus the boundary conditions (14.21) take the forms

$$X(0)T(t) = 0, \qquad 0 \le t < \infty;$$

$$X(L)T(t) = 0, \qquad 0 \le t < \infty.$$

Since $T(t) = 0$, $0 \le t < \infty$, would reduce the assumed solution (14.23) to the trivial solution of (14.20), we must have

$$X(0) = 0 \quad \text{and} \quad X(L) = 0. \tag{14.27}$$

Thus the function X in the assumed solution (14.23) must satisfy both the ordinary differential equation (14.25) and the boundary conditions (14.27). That is, the function X must be a nontrivial solution of the Sturm–Liouville problem

$$\frac{d^2X}{dx^2} - \frac{k}{\alpha^2} X = 0, \tag{14.25}$$

$$X(0) = 0, \qquad X(L) = 0. \tag{14.27}$$

We have already solved a special case of this problem in Example 12.3 of Chapter 12. Our procedure here will parallel the treatment in that example. We must first find the general solution of the differential equation (14.25). The form of this general solution depends upon whether $k = 0$, $k > 0$, or $k < 0$.

If $k = 0$, the general solution of (14.25) is of the form

$$X = c_1 + c_2 x. \tag{14.28}$$

We apply the boundary conditions (14.27) to the solution (14.28). The condition $X(0) = 0$ requires that $c_1 = 0$. The condition $X(L) = 0$ becomes $c_1 + c_2 L = 0$. Since $c_1 = 0$, this requires that $c_2 = 0$ also. Thus the solution (14.28) reduces to the trivial solution.

If $k > 0$, the general solution of (14.25) is of the form

$$X = c_1 e^{\sqrt{k}x/\alpha} + c_2 e^{-\sqrt{k}x/\alpha}. \tag{14.29}$$

Applying the boundary conditions (14.27) to the solution (14.29), we obtain the system of equations

$$c_1 + c_2 = 0,$$
$$c_1 e^{\sqrt{k}L/\alpha} + c_2 e^{-\sqrt{k}L/\alpha} = 0. \tag{14.30}$$

To obtain nontrivial solutions of this system, we must have

$$\begin{vmatrix} 1 & 1 \\ e^{\sqrt{k}L/\alpha} & e^{-\sqrt{k}L/\alpha} \end{vmatrix} = 0.$$

But this implies that $e^{\sqrt{k}L/\alpha} = e^{-\sqrt{k}L/\alpha}$ and hence that $k = 0$, contrary to our assumption in this case. Thus the system (14.30) has no nontrivial solutions, and so the solution (14.29) also reduces to the trivial solution.

Finally, if $k < 0$, the general solution of (14.25) is of the form

$$X = c_1 \sin \frac{\sqrt{-k}x}{\alpha} + c_2 \cos \frac{\sqrt{-k}x}{\alpha}. \qquad (14.31)$$

Applying the boundary conditions (14.27) to the solution (14.31), we obtain

$$c_2 = 0$$

and

$$c_1 \sin \frac{\sqrt{-k}L}{\alpha} + c_2 \cos \frac{\sqrt{-k}L}{\alpha} = 0.$$

Since $c_2 = 0$, the latter condition reduces to

$$c_1 \sin \frac{\sqrt{-k}L}{\alpha} = 0.$$

Thus to obtain nontrivial solutions of the form (14.31), we must have

$$\frac{\sqrt{-k}L}{\alpha} = n\pi \qquad (n = 1, 2, 3, \ldots),$$

and so

$$k = -\frac{n^2\pi^2\alpha^2}{L^2} \qquad (n = 1, 2, 3, \ldots). \qquad (14.32)$$

We thus find that the constant k must be a negative number of the form (14.32). We recognize these values of k as the characteristic values of the Sturm–Liouville problem under consideration. The corresponding nontrivial solutions (the characteristic functions) of the problem are then

$$X_n = c_n \sin \frac{n\pi x}{L} \qquad (n = 1, 2, 3, \ldots), \qquad (14.33)$$

where the c_n $(n = 1, 2, 3, \ldots)$, are arbitrary constants. We thus find that the function X in the assumed solution (14.23) must be of the form (14.33). That is, corresponding to each positive integral value of n, we obtain functions X_n of the form (14.33) which will serve as the function X in the product solution (14.23).

Let us now return to the differential equation (14.26) which the function T in (14.23) must satisfy. Since k must be of the form (14.32), the differential equation (14.26) becomes

$$\frac{d^2 T}{dt^2} + \frac{n^2\pi^2\alpha^2}{L^2} T = 0,$$

where $n = 1, 2, 3, \ldots$. For each such value of n, this differential equation has solutions of the form

$$T_n = c_{n,1} \sin \frac{n\pi\alpha t}{L} + c_{n,2} \cos \frac{n\pi\alpha t}{L} \qquad (n = 1, 2, 3, \ldots), \qquad (14.34)$$

where the $c_{n,1}$ and $c_{n,2}$ $(n = 1, 2, 3, \ldots)$, are arbitrary constants. Thus the function T in the assumed solution (14.23) must be of the form (14.34). That is, corresponding to each positive integral value of n, we obtain functions T_n of the form (14.34) which will serve as the function T in the product solution (14.23).

Therefore, corresponding to each positive integral value of n $(n = 1, 2, 3, \ldots)$, we obtain solutions

$$X_n T_n = \left[c_n \sin \frac{n\pi x}{L} \right] \left[c_{n,1} \sin \frac{n\pi\alpha t}{L} + c_{n,2} \cos \frac{n\pi\alpha t}{L} \right]$$

which have the product form (14.23).

We set $a_n = c_n c_{n,1}$ and $b_n = c_n c_{n,2}$ $(n = 1, 2, 3, \ldots)$, and write these solutions as

$$y_n(x, t) = \left[\sin \frac{n\pi x}{L} \right] \left[a_n \sin \frac{n\pi\alpha t}{L} + b_n \cos \frac{n\pi\alpha t}{L} \right] \qquad (n = 1, 2, 3, \ldots). \quad (14.35)$$

We point out that each of these solutions (14.35) satisfies both the partial differential equation (14.20) and the two boundary conditions (14.21) for all values of the constants a_n and b_n.

We must now try to satisfy the two initial conditions (14.22). In general no single one of the solutions (14.35) will satisfy these conditions. For example, if we apply the first initial condition (14.22) to a solution of the form (14.35) we must have

$$b_n \sin \frac{n\pi x}{L} = f(x), \qquad 0 \le x \le L,$$

where n is some positive integer; and this is clearly impossible unless f happens to be a sine function of the form $A \sin(n\pi x/L)$ for some positive integer n.

What can we do now? By Theorem 14.1 every finite linear combination of solutions of (14.20) is also a solution of (14.20); and by Theorem 14.2, assuming appropriate convergence, an infinite series of solutions of (14.20) is also a solution of (14.20). This suggests that we should form either a finite linear combination or an infinite series of the solutions (14.35) and attempt to apply the initial conditions (14.22) to the "more general" solutions thus obtained. In general no finite linear combination will satisfy these conditions, and we must resort to an infinite series.

We therefore form an infinite series

$$\sum_{n=1}^{\infty} y_n(x, t) = \sum_{n=1}^{\infty} \left[\sin \frac{n\pi x}{L} \right] \left[a_n \sin \frac{n\pi\alpha t}{L} + b_n \cos \frac{n\pi\alpha t}{L} \right]$$

of the solutions (14.35). Assuming appropriate convergence, Theorem 14.2 applies and assures us that the sum of this series is also a solution of the differential equation (14.20). Denoting this sum by $y(x, t)$, we write

$$y(x, t) = \sum_{n=1}^{\infty} \left[\sin \frac{n\pi x}{L} \right] \left[a_n \sin \frac{n\pi\alpha t}{L} + b_n \cos \frac{n\pi\alpha t}{L} \right]. \quad (14.36)$$

We note that $y(0, t) = 0$ and $y(L, t) = 0$. Thus, assuming appropriate convergence, the function y given by (14.36) satisfies both the differential equation (14.20) and the two boundary conditions (14.21).

Let us now apply the initial conditions (14.22) to the series solution (14.36). The first condition $y(x, 0) = f(x)$, $0 \leq x \leq L$, reduces (14.36) to

$$\sum_{n=1}^{\infty} b_n \sin \frac{n\pi x}{L} = f(x), \qquad 0 \leq x \leq L. \tag{14.37}$$

Thus to satisfy the first initial condition (14.22), we must determine the coefficients b_n so that (14.37) is satisfied. We recognize this as a problem in Fourier sine series (see Section 12.4C). Using (12.54) we find that the coefficients b_n are given by

$$b_n = \frac{2}{L} \int_0^L f(x) \sin \frac{n\pi x}{L} dx \qquad (n = 1, 2, 3, \ldots). \tag{14.38}$$

Thus in order for the series solution (14.36) to satisfy the initial condition $y(x, 0) = f(x)$, $0 \leq x \leq L$, the coefficients b_n in the series must be given by (14.38).

The only condition which remains to be satisfied is the second initial condition (14.22), which is

$$\frac{\partial y(x, 0)}{\partial t} = g(x), \qquad 0 \leq x \leq L.$$

From (14.36), we find that

$$\frac{\partial y(x, t)}{\partial t} = \sum_{n=1}^{\infty} \left[\frac{n\pi\alpha}{L} \right] \left[\sin \frac{n\pi x}{L} \right] \left[a_n \cos \frac{n\pi\alpha t}{L} - b_n \sin \frac{n\pi\alpha t}{L} \right].$$

The second initial condition reduces this to

$$\sum_{n=1}^{\infty} \frac{a_n n\pi\alpha}{L} \sin \frac{n\pi x}{L} = g(x), \qquad 0 \leq x \leq L.$$

Letting $A_n = a_n n\pi\alpha/L$ $(n = 1, 2, 3, \ldots)$, this takes the form

$$\sum_{n=1}^{\infty} A_n \sin \frac{n\pi x}{L} = g(x), \qquad 0 \leq x \leq L. \tag{14.39}$$

Thus to satisfy the second initial condition (14.22), we must determine the coefficients A_n so that (14.39) is satisfied. This is another problem in Fourier sine series. Using (12.54) again, we find that the coefficients A_n are given by

$$A_n = \frac{2}{L} \int_0^L g(x) \sin \frac{n\pi x}{L} dx \qquad (n = 1, 2, 3, \ldots).$$

Since $A_n = a_n n\pi\alpha/L$ $(n = 1, 2, 3, \ldots)$, we find that

$$a_n = \frac{L}{n\pi\alpha} A_n = \frac{2}{n\pi\alpha} \int_0^L g(x) \sin \frac{n\pi x}{L} dx \qquad (n = 1, 2, 3, \ldots). \tag{14.40}$$

Thus in order for the series solution (14.36) to satisfy the second initial condition (14.22), the coefficients a_n in the series must be given by (14.40).

Therefore, the formal solution of the problem consisting of the partial differential equation (14.20), the two boundary conditions (14.21), and the two initial conditions (14.22) is

$$y(x, t) = \sum_{n=1}^{\infty} \left[\sin \frac{n\pi x}{L} \right] \left[a_n \sin \frac{n\pi\alpha t}{L} + b_n \cos \frac{n\pi\alpha t}{L} \right], \tag{14.36}$$

where

$$a_n = \frac{2}{n\pi\alpha} \int_0^L g(x)\sin\frac{n\pi x}{L}\,dx \qquad (n = 1, 2, 3,\dots). \tag{14.40}$$

and

$$b_n = \frac{2}{L} \int_0^L f(x)\sin\frac{n\pi x}{L}\,dx \qquad (n = 1, 2, 3,\dots). \tag{14.38}$$

Summary. We briefly summarize the principal steps in the solution of this problem. The initial step was to assume the product solution XT given by (14.23). This led to the ordinary differential equation (14.25) for the function X and the ordinary differential equation (14.26) for the function T. We then considered the boundary conditions (14.21) and found that they reduced to the boundary conditions (14.27) on the function X. Thus the function X had to be a nontrivial solution of the Sturm–Liouville problem consisting of (14.25) and (14.27). The next step was to solve this Sturm–Liouville problem. We did this and obtained for solutions the functions X_n given by (14.33). We then returned to the differential equation (14.26) for the function T and obtained the solutions T_n given by (14.34). Thus, for each positive integral value of n, we found the product solutions $X_n T_n$ denoted by y_n and given by (14.35). Each of these solutions y_n satisfied both the partial differential equation (14.20) and the boundary conditions (14.21), but no one of them satisfied the initial conditions (14.22). In order to satisfy these initial conditions, we formed an infinite series of the solutions y_n. We thus obtained the formal solution y given by (14.36), in which the coefficients a_n and b_n were arbitrary. We applied the initial conditions to this series solution and thereby determined the coefficients a_n and b_n. We thus obtained the formal solution y given by (14.36), in which the coefficients a_n and b_n are given by (14.40) and (14.38), respectively. We emphasize that this solution is a formal one, for in the process of obtaining it we made assumptions of convergence which we did not justify.

A Special Case. As a particular case of the vibrating string problem, we consider the problem of the so-called *plucked string*. Let us suppose that the string is such that the constant $\alpha^2 = 2500$ and that the ends of the string are fixed on the x axis at $x = 0$ and $x = 1$. Suppose the midpoint of the string is displaced into the xy plane a distance 0.01 in the direction of the positive y axis (see Figure 14.2).

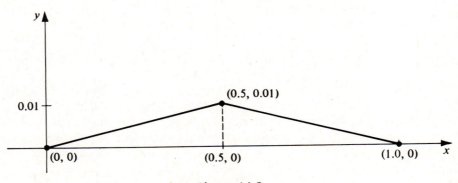

Figure 14.2

Then the displacement from the x axis on the interval $0 \le x \le 1$ is given by $f(x)$, where

$$f(x) = \begin{cases} \dfrac{x}{50}, & 0 \le x \le \dfrac{1}{2}; \\[2mm] -\dfrac{x}{50} + \dfrac{1}{50}, & \dfrac{1}{2} \le x \le 1. \end{cases} \tag{14.41}$$

Suppose that at $t = 0$ the string is released from rest from the initial position defined by $f(x), 0 \le x \le 1$. Let us find the formal expression (14.36) for the displacement $y(x, t)$ in the y direction, in this special case.

The coefficients a_n and b_n in the expression (14.36) are given by (14.40) and (14.38), respectively. In the special case under consideration we have $\alpha = 50$, $L = 1$, and $f(x)$ given by (14.41). Further, since the string is released from rest, the initial velocity is given by $g(x) = 0, 0 \le x \le 1$. Therefore from (14.40) we find that

$$a_n = 0 \qquad (n = 1, 2, 3, \ldots).$$

Using (14.38), we find that

$$b_n = 2 \int_0^1 f(x) \sin n\pi x \, dx$$

$$= 2 \int_0^{1/2} \frac{x}{50} \sin n\pi x \, dx + 2 \int_{1/2}^1 \left(-\frac{x}{50} + \frac{1}{50} \right) \sin n\pi x \, dx$$

$$= \frac{2}{25n^2\pi^2} \sin \frac{n\pi}{2} \qquad (n = 1, 2, 3, \ldots).$$

Hence the even coefficients $b_{2n} = 0$ ($n = 1, 2, 3, \ldots$), and the odd coefficients are given by

$$b_{2n-1} = \frac{(-1)^{n-1}2}{25\pi^2(2n-1)^2} \qquad (n = 1, 2, 3, \ldots).$$

Therefore in the special case under consideration the expression (14.36) for the displacement is

$$y(x, t) = \frac{2}{25\pi^2} \sum_{n=1}^{\infty} \frac{(-1)^{n-1}}{(2n-1)^2} \sin[(2n-1)\pi x] \cos[50(2n-1)\pi t].$$

C. An Example Involving the Laplace Equation

We now consider a second example of the application of the method of separation of variables to a partial differential equations problem. The problem which we shall consider originated from a problem of physics, but we shall not enter into a discussion of this related physical problem. Our sole purpose in presenting this second example is to help the reader to gain greater familiarity with the various details of the method under consideration.

Problem. Apply the method of separation of variables to obtain a formal solution $u(x, y)$ of the problem which consists of the two-dimensional Laplace equation

$$\frac{\partial^2 u}{\partial x^2} + \frac{\partial^2 u}{\partial y^2} = 0 \tag{14.42}$$

and the four boundary conditions

$$u(0, y) = 0, \qquad 0 \le y \le b; \tag{14.43}$$

$$u(a, y) = 0, \qquad 0 \le y \le b; \tag{14.44}$$

$$u(x, 0) = 0, \qquad 0 \le x \le a; \tag{14.45}$$

$$u(x, b) = f(x), \qquad 0 \le x \le a. \tag{14.46}$$

We point out that the numbers a and b are definite positive constants and the function f is a specified function of x, $0 \le x \le a$.

Formal Solution. We first make the basic assumption that the differential equation (14.42) has product solutions of the form XY, where X is a function of x only and Y is a function of y only. That is, we assume solutions

$$u(x, y) = X(x)Y(y). \tag{14.47}$$

We now differentiate (14.47) and substitute into the differential equation (14.42). Differentiating, we find

$$\frac{\partial^2 u}{\partial x^2} = Y \frac{d^2 X}{dx^2} \quad \text{and} \quad \frac{\partial^2 u}{\partial y^2} = X \frac{d^2 Y}{dy^2};$$

substituting, we obtain

$$Y \frac{d^2 X}{dx^2} + X \frac{d^2 Y}{dy^2} = 0.$$

From this we obtain at once

$$\frac{\dfrac{d^2 X}{dx^2}}{X} = -\frac{\dfrac{d^2 Y}{dy^2}}{Y}. \tag{14.48}$$

The left member of (14.48) is a function of x only and so is independent of y. The right member of (14.48) is a function of y only and so is independent of x. Therefore the two equal expressions in (14.48) must both be equal to a constant k. Setting each member of (14.48) equal to this constant k, we obtain the two ordinary differential equations

$$\frac{d^2 X}{dx^2} - kX = 0$$

and

$$\frac{d^2 Y}{dy^2} + kY = 0.$$

Let us now consider the four boundary conditions (14.43) through (14.46). The first three of these are homogeneous, but the fourth one is not. Let us attempt to satisfy the three homogeneous conditions first. Since $u(x, y) = X(x)Y(y)$, we see that the three homogeneous conditions (14.43), (14.44), and (14.45) reduce to

$$X(0)\,Y(y) = 0, \qquad 0 \le y \le b;$$

$$X(a)\,Y(y) = 0, \qquad 0 \le y \le b;$$

and

$$X(x)\,Y(0) = 0, \qquad 0 \le x \le a; \text{ respectively.}$$

Since either $X(x) = 0$, $0 \le x \le a$, or $Y(y) = 0$, $0 \le y \le b$, would reduce the assumed solution (14.47) to the trivial solution of (14.42), we must have

$$X(0) = 0, \qquad X(a) = 0, \quad \text{and} \quad Y(0) = 0.$$

Thus the function X in the assumed solution (14.47) must be a nontrivial solution of the Sturm–Liouville problem

$$\frac{d^2 X}{dx^2} - kX = 0, \tag{14.49}$$

$$X(0) = 0, \qquad X(a) = 0. \tag{14.50}$$

Further, the function Y in (14.47) must be a nontrivial solution of the problem

$$\frac{d^2 Y}{dy^2} + kY = 0, \tag{14.51}$$

$$Y(0) = 0. \tag{14.52}$$

The Sturm–Liouville problem (14.49) and (14.50) is essentially the same as the problem (14.25) and (14.27) which we encountered and solved in connection with the vibrating string problem in Part B of this section. Indeed, the present problem (14.49) and (14.50) is merely the special case of the problem (14.25) and (14.27) in which $\alpha^2 = 1$ and $L = a$. Thus if we set $\alpha^2 = 1$ and $L = a$ in the results (14.32) and (14.33) of the problem (14.25) and (14.27), we shall obtain the desired results for the present problem (14.49) and (14.50). Doing this, we first find from (14.32) that the constant k in (14.49) must be given by

$$k = -\frac{n^2 \pi^2}{a^2} \qquad (n = 1, 2, 3, \dots). \tag{14.53}$$

Then from (14.33) we find that the corresponding nontrivial solutions of the problem (14.49) and (14.50) are

$$X_n = c_n \sin \frac{n\pi x}{a} \qquad (n = 1, 2, 3, \dots), \tag{14.54}$$

where the c_n $(n = 1, 2, 3, \dots)$ are arbitrary constants. That is, corresponding to each positive integral value of n, we obtain functions X_n of the form (14.54) which will serve as the function X in the product solution (14.47).

We now return to the problem (14.51) and (14.52) involving the function Y. Since k must be of the form (14.53), the differential equation (14.51) becomes

$$\frac{d^2 Y}{dy^2} - \frac{n^2 \pi^2}{a^2} Y = 0,$$

where $n = 1, 2, 3, \dots$. For each such value of n, this differential equation has the general solution

$$Y_n = c_{n,1} e^{n\pi y/a} + c_{n,2} e^{-n\pi y/a} \qquad (n = 1, 2, 3, \dots),$$

where $c_{n,1}$ and $c_{n,2}$ $(n = 1, 2, 3, \dots)$, are arbitrary constants. In order to satisfy the condition (14.52), we must have

$$c_{n,1} + c_{n,2} = 0 \qquad (n = 1, 2, 3, \dots).$$

Thus nontrivial solutions of the problem (14.51) and (14.52) are

$$Y_n = c_{n,1}(e^{n\pi y/a} - e^{-n\pi y/a}) \qquad (n = 1, 2, 3, \dots),$$

where the $c_{n,1}$ $(n = 1, 2, 3, \dots)$, are arbitrary constants. Using the identity $e^{\theta} - e^{-\theta} = 2 \sinh \theta$, we may put these solutions in the form

$$Y_n = c'_{n,1} \sinh \frac{n\pi y}{a} \qquad (n = 1, 2, 3, \dots), \tag{14.55}$$

where the $c'_{n,1}$ $(n = 1, 2, 3, \dots)$, are arbitrary constants. Thus, corresponding to each positive integral value of n, we obtain functions Y_n of the form (14.55) which will serve as the function Y in the product solution (14.47).

Hence, corresponding to each positive integral value of $n(n = 1, 2, 3, \dots)$, we obtain solutions

$$X_n Y_n = \left[c_n \sin \frac{n\pi x}{a} \right]\left[c'_{n,1} \sinh \frac{n\pi y}{a} \right]$$

which have the product form (14.47). We set $C_n = c_n c'_{n,1}$ $(n = 1, 2, 3, \dots)$, and write these solutions as

$$u_n(x, y) = C_n \sin \frac{n\pi x}{a} \sinh \frac{n\pi y}{a} \qquad (n = 1, 2, 3, \dots). \tag{14.56}$$

Each one of these solutions (14.56) satisfies both the partial differential equation (14.42) and the three homogeneous boundary conditions (14.43), (14.44), and (14.45) for all values of the constant C_n.

We must now apply the nonhomogeneous boundary condition (14.46). In order to do this, we form an infinite series

$$\sum_{n=1}^{\infty} u_n(x, y) = \sum_{n=1}^{\infty} C_n \sin \frac{n\pi x}{a} \sinh \frac{n\pi y}{a}$$

of the solutions (14.56). Assuming appropriate convergence, Theorem 14.2 applies and assures us that the sum of this series is also a solution of the differential equation (14.42). Denoting this sum by $u(x, y)$, we write

$$u(x, y) = \sum_{n=1}^{\infty} C_n \sin \frac{n\pi x}{a} \sinh \frac{n\pi y}{a}. \tag{14.57}$$

We observe that $u(0, y) = 0$, $u(a, y) = 0$, and $u(x, 0) = 0$. Thus, assuming appropriate convergence, the function u given by (14.57) satisfies both the differential equation (14.42) and the three homogeneous boundary conditions (14.43), (14.44), and (14.45).

We now apply the nonhomogeneous boundary condition (14.46) to the series solution (14.57). Doing so, we obtain at once

$$\sum_{n=1}^{\infty} C_n \sin \frac{n\pi x}{a} \sinh \frac{n\pi b}{a} = f(x), \qquad 0 \le x \le a.$$

Letting $A_n = C_n \sinh(n\pi b/a)$ $(n = 1, 2, 3, \ldots)$, this takes the form

$$\sum_{n=1}^{\infty} A_n \sin \frac{n\pi x}{a} = f(x), \qquad 0 \le x \le a. \tag{14.58}$$

Thus in order to satisfy the condition (14.46), we must determine the coefficients A_n so that (14.58) is satisfied. This is a problem in Fourier sine series. Using (12.56), we find that the coefficients A_n are given by

$$A_n = \frac{2}{a} \int_0^a f(x) \sin \frac{n\pi x}{a} \, dx \qquad (n = 1, 2, 3, \ldots).$$

Since $A_n = C_n \sinh(n\pi b/a)$ $(n = 1, 2, 3, \ldots)$, we find that

$$C_n = \frac{A_n}{\sinh \dfrac{n\pi b}{a}} = \frac{2}{a \sinh \dfrac{n\pi b}{a}} \int_0^a f(x) \sin \frac{n\pi x}{a} \, dx \qquad (n = 1, 2, 3, \ldots). \tag{14.59}$$

Thus in order for the series solution (14.57) to satisfy the nonhomogeneous boundary condition (14.46), the coefficients C_n in the series must be given by (14.59).

Therefore the formal solution of the problem consisting of the partial differential equation (14.42) and the four boundary conditions (14.43) through (14.46) is

$$u(x, y) = \sum_{n=1}^{\infty} C_n \sin \frac{n\pi x}{a} \sinh \frac{n\pi y}{a}, \tag{14.57}$$

where

$$C_n = \frac{2}{a \sinh \dfrac{n\pi b}{a}} \int_0^a f(x) \sin \frac{n\pi x}{a} \, dx \qquad (n = 1, 2, 3, \ldots). \tag{14.59}$$

D. An Example Involving Bessel Functions

In this example we shall apply the method of separation of variables to a problem in which the partial differential equation has a variable coefficient. As a result of this variable coefficient we shall encounter certain difficulties which were not present in the two previous examples. Further, we shall need to know a few results which we have not yet proved. Whenever such a result is needed, we shall state it without proof.

▶ **Problem** Apply the method of separation of variables to obtain a formal solution $u(x, t)$ of the problem which consists of the partial differential equation

$$\frac{\partial^2 u}{\partial x^2} + \frac{1}{x} \frac{\partial u}{\partial x} = \frac{\partial u}{\partial t} \tag{14.60}$$

and the three conditions

$$\text{1.} \quad u(L, t) = 0, \qquad t > 0; \tag{14.61}$$

$$\text{2.} \quad u(x, 0) = f(x), \qquad 0 < x < L, \tag{14.62}$$

where f is a prescribed function of x, $0 < x < L$; and

$$\text{3.} \quad \lim_{t \to \infty} u(x, t) = 0 \tag{14.63}$$

for each x, $0 \le x \le L$.

Formal Solution. We begin by making the basic assumption that the differential equation (14.60) has product solutions of the form

$$u(x, t) = X(x)T(t), \tag{14.64}$$

where X is a function of x only and T is a function of t only. Upon differentiating (14.64) and substituting into the differential equation (14.60), we obtain

$$T\frac{d^2X}{dx^2} + \frac{1}{x}T\frac{dX}{dx} = X\frac{dT}{dt}.$$

From this we obtain at once

$$\frac{1}{X}\left(\frac{d^2X}{dx^2} + \frac{1}{x}\frac{dX}{dx}\right) = \frac{1}{T}\frac{dT}{dt}. \tag{14.65}$$

The left member of (14.65) is a function of x only and so is independent of t. The right member of (14.65) is a function of t only and so is independent of x. Therefore the two equal expressions in (14.65) must both be equal to a constant k. Setting each member of (14.65) equal to this constant k, we obtain the two ordinary differential equations

$$\frac{d^2X}{dx^2} + \frac{1}{x}\frac{dX}{dx} - kX = 0 \tag{14.66}$$

and

$$\frac{dT}{dt} - kT = 0. \tag{14.67}$$

The effect of the variable coefficient $(1/x)$ in the partial differential equation (14.60) appears here, for the ordinary differential equation (14.66) also has this same variable coefficient.

We shall need the general solutions of both of the ordinary differential equations (14.66) and (14.67). Equation (14.67) is the more promising of the two; let us work with it first. We find at once that the general solution of (14.67) is of the form

$$T = Ce^{kt}, \tag{14.68}$$

where C is an arbitrary constant.

Let us now examine the three conditions (14.61), (14.62), and (14.63) to see if any of them will lead to further information about the solution (14.68). The first two of these conditions lead to conditions on the function X. Let us therefore examine the third condition (14.63). Since $u(x, t) = X(x)T(t)$, this condition reduces to

$$X(x)[\lim_{t\to\infty} T(t)] = 0$$

for each $x, 0 \le x \le L$. Hence we require that

$$\lim_{t\to\infty} T(t) = 0.$$

From this we see that the constant k in (14.68) must be a negative number. Therefore we set $k = -\lambda^2$, where λ is real and positive. The general solution (14.68) of the differential equation (14.67) now takes the form

$$T = Ce^{-\lambda^2 t}, \tag{14.69}$$

where C is an arbitrary constant.

Let us now return to the differential equation (14.66) for X. Since $k = -\lambda^2$ it now takes the form

$$\frac{d^2 X}{dx^2} + \frac{1}{x}\frac{dX}{dx} + \lambda^2 X = 0$$

or equivalently,

$$x^2 \frac{d^2 X}{dx^2} + x \frac{dX}{dx} + \lambda^2 x^2 X = 0. \tag{14.70}$$

The transformation $\theta = \lambda x$ reduces (14.70) to the equation

$$\theta^2 \frac{d^2 X}{d\theta^2} + \theta \frac{dX}{d\theta} + \theta^2 X = 0.$$

We readily recognize this equation as the Bessel equation of order zero. Its general solution may be written

$$X = c_1 J_0(\theta) + c_2 Y_0(\theta),$$

where J_0 and Y_0 are the Bessel functions of order zero of the first and second kind, respectively, and c_1 and c_2 are arbitrary constants (see Section 6.3). Thus the general solution of (14.70) may be written

$$X = c_1 J_0(\lambda x) + c_2 Y_0(\lambda x), \tag{14.71}$$

where c_1 and c_2 are arbitrary constants.

Let us now return to the condition (14.63). This condition requires that

$$\lim_{t \to \infty} u(x, t) = 0$$

for *each* x, $0 \le x \le L$. For $x = 0$ this becomes

$$\lim_{t \to \infty} u(0, t) = 0$$

or

$$X(0)[\lim_{t \to \infty} T(t)] = 0.$$

In order to satisfy this condition we must require that $X(0)$ be finite. We recall that $J_0(0) = 1$. However, it can be shown that

$$\lim_{x \to \infty} Y_0(\lambda x) = -\infty.$$

Thus in order for $X(0)$ to be finite we must set $c_2 = 0$. Thus the solution (14.71) reduces to

$$X = c_1 J_0(\lambda x). \tag{14.72}$$

Let us now consider the condition (14.61). We have already noticed that this condition leads to a condition on X. Indeed, it reduces to

$$X(L)T(t) = 0, \qquad t > 0;$$

thus we must have $X(L) = 0$. Applying this to the solution (14.72), we see that λ must satisfy the equation

$$J_0(\lambda L) = 0. \tag{14.73}$$

In Section 6.3B we pointed out that the function J_0 has a damped oscillatory behavior as $x \to +\infty$. Thus the equation $J_0(x) = 0$ has an infinite sequence of positive roots x_n $(n = 1, 2, 3, \ldots)$. Let us arrange these positive roots such that $x_n < x_{n+1}$ $(n = 1, 2, 3, \ldots)$. Then there exists a monotonic increasing sequence of positive numbers

$$\lambda_n = \frac{x_n}{L} \qquad (n = 1, 2, 3, \ldots),$$

each of which satisfies Equation (14.73). Thus corresponding to each positive integer n, the differential equation (14.70) has solutions which satisfy the condition (14.61). These solutions are of the form

$$X_n = c_{1,n} J_0(\lambda_n x) \qquad (n = 1, 2, 3, \ldots), \tag{14.74}$$

where the $c_{1,n}$ $(n = 1, 2, 3, \ldots)$ are arbitrary constants and the λ_n $(n = 1, 2, 3, \ldots)$ are the positive roots of Equation (14.73). That is, corresponding to each positive integer n, we obtain functions X_n of the form (14.74) which will serve as the function X in the product solution (14.64).

Let us now return to the solution (14.69) of the differential equation (14.67). We see that, corresponding to each positive integer n, the differential equation (14.67) has solutions of the form

$$T_n = c_{2,n} e^{-\lambda_n^2 t} \qquad (n = 1, 2, 3, \ldots), \tag{14.75}$$

where the $c_{2,n}$ $(n = 1, 2, 3, \ldots)$ are arbitrary constants and the λ_n $(n = 1, 2, 3, \ldots)$ are the positive roots of Equation (14.73). That is, corresponding to each positive integer n, we obtain functions T_n of the form (14.75) which will serve as the function T in the product solution (14.64).

Hence, corresponding to each positive integral value of n $(n = 1, 2, 3, \ldots)$, we obtain product solutions of the form

$$u_n(x, t) = A_n J_0(\lambda_n x) e^{-\lambda_n^2 t} \qquad (n = 1, 2, 3, \ldots), \tag{14.76}$$

where the $A_n = c_{1,n} c_{2,n}$ $(n = 1, 2, 3, \ldots)$ are arbitrary constants.. Each one of these solutions (14.76) satisfies the partial differential equation (14.60) and the conditions (14.61) and (14.63) for all values of the constant A_n.

We must now apply the initial condition (14.62). In order to do this, we form an infinite series of the solutions (14.76). Assuming appropriate convergence, the sum of this series is also a solution of the partial differential equation (14.60). We denote this sum by $u(x, t)$ and thus write

$$u(x, t) = \sum_{n=1}^{\infty} A_n J_0(\lambda_n x) e^{-\lambda_n^2 t}. \tag{14.77}$$

Applying the initial condition (14.62) to the series solution (14.77), we obtain

$$\sum_{n=1}^{\infty} A_n J_0(\lambda_n x) = f(x), \qquad 0 < x < L. \tag{14.78}$$

Thus in order to satisfy the initial condition (14.62), the coefficients A_n must be determined so that (14.78) is satisfied. In other words, we must expand the function f in a series of Bessel functions of the first kind of order zero, valid on the interval $0 < x < L$.

Here we have encountered a new difficulty, and this difficulty leads to matters which are outside the scope of this book. Nevertheless, we shall indicate briefly what can be done. This is one place where we need to know certain of the results which we referred to at the beginning of this section. Let us state them and get on with the problem!

In the first place, it can be shown that if the numbers λ_n $(n = 1, 2, 3, \ldots)$ are the positive roots of the equation $J_0(\lambda L) = 0$, then the set of functions defined by $\{J_0(\lambda_n x)\}$ $(n = 1, 2, 3, \ldots)$ is an orthogonal system with respect to the weight function r such that $r(x) = x$ on the interval $0 \leq x \leq L$. Therefore,

$$\int_0^L x J_0(\lambda_m x) J_0(\lambda_n x)\, dx = 0 \qquad (m = 1, 2, 3, \ldots; n = 1, 2, 3, \ldots; m \neq n).$$

Further, if $m = n$, we have

$$\int_0^L x [J_0(\lambda_n x)]^2\, dx = \Gamma_n > 0 \qquad (n = 1, 2, 3, \ldots). \tag{14.79}$$

In Section 12.3A we learned how to form a set of *orthonormal* functions from a set of orthogonal characteristic functions of a Sturm–Liouville problem. Applying this procedure to the orthogonal set defined by $\{J_0(\lambda_n x)\}$, we obtain the corresponding orthonormal system $\{\phi_n\}$, where

$$\phi_n(x) = \frac{J_0(\lambda_n x)}{\sqrt{\Gamma_n}} \qquad (n = 1, 2, 3, \ldots), \tag{14.80}$$

and Γ_n $(n = 1, 2, 3, \ldots)$ is given by (14.79). Let us now recall the results of Section 12.3B concerning the formal expansion of a function f in a series

$$\sum_{n=1}^{\infty} c_n \phi_n$$

of orthonormal functions $\{\phi_n\}$. According to (12.37) the coefficients c_n in the expansion of f in the series of orthonormal functions ϕ_n defined by (14.80) are given by

$$c_n = \frac{1}{\sqrt{\Gamma_n}} \int_0^L x f(x) J_0(\lambda_n x)\, dx \qquad (n = 1, 2, 3, \ldots).$$

Thus this expansion takes the form

$$\sum_{n=1}^{\infty} \left[\frac{1}{\sqrt{\Gamma_n}} \int_0^L x f(x) J_0(\lambda_n x)\, dx \right] \frac{J_0(\lambda_n x)}{\sqrt{\Gamma_n}},$$

and we write formally

$$f(x) = \sum_{n=1}^{\infty} \left[\frac{1}{\Gamma_n} \int_0^L x f(x) J_0(\lambda_n x)\, dx \right] J_0(\lambda_n x), \qquad 0 < x < L. \tag{14.81}$$

Comparing (14.78) and (14.81), we see that if the coefficients A_n in (14.78) are given by

$$A_n = \frac{1}{\Gamma_n} \int_0^L x f(x) J_0(\lambda_n x)\, dx \qquad (n = 1, 2, 3, \ldots), \tag{14.82}$$

then the requirement (14.78) will be formally satisfied. We note that the constants Γ_n in (14.82) are given by (14.79). That is,

$$\Gamma_n = \int_0^L x [J_0(\lambda_n x)]^2\, dx \qquad (n = 1, 2, 3, \ldots).$$

This integral can be evaluated in terms of values of the Bessel function of the first kind of order one, J_1. Indeed, it can be shown that

$$\int_0^L x[J_0(\lambda_n x)]^2\,dx = \frac{L^2}{2}[J_1(\lambda_n L)]^2 \qquad (n = 1, 2, 3, \ldots),$$

and thus

$$\Gamma_n = \frac{L^2}{2}[J_1(\lambda_n L)]^2 \qquad (n = 1, 2, 3, \ldots).$$

Thus the coefficients A_n in (14.78) are given by

$$A_n = \frac{2}{L^2[J_1(\lambda_n L)]^2}\int_0^L xf(x)J_0(\lambda_n x)\,dx \qquad (n = 1, 2, 3, \ldots). \qquad (14.83)$$

Finally, then, we obtain the formal solution of the problem consisting of the partial differential equation (14.60) and the conditions (14.61), (14.62), and (14.63). The formal solution is

$$u(x, t) = \sum_{n=1}^{\infty} A_n J_0(\lambda_n x)e^{-\lambda_n^2 t},$$

where

$$A_n = \frac{2}{L^2[J_1(\lambda_n L)]^2}\int_0^L xf(x)J_0(\lambda_n x)\,dx \qquad (n = 1, 2, 3, \ldots),$$

the λ_n $(n = 1, 2, 3, \ldots)$, are the positive roots of the equation $J_0(\lambda L) = 0$, and J_0 and J_1 denote the Bessel functions of the first kind of orders zero and one, respectively.

Remarks and Observations. At the risk of being unduly repetitious, but with the good intentions of promoting a cautious attitude, we emphasize that the results which we have obtained are strictly formal results. We *assumed* "appropriate convergence" of the series in (14.77), and we have said nothing concerning the convergence of the Bessel function expansion which we obtained for the function f. In order to relieve our consciences concerning the latter point, we state that there do exist classes of functions f such that the expansions (14.78), in which the coefficients A_n are given by (14.83), is valid on the interval $0 < x < L$. The study of these classes is definitely beyond the scope of this book, and we refer the reader to more advanced works for a discussion of this and other pertinent problems of convergence.

Finally, we point out that the problem which we have considered here gives some indication of the types of difficulties which may be encountered if the method of separation of variables is applied to a problem in which the partial differential equation has variable coefficients. The variable coefficient $(1/x)$ in the partial differential equation (14.60) led to the variable coefficient $(1/x)$ in the ordinary differential equation (14.66) which resulted from the separation of the variables. A similar situation occurs in other problems in which the partial differential equation has variable coefficients. In such problems, one or more of the ordinary differential equations which result from the separation process will also contain variable coefficients. Obtaining the general solutions of these ordinary differential equations can then be a formidable task in its own right. But even if these general solutions can be obtained, they may involve functions which will lead to further difficulties when one attempts to apply certain of the supplementary conditions. This sort of difficulty occurred in the problem of this section when we tried to apply the initial condition (14.62). We were forced to consider the problem of expanding the function f in a series of Bessel functions, valid on the interval $0 < x < L$. A similar situation often occurs in other problems which involve a partial differential equation with variable coefficients. In such problems one is faced with the task of expanding a prescribed function f in a series of nonelementary orthonormal functions $\{\phi_n\}$, valid on a certain interval. The set of orthonormal functions $\{\phi_n\}$ might happen to be one of the many such sets which have been carefully studied, or it might turn out to be a set about which little is known. In any case, additional difficulties may occur which will necessitate further study and possibly some research.

Exercises

Use the method of separation of variables to find a formal solution $y(x, t)$ of each of the problems stated in Exercises 1–4

1. $\dfrac{\partial^2 y}{\partial x^2} = \dfrac{\partial^2 y}{\partial t^2}$,

   $y(0, t) = 0, \quad 0 \le t < \infty,$

   $y(\pi, t) = 0, \quad 0 \le t < \infty,$

   $y(x, 0) = \sin 2x, \quad 0 \le x \le \pi,$

   $\dfrac{\partial y(x, 0)}{\partial t} = 0, \quad 0 \le x \le \pi.$

2. $\dfrac{\partial^2 y}{\partial x^2} = \dfrac{\partial^2 y}{\partial t^2}$,

   $y(0, t) = 0, \quad 0 \le t < \infty,$

   $y(3\pi, t) = 0, \quad 0 \le t < \infty,$

   $y(x, 0) = 2 \sin^3 x, \quad 0 \le x \le 3\pi,$

   $\dfrac{\partial y(x, 0)}{\partial t} = 0, \quad 0 \le x \le 3\pi.$

3. $4\dfrac{\partial^2 y}{\partial x^2} = \dfrac{\partial^2 y}{\partial t^2}$,

   $y(0, t) = 0, \quad 0 \le t < \infty,$

   $y(3, t) = 0, \quad 0 \le t < \infty,$

   $y(x, 0) = \begin{cases} x, & 0 \le x \le 1, \\ 1, & 1 \le x \le 2, \\ 3 - x, & 2 \le x \le 3, \end{cases}$

   $\dfrac{\partial y(x, 0)}{\partial t} = 0, \quad 0 \le x \le 3.$

4. $4\dfrac{\partial^2 y}{\partial x^2} = 9\dfrac{\partial^2 y}{\partial t^2}$,

$y(0, t) = 0, \quad 0 \le t < \infty,$

$y(\pi, t) = 0, \quad 0 \le t < \infty,$

$y(x, 0) = \sin^2 x, \quad 0 \le x \le \pi,$

$\dfrac{\partial y(x, 0)}{\partial t} = \sin x, \quad 0 \le x \le \pi.$

5. Apply the method of separation of variables to obtain a formal solution $u(x, t)$ of the problem which consists of the heat equation

$$\frac{\partial^2 u}{\partial x^2} = \frac{\partial u}{\partial t},$$

the boundary conditions

$$u(0, t) = 0, \quad t > 0,$$
$$u(L, t) = 0, \quad t > 0,$$

and the initial condition

$$u(x, 0) = f(x), \qquad 0 \le x \le L,$$

where $L > 0$, and f is a specified function of $x, 0 \le x \le L$.

6. Use the method of separation of variables to find a formal solution $u(x, y)$ of the problem which consists of the partial differential equation

$$\frac{\partial^2 u}{\partial x^2} - \frac{\partial u}{\partial y} = -Ae^{-\alpha x},$$

where $A \ge 0$ and $\alpha > 0$, and the conditions

$$u(0, y) = 0, \qquad y > 0,$$
$$u(L, y) = 0, \qquad y > 0,$$
$$u(x, 0) = f(x), \qquad 0 < x < L.$$

[*Hint*: Let $u(x, y) = v(x, y) + \psi(x)$, where ψ is such that v satisfies the "homogeneous" equation

$$\frac{\partial^2 v}{\partial x^2} - \frac{\partial v}{\partial y} = 0$$

and the homogeneous boundary conditions

$$v(0, y) = 0, \, v(L, y) = 0.]$$

7. Use the method of separation of variables to find a formal solution $u(x, y)$ of the problem which consists of Laplace's equation

$$\frac{\partial^2 u}{\partial x^2} + \frac{\partial^2 u}{\partial y^2} = 0$$

and the boundary conditions

$$u(0, y) = 0, \qquad 0 \le y \le \pi,$$
$$u(\pi, y) = 0, \qquad 0 \le y \le \pi,$$
$$u(x, \pi) = 0, \qquad 0 \le x \le \pi,$$
$$u(x, 0) = f(x), \qquad 0 \le x \le \pi,$$

where f is a specified function of x, $0 \le x \le \pi$.

8. Use the method of separation of variables to obtain a formal solution $u(r, \theta)$ of the problem which consists of the equation

$$\frac{\partial^2 u}{\partial r^2} + \frac{1}{r}\frac{\partial u}{\partial r} + \frac{1}{r^2}\frac{\partial^2 u}{\partial \theta^2} = 0,$$

the periodicity condition

$$u(r, \theta) = u(r, \theta + 2\pi), \qquad 0 \le r \le L, \quad \text{for all values of } \theta,$$

and the boundary conditions

$$u(0, \theta) = \alpha, \qquad \text{where } \alpha \text{ is finite,}$$
$$u(L, \theta) = f(\theta), \qquad \text{where } f \text{ is a prescribed function of } \theta, \text{ for all values of } \theta.$$

14.3 CANONICAL FORMS OF SECOND-ORDER LINEAR EQUATIONS WITH CONSTANT COEFFICIENTS

A. Canonical Forms

In this section we restrict our attention to second-order linear partial differential equations of the form

$$A\frac{\partial^2 u}{\partial x^2} + B\frac{\partial^2 u}{\partial x \, \partial y} + C\frac{\partial^2 u}{\partial y^2} + D\frac{\partial u}{\partial x} + E\frac{\partial u}{\partial y} + Fu = 0, \qquad (14.84)$$

in which the coefficients A, B, C, D, E, and F are real *constants*. This equation is a special case of the more general equation (14.6) in which these coefficients are functions of x and y. In Section 14.1 we classified such equations according to the sign of $B^2 - 4AC$. Using this classification, Equation (14.84) is said to be

1. *hyperbolic* if $B^2 - 4AC > 0$;
2. *parabolic* if $B^2 - 4AC = 0$;
3. *elliptic* if $B^2 - 4AC < 0$.

We shall now show that in each of these three cases Equation (14.84) can be reduced to a more simple form by a suitable change of the independent variables. The simpler forms which result in this way are called *canonical forms* of Equation (14.84). We therefore introduce new independent variables ξ, η by means of the transformation

$$\xi = \xi(x, y), \qquad \eta = \eta(x, y). \qquad (14.85)$$

We now compute derivatives of u, regarding ξ and η as intermediate variables, so that $u = u(\xi, \eta)$, where ξ and η are given by (14.85). We first find

$$\frac{\partial u}{\partial x} = \frac{\partial u}{\partial \xi}\frac{\partial \xi}{\partial x} + \frac{\partial u}{\partial \eta}\frac{\partial \eta}{\partial x} \tag{14.86}$$

and

$$\frac{\partial u}{\partial y} = \frac{\partial u}{\partial \xi}\frac{\partial \xi}{\partial y} + \frac{\partial u}{\partial \eta}\frac{\partial \eta}{\partial y}. \tag{14.87}$$

Using (14.86), we next determine $\dfrac{\partial^2 u}{\partial x^2}$.

We find

$$\frac{\partial^2 u}{\partial x^2} = \frac{\partial}{\partial x}\left(\frac{\partial u}{\partial x}\right)$$

$$= \frac{\partial u}{\partial \xi}\frac{\partial}{\partial x}\left(\frac{\partial \xi}{\partial x}\right) + \frac{\partial \xi}{\partial x}\frac{\partial}{\partial x}\left(\frac{\partial u}{\partial \xi}\right) + \frac{\partial u}{\partial \eta}\frac{\partial}{\partial x}\left(\frac{\partial \eta}{\partial x}\right) + \frac{\partial \eta}{\partial x}\frac{\partial}{\partial x}\left(\frac{\partial u}{\partial \eta}\right). \tag{14.88}$$

Since $\xi = \xi(x, y)$ and $\eta = \eta(x, y)$,

$$\frac{\partial}{\partial x}\left(\frac{\partial \xi}{\partial x}\right) \quad \text{is simply} \quad \frac{\partial^2 \xi}{\partial x^2}$$

and

$$\frac{\partial}{\partial x}\left(\frac{\partial \eta}{\partial x}\right) \quad \text{is simply} \quad \frac{\partial^2 \eta}{\partial x^2}.$$

However, in computing $\dfrac{\partial}{\partial x}\left(\dfrac{\partial u}{\partial \xi}\right)$ and $\dfrac{\partial}{\partial x}\left(\dfrac{\partial u}{\partial \eta}\right)$, the situation is somewhat more complicated and we must be careful to remember what is involved. We are regarding u as a function of ξ and η, where ξ and η are themselves functions of x and y. That is, $u = u(\xi, \eta)$, where $\xi = \xi(x, y)$ and $\eta = \eta(x, y)$. Thus $\dfrac{\partial u}{\partial \xi}$ and $\dfrac{\partial u}{\partial \eta}$ are regarded as functions of ξ and η, where ξ and η are themselves functions of x and y. That is, $\dfrac{\partial u}{\partial \xi} = u_1(\xi, \eta)$ and $\dfrac{\partial u}{\partial \eta} = u_2(\xi, \eta)$, where in each case $\xi = \xi(x, y)$ and $\eta = \eta(x, y)$. With this in mind, we compute $\dfrac{\partial}{\partial x}\left(\dfrac{\partial u}{\partial \xi}\right)$. We have

$$\frac{\partial}{\partial x}\left(\frac{\partial u}{\partial \xi}\right) = \frac{\partial}{\partial x}[u_1(\xi, \eta)] = \frac{\partial u_1}{\partial \xi}\frac{\partial \xi}{\partial x} + \frac{\partial u_1}{\partial \eta}\frac{\partial \eta}{\partial x},$$

and since $u_1 = \dfrac{\partial u}{\partial \xi}$, we thus find

$$\frac{\partial}{\partial x}\left(\frac{\partial u}{\partial \xi}\right) = \frac{\partial^2 u}{\partial \xi^2}\frac{\partial \xi}{\partial x} + \frac{\partial^2 u}{\partial \eta\, \partial \xi}\frac{\partial \eta}{\partial x}.$$

In like manner, we obtain

$$\frac{\partial}{\partial x}\left(\frac{\partial u}{\partial \eta}\right) = \frac{\partial^2 u}{\partial \xi \, \partial \eta}\frac{\partial \xi}{\partial x} + \frac{\partial^2 u}{\partial \eta^2}\frac{\partial \eta}{\partial x}.$$

Substituting these results into (14.88), we thus obtain

$$\frac{\partial^2 u}{\partial x^2} = \frac{\partial u}{\partial \xi}\frac{\partial^2 \xi}{\partial x^2} + \frac{\partial \xi}{\partial x}\left(\frac{\partial^2 u}{\partial \xi^2}\frac{\partial \xi}{\partial x} + \frac{\partial^2 u}{\partial \eta \, \partial \xi}\frac{\partial \eta}{\partial x}\right)$$

$$+ \frac{\partial u}{\partial \eta}\frac{\partial^2 \eta}{\partial x^2} + \frac{\partial \eta}{\partial x}\left(\frac{\partial^2 u}{\partial \xi \, \partial \eta}\frac{\partial \xi}{\partial x} + \frac{\partial^2 u}{\partial \eta^2}\frac{\partial \eta}{\partial x}\right).$$

Assuming $u(\xi, \eta)$ has continuous second derivatives with respect to ξ and η, the so-called cross derivatives are equal, and we have

$$\frac{\partial^2 u}{\partial x^2} = \frac{\partial^2 u}{\partial \xi^2}\left(\frac{\partial \xi}{\partial x}\right)^2 + 2\frac{\partial^2 u}{\partial \xi \, \partial \eta}\frac{\partial \xi}{\partial x}\frac{\partial \eta}{\partial x} + \frac{\partial^2 u}{\partial \eta^2}\left(\frac{\partial \eta}{\partial x}\right)^2$$

$$+ \frac{\partial u}{\partial \xi}\frac{\partial^2 \xi}{\partial x^2} + \frac{\partial u}{\partial \eta}\frac{\partial^2 \eta}{\partial x^2}. \tag{14.89}$$

In like manner, we find

$$\frac{\partial^2 u}{\partial x \, \partial y} = \frac{\partial^2 u}{\partial \xi^2}\frac{\partial \xi}{\partial x}\frac{\partial \xi}{\partial y} + \frac{\partial^2 u}{\partial \xi \, \partial \eta}\left[\frac{\partial \xi}{\partial x}\frac{\partial \eta}{\partial y} + \frac{\partial \xi}{\partial y}\frac{\partial \eta}{\partial x}\right]$$

$$+ \frac{\partial^2 u}{\partial \eta^2}\frac{\partial \eta}{\partial x}\frac{\partial \eta}{\partial y} + \frac{\partial u}{\partial \xi}\frac{\partial^2 \xi}{\partial x \, \partial y} + \frac{\partial u}{\partial \eta}\frac{\partial^2 \eta}{\partial x \, \partial y} \tag{14.90}$$

and

$$\frac{\partial^2 u}{\partial y^2} = \frac{\partial^2 u}{\partial \xi^2}\left(\frac{\partial \xi}{\partial y}\right)^2 + 2\frac{\partial^2 u}{\partial \xi \, \partial \eta}\frac{\partial \xi}{\partial y}\frac{\partial \eta}{\partial y} + \frac{\partial^2 u}{\partial \eta^2}\left(\frac{\partial \eta}{\partial y}\right)^2$$

$$+ \frac{\partial u}{\partial \xi}\frac{\partial^2 \xi}{\partial y^2} + \frac{\partial u}{\partial \eta}\frac{\partial^2 \eta}{\partial y^2}. \tag{14.91}$$

We now substitute (14.86), (14.87), (14.89), (14.90), and (14.91) into the partial differential equation (14.84), to obtain

$$A\left[\frac{\partial^2 u}{\partial \xi^2}\left(\frac{\partial \xi}{\partial x}\right)^2 + 2\frac{\partial^2 u}{\partial \xi \, \partial \eta}\frac{\partial \xi}{\partial x}\frac{\partial \eta}{\partial x} + \frac{\partial^2 u}{\partial \eta^2}\left(\frac{\partial \eta}{\partial x}\right)^2 + \frac{\partial u}{\partial \xi}\frac{\partial^2 \xi}{\partial x^2} + \frac{\partial u}{\partial \eta}\frac{\partial^2 \eta}{\partial x^2}\right]$$

$$+ B\left[\frac{\partial^2 u}{\partial \xi^2}\frac{\partial \xi}{\partial x}\frac{\partial \xi}{\partial y} + \frac{\partial^2 u}{\partial \xi \, \partial \eta}\left(\frac{\partial \xi}{\partial x}\frac{\partial \eta}{\partial y} + \frac{\partial \xi}{\partial y}\frac{\partial \eta}{\partial x}\right) + \frac{\partial^2 u}{\partial \eta^2}\frac{\partial \eta}{\partial x}\frac{\partial \eta}{\partial y} + \frac{\partial u}{\partial \xi}\frac{\partial^2 \xi}{\partial x \, \partial y} + \frac{\partial u}{\partial \eta}\frac{\partial^2 \eta}{\partial x \, \partial y}\right]$$

$$+ C\left[\frac{\partial^2 u}{\partial \xi^2}\left(\frac{\partial \xi}{\partial y}\right)^2 + 2\frac{\partial^2 u}{\partial \xi \, \partial \eta}\frac{\partial \xi}{\partial y}\frac{\partial \eta}{\partial y} + \frac{\partial^2 u}{\partial \eta^2}\left(\frac{\partial \eta}{\partial y}\right)^2 + \frac{\partial u}{\partial \xi}\frac{\partial^2 \xi}{\partial y^2} + \frac{\partial u}{\partial \eta}\frac{\partial^2 \eta}{\partial y^2}\right]$$

$$+ D\left[\frac{\partial u}{\partial \xi}\frac{\partial \xi}{\partial x} + \frac{\partial u}{\partial \eta}\frac{\partial \eta}{\partial x}\right] + E\left[\frac{\partial u}{\partial \xi}\frac{\partial \xi}{\partial y} + \frac{\partial u}{\partial \eta}\frac{\partial \eta}{\partial y}\right] + Fu = 0.$$

Rearranging terms, this becomes

$$\left[A\left(\frac{\partial \xi}{\partial x}\right)^2 + B\frac{\partial \xi}{\partial x}\frac{\partial \xi}{\partial y} + C\left(\frac{\partial \xi}{\partial y}\right)^2\right]\frac{\partial^2 u}{\partial \xi^2} + \left[2A\frac{\partial \xi}{\partial x}\frac{\partial \eta}{\partial x} + B\left(\frac{\partial \xi}{\partial x}\frac{\partial \eta}{\partial y} + \frac{\partial \xi}{\partial y}\frac{\partial \eta}{\partial x}\right)\right.$$

$$+ 2C \frac{\partial \xi}{\partial y} \frac{\partial \eta}{\partial y} \bigg] \frac{\partial^2 u}{\partial \xi \, \partial \eta} + \left[A \left(\frac{\partial \eta}{\partial x} \right)^2 + B \frac{\partial \eta}{\partial x} \frac{\partial \eta}{\partial y} + C \left(\frac{\partial \eta}{\partial y} \right)^2 \right] \frac{\partial^2 u}{\partial \eta^2}$$

$$+ \left[A \frac{\partial^2 \xi}{\partial x^2} + B \frac{\partial^2 \xi}{\partial x \, \partial y} + C \frac{\partial^2 \xi}{\partial y^2} + D \frac{\partial \xi}{\partial x} + E \frac{\partial \xi}{\partial y} \right] \frac{\partial u}{\partial \xi}$$

$$+ \left[A \frac{\partial^2 \eta}{\partial x^2} + B \frac{\partial^2 \eta}{\partial x \, \partial y} + C \frac{\partial^2 \eta}{\partial y^2} + D \frac{\partial \eta}{\partial x} + E \frac{\partial \eta}{\partial y} \right] \frac{\partial u}{\partial \eta} + Fu = 0.$$

Thus using the transformation (14.85), the equation (14.84) is reduced to the form

$$A_1 \frac{\partial^2 u}{\partial \xi^2} + B_1 \frac{\partial^2 u}{\partial \xi \, \partial \eta} + C_1 \frac{\partial^2 u}{\partial \eta^2} + D_1 \frac{\partial u}{\partial \xi} + E_1 \frac{\partial u}{\partial \eta} + F_1 u = 0, \qquad (14.92)$$

where

$$A_1 = A \left(\frac{\partial \xi}{\partial x} \right)^2 + B \frac{\partial \xi}{\partial x} \frac{\partial \xi}{\partial y} + C \left(\frac{\partial \xi}{\partial y} \right)^2,$$

$$B_1 = 2A \frac{\partial \xi}{\partial x} \frac{\partial \eta}{\partial x} + B \left(\frac{\partial \xi}{\partial x} \frac{\partial \eta}{\partial y} + \frac{\partial \xi}{\partial y} \frac{\partial \eta}{\partial x} \right) + 2C \frac{\partial \xi}{\partial y} \frac{\partial \eta}{\partial y},$$

$$C_1 = A \left(\frac{\partial \eta}{\partial x} \right)^2 + B \frac{\partial \eta}{\partial x} \frac{\partial \eta}{\partial y} + C \left(\frac{\partial \eta}{\partial y} \right)^2, \qquad (14.93)$$

$$D_1 = A \frac{\partial^2 \xi}{\partial x^2} + B \frac{\partial^2 \xi}{\partial x \, \partial y} + C \frac{\partial^2 \xi}{\partial y^2} + D \frac{\partial \xi}{\partial x} + E \frac{\partial \xi}{\partial y},$$

$$E_1 = A \frac{\partial^2 \eta}{\partial x^2} + B \frac{\partial^2 \eta}{\partial x \, \partial y} + C \frac{\partial^2 \eta}{\partial y^2} + D \frac{\partial \eta}{\partial x} + E \frac{\partial \eta}{\partial y},$$

and

$$F_1 = F.$$

We now show that the new equation (14.92) can be simplified by a suitable choice of $\xi(x, y)$ and $\eta(x, y)$ in the transformation (14.85). The choice of these functions ξ and η and the form of the resulting simplified equation (the canonical form) depend upon whether the original partial differential equation (14.84) is hyperbolic, parabolic, or elliptic.

B. The Hyperbolic Equation

Concerning the canonical form in the hyperbolic case, we state and prove the following theorem.

THEOREM 14.3

Hypothesis. *Consider the second-order linear partial differential equation*

$$A \frac{\partial^2 u}{\partial x^2} + B \frac{\partial^2 u}{\partial x \, \partial y} + C \frac{\partial^2 u}{\partial y^2} + D \frac{\partial u}{\partial x} + E \frac{\partial u}{\partial y} + Fu = 0, \qquad (14.84)$$

where the coefficients A, B, C, D, E, and F are real constants and $B^2 - 4AC > 0$ so that the equation is hyperbolic.

Conclusion.   *There exists a transformation*

$$\xi = \xi(x, y), \qquad \eta = \eta(x, y) \tag{14.85}$$

of the independent variables in (14.84) so that the transformed equation in the independent variables (ξ, η) may be written in the canonical form

$$\frac{\partial^2 u}{\partial \xi \, \partial \eta} = d \frac{\partial u}{\partial \xi} + e \frac{\partial u}{\partial \eta} + fu, \tag{14.94}$$

where d, e, and f are real constants.
  If $A \neq 0$, such a transformation is given by

$$\xi = \lambda_1 x + y,$$
$$\eta = \lambda_2 x + y, \tag{14.95}$$

where λ_1 and λ_2 are the roots of the quadratic equation

$$A\lambda^2 + B\lambda + C = 0. \tag{14.96}$$

If $A = 0, B \neq 0, C \neq 0$, such a transformation is given by

$$\xi = x,$$
$$\eta = x - \frac{B}{C} y. \tag{14.97}$$

If $A = 0, B \neq 0, C = 0$, such a transformation is merely the identity transformation

$$\xi = x,$$
$$\eta = y. \tag{14.98}$$

Proof.   We shall first show that the transformations given by (14.95), (14.97), and (14.98) actually do reduce Equation (14.84) so that it may be written in the form (14.94) in the three respective cases described in the conclusion. We shall then observe that these three cases cover all possibilities for the hyperbolic equation, thereby completing the proof.
   We have seen that a transformation of the form (14.85) reduces Equation (14.84) to the form (14.92), where the coefficients are given by (14.93). In the case $A \neq 0$, we apply the special case of (14.85) given by

$$\xi = \lambda_1 x + y,$$
$$\eta = \lambda_2 x + y, \tag{14.95}$$

where λ_1 and λ_2 are the roots of the quadratic equation

$$A\lambda^2 + B\lambda + C = 0. \tag{14.96}$$

Then the coefficients in the transformed equation (14.92) are given by (14.93), where $\xi(x, y)$ and $\eta(x, y)$ are given by (14.95). Evaluating these coefficients in this case we find that

$$A_1 = A\lambda_1^2 + B\lambda_1 + C,$$

$$B_1 = 2A\lambda_1\lambda_2 + B(\lambda_1 + \lambda_2) + 2C$$

$$= 2A\left(\frac{C}{A}\right) + B\left(-\frac{B}{A}\right) + 2C = \frac{B^2 - 4AC}{-A},$$

$$C_1 = A\lambda_2^2 + B\lambda_2 + C,$$

$$D_1 = D\lambda_1 + E,$$

$$E_1 = D\lambda_2 + E,$$

$$F_1 = F.$$

Since λ_1 and λ_2 satisfy the quadratic equation (14.96), we see that $A_1 = 0$ and $C_1 = 0$. Therefore in this case the transformed equation (14.92) is

$$\left(\frac{B^2 - 4AC}{-A}\right)\frac{\partial^2 u}{\partial\xi\,\partial\eta} + (D\lambda_1 + E)\frac{\partial u}{\partial\xi} + (D\lambda_2 + E)\frac{\partial u}{\partial\eta} + Fu = 0. \qquad (14.99)$$

Since $B^2 - 4AC > 0$ [Equation (14.84) is hyperbolic], the roots λ_1 and λ_2 of (14.96) are real and distinct. Therefore the coefficients in (14.99) are all real. Furthermore, the leading coefficient $(B^2 - 4AC)/(-A)$ is unequal to zero. Therefore we may write the transformed equation (14.99) in the form

$$\frac{\partial^2 u}{\partial\xi\,\partial\eta} = d\frac{\partial u}{\partial\xi} + e\frac{\partial u}{\partial\eta} + fu,$$

where the coefficients

$$d = \frac{A(D\lambda_1 + E)}{B^2 - 4AC}, \qquad e = \frac{A(D\lambda_2 + E)}{B^2 - 4AC}, \qquad f = \frac{AF}{B^2 - 4AC}$$

are all real. This is the canonical form (14.94).

Now consider the case in which $A = 0$, $B \neq 0$, $C \neq 0$, and apply the special case of (14.85) given by

$$\xi = x,$$

$$\eta = x - \frac{B}{C}y. \qquad (14.97)$$

In this case the coefficients in the transformed equation (14.92) are given by (14.93), where $\xi(x, y)$ and $\eta(x, y)$ are given by (14.97). Evaluating these coefficients (recall $A = 0$ here), we find that

$$A_1 = 0,$$

$$B_1 = -\frac{B^2}{C} \neq 0,$$

$$C_1 = B\left(-\frac{B}{C}\right) + C\left(-\frac{B}{C}\right)^2 = 0,$$

$$D_1 = D, \qquad E_1 = D - \frac{EB}{C}, \qquad F_1 = F.$$

Therefore in this case the transformed equation is

$$\left(-\frac{B^2}{C}\right)\frac{\partial^2 u}{\partial\xi\,\partial\eta} + D\frac{\partial u}{\partial\xi} + \left(\frac{DC-EB}{C}\right)\frac{\partial u}{\partial\eta} + Fu = 0.$$

Since $(-B^2)/C \neq 0$, we may write this in the form

$$\frac{\partial^2 u}{\partial\xi\,\partial\eta} = d\frac{\partial u}{\partial\xi} + e\frac{\partial u}{\partial\eta} + fu,$$

where the coefficients

$$d = \frac{CD}{B^2}, \qquad e = \frac{DC-EB}{B^2}, \qquad f = \frac{CF}{B^2}$$

are all real. This is again the canonical form (14.94).

Finally, consider the case in which $A = 0, B \neq 0, C = 0$. In this case Equation (14.84) is simply

$$B\frac{\partial^2 u}{\partial x\,\partial y} + D\frac{\partial u}{\partial x} + E\frac{\partial u}{\partial y} + Fu = 0,$$

and the identity transformation (14.98) reduces it to

$$B\frac{\partial^2 u}{\partial\xi\,\partial\eta} + D\frac{\partial u}{\partial\xi} + E\frac{\partial u}{\partial\eta} + Fu = 0.$$

Since $B \neq 0$, we may write this in the form

$$\frac{\partial^2 u}{\partial\xi\,\partial\eta} = d\frac{\partial u}{\partial\xi} + e\frac{\partial u}{\partial\eta} + fu,$$

where

$$d = -\frac{D}{B}, \qquad e = -\frac{E}{B}, \qquad f = -\frac{F}{B}.$$

This is again the canonical form (14.94). In effect, in this special case ($A = 0, B \neq 0$, $C = 0$) Equation (14.84) may be put in the canonical form (14.94) simply by transposing the appropriate terms and dividing by B.

We now observe that the three special cases considered exhaust all possibilities for the hyperbolic equation (14.84). We first note that either $A = 0$ or $A \neq 0$. All cases in which $A \neq 0$ are covered by the first of the three special cases which we have considered. Turning to the cases in which $A = 0$, it would appear that the following four distinct possibilities deserve consideration: (a) $B \neq 0, C \neq 0$; (b) $B \neq 0, C = 0$; (c) $B = 0$, $C \neq 0$; and (d) $B = 0, C = 0$. We note that (a) and (b) are covered by the second and third of the three special cases which we have considered. Concerning (c) and (d), in both cases $B^2 - 4AC = 0$, contrary to hypothesis. In particular, if (c) holds, Equation (14.84) is parabolic (not hyperbolic); and if (d) holds, Equation (14.84) is of the first order.

We thus observe that the three special cases considered cover all possibilities for the hyperbolic equation (14.84). Thus there always exists a transformation (14.85) which transforms the hyperbolic equation (14.84) into one which may be written in the canonical form (14.94). Q.E.D

▶ **Example 14.6**

Consider the equation

$$\frac{\partial^2 u}{\partial x^2} + 4\frac{\partial^2 u}{\partial x\,\partial y} - 5\frac{\partial^2 u}{\partial y^2} + 6\frac{\partial u}{\partial x} + 3\frac{\partial u}{\partial y} - 9u = 0. \qquad (14.100)$$

We first observe that $B^2 - 4AC = 36 > 0$ and so Equation (14.100) is hyperbolic. Since $A \neq 0$, we consider the transformation

$$\xi = \lambda_1 x + y,$$
$$\eta = \lambda_2 x + y, \qquad (14.95)$$

where λ_1 and λ_2 are the roots of the quadratic equation $\lambda^2 + 4\lambda - 5 = 0$. We find that $\lambda_1 = 1$ and $\lambda_2 = -5$, and so the transformation (14.95) is

$$\xi = x + y,$$
$$\eta = -5x + y. \qquad (14.101)$$

Applying (14.101) to Equation (14.100), we see that this equation transforms into

$$-36\frac{\partial^2 u}{\partial \xi\,\partial \eta} + 9\frac{\partial u}{\partial \xi} - 27\frac{\partial u}{\partial \eta} - 9u = 0.$$

Dividing by -36 and transposing terms, we obtain the canonical form

$$\frac{\partial^2 u}{\partial \xi\,\partial \eta} = \frac{1}{4}\frac{\partial u}{\partial \xi} - \frac{3}{4}\frac{\partial u}{\partial \eta} - \frac{1}{4}u.$$

C. The Parabolic Equation

We now investigate the canonical form in the parabolic case and obtain the following theorem.

THEOREM 14.4

Hypothesis. *Consider the second-order linear partial differential equation*

$$A\frac{\partial^2 u}{\partial x^2} + B\frac{\partial^2 u}{\partial x\,\partial y} + C\frac{\partial^2 u}{\partial y^2} + D\frac{\partial u}{\partial x} + E\frac{\partial u}{\partial y} + Fu = 0, \qquad (14.84)$$

where the coefficients A, B, C, D, E, and F are real constants and $B^2 - 4AC = 0$ so that the equation is parabolic.

Conclusion. *There exists a transformation*

$$\xi = \xi(x, y), \qquad \eta = \eta(x, y) \qquad (14.85)$$

of the independent variables in (14.84) so that the transformed equation in the independent variables (ξ, η) *may be written in the canonical form*

$$\frac{\partial^2 u}{\partial \eta^2} = d \frac{\partial u}{\partial \xi} + e \frac{\partial u}{\partial \eta} + fu, \tag{14.102}$$

where d, e, and f are real constants.

If $A \neq 0$ and $C \neq 0$, *such a transformation is given by*

$$\xi = \lambda x + y,$$
$$\eta = y, \tag{14.103}$$

where λ is the repeated real root of the quadratic equation

$$A\lambda^2 + B\lambda + C = 0. \tag{14.104}$$

If $A \neq 0$ and $C = 0$, *such a transformation is given by*

$$\xi = y,$$
$$\eta = x. \tag{14.105}$$

If $A = 0$ and $C \neq 0$, *such a transformation is merely the identity transformation*

$$\xi = x,$$
$$\eta = y. \tag{14.106}$$

Proof. We shall proceed in a manner similar to that by which we proved Theorem 14.3.

If $A \neq 0$ and $C \neq 0$, we apply the transformation (14.103) to obtain the transformed equation (14.92) with coefficients (14.93), where in this case $\xi(x, y)$ and $\eta(x, y)$ are given by (14.103). Evaluating these coefficients we find that

$$A_1 = A\lambda^2 + B\lambda + C,$$
$$B_1 = B\lambda + 2C,$$
$$C_1 = C,$$
$$D_1 = D\lambda + E, \qquad E_1 = E, \qquad F_1 = F.$$

Since λ satisfies the quadratic equation (14.104), we see at once that $A_1 = 0$. Also, since $B^2 - 4AC = 0$, $\lambda = -B/2A$ and so

$$B_1 = -\frac{B^2}{2A} + 2C = \frac{4AC - B^2}{2A} = 0.$$

Thus in the present case the transformed equation (14.92) is

$$C \frac{\partial^2 u}{\partial \eta^2} + (D\lambda + E) \frac{\partial u}{\partial \xi} + E \frac{\partial u}{\partial \eta} + Fu = 0. \tag{14.107}$$

Since λ is real, all coefficients in (14.107) are real; and since $C \neq 0$, we may write equation (14.107) in the form

$$\frac{\partial^2 u}{\partial \eta^2} = d \frac{\partial u}{\partial \xi} + e \frac{\partial u}{\partial \eta} + fu,$$

where the coefficients

$$d = -\frac{D\lambda + E}{C}, \qquad e = -\frac{E}{C}, \qquad f = -\frac{F}{C}$$

are all real. This is the canonical form (14.102).

If $A \neq 0$ and $C = 0$, we apply the transformation (14.105) to obtain the transformed equation (14.92) with coefficients (14.93), where in this case $\xi(x, y)$ and $\eta(x, y)$ are given by (14.105). Evaluating these coefficients, we obtain

$$A_1 = C = 0,$$
$$B_1 = B = 0 \qquad \text{(since } B^2 - 4AC = 0 \quad \text{and} \quad C = 0),$$
$$C_1 = A \neq 0,$$
$$D_1 = E, \qquad E_1 = D, \qquad F_1 = F.$$

Thus in the case under consideration the transformed equation (14.92) is

$$A\frac{\partial^2 u}{\partial \eta^2} + E\frac{\partial u}{\partial \xi} + D\frac{\partial u}{\partial \eta} + Fu = 0.$$

Since $A \neq 0$, we may write this in the form

$$\frac{\partial^2 u}{\partial \eta^2} = d\frac{\partial u}{\partial \xi} + e\frac{\partial u}{\partial \eta} + fu,$$

where the coefficients $d = -E/A$, $e = -D/A$, and $f = -F/A$ are all real. This is again the canonical form (14.102).

Finally, consider the case in which $A = 0$ and $C \neq 0$. Since $B^2 - 4AC = 0$, we must also have $B = 0$. Therefore in this case Equation (14.84) is simply

$$C\frac{\partial^2 u}{\partial y^2} + D\frac{\partial u}{\partial x} + E\frac{\partial u}{\partial y} + Fu = 0,$$

and the identity transformation (14.106) reduces it to

$$C\frac{\partial^2 u}{\partial \eta^2} + D\frac{\partial u}{\partial \xi} + E\frac{\partial u}{\partial \eta} + Fu = 0.$$

Since $C \neq 0$, we may write this in the form

$$\frac{\partial^2 u}{\partial \eta^2} = d\frac{\partial u}{\partial \xi} + e\frac{\partial u}{\partial \eta} + fu,$$

where $d = -D/C$, $e = -E/C$, $f = -F/C$. This is again the canonical form (14.102). We thus see that in this special case ($A = 0$, $B = 0$, $C \neq 0$) Equation (14.84) may be put in the canonical form (14.102) simply by transposing the appropriate terms and dividing by C.

Finally, we observe that the three special cases considered exhaust all possibilities for the parabolic equation (14.84). For, either $A = 0$ or $A \neq 0$. If $A \neq 0$, either $C \neq 0$ or $C = 0$. These two possibilities are, respectively, the first and second special cases considered, and so all cases in which $A \neq 0$ are thus covered. If $A = 0$, either $C \neq 0$ or $C = 0$. The first of these two possibilities is the third special case considered. Finally, consider the situation in which $A = 0$ and $C = 0$. Since $B^2 - 4AC = 0$, we must also have $B = 0$ and so Equation (14.84) reduces to a first-order equation.

Therefore there always exists a transformation (14.85) which transforms the parabolic equation (14.84) into one which may be written in the canonical form (14.102).

Q.E.D.

▶ **Example 14.7**

Consider the equation

$$\frac{\partial^2 u}{\partial x^2} - 6\frac{\partial^2 u}{\partial x\,\partial y} + 9\frac{\partial^2 u}{\partial y^2} + 2\frac{\partial u}{\partial x} + 3\frac{\partial u}{\partial y} - u = 0. \tag{14.108}$$

We observe that $B^2 - 4AC = 0$, and so Equation (14.108) is parabolic. Since $A \neq 0$ and $C \neq 0$, we consider the transformation

$$\xi = \lambda x + y,$$
$$\eta = y, \tag{14.103}$$

where λ is the repeated real root of the quadratic equation $\lambda^2 - 6\lambda + 9 = 0$. We find that $\lambda = 3$, and so the transformation (14.103) is

$$\xi = 3x + y,$$
$$\eta = y. \tag{14.109}$$

Applying (14.109) to Equation (14.108), we see that this equation transforms into

$$9\frac{\partial^2 u}{\partial \eta^2} + 9\frac{\partial u}{\partial \xi} + 3\frac{\partial u}{\partial \eta} - u = 0.$$

Dividing by 9 and transposing terms, we obtain the canonical form

$$\frac{\partial^2 u}{\partial \eta^2} = -\frac{\partial u}{\partial \xi} - \frac{1}{3}\frac{\partial u}{\partial \eta} + \frac{1}{9}u.$$

D. The Elliptic Equation

Finally, we prove the following theorem concerning the canonical form in the elliptic case.

THEOREM 14.5

Hypothesis. *Consider the second-order linear partial differential equation*

$$A\frac{\partial^2 u}{\partial x^2} + B\frac{\partial^2 u}{\partial x\,\partial y} + C\frac{\partial^2 u}{\partial y^2} + D\frac{\partial u}{\partial x} + E\frac{\partial u}{\partial y} + Fu = 0, \tag{14.84}$$

where the coefficients A, B, C, D, E, and F are real constants and $B^2 - 4AC < 0$ so that the equation is elliptic.

Conclusion. *There exists a transformation*

$$\xi = \xi(x, y), \qquad \eta = \eta(x, y) \tag{14.85}$$

of the independent variables in (14.84) so that the transformed equation in the independent variables (ξ, η) may be written in the canonical form

$$\frac{\partial^2 u}{\partial \xi^2} + \frac{\partial^2 u}{\partial \eta^2} = d\frac{\partial u}{\partial \xi} + e\frac{\partial u}{\partial \eta} + fu, \tag{14.110}$$

where d, e, and f are real constants.
    Such a transformation is given by

$$\xi = ax + y,$$
$$\eta = bx, \tag{14.111}$$

where a ± bi (a and b real, b ≠ 0) are the conjugate complex roots of the quadratic equation

$$A\lambda^2 + B\lambda + C = 0 \tag{14.112}$$

Proof. Since $B^2 - 4AC < 0$, we cannot have $A = 0$ in the elliptic case. Thus Equation (14.112) is a full-fledged quadratic equation with two roots. The condition $B^2 - 4AC < 0$ further shows that these two roots must indeed be conjugate complex.

We apply the transformation (14.111) to obtain the transformed equation (14.92) with coefficients (14.93), where in this case $\xi(x, y)$ and $\eta(x, y)$ are given by (14.111). Evaluating these coefficients we find that

$$A_1 = Aa^2 + Ba + C,$$
$$B_1 = 2Aab + Bb = b(2Aa + B)$$
$$C_1 = Ab^2 \neq 0 \quad \text{(since } A \neq 0, \quad b \neq 0\text{)}$$
$$D_1 = Da + E, \quad E_1 = Db, \quad F_1 = F.$$

Since $a + bi$ satisfies the quadratic equation (14.112), we have

$$A(a + bi)^2 + B(a + bi) + C = 0$$

or

$$[A(a^2 - b^2) + Ba + C] + [b(2Aa + B)]i = 0.$$

Therefore

$$A(a^2 - b^2) + Ba + C = 0$$

and

$$b(2Aa + B) = 0.$$

Thus

$$A_1 = Aa^2 + Ba + C = Ab^2$$

and

$$B_1 = 0.$$

Hence the transformed equation (14.92) is

$$Ab^2\left(\frac{\partial^2 u}{\partial \xi^2} + \frac{\partial^2 u}{\partial \eta^2}\right) + (Da + E)\frac{\partial u}{\partial \xi} + Db\frac{\partial u}{\partial \eta} + Fu = 0. \tag{14.113}$$

Since a and b are real, all coefficients in (14.113) are real; and since $Ab^2 \neq 0$, we may write Equation (14.113) in the form

$$\frac{\partial^2 u}{\partial \xi^2} + \frac{\partial^2 u}{\partial \eta^2} = d \frac{\partial u}{\partial \xi} + e \frac{\partial u}{\partial \eta} + fu,$$

where the coefficients

$$d = -\frac{Da + E}{Ab^2}, \qquad e = -\frac{D}{Ab}, \qquad f = -\frac{F}{Ab^2}$$

are all real. This is the canonical form (14.110).   *Q.E.D.*

▶ **Example 14.8**

Consider the equation

$$\frac{\partial^2 u}{\partial x^2} + 2 \frac{\partial^2 u}{\partial x \, \partial y} + 5 \frac{\partial^2 u}{\partial y^2} + \frac{\partial u}{\partial x} - 2 \frac{\partial u}{\partial y} - 3u = 0. \tag{14.114}$$

We observe that $B^2 - 4AC = -16 < 0$ and so Equation (14.114) is elliptic. We consider the transformation

$$\xi = ax + y,$$
$$\eta = bx, \tag{14.111}$$

where $a \pm bi$ are the conjugate complex roots of the quadratic equation $\lambda^2 + 2\lambda + 5 = 0$. We find that these roots are $-1 \pm 2i$, and so the transformation (14.111) is

$$\xi = -x + y,$$
$$\eta = 2x. \tag{14.115}$$

Applying (14.115) to Equation (14.114), we see that this equation transforms into

$$4 \frac{\partial^2 u}{\partial \xi^2} + 4 \frac{\partial^2 u}{\partial \eta^2} - 3 \frac{\partial u}{\partial \xi} + 2 \frac{\partial u}{\partial \eta} - 3u = 0.$$

Dividing by 4 and transposing terms, we obtain the canonical form

$$\frac{\partial^2 u}{\partial \xi^2} + \frac{\partial^2 u}{\partial \eta^2} = \frac{3}{4} \frac{\partial u}{\partial \xi} - \frac{1}{2} \frac{\partial u}{\partial \eta} + \frac{3}{4} u.$$

E. Summary

Summarizing, we list in Table 14.1 the canonical forms which we have obtained for the second-order linear partial differential equation

$$A \frac{\partial^2 u}{\partial x^2} + B \frac{\partial^2 u}{\partial x \, \partial y} + C \frac{\partial^2 u}{\partial y^2} + D \frac{\partial u}{\partial x} + E \frac{\partial u}{\partial y} + Fu = 0, \tag{14.84}$$

where $A, B, C, D, E,$ and F are real constants.

TABLE 14.1

Type of equation (14.84)	Canonical form (where d, e, and f are real constants)
hyperbolic: $B^2 - 4AC > 0$	$\dfrac{\partial^2 u}{\partial \xi \partial \eta} = d\dfrac{\partial u}{\partial \xi} + e\dfrac{\partial u}{\partial \eta} + fu$
parabolic: $B^2 - 4AC = 0$	$\dfrac{\partial^2 u}{\partial \eta^2} = d\dfrac{\partial u}{\partial \xi} + e\dfrac{\partial u}{\partial \eta} + fu$
elliptic: $B^2 - 4AC < 0$	$\dfrac{\partial^2 u}{\partial \xi^2} + \dfrac{\partial^2 u}{\partial \eta^2} = d\dfrac{\partial u}{\partial \xi} + e\dfrac{\partial u}{\partial \eta} + fu$

Exercises

Transform each of the partial differential equations in Exercises 1–10 into canonical form.

1. $\dfrac{\partial^2 u}{\partial x^2} - 5\dfrac{\partial^2 u}{\partial x\, \partial y} + 6\dfrac{\partial^2 u}{\partial y^2} = 0.$

2. $\dfrac{\partial^2 u}{\partial x^2} - 4\dfrac{\partial^2 u}{\partial x\, \partial y} + 4\dfrac{\partial^2 u}{\partial y^2} = 0.$

3. $\dfrac{\partial^2 u}{\partial x^2} - 2\dfrac{\partial^2 u}{\partial x\, \partial y} - 8\dfrac{\partial^2 u}{\partial y^2} + 9\dfrac{\partial u}{\partial x} = 0.$

4. $2\dfrac{\partial^2 u}{\partial x^2} + 3\dfrac{\partial^2 u}{\partial x\, \partial y} - 9\dfrac{\partial^2 u}{\partial y^2} + 4\dfrac{\partial u}{\partial x} = 0.$

5. $\dfrac{\partial^2 u}{\partial x^2} - 4\dfrac{\partial^2 u}{\partial x\, \partial y} + 13\dfrac{\partial^2 u}{\partial y^2} - 9\dfrac{\partial u}{\partial y} = 0.$

6. $\dfrac{\partial^2 u}{\partial x^2} + 2\dfrac{\partial^2 u}{\partial x\, \partial y} + \dfrac{\partial^2 u}{\partial y^2} + 3\dfrac{\partial u}{\partial x} + 9u = 0.$

7. $6\dfrac{\partial^2 u}{\partial x\, \partial y} + 3\dfrac{\partial^2 u}{\partial y^2} + 2\dfrac{\partial u}{\partial x} = 0.$

8. $\dfrac{\partial^2 u}{\partial x^2} + 4\dfrac{\partial^2 u}{\partial x\, \partial y} + 3\dfrac{\partial u}{\partial x} + 5\dfrac{\partial u}{\partial y} = 0.$

9. $2\dfrac{\partial^2 u}{\partial x^2} - 2\dfrac{\partial^2 u}{\partial x\, \partial y} + 5\dfrac{\partial^2 u}{\partial y^2} + u = 0.$

10. $\dfrac{\partial^2 u}{\partial x^2} - 5\dfrac{\partial^2 u}{\partial x\, \partial y} + 5\dfrac{\partial u}{\partial x} - \dfrac{\partial u}{\partial y} + 3u = 0.$

11. Show that the transformation

$$\xi = y - \frac{x^2}{2},$$

$$\eta = x,$$

reduces the equation

$$\frac{\partial^2 u}{\partial x^2} + 2x\frac{\partial^2 u}{\partial x\,\partial y} + x^2\frac{\partial^2 u}{\partial y^2} = 0$$

to

$$\frac{\partial^2 u}{\partial \eta^2} = \frac{\partial u}{\partial \xi}.$$

12.   Consider the equation

$$\frac{\partial^2 u}{\partial x^2} + (2x + 3)\frac{\partial^2 u}{\partial x\,\partial y} + 6x\frac{\partial^2 u}{\partial y^2} = 0. \tag{A}$$

(a)   Show that for $x = \frac{3}{2}$, Equation (A) reduces to a parabolic equation, and reduce this parabolic equation to canonical form.

(b)   Show that for $x \neq \frac{3}{2}$, the transformation

$$\xi = y - 3x,$$

$$\eta = y - x^2$$

reduces Equation (A) to

$$\frac{\partial^2 u}{\partial \xi\,\partial \eta} = \frac{2\dfrac{\partial u}{\partial \eta}}{4(\eta - \xi) - 9}.$$

14.4   AN INITIAL-VALUE PROBLEM; CHARACTERISTICS

A. An Initial-Value Problem

We shall now consider an initial-value problem for the second-order linear partial differential equation

$$A\frac{\partial^2 u}{\partial x^2} + B\frac{\partial^2 u}{\partial x\,\partial y} + C\frac{\partial^2 u}{\partial y^2} + D\frac{\partial u}{\partial x} + E\frac{\partial u}{\partial y} + Fu = 0, \tag{14.84}$$

where A, B, C, D, E, and F are real constants. In Chapter 4 we considered an initial-value problem for linear *ordinary* differential equations. Let us begin by recalling this problem for the second-order homogeneous linear ordinary differential equation with constant coefficients,

$$a\frac{d^2 y}{dx^2} + b\frac{dy}{dx} + cy = 0. \tag{14.116}$$

The problem is to find a solution f of (14.116) such that $f(x_0) = c_0$ and $f'(x_0) = c_1$, where x_0 is some definite real number and c_0 and c_1 are arbitrary real constants. Interpreting this geometrically, the problem is to find a solution curve $y = f(x)$ of (14.116) which passes through the point (x_0, c_0) and whose tangent line has the slope c_1 at this point.

Let us now attempt to formulate an analogous initial-value problem for the *partial* differential equation (14.84). In doing this we shall let ourselves be guided by the geometric interpretation of the initial-value problem for the ordinary differential equation (14.116). In this problem we must find a solution of Equation (14.116) which satisfies certain supplementary conditions (the initial conditions). A solution of Equation (14.116) is a function f of the one real variable x, and this defines a curve $y = f(x)$ (a solution curve) in the xy plane. In the analogous problem for the partial differential equation (14.84), we shall also seek a solution which satisfies certain supplementary conditions. But a solution of Equation (14.84) is a function ϕ of the two real variables x and y, and this defines a surface $u = \phi(x, y)$ (a solution surface) in three-dimensional x, y, u space. Geometrically speaking, then, for the partial differential equation (14.84) we must find a particular solution surface $u = \phi(x, y)$ in x, y, u space. Now in the initial-value problem for the ordinary differential equation (14.116), there are two supplementary requirements which the solution curve $y = f(x)$ in the xy plane must satisfy. First, the solution curve $y = f(x)$ must pass through a prescribed point (x_0, c_0) in the xy plane; and, second, this solution curve must be such that its tangent line has a prescribed slope c_1 at this point. What sort of analogous requirements might be imposed in an initial-value problem for the partial differential equation (14.84)? Instead of prescribing a point in the xy plane through which the solution curve must pass, we would prescribe a curve in x, y, u space through which the solution surface must pass. Instead of prescribing a slope for the tangent line to the solution curve at the prescribed point, we would prescribe a normal direction for the tangent plane to the solution surface along the prescribed curve. We thus formulate an initial-value problem for the partial differential equation (14.84) in the following geometric language.

We seek a solution surface $u = \phi(x, y)$ of the partial differential equation (14.84) which (a) passes through a prescribed curve Γ (the initial curve) in x, y, u space, and (b) is such that its tangent plane has a prescribed normal direction at all points of the initial curve Γ.

Let us now proceed to formulate this initial-value problem analytically. To do so, let us assume that the prescribed curve Γ has the parametric representation

$$x = x(t), \qquad y = y(t), \qquad u = u(t)$$

for all t on some real interval I. Now let $t_0 \in I$ and consider the corresponding point on Γ. This point has coordinates (x_0, y_0, u_0), where $x_0 = x(t_0)$, $y_0 = y(t_0)$, $u_0 = u(t_0)$. If the solution surface $u = \phi(x, y)$ is to pass through Γ at (x_0, y_0, u_0), then we must have $u_0 = \phi(x_0, y_0)$ or, equivalently, $u(t_0) = \phi[x(t_0), y(t_0)]$. Therefore the requirement that the solution surface $u = \phi(x, y)$ pass through the entire curve Γ is expressed by the condition

$$u(t) = \phi[x(t), y(t)] \tag{14.117}$$

for all $t \in I$.

We now consider the requirement that the tangent plane to the solution surface $u = \phi(x, y)$ have a prescribed normal direction at all points of Γ. Let us assume that this prescribed normal direction is given by $[p(t), q(t), -1]$ for all $t \in I$. Recalling that the normal direction to the tangent plane to $u = \phi(x, y)$ is given by

$$[\phi_x(x, y), \phi_y(x, y), -1],\!^*$$

---

* Here and throughout the remainder of the chapter it is convenient to employ subscript notation for the various partial derivatives of ϕ. Thus we denote

$$\frac{\partial \phi}{\partial x} \text{ by } \phi_x, \quad \frac{\partial \phi}{\partial y} \text{ by } \phi_y, \quad \frac{\partial^2 \phi}{\partial x^2} \text{ by } \phi_{xx},$$

$$\frac{\partial^2 \phi}{\partial y\, \partial x} \text{ by } \phi_{xy}, \text{ and } \frac{\partial^2 \phi}{\partial y^2} \text{ by } \phi_{yy}.$$

we see that the requirement under consideration is expressed by the conditions that

$$p(t) = \phi_x[x(t), y(t)],$$
$$q(t) = \phi_y[x(t), y(t)], \tag{14.118}$$

for all $t \in I$.

We have thus far said nothing concerning the nature of the functions x, y, u which define Γ or the functions p, q which prescribe the normal direction. Let us assume that each of these five functions is analytic for all $t \in I$. We must now observe that the functions p and q cannot be chosen arbitrarily. For, from (14.117) we must have

$$\frac{du(t)}{dt} = \phi_x[x(t), y(t)] \frac{dx(t)}{dt} + \phi_y[x(t), y(t)] \frac{dy(t)}{dt}$$

for all $t \in I$; and since (14.118) must hold for all $t \in I$, this reduces to

$$\frac{du(t)}{dt} = p(t) \frac{dx(t)}{dt} + q(t) \frac{dy(t)}{dt} \tag{14.119}$$

for all $t \in I$. Thus the functions p and q are not arbitrary; rather they must satisfy the identity (14.119).

In our attempt to formulate the initial-value problem analytically, we have thus introduced a set of five functions defined by $\{x(t), y(t), u(t), p(t), q(t)\}$, where (1) $x = x(t)$, $y = y(t)$, $u = u(t)$ is the analytic representation of a prescribed curve Γ in x, y, u space, (2) $[p(t), q(t), -1]$ is the analytic representation of a prescribed normal direction along Γ, and (3) these five functions satisfy the identity (14.119). Such a set of five functions is called a *strip*, and the identity (14.119) is called the *strip condition*.

We may now state the initial-value problem in the following way.

Initial-Value Problem. Consider the second-order linear partial differential equation

$$A \frac{\partial^2 u}{\partial x^2} + B \frac{\partial^2 u}{\partial x \, \partial y} + C \frac{\partial^2 u}{\partial y^2} + D \frac{\partial u}{\partial x} + E \frac{\partial u}{\partial y} + Fu = 0, \tag{14.84}$$

where A, B, C, D, E, and F are real constants. Let $x(t)$, $y(t)$, $u(t)$, $p(t)$, and $q(t)$ denote five real functions, each of which is analytic for all t on a real interval I. Further, let these five functions be such that the condition

$$\frac{du(t)}{dt} = p(t) \frac{dx(t)}{dt} + q(t) \frac{dy(t)}{dt} \tag{14.119}$$

holds for all $t \in I$. Let R be a region of the xy plane such that $[x(t), y(t)] \in R$ for all $t \in I$.

We seek a solution $\phi(x, y)$ for the partial differential equation (14.84), defined for all $(x, y) \in R$, such that

$$\phi[x(t), y(t)] = u(t), \tag{14.117}$$

and

$$\phi_x[x(t), y(t)] = p(t),$$
$$\phi_y[x(t), y(t)] = q(t), \tag{14.118}$$

for all $t \in I$.

▶ **Example 14.9**

Consider the partial differential equation

$$\frac{\partial^2 u}{\partial x^2} + \frac{\partial^2 u}{\partial x\,\partial y} - 2\frac{\partial^2 u}{\partial y^2} = 0 \tag{14.120}$$

and the five real functions defined by

$$x(t) = t, \quad y(t) = t^2, \quad u(t) = t^4 + 2t^2, \quad p(t) = 4t, \quad q(t) = 2t^2 \tag{14.121}$$

for all real t. We observe that these five functions satisfy the strip condition (14.119); indeed we have

$$4t^3 + 4t = (4t)(1) + (2t^2)(2t)$$

for all real t. We may therefore interpret the set of five functions defined by (14.121) as a strip and state the following initial-value problem.

   We seek a solution $\phi(x, y)$ of the partial differential equation (14.120), defined for all (x, y), such that

$$\phi(t, t^2) = t^4 + 2t^2,$$

$$\phi_x(t, t^2) = 4t, \tag{14.122}$$

$$\phi_y(t, t^2) = 2t^2,$$

for all real t.

   Geometrically, we seek a solution surface $u = \phi(x, y)$ of the partial differential equation (14.120) which (a) passes through the space curve Γ defined by

$$x = t, \quad y = t^2, \quad u = t^4 + 2t^2$$

for all real t, and (b) is such that its tangent plane has the normal direction $(4t, 2t^2, -1)$ at all points of Γ.

   We now observe that a solution of this problem is the function ϕ defined by

$$\phi(x, y) = 2x^2 + y^2$$

for all (x, y). In the first place, this function ϕ satisfies the partial differential equation (14.120). For we find that

$$\phi_{xx}(x, y) = 4, \quad \phi_{yx}(x, y) = 0, \quad \text{and} \quad \phi_{yy}(x, y) = 2,$$

so that

$$\phi_{xx}(x, y) + \phi_{yx}(x, y) - 2\phi_{yy}(x, y) = 0$$

for all (x, y). In the second place, the conditions (14.122) are satisfied. Clearly $\phi(t, t^2) = 2t^2 + t^4$ for all real t. From $\phi_x(x, y) = 4x$, we see that $\phi_x(t, t^2) = 4t$ for all real t; and from $\phi_y(x, y) = 2y$, we see that $\phi_y(t, t^2) = 2t^2$ for all real t.

   We shall now examine the initial-value problem from a slightly different viewpoint, again employing useful geometric terminology in our discussion. Suppose, as before, that the curve Γ through which the solution surface $u = \phi(x, y)$ must pass is defined by

$$x = x(t), \quad y = y(t), \quad u = u(t) \tag{14.123}$$

for all t on some real interval I. Then the projection of Γ on the xy plane is the curve Γ_0

defined by

$$x = x(t), \qquad y = y(t) \tag{14.124}$$

for all $t \in I$. Employing this curve Γ_0, we may express the initial-value problem in the following form.

We seek a solution $\phi(x, y)$ of the partial differential equation (14.84) such that ϕ, ϕ_x, and ϕ_y assume prescribed values given by $u(t)$, $p(t)$, and $q(t)$, respectively, for each $t \in I$, or, in other words, at each point of Γ_0 [where we assume that the five functions given by $x(t)$, $y(t)$, $u(t)$, $p(t)$, and $q(t)$ satisfy the strip condition (14.119)].

We shall find that this interpretation of the initial-value problem will be useful in the discussion which follows.

B. Characteristics

Continuing our discussion of the initial-value problem, let us now assume that the problem has a unique solution $\phi(x, y)$ defined on a region R (of the xy plane) which includes the curve Γ_0. Let us also assume that at each point of Γ_0 the solution $\phi(x, y)$ has a power series expansion which is valid in some circle about this point. Now let $t_0 \in I$ and let $x_0 = x(t_0)$, $y_0 = y(t_0)$, so that (x_0, y_0) is a point of Γ_0. Then for all (x, y) in some circle K about (x_0, y_0) we have the valid power series expansion

$$\phi(x, y) = \phi(x_0, y_0) + [\phi_x(x_0, y_0)(x - x_0) + \phi_y(x_0, y_0)(y - y_0)]$$

$$+ \frac{1}{2!} [\phi_{xx}(x_0, y_0)(x - x_0)^2 + 2\phi_{xy}(x_0, y_0)(x - x_0)(y - y_0) \tag{14.125}$$

$$+ \phi_{yy}(x_0, y_0)(y - y_0)^2] + \cdots.$$

The solution $\phi(x, y)$ will be determined in the circle K if we can determine the coefficients

$$\phi(x_0, y_0), \qquad \phi_x(x_0, y_0), \quad \phi_y(x_0, y_0), \quad \phi_{xx}(x_0, y_0), \ldots$$

in the expansion (14.125).

Now we already know the first three of these coefficients. For the initial-value problem requires that ϕ, ϕ_x, and ϕ_y assume prescribed values given by $u(t)$, $p(t)$, and $q(t)$, respectively, at each point of Γ_0. Therefore at the point (x_0, y_0) at which $t = t_0$, we have

$$\phi(x_0, y_0) = u(t_0),$$

$$\phi_x(x_0, y_0) = p(t_0),$$

and

$$\phi_y(x_0, y_0) = q(t_0).$$

Let us now attempt to calculate the next three coefficients in (14.125). That is, let us see if we can determine

$$\phi_{xx}(x_0, y_0), \quad \phi_{xy}(x_0, y_0), \quad \phi_{yy}(x_0, y_0). \tag{14.126}$$

To do this, we need to know conditions which these numbers must satisfy. One such condition is readily available: it can be obtained at once from the partial differential equation (14.84)! For, since $\phi(x, y)$ is a solution of (14.84) in the region R, we have

$$A\phi_{xx}(x, y) + B\phi_{xy}(x, y) + C\phi_{yy}(x, y) = -D\phi_x(x, y) - E\phi_y(x, y) - F\phi(x, y)$$

for all $(x, y) \in R$. Thus, since $(x_0, y_0) \in R$, we have the condition

$$A\phi_{xx}(x_0, y_0) + B\phi_{xy}(x_0, y_0) + C\phi_{yy}(x_0, y_0)$$
$$= -D\phi_x(x_0, y_0) - E\phi_y(x_0, y_0) - F\phi(x_0, y_0). \qquad (14.127)$$

Now along the curve Γ_0 we know that

$$\phi_x[x(t), y(t)] = p(t)$$

and

$$\phi_y[x(t), y(t)] = q(t).$$

Differentiating these identities with respect to t, we find that along the curve Γ_0 we have

$$\phi_{xx}[x(t), y(t)] \frac{dx}{dt} + \phi_{xy}[x(t), y(t)] \frac{dy}{dt} = \frac{dp}{dt},$$

$$\phi_{xy}[x(t), y(t)] \frac{dx}{dt} + \phi_{yy}[x(t), y(t)] \frac{dy}{dt} = \frac{dq}{dt}.$$

Since (x_0, y_0) is a point of Γ_0, we thus obtain the two additional conditions

$$x'(t_0)\phi_{xx}(x_0, y_0) + y'(t_0)\phi_{xy}(x_0, y_0) = p'(t_0) \qquad (14.128)$$

and

$$x'(t_0)\phi_{xy}(x_0, y_0) + y'(t_0)\phi_{yy}(x_0, y_0) = q'(t_0), \qquad (14.129)$$

where $x'(t_0), y'(t_0), p'(t_0),$ and $q'(t_0)$ denote the values of $\dfrac{dx}{dt}, \dfrac{dy}{dt}, \dfrac{dp}{dt},$ and $\dfrac{dq}{dt}$, respectively, at $t = t_0$.

Let us examine more closely the three conditions (14.127), (14.128), and (14.129). They are actually a system of three linear algebraic equations in the three unknown numbers (14.126); all other quantities involved are known numbers! A necessary and sufficient condition that this system have a unique solution is that the determinant of coefficients be unequal to zero (Section 7.5C, Theorem B). Therefore in order to have a unique solution for the three coefficients (14.126), we must have

$$\Delta(t_0) = \begin{vmatrix} A & B & C \\ x'(t_0) & y'(t_0) & 0 \\ 0 & x'(t_0) & y'(t_0) \end{vmatrix} \neq 0.$$

Conversely, if $\Delta(t_0) \neq 0$, we can determine the three coefficients (14.126) from the linear system consisting of Equations (14.127), (14.128), and (14.129).

We have assumed that the solution $\phi(x, y)$ has a power series expansion of the form (14.125) at *each* point of the curve Γ_0. Thus if the determinant

$$\Delta(t) = \begin{vmatrix} A & B & C \\ \dfrac{dx}{dt} & \dfrac{dy}{dt} & 0 \\ 0 & \dfrac{dx}{dt} & \dfrac{dy}{dt} \end{vmatrix} \neq 0$$

at all points of Γ_0, we can determine the values of ϕ_{xx}, ϕ_{xy}, and ϕ_{yy} all along Γ_0. We can then attempt to find the values of the higher derivatives of $\phi(x, y)$ along Γ_0, employing a procedure similar to that by which we found the values of the three second derivatives. One can show that if $\Delta(t) \neq 0$ at all points of Γ_0, then the values of these higher derivatives can be determined all along Γ_0. In this manner we find the coefficients in the series expansion of $\phi(x, y)$ along Γ_0.

It can be shown that if $\Delta(t) \neq 0$ at all points of Γ_0, then there exists a unique solution of the initial-value problem. On the other hand, if $\Delta(t) = 0$ at all points of Γ_0, then there is either no solution or infinitely many solutions.

Let us now consider a curve Γ_0 at all points of which

$$\Delta(t) = \begin{vmatrix} A & B & C \\ \dfrac{dx}{dt} & \dfrac{dy}{dt} & 0 \\ 0 & \dfrac{dx}{dt} & \dfrac{dy}{dt} \end{vmatrix} = 0.$$

Expanding this determinant, we thus see that

$$A\left(\frac{dy}{dt}\right)^2 - B\frac{dx}{dt}\frac{dy}{dt} + C\left(\frac{dx}{dt}\right)^2 = 0$$

at all points of Γ_0. Therefore Γ_0 must be a curve having an equation $\zeta(x, y) = c_0$ which satisfies

$$A\,dy^2 - B\,dx\,dy + C\,dx^2 = 0. \tag{14.130}$$

We thus investigate the solutions of Equation (14.130). We shall divide this discussion into the following three subcases: (a) $A \neq 0$; (b) $A = 0$, $C \neq 0$; and (c) $A = 0$, $C = 0$.

Subcase (a): $A \neq 0$. In this case if follows from (14.130) that the curve Γ_0 is defined by $y = \phi(x)$, where $\phi(x)$ satisfies the ordinary differential equation

$$A\left(\frac{dy}{dx}\right)^2 - B\frac{dy}{dx} + C = 0.$$

From this equation we see that

$$\frac{dy}{dx} = \frac{B + \sqrt{B^2 - 4AC}}{2A} \quad \text{or} \quad \frac{dy}{dx} = \frac{B - \sqrt{B^2 - 4AC}}{2A}.$$

Thus if $A \neq 0$ the curve Γ_0 must be a member of one of the families of straight lines defined by

$$y = \left[\frac{B + \sqrt{B^2 - 4AC}}{2A}\right]x + c_1 \quad \text{or} \quad y = \left[\frac{B - \sqrt{B^2 - 4AC}}{2A}\right]x + c_2. \tag{14.131}$$

where c_1 and c_2 are arbitrary constants.

Subcase (b): $A = 0$, $C \neq 0$. In this case it follows from (14.130) that Γ_0 is defined by $x = \theta(y)$, where $\theta(y)$ satisfies the ordinary differential equation

$$C\left(\frac{dx}{dy}\right)^2 - B\frac{dx}{dy} = 0.$$

From this equation we see that

$$\frac{dx}{dy} = \frac{B}{C} \quad \text{or} \quad \frac{dx}{dy} = 0.$$

Thus if $A = 0$ and $C \neq 0$, the curve Γ_0 must be a member of one of the families of straight lines defined by

$$x = \frac{B}{C} y + c_1 \quad \text{or} \quad x = c_2, \tag{14.132}$$

where c_1 and c_2 are arbitrary constants.

Subcase (c): $A = 0$, $C = 0$. In this case it follows at once from (14.130) that the curve Γ_0 must be a member of one of the families of straight lines defined by

$$y = c_1 \quad \text{or} \quad x = c_2, \tag{14.133}$$

where c_1 and c_2 are arbitrary constants.

Thus if Γ_0 is a curve at all points of which $\Delta(t) = 0$, then Γ_0 must be one of the straight lines defined by (14.131), (14.132), or (14.133). Any such curve Γ_0 in the xy plane along which $\Delta(t) = 0$ is called a *characteristic* (or *characteristic base curve*) of the partial differential equation (14.84). In other words, a characteristic of (14.84) is any curve Γ_0 in the xy plane having an equation $\zeta(x, y) = c_0$ which satisfies

$$A \, dy^2 - B \, dx \, dy + C \, dx^2 = 0. \tag{14.130}$$

We have shown that the characteristics of the partial differential equation (14.84) are straight lines. Specifically, if $A \neq 0$, they are the straight lines defined by (14.131); if $A = 0$ and $C \neq 0$, they are those defined by (14.132); and if $A = 0$ and $C = 0$, they are given by (14.133).

Let us consider the connection between the characteristics of the partial differential equation (14.84) and the initial-value problem associated with this equation. Recall that in this problem we seek a solution $\phi(x, y)$ of (14.84) such that ϕ, ϕ_x, and ϕ_y assume prescribed values given by $u(t)$, $p(t)$, and $q(t)$, respectively, at each point of a curve Γ_0 (in the xy plane) defined by $x = x(t)$, $y = y(t)$ for all t on some real interval I.

Further recall that if

$$\Delta(t) = \begin{vmatrix} A & B & C \\ \dfrac{dx}{dt} & \dfrac{dy}{dt} & 0 \\ 0 & \dfrac{dx}{dt} & \dfrac{dy}{dt} \end{vmatrix}$$

is unequal to zero at all points of Γ_0, then this initial-value problem has a unique solution; but if $\Delta(t) = 0$ at all points of Γ_0, then there is either no solution or infinitely many solutions. Now we have defined a characteristic of (14.84) as a curve in the xy plane along which $\Delta(t) = 0$. Thus we may say that if the curve Γ_0 is nowhere tangent to a characteristic, then the initial-value problem has a unique solution; but if Γ_0 is a characteristic, then the problem has either no solution or infinitely many solutions.

We now examine the characteristics of Equation (14.84) in the hyperbolic, parabolic. and elliptic cases. We shall find it convenient to consider once again the three subcases (a) $A \neq 0$, (b) $A = 0$, $C \neq 0$, and (c) $A = 0$, $C = 0$.

Subcase (a): $A \neq 0$. In this case the characteristics of (14.84) are the families of straight lines defined by (14.131). If Equation (14.84) is hyperbolic, then $B^2 - 4AC > 0$ and (14.131) defines two distinct families of real straight lines. If (14.84) is parabolic, then $B^2 - 4AC = 0$, the two families of straight lines coincide, and (14.131) defines only one family of real straight lines. If (14.84) is elliptic, then $B^2 - 4AC < 0$, the lines defined by (14.131) are not real, and there are no real characteristics.

Subcase (b): $A = 0, C \neq 0$. In this case the characteristics are defined by (14.132). Since $A = 0$, $B^2 - 4AC = B^2 \geq 0$ and so the elliptic case cannot occur. In order for Equation (14.84) to be hyperbolic, we must have $B \neq 0$ and then (14.132) defines two distinct families of real straight lines. In the parabolic case, we must have $B = 0$ and then the two families coincide, so that (14.132) defines only one family of real straight lines.

Subcase (c): $A = 0, C = 0$. In this case we must have $B \neq 0$ in order for Equation (14.84) to be of the second order. But then $B^2 - 4AC = B^2 > 0$, and so Equation (14.84) must be hyperbolic. The characteristics are given by (14.133) and hence are two distinct families of real straight lines.

Let us summarize what we have thus found concerning the nature of the characteristics of Equation (14.84). If Equation (14.84) is hyperbolic, then the characteristics consist of two distinct families of real straight lines. If Equation (14.84) is parabolic, then the characteristics consist of a single family of real straight lines. If Equation (14.84) is elliptic, then there are no real characteristics.

We observe the significance of these results as they relate to the initial-value problem associated with Equation (14.84). Recall that if the curve Γ_0 is nowhere tangent to a characteristic, then the initial-value problem has a unique solution. In the elliptic case there are no real characteristics and so this requirement concerning Γ_0 is automatically satisfied. Thus if Equation (14.84) is elliptic, the associated initial-value problem has a unique solution. In the hyperbolic and parabolic cases real characteristics exist; they are certain straight lines. Thus if Equation (14.84) is hyperbolic or parabolic, the associated initial-value problem has a unique solution if the curve Γ_0 is nowhere tangent to one of the straight lines which is a characteristic.

▶ **Example 14.10**

Find the characteristics of the equation

$$\frac{\partial^2 u}{\partial x^2} + \frac{\partial^2 u}{\partial x\, \partial y} - 2\frac{\partial^2 u}{\partial y^2} = 0. \tag{14.120}$$

We first note that $A = 1$, $B = 1$, $C = -2$, and $B^2 - 4AC = 9 > 0$. Therefore equation (14.120) is hyperbolic and so the characteristics consist of two distinct families of real straight lines. Let us proceed to find them. We have observed that they satisfy the equation

$$A\, dy^2 - B\, dx\, dy + C\, dx^2 = 0. \tag{14.130}$$

For Equation (14.120), Equation (14.130) reduces to

$$dy^2 - dx\, dy - 2\, dx^2 = 0,$$

and so the characteristics of Equation (14.120) are the solutions of

$$\left(\frac{dy}{dx}\right)^2 - \frac{dy}{dx} - 2 = 0.$$

From this we see that

$$\frac{dy}{dx} = 2 \quad \text{or} \quad \frac{dy}{dx} = -1.$$

Thus the characteristics of Equation (14.120) are the straight lines of the two real families defined by

$$y = 2x + c_1 \quad \text{and} \quad y = -x + c_2, \tag{14.134}$$

where c_1 and c_2 are arbitrary constants (see Figure 14.3).

In Example 14.9 we considered an initial-value problem associated with Equation (14.120). We sought a solution $\phi(x, y)$ of Equation (14.120), defined for all (x, y), such that

$$\phi(t, t^2) = t^4 + 2t^2,$$
$$\phi_x(t, t^2) = 4t, \tag{14.122}$$
$$\phi_y(t, t^2) = 2t^2,$$

for all real t. The curve Γ_0 defined for all real t by $x = t$, $y = t^2$ is tangent to a characteristic (14.134) of Equation (14.120) only at $t = 1$ and $t = -\frac{1}{2}$. Thus on any interval $a \le t \le b$ *not* including either of these two values, we are assured that this initial-value problem has a unique solution. Recall that in Example 14.9 we observed that a solution valid for all (x, y) is the function ϕ defined by $\phi(x, y) = 2x^2 + y^2$.

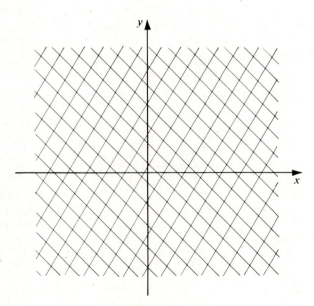

Figure 14.3

In Section 14.3 we proved that Equation (14.84) can always be transformed into an equation having a so-called canonical form. This reduction involved a transformation

$$\xi = \xi(x, y), \qquad \eta = \eta(x, y) \tag{14.85}$$

of the independent variables in (14.84). In Theorems 14.3, 14.4, and 14.5 we listed specific transformations of this form which led to the desired reduction in the hyperbolic, parabolic, and elliptic cases, respectively. The resulting canonical forms were given by (14.94), (14.102), and (14.110) in these three respective cases. The specific transformations listed for the various cases are directly related to the characteristics of the Equation (14.84). We shall not attempt to determine the exact nature of this relation in all cases; rather we shall restrict our investigation to the case in which Equation (14.84) is hyperbolic and $A \neq 0$.

Theorem 14.3 states that if Equation (14.84) is hyperbolic and $A \neq 0$, then a specific transformation which leads to the desired canonical form is given by

$$\begin{aligned} \xi &= \lambda_1 x + y, \\ \eta &= \lambda_2 x + y, \end{aligned} \tag{14.95}$$

where λ_1 and λ_2 are the roots of the quadratic equation

$$A\lambda^2 + B\lambda + C = 0. \tag{14.96}$$

More explicitly, this transformation may be written

$$\begin{aligned} \xi &= \left[\frac{-B - \sqrt{B^2 - 4AC}}{2A} \right] x + y, \\ \eta &= \left[\frac{-B + \sqrt{B^2 - 4AC}}{2A} \right] x + y. \end{aligned} \tag{14.135}$$

Turning to the characteristics of Equation (14.84) in the case under consideration, we note that they are the two distinct families of straight lines defined by

$$y = \left[\frac{B + \sqrt{B^2 - 4AC}}{2A} \right] x + c_1$$

and

$$\tag{14.131}$$

$$y = \left[\frac{B - \sqrt{B^2 - 4AC}}{2A} \right] x + c_2,$$

where c_1 and c_2 are arbitrary constants. Comparison of (14.135) and (14.131) clearly indicates the relation between the canonical form-producing transformation and the characteristics in this case. For here the transformation is given by $\xi = \xi(x, y)$, $\eta = \eta(x, y)$, where $\xi(x, y) = c_1, \eta(x, y) = c_2$ are the two families of characteristics of Equation (14.84).

▶ **Example 14.11**

Let us again consider the equation

$$\frac{\partial^2 u}{\partial x^2} + \frac{\partial^2 u}{\partial x \, \partial y} - 2 \frac{\partial^2 u}{\partial y^2} = 0. \tag{14.120}$$

Since Equation (14.120) is hyperbolic and $A \neq 0$, a canonical form-producing transformation is given by $\xi = \xi(x, y)$, $\eta = \eta(x, y)$, where $\xi(x, y) = c_1$, $\eta(x, y) = c_2$ are the two families of characteristics of Equation (14.120). In Example 14.10 we found that the characteristics of Equation (14.120) are the straight lines defined by (14.134). Therefore the two families of characteristics are given by

$$y - 2x = c_1 \quad \text{and} \quad y + x = c_2,$$

and a canonical form-producing transformation of Equation (14.120) is given by

$$\xi = y - 2x,$$
$$\eta = y + x. \tag{14.136}$$

Applying (14.136) to Equation (14.120), we see that this equation transforms into

$$-7\frac{\partial^2 u}{\partial \xi \, \partial \eta} = 0 \quad \text{and hence into} \quad \frac{\partial^2 u}{\partial \xi \, \partial \eta} = 0, \tag{14.137}$$

which is the canonical form in the hyperbolic case. We observe that the particular characteristics $y = 2x$ and $y = -x$ transform into the coordinate axes $\xi = 0$ and $\eta = 0$, respectively, in the $\xi\eta$ plane.

We note that we may obtain a solution of Equation (14.137) involving two arbitrary functions merely by integrating first with respect to ξ and then with respect to η. We thus obtain the solution

$$u = f(\xi) + g(\eta),$$

where f and g are arbitrary functions of their respective arguments. Using (14.136) to return to the original independent variables x and y, we obtain the solution

$$u = f(y - 2x) + g(y + x)$$

of Equation (14.120) involving the two arbitrary functions f and g. We point out that this solution could be obtained directly by the procedure outlined in Section 14.1.

Exercises

1. Consider the partial differential equation

$$\frac{\partial^2 u}{\partial x \, \partial y} = 0 \tag{A}$$

and the five real functions defined for all real t by

$$x(t) = t, \qquad y(t) = 3 - t, \qquad u(t) = t,$$
$$p(t) = t, \qquad q(t) = t - 1.$$

(a) Show that these five functions satisfy the strip condition (14.119). Then consider the initial-value problem in which one seeks a solution $\phi(x, y)$ of partial differential equation (A), defined for all (x, y) such that

$$\phi(t, 3 - t) = t,$$
$$\phi_x(t, 3 - t) = t,$$
$$\phi_y(t, 3 - t) = t - 1,$$

for all real t. Interpret this problem geometrically.

(b)   Show that the function ϕ defined for all (x, y) by

$$\phi(x, y) = \frac{x^2}{2} - \frac{y^2}{2} + 2y - \frac{3}{2}$$

is a solution of the initial-value problem considered in part (a).

2.   Consider the partial differential equation

$$\frac{\partial^2 u}{\partial x^2} - \frac{\partial^2 u}{\partial y^2} = 0 \qquad\qquad (B)$$

and the five real functions defined for all real t by

$$x(t) = t, \qquad y(t) = 0, \qquad u(t) = t^4 + t^2,$$
$$p(t) = 4t^3 + 2t, \qquad q(t) = 4t.$$

(a)   Show that these five functions satisfy the strip condition (14.119). Then consider the initial-value problem in which one seeks a solution $\phi(x, y)$ of partial differential equation (B), defined for all (x, y), such that

$$\phi(t, 0) = t^4 + t^2,$$
$$\phi_x(t, 0) = 4t^3 + 2t,$$
$$\phi_y(t, 0) = 4t$$

for all real t. Interpret this problem geometrically.

(b)   Show that the function ϕ defined for all (x, y) by

$$\phi(x, y) = x^4 + 6x^2y^2 + y^4 + x^2 + y^2 + 4xy$$

is a solution of the initial-value problem considered in part (a).

For each of the partial differential equations in Exercises 3 through 6, determine whether or not real characteristics exist. For those equations for which families of real characteristics exist, find these families and sketch several members of each family.

3. $\dfrac{\partial^2 u}{\partial x^2} - 2 \dfrac{\partial^2 u}{\partial x\,\partial y} - 8 \dfrac{\partial^2 u}{\partial y^2} = 0.$

4. $\dfrac{\partial^2 u}{\partial x^2} + 4 \dfrac{\partial^2 u}{\partial x\,\partial y} + 4 \dfrac{\partial^2 u}{\partial y^2} = 0.$

5. $2 \dfrac{\partial^2 u}{\partial x^2} - 2 \dfrac{\partial^2 u}{\partial x\,\partial y} + 5 \dfrac{\partial^2 u}{\partial y^2} = 0.$

6. $2 \dfrac{\partial^2 u}{\partial x^2} + 7 \dfrac{\partial^2 u}{\partial x\,\partial y} + 6 \dfrac{\partial^2 u}{\partial y^2} = 0.$

APPENDIX ONE
Second- and Third-Order Determinants

Determinants played an important role on a number of occasions in several chapters. Although certain definitions and theorems involved nth-order determinants, the relevant illustrative examples and exercises were concerned almost entirely with second- and third-order determinants. In this brief appendix, we discuss the evaluation of these two simple cases.

The second-order determinant denoted by

$$\begin{vmatrix} a_{11} & a_{12} \\ a_{21} & a_{22} \end{vmatrix}$$

is defined to be

$$a_{11}a_{22} - a_{12}a_{21}.$$

Thus, for example, the second-order determinant

$$\begin{vmatrix} 6 & 5 \\ -3 & 2 \end{vmatrix}$$

is given by

$$(6)(2) - (5)(-3) = 27.$$

The third-order determinant denoted by

$$\begin{vmatrix} a_{11} & a_{12} & a_{13} \\ a_{21} & a_{22} & a_{23} \\ a_{31} & a_{32} & a_{33} \end{vmatrix} \tag{1}$$

may be defined as

$$a_{11}\begin{vmatrix} a_{22} & a_{23} \\ a_{32} & a_{33} \end{vmatrix} - a_{12}\begin{vmatrix} a_{21} & a_{23} \\ a_{31} & a_{33} \end{vmatrix} + a_{13}\begin{vmatrix} a_{21} & a_{22} \\ a_{31} & a_{32} \end{vmatrix}, \tag{2}$$

where each of the three second-order determinants involved is evaluated as explained in the preceding paragraph. Thus, for example, the third-order determinant

$$\begin{vmatrix} 1 & 2 & -3 \\ 3 & 4 & -2 \\ -1 & 5 & 6 \end{vmatrix} \tag{3}$$

is given by

$$(1)\begin{vmatrix} 4 & -2 \\ 5 & 6 \end{vmatrix} - (2)\begin{vmatrix} 3 & -2 \\ -1 & 6 \end{vmatrix} + (-3)\begin{vmatrix} 3 & 4 \\ -1 & 5 \end{vmatrix}$$

$$= (1)[(4)(6) - (-2)(5)] - (2)[(3)(6) - (-2)(-1)] + (-3)[(3)(5) - (4)(-1)]$$

$$= (24 + 10) - 2(18 - 2) - 3(15 + 4) = -55.$$

The expression (2) for the third-order determinant (1) is sometimes called its expansion by cofactors along its first row. Similar expansions exist along each remaining row and along each column, and each of these also gives the value defined by (1). We list these other five expansions for evaluating (1):

$$-a_{21}\begin{vmatrix} a_{12} & a_{13} \\ a_{32} & a_{33} \end{vmatrix} + a_{22}\begin{vmatrix} a_{11} & a_{13} \\ a_{31} & a_{33} \end{vmatrix} - a_{23}\begin{vmatrix} a_{11} & a_{12} \\ a_{31} & a_{32} \end{vmatrix}, \tag{4}$$

$$a_{31}\begin{vmatrix} a_{12} & a_{13} \\ a_{22} & a_{23} \end{vmatrix} - a_{32}\begin{vmatrix} a_{11} & a_{13} \\ a_{21} & a_{23} \end{vmatrix} + a_{33}\begin{vmatrix} a_{11} & a_{12} \\ a_{21} & a_{22} \end{vmatrix}, \tag{5}$$

$$a_{11}\begin{vmatrix} a_{22} & a_{23} \\ a_{32} & a_{33} \end{vmatrix} - a_{21}\begin{vmatrix} a_{12} & a_{13} \\ a_{32} & a_{33} \end{vmatrix} + a_{31}\begin{vmatrix} a_{12} & a_{13} \\ a_{22} & a_{23} \end{vmatrix}, \tag{6}$$

$$-a_{12}\begin{vmatrix} a_{21} & a_{23} \\ a_{31} & a_{33} \end{vmatrix} + a_{22}\begin{vmatrix} a_{11} & a_{13} \\ a_{31} & a_{33} \end{vmatrix} - a_{32}\begin{vmatrix} a_{11} & a_{13} \\ a_{21} & a_{23} \end{vmatrix}, \tag{7}$$

$$a_{13}\begin{vmatrix} a_{21} & a_{22} \\ a_{31} & a_{32} \end{vmatrix} - a_{23}\begin{vmatrix} a_{11} & a_{12} \\ a_{31} & a_{32} \end{vmatrix} + a_{33}\begin{vmatrix} a_{11} & a_{12} \\ a_{21} & a_{22} \end{vmatrix}. \tag{8}$$

For example, we evaluate the determinant (3) using the expansion (7), its so-called expansion along its second column. We have

$$-(2)\begin{vmatrix} 3 & -2 \\ -1 & 6 \end{vmatrix} + (4)\begin{vmatrix} 1 & -3 \\ -1 & 6 \end{vmatrix} - (5)\begin{vmatrix} 1 & -3 \\ 3 & -2 \end{vmatrix}$$

$$= -(2)[(3)(6) - (-2)(-1)] + (4)[(1)(6) - (-3)(-1)] - (5)[(1)(-2) - (-3)(3)]$$

$$= -2(18 - 2) + 4(6 - 3) - 5(-2 + 9) = -55.$$

Of course, one does not need all six of the expansions (2), (4), (5), (6), (7), and (8) to evaluate the determinant (1). Any one will do. In particular, the expansion (2) which we gave initially is quite sufficient for the evaluation. However, we shall see that under certain conditions it may happen that a particular one of these six expansions involves considerably less calculation than the other five.

We now state and illustrate certain results that facilitate the evaluation of determinants. We first have:

Result A. If each element in a row (or column) of a determinant is zero, then the determinant itself is zero.

Result B. If each element in a row (or column) of a determinant is multiplied by the same quantity c, then the value of the determinant itself is also multiplied by c.

Thus, for the third-order determinant (1), if each element in its second row is multiplied by c, we have

$$\begin{vmatrix} a_{11} & a_{12} & a_{13} \\ ca_{21} & ca_{22} & ca_{23} \\ a_{31} & a_{32} & a_{33} \end{vmatrix} = c \begin{vmatrix} a_{11} & a_{12} & a_{13} \\ a_{21} & a_{22} & a_{23} \\ a_{31} & a_{32} & a_{33} \end{vmatrix}.$$

For example, we have

$$\begin{vmatrix} 1 & 2 & -3 \\ 12 & 16 & -8 \\ -1 & 5 & 6 \end{vmatrix} = \begin{vmatrix} 1 & 2 & -3 \\ 4(3) & 4(4) & 4(-2) \\ -1 & 5 & 6 \end{vmatrix} = 4 \begin{vmatrix} 1 & 2 & -3 \\ 3 & 4 & -2 \\ -1 & 5 & 6 \end{vmatrix}.$$

Result C. If two rows (or columns) of a determinant are identical or proportional, then the determinant is zero.

For example,

$$\begin{vmatrix} 1 & 2 & 4 \\ -1 & 2 & 5 \\ 3 & 6 & 12 \end{vmatrix} = 0,$$

because the first and third rows of this determinant are proportional.

Result D. (1) Suppose each element of the first row of a third-order determinant is expressed as a sum of two terms, thus

$$a_{11} = b_{11} + c_{11}, \quad a_{12} = b_{12} + c_{12}, \quad a_{13} = b_{13} + c_{13}.$$

Then the given determinant can be expressed as the sum of two determinants, where the first is obtained from the given determinant by replacing a_{11} by b_{11}, a_{12} by b_{12}, and a_{13} by b_{13}, and the second is obtained from the given determinant by replacing a_{11} by c_{11}, a_{12} by c_{12}, and a_{13} by c_{13}. That is,

$$\begin{vmatrix} b_{11}+c_{11} & b_{12}+c_{12} & b_{13}+c_{13} \\ a_{21} & a_{22} & a_{23} \\ a_{31} & a_{32} & a_{33} \end{vmatrix} = \begin{vmatrix} b_{11} & b_{12} & b_{13} \\ a_{21} & a_{22} & a_{23} \\ a_{31} & a_{32} & a_{33} \end{vmatrix} + \begin{vmatrix} c_{11} & c_{12} & c_{13} \\ a_{21} & a_{22} & a_{23} \\ a_{31} & a_{32} & a_{33} \end{vmatrix}.$$

(2) Analogous results hold for each of the other rows and for each column.

For example, we see that

$$\begin{vmatrix} 3+5 & 6+2 & 7+1 \\ 4 & 8 & -2 \\ 9 & -4 & -7 \end{vmatrix} = \begin{vmatrix} 3 & 6 & 7 \\ 4 & 8 & -2 \\ 9 & -4 & -7 \end{vmatrix} + \begin{vmatrix} 5 & 2 & 1 \\ 4 & 8 & -2 \\ 9 & -4 & -7 \end{vmatrix},$$

and similarly that

$$\begin{vmatrix} -2 & 1+8 & 7 \\ 3 & 5-3 & -1 \\ 4 & 6+9 & 5 \end{vmatrix} = \begin{vmatrix} -2 & 1 & 7 \\ 3 & 5 & -1 \\ 4 & 6 & 5 \end{vmatrix} + \begin{vmatrix} -2 & 8 & 7 \\ 3 & -3 & -1 \\ 4 & 9 & 5 \end{vmatrix}.$$

Result E. The value of a determinant is unchanged if each element of a row (or column) is multiplied by the same quantity c and then added to the corresponding element of another row (or column).

Thus, for the third-order determinant (1), if each element of the second row is multiplied by c and then added to the corresponding element of the third row, we have

$$\begin{vmatrix} a_{11} & a_{12} & a_{13} \\ a_{21} & a_{22} & a_{23} \\ a_{31} & a_{32} & a_{33} \end{vmatrix} = \begin{vmatrix} a_{11} & a_{12} & a_{13} \\ a_{21} & a_{22} & a_{23} \\ a_{31}+ca_{21} & a_{32}+ca_{22} & a_{33}+ca_{23} \end{vmatrix}.$$

For example, we have

$$\begin{vmatrix} 5 & 7 & -1 \\ 1 & 4 & -2 \\ 6 & -5 & 8 \end{vmatrix} = \begin{vmatrix} 5 & 7 & -1 \\ 1 & 4 & -2 \\ 6+(3)(1) & -5+(3)(4) & 8+(3)(-2) \end{vmatrix} = \begin{vmatrix} 5 & 7 & -1 \\ 1 & 4 & -2 \\ 9 & 7 & 2 \end{vmatrix}.$$

Result E is very useful in evaluating third-order determinants. It is applied twice to introduce two zeros, one after the other, in a certain row (or column). Then when one employs the expansion of the determinant along that particular row (or column), two of the second-order determinants in the expansion are multiplied by zero and hence drop out. Thus the given third-order determinant is reduced to a multiple of a single second-order determinant. We illustrate by evaluating determinant (3) in this manner, introducing two zeros in the first column and then using expansion (6) along this column:

$$\begin{vmatrix} 1 & 2 & -3 \\ 3 & 4 & -2 \\ -1 & 5 & 6 \end{vmatrix} = \begin{vmatrix} 1 & 2 & -3 \\ 3+(-3)(1) & 4+(-3)(2) & -2+(-3)(-3) \\ -1+(1)(1) & 5+(1)(2) & 6+(1)(-3) \end{vmatrix}$$

$$= \begin{vmatrix} 1 & 2 & -3 \\ 0 & -2 & 7 \\ 0 & 7 & 3 \end{vmatrix}$$

$$= (1)\begin{vmatrix} -2 & 7 \\ 7 & 3 \end{vmatrix} - (0)\begin{vmatrix} 2 & -3 \\ 7 & 3 \end{vmatrix} + (0)\begin{vmatrix} 2 & -3 \\ -2 & 7 \end{vmatrix}$$

$$= \begin{vmatrix} -2 & 7 \\ 7 & 3 \end{vmatrix} = (-2)(3) - (7)(7) = -55.$$

APPENDIX TWO

x	$\ln x$	x	$\ln x$	x	$\ln x$	x	$\ln x$
		3.0	1.099	6.0	1.792	9.0	2.197
0.1	-2.303	3.1	1.131	6.1	1.808	9.1	2.208
0.2	-1.609	3.2	1.163	6.2	1.825	9.2	2.219
0.3	-1.204	3.3	1.194	6.3	1.841	9.3	2.230
0.4	-0.916	3.4	1.224	6.4	1.856	9.4	2.241
0.5	-0.693	3.5	1.253	6.5	1.872	9.5	2.251
0.6	-0.511	3.6	1.281	6.6	1.887	9.6	2.262
0.7	-0.357	3.7	1.308	6.7	1.902	9.7	2.272
0.8	-0.223	3.8	1.335	6.8	1.917	9.8	2.282
0.9	-0.105	3.9	1.361	6.9	1.932	9.9	2.293
1.0	0.000	4.0	1.386	7.0	1.946	10	2.303
1.1	0.095	4.1	1.411	7.1	1.960	20	2.996
1.2	0.182	4.2	1.435	7.2	1.974	30	3.401
1.3	0.262	4.3	1.459	7.3	1.988	40	3.689
1.4	0.336	4.4	1.482	7.4	2.001	50	3.912
1.5	0.405	4.5	1.504	7.5	2.015	60	4.094
1.6	0.470	4.6	1.526	7.6	2.028	70	4.248
1.7	0.531	4.7	1.548	7.7	2.041	80	4.382
1.8	0.588	4.8	1.569	7.8	2.054	90	4.500
1.9	0.642	4.9	1.589	7.9	2.067	100	4.605
2.0	0.693	5.0	1.609	8.0	2.079		
2.1	0.742	5.1	1.629	8.1	2.092		
2.2	0.788	5.2	1.649	8.2	2.105		
2.3	0.833	5.3	1.668	8.3	2.116		
2.4	0.875	5.4	1.686	8.4	2.128		
2.5	0.916	5.5	1.705	8.5	2.140		
2.6	0.956	5.6	1.723	8.6	2.152		
2.7	0.993	5.7	1.740	8.7	2.163		
2.8	1.030	5.8	1.758	8.8	2.175		
2.9	1.065	5.9	1.775	8.9	2.186		

x	e^x	e^{-x}	x	e^x	e^{-x}
0.0	1.000	1.0000	3.0	20.086	0.0498
0.1	1.105	0.9048	3.1	22.198	0.0450
0.2	1.221	0.8187	3.2	24.533	0.0408
0.3	1.350	0.7408	3.3	27.113	0.0369
0.4	1.492	0.6703	3.4	29.964	0.0334
0.5	1.649	0.6065	3.5	33.115	0.0302
0.6	1.822	0.5488	3.6	36.598	0.0273
0.7	2.014	0.4966	3.7	40.447	0.0247
0.8	2.226	0.4493	3.8	44.701	0.0224
0.9	2.460	0.4066	3.9	49.402	0.0202
1.0	2.718	0.3679	4.0	54.598	0.0183
1.1	3.004	0.3329	4.1	60.340	0.0166
1.2	3.320	0.3012	4.2	66.686	0.0150
1.3	3.669	0.2725	4.3	73.700	0.0136
1.4	4.055	0.2466	4.4	81.451	0.0123
1.5	4.482	0.2231	4.5	90.017	0.0111
1.6	4.953	0.2019	4.6	99.484	0.0101
1.7	5.474	0.1827	4.7	109.947	0.0091
1.8	6.050	0.1653	4.8	121.510	0.0082
1.9	6.686	0.1496	4.9	134.290	0.0074
2.0	7.389	0.1353	5	148.4	0.00674
2.1	8.166	0.1225	6	403.4	0.00248
2.2	9.025	0.1108	7	1096.6	0.00091
2.3	9.974	0.1003	8	2981.0	0.00034
2.4	11.023	0.0907	9	8103.1	0.00012
2.5	12.182	0.0821	10	22026.5	0.00005
2.6	13.464	0.0743			
2.7	14.880	0.0672			
2.8	16.445	0.0608			
2.9	18.174	0.0550			

ANSWERS TO
ODD-NUMBERED EXERCISES

Section 1.1
1. Ordinary; first; linear.
3. Partial; second; linear.
5. Ordinary; fourth; nonlinear.
7. Ordinary; second; linear.
9. Ordinary; sixth; nonlinear.

Section 1.2
5. (a) $2, 3, -2$. (b) $-1, -2, 4$.

Section 1.3
1. No; one of the supplementary conditions is not satisfied.
3. (a) $y = 3e^{4x} + 2e^{-3x}$. (b) $y = -2e^{-3x}$.
5. $y = 2x - 3x^2 + x^3$.
7. Yes.

Section 2.1
1. $3x^2 + 4xy + y^2 = c$.
3. $x + x^2y + 2y^2 = c$.
5. $3x^2y + 2y^2x - 5x - 6y = c$.
7. $y \tan x + \sec x + y^2 = c$.
9. $s^2 - s = ct$.
11. $x^2y - 3x + 2y^2 = 7$.
13. $y^2 \cos x - y \sin^2 x = 9$.
15. $-3y + 2x + y^2 = 2xy$.
17. (a) $A = \frac{3}{2}$; $2x^3 + 9x^2y + 12y^2 = c$.
    (b) $A = -2$; $2x^2 - 2y^2 - x = cxy^2$.

19. (a) $x^2y + c.$ (b) $2x^{-1}y^{-3} - \frac{3}{2}x^2y^{-4} + c.$
21. (b) $x^2.$ (c) $x^4 + x^3y^2 = c.$

Section 2.2

1. $(x^2 + 1)^2y = c.$
3. $r^2 + s = c(1 - r^2s).$
5. $r \sin^2 \theta = c.$
7. $(x + 1)^6(y^2 + 1) = c(x + 2)^4.$
9. $y^2 + xy = cx^3.$

11. $\sin \dfrac{y}{x} = cx.$

13. $(x^2 + y^2)^{3/2} = x^3 \ln cx^3.$
15. $x + 4 = (y + 2)^2e^{-(y+1)}.$
17. $16(x + 3)(x + 2)^2 = 9(y^2 + 4)^2.$
19. $(2x + y)^2 = 12(y - x).$
23. (a) $x^3 - y^3 + 6xy^2 = c.$
    (b) $2x^3 + 3x^2y + 3xy^2 = c.$

Section 2.3

1. $y = x^3 + cx^{-3}.$
3. $y = (x^3 + c)e^{-3x}.$
5. $x = 1 + ce^{1/t}.$
7. $3(x^2 + x)y = x^3 - 3x + c.$
9. $y = x^{-1}(1 + ce^{-x}).$
11. $r = (\theta + c)\cos \theta.$
13. $2(1 + \sin x)y = x + \sin x \cos x + c.$
15. $y = (1 + cx^{-1})^{-1}.$
17. $y = (2 + ce^{-8x^2})^{1/4}.$
19. $y = x^4 - 2x^2.$
21. $y = (e^x + 1)^2.$
23. $2r = \sin 2\theta + \sqrt{2} \cos \theta.$
25. $x^2y^4 = x^4 + 15.$

27. $y = \begin{cases} 2(1 - e^{-x}), & 0 \le x < 1, \\ 2(e - 1)e^{-x}, & x \ge 1. \end{cases}$

29. $y = \begin{cases} e^{-x}(x + 1), & 0 \le x < 2, \\ 2e^{-x} + e^{-2}, & x \ge 2. \end{cases}$

31. (a) $y = \dfrac{ke^{-\lambda x}}{b - a\lambda} + ce^{-bx/a}$ if $\lambda \ne b/a;$

    $y = \dfrac{kxe^{-bx/a}}{a} + ce^{-bx/a}$ if $\lambda = b/a.$

35. (b) $y = \sin x - \cos x + \sin 2x - 2 \cos 2x + ce^{-x}.$
37. (a) $2x \sin y - x^2 = c.$
    (b) $y^2 + 2y + ce^{-x^2} - 1 = 0.$
39. $y = (x - 2 + ce^{-x})^{-1} + 1.$
41. $y = (2 + ce^{-2x^2})^{-1} + x.$

Section 2.3. Miscellaneous Exercises

1. $(x^3 + 1)^2 = |cy|$.
3. $xy + 1 = c(x + 1)$.
5. $(3x - y)^2 = |c(y - x)|$.
7. $y = \frac{3}{2} + c(x^4 + 1)^{-2}$.
9. $x^4y^2 - x^3y = c$.
11. $(2x + 3y)^2 = |c(y - x)|$.
13. $y = 1 + c(x^3 + 1)^{-2}$.
15. $y = (x^2 + 3x)^{1/2}$.
17. $y = e^{-x}(2x^2 + 4)^{1/2}$.
19. $(y^2 + 1)^2 = 2x$.
21. $y = (x^2 + 1)^{1/2}/2$.
23. $(x + 2)y = x^2 + 8, 0 \le x \le 2$,
    $(x + 2)y = 4x + 4, x > 2$.

Section 2.4

1. $4x^5y + 4x^4y^2 + x^4 = c$.
3. $xy^2e^x + ye^x = c$.
5. $x^3y^4(xy + 2) = c$.
7. $5x^2 + 4xy + y^2 + 2x + 2y = c$.

9. $\ln[c(x^2 + y^2 - 2x + 2y + 2)] + 4 \arctan\left(\dfrac{y + 1}{x - 1}\right) = 0$.

11. $12x^2 + 16xy + 4y^2 + 4x + 8y - 89 = 0$.
13. $x + 2y - \ln|2x + 3y - 1| - 2 = 0$.
21. (a) $y = cx + c^2$. (b) $y = -x^2/4$.

Section 3.1

1. $x^2 + 3y^2 = k^2$.
3. $x^2 + y^2 - \ln y^2 = k$.
5. $x = y - 1 + ke^{-y}$.
7. $x^2y + y^3 = k$.
9. $y^2 = 2x - 1 + ke^{-2x}$.
11. $x^2 + y^2 = ky$.
13. $n = 3$.
15. $\ln(x^2 + y^2) + 2 \arctan(y/x) = c$.
17. $\ln|3x^2 + 3xy + 4y^2| - (2/\sqrt{39})\arctan[(3x + 8y)/\sqrt{39}x] = c$.

Section 3.2

1. (a) $v = 8(1 - e^{-4t})$, $x = 2(4t - 1 + e^{-4t})$.
   (b) 8 ft/sec; 38 feet.
3. Rises 10.92 feet.
5. (a) $v = 245(1 - e^{-t/25})$, $x = 245[t + 25(e^{-t/25} - 1)]$.
   (b) $t = 5.58$ sec.
7. (a) 10.36 ft/sec. (b) 13.19 ft/sec.

9. $v = \dfrac{256[91 + 59e^{-t/4}]}{91 - 59e^{-t/4}}$.

11. $v = \dfrac{\sqrt{5}(1 - e^{-8\sqrt{5}t})}{1 + e^{-8\sqrt{5}t}}.$

13. 16.18 feet.

15. (a) 10.96 ft/sec. (b) 20.46 ft/sec.

17. 4.03 ft/sec.

19. $v^2 = -(k/m)x^2 + v_0^2 + (k/m)x_0^2.$

Section 3.3

1. (a) 59.05%. (b) 2631 years.

3. (a) $50\sqrt{2}$ grams. (b) 12.5 grams. (c) 10.29 hours.

5. 63.5 years.

7. (a) 9 times original number; (b) 10.48 hours.

9. (a) $x = \dfrac{(10)^6}{1 + 4e^{59.4 - 3t/100}};$

(b) 312,966; (c) 1,000,000.

11. $x = \dfrac{(10)^5[3e^{15(1980 - t)/(10)^4} - 2]}{6e^{15(1980 - t)/(10)^4} - 1}.$

13. (a) \$1822.10. (b) approx. 11.52 years.

15. (a) 112.31 lb. (b) 17.33 minutes.

17. (a) 318.53 lb. (b) 2.74 minutes.

19. 11,179.96 grams.

21. (a) 0.072%. (b) 1.39 minutes.

23. (a) 63.33 °F. (b) 13.55 minutes.

25. 7.14 grams.

Section 4.1B

7. (c) $y = 3e^{2x} - e^{3x}.$

9. (c) $y = 4x - x^2.$

Section 4.1D

1. $y = c_1 x + c_2 x^4.$

3. $y = c_1 x + c_2(x^2 + 1).$

5. $y = c_1 e^{2x} + c_2(x + 1).$

9. $y_p = -\frac{4}{3} - 2x + 3e^x.$

Section 4.2

1. $y = c_1 e^{2x} + c_2 e^{3x}.$

3. $y = c_1 e^{x/2} + c_2 e^{5x/2}.$

5. $y = c_1 e^x + c_2 e^{-x} + c_3 e^{3x}.$

7. $y = (c_1 + c_2 x)e^{4x}.$

9. $y = e^{2x}(c_1 \sin 3x + c_2 \cos 3x).$

11. $y = c_1 \sin 3x + c_2 \cos 3x.$

13. $y = (c_1 + c_2 x)e^x + c_3 e^{3x}.$

15. $y = (c_1 + c_2 x + c_3 x^2)e^{2x}.$

17. $y = c_1 e^x + c_2 \sin x + c_3 \cos x$.

19. $y = c_1 + c_2 x + c_3 x^2 + (c_4 + c_5 x)e^x$.

21. $y = c_1 e^{2x} + c_2 e^{3x} + e^{-x}(c_3 \sin x + c_4 \cos x)$.

23. $y = e^{\sqrt{2}x/2}\left(c_1 \sin \dfrac{\sqrt{2}x}{2} + c_2 \cos \dfrac{\sqrt{2}x}{2}\right)$

$\qquad + e^{-\sqrt{2}x/2}\left(c_3 \sin \dfrac{\sqrt{2}x}{2} + c_4 \cos \dfrac{\sqrt{2}x}{2}\right).$

25. $y = 2e^{4x} + e^{-3x}$.

27. $y = -e^{2x} + 2e^{4x}$.

29. $y = (3x + 2)e^{-3x}$.

31. $y = (13x + 3)e^{-2x}$.

33. $y = e^{2x} \sin 5x$.

35. $y = e^{-3x}(4 \sin 2x + 3 \cos 2x)$.

37. $y = 3e^{-(1/3)x}[\sin \tfrac{2}{3}x + 2 \cos \tfrac{2}{3}x]$.

39. $y = e^x - 2e^{2x} + e^{3x}$.

41. $y = \tfrac{32}{9}e^{-x} - \tfrac{23}{9}e^{2x} + \tfrac{2}{3}xe^{2x}$.

45. $y = c_1 \sin x + c_2 \cos x + e^{-x}(c_3 \sin 2x + c_4 \cos 2x)$.

Section 4.3

1. $y = c_1 e^x + c_2 e^{2x} + 2x^2 + 6x + 7$.

3. $y = e^{-x}(c_1 \sin 2x + c_2 \cos 2x) + 2 \sin 2x - \cos 2x$.

5. $y = e^{-x}(c_1 \sin \sqrt{3}x + c_2 \cos \sqrt{3}x) + \dfrac{\sin 4x}{26} - \dfrac{3 \cos 4x}{52}$.

7. $y = c_1 e^{-x} + c_2 e^{-5x} + \tfrac{1}{6}(e^x + e^{5x})$.

9. $y = c_1 e^x + c_2 e^{-2x} + c_3 e^{-3x} + 3x^2 + x + 4$.

11. $y = c_1 e^x + e^{-x}(c_2 \sin 2x + c_3 \cos 2x) - \dfrac{9 \sin 2x}{17} + \dfrac{2 \cos 2x}{17} - 2x^2$

$\qquad - \dfrac{9x}{5} - \dfrac{82}{25}.$

13. $y = c_1 e^{2x} + c_2 e^{-3x} + 2xe^{2x} - 3e^{3x} + x + 2$.

15. $y = c_1 e^{-x} + (c_2 + c_3 x)e^{2x} + 2e^x - 2xe^{-x}$.

17. $y = c_1 + c_2 \sin x + c_3 \cos x + \dfrac{2x^3}{3} - 4x - 2x \sin x$.

19. $y = c_1 e^x + c_2 e^{2x} + c_3 e^{3x} + \dfrac{x^2 e^x}{4} + \dfrac{3xe^x}{4} + 4xe^{2x} + e^{4x}$.

21. $y = c_1 \sin x + c_2 \cos x - \tfrac{1}{4}x^2 \cos x + \tfrac{1}{4}x \sin x$.

23. $y = c_1 + c_2 x + c_3 e^x + c_4 e^{-3x} - \dfrac{x^4}{2} - \dfrac{4x^3}{3} - \dfrac{19}{6}x^2 + 2x^2 e^x - 9xe^x + \dfrac{e^{3x}}{27}$.

25. $y = -6e^x + 2e^{3x} + 3x^2 + 8x + 10$.

27. $y = 3e^{3x} - 2e^{5x} + 3xe^{2x} + 4e^{2x}$.

29. $y = 4xe^{-4x} + 2e^{-2x}$.

31. $y = e^{-2x}(\frac{4}{3}\sin 3x - 2\cos 3x) + 2e^{-2x}$.

33. $y = \frac{1}{5}[e^{2x}(\sin 3x + 2\cos 3x) + \sin 3x + 3\cos 3x]$.

35. $y = (x + 5)e^x + 3x^2 e^x + 2xe^{2x} - 4e^{2x}$.

37. $y = 6\cos x - \sin x + 3x^2 - 6 + 2x\cos x$.

39. $y = \dfrac{7e^{-x}}{20} - \dfrac{31e^{2x}}{40} + \dfrac{3xe^x}{4} + \dfrac{5e^x}{4} - \dfrac{\sin x}{10}$.

41. $y_p = Ax^3 + Bx^2 + Cx + D + Ee^{-2x}$.

43. $y_p = Ae^{-2x} + Bxe^{-2x}\sin x + Cxe^{-2x}\cos x$.

45. $y_p = Ax^2 e^{-3x}\sin 2x + Bx^2 e^{-3x}\cos 2x + Cxe^{-3x}\sin 2x$
$+ Dxe^{-3x}\cos 2x + Ex^2 e^{-2x}\sin 3x + Fx^2 e^{-2x}\cos 3x$
$+ Gxe^{-2x}\sin 3x + Hxe^{-2x}\cos 3x + Ie^{-2x}\sin 3x + Je^{-2x}\cos 3x$.

47. $y_p = Ax^4 e^{2x} + Bx^3 e^{2x} + Cx^2 e^{3x} + Dxe^{3x} + Ee^{3x}$.

49. $y_p = Ax^3\sin 2x + Bx^3\cos 2x + Cx^2\sin 2x + Dx^2\cos 2x + Ex\sin 2x$
$Fx\cos 2x + Gx^5 e^{2x} + Hx^4 e^{2x} + Ix^3 e^{2x} + Jx^2 e^{2x} + Kxe^{2x}$.

51. $y_p = Ax^4\sin x + Bx^4\cos x + Cx^3\sin x + Dx^3\cos x + Ex^2\sin x$
$+ Fx^2\cos x$.

53. $y_p = A + Bx\sin 2x + Cx\cos 2x + Dxe^x + Exe^{-x}$
or $y_p = Ax\sin^2 x + Bx\cos^2 x + Cx\sin x\cos x + Dx\sinh x + Ex\cosh x$.

Section 4.4

1. $y = c_1\sin x + c_2\cos x + (\sin x)[\ln|\csc x - \cot x|]$.

3. $y = c_1\sin x + c_2\cos x + (\cos x)[\ln|\cos x|] + x\sin x$.

5. $y = c_1\sin 2x + c_2\cos 2x + \dfrac{\sin 2x}{4}[\ln|\sec 2x + \tan 2x|] - \dfrac{1}{4}$.

7. $y = e^{-2x}(c_1\sin x + c_2\cos x) + xe^{-2x}\sin x + (\ln|\cos x|)e^{-2x}\cos x$.

9. $y = \left(c_1 + c_2 x + \dfrac{1}{2x}\right)e^{-3x}$.

11. $y = c_1\sin x + c_2\cos x + [\sin x][\ln|\csc x - \cot x|]$
$- [\cos x][\ln|\sec x + \tan x|]$.

13. $y = c_1 e^{-x} + c_2 e^{-2x} + (e^{-x} + e^{-2x})[\ln(1 + e^x)]$.

15. $y = c_1\sin x + c_2\cos x + (\sin x)[\ln(1 + \sin x)] - x\cos x - \dfrac{\cos^2 x}{1 + \sin x}$.

17. $y = c_1 e^{-x} + c_2 e^{-2x} + e^{-x}\ln|x| - e^{-2x}\displaystyle\int\dfrac{e^x}{x}\,dx$.

19. $y = c_1 x^2 + c_2 x^5 - 3x^3 - \frac{3}{2}x^4$.

21. $y = c_1(x + 1) + c_2 x^2 - x^2 - 2x + x^2\ln|x|$.

23. $y = c_1 e^x + c_2 x^2 + x^3 e^x - 3x^2 e^x$.

25. $y = c_1\sin x + c_2 x\sin x + \dfrac{x^2}{2}\sin x$.

Section 4.5

1. $y = c_1 x + c_2 x^3$.

3. $y = c_1 x^{1/2} + c_2 x^{3/2}$.

5. $y = c_1 \sin(\ln x^2) + c_2 \cos(\ln x^2)$.
7. $y = c_1 x^2 + c_2 x^{1/3}$.
9. $y = (c_1 + c_2 \ln x) x^{1/3}$.
11. $y = c_1 x + c_2 x^2 + c_3 x^3$.
13. $y = (c_1 + c_2 \ln x) x^3 + c_3 x^{-2}$.
15. $y = c_1 x^2 + c_2 x^4 - 2x^3$.

17. $y = c_1 \sin(\ln x^2) + c_2 \cos(\ln x^2) + \dfrac{x \ln x^2}{5} - \dfrac{4x}{25}$.

19. $y = (c_1 + c_2 \ln x) x + c_3 x^2 + \dfrac{x^3}{4}$.

21. $y = -2x^2 + x^3$.
23. $y = x^{-1} + x^2 - 2x + 4$.
25. $y = -x^2 + 2x^{-3} + 2x^2 \ln x$.

27. $y = \dfrac{1}{6}\left(\dfrac{x^{-2}}{2} + \dfrac{x^3}{3} - \ln x + \dfrac{1}{6}\right)$.

29. $y = c_1(2x - 3) + c_2(2x - 3)^3$.

Section 4.6

1. (a) $f_1(x) = 2e^x - e^{2x}$, $f_2(x) = -e^x + e^{2x}$. (b) $5f_1(x) + 7f_2(x)$.

Section 5.2

1. $x = \dfrac{\cos 16t}{6}$; $\frac{1}{6}$ (ft), $\pi/8$ (sec), $8/\pi$ oscillations/sec.

3. (a) $x = \dfrac{\sin 8t}{4}$. (b) $\frac{1}{4}$ (ft), $\pi/4$ (sec), $4/\pi$ oscillations/sec.

(c) $t = \dfrac{\pi}{48} + \dfrac{n\pi}{4}$ $(n = 0, 1, 2, \dots)$. (d) $t = \dfrac{5\pi}{48} + \dfrac{n\pi}{4}$ $(n = 0, 1, 2, \dots)$.

5. (a) $x = -\dfrac{\sin 10t}{5} + \dfrac{\cos 10t}{3}$. (b) $\dfrac{\sqrt{34}}{15}$ (ft); $\pi/5$ (sec); $5/\pi$ oscillations/sec.
(c) 0.103 (sec); -3.888 ft/sec.
7. 18 lb.

Section 5.3

1. (a) $\dfrac{1}{4}\dfrac{d^2 x}{dt^2} + 2\dfrac{dx}{dt} + 20x = 0$, $x(0) = \frac{1}{2}$, $x'(0) = 0$.

(b) $x = e^{-4t}\left(\dfrac{\sin 8t}{4} + \dfrac{\cos 8t}{2}\right)$.

(c) $x = \dfrac{\sqrt{5}}{4} e^{-4t} \cos(8t - \phi)$, where $\phi \approx 0.46$.

(d) $\pi/4$ (sec).

3. $x = (6t + \frac{3}{4})e^{-8t}$.

5. (a) $x = e^{-8t}\left(\dfrac{\sin 16t}{3} + \dfrac{2\cos 16t}{3}\right)$; $x = \dfrac{\sqrt{5}}{3}e^{-8t}\cos(16t - \phi)$, where
$\phi \approx 0.46$.
 (b) $\pi/8$ (sec); π. (c) 0.127.

7. (a) $x = (\frac{1}{4} + 2t)e^{-8t}$.

 (b) $x = e^{-4t}\left(\dfrac{\sqrt{3}}{12}\sin 4\sqrt{3}t + \dfrac{1}{4}\cos 4\sqrt{3}t\right)$.

 (c) $x = \left(\dfrac{3 + 2\sqrt{3}}{24}\right)e^{(-16+8\sqrt{3})t} + \left(\dfrac{3 - 2\sqrt{3}}{24}\right)e^{(-16-8\sqrt{3})t}$.

9. (a) 64. (b) $8\sqrt{3}$.

Section 5.4

1. $x = \dfrac{\cos 12t - \cos 20t}{4}$.

3. $x = -2te^{-8t} + \dfrac{\sin 8t}{4}$.

5. $x = e^{-8t}\left(\dfrac{\sqrt{2}}{2}\sin 4\sqrt{2}t + \cos 4\sqrt{2}t\right) + \sin 4t - \cos 4t$.

7. $x = e^{-2t}\left(-\dfrac{3\sin 4t}{2} - 2\cos 4t\right) + \sin 2t + 2\cos 2t, 0 \le t \le \pi$;

 $x = (e^{2\pi} - 1)e^{-2t}\left(\dfrac{3\sin 4t}{2} + 2\cos 4t\right), t > \pi$.

9. (a) $x = \cos 7t - \cos 8t$.

Section 5.5

1. (a) $\dfrac{2\sqrt{2}}{\pi}$; $x = \dfrac{e^{-4t}(-\sqrt{3}\sin 4\sqrt{3}t - \cos 4\sqrt{3}t)}{18}$
 $+ \dfrac{\sqrt{2}\sin 4\sqrt{2}t + \cos 4\sqrt{2}t}{18}$

 (b) 8; $x = \dfrac{t\sin 8t}{3}$.

3. (b) $2/\pi$; $3\sqrt{5}/4$. (c) $\sqrt{22}/2\pi$; $15\sqrt{23}/23$.

Section 5.6

1. $i = 4(1 - e^{-50t})$.

3. $q = \dfrac{1 - e^{-500t}}{50}$; $i = 10e^{-500t}$.

5. $i = e^{-80t}(-4.588\sin 60t + 1.247\cos 60t) - 1.247\cos 200t + 1.331\sin 200t$.

7. $q = e^{-Rt/2L} \left[\dfrac{Q_0 \sqrt{c}\, R}{\sqrt{4L - R^2 c}} \sin\left(\dfrac{\sqrt{4L - R^2 c}}{2\sqrt{c}\, L} t \right) + Q_0 \cos\left(\dfrac{\sqrt{4L - R^2 c}}{2\sqrt{c}\, L} t \right) \right]$,

$i = -\dfrac{2Q_0}{\sqrt{4Lc - R^2 c^2}} e^{-Rt/2L} \sin\left(\dfrac{\sqrt{4L - R^2 c}}{2\sqrt{c}\, L} t \right)$.

Section 6.1

1. $y = c_0 \left[1 + \displaystyle\sum_{n=1}^{\infty} \dfrac{(-1)^n x^{2n}}{[2 \cdot 4 \cdot 6 \cdots (2n)]} \right] + c_1 \left[x + \displaystyle\sum_{n=1}^{\infty} \dfrac{(-1)^n x^{2n+1}}{[3 \cdot 5 \cdot 7 \cdots (2n+1)]} \right]$.

3. $y = c_0 \left(1 - \dfrac{x^2}{2} - \dfrac{x^4}{24} + \cdots \right) + c_1 \left(x - \dfrac{x^3}{3} - \dfrac{x^5}{30} + \cdots \right)$.

5. $y = c_0 \left(1 - x^2 - \dfrac{x^3}{2} + \dfrac{x^4}{3} + \dfrac{11x^5}{40} + \cdots \right)$

$\quad + c_1 \left(x - \dfrac{x^3}{2} - \dfrac{x^4}{4} + \dfrac{x^5}{8} + \cdots \right)$.

7. $y = c_0 \left(1 - \dfrac{x^3}{6} + \dfrac{3x^5}{40} + \cdots \right) + c_1 \left(x - \dfrac{x^3}{6} - \dfrac{x^4}{12} + \dfrac{3x^5}{40} + \cdots \right)$.

9. $y = c_0 \left(1 + \dfrac{x^3}{6} + \dfrac{x^6}{18} + \cdots \right) + c_1 \left(x + \dfrac{x^4}{6} + \dfrac{17x^7}{252} + \cdots \right)$.

11. $y = 1 + \displaystyle\sum_{n=1}^{\infty} \dfrac{x^{2n}}{[2 \cdot 4 \cdot 6 \cdots (2n)]} = \displaystyle\sum_{n=0}^{\infty} \dfrac{x^{2n}}{2^n n!}$.

13. $y = 2 + 3x - \dfrac{7x^3}{6} - \dfrac{x^4}{2} + \dfrac{21x^5}{40} + \cdots$.

15. $y = c_0 \left[1 - \dfrac{(x-1)^2}{2} + \dfrac{(x-1)^3}{2} - \dfrac{5(x-1)^4}{12} + \dfrac{(x-1)^5}{3} + \cdots \right]$

$\quad + c_1 \left[(x-1) - \dfrac{(x-1)^2}{2} + \dfrac{(x-1)^3}{6} - \dfrac{(x-1)^5}{12} + \cdots \right]$.

17. $y = 2 + 4(x-1) - 4(x-1)^2 + \dfrac{4(x-1)^3}{3} - \dfrac{(x-1)^4}{3} + \dfrac{2(x-1)^5}{15} + \cdots$.

Section 6.2

1. $x = 0$ and $x = 3$ are regular singular points.

3. $x = 1$ is a regular singular point; $x = 0$ is an irregular singular point.

5. $y = C_1 x \left(1 - \dfrac{x^2}{14} + \dfrac{x^4}{616} - \cdots \right) + C_2 x^{-1/2} \left(1 - \dfrac{x^2}{2} + \dfrac{x^4}{40} - \cdots \right)$.

7. $y = C_1 x^{4/3} \left(1 - \dfrac{3x^2}{16} + \dfrac{9x^4}{896} - \cdots \right) + C_2 x^{2/3} \left(1 - \dfrac{3x^2}{8} + \dfrac{9x^4}{320} - \cdots \right)$.

9. $y = C_1 x^{1/3} \left(1 - \dfrac{3x^2}{16} + \dfrac{9x^4}{896} - \cdots \right) + C_2 x^{-1/3} \left(1 - \dfrac{3x^2}{8} + \dfrac{9x^4}{320} - \cdots \right)$.

11. $y = C_1\left(1 + x + \dfrac{3x^2}{10} + \cdots\right) + C_2 x^{1/3}\left(1 + \dfrac{7x}{12} + \dfrac{5x^2}{36} + \cdots\right).$

13. $y = C_1 x^{1/2}\left(1 - \dfrac{x^2}{6} + \dfrac{x^4}{120} - \cdots\right) + C_2 x^{-1/2}\left(1 - \dfrac{x^2}{2} + \dfrac{x^4}{24} - \cdots\right).$

15. $y = C_1\left[1 + \displaystyle\sum_{n=1}^{\infty} \dfrac{x^{2n}}{2^n n!}\right] + C_2 x^3\left[1 + \displaystyle\sum_{n=1}^{\infty} \dfrac{x^{2n}}{[5 \cdot 7 \cdot 9 \cdots (2n+3)]}\right]$

$\quad = C_1\left(1 + \dfrac{x^2}{2} + \dfrac{x^4}{8} + \cdots\right) + C_2 x^3\left(1 + \dfrac{x^2}{5} + \dfrac{x^4}{35} + \cdots\right).$

17. $y = C_1\left(1 - x + \dfrac{x^2}{2} - \dfrac{x^3}{6} + \cdots\right) + C_2 x^{-1}\left(1 + \dfrac{3x^2}{2} - \dfrac{x^3}{3} + \cdots\right)$

$\quad = Ce^{-x} + C_2 x^{-1} e^x,$ where $C = C_1 - C_2.$

19. $y = C_1 x\left[1 + 2\displaystyle\sum_{n=1}^{\infty} \dfrac{(-1)^n x^n}{n!(n+2)!}\right]$

$\quad + C_2\left[x^{-1}\left(-\dfrac{1}{2} - \dfrac{x}{2} + \dfrac{29x^2}{144} + \cdots\right) + \dfrac{1}{4} y_1(x) \ln|x|\right],$

where $y_1(x)$ denotes the solution of which C_1 is the coefficient.

21. $y = C_1 x^4\left(1 - \dfrac{x^2}{2} + \dfrac{x^4}{10} - \cdots\right)$

$\quad + C_2\left[x^{-2}\left(-\dfrac{1}{6} - \dfrac{x^2}{6} - \dfrac{x^4}{6} + \cdots\right) + \dfrac{2}{9} y_1(x) \ln|x|\right],$

where $y_1(x)$ denotes the solution of which C_1 is the coefficient.

23. $y = C_1\left[1 + \displaystyle\sum_{n=1}^{\infty} \dfrac{(-1)^n 2^n x^n}{(n!)^2}\right] + C_2\left[4x - 3x^2 + \dfrac{22x^3}{7} + \cdots + y_1(x) \ln|x|\right],$

where $y_1(x)$ denotes the solution of which C_1 is the coefficient.

25. $y = C_1 x\left[1 + \displaystyle\sum_{n=1}^{\infty} \dfrac{(-1)^n x^{2n}}{[2 \cdot 4 \cdot 6 \cdots (2n)]^2}\right]$

$\quad + C_2\left[\dfrac{x^3}{4} - \dfrac{3x^5}{128} + \dfrac{11x^7}{13824} + \cdots + y_1(x) \ln|x|\right],$

where $y_1(x)$ denotes the solution of which C_1 is the coefficient.

Section 6.3

3. $y = \dfrac{c_1 \sin x + c_2 \cos x}{\sqrt{x}}.$

Section 7.1

1. $x = ce^{-2t},\ y = -\tfrac{2}{3}ce^{-2t} + \tfrac{1}{3}e^{4t} - \tfrac{1}{3}e^t.$

3. $x = ce^{-3t} + \dfrac{e^t}{4},\ y = -\dfrac{2ce^{-3t}}{3} + \dfrac{e^{3t}}{3} - \dfrac{e^t}{2}.$

5. $x = c_1 \sin t + c_2 \cos t,$

$$y = -\left(\frac{3c_1 + c_2}{2}\right)\sin t + \left(\frac{c_1 - 3c_2}{2}\right)\cos t + \frac{e^t}{2} - \frac{e^{-t}}{2}.$$

7. $x = c_1 e^{\sqrt{6}t} + c_2 e^{-\sqrt{6}t} - t + \frac{1}{6},$

$$y = \frac{\sqrt{6}c_1 e^{\sqrt{6}t}}{6} - \frac{\sqrt{6}c_2 e^{-\sqrt{6}t}}{6} + \frac{t}{6} - \frac{1}{6} - \frac{e^{3t}}{3}.$$

9. $x = c_1 e^t - \dfrac{\sin t}{2},\ y = -\dfrac{c_1 e^t}{3} + \dfrac{\sin t}{2}.$

11. $x = c_1 e^{4t} + c_2 e^{-2t} - t + 1,\ y = -c_1 e^{4t} + c_2 e^{-2t} + t.$

13. $x = c_1 + c_2 e^{-2t} + 2t^2 + t,\ y = (1 - c_1) - 3c_2 e^{-2t} - t^2 - 3t.$

15. $x = c_1 e^t + c_2 e^{-2t} - te^t;$
    $y = (\frac{1}{3} - c_1)e^t - \frac{1}{3}c_2 e^{-2t} + te^t.$

17. $x = c_1 e^{3t} + \dfrac{t}{3} - \dfrac{2}{9},\ y = c_2 e^t - \dfrac{5c_1 e^{3t}}{2} - \dfrac{t}{3} - \dfrac{4}{9}.$

19. $x = c_1 + c_2 t + c_3 e^t + c_4 e^{-t} - \dfrac{t^2}{2},$

$$y = (c_1 - c_2 + 1) + (c_2 + 1)t + c_4 e^{-t} - \frac{t^2}{2}.$$

21. $x = c_1 + c_2 e^t + c_3 e^{-3t} - \dfrac{t^2}{6} - \dfrac{14t}{9},$

$$y = \left(3c_1 + \frac{17}{9}\right) + c_2 e^t - 3c_3 e^{-3t} - \frac{t^2}{2} - \frac{4t}{3}.$$

23. $\dfrac{dx_1}{dt} = x_2,$

$$\frac{dx_2}{dt} = -2x_1 + 3x_2 + t^2.$$

25. $\dfrac{dx_1}{dt} = x_2,$

$$\frac{dx_2}{dt} = x_3,$$

$$\frac{dx_3}{dt} = -2t^3 x_2 - tx_3 + 5t^4.$$

Section 7.2

1. $x_1 = 2 \cos t - \cos 2t,\ x_2 = 4 \cos t + \cos 2t.$

3. $i_1 = -\dfrac{10e^{-1000t}}{3} - \dfrac{5e^{-4000t}}{3} + 5,\qquad i_2 = -\dfrac{10e^{-1000t}}{3} + \dfrac{5e^{-4000t}}{6} + \dfrac{5}{2}.$

5. $x = e^{-t/10} + 2e^{-t/30},\ y = -2e^{-t/10} + 4e^{-t/30}.$

Section 7.3

1. (c) $x = 2e^{5t} - e^{-t}$, $y = e^{5t} + e^{-t}$.

Section 7.4

1. $x = c_1 e^t + c_2 e^{3t}$, $y = 2c_1 e^t + c_2 e^{3t}$.
3. $x = 2c_1 e^{4t} + c_2 e^{-t}$, $y = 3c_1 e^{4t} - c_2 e^{-t}$.
5. $x = c_1 e^t + c_2 e^{5t}$, $y = -2c_1 e^t + 2c_2 e^{5t}$.
7. $x = 2c_1 e^t + c_2 e^{-t}$, $y = c_1 e^t + c_2 e^{-t}$.
9. $x = c_1 e^{4t} + c_2 e^{-2t}$, $y = c_1 e^{4t} - c_2 e^{-2t}$.
11. $x = 2e^t(-c_1 \sin 2t + c_2 \cos 2t)$, $y = e^t(c_1 \cos 2t + c_2 \sin 2t)$.
13. $x = e^t(c_1 \cos 3t + c_2 \sin 3t)$, $y = e^t(c_1 \sin 3t - c_2 \cos 3t)$.
15. $x = 2e^{3t}(c_1 \cos 3t + c_2 \sin 3t)$,
    $y = e^{3t}[c_1(\cos 3t + 3 \sin 3t) + c_2(\sin 3t - 3 \cos 3t)]$.
17. $x = e^{3t}(c_1 \cos 2t + c_2 \sin 2t)$, $y = e^{3t}(c_1 \sin 2t - c_2 \cos 2t)$.
19. $x = c_1 e^t + c_2 t e^t$, $y = 2c_1 e^t + c_2(2t - 1)e^t$.
21. $x = -2c_1 e^{3t} + c_2(2t + 1)e^{3t}$, $y = c_1 e^{3t} - c_2 t e^{3t}$.
23. $x = 2e^{5t} + 7e^{-5t}$, $y = 2e^{5t} - 3e^{-5t}$.
25. $x = 4e^{4t}[\cos 2t - 2 \sin 2t]$, $y = e^{4t}[\cos 2t + 3 \sin 2t]$.
27. $x = 2e^{4t} - 8t e^{4t}$, $y = 3e^{4t} - 4t e^{4t}$.
31. $x = 3c_1 t^4 + c_2 t^{-1}$, $y = 2c_1 t^4 - c_2 t^{-1}$.

Section 7.5A

1. (b) $\begin{pmatrix} 9 & 0 & 9 \\ 1 & 4 & 2 \\ 1 & -2 & -1 \end{pmatrix}$.

2. (b) $\begin{pmatrix} -4 & 12 & -20 \\ -24 & 8 & 0 \\ 12 & -4 & -8 \end{pmatrix}$.

3. (b) $\begin{pmatrix} 7 \\ 8 \\ -25 \\ 36 \end{pmatrix}$.

4. (b) $\begin{pmatrix} -35 \\ 10 \\ -7 \end{pmatrix}$.

6. (b) (i) $\begin{pmatrix} 3e^{3t} \\ (6t + 11)e^{3t} \\ (3t^2 + 2t)e^{3t} \end{pmatrix}$;

   (ii) $\begin{pmatrix} (e^{3t} - 1)/3 \\ [e^{3t}(6t + 7) - 7]/9 \\ [e^{3t}(9t^2 - 6t + 2) - 2]/27 \end{pmatrix}$.

Section 7.5B

1. $\mathbf{AB} = \begin{pmatrix} 22 & 23 \\ 18 & 13 \end{pmatrix}, \quad \mathbf{BA} = \begin{pmatrix} 18 & 62 \\ 7 & 17 \end{pmatrix}.$

3. $\mathbf{AB} = \begin{pmatrix} 7 & 4 & 0 \\ -7 & 8 & -14 \\ 17 & -4 & 16 \end{pmatrix}, \quad \mathbf{BA} = \begin{pmatrix} 19 & 2 \\ -20 & 12 \end{pmatrix}$

5. $\mathbf{AB} = \begin{pmatrix} 7 & 5 \\ 9 & 1 \\ 10 & 4 \end{pmatrix}, \quad \mathbf{BA}$ not defined.

7. $\mathbf{AB} = \begin{pmatrix} 42 & 14 & 4 \\ 34 & 15 & 9 \\ 6 & -4 & -2 \end{pmatrix}, \quad \mathbf{BA} = \begin{pmatrix} 1 & 7 & 5 \\ -8 & 4 & 8 \\ -6 & 36 & 50 \end{pmatrix}.$

9. $\mathbf{AB} = \begin{pmatrix} 3 & 5 \\ -4 & 8 \\ 0 & -4 \end{pmatrix}, \quad \mathbf{BA}$ not defined.

11. $\mathbf{A}^2 = \begin{pmatrix} 1 & 4 & 4 \\ 2 & -3 & 8 \\ 4 & 4 & -1 \end{pmatrix}, \quad \mathbf{A}^3 = \begin{pmatrix} 1 & -6 & 21 \\ 12 & 1 & -3 \\ 6 & 12 & 7 \end{pmatrix}.$

13. $\mathbf{A}^{-1} = \begin{pmatrix} -5 & 3 \\ 2 & -1 \end{pmatrix}.$

15. $\mathbf{A}^{-1} = \begin{pmatrix} 1/16 & -1/8 \\ -3/8 & -1/4 \end{pmatrix}.$

17. $\mathbf{A}^{-1} = \begin{pmatrix} -1/5 & 2/5 & -3/5 \\ 1 & -1 & 1 \\ -6/5 & 7/5 & -3/5 \end{pmatrix}.$

19. $\mathbf{A}^{-1} = \begin{pmatrix} -2 & 19/3 & 1/3 \\ 0 & -2/3 & 1/3 \\ 1 & -7/3 & -1/3 \end{pmatrix}.$

21. $\mathbf{A}^{-1} = \begin{pmatrix} 4 & 1 & -6 \\ 1/2 & 1/2 & -1 \\ -3 & -1 & 5 \end{pmatrix}.$

23. $\mathbf{A}^{-1} = \begin{pmatrix} 1 & -1/5 & 2/5 \\ -2 & 1 & -1 \\ 0 & -1/5 & 2/5 \end{pmatrix},$

Section 7.5C

3. (a) $k = 3$, (b) $k = 2$.

Section 7.5D

1. Characteristic values: -1 and 4;
    Respective corresponding characteristic vectors:

    $$\begin{pmatrix} k \\ -k \end{pmatrix} \quad \text{and} \quad \begin{pmatrix} 2k \\ 3k \end{pmatrix},$$

    where in each vector k is an arbitrary nonzero real number.

3. Characteristic values: -1 and 6;
    Respective corresponding characteristic vectors:

    $$\begin{pmatrix} k \\ -4k \end{pmatrix} \quad \text{and} \quad \begin{pmatrix} k \\ 3k \end{pmatrix},$$

    where in each vector k is an arbitrary nonzero real number.

5. Characteristic values: 7 and -2;
    Respective corresponding characteristic vectors:

    $$\begin{pmatrix} k \\ k \end{pmatrix} \quad \text{and} \quad \begin{pmatrix} 4k \\ -5k \end{pmatrix},$$

    where in each vector k is an arbitrary nonzero real number.

7. Characteristic values: 1, 2, and -3;
    Respective corresponding characteristic vectors:

    $$\begin{pmatrix} k \\ k \\ k \end{pmatrix}, \begin{pmatrix} k \\ 2k \\ k \end{pmatrix}, \quad \text{and} \quad \begin{pmatrix} k \\ 7k \\ 11k \end{pmatrix},$$

    where in each vector k is an arbitrary nonzero real number.

9. Characteristic values: 2, 3, and -2;
    Respective corresponding characteristic vectors:

    $$\begin{pmatrix} k \\ 0 \\ -k \end{pmatrix}, \begin{pmatrix} k \\ -k \\ -k \end{pmatrix}, \quad \text{and} \quad \begin{pmatrix} k \\ -k \\ 4k \end{pmatrix},$$

    where in each vector k is an arbitrary nonzero real number.

11. Characteristic values: 1, 3, and 4;
    Respective corresponding characteristic vectors:

    $$\begin{pmatrix} k \\ 0 \\ 0 \end{pmatrix}, \begin{pmatrix} 0 \\ 2k \\ k \end{pmatrix}, \quad \text{and} \quad \begin{pmatrix} -k \\ k \\ k \end{pmatrix}.$$

    where in each vector k is an arbitrary nonzero real number.

13. Characteristic values: -1, 2, and 3;
    Respective corresponding characteristic vectors:

    $$\begin{pmatrix} 5k \\ 3k \\ -2k \end{pmatrix}, \begin{pmatrix} 5k \\ 5k \\ -k \end{pmatrix}, \quad \text{and} \quad \begin{pmatrix} k \\ k \\ 0 \end{pmatrix}.$$

    where in each vector k is an arbitrary nonzero real number.

Section 7.6

1. $x_1 = c_1 e^t + c_2 e^{3t}, x_2 = 2c_1 e^t + c_2 e^{3t}$.
3. $x_1 = 2c_1 e^{4t} + c_2 e^{-t}, x_2 = 3c_1 e^{4t} - c_2 e^{-t}$.
5. $x_1 = c_1 e^t + c_2 e^{5t}, x_2 = -2c_1 e^t + 2c_2 e^{5t}$.
7. $x_1 = c_1 e^{4t} + c_2 e^{-2t}, x_2 = c_1 e^{4t} - c_2 e^{-2t}$.
9. $x_1 = 2e^t(-c_1 \sin 2t + c_2 \cos 2t), x_2 = e^t(c_1 \cos 2t + c_2 \sin 2t)$.
11. $x_1 = e^t(c_1 \cos 3t + c_2 \sin 3t), x_2 = e^t(c_1 \sin 3t - c_2 \cos 3t)$.
13. $x_1 = c_1 e^t + c_2 te^t, x_2 = 2c_1 e^t + c_2(2t - 1)e^t$.
15. $x_1 = -2c_1 e^{3t} + c_2(2t + 1)e^{3t}, x_2 = c_1 e^{3t} - c_2 te^{3t}$.
17. $x_1 = 2c_1 e^{4t} + c_2(2t + 1)e^{4t}, x_2 = c_1 e^{4t} + c_2 te^{4t}$.

Section 7.7

1. $x_1 = c_1 e^t + c_2 e^{2t} + c_3 e^{-3t}$,
   $x_2 = c_1 e^t + 2c_2 e^{2t} + 7c_3 e^{-3t}$,
   $x_3 = c_1 e^t + c_2 e^{2t} + 11c_3 e^{-3t}$.
3. $x_1 = c_1 e^{2t} + c_2 e^{3t} + c_3 e^{-2t}$,
   $x_2 = -c_2 e^{3t} - c_3 e^{-2t}$,
   $x_3 = -c_1 e^{2t} - c_2 e^{3t} + 4c_3 e^{-2t}$.
5. $x_1 = c_1 e^{5t} + 2c_2 e^{-t}$,
   $x_2 = c_1 e^{5t} - c_2 e^{-t} + c_3 e^{-3t}$,
   $x_3 = c_1 e^{5t} - c_2 e^{-t} - c_3 e^{-3t}$.
7. $x_1 = -2c_1 e^{(2+\sqrt{5})t} + 2c_2 e^{(2-\sqrt{5})t}$,
   $x_2 = (1 + \sqrt{5})c_1 e^{(2+\sqrt{5})t} + (-1 + \sqrt{5})c_2 e^{(2-\sqrt{5})t}$,
   $x_3 = c_3 e^{2t}$.
9. $x_1 = c_1 e^{-t} + c_3 e^{3t}$,
   $x_2 = 2c_1 e^{-t} + c_2 e^{3t}$,
   $x_3 = -2c_1 e^{-t} - c_3 e^{3t}$.
11. $x_1 = c_1 e^{4t} + c_2 e^t$,
    $x_2 = 2c_1 e^{4t} + 3c_2 e^t + 3c_3 e^t$,
    $x_3 = c_1 e^{4t} + c_2 e^t + c_3 e^t$.
13. $x_1 = c_1 e^{-t} + 2c_2 e^{2t}$,
    $x_2 = c_1 e^{-t} + 3c_3 e^{2t}$,
    $x_3 = -c_1 e^{-t} - c_2 e^{2t} - c_3 e^{2t}$.
15. $x_1 = c_1 e^t + 2c_3 e^{4t}$,
    $x_2 = c_2 e^{3t} - c_3 e^{4t}$,
    $x_3 = c_1 e^t + 2c_2 e^{3t} + c_3 e^{4t}$.
17. $x_1 = c_1 e^t + c_3 e^{-2t}$,
    $x_2 = -2c_1 e^t + c_2 e^{-2t} + c_3(t - 1)e^{-2t}$,
    $x_3 = -c_2 e^{-2t} - c_3 te^{-2t}$.
19. $x_1 = c_1 e^{-t} + c_3 e^{3t}$,
    $x_2 = -2c_1 e^{-t} + c_2 e^{3t} + c_3(t - 1)e^{3t}$,
    $x_3 = -c_2 e^{3t} - c_3 te^{3t}$.
21. $x_1 = c_1 e^{2t} + c_3 te^{2t}$,
    $x_2 = c_2 e^{2t} - 2c_3 te^{2t}$,
    $x_3 = -2c_1 e^{2t} - 3c_2 e^{2t} + c_3(4t + 1)e^{2t}$.
23. $x_1 = 2c_1 e^{2t} + 2c_2 te^{2t} + c_3(t^2 + 1)e^{2t}$,
    $x_2 = -c_1 e^{2t} - c_2 te^{2t} - \frac{1}{2}c_3 t^2 e^{2t}$,
    $x_3 = -c_2 e^{2t} - c_3(t - 3)e^{2t}$.

Section 8.2

1. $y = 1 + x + 2 \sum_{n=2}^{\infty} \dfrac{x^n}{n!} = -x - 1 + 2e^x$.

3. $y = 2 + x + 2x^2 + \dfrac{4x^3}{3} + \dfrac{9x^4}{4} + \cdots$.

5. $y = \dfrac{x^2}{2} + \dfrac{x^3}{6} + \dfrac{x^4}{24} + \dfrac{x^5}{120} + \cdots$.

7. $y = x + x^2 + \dfrac{x^3}{6} - \dfrac{x^4}{12} + \cdots$.

9. $y = 1 + 2(x - 1) + \dfrac{7(x - 1)^2}{2} + \dfrac{14(x - 1)^3}{3} + \dfrac{73(x - 1)^4}{12} + \cdots$.

11. $y = \pi + \dfrac{(x - 1)^2}{2} + \dfrac{(x - 1)^5}{40} + \cdots$.

Section 8.3

1. $\phi_1(x) = 1 + \dfrac{x^2}{2}$, $\phi_2(x) = 1 + \dfrac{x^2}{2} + \dfrac{x^4}{8}$, $\phi_3(x) = 1 + \dfrac{x^2}{2} + \dfrac{x^4}{8} + \dfrac{x^6}{48}$.

3. $\phi_1(x) = \dfrac{x^2}{2}$, $\phi_2(x) = \dfrac{x^2}{2} + \dfrac{x^5}{20}$, $\phi_3(x) = \dfrac{x^2}{2} + \dfrac{x^5}{20} + \dfrac{x^8}{160} + \dfrac{x^{11}}{4400}$.

5. $\phi_1(x) = e^x - 1$, $\phi_2(x) = \dfrac{e^{2x}}{2} - e^x + x + \dfrac{1}{2}$,

$\phi_3(x) = \dfrac{e^{4x}}{16} - \dfrac{e^{3x}}{3} + \dfrac{xe^{2x}}{2} + \dfrac{e^{2x}}{2} - 2xe^x + 2e^x + \dfrac{x^3}{3} + \dfrac{x^2}{2} + \dfrac{x}{4} - \dfrac{107}{48}$.

7. $\phi_1(x) = x^2$, $\phi_2(x) = x^2 + \dfrac{x^7}{7}$, $\phi_3(x) = x^2 + \dfrac{x^7}{7} + \dfrac{x^{12}}{28} + \dfrac{3x^{17}}{833} + \dfrac{x^{22}}{7546}$.

Section 8.4B

1. (a) 0.800, 0.650, 0.540, 0.462. (b) 0.812, 0.670, 0.564, 0.488.
   (c) 0.823, 0.688, 0.586, 0.512.

Section 8.4C

1. (a) 1.239, 1.564, 1.995. (b) 1.238, 1.562, 1.991. (c) 1.237, 1.561, 1.989.
3. (a) 1.112, 1.255, 1.445. (b) 1.100, 1.222, 1.375.

Section 8.4D

1. (a) 1.2374, 1.5606, 1.9886. (b) 1.2374, 1.5606, 1.9887.
3. 1.1115, 1.2531, 1.4398.

Section 8.4E

1. 2.5447.

Section 9.1A

1. $\dfrac{2}{s^3}$.

3. $\dfrac{5}{s}(1 - e^{-2s})$.

5. $\dfrac{1}{s^2} + \dfrac{e^{-2s}}{s} - \dfrac{e^{-2s}}{s^2}$

Section 9.1B

1. $\dfrac{s^2 + 2a^2}{s(s^2 + 4a^2)}$.

3. $\mathscr{L}\{\sin^3 at\} = \dfrac{6a^3}{(s^2 + a^2)(s^2 + 9a^2)}$,

 $\mathscr{L}\{\sin^2 at \cos at\} = \dfrac{2a^2 s}{(s^2 + a^2)(s^2 + 9a^2)}$.

5. $\dfrac{6}{s^4}$.

7. $\dfrac{s + 5}{s^2 + 3s + 2}$.

9. $\dfrac{2}{(s - a)^3}$.

11. $\dfrac{2s(s^2 - 3b^2)}{(s^2 + b^2)^3}$.

13. $\dfrac{6}{(s - a)^4}$.

Section 9.1C

1. $\dfrac{5e^{-6s}}{s}$.

3. $\dfrac{4}{s}(1 - e^{-6s})$.

5. $\dfrac{2}{s}(e^{-5s} - e^{-7s})$.

7. $\dfrac{1 + e^{-2s} + e^{-4s} - 3e^{-6s}}{s}$.

9. $\dfrac{2(1 - e^{-3s} + e^{-6s})}{s}$.

11. $e^{-2s}\left(\dfrac{1}{s^2} + \dfrac{2}{s}\right).$

13. $\dfrac{1}{s^2}(1 - e^{-3s}).$

15. $\dfrac{-e^{-\pi s/2}}{s^2 + 1}.$

17. $\dfrac{e^{-4s} - e^{-7s}}{s^2}.$

Section 9.2A

1. $\dfrac{2 \sin 3t}{3}.$

3. $\dfrac{5t^3 e^{2t}}{6}.$

5. $e^{-2t} \cos \sqrt{3}t.$

7. $\dfrac{1}{3} - \dfrac{e^{-t}}{2} + \dfrac{e^{-3t}}{6}.$

9. $\dfrac{4t \sin 2t + 3 \sin 2t - 6t \cos 2t}{16}.$

11. $\frac{5}{24}t^4 e^{-2t}.$

13. $t^2 e^{-3t}\left(1 + \dfrac{t}{6}\right).$

15. $-4e^{-2t} + 5e^{-3t}.$
17. $7 \cos 3t + 4 \sin 3t.$

19. $f(t) = \begin{cases} 0, & 0 < t < \pi, \\ -5 \cos 3t - 2 \sin 3t, & t > \pi. \end{cases}$

21. $f(t) = \begin{cases} 0, & 0 < t < \pi/2, \\ e^{-2(t-\pi/2)}[2 \cos 3t - \sin 3t], & t > \pi/2. \end{cases}$

23. $f(t) = \begin{cases} 0, & 0 < t < 4, \\ t - 4, & 4 < t < 7, \\ 3, & t > 7. \end{cases}$

25. $f(t) = \begin{cases} \frac{1}{2} \sin 2t, & 0 < t < \pi. \\ \sin 2t, & t > \pi. \end{cases}$

27. $f(t) = \begin{cases} e^t \sin 2t, & 0 < t < \pi/2, \\ (1 - e^{-\pi/2})e^t \sin 2t, & t > \pi/2. \end{cases}$

Section 9.2B
1. $e^{-2t} - e^{-3t}.$

3. $(1 - \cos 3t)/9.$

5. $(-1 + 3t + e^{-3t})/9.$

Section 9.3

1. $y = \dfrac{3e^t + e^{3t}}{2}.$

3. $y = e^{2t}.$

5. $y = 2 - 2\cos 2t + 3\sin 2t.$

7. $y = 2e^{2t} - 3e^{-t} - e^{-t}\sin 3t + e^{-t}\cos 3t.$

9. $y = e^{-2t} - e^{-3t} - 2te^{-3t}.$

11. $y = (3 - 4t)e^t + \sin t - 3\cos t.$

13. $y = -2e^t + e^{2t} + 1, 0 < t < 4;$
$y = 2(e^{-4} - 1)e^t + (1 - e^{-8})e^{2t}, t > 4.$

15. $y = \frac{1}{5}[1 + e^{-2t}(3\sin t - \cos t)], 0 < t < \pi/2,$

$y = \dfrac{e^{-2t}}{5}[(e^\pi + 3)\sin t - (2e^\pi + 1)\cos t], t > \pi/2.$

17. $y = -t + 2\pi + \frac{1}{2}\sin 2t + (2 - 2\pi)\cos 2t, 0 < t < 2\pi,$
$y = (2 - 2\pi)\cos 2t, t > 2\pi.$

Section 9.4

1. $x = -\dfrac{e^t}{2} + \dfrac{e^{-t}}{2} + 2e^{2t}, y = \dfrac{e^t}{2} + \dfrac{e^{-t}}{2} - e^{2t}.$

3. $-2e^t + 5e^{4t}, y = -4e^t + 4e^{4t}.$

5. $x = 3 + 2t + \frac{4}{3}t^3, y = 5 + 5t - 2t^2 + \frac{8}{3}t^3.$

7. $x = 8\sin t + 2\cos t,$

$y = -13\sin t + \cos t + \dfrac{e^t}{2} - \dfrac{e^{-t}}{2}.$

9. $x = e^{-2t} - te^t, y = \frac{1}{3}e^t - \frac{1}{3}e^{-2t} + te^t.$

Section 10.2

1. (a) There exists a solution ϕ on some interval $|x - x_0| \le h$ about $x = x_0.$
(b) There exists a *unique* solution on $|x - x_0| \le h.$

3. (a) $|x| \le \dfrac{1}{2}; y = \tan x, |x| < \dfrac{\pi}{2}.$

(b) $|x| \le \dfrac{1}{2e}; y = -\dfrac{\ln(1 - 2x)}{2}, -\infty < x < \dfrac{1}{2}.$

Section 10.4

1. (a) Yes. (b) $\dfrac{dy_1}{dx} = y_2, \dfrac{dy_2}{dx} = y_3, \dfrac{dy_3}{dx} = x^2 + y_1y_2 + y_3^2.$

3. Theorem 10.8

Section 11.2
1. Yes.
3. No.
5. Yes.

Section 11.3
1. See Example 11.10 in the text.
3. (b) $x_1 = (11/4)e^{3t} + (9/4)e^{-3t}$, $x_2 = (55/4)e^{3t} - (27/4)e^{-3t}$.
5. (b) $x_1 = 6e^t$, $x_2 = 6e^t - 3e^{5t}$, $x_3 = 3e^t - e^{5t}$.
7. (c) $\mathbf{B} = \mathbf{CAC}^{-1}$.

Section 11.4

1. (a) $\begin{pmatrix} e^{3t} & 3e^{4t} \\ e^{3t} & 2e^{4t} \end{pmatrix}$,   (b) $\begin{pmatrix} 2e^{2t} \\ 3e^{2t} \end{pmatrix}$.

3. (a) $\begin{pmatrix} e^{-t} & 2e^{4t} \\ -e^{-t} & 3e^{4t} \end{pmatrix}$,   (b) $\begin{pmatrix} e^t + 2e^{2t} \\ -3e^t + e^{2t} \end{pmatrix}$.

5. (a) $\begin{pmatrix} e^t & e^{5t} \\ -2e^t & 2e^{5t} \end{pmatrix}$,   (b) $\begin{pmatrix} \cos t \\ \sin t - 3\cos t \end{pmatrix}$.

7. $\begin{pmatrix} -6e^{-t} + 3e^{-2t} + 4e^{-3t} \\ -9e^{-t} + 4e^{-2t} + 3e^{-3t} \end{pmatrix}$.

Section 11.5

1. (a) $f_1(t) = \dfrac{e^t}{2} - \dfrac{\sin t}{2} + \dfrac{\cos t}{2}$,

   $f_2(t) = \sin t$,

   $f_3(t) = \dfrac{e^t}{2} - \dfrac{\sin t}{2} - \dfrac{\cos t}{2}$.

   (b) $8f_1(t) + 7f_2(t) + 2f_3(t)$.

9. $\dfrac{d^2 y}{dt^2} - y = 0$; $y = e^t$, $u = e^{-t}$; $x = \exp\left(-\dfrac{t^2 - t}{2}\right)$,

   $x = \exp\left(-\dfrac{t^2 + 3t}{2}\right)$.

Section 11.6
5. (a) $x = t^3$. (b) $x = c_1 t^3 + c_2 t e^t$.

Section 11.8

1. (a) $t^2 \dfrac{d^2 x}{dt^2} + t \dfrac{dx}{dt} + 2x = 0$.

(b) $(2t + 1)\dfrac{d^2x}{dt^2} + (4 - t^3)\dfrac{dx}{dt} + (1 - 3t^2)x = 0.$

(c) $t^2\dfrac{d^2x}{dt^2} - (2t^3 + 3t)\dfrac{dx}{dt} + (2t^2 + 3)x = 0.$

(d) $t^3\dfrac{d^2x}{dt^2} + (t^3 + 8t^2 - t)\dfrac{dx}{dt} + (4t^2 + 11t - 2)x = 0.$

5. (b) $u = -t, \; x = e^{-t}, \; x = c_1 e^{-t} + c_2 e^{-t}\displaystyle\int e^{2t}t^{-1}\,dt.$

Section 12.1

1. $\lambda = 4n^2 \; (n = 1, 2, 3, \ldots); \; y = c_n \sin 2nx \; (n = 1, 2, 3, \ldots).$

3. $\lambda = \left(\dfrac{n\pi}{L}\right)^2 (n = 1, 2, 3, \ldots); \; y = c_n \sin \dfrac{n\pi x}{L} \; (n = 1, 2, 3, \ldots).$

5. $\lambda = \alpha_n^2$, where $\alpha_n \; (n = 1, 2, 3, \ldots)$ are the positive roots of the equation $\alpha = \tan \pi\alpha; \; y = c_n \sin \alpha_n x \; (n = 1, 2, 3, \ldots).$

7. $\lambda = n^2 \; (n = 1, 2, 3, \ldots); \; y = c_n \sin(n \ln|x|) \; (n = 1, 2, 3, \ldots).$

9. $\lambda = 16n^2 \; (n = 1, 2, 3, \ldots); \; y = c_n \sin(4n \arctan x) \; (n = 1, 2, 3, \ldots).$

Section 12.3

3. $\dfrac{2}{\pi}\displaystyle\sum_{n=1}^{\infty}\dfrac{1 - (-1)^n}{n}\sin(n \ln|x|) = \dfrac{4}{\pi}\sum_{n=1}^{\infty}\dfrac{\sin[(2n-1)\ln|x|]}{2n-1}.$

Section 12.4B

1. $2\displaystyle\sum_{n=1}^{\infty}\dfrac{(-1)^{n+1}\sin nx}{n} = 2\left[\sin x - \dfrac{\sin 2x}{2} + \dfrac{\sin 3x}{3} - \cdots\right].$

3. $\dfrac{4}{\pi}\displaystyle\sum_{n=1}^{\infty}\left(\dfrac{1 - \cos n\pi}{n}\right)\sin nx = \dfrac{8}{\pi}\sum_{n=1}^{\infty}\dfrac{\sin(2n-1)x}{2n-1}.$

5. $\dfrac{\pi^2}{6} + \displaystyle\sum_{n=1}^{\infty}\left\{\left[\dfrac{2(-1)^n}{n^2}\right]\cos nx + \dfrac{1}{\pi}\left[\left(\dfrac{2}{n^3} - \dfrac{\pi^2}{n}\right)(-1)^n - \dfrac{2}{n^3}\right]\sin nx\right\}$

$= \dfrac{\pi^2}{6} + \left[-2\cos x + \left(\pi - \dfrac{4}{\pi}\right)\sin x\right] + \left[\dfrac{\cos 2x}{2} - \dfrac{\pi \sin 2x}{2}\right]$

$+ \left[-\dfrac{2\cos 3x}{9} + \dfrac{1}{3}\left(\pi - \dfrac{4}{9\pi}\right)\sin 3x\right] + \left[\dfrac{\cos 4x}{8} - \dfrac{\pi \sin 4x}{4}\right] + \cdots.$

7. $\dfrac{2}{\pi}\displaystyle\sum_{n=1}^{\infty}\left[\left(\dfrac{2}{n^3} - \dfrac{\pi^2}{n}\right)(-1)^n - \dfrac{2}{n^3}\right]\sin nx.$

9. $3 + \dfrac{12}{\pi^2}\displaystyle\sum_{n=1}^{\infty}\left[\dfrac{(-1)^n - 1}{n^2}\right]\cos\dfrac{n\pi x}{6}$

$= 3 - \dfrac{24}{\pi^2}\displaystyle\sum_{n=1}^{\infty}\dfrac{1}{(2n-1)^2}\cos\dfrac{(2n-1)\pi x}{6}.$

11. $\dfrac{1}{2} + \dfrac{1}{\pi} \displaystyle\sum_{n=1}^{\infty} \left[\dfrac{1-(-1)^n}{n}\right] \sin \dfrac{n\pi x}{3} = \dfrac{1}{2} + \dfrac{2}{\pi} \displaystyle\sum_{n=1}^{\infty} \dfrac{1}{2n-1} \sin \dfrac{(2n-1)\pi x}{3}$.

13. $\dfrac{L}{2} + \dfrac{L}{\pi} \displaystyle\sum_{n=1}^{\infty} \left[\dfrac{1-(-1)^n}{n}\right] \sin \dfrac{n\pi x}{L} = \dfrac{L}{2} + \dfrac{2L}{\pi} \displaystyle\sum_{n=1}^{\infty} \dfrac{1}{2n-1} \sin \dfrac{(2n-1)\pi x}{L}$.

15. $b + \dfrac{2aL}{\pi} \displaystyle\sum_{n=1}^{\infty} \dfrac{(-1)^{n+1}}{n} \sin \dfrac{n\pi x}{L}$.

Section 12.4C

1. (a) $\dfrac{2}{\pi} \displaystyle\sum_{n=1}^{\infty} \left[\dfrac{1-(-1)^n}{n}\right] \sin nx = \dfrac{4}{\pi} \displaystyle\sum_{n=1}^{\infty} \dfrac{\sin(2n-1)x}{2n-1}$. (b) 1.

3. (a) $\dfrac{4}{\pi} \displaystyle\sum_{n=1}^{\infty} \dfrac{\cos(n\pi/2) - \cos n\pi}{n} \sin nx$.

   (b) $1 - \dfrac{4}{\pi} \displaystyle\sum_{n=1}^{\infty} \dfrac{\sin(n\pi/2)}{n} \cos nx = 1 + \dfrac{4}{\pi} \displaystyle\sum_{n=1}^{\infty} \dfrac{(-1)^n}{2n-1} \cos(2n-1)x$.

5. (a) $\sin x$. (b) $\dfrac{2}{\pi} - \dfrac{4}{\pi} \displaystyle\sum_{n=1}^{\infty} \dfrac{\cos 2nx}{4n^2-1}$.

7. (a) $\dfrac{4L}{\pi} \displaystyle\sum_{n=1}^{\infty} \left[\dfrac{1-(-1)^n}{n}\right] \sin \dfrac{n\pi x}{L} = \dfrac{8L}{\pi} \displaystyle\sum_{n=1}^{\infty} \dfrac{1}{2n-1} \sin \dfrac{(2n-1)\pi x}{L}$.

   (b) $2L$.

9. (a) $\dfrac{2L^2}{\pi} \displaystyle\sum_{n=1}^{\infty} \dfrac{\cos(n\pi/2) - \cos n\pi}{n} \sin \dfrac{n\pi x}{L}$.

   (b) $\dfrac{L^2}{2} - \dfrac{2L^2}{\pi} \displaystyle\sum_{n=1}^{\infty} \dfrac{\sin(n\pi/2)}{n} \cos \dfrac{n\pi x}{L}$

   $= \dfrac{L^2}{2} + \dfrac{2L^2}{\pi} \displaystyle\sum_{n=1}^{\infty} \dfrac{(-1)^n}{2n-1} \cos \dfrac{(2n-1)\pi x}{L}$.

11. (a) $\dfrac{4L^2}{\pi^3} \displaystyle\sum_{n=1}^{\infty} \left[\dfrac{1-(-1)^n}{n^3}\right] \sin \dfrac{n\pi x}{L} = \dfrac{8L^2}{\pi^3} \displaystyle\sum_{n=1}^{\infty} \dfrac{1}{(2n-1)^3} \sin \dfrac{(2n-1)\pi x}{L}$.

   (b) $\dfrac{L^2}{6} - \dfrac{2L^2}{\pi^2} \displaystyle\sum_{n=1}^{\infty} \left[\dfrac{1+(-1)^n}{n^2}\right] \cos \dfrac{n\pi x}{L} = \dfrac{L^2}{6} - \dfrac{L^2}{\pi^2} \displaystyle\sum_{n=1}^{\infty} \dfrac{1}{n^2} \cos \dfrac{2n\pi x}{L}$.

Section 12.4D

1. At $x = 0$, series converges to $\frac{1}{2}$; at $x = \pi/2$, to $(3\pi + 2)/4$; and at $x = \pi$, to π.

Section 13.1

1. (a) $(0, 0), (1, 1)$; (b) $\dfrac{dy}{dx} = \dfrac{x^2 - y}{x - y^2}$; (c) $x^3 - 3xy + y^3 = c$.

3. (a) $x = e^t$, $y = 3e^t + te^t$; (b) $x = e^{t-3}$, $y = te^{t-3}$;
   (c) The equation of the common path is $y = x(\ln |x| + 3)$;
   (d) The family of paths is $y = x(\ln |x| + c)$.

Section 13.2
1. Saddle point, unstable.
3. Saddle point, unstable.
5. Center, stable.
7. Node, unstable.
9. (a) Saddle point; (b) $x = c_1 e^{2t} + c_2 e^{-2t}$, $y = c_1 e^{2t} - 3c_2 e^{-2t}$.
   (c) The general solution is $3x^2 - 2xy - y^2 = c$.

Section 13.3
1. Node, unstable.
3. Saddle point, unstable.
5. Saddle point, unstable.
7. (a) Saddle point. (b) $y^2 = 2(x + e^{-x} + c)$. (d) $x^2 - y^2 = c$.
9. (a) Critical points are $(-1, 0)$, $(0, 0)$, $(1, 0)$, $(2, 0)$, and $(3, 0)$.
   (d) $(0, 0)$ and $(2, 0)$ are centers; the other three are saddle points.
11. (a) Critical points are $(-1, 0)$, $(0, 0)$, $(1, 0)$, and $(2, 0)$.
    (d) $(0, 0)$ is a center; $(-1, 0)$ and $(2, 0)$ are saddle points; $(1, 0)$ is a "cusp."
13. Saddle point if $\alpha < -12$; node if $-12 < \alpha \le 4$; spiral point if $\alpha > 4$.
15. (a) $(0, 0)$, saddle point, unstable; $(\frac{3}{4}, -\frac{3}{16})$, spiral point, unstable.
    (b) $(0, 0)$, node, asymptotically stable; $(2, 6)$, saddle point, unstable; $(3, 12)$, node, unstable.
17. (a) $\lambda_0 = 9$.
    (b) Center at $(1, 0)$; saddle point at $(5, 0)$.
    (c) "Cusp" at $(3, 0)$.
19. (a) $E(x, y) = x^2 + y^2$; asymptotically stable.
    (c) $E(x, y) = x^2 + y^2$; asymptotically stable.

Section 13.4
1. (a) $\dfrac{dr}{dt} = r(4 - r^2)$, $\dfrac{d\theta}{dt} = 4$.

   (c) $x = \dfrac{2 \cos(4t - t_0)}{\sqrt{1 + ce^{-8t}}}$, $y = \dfrac{2 \sin(4t - t_0)}{\sqrt{1 + ce^{-8t}}}$;

   $x = 2 \cos(4t - t_0)$, $y = 2 \sin(4t - t_0)$.
3. Equation has periodic solutions.
5. Equation has no nontrivial periodic solutions.
7. Equation has no nontrivial periodic solutions.

Section 13.5
1. $x = a_0 \sin\left\{\left[\dfrac{5\mu a_0^4 + 16\omega^2}{16\omega}\right] t + \phi_0 \right\}$.

3. $x = \dfrac{2a_0}{\sqrt{3a_0^2 \mu \omega^2 t + 4}} \sin\left[\omega t + \dfrac{\ln(3a_0^2 \mu \omega^2 t + 4)}{2\omega^3} + \phi_0 \right]$.

5. $x = a_0 e^{-\mu t/2} \sin\left[\omega t - \dfrac{3a_0^2 e^{-\mu t}}{8\omega} + \phi_0 \right]$.

Section 14.1

1. $u = f(y + x) + g(y + 6x)$; hyperbolic.
3. $u = f(y - 3x) + xg(y - 3x)$; parabolic.
5. $u = f(y + ix) + g(y - ix)$; elliptic.
7. $u = f(y) + g(y + 2x)$; hyperbolic.

9. $u = f\left(y - \dfrac{x}{2}\right) + g\left(y + \dfrac{3x}{4}\right)$; hyperbolic.

11. (b) Hyperbolic at all (x, y) such that $y < x^2/4$.

Section 14.2

1. $y(x, t) = \sin 2x \cos 2t$.

3. $y(x, t) = \displaystyle\sum_{n=1}^{\infty} b_n \sin \dfrac{n\pi x}{3} \cos \dfrac{2n\pi t}{3}$, where

$$b_n = \dfrac{12}{n^2\pi^2} \sin \dfrac{n\pi}{2} \cos \dfrac{n\pi}{6} \ (n = 1, 2, 3, \dots)$$

$$= 6\sqrt{3}/n^2\pi^2 \ (n = 1, 7, 13, \dots); \ -6\sqrt{3}/n^2\pi^2 \ (n = 5, 11, 17, \dots);$$

0 (n even or $n = 3, 9, 15, \dots$).

5. $u(x, t) = \displaystyle\sum_{n=1}^{\infty} c_n \sin \dfrac{n\pi x}{L} e^{-n^2\pi^2 t/L^2}$, where

$$c_n = \dfrac{2}{L} \int_0^L f(x) \sin \dfrac{n\pi x}{L} \, dx \ (n = 1, 2, 3, \dots).$$

7. $u(x, y) = \displaystyle\sum_{n=1}^{\infty} a_n \sin nx \sinh n(y - \pi)$, where

$$a_n = -\dfrac{2}{\pi \sinh n\pi} \int_0^\pi f(x) \sin nx \, dx \ (n = 1, 2, 3, \dots).$$

Section 14.3

1. $\dfrac{\partial^2 u}{\partial \xi \, \partial \eta} = 0$.

3. $\dfrac{\partial^2 u}{\partial \xi \, \partial \eta} = \dfrac{\partial u}{\partial \xi} - \dfrac{1}{2} \dfrac{\partial u}{\partial \eta}$.

5. $\dfrac{\partial^2 u}{\partial \xi^2} + \dfrac{\partial^2 u}{\partial \eta^2} = \dfrac{\partial u}{\partial \xi}$.

7. $\dfrac{\partial^2 u}{\partial \xi \, \partial \eta} = \dfrac{1}{6}\left(\dfrac{\partial u}{\partial \xi} + \dfrac{\partial u}{\partial \eta}\right)$.

9. $\dfrac{\partial^2 u}{\partial \xi^2} + \dfrac{\partial^2 u}{\partial \eta^2} = -\dfrac{2}{9} u$.

Section 14.4

3. $y = 2x + c_1$, $y = -4x + c_2$.
5. No real characteristics exist.

Suggested Reading

The following are other introductions to differential equations:

BOYCE, W., and R. DIPRIMA. *Elementary Differential Equations and Boundary Value Problems*, 3rd ed. (Wiley, New York, 1977).

BRAUER, F., and J. NOHEL. *Ordinary Differential Equations*, 2nd ed. (Benjamin, New York, 1973).

BRAUN, M. *Differential Equations and Their Applications: An Introduction to Applied Mathematics*, 3rd ed. (Springer-Verlag, New York, 1983).

DERRICK, W., and S. GROSSMAN. *Elementary Differential Equations with Applications*, 2nd ed. (Addison-Wesley, Reading, Mass., 1981).

FINIZIO, N., and G. LADAS. *Ordinary Differential Equations with Modern Applications*, 2nd ed. (Wadsworth, Belmont, Calif., 1982).

FINNEY, R., and D. OSTBERG. *Elementary Differential Equations with Linear Algebra* (Addison-Wesley, Reading, Mass., 1976)

RAINVILLE, E., and P. BEDIENT. *Elementary Differential Equations*, 6th ed. (Macmillan, New York, 1981).

REISS, E., A. CALLEGRI, and D. AHLUWALIA. *Ordinary Differential Equations with Applications* (Holt, Rinehart, & Winston, New York, 1976).

SANCHEZ, D., R. ALLEN, and W. KYNER. *Differential Equations, An Introduction* (Addison-Wesley, Reading, Mass., 1983).

SPIEGEL, M. *Applied Differential Equations*, 3rd ed. (Prentice-Hall, Englewood Cliffs, N.J., 1981).

WYLIE, C. R. *Differential Equations* (McGraw-Hill, New York, 1979).

ZILL, D. *A First Course in Differential Equations with Applications*, 2nd ed. (Prindle, Weber, & Schmidt, Boston, 1982).

An introduction to matrices and vectors is presented in the following text:

CAMPBELL, H. *An Introduction to Matrices, Vectors,* and Linear Programming, 2nd ed. (Prentice-Hall, Englewood Cliffs, N.J., 1977).

An introduction to uniform convergence is contained in the following text:

ROSS, K. *Elementary Analysis: The Theory of Calculus* (Springer-Verlag, New York, 1980).

The following are more advanced works on ordinary differential equations:

ARNOLD, V. *Ordinary Differential Equations* (MIT Press, Cambridge, Mass., 1973).

ARROWSMITH, D., and C. PLACE. *Ordinary Differential Equations* (Chapman & Hall, London, 1982).

BIRKHOFF, G., and G.-C. ROTA. *Ordinary Differential Equations,* 3rd ed. (Wiley, New York, 1978).

BRAUER, F., and J. NOHEL. *Qualitative Theory of Ordinary Differential Equations* (Benjamin, New York, 1969).

CODDINGTON, E., and N. LEVINSON. *Theory of Ordinary Differential Equations* (McGraw-Hill, New York, 1955).

COLE, R. *Theory of Ordinary Differential Equations (Appleton-Century-Crofts, New York, 1968).*

CORDUNEANU, C. *Principles of Differential and Integral Equations* (Allyn & Bacon, Boston, 1971).

HALE, J. *Ordinary Differential Equations* (Wiley-Interscience, New York, 1969).

HARTMAN, P. *Ordinary Differential Equations* (Wiley, New York, 1964).

HIRSCH, M., and S. SMALE. *Differential Equations, Dynamical Systems, and Linear Algebra* (Academic Press, New York, 1974).

HUREWICZ, W. *Lectures on Ordinary Differential Equations* (Technology Press and Wiley, New York, 1958).

JORDAN, D., and P. SMITH. *Nonlinear Ordinary Differential Equations* (Clarendon Press, Oxford, 1977).

LAKIN, W., and D. SANCHEZ. *Topics in Ordinary Differential Equations: A Potpourri* (Prindle, Weber & Schmidt, Boston, 1970).

MILLER. R., and A. MICHEL. *Ordinary Differential Equations* (Academic Press, New York, 1982).

SANCHEZ, D. *Ordinary Differential Equations and Stability Theory: An Introduction* (Freeman, San Francisco, 1968).

The following are works dealing with partial differential equations:

BERG, P., and J. MCGREGOR. *Elementary Partial Differential Equations* (Holden-Day, San Francisco, 1966).

CHURCHILL, R., and J. BROWN. *Fourier Series and Boundary Value Problems*, 3rd ed. (McGraw-Hill, New York, 1978).

DENNEMEYER, R. *Introduction to Partial Differential Equations and Boundary Value Problems* (McGraw-Hill, New York, 1968).

MYINT-U, T. *Partial Differential Equations of Mathematical Physics* (Elsevier, New York, 1973).

POWERS, D. *Boundary Value Problems*, 2nd ed. (Academic Press, New York, 1979).

SAGAN, H. *Boundary and Eigenvalue Problems in Mathematical Physics* (Wiley, New York, 1961).

WEINBERGER, H. *A First Course in Partial Differential Equations* (Blaisdell, New York, 1965).

INDEX

Abel-Liouville formula:
  for linear systems, 528
  for nth order linear equations, 558
Abel's formula, 578
Adjoint equation, 573
Amplitude, 183
Analytic function, 222
Applications:
  to electric circuit problems, 211, 282
  to geometric problems, 70
  to mechanics problems, 77, 179, 278, 655, 668,
    674, 724
  to mixture problems, 94, 285
  to population growth, 91
  to rate problems, 89
Approach, of path to critical point, 637
Approximate methods, first-order equations:
  graphical, 377
  numerical, 394
  power series, 384
  successive approximations, 390
Asymptotic stability of critical point, 643, 653,
    661, 682, 686
Autonomous system, 633
Auxiliary equation, 126

Beats, 205 (Exercise 9)
Bendixson's nonexistence criterion, 695
Bernoulli differential equation, 54
Bessel functions:

  of first kind, order one, 259
  of first kind, order p, 259
  of first kind, order zero, 254
  of second kind, order p, 262
  of second kind, order zero, 256
Bessel's differential equation:
  applications to partial differential equations, 737
  of order p, 252, 256, 586 (Exercise 11)
  of order zero, 253
Boundary-value problem, 16, 588

Canonical forms, of partial differential equations,
    743, 746, 750, 753, 756
Cauchy-Euler differential equation, 164
Center, 638, 652, 660
Characteristic equation, 301, 340, 360, 645
Characteristic function:
  definition of, 592
  existence of, 595
  expansion in series of, 603
  orthogonality of, 598
  orthonormal system of, 602
Characteristic roots, 301
Characteristics, of partial differential equations,
    761, 764
Characteristic values:
  of matrix, 337, 340, 360
  of Sturm-Liouville boundary-value problem, 592
Characteristic vectors, 337, 340
Clairaut's equation, 68 (Exercise 20)

Companion vector equation, 545
Comparison theorem of Sturm, 581
Complementary function, 121
Continuation of solutions, 484
Continuing methods, 395, 407
Convergence:
    of sequence, 461, 464
    of series, uniform, 468
    of trigonometric Fourier series, 624
Convolution, 437
Critical damping, 191
Critical point:
    approached by path, 637
    criteria for determination of type of, 645, 654
        (Table 13.1), 660
    definition of, 634
    entered by path, 637
    index of, 700
    isolated, 637
    of linear system, 644
    of nonlinear system, 658
    stability of, 642, 653, 661, 682, 684
    types of, 638

Damped oscillatory motion, 189
Damping, 187
Damping factor, 190
Damping force, 181
Dependence of solutions:
    on initial conditions, 488
    on a parameter, 678
Determinants, fundamentals of, 771
Differential equation:
    analytic significance of, 11
    definition, 3
    geometric significance of, 11
Differential operator, 268
Direction field, 378
Domain, 469

Entering, of path to critical point, 637
Equidimensional differential equation, 164
Euler method, 395
Euler's constant, 256
Even function, 611
Exact differential, 27
Exact differential equation, 27
Existence theorems:
    for characteristic values and functions of
        Sturm-Liouville Problems, 595
    for closed paths of nonlinear systems, 698
    for first-order initial-value problems, 19, 475
    for Laplace transforms, 416
    for linear differential equation with given
        fundamental set, 563
    for linearly independent solutions:
        linear systems, 295, 357, 361
        nth-order linear equations, 109, 553
        second-order linear equations, 112, 223, 240

    for linear systems, initial-value problems, 292,
        500, 509
    for nonlinear systems, initial-value problems, 496
    for nth-order linear equations, initial-value
        problems, 103, 501, 543
    for nth-order nonlinear equations, initial-value
        problems, 498
    for periodic solutions of nonlinear equations, 703
Expansion in series of orthonormal functions, 603
Exponential order, functions of, 415
Exponents of differential equation, 237

Focal point, 640
Forced motion, 199
Fourier coefficients, 608, 611
Fourier cosine series, 620
Fourier series, relative to orthonormal system, 608
Fourier series, trigonometric:
    convergence, 624
    definition of, 611
    of even function, 611
    examples illustrating, 613
    of odd function, 612
Fourier sine series:
    applied to partial differential equations, 729, 735
    definition of, 619
Free damped motion, 189
Free undamped motion, 182
Frequency, natural, 184
Frobenius, method of, 236
Fundamental matrix, 517
Fundamental set of solutions, 110, 517, 553

Gamma function, 258
Graphical methods, 377

Half-path, 697
Heat equation, 721, 742 (Exercise 5)
Homogeneous differential equation:
    first-order, 42
    linear, higher, 102, 106, 125, 543, 558
    linear partial of second-order, 717
    linear systems, 290, 292, 505, 510
Homogeneous function of degree n, 43
Hooke's law, 180

Improper integrals, 416
Index:
    of critical point, 700
    of curve, 700
Indicial equation, 237
Initial-valve problem:
    for first-order ordinary differential equations,
        17, 473
    for linear partial differential equations, 757, 759
    for nth-order linear ordinary differential
        equations, 1, 3, 501
    for nth-order ordinary differential equations, 498

for systems of linear ordinary differential
  equations, 292, 500, 510
for systems of nonlinear ordinary differential
  equations, 496
Integral curve, 11, 377
Integral equation, 474
Integrating factor:
  definition of, 35
  for first-order linear differential equations, 50
  methods of finding, 61
  special types of, 63
Inverse Laplace transform, 431
Irregular singular point, 234
Isocline, 380

Kirchoff's laws, 212, 282
Kryloff and Bogoliuboff, method of, 707

Laplace differential equation, 721, 731, 742
  (Exercise 7)
Laplace transform:
  convolution, 437
  definition of, 411
  of derivatives, 419
  existence of, 416
  inverse transform, 431
  linear property, 418
  of periodic functions, 428
  properties of, 418
  solution of linear differential equations by, 441
  solution of systems of linear differential
    equations by, 453
  of step functions, 424
  table of, 434
  of translated functions, 426
  translation property, 421
Legendre's differential equation, 233 (Exercise 18),
  575
Liapunov function, 684
Liapunov's direct method, 682
Lienard-Levinson-Smith Theorem, 703
Limit cycle, 692
Limit set, of half-path, 698
Linear algebraic systems, theorems on, 171, 332,
  559
Linear combination, 106, 293, 316
Linear dependence:
  of constant vectors, 332
  of functions, 107
  of solutions, 294, 549
  of vector functions, 334
Linear independence:
  of constant vectors, 333
  of functions, 108
  of solutions, 109, 294, 549
  of vector functions, 334
Linear ordinary differential equations:
  first-order, 49
  higher-order:

companion vector equation of, 545
constant coefficients, 125, 441
definition of, 5, 102
existence theorems for, 103, 501, 543
general solution of, 110, 122, 501
homogeneous, 102, 106, 125, 510, 522, 543
Laplace transform solution of, 441
nonhomogeneous, 120, 137, 155, 569
reduction of order, 116, 565
relation to linear system, 267, 544
theory of, 102, 543, 558, 569
second-order:
  applications of, 179
  general, 103, 107, 112, 116, 221
  self-adjoint, 574
  Sturm theory, 577
  theory of, 170
Linear partial differential equation, 717
Line element, 378
Lipschitz condition, 470
Logarithmic decrement, 190

Matrix:
  adjoint of, 327
  characteristic equation of, 340, 347
  characteristic values of, 337, 340, 347
  characteristic vectors of, 337, 340
  coefficient matrix, 347, 356
  cofactor of an element of, 326
  definitions, basic, 312–315
  derivative of, 329
  fundamental matrix, 517
  identity matrix, 324
  inverse of, 325
  minor of an element of, 326
  multiplication of, 322
  nonsingular, 325
  product of, by vector, 317
  symmetric matrix, 341
  transpose of, 327
Method:
  of Euler, 395
  of Euler, modified, 399
  of Frobenius, 236
  of isoclines, 380
  of Kryloff and Bogoliuboff, 707
  of Liapunov, direct, 682
  of Milne, 407
  of Runge-Kutta, 403
  of separation of variables, 722
  of successive approximations, 390
  of power series, 384
  of Taylor series, 384
  of undetermined coefficients, 137, 386
  of variation of parameters, 155, 537, 570
Milne, method of, 407
Modified Euler method, 399

Negative definite function, 682

Negative semidefinite function, 682
Newton's second law, 77, 182
Node, 641, 645, 648, 660
Nonhomogeneous differential equation,
    higher-order linear, 102, 120, 137, 155, 569
Nonlinear differential equation, 5, 632
Normalized function, 601
Normalized linear differential equation, 543, 563
Numerical methods, 394

Oblique trajectories, 74
Odd function, 611
Operator method for linear systems, 270
Order, of a differential equation, 5
Ordinary point, 222
Orthogonal functions, 597
Orthogonal trajectory, 70
Orthonormal system, 601
Overcritical damping, 192

Partial differential equations:
    canonical forms of, 743, 746, 750, 753, 756
    characteristics of, 761, 764
    classification of, 720
    elliptic, 720, 753
    homogeneous linear, 717
    hyperbolic, 720, 746
    infinite series of solutions of, 717
    initial-value problem for, 757, 759
    parabolic, 720, 750
    separation of variables in, 722
Particular integral:
    definition of, 121
    methods of finding, 137, 155, 570
    superposition principle for, 572
Path, 633
Pendulum problem, 187 (Exercise 8), 674
Periodic function, 624
Periodic motion, 183
Periodic solutions, 692, 703
Period of motion, 184, 190
Phase constant, 184
Phase plane, 633
Picard method of successive approximations, 390
Piecewise continuous function, 414
Piecewise smooth function, 625
Plucked string problem, 730
Poincare-Bendixson theorem, 698
Pointwise convergence, 462
Positive definite function, 682
Positive semidefinite function, 682
Power series solutions, 221, 384

Recurrence formula, 228, 239
Reduction of order, 116, 565
Regular singular point, 324
Resonance, 206
Restoring force, 180

Riccati equation, 59 (Exercise 38), 584
    (Exercise 5)
Runge-Kutta method, 403

Saddle point, 640, 646, 660
Self-adjoint equation, 574
Separable differential equation, 39
Separation of variables, method of, 722
Separation theorem, of Sturm, 580
Separatrix, 674
Series solutions:
    of first-order differential equations, 384
    about ordinary point, 221, 223
    about regular singular point, 233, 235, 240
    of second-order linear differential equations, 221
Simple harmonic motion, 183
Singular point, 222
Solution:
    definition, 8
    explicit, 8
    of first-order initial-value problem, 17, 473
    formal, 10
    fundamental set of, 110, 517, 553
    general, nth-order linear differential equations,
        110, 121
    general, system of linear differential equations,
        273, 296, 298, 520, 535
    geometric significance of, 11
    implicit, 8
    of nth-order ordinary differential equation, 8, 497
    one-parameter family of, 10
    of partial differential equation, 715
    of system of linear ordinary differential
        equations, 265, 272, 290, 356, 508
    of system of ordinary differential equations, 495
Spiral point, 640, 650, 660
Spring constant, 180
Stability of critical point, 642, 653, 661, 682
Starting methods, 394
Steady state term, 201
Step function, 424
Strip condition, 759
Sturm comparison theorem, 581
Sturm-Liouville problem:
    application of, to partial differential
        equations, 726, 733
    definition of, 588
    properties of solutions of, 595, 598
Sturm separation theorem, 580
Sturm theory, 577
Superposition principle, 572
Systems of linear differential equations:
    applications of, 278
    coefficient matrix of, 347, 356
    constant coefficients, 270, 301, 346, 355
    critical points of, 644
    existence theorems for solutions of, 292, 295,
        349, 352, 362, 500, 509, 518

fundamental matrix of, 517
fundamental set of solutions of, 517
general solution of, 273, 296, 298, 520, 535
homogeneous, normal form, 290, 292, 505, 510
Laplace transform solution of, 453
matrix method for solving, 346, 355
nonhomogeneous, normal form, 290, 297, 505, 533
normal form, 266, 505
operator methods for, 270, 275
relation to single higher-order linear equation, 267, 545
scalar form, 507
solution of, 272, 290, 508
theory, normal form, 297, 499, 505
types of, 264
vector form, 346, 356, 507, 545
Systems of nonlinear differential equations:
autonomous, 633, 658
conservative, 668
critical points of, 634, 658, 673
derivative of a function with respect to, 682
examples of, 662
for n first-order, 495
without linear terms, 667

Taylor series method, 384
Total differential, 26
Transient term, 201
Trivial solution, 105
Two-point boundary-value problem, 16, 588

UC function, 141
UC set, 142
Undetermined coefficients, method of:
for first-order differential equations, 386
for nth-order linear differential equations, 137
table for, 143
Uniform convergence, 464, 468
Uniqueness of solutions, 19, 103, 475, 482
Unit step function, 424

Van der Pol equation, 704, 712
Variables separable, equations with, 39
Variation of parameters, 155, 537, 570
Vector:
constant vector, 319
definitions, basic, 313, 314
derivative of vector function, 319
integral of, 319
linearly dependent, 332, 334
linearly independent, 333, 334
product of matrix by, 317
vector differential equation, 346, 356, 507, 545
vector function, 319
Vibrating string problem, 724
Vibrations of mass on spring, 179

Wave equation, 720, 724
Weierstrass M-test, 468
Wronskian, 111, 296, 358, 513, 551

Zeros of solutions, 260, 579, 595